CAMBRIDGE LIBRARY COLLECTION

Books of enduring scholarly value

Mathematical Sciences

From its pre-historic roots in simple counting to the algorithms powering modern desktop computers, from the genius of Archimedes to the genius of Einstein, advances in mathematical understanding and numerical techniques have been directly responsible for creating the modern world as we know it. This series will provide a library of the most influential publications and writers on mathematics in its broadest sense. As such, it will show not only the deep roots from which modern science and technology have grown, but also the astonishing breadth of application of mathematical techniques in the humanities and social sciences, and in everyday life.

The Collected Mathematical Papers

Arthur Cayley (1821-1895) was a key figure in the creation of modern algebra. He studied mathematics at Cambridge and published three papers while still an undergraduate. He then qualified as a lawyer and published about 250 mathematical papers during his fourteen years at the Bar. In 1863 he took a significant salary cut to become the first Sadleirian Professor of Pure Mathematics at Cambridge, where he continued to publish at a phenomenal rate on nearly every aspect of the subject, his most important work being in matrices, geometry and abstract groups. In 1882 he spent five months at Johns Hopkins University, and in 1883 he became president of the British Association for the Advancement of Science. Publication of his Collected Papers - 967 papers in 13 volumes plus an index volume - began in 1889 and was completed after his death (while this volume was in the press) under the editorship of his successor in the Sadleirian Chair. This volume contains 70 papers published mostly between 1871 to 1873, as well as a 36-page obituary of the author.

The Collected Mathematical Papers

VOLUME 8

ARTHUR CAYLEY

CAMBRIDGE
UNIVERSITY PRESS

CAMBRIDGE UNIVERSITY PRESS

Cambridge New York Melbourne Madrid Cape Town Singapore São Paolo Delhi

Published in the United States of America by Cambridge University Press, New York

www.cambridge.org
Information on this title: www.cambridge.org/9781108005005

This edition first published 1895
This digitally printed version 2009

ISBN 978-1-108-00500-5

MATHEMATICAL PAPERS.

London: C. J. CLAY AND SONS,
CAMBRIDGE UNIVERSITY PRESS WAREHOUSE,
AVE MARIA LANE.
Glasgow: 263, ARGYLE STREET.

Leipzig: F. A. BROCKHAUS.
New York: MACMILLAN AND CO.

Vol. 8. Notes & References.

Ribaucour C. R. t. 75. 1872 pp. 533–536. Referring to my Note remarks that the condition (by means of the imaginary coordinates of M. Ossian Bonnet) expressed can be ~~expressed~~ in a simple form communicated by him to the Philomathic Society, May 1870. I reproduce this investigation, altho' it is not easy to present it in a quite intelligible form: ~~we take~~ ~~the investigation is as follows: he takes~~ $p = f(x, y, z)$ to represent a family of surfaces belonging to a triply orthotomic system. Considering two neighbouring surfaces (A) & (A') corresponding to the values z & $z + dz$; A & A' the two points where they meet the trajectories of the surfaces; AT, A'T' the tangents to the curves of curvature of the same system at A, A' respectively. Then according to the remark of M. Lévy, it is to be expressed that these lines meet, and this is done by expressing that along the trajectory AA', the variation of the angle of AT with the osculating plane at A is equal to the angle of the osculating planes at A, A' respectively.

Let B' be the spherical image of A', the plane OBB' is parallel to the osculating plane at A of the trajectory, and the angle of the two osculating planes measures the geodesic curvature of BB': denote this by $d\gamma$

Let β be the angle of BB' with BX, θ the angle of AT with BX, $\beta - \theta$ is the angle of AT with the osculating plane at A of the trajectory: $d\beta - d\theta = d\gamma$, corresponding to dz in the passage from

But dx & dy being the increments of x, y

of M. Liouville

$dy =$

The condition thus is $d\theta = i\left(\frac{d\lambda}{\lambda dx}dx - \frac{d\lambda}{\lambda dy}dy\right)$

write

$a = \frac{dp}{\lambda^2 dx},\ b = \frac{dp}{\lambda^2 dy},\ c = \frac{1}{\lambda^2}\frac{d^2p}{dx\,dy}$

$ds^2 = 4\lambda^2 \frac{da}{dx}\frac{db}{dy}\,dx\,dy$

and the formula $e^{-2i\theta} = \pm\sqrt{\frac{da}{dx}} \div \sqrt{\frac{db}{dx}}$

enables this to be written in the definitive form

$$dx\,\frac{d}{dx}\,l\left(\frac{db}{dy} \div \lambda^4\frac{da}{dx}\right) - dy\,\frac{d}{dy}\,l\left(\frac{da}{dx} \div \lambda^4\frac{db}{dy}\right) + dz\,\frac{d}{dz}\left(l\frac{db}{dy} - \frac{d}{dz}\,l\frac{da}{dx}\right) = 0.$$

We have

$$dx\left(\tfrac{1}{2}p + c\right) + dy\,\frac{db}{dy} + dz\,\frac{db}{dz} = 0$$

$$dx\,\frac{da}{dx} + dy\left(\tfrac{1}{2}p + c\right) + dz\,\frac{da}{dz} = 0$$

and thence eliminating dx, dy, dz we have

$$\begin{vmatrix} \frac{d}{dx}\,l\left(\frac{db}{dy} \div \lambda^4\frac{da}{dx}\right), & \frac{d}{dy}\,l\left(\frac{db}{dy} \div \lambda^4\frac{da}{dx}\right), & \frac{d}{dz}\,l\left(\frac{db}{dy} \div \frac{da}{dx}\right) \\ \tfrac{1}{2}p + c, & \frac{db}{dy} & ; \frac{db}{dz} \\ \frac{da}{dx} & \tfrac{1}{2}p + c & \frac{da}{dz} \end{vmatrix} = 0$$

which defines the triply orthotomic system.

THE COLLECTED

MATHEMATICAL PAPERS

OF

ARTHUR CAYLEY, Sc.D., F.R.S.,

LATE SADLERIAN PROFESSOR OF PURE MATHEMATICS IN THE UNIVERSITY OF CAMBRIDGE.

VOL. VIII.

CAMBRIDGE:
AT THE UNIVERSITY PRESS.
1895

CAMBRIDGE:
PRINTED BY J. AND C. F. CLAY,
AT THE UNIVERSITY PRESS.

ADVERTISEMENT.

THE present volume contains 70 papers, numbered 486 to 555, published for the most part in the years 1871 to 1873.

The Table for the eight volumes is

Vol.	I.	Numbers	1	to	100.
,,	II.	,,	101	,,	158.
,,	III.	,,	159	,,	222.
,,	IV.	,,	223	,,	299.
,,	V.	,,	300	,,	383.
,,	VI.	,,	384	,,	416.
,,	VII.	,,	417	,,	485.
,,	VIII.	,,	486	,,	555.

PREFATORY NOTE.

THE death of Professor Cayley, which occurred on the 26th of January, 1895, has deprived the later part of this volume, as it will deprive the succeeding volumes, of the advantage of his supervision. The Syndics of the Press desired that the collection of the papers should be completed; and on the 15th of February, they asked me to undertake the duty of editing the remaining volumes. I willingly acceded to their request.

Professor Cayley had himself passed the first thirty-eight sheets of this volume for press; his illness prevented him from even revising any succeeding sheets. He had prepared one Note for the volume: it is printed at the end. The remaining volumes must appear without Notes and References: the reason being that he did not prepare these Notes in advance but only when the corresponding papers came before him in the proof-sheets.

The actual manuscript of the Note has been reproduced in facsimile upon the kind of paper which he regularly used during his mathematical investigations. As it refers to the memoir that ends only in the last sheet but one which he passed for press, it is one of the last pieces of his writing.

He left no instructions as to the Collection of his Mathematical Papers; the statement, prefixed to the first volume, is the only account of his method of arrangement. A comparison of the contents of the first seven volumes with the list of his papers in the Royal Society's *Catalogue of Scientific Papers* has enabled me to make out the detailed course of the method which will, of course, be followed in the remaining volumes.

The Syndics expressed their desire that I should insert some biographical notice of Professor Cayley in a volume of the series. Accordingly, one is inserted in the present volume; it is a reprint (with only slight verbal changes) of the notice which was written for the *Proceedings of the Royal Society.* And I have ventured to add a complete list of the lectures which he announced from year to year after his return to Cambridge in 1863 as Sadlerian Professor.

A. R. FORSYTH.

8 *June,* 1895.

ARTHUR CAYLEY.

[From the Obituary Notices in the *Proceedings of the Royal Society*, vol. LVIII. 1895.]

ARTHUR CAYLEY was the second son of Henry Cayley and Maria Antonia Doughty; he was born at Richmond, in Surrey, on 16 August, 1821.

The family, to whose fame so much honour has been added by one of the greatest mathematicians of all time, is of old origin and illustrious descent. Its name, like not a few English names, is derived from a locality in Normandy; there was a Castellum Cailleii, near Rouen, held by baronial tenure. The head of the house appears to have come to England with William the Conqueror and to have settled in Norfolk, becoming Lord of Massingham, Cranwich, Brodercross, and Hiburgh in that county. The influence of the family increased and, by the time of Edward II., Sir Thomas de Cailli possessed estates also in Yorkshire. On his decease without issue, the Yorkshire property was transferred to a younger branch of the family and was inherited by a long succession of Cayleys who made their home at Thormanby. One of these was knighted, as Sir William Cayley, in 1641; in 1661 he was created a baronet in recognition of his services during the Civil Wars, the title surviving to the present day. The fourth son of Sir William, Cornelius, settled at York; and the eldest son of the latter, also Cornelius, born in 1692, was a barrister and in 1725 was appointed Recorder of Kingston-upon-Hull, an office which he held until a few years before his death in 1779. Probably the advantages offered by Hull, then, as now, the greatest port on the northern coast of England, suggested commerce as an occupation for some members of the Recorder's large family; two of his sons became Russia merchants, settling in St Petersburg. The younger of these, being the fifth son of the Recorder, was Henry Cayley, born in 1768; he married, in 1814, Maria Antonia Doughty, a daughter of William Doughty. The eldest son of this marriage died in infancy. The youngest son, Charles Bagot, was a scholar, possessed of linguistic genius; he was particularly interested in the Romance Languages and he made verse-translations of Homer's *Iliad*, Dante's *Divine Comedy*, and the *Sonnets* of Petrarch. The second son was Arthur, the subject of the present sketch; he was born during a visit of his parents to England. Before passing to the details of his life, it may be added that the second of his father's sisters married Edward Moberly—also a Russia merchant living in St Petersburg—and was the mother of the late Dr. George Moberly, Bishop of Salisbury.

Mr. Henry Cayley took his young family to Russia and remained there for a few years. On retiring from business in 1829, he returned to England and settled into residence at Blackheath. Arthur was sent soon afterwards to a private school there, kept by the Rev. G. B. F. Potticary; and when he was fourteen he was transferred to King's College School, London. At a very early age he had begun to show some of those preferences by which the existence of mathematical ability is wont to reveal itself; he had a great liking for numerical calculations and he developed a great aptitude for them.

In his new school the boy showed himself to be possessed of remarkable ability: his power of grasping a new subject very rapidly and of seizing its central principles was certainly unusual. An old friend tells of an examination in chemistry: the subject had not been studied by Cayley before, but he soon acquired sufficient knowledge to carry off the medal from the professedly chemical students, to their surprise and mortification *. But it was most of all by the indications of mathematical genius that he astonished his teachers. It had been Mr. Cayley's intention to educate his son with the view of placing him in his former business—an intention not abandoned without reluctance. The impression, however, produced upon his teachers could not lightly be set aside; and the advice of the Principal to send him to Cambridge, where his abilities promised to secure brilliant distinction, was adopted.

Accordingly, he went to Cambridge. He was entered at Trinity College on 2nd May, 1838, as a pensioner, and he began residence in the succeeding October at the unusually early age of seventeen. He passed through the ordinary stages in the career of a successful student of mathematics. Like the other able undergraduates of his period, he "coached" with William Hopkins of Peterhouse, who has been described as a great and stimulating teacher—a description justified by the high achievements of a long line of distinguished and grateful pupils.

Cayley's fame grew rapidly: and, as is the way of Cambridge undergraduates, he soon was pointed out as the future Senior Wrangler of the year. It is interesting to find a record of him written about this time and published not long afterwards by an acquaintance†, who says that:—

> "As an undergraduate he had generally the reputation of a mere mathematician, which did him great injustice, for he was really a man of much varied information, and that on some subjects the very opposite of scientific—for instance, he was well up in all the current novels, an uncommon thing at Cambridge where novel-reading is not one of the popular weaknesses."

* It may be added that he maintained his interest in chemistry throughout his life, and acquired a considerable knowledge of it. When he was at Baltimore, in 1882, lecturing at the Johns Hopkins University by special invitation, he attended Professor Remsen's lectures with a pleasure which found expression in his letters home to his children in England. And on one occasion, at Professor Remsen's request, he lectured to the chemistry class on the hydrocarbon "trees" (*Brit. Assoc. Report*, 1875, pp. 257—305).

† Bristed, *Five Years in an English University* (second edition, 1852), p. 95.

It may be added that Cayley declared the story about him in the tripos, recorded by Bristed, to be quite apocryphal.

So also was another story, belonging to a later part of his life, according to which he is reported to have said that "the object of law was to say a thing in the greatest number of words, and of mathematics to say it in the fewest": this view, and the possibility of his ever having held it, he repudiated entirely.

Novel-readers are more frequent in Cambridge now than they appear to have been in 1842, and Cayley in his later days avoided reading some of the modern novels; but it is worth noting, as will subsequently be seen more in detail, that he had this "popular weakness" all his life.

He was admitted a scholar of the College on 1st May, 1840, winning his scholarship at the earliest time when it was possible to do so: and he secured a first class in each of the annual examinations of the College. No record of marks for the first and the second years is given in the Trinity Head Examiner's Book; but in the third year the marks are given and, as he then scored more than twice the marks of the second candidate, the Head Examiner separated him from the rest of the first class by drawing a line under his name. This presage of his powers was confirmed in the following year, 1842, when he graduated as Senior Wrangler; the Examiners were so definitely satisfied that he was first as to dispense in his case with the viva voce tests which at that time were a customary part of the Tripos. And in due course the first Smith's Prize was awarded to him in the succeeding examination.

Cayley's own "year" at Trinity was a distinguished one; for, in addition to himself, it contained Mr. (now the Right Honourable) George Denman, for many years a Judge of the High Court of Justice, and Mr. Hugh Andrew Johnstone Munro, one of the foremost of Latin Scholars of any period. And the distinction of Cayley's contemporaries in neighbouring years is marked: it is impossible to avoid noticing the names of some of the graduates in the Mathematical Tripos about that time. Sylvester and Green (second and fourth wranglers respectively in 1837), Leslie Ellis (senior in 1840), Stokes (senior in 1841), Cayley (senior in 1842), Adams (senior in 1843), Thomson—now Lord Kelvin—(second in 1845), constitute an extraordinary succession of mathematicians of whom England is justly proud. Their achievements in mathematical science have done much to render their University one of the acknowledged chief mathematical schools of the world.

Cayley was elected a Fellow of Trinity and admitted to fellowship on 3rd October, 1842, at an age younger than any other Fellow of the College, at least in the present century; and he was promoted from the position of Minor Fellow to that of Major Fellow on 2nd July, 1845, the year in which he proceeded to his M.A. degree. He was an Assistant Tutor of the College for three years; but such a post was then of an almost nominal character, and there appears to be no indication that any of the mathematical teaching of the College fell to him. He did, indeed, accept some private pupils: his lifelong friend, Canon Venables, has given a pleasant account* of a reading-party which Cayley took to Aberfeldie in 1842.

His pupils, however, did not tie him strictly to Cambridge, for it appears that the latter half of the year 1843 was devoted to continental rambles. The summer was spent in Switzerland, where his zest for walking and for mountain-climbing, a pleasure that never failed while his health lasted, found an active outlet: he had become a member of the Alpine Club in its comparatively early days. The last four months of the year were spent in Italy, partly in the North and in Florence, partly in Rome and Naples. It may have been on this tour that he acquired his love for

* *Guardian*, 6th Feb. 1895, p. 201.

both painting and architecture. The works of painters such as Masaccio, Giovanni Bellini, Perugino, and Luini, then first became known to him; they proved a delight at the time and remained a happy remembrance with him.

These and other continental journeys from time to time, while he remained in residence as a Fellow of his College, were his relaxations. He had no formal lecturing and he did not attempt to obtain a large number of private pupils. The leisure that he thus secured was turned to the best, and to him the most pleasant, of uses, in carrying out mathematical researches. It was, indeed, as an undergraduate that Cayley began the marvellous series of publications which, extending over more than fifty years of his life, have been concerned with practically every branch of pure mathematics as well as with theoretical dynamics and physical astronomy.

The time seemed ripe for the outburst of some mathematical activity. By the efforts of Herschel, Peacock, and Whewell, Cambridge teaching had been set free from the bonds that restricted methods of procedure to those which had proved effective in Newton's days; and the struggle to secure the admittance of analytical methods had been successfully completed. One sign of the new freedom was the foundation of the *Cambridge Mathematical Journal*, in 1837, by D. F. Gregory and Leslie Ellis. Before that time, practically the only English means of publication open to mathematicians was in the *Philosophical Transactions of the Royal Society;* and young writers, whether modest or not about the value of their researches, might well have hesitated before seeking publication in a quarter that exacts so high a standard. The new journal then founded was open to young students and gave them an opportunity, previously difficult to obtain, of making their researches known; and it proved a great stimulus to the intellectual activity of those members of the University. Only four volumes of the journal appeared; but it was continued, first under the name of the *Cambridge and Dublin Mathematical Journal*, and, subsequently down to the present time, under that of the *Quarterly Journal of Pure and Applied Mathematics*. Though the opportunities of publication, which now are afforded to mathematicians both in England and abroad, are vastly more numerous than they were half a century ago, the undoubted service rendered to English mathematics by the initial venture of the two young Cambridge men should not be forgotten.

It was in the second volume of this journal that Cayley's earliest paper, written in 1841, was printed: and two other papers bearing the same date—it was the year before his degree—are included in the third volume. Though the results are not remarkable, the freshness and the independence of these early investigations are worthy of notice. Cayley had evidently read with enquiring and critical care the *Mécanique Analytique* of Lagrange, some of the work of Laplace, and several memoirs in the two continental journals of the time, those of Liouville and Crelle. These achievements of an undergraduate of nineteen or twenty, which are rarely accomplished now and were still rarer in his day, recall Abel's dictum*:—

"Si l'on veut faire des progrès dans les mathématiques, il faut étudier les maîtres et non pas les écoliers."

* *Niels-Henrik Abel* (par Bjerknes, Paris, 1885), p. 173.

It was as certainly one of the characteristics of Cayley to find a stimulus to new developments in the main ideas of other writers as it was one of his characteristics to be able to follow out his own ideas with the insistent unwearying patience of an investigator creating a new work complete. And it is interesting to see how this faculty of receiving inspiration reveals itself from the beginning of his career.

Once free from the necessity of preparing for his Tripos and his Fellowship examination, he was able to throw himself into the work of production. His activity may be estimated from the fact that he produced three papers in 1842, eight in 1843, four in 1844, and thirteen in 1845. Moreover, these papers deal with a great variety of subjects. Thus he makes his first investigations in the numerative calculus of plane curves: he initiates his discussions about geometry of n dimensions: he founds the theory of invariants and covariants: and he elucidates the connexion between doubly-infinite products and elliptic functions. Some of these early papers are now classical; and the briefest inspection of them is sufficient to reveal the suggestiveness and the easy strength of the young mathematician who was not yet in his twenty-fifth year.

Even by this date the opportunities of publication in England had become inadequate to his needs. Curiously enough, he does not appear to have sent any paper to the Royal Society until the year 1852, when Sylvester communicated the "Analytical Researches connected with Steiner's Extension of Malfatti's Problem*" to that Society. Later in the same year, Cayley was elected a Fellow of the Society, and thereafter many of his papers appear in its *Philosophical Transactions*. Before 1852, there were few journals either at home or abroad which did not receive communications from him: and even in the quite early years of his researches, several of his papers, written in French, appeared in Liouville's journal and in Crelle's journal. As societies and journals grew in number, so the area over which his papers spread became ever wider.

At first, after winning his Trinity Fellowship, he remained at Cambridge, and his time must then have been largely at his own disposal. This freedom, in his circumstances, could last for only a limited time because, unless he either entered holy orders or devoted himself to teaching in some permanent post (if obtainable) in the College, the Fellowship could be held for not more than seven years after his M.A. degree—a period that would expire in 1852. He was unwilling to take holy orders—not that there was any religious obstacle in his way, for he was not harassed either by philosophical doubts or critical difficulties. His simple reason for remaining a layman was that, though devout in spirit and an active Churchman, he felt no vocation for the sacred office.

In consequence, it became necessary to choose some profession. Cayley selected the law, left Cambridge in 1846, entered at Lincoln's Inn, and became a pupil of the famous conveyancer, Mr. Christie. A story of their first interview, that Mr. Christie used to tell in after years, is an illustration of the modesty and the lack of self-

* Cayley's *Collected Mathematical Papers*, vol. II. No. 114. Subsequent references to this series will be made in the form *C. M. P.*

assertiveness which were leading features of Cayley's character: and this impression is confirmed by the recollections of a fellow-pupil, Mr. T. C. Wright, who says:—

> "... We fellow-pupils knew that Arthur Cayley had been the Senior Wrangler of his year, and that he possessed extraordinary abilities; but they were not indicated by his personal bearing, and the retiring modesty of his disposition prevented him from ever alluding to the honours he had won at Cambridge. He had one of the most unsophisticated minds I have ever known; jokes, and the badinage of the pupil-room, seemed to be delightful novelties to him, and his face beamed with amusement as he listened to them without taking much part in the conversation, being content to devote his time assiduously to work which I suspect was not altogether congenial to his taste...."

But if the modest, almost shy, man did not display his honours, he could not conceal his powers; and very soon his clearness of head, his almost intuitive grasp of the principles of any subject that came before him, his capacity for work and his power of concentration, made him a favourite pupil. He was called to the Bar on 3rd May, 1849, and thereafter he had no occasion to wait for business. Mr. Christie was always ready to supply him with at least as much conveyancing work as he was willing to undertake: but no advice, no encouragement, no opening however favourable, least of all any wish for fame or fortune, could tempt him to subside into a large practice. He restricted himself to "devilling" for Mr. Christie, and he limited the amount of work he would undertake in this way, always refusing work that came to him at first hand. There is no doubt that, had he remained at the Bar and devoted himself to its business, he could have made a great legal reputation and a substantial fortune: even as it was, some of his drafts* have been made to serve as models. But the spirit of research possessed him; it was not merely will but an irresistible impulse that made the pursuit of mathematics, not the practice of law, his chief desire. To achieve this desire, he reserved with jealous care a due portion of his time; and he regarded his legal occupations mainly as the means of providing a livelihood.

He remained at the Bar for fourteen years. Between two and three hundred papers are the mathematical outcome of that period; and they include some of the most brilliant of his discoveries. Among these papers are to be found the majority of his famous memoirs on quantics (particularly the sixth memoir, in which he develops his theory of geometry, and shows that all geometry can be made entirely descriptive), his work upon matrices, numerous contributions to the theory of symmetric functions of the roots of an equation, the elaborate calculations connected with the development of functions arising in the planetary and the lunar theories, and his valuable reports on theoretical dynamics. The enormous range over which his papers of these fourteen years extend is not more remarkable than the vigour of his contributions to knowledge; and a reference to them will show that he frequently recurs to some given problem, always adding something to the development.

* In Davidson's *Precedents and Forms in Conveyancing* (third edition, 1873), vol. III. Part II. p. 1067, the author adds a footnote, calling "attention to the remarkable skill exhibited in [a] settlement, the work of Mr. Arthur Cayley."

In judging of this persistent and unflagging activity, some account ought to be taken of his surroundings. It can hardly be that 2, Stone Court, from which many of his papers are dated, proved an inspiration to mathematical research. For part of the time, his friend Sylvester was in London—then as an actuary; and I have heard Cayley describe how Sylvester and he walked round the Courts of Lincoln's Inn discussing the theory of invariants and covariants which occupied (and occasionally absorbed) the attention of both of them during the fifties. And on matters which related to analytical geometry he was in frequent (but formal) correspondence with Salmon; indeed, the relation that existed between the two men developed ultimately into one of warm friendship and deep mutual regard: its sincerity can be gathered from the spirit animating Salmon's notice of Cayley, published in *Nature* in 1883, at the time when the latter was President of the British Association. But, with special exceptions of the types indicated, his work was so largely of the kind that is called path-breaking that he was bound to do it alone: he did it with a simple unconscious courage and with unfailing resolution.

It may easily be imagined that his links with life at Cambridge had now become slight. During the earliest of the years spent at the bar, he had returned on a few occasions. In 1848, the year before his call, he was the junior mathematical examiner in the regular annual examinations of Trinity; in 1849, and also in 1850, he was the senior mathematical examiner in the same examinations. In 1851 he was Senior Moderator for the Mathematical Tripos; one of the wranglers, Lightfoot, becoming subsequently his friend, and his colleague in the University, before going to his great work in the diocese of Durham as Bishop. In 1852 he was Senior Examiner for the Tripos, the senior wrangler of the year being Tait (also afterwards one of his intimate friends), now Professor of Natural Philosophy at Edinburgh. These seem to have been the only occasions when he was recalled to Cambridge; and they did not require any permanent connexion with the College or the University. He was settled in London, his allegiance divided between law and mathematics.

A change, however, in the statutes of the University offered an opportunity for his return to Cambridge; a professorship of pure mathematics was established upon an old foundation. Lady Mary Sadleir (who endowed the Croonian Lecture Fund of the Royal College of Physicians of London and also that of the Royal Society in memory of her first husband, Dr. William Croone, a physician and one of the earliest Fellows of the Royal Society) had, by her will, dated 25th September, 1701, and proved 6th November, 1706, given to the University an estate, which was to be used as an endowment of lectureships in algebra at nine of the colleges in Cambridge. These posts were duly established. The great developments of analysis, which took place at the end of the last century and during the first half of the present century, gradually proved that the restriction to algebra prevented the lectureships from being as adequate an encouragement to the advancement of mathematics as they were designed to be at the time of their establishment. Moreover, the lecturers had ceased to attract undergraduates to their lectures: so that the purpose of the foundation was not being fulfilled. Consequently, in 1857, a proposal was made by the Council of the Senate of the University that a new direction should be given to the endowment by the establishment of a professorship, to be called the Sadlerian Professorship of Pure Mathematics: the duty of the professor was "to explain and teach the principles of pure mathematics, and to apply himself to the advancement of that science." The proposal was approved by the Senate on 3rd December, 1857, and the new statute was sanctioned by an Order of the Queen in Council on 7th March, 1860. Some time had to elapse before certain provisional arrangements could be completed, and it was not until after three years that the University was in a position to act.

On 10th June, 1863, Cayley was elected Sadlerian professor: he held the chair for the rest of his life. The stipend attached to the professorship was modest, though it was improved in the course of subsequent legislation; these changes, however, could not have been foreseen at the time when Cayley was elected. Yet he had no hesitation about returning to Cambridge: for the post enabled him to devote his life to the pursuit he liked best. He never showed the slightest regret at having neglected the prospects of distinction at the bar, or at having chosen to return to his University; and he always expressed perfect satisfaction and content with his life in Cambridge, which was one of great happiness.

His appointment as Sadlerian professor marks a turning point in his life. Henceforward he lived, for the most part, in the quiet of the University; yet it was by no means in seclusion, for he took his share in administration, which claims a part (often too large a part) of the leisure of men fitted for this necessary duty. But he was not burdened by heavy claims arising out of his official position: and he was directed by the statutes governing him to do what was, as a matter of fact, his ideal in life. No man could have been better suited than Cayley was to fulfil the charge of the statutes: his knowledge and his power of research pointed him out as the obvious choice of the electors.

He settled in Cambridge at once. On 8th September, 1863, he married Susan, daughter of Robert Moline, of Greenwich. This is not the place to dwell upon his domestic life; but it is impossible to omit in silence all reference to its singular happiness, based upon the affection felt by its members for one another. Friends and visitors who have been in that home will not soon forget the kindness and the gracious courtesy of the welcome they received, or the atmosphere of peace into which they were raised. Sometimes in the old garden by the river-side, more often in the drawing-room, the talk went on; the professor himself listening, attentive and watchful, frequently taking only a slight share, but ever ready to join in. No cynicism or paradox in speech was ventured upon in his presence; no harshness of judgment was tolerated without a quiet protest; no sense of bustle or ambition was felt there; in all things the charm of an old-world home, centred round him. His widow and their two children, Mary and Henry, remain to mourn their loss.

His teaching duty was limited to the delivery of one course of lectures in the academic year, and he usually chose the Michaelmas term. This practice was maintained for twenty-three years until he was placed under the new statutes, which in 1882 had come into operation so far as concerned all future appointments. After that change, he delivered two courses of lectures, one in the Michaelmas term, the other in the Lent term. An inspection of the list of his lectures* shows that he chose his

* The list is given on pp. xlv, xlvi.

subjects by preference from analytical geometry, dynamics (in his view, theoretical dynamics is a portion of pure mathematics), differential equations, theory of equations, Abelian functions, elliptic functions, and modern algebra. The titles of the lectures, as announced, were sometimes vague, nor were they intended to limit his range; in all cases he went far beyond the boundary that so frequently limits Cambridge studies. Thus a course of lectures on differential equations, announced for the Michaelmas term in 1879, was chiefly concerned with conformal representation, polyhedral functions, and Schwarz's investigations on the hypergeometric series.

For many years he dispensed with the use of blackboard and chalk in his class-room; this was possible because his class usually was small. He brought his work written out upon the blue draft-paper,* which was regularly used by him in all his writing of mathematics; the exposition consisted partly of verbal explanations made as he showed the manuscript, partly of details written out at the moment. A change came in 1881, when his class amounted to fifteen or sixteen: he was then obliged to use the blackboard, and he subsequently maintained the new practice. Occasionally his older habit of explaining his manuscript recurred—he then placed it upon the board. This was especially the case when he brought carefully prepared diagrams, such as those used in the modular-function division of the plane: these diagrams were made much clearer by the use of water-colours to distinguish different sets of regions, and their preparation evidently gave him pleasure.

But, as may be surmised, his influence as a teacher was overshadowed by his influence as an investigator. Those whom he affected by his lectures belonged for the most part to the mathematical teachers in Cambridge: the number of undergraduates whom he influenced was small, though, when any one of them did come under his influence, the effect was well marked. His starting point in any subject was usually beyond the range of all other than quite advanced students; but to any able undergraduate who was willing to devote time, not merely to the comprehension of the matter in the lectures but also to collateral reading, the lectures were stimulating and inspiring. This effect was partly due to the easy strength with which he worked, partly to the spirit in which he approached old and new subjects alike; an independent suggestiveness and a singular freshness marked his views, and gave an added interest to his exposition even of a well-known theory. One reason of this freshness may be found in the fact that his lectures consisted of the current researches upon which he was engaged at the time; sometimes, even, a lecture would be devoted to results which he had obtained since the preceding lecture. Though the titles of the courses occasionally recur from one year to another, the same course was never given twice. The new matter in any course, once given, was usually incorporated in a paper or memoir; and when the same subject was nominally lectured upon again, it was a distinct part of the subject—old notes were never used a second time.

It was not alone by his lectures that he acted as professor. Students, seeking help or desiring to interest him in their work, found him always willing to give them the benefit of his advice, his criticism, and his knowledge. Nor was it merely mathematicians in Cambridge whom he helped in this way. He was continually consulted by

* It was the customary "scribbling paper" of his undergraduate days.

foreigners, who appreciated the promptness no less than the fulness of information in his replies.

It frequently happens that a man of genius, great enough to leave a distinct impress of his originality upon his science, finds it irksome to study what others have written. With the growth of all sciences during the last fifty years, especially—it may be said—with the growth of pure mathematics in that time, the tendency of workers is to become specialists in their own subject and, perhaps, in subjects immediately cognate with it, and to acquire only a slight acquaintance with what is being done outside the circle of their limited interests. Not so was Cayley: he was singularly learned in the work of other men, and catholic in his range of knowledge. Yet he did not read a memoir completely through: his custom was to read only so much as would enable him to grasp the meaning of the symbols and understand its scope. The main result would then become to him a subject of investigation: he would establish it (or test it) by algebraical analysis and, not infrequently, develop it so as to obtain other results. This faculty of grasping and testing rapidly the work of others, together with his great knowledge, made him an invaluable referee; his services in this capacity were used through a long series of years by a number of societies to which he almost was in the position of standing mathematical adviser.

Concurrently with his teaching, he continued his investigations. He wrote only one book—a *Treatise on Elliptic Functions*, published in 1876, which was intended to bridge over the gap from Legendre's *Traité des Fonctions Elliptiques* to Jacobi's *Fundamenta Nova*; it contains a considerable amount of new matter. But paper after paper was published in a long unfailing succession almost until his death; their tale amounts to more than 800. Happily for the convenience of mathematicians, the republication of his papers in collected form was undertaken by the Cambridge University Press,—perhaps the most enduring, certainly not the least fitting, monument of his fame. The request was made to him in 1889 by the Syndics of the Press; he willingly acceded to it and deeply appreciated, both then and afterwards, what he regarded as a great compliment to himself. Seven large quarto volumes, under his own editorship, have already appeared. The preparation of them was always a great happiness to him; and, especially in the later years of his life, it gave him an occupation in his science which was still within the range of his failing strength. At the time when the collection was begun it was estimated that ten volumes would suffice for the purpose, but it is now evident that ten will be certainly insufficient. The Syndics of the Press intend to complete the series of volumes; it is a matter of regret that the illustrious author of the papers has not lived to complete it himself.

Even his teaching and investigations did not fully occupy his time. For the first few years after his return he was left comparatively free from a large share in administration, but gradually it was assigned to him. As he became better known for his effective business capacity, his share in administration grew until he came to be regarded as an indispensable member of the Council of the Senate. He was elected a member of that body on 7th November, 1876, and with the exception of some six months when he was absent in America, he continued a member of it until 1892, when failing health compelled him to resign. During this period of service he was re-elected three times. Party feeling ran rather strongly at times during the discussions that led to the new statutes; but both parties included his name among their lists of nominations—an adequate proof that he possessed the confidence of the Senate. He was free from party bias, and he became established in his position of strength by his fairmindedness, his sound judgment, and his calm temperament. He would listen to a discussion, speaking only when he had something of importance to add; when speaking he was listened to with full attention. More frequently he would take no part in the discussion until his opinion was asked, as was usually the case in difficult questions; his opinion was always valued and sometimes final. Similarly, on syndicates, his co-operation was much sought, and in particular the services which he rendered to the Library Syndicate and the Press Syndicate were of substantial importance. He also took great interest in the movement for the higher education of women. In the early days of Girton College he gave direct help in teaching, and for some years he was Chairman of the Council of Newnham College, in the progress of which he took the keenest interest even to the last.

But, with all his general aptitude for business, he was perhaps most specially helpful by his legal knowledge. The training he had undergone and the knowledge he had acquired at the bar ultimately proved invaluable. His opinion on legal matters was sought by the University, by his own college, and by the scientific societies with which he was connected; when given, it frequently had the effect of a judicial decision. His powers of drafting were constantly being called into requisition; he responded to the calls upon him and, with unstinted generosity, placed his time and skill at the disposal of these bodies, so that the new statutes of Trinity College, and not a few of the statutes and ordinances of the University, owe much to him.

One other illustration, at once of his general business capacity and of the confidence reposed in him, may be given. The elections for representatives of the Universities in the House of Commons are still conducted openly and by means of voting papers, delivered either by the elector himself or by another elector whom he has nominated; objections may be raised against any voting paper, but they must be decided at once. In Cambridge the Vice-Chancellor, being the returning officer, nominates a number of assessors to act with him in the case of a contested election. At a bye-election in 1882, when the candidates were Mr. H. C. Raikes and Professor James Stuart, Cayley was nominated as presiding officer at one of the polling places. His imperturbable firmness, his calm courtesy, and the justice of his decisions secured for his effectiveness in this capacity the admiration of the University.

This brief account of his participation in business affairs is necessary; without some such indication a proper estimate of his position in Cambridge cannot be framed. And it also may help to show that his supremacy in the subjects of his investigations neither made him a recluse, nor limited his other interests, nor restricted his practical usefulness.

The merits of such a man were recognised by the only means at the disposal of a grateful and appreciative University. He was elected an honorary Fellow of Trinity College on 22nd May, 1872, at the same time as Dr. Lightfoot, Mr. James

Spedding, and Professor Clerk Maxwell; and on 11th October, 1875, he was made an ordinary Fellow, a position which he retained for the rest of his life. His friends subscribed for a presentation portrait,* painted by Lowes Dickenson in 1874; it now hangs in the College Hall. The simplest of inscriptions is on its frame, but the humorous lines which Clerk Maxwell† wrote at the time should not readily be forgotten. The graver element, seldom absent from his verses, is not entirely repressed even by his wit, and the lines were based upon a deep admiration of the man

> "Whose soul, too large for vulgar space,
> In *n* dimensions flourished unrestricted."

His bust, by Mr. Henry Wiles, was given to Trinity College by a donor who wished to remain anonymous. It was placed in the beautiful library of the College on 3rd December, 1888, an honour that has been conferred during life in only two other cases—Tennyson and Sedgwick.

After the new statutes came into operation, the Senate on 27th May, 1886, decided that the Sadlerian Professorship should at once be made subject to the improved provisions, a decision which, though it increased the amount of lecturing required, gave him the benefit of the full stipend. At the same time the Lucasian Professorship, held by Professor Stokes, was also made subject to the new statutes; and it was currently believed that the Lowndean Professorship would have been included in the proposal had Professor Adams been willing to have the change made. There was a wish on the part of members of the University to give some recognition to the glory conferred upon the mathematical school by Stokes, Adams, and Cayley; one possibility remained. The opportunity came in 1888 when Prince Edward (as he was known in Cambridge), afterwards Duke of Clarence, received the degree of LL.D. Such an occasion is customarily marked by the conferment of a number of honorary degrees upon distinguished men; among them, on this particular occasion, were the three professors who had been colleagues for a quarter of a century. On the 9th of June in that year a great assembly gathered to see these degrees conferred upon the recipients. It need hardly be said that the men singled out for honour received ovations on being presented; among the most enthusiastic ovations were those accorded to the three professors.

Nor were external bodies and learned societies, both at home and abroad, backward in recognising the merits of his work; the honours he received were numerous and came from all quarters. Honorary degrees were conferred upon him by several universities as well as his own, among them being Oxford, Dublin, Edinburgh, Göttingen, Heidelberg, Leyden, and Bologna. President Carnot nominated him an Officer of the Legion of Honour. He was either a Fellow or a foreign corresponding member of most of the scientific societies of the Continent, among them being the French Institute, the Academies of Berlin, Göttingen, St. Petersburg, Milan, Rome, Leyden, Upsala, and Hungary. He was also a Fellow of the Royal Society of Edinburgh, of the Royal Irish

* A photographic reproduction of the portrait is prefixed to vol. VI. of the *C. M. P.*

† See Campbell and Garnett's *Life of James Clerk Maxwell*, p. 636.

Academy, and of the Royal Astronomical Society. He had been President of the Cambridge Philosophical Society, and he sat on its Council for many years; also President of the London Mathematical Society and of the Royal Astronomical Society. He was elected a Fellow of the Royal Society on 3rd June, 1852, and he served as a member of its Council for six periods of office. In 1859 he received from the Royal Society a Royal medal, and in 1882 the Copley Medal, the highest scientific distinction it is in its power to bestow. When the De Morgan Medal was instituted in connexion with the London Mathematical Society, the first award was fitly made to Cayley. And from Leyden he received the Huyghens Medal.

Mention should be made of one other honour which he received: it is of a kind seldom conferred. The high opinion of his work which was held in America was indicated by an invitation in 1881 to deliver a course of lectures in the Johns Hopkins University, Baltimore, where his friend and fellow investigator, Sylvester, was then professor. He accepted the invitation, and left England in December of that year. During the next five months he lectured on Abelian and Theta Functions; the substance of these lectures was incorporated in a memoir subsequently published in the *American Journal of Mathematics**. He returned to England in June, 1882, bringing back pleasant remembrances of kindnesses and friendships.

His life, spent in mathematical research and in the quiet round of activity in the University, offered little of either interest or incident to make his name known by the outside world to the same extent or in the same way as the names of many scientific men, engaged in other lines of enquiry, are known. Once, however, in his life circumstances brought him prominently into notice. In 1883 he was President of the British Association for the Advancement of Science, the meeting being held at Southport; and, in that capacity at the opening of the meeting, he had to deliver a formal address, an abstract of which appeared as usual in the leading newspapers of the country.

In the early days of the Association, the President's address frequently reviewed the whole field of science; but as knowledge has developed, a tendency has set in, according to which each later President has confined himself more particularly to those matters within whose range he is an authority. And, subject to this restriction, it is hoped that the address may be legitimately popular. There have been critics of presidential addresses prepared to assert that science was sacrificed to popularity; there have been immense audiences convinced that popularity was sacrificed to science. Taken together, the presidential addresses, some severe and others popular, form an interesting series of reviews of the successive stages in scientific achievements.

Cayley's address belonged to the severely scientific class. From the nature of his subject—the progress of mathematics, more particularly of pure mathematics—it was bound to have this character. Few of the members of a regular Association audience have more than a slight acquaintance with pure mathematics; and, consequently, it is impossible to deliver to such a gathering an address which, in a reasonable time, can give them any real idea of the condition or the progress of the science. Cayley felt

* Vol. v. (1883), pp. 137—179; vol. vii. (1885), pp. 101—167.

this and confessed to the feeling in a passage which is perhaps the best known in the address:—

> "It is difficult to give an idea of the vast extent of modern mathematics. The word 'extent' is not the right one: I mean extent crowded with beautiful detail—not an extent of mere uniformity such as an objectless plain, but of a tract of beautiful country seen at first in the distance, but which will bear to be rambled through and studied in every detail of hillside and valley, stream, rock, wood, and flower. But, as for everything else, so for a mathematical theory—beauty can be perceived but not explained."

But he also felt that the respect due to the Association requires its President to deal with that branch of science about which, as he knows it best, he is best fitted to tell them, so that different subjects may thus in turn be brought before successive meetings.

> "So much the worse," he added, "it may be, for a particular meeting; but the meeting is the individual which on evolution principles must be sacrificed for the development of the race."

Granting then the inevitably stern character (as popularly estimated) that must mark any proper exposition of his subject, the address is one of singular interest. It undoubtedly made a great impression. Parts of it were incomprehensible to all but mathematicians: still, there was much which others could understand and, understanding, found excellent. Even leader-writers at the time recognised its lucidity, its finish, its native elegance, and its instructive and stimulating essence. To mathematicians it counts for much. Not merely is it a valuable historical review of various mathematical theories; but the exposition possesses all the freshness, the independence of view, the suggestiveness and the amazing knowledge that were so characteristic of Cayley. And, consequently, it can often be recurred to with unfailing profit.

After this event, his life pursued the unbroken tenor of its scientific course. Ever thinking, working, writing, he maintained the flow of his papers with the same unslackening vigour, and he showed the same sympathetic encouragement of others, as had marked him before the scientific world had tried to acknowledge his genius by showering its honours upon him.

It is now some years since the painful internal malady, which ultimately was the cause of death, began to show itself. At first, its action was slow; and there was reasonable hope that his naturally strong constitution would enable him to throw it off. Unfortunately these hopes were not realised; its growth was steady, its undermining influence persistent. Change of scene was tried once or twice, but without good effect; and it soon appeared that Cambridge itself troubled him least. Three years ago his friends saw that his health began to fail: he had occasional attacks of severe illness which confined him to his bed for weeks together, each of them leaving him gravely frailer than before. Gradually he became confined to his house and his garden; he could see only very few friends, and usually even them only for a short time. When they did see him, they found only too clearly how rare and brief were his intervals of relief from pain, though occasionally his gentleness and his patience would almost delude them into hope.

The last of the severe attacks began on the 8th of January; he seemed to be getting better when, on the 21st, his strength suddenly began to collapse. He died about six o'clock on the evening of Saturday, 26th January, 1895. The funeral took place on the succeeding Friday when, in Trinity Chapel, a great assemblage, composed of members of the University, of representatives of the embassies of Russia and America, as well as of various learned societies and of personal friends, gathered to pay him their last homage of respect and reverence.

Sufficient has been said to show that Cayley was a man of general activities; but his scientific work and his public duties by no means exhausted or limited his general interests.

It has already been stated that, as an undergraduate, he was fond of reading novels; this practice remained with him all his days. He preferred a novel of the old orthodox type with a "happy ending"; and though his greatest delight was in the older novels, a modern book, such as *Beside the Bonnie Briar Bush* (which he read quite late in 1894), met with words of warm praise. He had a good memory, and used to discuss plots and characters with considerable animation. The two novelists, by whose works many English people are divided into one or other of two classes, did not affect him much; Thackeray he read but did not like, and he would not read Dickens. His favourite authors were Scott and Jane Austen; all their works had been read by him many times, and they were read aloud to him during the long period of his illness. *Guy Mannering* and *The Heart of Midlothian*, among Scott's, and *Persuasion*, among Jane Austen's, were the books he liked the best. He also was fond of George Eliot's novels, particularly of *Romola*. Indeed, though he had aversions, his taste was somewhat general. Commendation of a book was enough to make him willing to try it; and there was only one limitation to his range of novel-reading—he had an instinctive abhorrence of anything that suggested either coarseness or vulgarity.

His English reading was not confined to novels. He had a keen liking for many of Shakespeare's plays, notably *Much Ado About Nothing*, and some of the historical dramas. He delighted in Milton's shorter poems, though he would not tolerate *Paradise Lost*. Scott's poems were frequently read; and he had a great appreciation of Byron's *Tales* and of Coleridge's *Ancient Mariner*. Grote's *History of Greece* and Macaulay's *History of England* he read repeatedly and with zest; and he never seemed tired of Lockhart's *Life of Scott*.

He was also a good linguist. He knew French well; it was a second writing-language to him, as will be seen from the large number of papers, written in French, which occur in his collected mathematical papers. He read (but he did not talk) German and Italian with ease, and his Greek remained fresh throughout his life. This last power may have been due to the admiration he felt for Plato; he referred to the *Republic* and the *Theætetus* in his Presidential Address; and, on the afternoon of the day of the "Greek division"* in the Senate House, I remember finding him at home reading the *Gorgias*.

He had the keenest interest, amounting almost to a passionate delight, in travelling; cities of historic or artistic fame delighted him equally with beautiful scenery. Long after he had become an invalid, he found a fascination in guide-books and maps; and all his younger friends will recall the sympathetic zeal with which he entered into their projected journeys, and the happy pleasure he took in hearing them speak of recent journeyings and in recalling, with a wonderful vivid memory, his own experiences and ideas about places they had visited.

Reference has been made to his early pleasure in the old Italian masters. Yet, if any inferences can be drawn from the likings of his later years, architecture attracted him even as much as pictures. He had a true feeling and a clear judgment as to genuine excellence: he sketched well, and had a quick eye for proportions, perspective, light and shade. One of his relaxations was to make coloured sketches of buildings that he liked, notably sepia drawings of some of the great Gothic cathedrals and churches of northern France. He kept up his practice of water-colour painting all his life, and in his closing years it proved a great solace to him at times when his strength was so far reduced that he could not work. He had great happiness in looking at architectural pictures and at books on architecture, one of his favourites among the latter being Street's *Brick and Marble in the Middle Ages*.

Financial matters and accounts also interested him; and only a few months before his death he published a brief pamphlet on book-keeping by double entry, which he has been known to declare one of the two perfect sciences. He could not resist some reference to the subject in his Presidential Address, making the remark that the notion of a negative magnitude "is used in a very refined manner in book-keeping by double-entry."

His bearing was gentle, and it was marked by a courtesy that was unfailing. On questions of administration and in discussions, his opinions were stated clearly and quietly. Not that he did not hold decided views or that he would abate one jot of his firm, even chivalrous, defence of what he held to be right; but there was a judicial temper in his mind which prevented the subjective element in a discussion from disturbing his equanimity. The even balance of his mind enabled him to recognise and appreciate the position of one who differed from him, and his quiet "I do not think so" was all the more effective because its very calmness excluded the slightest suggestion of hostile spirit.

His figure was spare: until his illness, he could easily endure the fatigue of long walks, in which he delighted, especially in hill country. In later years it became rather bent, and he had the appearance of being frail. His head was very impressive,

* In 1891 a proposal was made by the Council of the Senate for the appointment of a Syndicate to enquire, among other things, into the expediency of allowing alternatives for one of the two classical languages in the Previous Examination. Many members of the Senate were convinced that the adoption of an alternative would lead to the extinction of the study of Greek except in the greater public schools; they consequently opposed the proposal, which, on 29th October, 1891, was rejected by a great majority (525 to 185).

It may be added that Cayley was in the minority. He allowed his signature to be added to a letter which was sent to the London newspapers as an appeal for assistance in defeating the attempt to resist enquiry.

as may be seen from his portrait and from photographs. In repose, and when his attention was not concentrated upon what was passing, his face had a grave air and the blue-grey eyes suggested that he was far away in thought; but when attentive or amused, and when expressing pleasure, the eyes became singularly keen and a peculiar charm lightened up the whole face.

He was absolutely modest. The honours conferred on him in full profusion never injured in the least degree the grand simplicity of his character, never gave rise to the slightest trace of vanity, which was alien to his nature. He rarely spoke of them, and, when he did, it never was as of honours: they pleased him, but, perhaps, rather as recognition of his work than as tributes to the worker. If any one expressed appreciation of any of his papers, owing to the help it had given, he would reply very quietly: but he did not stint the expression of his pleasure at advances beyond his own results when they were made by others. Public appearances were rather distressing to him at first, for his disposition was retiring and he could be reserved; but as time wore on, duty often compelled him to take part in them. In such cases he accepted the claim and discharged it with a straightforward simplicity that was entirely devoid of self-consciousness; but he gladly avoided demonstrations whenever it was possible.

In the spirit of his work one great quality was his generosity to others, particularly to young men, whose work he was always willing to recognise. He ignored the fact that he was a great mathematician—probably it never occurred to him to think of his doings: but it may be doubted whether this unconsciousness of his greatness ever proved at once more fascinating or more bewildering than when he was discussing scientific results with young men. He so evidently had his wishes centred on a single-hearted desire for the right result that it was difficult to conceive him approaching a question merely as a learner: yet he was ever a learner. There are few men, if any, with not even a tithe of his scientific achievements, who have had less of controversy or have had such immunity from questions as to priority of discovery. This arose not merely from the indisputable priority of his results: it was partly owing to his nature. Salmon says of him:—

> "His motto has always been 'esse quam videri,' and I do not know any one to whom it would be more repulsive to engage in a personal contest by claiming for himself a particle of honour or of money more than was spontaneously conceded. He would be apt to take for his model the patriarch Isaac, who, when the Philistines claimed a well which he had dug, went on and dug another, and when they claimed that, too, went on and dug a third":

an exceedingly happy description of the man the tide of whose genius was

"Too full for sound or foam."

Some account of his work, some estimate of its character, some indication of the original contributions made by him to his science, may not improperly be given here. It is, of course, impossible to predict what his permanent influence will be upon mathematics, or what opinion coming generations of workers will hold of him: certainly, by his own contemporaries, he was deemed one of the greatest mathematicians the world has seen. Bertrand, Darboux, and Glaisher have compared him to Euler, alike for his range, his analytical power, and, not least, for his prolific production of new views and fertile theories. There is hardly a subject in the whole of pure mathematics at which he has not worked. Some new subjects owe their existence to him; to others he has made very definite contributions, so that their boundaries have been enlarged often to an enormous extent; there are few upon which he has not left the mark of his genius.

In several of the notices that appeared at his death he was described as a great explorer. Such he undoubtedly was, but he was more. He not merely discovered new countries but he also opened them up, so that others were able to enter into some possession of those regions without undergoing the difficulties that he had overcome. And if the metaphor may be carried further, he had the restlessness of the explorer: he could not long remain satisfied with an achievement concluded, but must try his fortune again and elsewhere.

Varying opinions have been expressed as to Cayley's style; the variations are largely due to preconceived views of what a mathematical paper should be. It certainly is not easy to skim one of his papers; any attempt to do so leads to an inadequate estimate of what it usually establishes. It is not difficult to read one of his papers, even to grasp the contents well, provided proper care be devoted to it, because difficulties that occur are completely solved, and nothing lies in the background to cause doubt or suggest incompleteness. He has been well described by Glaisher as an unequalled master of analytical processes; it is especially in algebraical manipulation that his strength and his facility stand out in clear view. His success in this direction was achieved by a skill that cannot be explained by describing it as due to acquired knowledge, or to practice, or to long consideration and patient selection. It was rather an instinct for the management of the most complicated processes, and the way in which he controls the most elaborate calculations is sometimes little short of extraordinary.

As regards his methods, he does not seem to have cast about so as to choose one rather than another. As soon as he had thought of any method the possible effectiveness of which he could settle almost intuitively ("one's best things are done in five minutes," he once said to me, in confirmation of the satisfaction I was expressing at the fruitfulness of an idea that had occurred to me unexpectedly), the rest was the exercise of his powers. Among the methods he preferred, especially during the last twenty-five years of his life, was that of verification; in his hands it proved a weapon of great force. Indeed, only less remarkable than his algebraical skill, was the insight which enabled him to preserve the exact equivalence of all the equations in any particular process, so that he could have reversed each process merely by reversing the steps as they were made, and could have proceeded to the required theorem from the initial expression of an algebraical fact. Numerous instances of this quality in his work could be adduced; it will be sufficient to refer to some parts of his paper* "On the centro-surface of an ellipsoid."

But though Cayley was specially happy in the treatment of algebraical developments, an inadequate estimate of his genius would be obtained by supposing that he

* *C. M. P.* vol. VIII. No. 520: *Camb. Phil. Trans.* vol. XII. (1873), pp. 319—365.

was almost entirely an analyst. Much of his thinking, not a little of his writings, is completely geometrical; and his contributions to line geometry, his introduction of the Absolute into geometry, his continued recurrence to the methods in pure geometry invented by Poncelet and Chasles, should be sufficient to range him among geometricians.

Moreover, even in strictly analytical work, the synthetic element is often not far away though it does not always appear on the surface. In this connexion an acute suggestion, made by Salmon and perhaps based upon his remembrance of their mathematical correspondence that lasted through many years, is confirmed by one of the Notes Cayley himself added at the end of the second volume of his *Collected Mathematical Papers.* An enquiry sometimes begins by a comparatively easy problem which, when solved, leads to wider inferences; so that, ultimately in the development, considerable generalisations are effected. Now the usual writer, in publishing the results of such an enquiry, draws them up in a sequence that partly marks the order of their connected discovery: and, in doing so, he makes his work easier for his readers. But Cayley was not the usual writer. When he had reached his most advanced generalisations he proceeded to establish them directly by some method or other, though he seldom gave the clue by which they had first been obtained: a proceeding which does not tend to make his papers easy reading. An instance of the fact occurs* in his "Memoir on the Theory of Matrices," where he proves that a matrix satisfies an algebraical equation of his own order; he proves it by verification in simple cases, but he gives no clue as to his line of discovery. An instance of the method occurs in a note† added to one of his papers, where he says that the general equations

$$\{yd_x\} - yd_x = 0, \quad \{xd_y\} - xd_y = 0,$$

characteristic of covariants and invariants of binary quantics, were initially suggested by considering the relation of the quadratic $ax^2 + 2bxy + cy^2$ and its discriminant $ac - b^2$ to these equations. In the paper he drops linear transformation as connected with the covariantive property and defines a covariant as a function satisfying these two equations.

His literary style is direct, simple and clear. His legal training had an influence, not merely upon his mode of arrangement but also upon his expression; the result is that his papers are severe and present a curious contrast to the luxuriant enthusiasm which pervades so many of Sylvester's papers. He used to prepare his work for publication as soon as he had carried his investigations in any subject far enough for his immediate purpose. He found it an easy matter to do this part of his work, and thus differed widely in experience from those to whom the preparation of a paper is laborious even when the results to be incorporated have been obtained. As a matter of fact, he took the straightforward course of saying what he had to say in a clear and simple manner, fixing his mind upon the substance and never going out of his way in order to secure beautiful form for the presentation of results. Yet not infrequently his papers are so admirably written that they satisfy the exacting critics; thus it is perhaps not too much to affirm that his "Sixth Memoir on Quantics‡" could not be presented in more attractive form—a character due, however, to the tendency of

* *C. M. P.* vol. II. No. 152, pp. 482, 483; *Phil. Trans.* (1858), pp. 24, 25.

† *C. M. P.* vol. II. p. 600.

‡ *C. M. P.* vol. II. No. 158; *Phil. Trans.* (1859), pp. 61—90.

his method and to his results, but not acquired by any effort specially devoted to elaboration of clear expression. Again, a paper once written out was promptly sent for publication; this practice he maintained throughout his life. He undoubtedly formed projects for the immediate future; thus to the second edition* of his *Treatise on Elliptic Functions* he intended to add a couple of chapters, which, however, remained unwritten solely for the reason that all such projects were carried into effect only about the time when the need arose. The consequence is that he has left few arrears of unfinished or unpublished papers; his work has been given by himself to the world.

Only one other remark as to the form of his papers need be made. Readers must be struck with the number of exact references he makes to other writers. It was a practice about which he had very decided opinions: he wished not merely to make honourable acknowledgment of indebtedness but also to give indications of the history of the subject. In the latter particular he was always careful to insert in the reference the year in which the book or the paper had appeared; and he steadily urged others to insert dates in their references.

Cayley made additions to every important subject that lies within the range of pure mathematics. Their importance and their amount have varied in different subjects; thus on geometry his writings have a dominating influence: while on the general theory of functions, though he knew the subject well, he has left little mark, for he concerned himself chiefly with details such as the solution of more or less special problems in conformal representation. His papers in general have such value that he is the author most frequently quoted by the great body of current mathematicians. A full record of what he has done in pure mathematics could be made only by writing its history during the last half century; all that is attempted here consists of some brief indications of a selection among his more obviously important contributions to mathematical knowledge.

One of the subjects with which Cayley's name will probably be most closely associated is the theory of invariance. It is easy to cite simple cases of what is implied by an invariantive function: two will suffice.

It is known that, in solving an ordinary algebraical equation with literal coefficients, a certain functional combination of these coefficients (called the discriminant) must vanish in order that two roots of the equation may be equal; for example, the equation $ax^2 + 2bx + c = 0$, has equal roots if (and only if) the quantity $ac - b^2$ vanishes. When the variable is transformed from x to y by a relation $(l'x + m')y = lx + m$, where l, m, l', m' are constants, then evidently two values of y, corresponding to the two equal values of x, are equal. When x is eliminated from the equation by means of the assumed relation, a new quadratic arises having y for its variable; let it be $a'y^2 + 2b'y + c' = 0$, where a', b', c' depend upon a, b, c and l, m, l', m'. The two values of y determined by this equation are equal if (and only if) the quantity $a'c' - b'^2$ vanishes. But the equality of the two values of y depends upon and is determined by the equality of the two values of x, the latter equality being secured if the quantity $ac - b^2$ vanishes. It follows that the vanishing of either of the quantities $a'c' - b'^2$ and $ac - b^2$ requires

* It was published four months after his death; only the earlier sheets had the benefit of his revision.

the vanishing of the other; and it is therefore inferred that, when neither of them vanishes, one of them contains the other as a factor. When the actual calculation is made, it is found that $a'c' - b'^2$ is the product of $ac - b^2$ and $(lm' - l'm)^2$, the latter being a quantity that depends solely upon the transforming relation. Consequently it appears that a combination of the coefficients in the original equation exists, such that when the equation is transformed by any relation of the type indicated and exactly the same combination of the new coefficients is constructed, the two combinations are equal to one another save as to a factor depending solely upon the transforming relation. Such a combination of the coefficients is called an invariant.

Again, it is known that every curve (of degree higher than two) possesses a number of points where a tangent to the curve not merely touches it but, having contact of one degree closer, crosses it; and it is found that all these points, called points of inflexion, also lie upon another curve uniquely derived from the first. When the curves are represented by means of equations, the statement is that the points of inflexion of a curve $U = 0$ are given as the intersections of this curve with a curve $H = 0$, the latter equation being uniquely derived from $U = 0$. Now suppose that the axes, to which the curves have been referred, are changed to another system, so that new co-ordinates x', y' are connected with the former co-ordinates by relations

$$\frac{x'}{a_1x + b_1y + c_1} = \frac{y'}{a_2x + b_2y + c_2} = \frac{1}{ax + by + c}.$$

A new equation $U' = 0$, obtained by eliminating x and y between these relations and $U = 0$, will now represent the curve. The change thus made does not affect the geometrical properties of the curve; its points of inflexion are still given as its intersections with the curve $H = 0$. But the points of inflexion of the curve represented by $U' = 0$ are the intersections of this curve with another curve represented by $H' = 0$, an equation derived from $U' = 0$ in exactly the same way as $H = 0$ is derived from $U = 0$. It therefore appears that the associated curve $H' = 0$ cuts the given curve in precisely the same points as the associated curve $H = 0$, a result which suggests that the associated curves are the same. Now $H' = 0$ has been derived from $U' = 0$; but actual calculation shows that, if the relations between x', y' and x, y be used to eliminate x, y from $H = 0$, the resulting equation is $H' = 0$; in other words, the relations between x', y' and x, y transform the equation $H = 0$, derived from $U = 0$, into the equation $H' = 0$, derived in the same way from $U' = 0$. Moreover, as in the case of the invariant, it is found that H', a specially constructed function of x', y' and the coefficients in U', is divisible by H, the same function of x, y and the coefficients in U; the quotient being a quantity depending again only upon the constants in the transforming relations. Consequently it appears that a combination of the coefficients and the variables in the original equation exists such that, when the equation is transformed by means of relations of the type indicated, and exactly the same combination of the new coefficients and the new variables is constructed, the two combinations are equal to one another save as to a factor dependent solely upon the transforming relations. Such a combination of the coefficients and the variables is called a covariant.

The first notice of such a property appears to have been made by Lagrange. And Gauss discussed the invariance of the discriminants of certain expressions when the latter are subjected to linear transformations. Again, Boole in 1841 had shown that this invariantive property belongs to all discriminants, and he gave a method of deducing some other functions of this kind. Boole's paper suggested to Cayley a much more general subject—the permanence of invariantive form—so that he set himself the question of finding "all the derivatives of any number of functions which have the property of preserving their form unaltered after any linear transformation of the variables." The first set of results obtained by his investigations related to invariants; they appeared in his famous paper,* "On the Theory of Linear Transformations," published half a century ago. The second set of results related to covariants; they appeared in the paper,† "On Linear Transformations," published in the succeeding year. In these two papers Cayley demonstrated the general existence of a number of functions, both invariants and covariants (at first he called them hyperdeterminants), which preserve their form under linear transformation.

These discoveries of Cayley establish him as the founder of what is called sometimes modern algebra, sometimes invariants and covariants, sometimes theory of forms; the origination of the theory is incontestably his, and it is universally ascribed to him.

A discovery of this general importance and complete novelty soon attracted the attention of other workers. It is not too much to say that the subsequent investigations long absorbed the active interest of many mathematicians, and, as a result, the theory has influenced all that domain of mathematical science which is in any way connected with algebraical form. Among the first to enter the field was Sylvester, then living in London; he and Cayley were in constant communication, alike oral and written, and carried on their work in the most friendly relations with one another. Boole also resumed his investigations, and both he and Salmon made substantial additions to the theory. The continental mathematicians also had begun their important contributions, chief among them being Aronhold, Hesse, and, at a later date, Hermite. Aronhold, indeed, devised the so-called symbolical method, now the favourite method with German workers; in its origin it is nearly the same as the symbolical method introduced by Cayley, but the subsequent developments—due largely also to Clebsch and to Gordan—run on lines entirely different from Cayley's.

After a time, Cayley began his series of ten memoirs on quantics; they must rank among the most wonderful combinations of original researches and papers upon a single theory ever produced. They contain a splendid exposition of the theory as already established; they are full of original contributions to the subject, and as they take account of the work done by other authors, they have the further interest of showing how the subject grew between the appearance of the "Introductory Memoir" in 1854 and the appearance of the "Tenth Memoir" in 1878. This is hardly the

* *C. M. P.* vol. I. No. 13; *Camb. Math. Jour.* vol. IV. (1845), pp. 193—209.

† *C. M. P.* vol. I. No. 14; *Camb. and Dubl. Math. Jour.* vol. I. (1846), pp. 104—122. The two papers were rewritten, and appeared in *Crelle*, vol. XXX. (1846), pp. 1—37, under the title "Mémoire sur les Hyperdéterminants."

opportunity to write a history of the subject by apportioning among the various investigators the sections which they respectively originated;* yet reference should be made to two matters.

First, one of the problems that greatly interested Cayley was the determination of the complete asyzygetic system of irreducible invariants and covariants appertaining to a binary form, that is, the system such that every invariant and every covariant of the form can be expressed as a rational integral algebraical function of the members of the system, the coefficients in the function being numerical only. In his "Second Memoir on Quantics"† he had accurately determined the number (and their degrees) of the asyzygetic invariants for binary forms of orders 2, 3, 4, 5, 6; he had also accurately inferred the number (together with their degrees and their orders) of the asyzygetic covariants for binary forms of orders 2, 3, 4, all these concomitants being subsequently tabulated. But, in regard to the invariants of forms of order higher than 6 and the covariants of forms of order higher than 4, he came to the erroneous conclusion that the respective numbers are infinite. The error was not corrected until Gordan in his memoir‡, dated 8th June, 1868, and entitled "Beweis dass jede Covariante und Invariante einer binären Form eine ganze Function mit numerischen Coefficienten einer endlichen Anzahl solcher Formen ist," showed that the complete system for a binary quantic of any order contains only a limited number of members. Cayley at once returned to the question, and having found a source of error (it was the neglected interdependence of certain syzygies, reducing the numbers of invariants and covariants; the interdependence had not previously been suspected), he dedicated his "Ninth Memoir on Quantics§" (dated 7th April, 1870), to the correction of the error and a further development of the theory in the light of Gordan's results. His promptness in recognising and giving immediate prominence to the work of the younger author possibly prevented some controversy among unwise partisans; it was characteristic of the man.

And, secondly, though his series of memoirs was brought to an end with the tenth, his interest in the subject did not cease, and he frequently wrote upon parts of it under other titles. In particular, Captain P. A. MacMahon's discovery of a relation of a new character between seminvariants and symmetric functions (viz., that the leading coefficients of the covariants of a binary quantic are the same as the non-unitary partition symmetric functions of the roots of an equation connected with a modified quantic) proved of the keenest satisfaction to him. From time to time he wrote in the *American Journal of Mathematics* upon this subject and upon symmetric functions generally in this connexion, always sympathetic and appreciative of the advances made by others, able to grasp and assimilate

* Some information will be found in an appendix to Salmon's *Lessons on Higher Algebra*; also in the notes and references at the end of the second volume (pp. 598—601) of the *Collected Mathematical Papers*. A valuable and exhaustive report, containing a full history of the subject, was drawn up by Prof. Dr. Franz Meyer, and published under the title "Bericht über den gegenwärtigen Stand der Invariantentheorie" (*Jahresber. d. Deutschen Mathem.-Vereinigung*, I. 1892).

† *C. M. P.* vol. II. No. 250; *Phil. Trans.* (1856), pp. 101—126.

‡ *Crelle*, vol. LXIX. (1869), pp. 323—354.

§ *C. M. P.* vol. VII. No. 462; *Phil. Trans.* (1871), pp. 17—50.

their ideas, but using them as a master and not as a follower. It was not alone, however, to symmetric functions, upon which he had written long and important memoirs as early as 1857, but to many other cognate subjects that he extended his researches upon invariants and covariants. The theory of equations of the fifth and higher degrees, Sturm's functions, Tschirnhausen's transformation, partition of numbers, Arbogast's method of derivation, skew determinants*—to quote no others—are titles and subjects of papers, all of which contain investigations of great value. The reason that they are less known (if such be the case) than his other work in the same line of ideas is perhaps due to the fact that the direct theory of invariants and covariants was rapidly brought within the range of students through Salmon's *Lessons on Higher Algebra*, dedicated by the author to Cayley and Sylvester.

Another subject, of which he must be regarded as the creator, is the theory of matrices. His first memoir† upon this theory, "wherein," to quote Sylvester,‡ "he may be said to have laid the foundation-stone of multiple quantity," was published in 1858. A couple of isolated results had been obtained by Hamilton in 1852 through the methods of quaternions; but they were unknown to Cayley at the time of his memoir, and, owing to the connexion in which they occur, they have an entirely detached aspect.

A matrix may initially be defined as a symbol of linear operation; thus, when the equations

$$X = ax + by + cz, \quad Y = a'x + b'y + c'z, \quad Z = a''z + b''y + c''z$$

are expressed in the form

$$(X, Y, Z) = \left(\begin{array}{ccc} a, & b, & c \\ a', & b', & c' \\ a'', & b'', & c'' \end{array} \right\rangle (x, y, z) = M(x, y, z),$$

the symbol M is a matrix. Cayley was the first to discuss the theory of such symbols as subjects of functional operation and to dispense with the hitherto regular return at each stage to the equations of substitution in which the symbol first arises; in fact, he replaces the notion of substitutional operation by the notion of a new class of quantity.

Matrices (being of the same order or dimension) can be added like ordinary algebraical quantities; as regards multiplication, they are subject to the associative law, but not to the commutative law. Hence powers of a matrix (positive and negative, integral and fractional) can be obtained, and likewise algebraical functions of a matrix. It also follows that two general matrices are not convertible, that is, LM is not the same as ML save under special conditions; and it is a part of the theory to find the most general matrix convertible with a given matrix. The expression of this convertible

* His discoveries in this subject alone have done much to simplify the analytical investigations connected with Pfaff's problem and the allied theory.

† *C. M. P.* vol. II. No. 152; *Phil. Trans.* (1858), pp. 17—37.

‡ *Amer. Jour. Math.* vol. VI. (1884), p. 271.

matrix can be deduced by means of the fundamental equation which every matrix satisfies, viz., an algebraical equation of its own order, the coefficient of the highest term being unity, and the last term being the determinant of the constants in the matrix. All these results were given by Cayley in his initial memoir; and, at the same time, they were applied by him to obtain the most general automorphic linear transformation of a bipartite quadric function, an extension of the problem which requires the most general (orthogonal) substitution transforming the function $x^2+y^2+z^2+\ldots\ldots$ into the function $x'^2+y'^2+z'^2+\ldots\ldots$

How fruitful the subject has proved may be inferred by noting the subsequent investigations of Sylvester, who has developed it on Cayley's lines, and has added to it many new ideas; of Tait, who developed the theory of quaternions on parallel lines: of the Peirces, father and son, whose researches on linear associate algebra* gave rise to the notion of matrices from a different source; of Clifford and Buchheim, who connected the theory with Grassmann's methods; of Laguerre, in whose memoir† the treatment of a "linear system" (the same as a Cayley matrix) is similar to Cayley's; and of many other writers, among whom Taber should be mentioned.

Connected with non-commutative algebraical quantities, Cayley's researches on the theory of groups require a passing notice. He devoted several papers to questions in this theory. Some of them relate to those groups of substitutions, the introduction of which by Galois made an epoch in the theory of equations, others of them relate to groups of homographic transformations, particularly those related to the polyhedral functions. But, so far as can be seen, he limited his published investigations to those groups which are finite and discontinuous.

Abstract geometry—the ideal geometry of n dimensions—is a subject that he may almost be said to have created; no other name than his has been associated with its origin. More than anything else, it marks the line of difference between the kinds of homage accorded to him. Experts regard it as an illustration of his imaginative power: the unlearned regard it as an incomprehensible mystery.

It finds a place among his earliest investigations,‡ it was steadily present to his mind, illuminating many of his researches; and occasionally it found explicit treatment, e.g., in his "Memoir on Abstract Geometry,"§ and in his Presidential Address at Southport. The theory presents itself in two connexions: one, as a need in analysis, the other as a generalisation of the ordinary geometries of two dimensions and of three dimensions.

The former origin can be indicated in a brief statement. When an occasion arises for dealing with a number of variables, connected in any manner and regarded as either variable or determinate (wholly or partially), the nature of the relations among them is frequently indicated, and often is made more easily intelligible, by associating some geometrical interpretation with the given system of relations. Thus

* *Amer. Jour. Math.* vol. IV. (1881), pp. 97—229.

† "Sur le Calcul des Systèmes Linéaires (*Journal de l'Éc. Poly.* vol. XXV. 1867, pp. 215—264).

‡ *C. M. P.* vol. I. No, 11; *Camb. Math. Jour.* vol. IV. (1845), pp. 119—127.

§ *C. M. P.* vol. VI. No. 413; *Phil. Trans.* (1870), pp. 51—63.

the momental ellipsoid is of great use in the discussion of moments of inertia, in representing the motion of a body round a fixed point when there are no impressed forces, and in other questions in dynamics. Again, two non-homogeneous (or three homogeneous) variables can be regarded as the co-ordinates of a point in a two-dimensional geometry, such as that of a plane or the surface of a sphere or any analytical surface; and any equation among the co-ordinates is then interpreted as representing a curve (or curves, or portion of a curve or curves) upon the surface. Similarly, when there are three non-homogeneous (or four homogeneous) variables, they can be regarded as the co-ordinates of a point in a three-dimensional geometry, such as that of ordinary space; corresponding to an equation among the variables, there is a surface (or surfaces) in space; corresponding to two independent equations among the variables, there is a curve (or curves) in space; and corresponding to three independent equations, there is a point (or points) in space. In such cases the analytical relations can often, with great advantage, be exhibited as geometrical properties. When the number of non-homogeneous co-ordinates is greater than three (or the number of homogeneous co-ordinates is greater than four), the circumstances have greater need of such a representation, while there is a greater difficulty in constructing some geometrical illustration; and then it can be obtained in a corresponding form only by the idea of a space of the proper number of dimensions. To secure the possibility of such a representation, it is necessary to evolve the geometry of multiple space.

For example, there are four single theta-functions, and their squares are connected by linear homogeneous relations. In order to obtain other properties of the functions themselves, it is convenient to regard them as homogeneous co-ordinates of a point in (ordinary) space; the amplitude in space that then is to be selected is the quadri-quadric tortuous curve represented by those linear relations, viz., the curve which is common to two quadric cylinders with intersecting axes. Similarly there are sixteen double theta-functions, with corresponding linear relations among their squares. The associated geometry is fifteen-dimensional; the manifoldness in this space to be selected for the discussion of the properties is the quadri-quadric two-dimensional amplitude common to thirteen quadric hyper-cylinders.

An initial difficulty in the construction of an analytical geometry of n dimensions is the expression of an amplitude of less than $n-1$ dimensions by means of equations that shall represent the complete amplitude, and nothing besides the amplitude. It occurs in ordinary solid geometry, the difficulty there being to obtain the expression of a tortuous curve in space by means of equations that represent it alone. For instance, a twisted cubic is frequently taken as the intersection of two quadrics having one common generator; but the equations of the quadrics taken together represent not the cubic curve alone but also the common generator. And the like for other cases.

Cayley's purpose in his "Memoir on Abstract Geometry," already referred to, was the exposition of some of the elementary principles of the subject. The paper is a remarkable instance of his power of presentation of abstract ideas, and of his clear precision of statement. Moreover, he makes it an explanatory paper; and, in view of the prevailing estimate of him as an analyst, it is worthy of notice that the paper does not contain a single equation, and contains only a few symbols. It is unnecessary to summarise its contents; the furthest stage reached is the establishment of the notion that underlies the principle of duality in geometry.

But though the necessity for hyperdimensional geometry can thus be met so far as it arises in connexion with analysis, it is a different matter when the geometry is to be regarded as the generalisation of the geometries of two-dimensional space and of three-dimensional space. Cayley's reply to his own question as to the meaning to be attached to hyperdimensional space is* that

> "It may be at once admitted that we cannot conceive of a fourth dimension of space; that space as we conceive of it, and the physical space of our experience, are alike three-dimensional; but we can, I think, conceive of space as being two- or even one-dimensional; we can imagine rational beings living in a one-dimensional space (a line) or in a two-dimensional space (a surface), and conceiving of space accordingly, and to whom, therefore, a two-dimensional space, or (as the case may be) a three-dimensional space would be as inconceivable as a four-dimensional space is to us."

By not a few people the first clause in this passage has been neglected and the later clauses have not always been read rightly; and his further remark, "I need hardly say that the first step is the difficulty, and that granting a fourth dimension we may assume as many more dimensions as we please," has left some readers rather puzzled as to whether Cayley had not, after all, some mysterious incommunicable conception of a fourth dimension. His position is stated in the first clause of the former passage; his conclusion is that hypergeometry is, and is only, a branch of mathematics.

Before passing from the consideration of his larger contributions to hypergeometry, it is proper to mention his introduction of the six co-ordinates of a line. These are six quantities connected by a homogeneous equation $af + bg + ch = 0$; and as only their ratios are used, they are thus equivalent to only four independent magnitudes, sufficient for the unique specification of a right line. They were first established, and primarily used by him, in connexion with his new analytical representation of curves in space;† and he often recurred to the subject, devoting in particular one paper‡ to the calculus of the six co-ordinates and to a discussion of Sylvester's involution of six lines. It should, however, be stated that these co-ordinates presented themselves independently to Plücker; the development of Plücker's theory as set forth in his memoir§ *On a New Geometry of Space*, and in his book‖ *Neue Geometrie des Raumes*, is entirely different from that obtained by Cayley, and it ought to be regarded as a separate creation. And it need hardly be remarked that while the introduction of a line, as an entity represented by a set of co-ordinates, leads to a new geometry of space, it is also clear that line-geometry can be regarded as a geometry of four dimensions.

* *Brit. Assoc. Report*, 1883, President's Address, p. 9.

† *C. M. P.* vol. IV. Nos. 284, 294.

‡ *C. M. P.* vol. VII. No. 435.

§ *Phil. Trans.* 1865, pp. 725—791.

‖ Leipzig, *Teubner*, 1868.

Another notion, entirely due to Cayley in its first form, is that of the Absolute; it was first introduced in his *Sixth Memoir on Quantics*,* which was devoted chiefly to his investigations on the generalised theory of metrical geometry.

It is a known property that the angle between two lines AB, AC, when multiplied by $2\sqrt{-1}$, is equal to the logarithm of the cross-ratio of the pencil made up of the lines AB, AC and (conjugate imaginary) lines joining A to the circular points at infinity; and the measure of the angle between two lines can thus be replaced by the consideration of a projective property of an extended system of lines. Other examples of similar changes could easily be quoted. The purpose of Cayley's theory was to replace metrical properties of a figure or figures by projective properties of an extended system composed of a given figure or figures and of an added figure.

But it is not solely owing to the generalisation of distance that the memoir is famous. It has revolutionised the theory of the so-called non-Euclidian geometry; and it has important bearings on the logical and philosophical analysis of the axioms of space-intuition. The independence and the importance of the ideas, originated by Cayley in this memoir, have never been questioned; but, as is often (and naturally) the case with the discoverer of a fertile subject, Cayley himself did not explain or foresee the full range of application of his new ideas. He did not recognise, at the time when his memoir was first published, the beautiful identification of his generalised theory of metrical geometry with the non-Euclidian geometry of Lobatchewsky and Bolyai. This fundamental step was taken by Klein in his admirable memoir†, *Ueber die sogenannte Nicht-Euklidische Geometrie*, which contains a considerable simplification in statement of Cayley's original point of view, and contributes one of the most important results of the whole theory. The work of the two mathematicians now being an organic whole, there is no advantage—at least here—in attempting to subdivide the subject for the purpose of specifying the exact share of each in its construction.

The scope of the Cayley-Klein ideas may briefly be gathered from the following sketch. Let A_1 and A_2 be two points, often called a point-pair; they are to be either both real or, if not both real, then conjugate imaginaries so far as their coordinates are concerned. Let P, Q, R be three other points on the line $A_1 A_2$; and let the symbol (PQ) denote

$$2\gamma \log \frac{A_1P \,.\, A_2Q}{A_1Q \,.\, A_2P} \text{ or } 2i\gamma \log \frac{A_1P \,.\, A_2Q}{A_1Q \,.\, A_2P},$$

according as A_1 and A_2 are a real point-pair, or an imaginary point-pair. Then it is manifest that

$$(PQ) + (QR) = (PR),$$

so that the functions (PQ), (QR), (PR) satisfy the fundamental property of the distances between P and Q, Q and R, and P and R. Consequently (PQ) may be taken as a generalised conception of the distance between the points P and Q.

* *C. M. P.* vol. II. No. 158; *Phil. Trans.* (1859), pp. 61—90.

† *Math. Ann.* vol. IV. (1871), pp. 573—625.

Now let a conic be described in a plane, either imaginary, say, of the form $x^2+y^2+z^2=0$ or real, say, of the form $x^2+y^2-z^2=0$. Choosing the latter case, let attention be confined to points lying within the conic, so that every straight line through a point cuts the conic in a real point-pair. Take two points, P and Q; and let the line joining them cut the conic in two points, A_1 and A_2. Then (PQ), as defined above (the constant γ being the same for all such lines), is the generalised distance between P and Q. This conic, which has been arbitrarily assumed, and upon which the generalised conception of distance depends, is termed by Cayley the Absolute.

Cayley, however, avoided the unsatisfactory procedure of using one conception of distance to define a more general conception. As he himself explains more fully,* he regarded the co-ordinates of points as some quantities which define the relative properties of points, considered without any reference to the idea of distance but conceived as ordered elements of a manifold. Thus if α_1, β_1, γ_1 and α_2, β_2, γ_2 be the co-ordinates of the point-pair A_1 and A_2, the co-ordinates of the points P and Q on the line A_1A_2 can be taken as $\lambda_1\alpha_1+\lambda_2\alpha_2$, $\lambda_1\beta_1+\lambda_2\beta_2$, $\lambda_1\gamma_1+\lambda_2\gamma_2$ and $\mu_1\alpha_1+\mu_2\alpha_2$, $\mu_1\beta_1+\mu_2\beta_2$, $\mu_1\gamma_1+\mu_2\gamma_2$ respectively. The function (PQ) can then be defined as

$$2\gamma \log \frac{\lambda_2\mu_1}{\lambda_1\mu_2} \text{ or } 2i\gamma \log \frac{\lambda_2\mu_1}{\lambda_1\mu_2};$$

the generalised idea of distance thus finds its definition without any antecedent use of the conception in its ordinary form. Cayley's view is summed up in his sentence†:—"......the theory in effect is, that the metrical properties of a figure are not the properties of the figure considered *per se* apart from everything else, but its properties when considered in connexion with another figure, viz. the conic termed the absolute."

The metrical formulæ obtained when the absolute is real are identical with those of Lobatchewsky's and Bolyai's "hyperbolic" geometry: when the absolute is imaginary the formulæ are identical with those of Riemann's "elliptic" geometry; the limiting case between the two being that of ordinary Euclidian ("parabolic") geometry.

Cayley's memoir leads inevitably to the question, as to how far projective geometry can be defined in terms of space perception without the introduction of distance. This has been discussed by von Staudt‡ (in 1847, previous to Cayley's memoir), by Klein§ and by Lindemann‖. The memoir thus points to a division of our space intuitions into two distinct parts: one, the more fundamental as not involving the idea of distance, the other, the more artificial as adding the idea of distance to the former. The consideration of the relation of these ideas to the philosophical account of space has not yet been brought to its ultimate issue.

* See the note which he added, *C. M. P.* vol. II. p. 604, to the Sixth Memoir; it contains some interesting historical and critical remarks.

† *Loc. cit.* § 230.

‡ *Geometrie der Lage*; also in his later *Beiträge zur Geometrie der Lage*, 1857.

§ *Math. Ann.* vol. VI. (1873), pp. 112—145.

‖ *Vorlesungen über Geometrie* (Clebsch-Lindemann), vol. II. part I.; the third section is devoted to the subject.

It is in analytical geometry, both of curves and of surfaces, that the greatest variety of Cayley's contributions is to be found. There is hardly an important question in the whole range of either subject in the solution of which he has not had some share; and there are many properties our acquaintance with which is due chiefly, if not entirely, to him. How widely he has advanced the boundaries of knowledge in analytical geometry can be inferred even from the amount of his researches already incorporated in treatises such as those by Salmon, Clebsch and Frost; and yet they represent only a portion of what he has done. In these circumstances only a selection among his contributions can be indicated: it must be understood that, here as elsewhere, the statement does not pretend to be a complete account.

It is an old-established property that two curves of degrees m and n cut in mn points, but that it is not possible to draw a curve of degree n through any mn arbitrarily selected points on a curve of degree m. As early as 1843, Cayley extended the property and showed that when a curve of degree r higher than either m or n is to be drawn through the mn points common to the two curves, they do not count for mn conditions in its determination, but only for a number of conditions smaller than mn by $\frac{1}{2}(m+n-r-1)(m+n-r-2)$. A single addition was made to the theorem by Bacharach* in 1886—taking account of the case when the undetermining points lie on a curve of degree $m+n-r-3$; with this exception the algebraical problem was completely solved by Cayley in his original paper†. The result is often called Cayley's intersection-theorem.

Another geometrical research of fundamental importance was embodied by him in a memoir‡, "On the higher singularities of a plane curve," published in 1866: it is there proved that any singularity whatever on a plane algebraical curve can be reckoned as equivalent to a definite number of the simple singularities constituted by the node, the ordinary cusp, the double tangent and the ordinary inflexional tangent. The theory has, since that date, been developed on lines different from Cayley's—owing to its importance in other theories, such as Abelian functions, variety in its development has proved both necessary and useful; but it was Cayley's investigations in continuation of Plücker's theory that have cleared the path for the later work of others.

The classification of cubic curves had been effected by Newton in his tract "Enumeratio linearum tertii ordinis," published in 1704: and six species had been added by Stirling and Cramer, the total then being 78. Plücker effected a new classification in his "System der analytischen Geometrie," published in 1835: his total number of species is 219, the division into species being more detailed than Newton's. Cayley re-examined the subject in his memoir§, "On the classification of cubic curves," expounding the principles of the two classifications and bringing them into comparison with one another; and entering into the discussion with full minuteness, he obtains the exact relation of the two classifications to one another—a result of great value in the theory.

* *Math. Ann.* vol. XXVI. (1886), pp. 275—299.

† *C. M. P.* vol. I. No. 5; *Camb. Math. Jour.* vol. III. (1843), pp. 211—213.

‡ *C. M. P.* vol. V. No. 374; *Quart. Math. Jour.* vol. VII. (1866), pp. 212—223.

§ *C. M. P.* vol. V. No. 350; *Camb. Phil. Trans.* vol. XI. (1864), pp. 81—128.

To the theories of rational transformation and correspondence he made considerable additions. Two figures are said to be rationally transformable into one another when to a variable point of one of them corresponds reciprocally one (and only one) variable point of the other. The figure may be a space or it may be a locus in a space. Rational transformations between two spaces give rational transformations between loci in those spaces; but it is not in general true that rational transformations between two loci necessarily give rational transformations between the spaces in which those loci exist. There is thus a distinction between the theory of transformation of spaces and the theory of correspondence of loci. Both theories have occupied many investigators, the latter in particular; and Cayley's work may fairly be claimed to have added much to the knowledge of the theory as due* to Riemann, Cremona and others.

Further, there may be singled out for special mention, his investigations on the bitangents of plane curves, and, in particular, on the 28 bitangents of a non-singular quartic; his developments of Plücker's conception of foci; his discussion of the osculating conics of curves, and of the sextactic points on a plane curve (these are the places where a conic can be drawn through six consecutive points); his contributions to the geometrical theory of the invariants and covariants of plane curves; and his memoirs on systems of curves subjected to specified conditions. Moreover, he was fond of making models and of constructing apparatus intended for the mechanical description of curves. The latter finds record in various of his papers; even so lately as 1893 he exhibited, at a meeting of the Cambridge Philosophical Society, a curve-tracing mechanism connected with three-bar motion.

All the preceding results belong to plane geometry; no less important or less numerous were the results he contributed to solid geometry. The twenty-seven lines that lie upon a cubic surface were first announced in his memoir†, "On the triple tangent planes of surfaces of the third order," published in 1849, after a correspondence between Salmon and himself. Cayley devised a new method for the analytical expression of curves in space by introducing into the representation the cone passing through the curve and having its vertex at an arbitrary point. Again, by using Plücker's equations that connect the ordinary (simple) singularities of plane curves, he deduced equations connecting the ordinary (simple) singularities of the developable surface that is generated by the osculating plane of a given tortuous curve, and, therefore, also of any developable surface. He greatly extended Salmon's theory of reciprocal surfaces; and resuming a subject already discussed by Schläfli he produced‡ in 1869 his "Memoir on cubic surfaces," in which he dealt with their complete classification. Many of his memoirs are devoted to the theory of skew ruled surfaces, or scrolls as he called them. Our knowledge of geodesics, of orthogonal systems of surfaces, of the centro-surface of an ellipsoid, of the wave-surface, of the 16-nodal quartic surface, not to mention more,

* In this connexion a report by Brill and Noether, "Bericht über die Entwicklung der Theorie der algebraischen Functionen in älterer und neuerer Zeit" (*Jahresber. d. Deutschen Mathem.-Vereinigung*, vol. III. 1894) will be found—particularly the sixth and the tenth sections—to give a very valuable *résumé* of the theory and its history.

† *C. M. P.* vol. I. No. 76; *Camb. and Dubl. Math. Jour.* vol. IV. (1849), pp. 118—132. See also Salmon's *Solid Geometry* (third edition, 1874), p. 464, note.

‡ *C. M. P.* vol. VI. p. 412; *Phil. Trans.* (1869), pp. 231—326.

is due in part to the extensions he achieved. It is difficult to indicate parts of the general theory of surfaces and of twisted curves that do not owe at least something and frequently much to his labours; a mere reference to the index of a book like Salmon's *Solid Geometry* will show how vast has been his influence.

One group of subjects interested him throughout his life, the theory of periodic functions, in particular, of elliptic functions: it was to the latter that his only book was devoted. But in a subject, the main lines of which were established so definitely before he began to write*, it is impossible, without entering into great detail, to mark out the contributions that are directly due to him. When a theory is in such a stage as was that of elliptic functions about 1842, the work of one writer sometimes helps to fill the gaps left by that of another, sometimes develops another writer's results from a different point of view; the composite theory depends, in part, upon the coordination of complementary results.

Abel's famous paper†, "Mémoire sur une propriété générale d'une classe très-étendue de fonctions transcendantes," presented to the French Academy of Sciences in 1826, and unfortunately delayed in publication‡ for nearly fifteen years, attracted Cayley's attention quite early in his scientific career. In 1845 Cayley published his "Mémoire sur les fonctions doublement périodiques,"§ in which he considered Abel's doubly-infinite products of the form

$$u(x) = x\Pi\Pi\left(1+\frac{x}{w}\right),$$

where $w=(m, n)=m\Omega+n\Upsilon$, the ratio $\Omega : \Upsilon$ is not real, and the product is taken for all positive and all negative integer values of m and of n between positive and negative infinity, except simultaneous zero values. He showed that such products can be used to obtain Jacobi's elliptic functions by constructing fractions such as

$$u(x+\tfrac{1}{2}\Omega) \div u(x);$$

and he also showed that the actual value of any product involves an exponential factor $e^{\frac{1}{2}Bx^2}$, where the value of the constant B depends upon the relation‖ between the infinities of m and of n. The results were of definite importance at the time of their discovery, and they still hold their place. But the form of the doubly-infinite product has been modified¶ by Weierstrass, who takes

$$\sigma(x) = x\Pi\Pi\left\{\left(1+\frac{x}{w}\right)e^{-\frac{x}{w}+\frac{1}{2}\frac{x^2}{w^2}}\right\},$$

* The history will be found in Casorati, *Teorica delle funzioni di variabili complesse*, 1868, and in Enneper, *Elliptische Functionen, Theorie und Geschichte*, second edition, 1890, where other references are given.

† *Œuvres complètes d'Abel* (Christiania, 1881), vol. I. pp. 145—211.

‡ The circumstances are recited in § 9 of the appendix to the volume, by Bjerknes, *Niels Henrik Abel, Tableau de sa vie et de son action scientifique* (Gauthier-Villars, Paris, 1885).

§ *C. M. P.* vol. I. No. 25; *Liouville*, vol. X. (1845), pp. 385—420.

‖ This is sometimes expressed differently, as follows. Points are taken having m and n for their Cartesian co-ordinates; those which occur for infinite values of m and of n lie at infinity, and may be considered to lie upon a curve altogether at infinity, the shape of which is determined by the relation between the infinities of m and of n.

The value of the constant B is said to depend upon the shape of this bounding curve.

¶ Weierstrass's investigations on infinite products are contained in his memoir "Zur Theorie der eindeutigen analytischen Functionen" (*Abh. d. K. Akad. d. Wiss. zu Berlin*, 1876); also in his book *Abhandlungen aus der Functionenlehre*, 1886.

a function the value of which is independent of any particular form of relation between the infinities of m and of n. Owing to the latter simplification, Cayley's results are, as he himself remarked*, partly superseded by those of Weierstrass.

Cayley had great admiration for the works of both Abel and Jacobi; he had begun to read the latter's *Fundamenta Nova* immediately after his degree. The prominent position occupied in that work by the theory of transformation naturally attracted his interest; and, even as early as 1844 and 1846, he wrote short memoirs upon the subject, obtaining in one of them a function, due to Abel and now known as the octahedral function. Further memoirs of a similar tenor appeared occasionally; they deal chiefly with transformation as concerned with the known differential relation of the form

$$\{(1-x^2)(1-k^2x^2)\}^{-\frac{1}{2}}\,dx = M\{(1-y^2)(1-\lambda^2y^2)\}^{-\frac{1}{2}}\,dy.$$

The contributions made to the transformation theory by Sohnke, Joubert, and Hermite, as well as Jacobi's original investigations, all depend upon the use of transcendental functions of the quantity $q\left(=e^{-\pi\frac{K'}{K}}\right)$: yet the results are such that they ought to be deducible by ordinary algebraical processes. It was Cayley's wish to deal with this theory by pure algebra; two simple cases had already thus been discussed by Jacobi, but the extension to the less simple cases proved difficult. Cayley's "Memoir on the transformation of elliptic functions†," carries on the algebraical theory and places it in a clearer light than before. But though he made a distinct advance in dealing with particular cases, he still found it necessary to use the q-transcendents for making any definite advance in the general case. And the same compulsion occurs in the chapters of his *Treatise on Elliptic Functions*, where transformation is discussed at considerable length.

He resumed his investigations in 1886, still dealing with the algebraical method, but applying it to a simplified form of elliptic integral due to Brioschi. Though the problem is not solved‡ completely for the general case, he has devised a method which is effective at least in part; it easily leads to new results connected with the modular equations in the known simpler cases previously solved.

The theta-functions are the subject of several of his papers. He began§ with a direct establishment of Jacobi's relation

$$\sqrt{k}\,snu = \mathrm{H}(u) \div \Theta(u),$$

obtained in the *Fundamenta Nova* by a long and cumbrous process; and he proceeded to the construction of the linear differential equations satisfied by the theta-functions. Except, however, in so far as they arise in the transformation theory, they do not appear to have occupied him until about 1877. In that year and in the succeeding

* *C. M. P.* vol. I. p. 586.

† *C. M. P.* vol. IX. No. 577; *Phil. Trans.* 1874, pp. 397—456.

‡ The memoirs of this period belonging to the transformation of elliptic functions were published in the *American Journal of Mathematics*, vol. IX. (1887), pp. 193—224; vol. X. (1888), pp. 71—93.

§ "On the Theory of Elliptic Functions," *C. M. P.* vol. I. No. 45; *Camb. and Dubl. Math. Jour.* vol. II. (1847), pp. 256—266.

years he wrote a number of papers dealing with the theta-functions as on an independent basis and not as a detail in elliptic functions. Though the investigations are concerned with p-tuple functions, yet, partly for simplicity, and partly in order to secure the greater detailed development of the theory, the papers deal chiefly with the cases $p=1$, $p=2$.

Previous to Cayley's investigations, the most valuable algebraical results in this subject were those of Rosenhain* and Göpel† which had connected the double theta-functions with the theory of the Abelian functions of two variables, and those of Weierstrass, developed by Königsberger‡ to give the "addition-theorem." Proceeding in his "Memoir on the single and double theta-functions"§ more by Göpel's method than by Rosenhain's, Cayley resumes the whole theory. He pays special attention to the relations among the squares of the functions and to the derivation of the biquadratic relation among four of the functions, which is the same as the equation of Kummer's sixteen-nodal quartic surface. To this relation and to the geometry of this associated surface he frequently recurred, both specifically in isolated papers and generally in researches upon quartic surfaces.

As connected, in part, with elliptic functions, his investigations on the porism of the in- and circumscribed polygon should be mentioned. The porismatic property of two conics, viz. that they may be related to each other so that one polygon (and, if one polygon, then an infinite number of polygons) can be inscribed in one and circumscribed about the other, is due to the geometrician Poncelet. The special case when the conics are two circles had been discussed analytically by Jacobi‖, using elliptic functions for the purpose. Cayley undertook, first in 1853, the analytical discussion of the most general case of two conics, also using elliptic functions; and he obtained¶ the relations, necessary for the porism, for the several polygons as far as the enneagon. And it may be remarked, as a characteristic instance of Cayley's habit of proceeding to general cases, that he did not leave the matter at this stage. In a memoir** "On the problem of the in- and circumscribed triangle" he raises the question as to the number of polygons which are such that their angular points lie on a given curve or given curves of any order and their sides touch another given curve or given curves of any class. Using the theory of correspondence, he solves the question completely in the case of a triangle—taking account of the fifty-two cases that arise through the possibility of two curves, or more than two curves, being one and the same curve.

From time to time Cayley turned his attention to questions in theoretical dynamics, choosing them as subjects of his lectures during his earlier years as professor. Among them may be mentioned his investigations on attractions, specially those on the attraction of ellipsoids, to which he devotes five memoirs††, discussing the methods of Legendre,

* *Mém. des Sav. Étr.* vol. XI. (1851), pp. 361—468; the paper is dated 1846.

† *Crelle*, vol. XXXV. (1847), pp. 277—312.

‡ *Crelle*, vol. LXIV. (1865), pp. 17—42.

§ *Phil. Trans.* 1880, pp. 897—1002.

‖ *Ges. Werke*, vol. I. pp. 277—293; this paper was published first in *Crelle*, vol. III. (1828), pp. 376—389.

¶ In a set of five papers, *C. M. P.* vol. II. Nos. 113, 115, 116, 128; *ibid.*, vol. IV. No. 267.

** *C. M. P.* vol. VIII. No. 514; *Phil. Trans.* (1871), pp. 369—412.

†† *C. M. P.* vol. I. Nos. 75, 89; vol. II. Nos. 164, 173, 193.

Jacobi, Gauss, Laplace, and Rodrigues; and his evaluations or reductions of multiple definite integrals connected with attractions and potentials in general, particularly his "Memoir on Prepotentials*," in which he discusses the reduction of the most general integral of the type that can occur in dealing with the potential-problem related to hyperspace. He also frequently recurred at intervals, before drawing up his reports about to be quoted, to the consideration of the motion of rotation of a solid body about a fixed point under no forces. By introducing Rodrigues's co-ordinates into the equations of motion he was able to reduce the solution of the problem to quadratures; but the final solution of this case, in the most elegant form, is due to Jacobi himself; it involves single theta-functions. It may be remarked that the next substantial advance made in the theory of motion of a body under the action of forces is due to the late Madame Sophie Kovalewsky, who, in a memoir†, to which the Bordin Prize of 1888 was awarded by the Paris Academy of Sciences, has shown that the motion can, in a particular case, be determined in terms of double theta-functions when the body rotating round a fixed point is subject to the force of gravity.

Sometimes, after reading widely upon a subject, Cayley would draw up a report recounting the chief researches in it made by the great writers. It occasionally happens in the development of a theory that periods come when the incorporation and the marshalling of created ideas seem almost necessary preliminaries to further progress. Cayley was admirably fitted for work of this kind, owing not only to his faculty of clear and concise exposition, but also to his wide and accurate knowledge. Among such reports, two are of particular importance; his "Report on the recent progress of theoretical dynamics‡" and his "Report on the progress of the solution of certain special problems of dynamics§" have proved of signal service to other writers and to students. His knowledge and his power of summarising are shown also in some interesting articles on mathematical topics, written by him for the *Encyclopædia Britannica.*

Cayley also had a great enthusiasm for some of the branches of physical astronomy. Some idea of the value and importance of his labours in this subject, particularly in connexion with the development of the disturbing function in both the lunar theory and the planetary theory, and with the general developments of the functions that arise in elliptic motion, may be gathered by consulting the series of memoirs‖ which he communicated to the Royal Astronomical Society.

Special reference should be made to one of Cayley's astronomical papers. In 1853 Adams had made a new investigation of the value of the secular acceleration of the moon's mean motion, and, taking account of the variation in the eccentricity of the earth's orbit, had obtained a value which differed from that given by Laplace. Unfortunately, Adams's result was disputed by some of the great school of French physical

* *Phil. Trans.* 1875, pp. 675—774.

† *Mém. des Sav. Étr.*, vol. XXXI. (1894), No. 1.

‡ *C. M. P.* vol. III. No. 195; *Brit. Assoc. Report* (1857), pp. 1—42.

§ *C. M. P.* vol. IV. No. 298; *Brit. Assoc. Report* (1862), pp. 184—252.

‖ They are included, with very few exceptions, in the third and the seventh volumes of the *Collected Mathematical Papers.*

astronomers, notably by Pontécoulant, and, in consequence, some hesitation about acceptance was felt by some English astronomers, perhaps not unnaturally in view of the severe criticisms expressed. Cayley made an independent investigation of the necessary approximations, and devised a new method for introducing the variation of the eccentricity in question—a method effective perhaps chiefly owing to the instinct and power with which he carried out the laborious analysis required. The memoir, in which he embodied his results and which was entitled "On the secular acceleration of the moon's mean motion*," completely confirmed the value obtained by Adams, and was of substantial help in settling the controversy.

And, in the last place, the preceding sketch of Cayley's contributions to mathematical science seems to refer, for the most part, only to long memoirs. Yet it must not therefore be supposed that his shorter papers (which are very numerous) can safely be neglected. Sometimes he wrote a simple note not so much to convey new results as to set out his view of some particular theorem; these notes were always fresh and often suggestive. He was specially gratified when he had obtained a brief solution of some question, and his quite short papers frequently contain most important results. For instance, in the brief paper †, "On the theory of the singular solutions of differential equations of the first order," he was the first to give a clear exposition of the theory which in Boole's book had been left in an imperfect state. He there obtained the broad essential results of the theory, and it is particularly on his work, and on the work of Darboux published very soon after Cayley's, that ulterior researches are based.

What has been said may be sufficient to point out Cayley's place among the mathematicians of his time, and to indicate the services he rendered to the science which he loved so well. But he was more than a mathematician. With a singleness of aim, which Wordsworth could have chosen for his "Happy Warrior," he persevered to the last in his nobly lived ideal. His life had a significant influence on those who knew him: they admired his character as much as they respected his genius: and they felt that, at his death, a great man had passed from the world.

A. R. F.

1 *June*, 1895.

* *C. M. P.* vol. III. No. 221; *Monthly Not. R. A. S.* vol. XXII. (1862), pp. 171—231.

† *C. M. P.* vol. VIII. No. 545; *Messenger of Math.* vol. II. (1873), pp. 6—12.

COURSES OF LECTURES DELIVERED BY PROFESSOR CAYLEY.

(M. denotes Michaelmas Term; L. denotes Lent Term.)

1863. M. Analytical Geometry.
1864. M. Analytical Geometry.
1865. M. Analytical Geometry and Mechanics.
1866. M. Dynamics.
1867. M. Miscellaneous Analysis.
1868. M. Dynamics and Differential Equations.
1869. M. Analytical Geometry.
1870. M. Theories of correspondence and transformation in analytical geometry.
1871. M. Graphical Geometry.
1872. M. Elliptic Functions.
1873. M. Theory of Equations and Miscellaneous Analysis.
1874. M. Integral Calculus.
1875. M. On a course of pure mathematics.
1876. M. Differential Equations.
1877. M. Algebra.
1878. M. Solid Geometry.
1879. M. Differential Equations.
1880. M. Theory of Equations.
1881. M. Abel's Theorem and the Theta Functions.
1882. M. Abelian and Theta Functions.
1883. M. Higher Algebra and the Theory of Numbers.
1884. M. Some recent developments in Analysis and Geometry.
1885. M. Higher Algebra.
1886. M. Differential Equations and Analytical Geometry.

1887.	L.	Differential Equations.
	M.	Quaternions and other non-commutative algebras.
1888.	L.	Analytical Geometry.
	M.	Elliptic Functions.
1889.	L.	Analytical Geometry.
	M.	Solid Geometry.
1890.	L.	Theory of Equations.
	M.	Elliptic Functions, in particular Transformation and the modular equations.
1891.	L.	Analytical Geometry.
	M.	Higher Algebra and Analytical Geometry.
1892.	L.	Higher Algebra and Analytical Geometry.
	M.	On a course of pure mathematics.
1893.	L.	On a course of pure mathematics.
	M.	Analytical Geometry.
1894.	L.	The known transcendental functions.
	M.	Analytical Geometry.
1895.	L.	Theory of Equations (announced).

CONTENTS.

[An Asterisk means that the paper is not printed in full.]

PAGE

486. *Note on Dr. Glaisher's paper on a theorem in definite integration* 1
Quart. Math. Jour. t. x. (1870), pp. 355, 356

487. *On the quartic surfaces* $(*\between U, V, W)^2=0$ 2
Quart. Math. Jour. t. xi. (1871), pp. 15—25

488. *Note on a relation between two circles* 12
Quart. Math. Jour. t. xi. (1871), pp. 82, 83

489. *On the porism of the in-and-circumscribed polygon, and the (2, 2) correspondence of points on a conic* 14
Quart. Math. Jour. t. xi. (1871), pp. 83—91

490. *On a problem of elimination* 22
Quart. Math. Jour. t. xi. (1871), pp. 99—101

491. *On the quartic surfaces* $(*\between U, V, W)^2=0$ 25
Quart. Math. Jour. t. xi. (1871), pp. 111—113

492. *Note on a system of algebraical equations* 29
Quart. Math. Jour. t. xi. (1871), pp. 132, 133

493. *On evolutes and parallel curves* 31
Quart. Math. Jour. t. xi. (1871), pp. 183—200

494. *Example of a special discriminant* 46
Quart. Math. Jour. t. xi. (1871), pp. 211—213

495. *On the envelope of a certain quadric surface* 48
Quart. Math. Jour. t. xi. (1871), pp. 244—246

PAGE

496. *Tables of the binary cubic forms for the negative determinants* $\equiv 0$ (mod. 4) *from* -4 *to* -400; *and* $\equiv 1$ (mod. 4) *from* -3 *to* -99; *and for five irregular negative determinants* . 51
Quart. Math. Jour. t. XI. (1871), pp. 246—261

497. *Note on the calculus of logic* 65
Quart. Math. Jour. t. XI. (1871), pp. 282, 283

498. *On the inversion of a quadric surface* 67
Quart. Math. Jour. t. XI. (1871), pp. 283—288

499. *On the theory of the curve and torse* 72
Quart. Math. Jour. t. XI. (1871), pp. 294—317

500. *On a theorem relating to eight points on a conic* . . . 92
Quart. Math. Jour. t. XI. (1871), pp. 344—346

501. *Review. Pineto's tables of logarithms* 95
Quart. Math. Jour. t. XI. (1871), pp. 375, 376

502. *On the surfaces divisible into squares by their curves of curvature* 97
Proc. Lond. Math. Society, t. IV. (1871—1873), pp. 8, 9

503. *On the surfaces each the locus of the vertex of a cone which passes through m given points and touches* $6-m$ *given lines* 99
Proc. Lond. Math. Society, t. IV. (1871—1873), pp. 11—47

504. *On the mechanical description of certain sextic curves* . . 138
Proc. Lond. Math. Society, t. IV. (1871—1873), pp. 105—111

505. *On the surfaces divisible into squares by their curves of curvature* 145
Proc. Lond. Math. Society, t. IV. (1871—1873), pp. 120, 121

506. *On the mechanical description of a cubic curve* . . . 147
Proc. Lond. Math. Society, t. IV. (1871—1873), pp. 175—178

507. *On the mechanical description of certain quartic curves by a modified oval chuck* 151
Proc. Lond. Math. Society, t. IV. (1871—1873), pp. 186—190

508. *On geodesic lines, in particular those of a quadric surface* . 156
Proc. Lond. Math. Society, t. IV. (1871—1873), pp. 191—211

PAGE

509. *Plan of a curve-tracing apparatus* 179
Proc. Lond. Math. Society, t. IV. (1871—1873), pp. 345—347

510. *On bicursal curves* 181
Proc. Lond. Math. Society, t. IV. (1871—1873), pp. 347—352

511. *Addition to the memoir on geodesic lines, in particular those of a quadric surface* 188
Proc. Lond. Math. Society, t. IV. (1871—1873), pp. 368—380

512. *On a correspondence of points in relation to two tetrahedra* . 200
Proc. Lond. Math. Society, t. IV. (1871—1873), pp. 396—404

513. *On a bicyclic chuck* 209
Phil. Mag. t. XLIII. (1872), pp. 365—367

514. *On the problem of the in-and-circumscribed triangle* . . 212
Phil. Trans. t. CLXI. (for 1871), pp. 369—412

515. *Sur les courbes aplaties* 258
Comptes Rendus, t. LXXIV. (1872), pp. 708—712

516. *Sur une surface quartique aplatie* 262
Comptes Rendus, t. LXXIV. (1872), pp. 1393—1395

517. *Sur les surfaces divisibles en carrés par leurs courbes de courbure et sur la théorie de Dupin* 264
Comptes Rendus, t. LXXIV. (1872), pp. 1445—1449

518. *Sur la condition pour qu'une famille de surfaces données puisse faire partie d'un système orthogonal* 269
Comptes Rendus, t. LXXV. (1872), pp. 177—185, 246—250, 324—330, 381—385, 1800—1803

519. *On curvature and orthogonal surfaces* 292
Phil. Trans. t. CLXIII. (for 1873), pp. 229—251

520. *On the centro-surface of an ellipsoid* 316
Camb. Phil. Trans. t. XII. Part I. (1873), pp. 319—365

521. *On Dr. Wiener's model of a cubic surface with 27 real lines; and on the construction of a double-sixer* 366
Camb. Phil. Trans. t. XII. Part I. (1873), pp. 366—383

PAGE

522. *Note on the theory of invariants* 385
Math. Ann. t. III. (1871), pp. 268—271

523. *On the transformation of unicursal surfaces* 388
Math. Ann. t. III. (1871), pp. 469—474

524. *On the deficiency of certain surfaces* 394
Math. Ann. t. III. (1871), pp. 526—529

525. *An example of the higher transformation of a binary form* . 398
Math. Ann. t. IV. (1871), pp. 359—361

526. *On a surface of the eighth order* 401
Math. Ann. t. IV. (1871), pp. 558—560

527. *On a theorem in covariants* 404
Math. Ann. t. V. (1872), pp. 625—629

528. *On the non-Euclidian geometry* 409
Math. Ann. t. V. (1872), pp. 630—634

529. *A "Smith's Prize" paper [1868]; solutions by Prof. Cayley* . 414
Oxford, Camb. and Dubl. Messenger of Mathematics, t. IV. (1868), pp. 201—226

530. *Solution of a Senate-House problem* 436
Oxford, Camb. and Dubl. Messenger of Mathematics, t. V. (1871), pp. 24—27

531. *A "Smith's Prize" paper [1869]; solutions by Prof. Cayley* . 439
Oxford, Camb. and Dubl. Messenger of Mathematics, t. V. (1871), pp. 41—64

532. *Note on the integration of certain differential equations by series* 458
Oxford, Camb. and Dubl. Messenger of Mathematics, t. V. (1871), pp. 77—82

533. *On the binomial theorem, factorials, and derivations* . . 463
Oxford, Camb. and Dubl. Messenger of Mathematics, t. V. (1871), pp. 102—114

534. *A "Smith's Prize" paper [1870]; solutions by Prof. Cayley*. 474
Oxford, Camb. and Dubl. Messenger of Mathematics, t. V. (1871), pp. 182—203

PAGE

535. *Note on the problem of envelopes* 491
Messenger of Mathematics, t. I. (1872), pp. 3, 4

536. *Note on Lagrange's demonstration of Taylor's theorem* . . 493
Messenger of Mathematics, t. I. (1872), pp. 22—24

537. *Solutions of a Smith's Prize paper for* 1871 496
Messenger of Mathematics, t. I. (1872), pp. 37—47, 71—77, 89—95

538. *Extract from a letter from Prof. Cayley to Mr. C. W. Merrifield* 517
Messenger of Mathematics, t. I. (1872), pp. 87, 88

*539. *Further note on Lagrange's demonstration of Taylor's theorem* 519
Messenger of Mathematics, t. I. (1872), pp. 105, 106

540. *On a property of the torse circumscribed about two quadric surfaces* 520
Messenger of Mathematics, t. I. (1872), pp. 111, 112

541. *On the reciprocal of a certain equation of a conic* . . . 522
Messenger of Mathematics, t. I. (1872), pp. 120, 121

*542. *Further note on Taylor's theorem* 524
Messenger of Mathematics, t. I. (1872), p. 137

543. *On an identity in spherical trigonometry* 525
Messenger of Mathematics, t. I. (1872), p. 145

544. *On a penultimate quartic curve* 526
Messenger of Mathematics, t. I. (1872), pp. 178—180

545. *On the theory of the singular solutions of differential equations of the first order* 529
Messenger of Mathematics, t. II. (1873), pp. 6—12

546. *Theorems in relation to certain sign-symbols* 535
Messenger of Mathematics, t. II. (1873), pp. 17—20

547. *On the representation of a spherical or other surface on a plane: a Smith's Prize dissertation* 538
Messenger of Mathematics, t. II. (1873), pp. 36, 37

PAGE

548. *On Listing's theorem.* 540

Messenger of Mathematics, t. II. (1873), pp. 81—89

549. *Note on the maxima of certain factorial functions* . . . 548

Messenger of Mathematics, t. II. (1873), pp. 129, 130

550. *Problem and hypothetical theorems in regard to two quadric surfaces* 550

Messenger of Mathematics, t. II. (1873), p. 137

551. *Two Smith's Prize dissertations* [1872] 551

Messenger of Mathematics, t. II. (1873), pp. 145—149

552. *On a differential formula connected with the theory of confocal conics* 556

Messenger of Mathematics, t. II. (1873), pp. 157, 158

553. *Two Smith's Prize dissertations* [1873] 558

Messenger of Mathematics, t. II. (1873), pp. 161—166

554. *An elliptic-transcendent identity* 564

Messenger of Mathematics, t. II. (1873), p. 179

555. *Notices of Communications to the British Association for the Advancement of Science* 565

Brit. Assoc. Reports, Notices and Abstracts of Communications to the Sections (1870, 1871, 1873).

Notes and References 569
Prefatory Note vii
Arthur Cayley: biographical notice by the Editor ix
List of courses of lectures delivered by Professor Cayley . . xlv
Facsimile of the manuscript of his note on p. 569 . . *Frontispiece*

CLASSIFICATION.

GEOMETRY :

Conics, 500, 541, 552

Mechanical description of curves, 504, 506, 507, 509, 513

Bicursal curves, 510

Penultimate forms of curves, 515, 516, 544

In-and-circumscribed figures, 489, 514

Evolutes and parallel curves, 493

Quadric surfaces, 495, 498, 550

Centro-surface of ellipsoid, 520

Cubic surfaces, 521

Quartic surfaces, 487, 491

Envelopes, 535

Curve and torse, 499, 540

Curves of curvature on surfaces, 502, 505, 517

Geodesics on quadrics, 508, 511

Correspondence, transformation, and deficiency of surfaces, 512, 523, 524

Cones satisfying six conditions, 503

Orthogonal surfaces, 518, 519

Surface of eighth order, 526

Non-Euclidian geometry, 528

Listing's theorem, 548

Miscellaneous geometry, 488, 555

ANALYSIS:

Pineto's table of logarithms, 501

Tables of binary cubic forms, 496

Calculus of logic, 497

Algebraical equations, 492

Elimination, 490

Discriminant, 494

Invariants and covariants, 522, 525, 527

Definite integrals, 486

Differential equations, 532, 545

Miscellaneous analysis, 533, 536, 538, 539, 542, 543, 546, 549, 554, 555

Smith's-Prize papers and solutions, 529, 530, 531, 534, 537, 547, 551, 553

486.

NOTE ON DR GLAISHER'S PAPER ON A THEOREM IN DEFINITE INTEGRATION.

[From the *Quarterly Journal of Pure and Applied Mathematics*, vol. X. (1870), pp. 355, 356.]

It is worth noticing how easily the case when $\phi = 1$ may be proved independently of the general formula with Θ; for (1) the equation

$$v = ax - \frac{a_1}{x-\lambda_1} - \frac{a_2}{x-\lambda_2} \dots - \frac{a_n}{x-\lambda_n}$$

is

$$(ax - v)(x-\lambda_1)(x-\lambda_2)\dots - a_1(x-\lambda_2)\dots - \dots = 0,$$

and has $n+1$ roots, say $x_1, x_2 \dots x_{n+1}$ where

$$x_1 + x_2 \dots + x_{n+1} = \lambda_1 + \lambda_2 \dots + \lambda_n + \frac{v}{a},$$

and (2) the equation

$$v = -\frac{a_1}{x-\lambda_1} - \frac{a_2}{x-\lambda_2} \dots - \frac{a_n}{x-\lambda_n}$$

is

$$-v(x-\lambda_1)(x-\lambda_2)\dots - a_1(x-\lambda_2)\dots - \dots = 0,$$

and has n roots $x_1, x_2 \dots x_n$ where

$$x_1 + x_2 \dots + x_n = \lambda_1 + \lambda_2 \dots + \lambda_n - \frac{a_1 + a_2 \dots + a_n}{v};$$

wherefore

$$\int v\, dx_1 + \int v\, dx_2 \dots = \int v\,(dx_1 + dx_2 \dots)$$

$$= \int v \frac{dv}{a} \quad \text{in the first case}$$

and

$$= \int v \frac{(a_1 + a_2 \dots + a_n)\, dv}{v^2} \quad \text{in the second}$$

[which are the two formulæ in question].

487.

ON THE QUARTIC SURFACES $(*\between U, V, W)^2=0$.

[From the *Quarterly Journal of Pure and Applied Mathematics*, vol. XI. (1871), pp. 15—25.]

AMONG the surfaces of the form in question are included the reciprocals of several interesting surfaces of the orders 6, 8, 9, 10, and 12, viz.

Order 6, parabolic ring.
„ 8, elliptic ring.
„ 9, centro-surface of a paraboloid.
„ 10, parallel surface of a paraboloid.
„ „ envelope of planes through the points of an ellipsoid at right angles to the radius vectors from the centre.
„ 12, centro-surface of an ellipsoid.
„ „ parallel surface of an ellipsoid.

I propose to consider these surfaces, not at present in any detail, but merely for the purpose of presenting them in connexion with each other and with the present theory. It will be convenient to use homogeneous equations, but for the metrical interpretation the coordinate W or w may be considered as equal to unity: I have not thought it necessary so to adjust the constants that the equations shall be homogeneous in regard to the constants; this can of course be done without difficulty, and in many cases it would be analytically advantageous to make the change.

I take throughout (X, Y, Z, W) for the coordinates of a point on the quartic surface (so that (U, V, W) in the equation $(*\between U, V, W)^2=0$ are to be considered as quadric functions of (X, Y, Z, W)), reserving (x, y, z, w) for the coordinates of a point on the reciprocal surface of the order 6, 8, 9, 10, or 12. The reciprocation is performed in regard to the imaginary sphere $x^2+y^2+z^2+w^2=0$: the relation between the coordinates (X, Y, Z, W) and (x, y, z, w) is then $Xx+Yy+Zz+Ww=0$, and the equation $(X, Y, Z, W)^4=0$ is the equation in point-coordinates of the quartic surface, or in line-coordinates of the reciprocal surface: and similarly the equation $(x, y, z, w)^n=0$ is the equation in point-coordinates of the reciprocal surface, or in line-coordinates of the quartic surface.

Parabolic ring, or envelope of a sphere of constant radius having its centre on a parabola.

Taking k for the radius of the sphere, and $z=0$, $y^2=4ax$ for the equations of the parabola, then the coordinates of a point on the parabola are $a\theta^2$, $2a\theta$, 0; where θ is a variable parameter. The equation of the sphere therefore is

$$(x-a\theta^2w)^2+(y-2a\theta w)^2+z^2-k^2w^2=0,$$

and the ring is the envelope of this sphere.

The reciprocal of the sphere is

$$k^2(X^2+Y^2+Z^2)-(a\theta^2X+2a\theta Y+W)^2=0;$$

writing this in the form

$$a\theta^2X+2a\theta Y+W+k\sqrt{(X^2+Y^2+Z^2)}=0,$$

and taking the envelope in regard to θ, we have

$$X\{W+k\sqrt{(X^2+Y^2+Z^2)}\}-aY^2=0,$$

or, what is the same thing,

$$(aY^2-XW)^2-k^2X^2(X^2+Y^2+Z^2)=0,$$

for the equation of the quartic surface. This has the line $X=0$, $Y=0$ for a tacnodal line, but I am not in possession of a theory enabling me thence to infer that the parabolic ring is of the order 6.

To show that it is so, I revert to the equation of the variable sphere

$$(x-a\theta^2w)^2+(y-2a\theta w)^2+z^2-k^2w^2=0,$$

or, what is the same thing,

$$(A, B, C, D, E\between\theta, 1)^4=0,$$

where

$$\begin{aligned} A &= 3a^2w^2, \\ B &= 0, \\ C &= a(2aw^2-xw), \\ D &= -3ayw, \\ E &= 3(x^2+y^2+z^2-k^2w^2). \end{aligned}$$

Then $I = 3a^2w^2I'$, $J = a^3w^3J'$, and the equation is $I'^3 - J'^2 = 0$, viz. this is

$$\{4x^2 + 3y^2 - 4axw + 4a^2w^2 + 3(z^2 - k^2w^2)\}^3$$
$$- \{(2aw - x)[8x^2 + 9y^2 + 4axw - 4a^2w^2 + 9(z^2 - k^2w^2)] - 27ay^2w\}^2 = 0,$$

or, as this may also be written,

$$\{4x^2 + 3y^3 - 4axw + 4a^2w^2 + 3(z^2 - k^2w^2)\}^3$$
$$- \{-8x^3 - 9xy^2 + 12ax^2w + 12a^2xw^2 - 8a^3w^3 - 9(x - 2aw)(z^2 - h^2w^2)\} = 0.$$

Developing, the whole divides by 27, and the equation of the ring finally is

$$(y^2 - 4axw)^2\{y^2 + (x - aw)^2\}$$
$$+ \{3y^4 + y^2(2x^2 - 2axw + 20a^2w^2) + 8ax^3w + 8a^2x^2w^2 - 32a^3xw^3 + 16a^4w^4\}(z^2 - k^2w^2)$$
$$+ (3y^2 + x^2 + 8axw - 8a^2w^2)(z^2 - k^2w^2)^2$$
$$+ (z^2 - k^2w^2)^3 = 0.$$

Elliptic ring, or envelope of a sphere of constant radius having its centre on an ellipse.

Taking k for the radius of the sphere, and $z = 0$, $\frac{x^2}{a^2} + \frac{y^2}{b^2} = 1$ for the equations of the ellipse, the coordinates of a point on the ellipse are $a\cos\theta$, $b\sin\theta$; hence the equation of the variable sphere is

$$(x - aw\cos\theta)^2 + (y - bw\sin\theta)^2 + z^2 - k^2w^2 = 0.$$

The reciprocal of this is

$$k^2(X^2 + Y^2 + Z^2) - (aX\cos\theta + bY\sin\theta + W)^2 = 0,$$

viz. writing this under the form

$$aX\cos\theta + bY\sin\theta + W + k\sqrt{(X^2 + Y^2 + Z^2)} = 0,$$

and taking the envelope in regard to θ, the equation of the reciprocal surface is

$$a^2X^2 + b^2Y^2 = \{W + k\sqrt{(X^2 + Y^2 + Z^2)}\}^2,$$

viz. this is

$$(a^2 - k^2)X^2 + (b^2 - k^2)Y^2 - k^2Z^2 - W^2 = 2kW\sqrt{(X^2 + Y^2 + Z^2)},$$

or

$$\{(a^2 - k^2)X^2 + (b^2 - k^2)Y^2 - k^2Z^2 - W^2\}^2 - 4k^2W^2(X^2 + Y^2 + Z^2) = 0,$$

that is

$$\{(a^2 - k^2)X^2 + (b^2 - k^2)Y^2 - k^2Z^2\}^2 - 2W^2\{(a^2 + k^2)X^2 + (b^2 + k^2)Y^2 + k^2Z^2\} + W^4 = 0,$$

which is a quartic surface having the nodal conic $W = 0$, $(a^2 - k^2)X^2 + (b^2 - k^2)Y^2 - k^2Z^2 = 0$. This singularity alone would only reduce the order of the reciprocal surface to 12; the reciprocal surface or elliptic ring is in fact (as I proceed to show) of the order 8.

For this purpose reverting to the equation

$$(x - aw\cos\theta)^2 + (y - bw\sin\theta)^2 + z^2 - k^2w^2 = 0,$$

this may be written

$$A\cos 2\theta + B\sin 2\theta + C\cos\theta + D\sin\theta + E = 0,$$

where

$$\begin{aligned} A &= \ (a^2 - b^2)\, w^2, \\ B &= \ 0, \\ C &= -\, 4axw, \\ D &= -\, 4byw, \\ E &= \ (a^2 + b^2)\, w^2 + 2\,(x^2 + y^2 + z^2 - k^2w^2), \end{aligned}$$

and the equation is

$$\{12\,(A^2 + B^2) - 3\,(C^2 + D^2) + 4E^2\}^3$$
$$- \{27A\,(C^2 - D^2) + 54BCD - [72\,(A^2 + B^2) + 9\,(C^2 + D^2)]\, E + 8E^3\}^2 = 0,$$

or say

$$\{12A^2 - 3\,(C^2 + D^2) + 4E^2\}^3 - \{27A\,(C^2 - D^2) - [72A^2 + 9\,(C^2 + D^2)]\, E + 8E^3\}^2 = 0.$$

This is of the order 12, but it is easy to see that the terms in E^6 and $E^4\,(C^2 + D^2)$ disappear from the equation, all the other terms divide by w^4; and the equation is thus of the order 8.

The equation may be obtained somewhat differently as follows. The equation of the variable sphere is

$$(x - \alpha w)^2 + (y - \beta w)^2 + z^2 - k^2w^2 = 0,$$

where (α, β) vary subject to the condition $\frac{\alpha^2}{a^2} + \frac{\beta^2}{b^2} = 1$. We have therefore

$$x - \alpha w - \lambda\,\frac{\alpha w}{a^2} = 0,$$

$$y - \beta w - \lambda\,\frac{\beta w}{b^2} = 0,$$

and thence

$$\alpha w = \frac{a^2x}{a^2 + \lambda}, \quad x - \alpha w = -\frac{\lambda x}{a^2 + \lambda},$$

$$\beta w = \frac{b^2y}{b^2 + \lambda}, \quad y - \beta w = -\frac{\lambda y}{b^2 + \lambda}.$$

Consequently

$$\frac{x^2}{(a^2 + \lambda)^2} + \frac{y^2}{(b^2 + \lambda)^2} + \frac{z^2 - k^2w^2}{\lambda^2} = 0,$$

$$\frac{a^2x^2}{(a^2 + \lambda)^2} + \frac{b^2y^2}{(b^2 + \lambda)^2} - \quad w^2 \quad = 0,$$

from which λ is to be eliminated. The second equation may be replaced by

$$\frac{x^2}{a^2+\lambda}+\frac{y^2}{b^2+\lambda}+\frac{z^2-k^2w^2}{\lambda}-w^2=0,$$

which has the first for its derived equation in regard to λ. Hence, writing this last equation in the form

$$w^2(a^2+\lambda)(b^2+\lambda)\lambda-(b^2+\lambda)\lambda x^2-(a^2+\lambda)\lambda y^2-(a^2+\lambda)(b^2+\lambda)(z^2-k^2w^2)=0,$$

we have to equate to zero the discriminant of this cubic function of λ. Calling the equation

$$(A, B, C, D\between\lambda, 1)^3=0,$$

we have

$$\begin{aligned} A &= 3w^2,\\ B &= (a^2+b^2)w^2-x^2-y^2-(z^2-k^2w^2),\\ C &= a^2b^2w^2-b^2x^2-a^2y^2-(a^2+b^2)(z^2-k^2w^2),\\ D &= -3a^2b^2(z^2-k^2w^2). \end{aligned}$$

The required equation then is

$$A^2D^2+4AC^3+4B^3D-3B^2C^2-6ABCD=0.$$

The developed equation (Salmon's *Conic Sections*, Ed. v., p. 325) is

$$\begin{aligned} &(b^2x^2+a^2y^2-a^2b^2w^2)^2\{(x-cw)^2+y^2\}\{(x+cw)^2+y^2\}\\ &-\left\{\begin{array}{l} 2b^2(a^2-2b^2)x^6-2(a^4-a^2b^2+3b^4)x^4y^2-2(3a^4-a^2b^2+b^4)x^2y^4+2a^2(b^2-2a^2)y^6\\ -b^2(6a^4-10a^2b^2+6b^4)x^4w^2+(4a^6-6a^4b^2-6a^2b^4+4b^6)x^2y^2w^2\\ \qquad -a^2(6a^4-10a^2b^2+6b^4)y^4w^2\\ +2c^2(3a^4-a^2b^2+b^4)x^2w^4-2c^2(a^4-3a^2b^2+3b^4)y^2w^4-2a^2b^2c^4(a^2+b^2)w^6 \end{array}\right\}(z^2-k^2w^2)\\ &+\left\{\begin{array}{l} (a^4-6a^2b^2+6b^4)x^4+(6a^4-10a^2b^2+6b^4)x^2y^2+(6a^4-6a^2b^2+b^4)y^4\\ -2c^2(a^4-a^2b^2+3b^4)x^2w^2+2c^2(3a^4-a^2b^2+b^4)y^2w^2+c^4(a^4+4a^2b^2+b^4)w^4 \end{array}\right\}(z^2-k^2w^2)^2\\ &+\left\{\begin{array}{l} (a^2-2b^2)x^2+(2a^2-b^2)y^2\\ \quad +c^2(a^2+b^2)w^2 \end{array}\right\}2c^2(z^2-k^2w^2)^3\\ &+c^4(z^2-k^2w^2)^4=0. \end{aligned}$$

I remark that the before-mentioned nodal conic $W=0$, $(a^2-k^2)X^2+(b^2-k^2)Y^2-k^2Z^2=0$ is the reciprocal of a quadric cone, which is a bitangent cone of the ring: this is a cone, vertex at the centre of the ring, and which is the envelope of the right cone, vertex the same point, circumscribed about the variable sphere which generates the ring.

Centro-surface of a paraboloid.

For the paraboloid $\dfrac{X^2}{a}+\dfrac{Y^2}{b}-2ZW=0$, it may be shown that the centro-surface is the envelope of the quadric

$$\frac{ax^2}{(a+\theta)^2}+\frac{by^2}{(b+\theta)^2}-2zw-2\theta w^2=0.$$

The quartic surface is consequently the envelope of the quadric

$$\frac{(a+\theta)^2}{a}X^2+\frac{(b+\theta)^2}{b}Y^2+2\theta Z^2-2ZW=0,$$

viz. this is

$$\theta^2\left(\frac{X^2}{a}+\frac{Y^2}{b}\right)+2\theta\,(X^2+Y^2+Z^2)+aX^2+bY^2-2ZW=0.$$

Hence the quartic surface is

$$\left(\frac{X^2}{a}+\frac{Y^2}{b}\right)(aX^2+bY^2-2ZW)-(X^2+Y^2+Z^2)^2=0,$$

or, what is the same thing,

$$X^2Y^2(a-b)^2-2ZW(bX^2+aY^2)-2abZ^2(X^2+Y^2)-abZ^4=0.$$

This has four conic nodes; viz. considering the equations

$$\frac{X^2}{a}+\frac{Y^2}{b}=0,\quad aX^2+bY^2-2ZW=0,\quad X^2+Y^2+Z^2=0,$$

these give the point $X=0$, $Y=0$, $Z=0$ four times, and four other points which are the nodes in question; the point $(X=0,\ Y=0,\ Z=0)$ is a singular point of a higher order; the reduction caused by these singularities should be $=8+19$, so as to make the order of the surface of centres $=9$; that is the reduction on account of the point $(X=0,\ Y=0,\ Z=0)$ must be $=19$; but it is not by any means obvious how this is so.

Parallel surface of the paraboloid.

This is given, Salmon's *Solid Geometry*, 2nd Edit., pp. 146 and 148, [Ed. 4, p. 180], for the paraboloid $aX^2+bY^2+2rZW=0$, as the envelope of the quadric surface

$$\frac{\theta ax^2}{\theta a+1}+\frac{\theta by^2}{\theta b+1}+2\theta rzw-(\theta^2r^2+k^2)\,w^2\quad=0.$$

The reciprocal quartic is thus the envelope of

$$\frac{\theta a+1}{\theta a}X^2+\frac{\theta b+1}{\theta b}Y^2+\frac{\theta^2r^2+k^2}{\theta^2r^2}Z^2+\frac{2}{\theta r}ZW=0,$$

that is

$$(X^2+Y^2+Z^2)+\frac{1}{\theta}\left(\frac{X^2}{a}+\frac{Y^2}{b}+\frac{2}{r}ZW\right)+\frac{1}{\theta^2}\frac{k^2}{r^2}Z^2=0,$$

whence the equation is

$$4\frac{k^2}{r^2}Z^2(X^2+Y^2+Z^2)-\left(\frac{X^2}{a}+\frac{Y^2}{b}+\frac{2}{r}ZW\right)^2=0,$$

viz. this is a quartic having the nodal line-pair $Z=0$, $\dfrac{X^2}{a}+\dfrac{Y^2}{b}=0$; and a further singularity at the point $X=0$, $Y=0$, $Z=0$. It would require some consideration to show that the order of the parallel surface is thence $=10$, as it should be.

Envelope of the planes through the points of an ellipsoid at right angles to the radius vectors from the centre.

This is given in my paper "Sur la surface &c." in the *Annali di Matematica*, t. II. (1859), [250], as the envelope of the quadric surface

$$\frac{x^2}{2-\frac{\theta}{a^2}}+\frac{y^2}{2-\frac{\theta}{b^2}}+\frac{z^2}{2-\frac{\theta}{c^2}}-\theta w^2=0.$$

The reciprocal quartic surface is thus the envelope of

$$\left(2-\frac{\theta}{a^2}\right)X^2+\left(2-\frac{\theta}{b^2}\right)Y^2+\left(2-\frac{\theta}{c^2}\right)Z^2-\frac{1}{\theta}W^2=0,$$

or, what is the same thing,

$$\theta\left(\frac{X^2}{a^2}+\frac{Y^2}{b^2}+\frac{Z^2}{c^2}\right)-2(X^2+Y^2+Z^2)+\frac{1}{\theta}W^2=0,$$

viz. this is

$$W^2\left(\frac{X^2}{a^2}+\frac{Y^2}{b^2}+\frac{Z^2}{c^2}\right)-(X^2+Y^2+Z^2)^2=0,$$

which is in fact the inverse surface

$$\left(\frac{X}{X^2+Y^2+Z^2},\ \frac{Y}{X^2+Y^2+Z^2},\ \frac{Z}{X^2+Y^2+Z^2}\text{ for } X, Y, Z\right)$$

of the ellipsoid $\frac{X^2}{a^2}+\frac{Y^2}{b^2}+\frac{Z^2}{c^2}=1$; this is obvious geometrically inasmuch as the reciprocal of the variable plane is the inverse of the point on the ellipsoid.

The quartic surface has the nodal conic

$$W=0,\quad X^2+Y^2+Z^2=0;$$

and also the node $X=0$, $Y=0$, $Z=0$; there is consequently in the order of the reciprocal surface a reduction $24+2=26$, or the order of the reciprocal surface is $=10$.

Centro-surface of the ellipsoid.

Writing the equation of the ellipsoid in the form $\frac{x^2}{a^2}+\frac{y^2}{b^2}+\frac{z^2}{c^2}-w^2=0$, the centro-surface is given as the envelope of the quadric surface

$$\frac{a^2x^2}{(\theta+a^2)^2}+\frac{b^2y^2}{(\theta+b^2)^2}+\frac{c^2z^2}{(\theta+c^2)^2}-w^2=0,$$

(Salmon, [Ed. 2], p. 400, [Ed. 4, p. 179]), and hence the reciprocal quartic surface is the envelope of

$$\left(a+\frac{\theta}{a}\right)^2X^2+\left(b+\frac{\theta}{b}\right)^2Y^2+\left(c+\frac{\theta}{c}\right)^2Z^2-W^2=0,$$

in regard to the variable parameter θ, viz. the equation is

$$\left(\frac{X^2}{a^2}+\frac{Y^2}{b^2}+\frac{Z^2}{c^2}\right)(a^2X^2+b^2Y^2+c^2Z^2-W^2)-(X^2+Y^2+Z^2)^2=0,$$

(see Salmon, [Ed. 2], p. 144 [Ed. 4, p. 172]). It hence at once appears, that the quartic surface has 12 nodes, viz. these are the four angles of the fundamental tetrahedron $(XYZW)$, and the eight points

$$\begin{cases}\dfrac{X^2}{a^2}+\dfrac{Y^2}{b^2}+\dfrac{Z^2}{c^2}=0,\\ X^2+Y^2+Z^2=0,\\ a^2X^2+b^2Y^2+c^2Z^2-W^2=0,\end{cases}$$

or writing as it is convenient to do

$$(\alpha,\ \beta,\ \gamma)=(b^2-c^2,\ c^2-a^2,\ a^2-b^2);$$

and therefore

$$\alpha+\beta+\gamma=0,\quad a^2\alpha+b^2\beta+c^2\gamma=0,\quad a^4\alpha+b^4\beta+c^4\gamma=-\alpha\beta\gamma;$$

these are the eight points

$$\frac{X^2}{W^2}=-\frac{a^2}{\beta\gamma},\quad \frac{Y^2}{W^2}=-\frac{b^2}{\gamma\alpha},\quad \frac{Z^2}{W^2}=-\frac{c^2}{\alpha\beta};$$

the order of the reciprocal of the quartic surface is thus $36-2.12$, $=12$, which is in fact the order of the surface of centres.

The equation of the centro-surface is given, Salmon, [Ed. 2], p. 151, and *Quart. Math. Jour.*, t. II. (1858), p. 220, in the form

$$(\alpha,\ \beta,\ \gamma)^6(\xi,\ \eta,\ \zeta,\ \omega)^{12}=0,$$

where ξ, η, ζ, ω stand for ax, by, cz, iw; it is therefore of the degree 18 in regard to a, b, c.

Parallel surface of the ellipsoid.

This is given, Salmon, [Ed. 2], p. 148 [Ed. 4, p. 176], as the envelope of the quadric surface

$$\frac{x^2}{a^2+\theta}+\frac{y^2}{b^2+\theta}+\frac{z^2}{c^2+\theta}-\left(1+\frac{k^2}{\theta}\right)w^2=0.$$

The reciprocal quartic is thus the envelope of

$$(a^2+\theta)X^2+(b^2+\theta)Y^2+(c^2+\theta)Z^2-\frac{\theta W^2}{k^2+\theta}=0,$$

or writing $k^2+\theta=\lambda$, this is

$$(a^2-k^2+\lambda)X^2+(b^2-k^2+\lambda)Y^2+(c^2-k^2+\lambda)Z^2-\left(1-\frac{k^2}{\lambda}\right)W^2=0,$$

or, what is the same thing,

$$\lambda^2(X^2+Y^2+Z^2)+\lambda\,[(a^2-k^2)X^2+(b^2-k^2)Y^2+(c^2-k^2)Z^2-W^2]+k^2W^2=0,$$

whence the equation is

$$\{(a^2-k^2)X^2+(b^2-k^2)Y^2+(c^2-k^2)Z^2-W^2\}^2-4k^2W^2(X^2+Y^2+Z^2)=0,$$

viz. this is a quartic having the nodal conic

$$W=0,\quad (a^2-k^2)X^2+(b^2-k^2)Y^2+(c^2-k^2)Z^2=0.$$

The order of the reciprocal or parallel surface is thus $36-24$, $=12$, as it should be. The nodal conic of the quartic surface is the reciprocal of a bitangent or node-couple quadric cone, vertex the centre, in the parallel surface: this cone is imaginary for the ellipsoid, but real for either of the hyperboloids, and its existence in the case of the hyperboloid is readily perceived.

Reverting to the equation

$$\frac{x^2}{a^2+\theta}+\frac{y^2}{b^2+\theta}+\frac{z^2}{c^2+\theta}-\left(1+\frac{k^2}{\theta}\right)w^2=0,$$

or say

$$(a^2+\theta)(b^2+\theta)(c^2+\theta)(k^2+\theta)\,w^2$$
$$-x^2(b^2+\theta)(c^2+\theta)\,\theta-y^2(c^2+\theta)(a^2+\theta)\,\theta-z^2(a^2+\theta)(b^2+\theta)\,\theta=0,$$

this is

$$(A,\ B,\ C,\ D,\ E\between\theta,\ 1)^4=0,$$

where putting for shortness

$$\begin{aligned}\alpha&=a^2+b^2+c^2+k^2,\\ \beta&=b^2c^2+c^2a^2+a^2b^2+k^2(a^2+b^2+c^2),\\ \gamma&=a^2b^2c^2+k^2(b^2c^2+c^2a^2+a^2b^2),\\ \delta&=a^2b^2c^2k^2,\end{aligned}$$

and

$$\begin{aligned}p&=x^2+y^2+z^2,\\ q&=(b^2+c^2)x^2+(c^2+a^2)y^2+(a^2+b^2)z^2,\\ r&=b^2c^2x^2+c^2a^2y^2+a^2b^2z^2,\end{aligned}$$

we have

$$\begin{aligned}A&=12w^2,\\ B&=3\alpha w^2-3p,\\ C&=2\beta w^2-2q,\\ D&=3\gamma w^2-3r,\\ E&=12\delta w^2.\end{aligned}$$

The equation of the parallel surface is of course

$$(AE-4BD+3C^2)^3-27\,(ACE-AD^2-B^2E+2BCD-C^3)^2=0.$$

It is remarked (Salmon, [Ed. 2], p. 148 [Ed. 4, p. 176]) that there is in the plane $z=0$, a nodal conic $\frac{x^2}{a-c}+\frac{y^2}{b-c}-\left(1+\frac{k^2}{c}\right)w^2=0$, the complete section being made up of this conic twice, and of the curve of the eighth order which is the parallel curve of the ellipse $\frac{x^2}{a^2}+\frac{y^2}{b^2}-w^2=0$; the like is of course the case as to the sections by the other two principal planes $x=0$ and $y=0$. For the section by the plane $w=0$ (or plane infinity) we have at once $p^2r^2(4pr-q^2)=0$, where observe that

$$\begin{aligned} q^2-4pr &= \{(b^2+c^2)x^2+(c^2+a^2)y^2+(a^2+b^2)z^2\}^2-4(x^2+y^2+z^2)(b^2c^2x^2+c^2a^2y^2+a^2b^2z^2), \\ &=(1,\ 1,\ 1,\ -1,\ -1,\ -1\between(b^2-c^2)x^2,\ (c^2-a^2)y^2,\ (a^2-b^2)z^2)^2 \\ &=\text{norm.}\ \{x\sqrt{(b^2-c^2)}+y\sqrt{(c^2-a^2)}+z\sqrt{(a^2-b^2)}\}. \end{aligned}$$

The section is thus made up of two conics, each twice, and of four right lines: viz. the conics are $x^2+y^2+z^2=0$, the circle at infinity and $\frac{x^2}{a^2}+\frac{y^2}{b^2}+\frac{z^2}{c^2}=0$, the section at infinity of the ellipsoid; and the lines are

$$x\sqrt{(b^2-c^2)}\pm y\sqrt{(c^2-a^2)}\pm z\sqrt{(a^2-b^2)}=0,$$

viz. these are the common tangents of the two conics. The circle at infinity is a nodal conic on the surface, which has thus 4 nodal conics.

488.

NOTE ON A RELATION BETWEEN TWO CIRCLES.

[From the *Quarterly Journal of Pure and Applied Mathematics*, vol. XI. (1871), pp. 82, 83.]

CONSIDER any two circles O, Q; and let AC, BD, $A'D'$, $B'C'$ be the common tangents touching the circles in the points A, A', B, B', C, C', D, D': the locus of a point P such that the pairs of tangents from it to the two circles respectively form a harmonic pencil, is a conic through the 8 points A, A', B, B', C, C', D, D'; but this conic may break up into two lines, viz. if (as in the figure) the points A, B', D', D are in a line, then the points C, C', A', B will be in a symmetrically situated line, and the conic breaks up into this pair of lines, meeting suppose in K. The condition

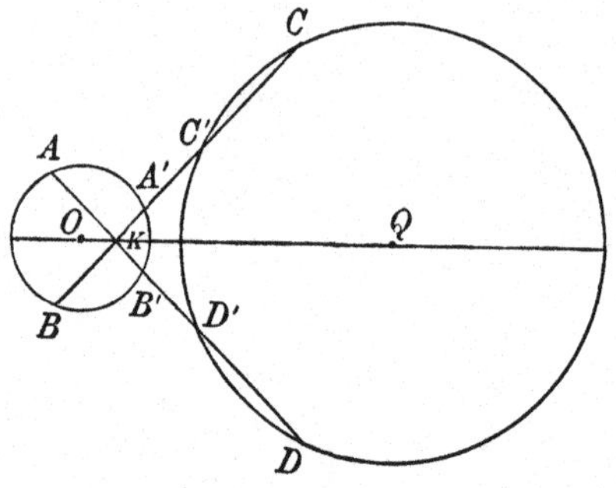

for this, if a, a' are the distances of the centres from a fixed point in the line of centres, and if the radii are c, c', is readily found to be

$$(a - a')^2 = 2 (c^2 + c'^2).$$

Suppose in general, that (given any two conics) the point P' is the intersection of the polars of P in regard to the two given conics respectively; then if P describes a line, the locus of P' is a conic passing through the three conjugate points of the given conics; if, however, the line which is the locus of P pass through one of the conjugate points, then the conic the locus of P' breaks up into a pair of lines, one of them a fixed line through the other two conjugate points, the other of them a line through the first-mentioned conjugate point. That is, if the locus of P be a line through a conjugate point, the locus of P' is a line through the same conjugate point; but in every other case the locus of P' is a conic.

Reverting to the figure of the two circles, in order that it may be possible that the two lines AD and BC may be loci of points P, P', related as above, it is necessary that K shall be a conjugate point of the two circles; that is, if the two circles intersect in points Λ, Λ' lying symmetrically in the radical axis, which meets, suppose, the line of centres in M, then it is necessary that K shall be one of the anti-points of Λ, Λ'; or, what is the same thing, the distance KM must be $=i$ into $M\Lambda$ or $M\Lambda'$; this condition, if as above $(a-a')^2=2(c^2+c'^2)$, implies $c^2=c'^2$, and we have then $(a-a')^2=4c^2$, that is, the circles must be equal, and the distance of the centres must be twice the radius, or, what is the same thing, the circles must be equal circles touching each other; when this is so, the two lines AD, BC (being then lines at right angles to each other intersecting in the point of contact), have, in fact, the above-mentioned relation. And it thus appears that given two circles, the necessary and sufficient conditions for the coexistence of the properties mentioned in the theorem are that they shall be equal circles touching each other.

489.

ON THE PORISM OF THE IN-AND-CIRCUMSCRIBED POLYGON, AND THE (2, 2) CORRESPONDENCE OF POINTS ON A CONIC.

[From the *Quarterly Journal of Pure and Applied Mathematics*, vol. XI. (1871), pp. 83—91.]

THE present paper includes, as will at once be seen, much that is perfectly well known; but the separate theories required, it seemed to me, to be put together; and there are, particularly as regards the unsymmetrical case afterwards referred to, some results which I believe to be new.

The porism of the in-and-circumscribed polygon has its foundation in the theory of the symmetrical (2, 2) correspondence of points on a conic; viz. a (2, 2) correspondence is such that to any given position of either point there correspond two positions of the other point; and in a symmetrical (2, 2) correspondence either point indifferently may be considered as the first point and the other of them will then be the second point of the correspondence. Or, what is the same thing, if x, y are the parameters which serve to determine the two points, then x, y are connected by an equation of the form $(*\between x, 1)^2(y, 1)^2=0$, which is symmetrical in regard to the two parameters (x, y). In the case of such symmetrical relation it is easy to show that the line joining the two points envelopes a conic. For the relation may be expressed in the form $(*\between 1, x+y, xy)^2=0$; we may imagine the coordinates (P, Q, R) fixed in such manner that for the point (x) on the first conic we have $P : Q : R = 1 : x : x^2$, and for the point (y), $P : Q : R = 1 : y : y^2$; the equation of the line joining the two points is then

$$\begin{vmatrix} P, & Q, & R \\ 1, & x, & x^2 \\ 1, & y, & y^2 \end{vmatrix} = 0;$$

that is

$$Pxy - Q(x+y) + R = 0,$$

or representing this by

$$P\xi + Q\eta + R\zeta = 0,$$

we have $\xi : \eta : \zeta = xy : -x-y : 1$; and consequently (ξ, η, ζ) are connected by a quadric equation; that is, the envelope is a conic.

The relation $(*\between x, 1)^2 (y, 1)^2 = 0$, whether symmetrical or not, leads as will be presently shown to a differential equation of the form

$$\frac{dx}{\sqrt{(X)}} \pm \frac{dy}{\sqrt{(Y)}} = 0,$$

where X, Y are quartic functions of x, y respectively; viz. these are unlike or like functions of the two variables according as the integral equation is not or is symmetrical in regard to the two variables. In the former case, however, the functions X, Y are so related to each other, that the two can be by a linear transformation converted into like functions of the variables: for instance, if y be changed into $ay_1 + b \div cy_1 + d$, then the constants may be determined in suchwise that Y is the same function of y_1, that X is of x; the original integral equation being hereby converted into a symmetrical equation $(*\between x, 1)^2 (y_1, 1)^2 = 0$ between x and y_1, so that in one point of view the unsymmetrical case is not really more general than the symmetrical one. It is to be added that the integral equation contains really one more constant than the differential equation (this is most readily seen in the symmetrical case, the differential equation depends only on the ratio of five constants a, b, c, d, e, whereas the integral equation depends on the ratio of six constants), so that the integral equation is really the complete integral of the differential equation.

Attending now to the symmetrical case; if A and B are corresponding points, then the corresponding points of B are A and a new point C; those of C are B and a new point D, and so on; so that the points form a series $A, B, C, D, \ldots$; and the porismatic property is that, if for a given position of A this series closes at a certain term, for instance, if $D = A$, then it will always thus close, whatever be the position of A. And this follows at once from the consideration of the differential equation $\frac{dx}{\sqrt{(X)}} = \frac{dy}{\sqrt{(Y)}}$; viz. as this is at once integrable *per se* in the form

$$\Pi(y) - \Pi(x) = \Pi(k),$$

this equation must be a transformation of the original equation $(*\between x, 1)^2 (y, 1)^2 = 0$, and equally with it represent the relation between the parameters x, y of the two points A, B; the constant of integration k is of course completely determined in terms of the coefficients of the last-mentioned equation, assumed to be given.

Hence forming the equations for the correspondences, B, C; C, D;... and assuming that the series closes F, A; we have

$$\Pi(z) - \Pi(y) = \Pi(k),$$

$$\vdots$$

$$\Pi(x) - \Pi(u) = \Pi(k);$$

where, however, the $\Pi(x)$ of the last equation must be regarded as differing from that of the first equation by a period, say Ω, of the integral; hence adding, we have

$$\Omega = n\Pi(k),$$

or

$$\Pi(k) = \frac{1}{n}\Omega,$$

which gives between the constants of the integral equation $(*\between x, 1)^2 (y, 1)^2 = 0$, a relation which must be satisfied when the series closes at the nth term (viz. when the term after this coincides with the first term); and this relation is independent of x, that is, of the position of the point A.

The analysis in regard to the differential equation is as follows:

Consider the equation

$$\begin{aligned} U = \quad & y^2 (ax^2 + 2bx + c) \\ & + 2y\,(a'x^2 + 2b'x + c') \\ & + \quad (a''x^2 + 2b''x + c'') = 0, \end{aligned}$$

say

$$U = (P, Q, R\between y, 1)^2 = (L, M, N\between x, 1)^2 = 0,$$

we have

$$dU = 0 = (Py + Q)\,dy + (Lx + M)\,dx.$$

But the equation $U = 0$ gives $(Py + Q)^2 = Q^2 - PR$, $(Lx + M)^2 = M^2 - NL$, and the differential equation therefore becomes

$$dy\,\sqrt{(Q^2 - PR)} \pm dx\,\sqrt{(M^2 - NL)} = 0,$$

viz. it is

$$\frac{dy}{\sqrt{\{(ay^2 + 2a'y + a'')(cy^2 + 2c'y + c'') - (bx^2 + 2b'y + b'')^2\}}}$$

$$\pm \frac{dx}{\sqrt{\{(ax^2 + 2bx + c)(a''x^2 + 2b''x + c'') - (a'x^2 + 2b'x + c')^2\}}} = 0.$$

Suppose the equation is

$$\begin{aligned} & y^2 (ax^2 + 2hx + g) \\ & + 2y\,(hx^2 + 2bx + f) \\ & + \quad (gx^2 + 2fx + c) = 0, \end{aligned}$$

then the differential equation is

$$\frac{dy}{\sqrt{\{(ay^2 + 2hy + g)(gy^2 + 2fy + c)^2 - (hy^2 + 2by + f)^2\}}}$$

$$\pm \frac{dx}{\sqrt{\{(ax^2 + 2hx + g)(gx^2 + 2fx + c) - (hx^2 + 2bx + f)^2\}}} = 0,$$

say

$$\frac{dy}{\sqrt{(Y)}} = \frac{\pm\, dx}{\sqrt{(X)}}.$$

Now starting from the differential equation

$$\frac{dx}{\sqrt{\{(a,\ b,\ c,\ d,\ e\between x,\ 1)^4\}}} = \pm\frac{dy}{\sqrt{\{(a,\ b,\ c,\ d,\ e\between y,\ 1)^4\}}},$$

the integral equation is known to be

$$\left[\frac{\sqrt{\{(a,\ b,\ c,\ d,\ e\between x,\ 1)^4\}} - \sqrt{\{(a,\ b,\ c,\ d,\ e\between y,\ 1)^4\}}}{x-y}\right]^2 = a(x+y)^2 + 4b(x+y) + 6\theta,$$

where θ is the constant of integration. Writing, for shortness, $X = (a,\ b,\ c,\ d,\ e\between x,\ 1)^4$, $Y = (a,\ b,\ c,\ d,\ e\between y,\ 1)^4$, this is

$$X + Y - 2\sqrt{(XY)} = a(x^2-y^2)^2 + 4b(x-y)(x^2-y^2) + 6\theta(x-y)^2;$$

or, what is the same thing,

$$\begin{aligned} a(x^4+y^4) - 2\sqrt{(XY)} &= a(x^2-y^2)^2 + 4b(x-y)(x^2-y^2) + 6\theta(x-y)^2, \\ + 4b(x^3+y^3) & \\ + 6c(x^2+y^2) & \\ + 4d(x+y) & \\ + 2e, & \end{aligned}$$

viz. this gives

$$\begin{aligned} \sqrt{(XY)} = {} & ax^2y^2 \\ & + 2b(x^2y + xy^2) \\ & + 3c(x^2 + y^2) \\ & + 3\theta(x - y)^2 \\ & + 2d(x + y) \\ & + e, \end{aligned}$$

and, rationalising, the integral equation becomes

$$\begin{aligned} & - 6a\theta x^2y^2 \\ & - 4adxy(x+y) \\ & - ae(x+y)^2 \\ & + 4b^2x^2y^2 \\ & + 12bcxy(x+y) - 12b\theta xy(x+y) \\ & - 8bdxy \\ & - 4be(x+y) \\ & + 9c^2(x+y)^2 - 18c\theta(x^2+y^2) \\ & - 12cd(x+y) \\ & + 9\theta^2(x-y)^2 - 12d\theta(x+y) - 6e\theta + 4d^2 = 0; \end{aligned}$$

or, as it may be written,

$$\begin{aligned}
&x^2y^2\,(4b^2-6a\theta)\\
&+(x^2y+xy^2)\,(-4ad+12bc-12b\theta)\\
&+(x^2\;+y^2)\;\;(-ae+9c^2-18c\theta+9\theta^2)\\
&+\qquad xy\;\;\;(-2ae-8bd+18c^2-18\theta^2)\\
&+(x\;\;+y)\;\;\;(-4be+12cd-12d\theta)\\
&+\qquad\qquad 4d^2-6e\theta=0.
\end{aligned}$$

Comparing this with the original integral equation $V=0$, and the form of differential equation deduced therefrom, we ought to have identically

$$\begin{aligned}
&[(4b^2-6a\theta)\,x^2+(-2ad+6bc-6b\theta)\,x+(-ae+9c^2-18c\theta+9\theta^2)]\\
&\qquad\times[(-ae+9c^2-18c\theta+9\theta^2)\,x^2+(-2be+6cd-6d\theta)\,x+(4d^2-6e\theta)]\\
&-[(-2ad+6bc-6b\theta)\,x^2+(-ae-4bd+9c^2-9\theta^2)\,x+(-2be+6cd-6d\theta)]^2\\
&=\text{multiple of } X,\\
&=\{(-4ad^2-4b^2e+24bcd)+(6ae-24bd-54c^2)\,\theta+108c\theta^2-54\theta^3\}\,(a,\ b,\ c,\ d,\ e\,\check{\curlyvee}\,x,\ 1)^4,
\end{aligned}$$

by comparing the coefficients of x^4.

I obtain this otherwise:

Write

$$V=\alpha U+6\beta H,$$

then, forming the Hessian of V, we have

$$\begin{aligned}
\tilde{H}V&=(\alpha^2-3I\beta^2)\,H+(I\alpha\beta+9J\beta^2)\,U,\\
&=\frac{(\alpha^2-3I\beta^2)}{6\beta}(V-\alpha U)+(I\alpha\beta+9J\beta^2)\,U,\\
&=\frac{\alpha^2-3I\beta^2}{6\beta}V+\frac{1}{6\beta}(-\alpha^3+9I\alpha\beta^2+54J\beta^3)\,U,
\end{aligned}$$

that is

$$d_x{}^2Vd_y{}^2V-(d_xd_yV)^2-\frac{2\,(\alpha^2-3I\beta^2)}{\beta}(x^2d_x{}^2V+2xyd_xd_yV+y^2d_y{}^2V)=\frac{24}{\beta}(-\alpha^3+9I\alpha\beta^2+54J\beta^3)\,U,$$

or writing

$$K=-\frac{2\,(\alpha^2-3I\beta^2)}{\beta},$$

this is

$$(d_x{}^2V+Ky^2)(d_y{}^2V+Kx^2)-(d_xd_yV-Kxy)^2=\frac{24}{\beta}(-\alpha^3+9I\alpha\beta^2+54J\beta^2)\,U,$$

so that the components are

$$d_x{}^2V+Ky^2,\quad d_xd_yV-Kxy,\quad d_y{}^2V+Kx^2,$$

$V=\alpha U+6\beta H=$

$$\alpha\,(a,\ b,\ c,\ d,\ e\,\check{\curlyvee}\,x,\ 1)^4+6\beta\,(ac-b^2,\ 2ad-2bc,\ ae+2bd-3c^2,\ 2be-2cd,\ ce-d^2\,\check{\curlyvee}\,x,\ 1)^4,$$

viz. the components are

$$\begin{aligned}
&(\alpha a + 6\beta(ac - b^2), && ab + 3\beta(ad - bc), && ac + \beta(ae + 2bd - 3c^2) + \tfrac{1}{12}K \between x,\ 1)^2,\\
&(\alpha b + 3\beta(ad - bc), && ac + \beta(ae + 2bd - 3c^2) + \tfrac{1}{24}K, && ad + 3\beta(be - cd) \between x,\ 1)^2,\\
&(\alpha c + \ \beta(ae + 2bd - 3c^2) + \tfrac{1}{12}K, && ad + 3\beta(be - cd), && ae + 6\beta(ce - d^2) \between x,\ 1)^2,
\end{aligned}$$

where as before

$$K = -\frac{2(\alpha^2 - 3I\beta^2)}{\beta}.$$

I assume

$$\beta = -\tfrac{2}{3},\quad \alpha = 4c - 6\theta,\quad K = 3\{(4c - 6\theta)^2 - \tfrac{4}{3}I\}.$$

$$\begin{aligned}
\alpha a + 6\beta(ac - b^2) &= a(4c - 6\theta) - 4(ac - b^2) = 4b^2 - 6a\theta,\\
\alpha b + 3\beta(ad - bc) &= b(4c - 6\theta) - 2(ad - bc) = -2ad + 6bc - 6b\theta,\\
\alpha d + 3\beta(be - cd) &= d(4c - 6\theta) - 2(be - cd) = -2be + 6cd - 6d\theta,\\
\alpha e + 6\beta(ce - d^2) &= e(4c - 6\theta) - 4(ce - d^2) = \ \ 4d^2 - 6e\theta,\\
\alpha c + \ \beta(ae + 2bd - 3c^2) - \tfrac{1}{24}K &= c(4c - 6\theta) - \tfrac{2}{3}(ae + 2bd - 3c^2) - \tfrac{1}{8}\{(4c - 6\theta)^2 - \tfrac{4}{3}I\}\\
&= -\tfrac{1}{2}ae - 2bd + \tfrac{9}{2}c^2 - \tfrac{9}{2}\theta^2,
\end{aligned}$$

$$\begin{aligned}
&\alpha c + \beta(ae + 2bd - 3c^2) + \tfrac{1}{12}K\\
&\qquad = c(4c - 6\theta) - \tfrac{2}{3}(ae + 2bd - 3c^2) + \tfrac{1}{4}\{(4c - 6\theta)^2 - \tfrac{4}{3}I\}\\
&\qquad = -ae + 9c^2 - 18c\theta + 9\theta^2,
\end{aligned}$$

agreeing with the former result.

I return to the general form

$$\begin{aligned}
&y^2(a\ ,\ b\ ,\ c\ \between x,\ 1)^2\\
+\ &2y(a',\ b',\ c' \between x,\ 1)^2\\
+\ &\ \ \ (a'',\ b'',\ c'' \between x,\ 1)^2 = 0,
\end{aligned}$$

giving

$$\frac{dx}{\sqrt{[(a,\ b,\ c \between x,\ 1)^2 (a'',\ b'',\ c'' \between x,\ 1)^2 - \{(a',\ b',\ c' \between x,\ 1)^2\}^2]}}$$

$$= \frac{dy}{\sqrt{[(a,\ a',\ a'' \between y,\ 1)^2 (c,\ c',\ c'' \between y,\ 1)^2 - \{(b,\ b',\ b'' \between y,\ 1)^2\}^2]}}.$$

Operate a linear transformation on the x, say

$$x = \frac{\lambda x' + \mu}{\nu x' + \rho};$$

the new coefficients are

$$(a\,,\,b\,,\,c\between\lambda,\,\nu)^2,\quad (a\,,\,b\,,\,c\between\lambda,\,\nu\between\mu,\,\rho),\quad (a\,,\,b\,,\,c\between\mu,\,\rho)^2,$$
$$(a',\,b',\,c'\between\lambda,\,\nu)^2,\quad (a',\,b',\,c'\between\lambda,\,\nu\between\mu,\,\rho),\quad (a',\,b',\,c'\between\mu,\,\rho)^2,$$
$$(a'',\,b'',\,c''\between\lambda,\,\nu)^2,\quad (a'',\,b'',\,c''\between\lambda,\,\nu\between\mu,\,\rho),\quad (a'',\,b'',\,c''\between\mu,\,\rho)^2:$$

assume now

$$\begin{aligned} a'\lambda + (b'-\theta)\,\nu - a\mu - b\rho &= 0,\\ (b'+\theta)\,\lambda + c'\nu - b\mu - c\rho &= 0,\\ a''\lambda + b''\nu - a'\mu - (b'+\theta)\,\rho &= 0,\\ b''\lambda + c''\nu - (b'-\theta)\,\mu - c'\rho &= 0,\end{aligned}$$

then it is easy to show that

$$\begin{aligned}(a,\,b,\,c\between\lambda,\,\nu\between\mu,\,\rho) &= (a',\,b',\,c'\between\lambda,\,\nu)^2,\\ (a',\,b',\,c'\between\mu,\,\rho)^2 &= (a'',\,b'',\,c''\between\lambda,\,\nu\between\mu,\,\rho),\\ (a,\,b,\,c\between\mu,\,\rho)^2 &= (a'',\,b'',\,c''\between\lambda,\,\nu)^2\\ &[= (a',\,b',\,c'\between\lambda,\,\nu\between\mu,\,\rho) + \theta\,(\lambda\rho-\mu\nu)],\end{aligned}$$

and the equations give

$$\begin{vmatrix} a' & b'-\theta, & a & b \\ b'+\theta, & c' & b & c \\ a'' & b'' & a' & b'+\theta \\ b'' & c'' & b'-\theta, & c' \end{vmatrix} = 0,$$

that is

$$\begin{aligned}&(a'c'-b'^2+\theta^2)^2 + (a''c''-b''^2)\,(ac-b^2)\\ &-(a'b''-a''b'+a''\theta)\,(bc'-b'c+c\theta)\\ &+(a'c''-b''b'+b''\theta)\,(bb'-a'c+b\theta)\\ &+(b'b''-a''c'+b''\theta)\,(ac'-b'b+b\theta)\\ &-(b'c''-b''c'+c''\theta)\,(ab'-a'b+a\theta)=0,\end{aligned}$$

which is

$$\begin{aligned}&(a'c'-b'^2)^2+(a''c''-b''^2)\,(ac-b^2) && -2\theta\begin{vmatrix} a, & b, & c\\ a', & b', & c'\\ a'', & b'', & c''\end{vmatrix}\\ &+a'^2\,(-cc'')\\ &+b'^2\,(-ac''-2bb''-a''c)\\ &+c'^2\,(-aa'')\\ &+2b'c'\,(ab''+a''b) && +\theta^2\big(2\,(a'c'-b'^2)-ac''+2bb''-a''c\big)\\ &+2c'a'\,(-bb'')\\ &+2a'b'\,(bc''+b''c) && +\theta^4=0.\end{aligned}$$

If the original matrix be symmetrical $= \begin{vmatrix} a, & h, & g \\ h, & b, & f \\ g, & f, & c \end{vmatrix}$, this is

$$\begin{aligned}
&(fh-b^2)^2+(ag-h^2)(cg-f^2) && -2\theta \begin{vmatrix} a, & h, & g \\ h, & b, & f \\ g, & f, & c \end{vmatrix} \\
&+h^2(-cg) && \\
&+b^2(-ac-g^2-2fh) && \\
&+f^2(-ag) && \\
&+2bf(af+gh) && +\theta^2[2(fh-b^2)-ac-g^2+2hf] \\
&+2fh(-fh) && \\
&+2bh(fg+ch) && +\theta^4=0,
\end{aligned}$$

that is

$$(b-g)\{(b^2-ac)(b+g)+2(af^2+ch^2-2bfh)\}$$
$$-2\theta(abc-af^2-bg^2-ch^2+2fgh)+\theta^2(4fh-2b^2-ac-g^2)+\theta^4=0,$$

satisfied by

$$\theta+b-g=0,$$

viz. the equation in θ is

$$(\theta+b-g)\{\theta^3-(b-g)\theta^2+(4fh-ac-2bg-b^2)\theta+(b^2-ac)(b+g)+2(af^2+ch^2-2bfh)\}=0.$$

490.

ON A PROBLEM OF ELIMINATION.

[From the *Quarterly Journal of Pure and Applied Mathematics*, vol. XI. (1871), pp. 99—101.]

I WRITE

$$P = (\alpha, \dots \between x, y, z)^k, \quad Q = (\alpha', \dots \between x, y, z)^k,$$
$$U = (a, \dots \between x, y, z)^m, \quad V = (b, \dots \between x, y, z)^n,$$

and I seek for the form of the relation between the coefficients $(\alpha, \dots)$, $(\alpha', \dots)$, $(a, \dots)$, $(b, \dots)$, in order that there may exist in the pencil

$$\lambda P + \mu Q = 0$$

a curve passing through *two* of the intersections of the curves $U = 0$, $V = 0$.

The ratio $\lambda : \mu$ may be determined so as that the curve $\lambda P + \mu Q = 0$ shall pass through one of the intersections of the curves $U = 0$, $V = 0$; or, what is the same thing, so as that the three curves shall have a common point; the condition for this is

$$\text{Reslt.} (\lambda P + \mu Q, U, V) = 0,$$

a condition of the form

$$(\lambda\alpha + \mu\alpha', \dots)^{mn} (a, \dots)^{kn} (b, \dots)^{km} = 0;$$

or, what is the same thing,

$$(\alpha, \dots, \alpha', \dots)^{mn} (a, \dots)^{kn} (b, \dots)^{km} (\lambda, \mu)^{mn} = 0,$$

which, for shortness, may be written

$$(A, \dots \between \lambda, \mu)^{mn} = 0.$$

Suppose this equation has equal roots, then we have

$$\text{Disct. Reslt. } (\lambda P + \mu Q, \ U, \ V) = 0,$$

the discriminant being taken in regard to λ, μ. This is of the form

$$(A, \ldots)^{2 \, (mn-1)} = 0;$$

that is

$$(a, \ldots, \ \alpha', \ldots)^{2mn \, (mn-1)} \, (a, \ldots)^{2kn \, (mn-1)} \, (b, \ldots)^{2km \, (mn-1)} = 0.$$

It is moreover clear that the nilfactum is a combinant of the functions P, Q; and the form of the equation is therefore

$$\left(\left\| \begin{array}{cc} \alpha, & \beta, \ldots \\ \alpha', & \beta', \ldots \end{array} \right\| \right)^{mn \, (mn-1)} (a, \ldots)^{2kn \, (mn-1)} \, (b, \ldots)^{2km \, (mn-1)} = 0.$$

Now the equation in question will be satisfied, 1°. if the curves $U = 0$, $V = 0$ touch each other; let the condition for this be $\nabla = 0$. 2°. If there exists a curve $\lambda P + \mu Q = 0$ passing through two of the intersections of the curves $U = 0$, $V = 0$; let the condition be $\Omega = 0$. There is reason to think that the equation contains the factor Ω^2, and that the form thereof is $\Omega^2 \nabla = 0$.

Assuming that this is so, and observing that ∇, the osculant or discriminant of the functions U, V, is of the form

$$\nabla = (a, \ldots)^{n \, (n+2m-3)} \, (b, \ldots)^{m \, (m+2n-3)},$$

we have

$$\Omega^2 = \left(\left\| \begin{array}{cc} \alpha, & \beta, \ldots \\ \alpha', & \beta', \end{array} \right\| \right)^{mn \, (mn-1)} (a, \ldots)^{kn \, (n-1) \, (2m-1) + (k-1) \, n \, (n+2m-3)} \times$$
$$(b, \ldots)^{km \, (m-1) \, (2n-1) + (k-1) \, m \, (m+2n-3)},$$

and consequently

$$\Omega = \left(\left\| \begin{array}{cc} \alpha, & \beta, \ldots \\ \alpha', & \beta', \end{array} \right\| \right)^{\frac{1}{2}mn \, (mn-1)}$$
$$(a, \ldots)^{\frac{1}{2}n \, (n-1) \, k \, (2m-1) + \frac{1}{2} \, (k-1) \, n \, (n+2m-3)} \times$$
$$(b, \ldots)^{\frac{1}{2}m \, (m-1) \, k \, (2n-1) + \frac{1}{2} \, (k-1) \, m \, (m+2n-3)},$$

which is the solution of the proposed question. Suppose for instance $n = 1$, then

$$\Omega = \left(\left\| \begin{array}{cc} \alpha, & \beta, \ldots \\ \alpha', & \beta', \end{array} \right\| \right)^{\frac{1}{2}m \, (m-1)} (a, \ldots)^{(k-1) \, (m-1)} \, (b, \ldots)^{\frac{1}{2}m \, (m-1) \, k + \frac{1}{2} \, (k-1) \, (m-1)}.$$

If moreover $k = 1$, then

$$\Omega = \left(\left\| \begin{array}{cc} \alpha, & \beta, \ldots \\ \alpha', & \beta', \end{array} \right\| \right)^{\frac{1}{2}m \, (m-1)} (b, \ldots)^{\frac{1}{2}m \, (m-1)};$$

this is right, for writing $P = \alpha x + \beta y + \gamma z$, $Q = \alpha' x + \beta' y + \gamma' z$, $V = bx + b'y + b''z$, then if two of the intersections of the curve $U = 0$ with the line $V = 0$ lie in a line with the point $P = 0$, $Q = 0$, then the point in question, that is the point $(\beta\gamma' - \beta'\gamma,\ \gamma\alpha' - \gamma'\alpha,\ \alpha\beta' - \alpha'\beta)$, must lie in the line $V = 0$; and the condition reduces itself to

$$\{(\beta\gamma' - \beta'\gamma,\ \gamma\alpha' - \gamma'\alpha,\ \alpha\beta' - \alpha'\beta \between b,\ b',\ b'')\}^{\frac{1}{2}m(m-1)} = 0,$$

where the index $\frac{1}{2}m(m-1)$ is accounted for as denoting the number of pairs of points out of the m intersections of the curve $U = 0$ with the line $V = 0$.

If in general $k = 1$, then writing as before $P = \alpha x + \beta y + \gamma z$, $Q = \alpha' x + \beta' y + \gamma' z$, we have

$$\Omega = (\beta\gamma' - \beta'\gamma, \ldots)^{\frac{1}{2}mn(mn-1)} (a, \ldots)^{\frac{1}{2}n(n-1)(2m-1)} (b, \ldots)^{\frac{1}{2}m(m-1)(2n-1)},$$

where $\Omega = 0$ is the condition in order that the point $(\beta\gamma' - \beta'\gamma, \ldots)$ may lie *in lineâ* with two of the intersections of the curves $U = 0$, $V = 0$. Or writing (X, Y, Z) for the coordinates of the given point, the condition is

$$\Omega = (a, \ldots)^{\frac{1}{2}n(n-1)(2m-1)} (b, \ldots)^{\frac{1}{2}m(m-1)(2n-1)} (X,\ Y,\ Z)^{\frac{1}{2}mn(mn-1)} = 0.$$

I have found that if

$$U = (a, \ldots \between x,\ y,\ z)^m, \quad V = (b, \ldots \between x,\ y,\ z)^n,$$
$$W = (c, \ldots \between x,\ y,\ z)^p, \quad T = (d, \ldots \between x,\ y,\ z)^q,$$

the condition in order that the point (X, Y, Z) may lie *in lineâ* with one of the intersections of the curves $U = 0$, $V = 0$, and one of the intersections of the curves $W = 0$, $T = 0$, is

$$\Omega = (a, \ldots)^{npq} (b, \ldots)^{mpq} (c, \ldots)^{mnq} (d, \ldots)^{mnp} (X,\ Y,\ Z)^{mnpq} = 0.$$

Supposing that the curves $W = 0$, $T = 0$ become identical with the curves $U = 0$, $V = 0$ respectively, this becomes

$$\Omega = (a, \ldots)^{n^2 \cdot 2m} (b, \ldots)^{m^2 \cdot 2n} (X,\ Y,\ Z)^{mn \cdot mn} = 0,$$

and the variation from the correct form given above is what might have been expected.

491.

ON THE QUARTIC SURFACES $(*\between U, V, W)^2 = 0$.

[From the *Quarterly Journal of Pure and Applied Mathematics*, vol. XI. (1871), pp. 111—113.]

THE general Torus, or surface generated by the rotation of a conic about a fixed axis anywise situate, has been investigated by M. De La Gournerie, *Jour. de l'École Polyt.*, t. XXIII. (1863), pp. 1—74. The surface is one of the fourth order, having a nodal circle; and with its equation of the form $V^2 - UW = 0$, consequently of the form in question. The leading points of the theory are as follows:

Consider (fig. 1) the plane of the conic in any particular position thereof; let this

Fig. 1.

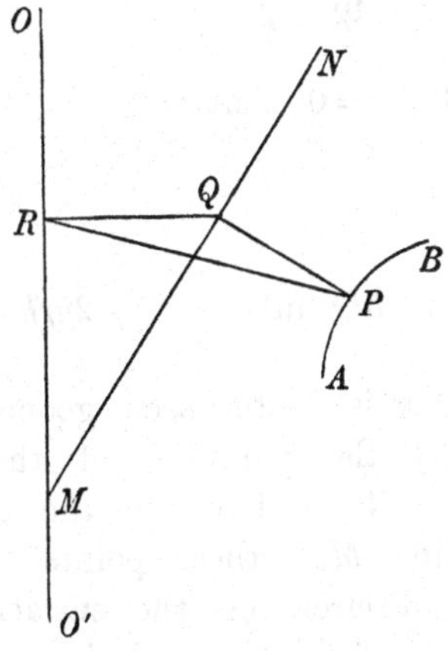

meet the axis of rotation OO' in the point M, and let the projection of OO' on the plane of the conic be MN. Take P any point of the conic; draw PQ in the plane of the conic, perpendicular to MN, and QR perpendicular to OO', and join PR: the point P of the conic describes a circle radius RP, $=\sqrt{(\overline{RQ}^2+\overline{QP}^2)}$. Hence if $\angle OMN=\alpha$, and if $MQ=\mathrm{x}$, $QP=\mathrm{y}$ are the coordinates in the plane of the conic of the point P; and if the coordinates x, y, z are measured, z upwards from M in the direction MO, and x, y in the plane at right angles to the axis OO': we have

$$z=\mathrm{x}\cos\alpha,\quad \sqrt{(x^2+y^2)}=\sqrt{(\mathrm{x}^2\sin^2\alpha+\mathrm{y}^2)};$$

or, what is the same thing,

$$\mathrm{x}=z\sec\alpha,\quad \mathrm{y}=\sqrt{(x^2+y^2-z^2\tan^2\alpha)}.$$

Hence the equation of the conic being $F(\mathrm{x}, \mathrm{y})=0$, that of the torus is

$$F\left(z\sec\alpha,\quad \sqrt{(x^2+y^2-z^2\tan^2\alpha)}\right)=0.$$

Thus taking the equation of the conic to be

$$(a, b, c, f, g, h\between \mathrm{x}, \mathrm{y}, 1)^2=0;$$

or, as this may be written,

$$(a\mathrm{x}^2+2g\mathrm{x}+c+b\mathrm{y}^2)^2 \qquad =4\mathrm{y}^2(h\mathrm{x}+f)^2,$$

we have at once the equation of the torus in the form

$$\{az^2\sec^2\alpha+2gz\sec\alpha+c+b(x^2+y^2-z^2\tan^2\alpha)\}^2=4(x^2+y^2-z^2\tan^2\alpha)(hz\sec\alpha+f)^2,$$

which is of the form $V^2-4UW=0$; or, as it is better to write it, $V^2-4UL^2=0$, where

$$V=az^2\sec^2\alpha+2gz\sec\alpha+c+b(x^2+y^2-z^2\tan^2\alpha),$$
$$U=x^2+y^2-z^2\tan^2\alpha,$$
$$L=hz\sec\alpha+f,\quad W=L^2.$$

There is thus a nodal circle $V=0$, $L=0$, that is

$$z=-\frac{f}{h}\cos\alpha,$$

$$bh^2(x^2+y^2)-bf^2\sin^2\alpha+af^2-2gfh+ch^2=0.$$

But the origin of this nodal circle is better seen geometrically. For observe that the radius of the circle described by the point P of the conic depends only on the *square* of the ordinate PQ: hence if we have on the conic two points S, S' situate symmetrically in regard to the line MN, these points S, S' will describe one and the same circle, which will be a nodal circle on the surface. And there is in fact one such pair of points S, S'; for (see fig. 2) considering in the plane of the conic the equal conic situate symmetrically thereto on the other side of the line MN, the two conics intersect in two points T, T' (real or imaginary) on the line MN, and in two other points S, S' (real or imaginary) situate symmetrically in regard to MN; we have thus the required pair of points which generate the nodal circle.

Fig. 2.

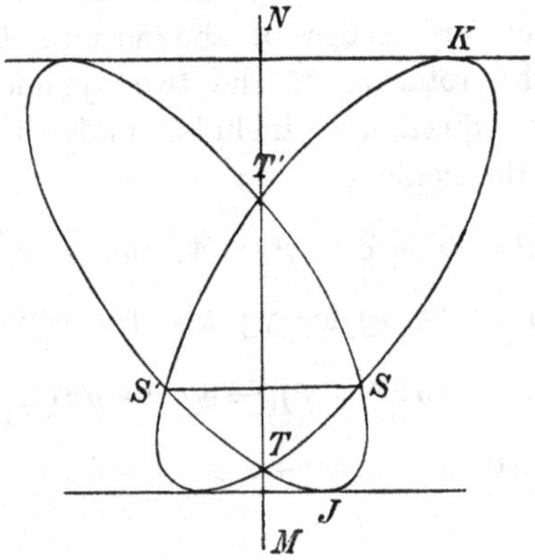

A meridian section of the torus (or section through the axis OO') is a quartic curve symmetrical in regard to this axis, and having two (real or imaginary) nodes the intersections of the plane by the nodal circle: see fig. 3, which shows the section for the surface generated by a conic such as in fig. 2. The quartic curve has 8 double tangents, 2 of them at right angles to the axis OO', the remaining 6 forming 3 pairs

Fig. 3.

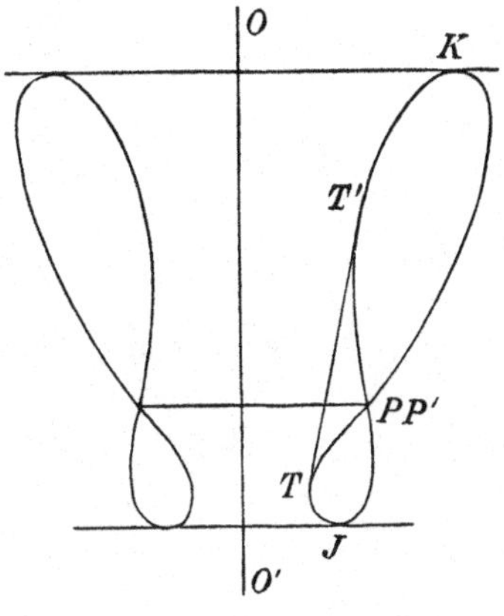

of tangents situate symmetrically in regard to this axis; so that attending only to one tangent of each pair, we may say that there are 3 oblique bitangents: one of these is the line TT'; and the section of the torus by a plane through this line at right angles to the plane of the meridian section is in fact the two conics of fig. 2, either of which by its rotation about OO' generates the torus. But taking either of the other two oblique bitangents, the section by a plane through the bitangent at right angles to the meridian plane is in like manner a pair of conics situate symmetrically in regard to the bitangent, and such that either of them by its rotation about the axis OO' generates the torus. It thus appears that the same torus may be generated in three different ways by the rotation of a conic about the axis OO'.

In the particular case where the plane of the conic passes through the axis, the meridian section consists it is clear of two symmetrically situate conics, intersecting the axis in the points T, T', which are nodes of the surface, the surface having as before a nodal circle generated by the rotation of the two symmetrically situate intersections S, S' of the two conics. The equation is included under the foregoing form, but it is at once obtained from that of the conic,

$$(a\text{x}^2 + 2g\text{x} + c + b\text{y}^2)^2 = 4\text{y}^2 (h\text{x} + f)^2,$$

by writing therein z for x and $\sqrt{(x^2+y^2)}$ for y; viz. the equation of the torus here is

$$\{az^2 + 2gz + c + b(x^2 + y^2)\}^2 = 4(x^2 + y^2)(hz + f)^2,$$

and the two nodes thus are $x=0$, $y=0$, $az^2 + 2gz + c = 0$.

492.

NOTE ON A SYSTEM OF ALGEBRAICAL EQUATIONS.

[From the *Quarterly Journal of Pure and Applied Mathematics*, vol. XI. (1871), pp. 132, 133.]

CONSIDER the system of equations

$$a + b(y+z)^2 + cy^2z^2 = 0,$$

$$a + b(z+x)^2 + cz^2x^2 = 0,$$

$$a + b(x+y)^2 + cx^2y^2 = 0,$$

which is a particular case of that belonging to the porism of the in-and-circumscribed triangle. We have y and z the roots of

$$a + bx^2 + 2u \,.\, bx + u^2(b + cx^2) = 0;$$

consequently

$$y + z = \frac{-2bx}{b + cx^2},$$

$$yz = \frac{a + bx^2}{b + cx^2},$$

or substituting in the equation between y and z, this becomes

$$(ac + b^2)(a + 4bx^2 + cx^4) = 0,$$

so that if $ac + b^2$ is not $= 0$, we have

$$a + 4bx^2 + cx^4 = 0,$$

and moreover

$$(x-y)(x-z) = x^2 + \frac{2bx^2}{b+cx^2} + \frac{a+bx^2}{b+cx^2}, = \frac{1}{b+cx^2}(a+4bx^2+cx^4) = 0,$$

so that $x=y$ or else $x=z$. If $x=z$, the three equations reduce themselves to the two

$$a + bx^2 + 2y \,.\, bx + y^2(b+cx^2) = 0,$$
$$a \;+\; 4bx^2 \;+\; cx^4 \;=0,$$

giving $y=x$, or else $y = -\dfrac{3bx+cx^3}{b+cx^2}$; and it hence appears that if from this last equation and $a+4bx^2+cx^4=0$ we eliminate x, the result must be $a+4by^2+cy^4=0$. For in the same way that the elimination of y, z from the original three equations gives $a+4bx^2+cx^4=0$, the elimination of x, z from the same three equations will give $a+4by^2+cy^4=0$, so that in any case y is a root of this equation.

493.

ON EVOLUTES AND PARALLEL CURVES.

[From the *Quarterly Journal of Pure and Applied Mathematics*, vol. XI. (1871), pp. 183—200.]

IN abstract geometry we have a conic called the Absolute; lines which are harmonics of each other in regard to the absolute, or, what is the same thing, which are such that each contains the pole of the other in regard to the absolute, are said to be at right angles. Similarly, points which are harmonics of each other in regard to the absolute, or, what is the same thing, which are such that each lies in the polar of the other, are said to be quadrantal.

A conic having double contact with the absolute is said to be a circle; the intersection of the two common tangents is the centre of the circle; the line joining the two points of contact, or chord of contact, is the axis of the circle.

Taking as a definition of equidistance that the points of a circle are equidistant from the centre, we arrive at the notion of distance generally, and we can thence pass down to that of equal circles; but the notion of equal circles may be established descriptively in a more simple manner:

Any two circles have an axis of symmetry, viz. this is the line joining their centres; and they have a centre of homology, viz. this is the intersection of their axes. They intersect in four points, lying in pairs on two lines through the centre of homology: they have also four tangents meeting in pairs in two points on the axis of symmetry. Now if the two lines through the centre of homology are harmonically related to the two axes, or, what is the same thing, if the two points on the axis of symmetry are harmonically related to the two centres, then the circles are *equal*.

Circles which are equal to the same circle are equal to each other, and the entire series of circles which are equal to a given circle, are said to be a system of circles of constant magnitude.

Starting from these general considerations, I pass to the question of evolutes and parallel curves: it will be understood that everything—lines at right angles, circles, poles, polars, reciprocal curves, &c.—refers to the absolute.

At any point of a curve we have a normal, viz. this is a line at right angles to the tangent; or, what is the same thing, it is the line joining the point with the pole of the tangent. The locus of the pole of the tangent is the reciprocal curve, and for any point of a given curve, the pole of the tangent at that point is the corresponding point of the reciprocal curve. Hence, also the normal is the line from the point to the corresponding point of the reciprocal curve. And the curve and its reciprocal have at corresponding points the same normal.

The envelope of the normals is the evolute; any curve having with the given curve the same normals (and therefore the same evolute) is a parallel curve; in other words, the parallel curve is any orthogonal trajectory of the normals of the given curve.

The parallel curve is also the envelope of a circle of constant radius having its centre on the given curve; or, again, it is the envelope of a circle of constant radius touching the given curve.

The theory in the above form is directly applicable to spherical, or rather conical, geometry; but in ordinary plane geometry the absolute degenerates into a point-pair, the two circular points at infinity, or say the points I, J; and this is a case that requires to be separately treated. The theory in the general case, the absolute a conic, is the more symmetrical and elegant, and it might appear advantageous to commence with this; but upon the whole I prefer the opposite course, and will commence with the case of plane geometry, the absolute a point-pair.

The subject connects itself with that of foci: I call to mind that a common tangent of the curve and the absolute is a focal tangent, and the intersection of two focal tangents a focus. In the case where the absolute is a point-pair, the focal tangents are the tangents from I to the curve, and the tangents from J to the curve, or say these are the I-tangents and the J-tangents; a focus is the intersection of an I-tangent and a J-tangent; the line IJ, when it touches the curve, and (when the curve passes through I and J or either of them) the tangents at I or J to the curve are usually not reckoned as focal tangents; and other singular tangents, for instance a double or stationary tangent through I or J, are also excluded from the focal tangents; and the number of foci is of course reckoned accordingly, viz. it is the product of the number of the I-tangents into that of the J-tangents. So when the absolute is a conic; if this is touched by the curve, the common tangent at the point of contact is not reckoned as a focal tangent; and we may also exclude any singular tangents which touch the absolute; and the number of foci is reckoned accordingly, viz. it is equal to the number of pairs of focal tangents.

Let the Plückerian numbers for the given curve be $(m, n, \delta, \kappa, \tau, \iota)$, viz. m the order, n the class, δ the number of nodes, κ of cusps, τ of bitangents, ι of inflexions; and suppose moreover that D is the deficiency, and α the statitude; viz.

$$\alpha = 3m + \iota, \quad = 3n + \kappa;$$

$$2D = (m-1)(m-2) - 2\delta - 2\kappa, \quad = (n-1)(n-2) - 2\tau - 2\iota, \quad = -2m - 2n + 2 + \alpha$$

$$= n - 2m + 2 + \kappa, \quad = m - 2n + 2 + \iota,$$

And let the corresponding numbers for the evolute be

$$(m'', n'', \delta'', \kappa'', \tau'', \iota''; D'', \alpha'').$$

These are most readily obtained, as in Clebsch's paper, "Ueber die Singularitäten algebraischer Curven," *Crelle*, t. LXIV. (1864), pp. 98—100, viz. it being assumed that we have

$$n'' = m + n, \ \iota'' = 0,$$

then by reason that the evolute has a (1, 1) correspondence with the original curve, the two curves have the same deficiency, or writing this relation under the form

$$m'' - 2n'' + \iota'' = m - 2n + \iota,$$

we have $m'' = 3m + \iota, = \alpha$; and the Plückerian relations then give the values of κ'', δ'', τ''.

In regard to these equations $n'' = m + n$, $\iota'' = 0$, I remark that if we have two curves of the orders m, m', and on these points P, Q having an (α, α') correspondence, the line PQ envelopes a curve of the class $m\alpha' + m'\alpha$, and the number of inflexions is in general $= 0$. Now in the present case, taking P on the given curve and Q the point of intersection with IJ of the normal (or harmonic of the tangent), the orders of the curves are $(m, 1)$, and the correspondence is $(n, 1)$; whence as stated $m'' = m + n$, $\iota'' = 0$.

The formulæ thus are

$$\begin{aligned} m'' &= \alpha, \\ n'' &= m + n, \\ \iota'' &= 0, \\ \kappa'' &= -3m - 3n + 3\alpha, \\ \alpha'' &= 3\alpha, \\ D'' &= D, \end{aligned}$$

in which formulæ it is assumed that the curve has no special relations to the points I, J; or, what is the same thing, that the line IJ intersects the curve in m points distinct from each other, and from the points I, J.

It is to be added (see Salmon's *Higher Plane Curves*, [Ed. 2], (1852), pp. 109 et seq.) that m of the κ'' cusps arise from the intersections of the curve with IJ, these cusps being situate on the line IJ, and each of them the harmonic of one of the intersections in question, and the cuspidal tangent being for each of them the line IJ. The intersections of the evolute by the line IJ are these m cusps each 3 times, and besides ι points arising from the ι inflexions of the curve; viz. at any inflexion the two consecutive normals intersect in a point on the line IJ, being in fact the harmonic of the intersection of IJ with the tangent at the inflexion. It was in this manner that Salmon obtained the number $3m + \iota$ of the points at infinity of the evolute, that is the expression $m'' = 3m + \iota$ $(= \alpha)$ for the order of the evolute.

The remaining $-4m - 3n + 3\alpha$ cusps arise from the points on the curve where there is a circle of 4-pointic intersection, or contact of the third order, and in this manner the number of them was found, Salmon, [Ed. 2], p. 113, in the particular case of a curve without nodes or cusps, and generally in Zeuthen's *Nyt Bidrag* &c., p. 91; the number of the points in question, in the foregoing form $-4m-3n+3\alpha$, is also obtained in my Memoir, "On the curves which satisfy given conditions," *Phil. Trans.* (1868), pp. 75—143, see p. 97, [406].

It is further to be noticed that the $m+n$ tangents to the evolute from either of the points I, J are made up of the line IJ counting m times (in respect that it is a tangent at each of the above-mentioned m cusps) and of the n tangents from the points in question to the original curve. Or taking the two points I, J conjointly, say the $2m+2n$ common tangents of the absolute and the evolute are made up of the line IJ (or axis of the absolute) counting m times, and of the $2n$ focal tangents of the original curve. The focal tangents of the original curve and of the evolute are thus the same $2n$ lines; and the two curves have the same foci.

The above are the ordinary values of m'', n'', ι'', κ'', but if the given curve touch the line IJ, then the evolute has at the point of contact an inflexion, the stationary tangent being the line IJ; and if the given curve pass through one or other of the points I, J, the evolute has in this case an inflexion on the tangent at the point in question, this tangent being the stationary tangent of the evolute: but observe that the inflexion is not at the point I or J in question: and for each inflexion there is a diminution $=1$ in the class, 3 in the order, and 5 in the number of cusps. Suppose that the point I is a f_1-tuple point on the given curve; then the evolute has f_1 inflexions; and similarly if the point J is a f_2-tuple point on the given curve, then the evolute has f_2 inflexions. Hence writing $f_1+f_2=f$, we have thus f inflexions; and if moreover the number of contacts with the line IJ be $=g$, then we have on this account g inflexions; or in all $f+g$ inflexions, and the formulæ become

$$\begin{aligned} m'' &= \alpha - 3f - 3g, \\ n'' &= m+n - f - g, \\ \iota'' &= f + g, \\ \kappa'' &= -3m - 3n + 3\alpha - 5f - 5g. \end{aligned}$$

It is to be noticed here that the number of the intersections of the given curve with the line IJ (other than the points I, J and the points of contact) is $=m-f_1-f_2-2g$, that is $m-f-2g$: each of these gives as before a cusp on the evolute, the cuspidal tangent being IJ; we have besides on the line IJ (in respect of the g contacts) g inflexions, the stationary tangent being the line IJ; and each of the ι inflexions gives for the evolute a point on the line IJ; hence the whole number of intersections with the line IJ is $3(m-f-2g)+3g+\iota$, $=3m+\iota-3f-3g$, which is thus the order of the evolute.

The tangents from the point I or J to the evolute are the line IJ counting $m-f-2g$ times in respect of the cusps on this line and $2g$ times in respect to the inflexions, that is $m-f$ times; the tangents at the point in question to the given curve each twice as touching the evolute at an inflexion, $2f_1$ or $2f_2$: and the remaining

$n-2f_1-g$, or $n-2f_2-g$ tangents from the point in question to the given curve; the whole number is thus $(m-f)+2f_1+(n-2f_1-g)$ or $(m-f)+2f_2+(n-2f_2-g)$, $=m+n-f-g$, the class of the evolute. The two values of n'' give

$$2n''=2m+(n-2f_1-g)+(n-2f_2-g),$$

viz. twice the class of the evolute = twice the order of the curve + the number of the focal I-tangents + that of the focal J-tangents; but this is not true for all relations whatever of the curve to the absolute.

The tangents from I to the given curve (excluding the line IJ and the tangents at I) are $n-2f_1-g$ tangents; and similarly the tangents from I to the evolute (excluding the line IJ and the stationary tangents through I) are the same $n-2f_1-g$ tangents; say the curve and the evolute have the same $n-2f_1-g$ I-tangents. Similarly they have the same $n-2f_2-g$ J-tangents; or together the same $2(n-f_1-f_2-g)$, $=2(n-f-g)$ focal tangents. And the curve and evolute have the same $(n-2f_1-g)(n-2f_2-g)$ foci.

The foregoing specialities f and g refer, g to the ordinary contacts of the line IJ with the curve, viz. the curve is supposed to have with the curve at an ordinary or non-singular point thereof a contact or 2-pointic intersection, and f, that is f_1 or f_2, to the multiple points having f_1 or f_2 ordinary branches, none of them touching the line IJ. Thus the formulæ do not apply to the cases of IJ passing through a node or a cusp of the given curve, or touching it at an inflexion; nor to the cases where at I or J the curve touches IJ, or where there is at I or J an ordinary double point with one of its branches touching IJ, or where there is at I or J a cusp, where the cuspidal tangent is or is not IJ.

It is easy to see that in the case of a multiple point of any kind whether situate on IJ or at I or J, each branch of the curve produces its own separate effect on the singularities of the evolute: thus if we have on IJ a double point neither branch touching IJ, then the separate effect of each branch is *nil*, therefore the effect of the double point is also *nil*: but if one branch touch the line IJ, then the whole effect is the same as if we had this branch only; viz. we have here the case $g=1$. And so if there is at I or J a double point with one branch touching the line IJ, then the effect of this branch is as if we had this branch only (a case not yet investigated) but the other branch is the case $f=1$. And so if we have at I or J a double point with two ordinary branches touching each other (tacnode or, if the two branches have a contact higher than the first order, oscnode), then if the branches do not touch the line IJ the case is $f=2$, but if they do, then the effect is twice that of an ordinary branch touching IJ. In support of these conclusions, observe that such multiple points, with ordinary branches, present themselves in the case of two or more curves which intersect or touch each other in any manner; and that the evolute of a system of two or more curves is simply the system of the evolutes of the several curves.

It follows as regards the relations of the given curve to the points I, J, and the effect thereby produced on the evolute, we only need to consider the case of a single branch; viz. the cases are

the given curve intersects the line IJ at a point other than I or J, and belonging thereto there is a branch ordinary or singular,

not touching IJ,

touching IJ;

and the given curve passes through the point I or J, and belonging thereto there is a branch ordinary or singular,

not touching IJ,

touching IJ.

I have succeeded in determining the effect, not for a singular branch of any kind whatever, but for branches of the form $y = x^k$, $y^{k-1} = x^k$; viz. $k = 2$, each of these is an ordinary branch, $k = 3$, the first $y = x^3$ is an inflexional branch and the second $y = x^{\frac{3}{2}}$ a cuspidal branch; and so $k > 3$ the two branches are respectively inflexional and cuspidal of a higher order. I do this very simply by consideration of the curve $x^{k-1}z = y^k$.

The curve in question $x^{k-1}z = y^k$, is a unicursal curve, and it has a reciprocal of the same form $X^{k-1}Z = Y^k$, hence

$$m = n = k; \quad 0 = n - 2m + 2 + \kappa,$$

whence

$$\iota = \kappa = k - 2, \quad \tau = \delta = \tfrac{1}{2}(k-2)(k-3);$$

viz. the point $x = 0$, $y = 0$ is a cusp equivalent to $k - 2$ cusps and $\frac{1}{2}(k-2)(k-3)$ nodes; and the point $z = 0$, $y = 0$ is an inflexion equivalent to $k - 2$ inflexions and $\frac{1}{2}(k-2)(k-3)$ bitangents.

The equation $U = x^{k-1}z - y^k = 0$ of the curve is satisfied by writing therein $x : y : z = 1 : \theta : \theta^k$; and these values give

$$d_xU : d_yU : d_zU = (k-1)x^{k-2}z : -ky^{k-1} : x^{k-1}, = (k-1)\theta^k : -k\theta^{k-1} : 1.$$

Taking the coordinates of I, J to be (α, β, γ) and $(\alpha', \beta', \gamma')$ respectively, and X, Y, Z as current coordinates, the equation of the normal at the point (x, y, z) of the curve $U = 0$ is readily found to be

$$(\alpha' d_xU + \beta' d_yU + \gamma' d_zU) \begin{vmatrix} X, & Y, & Z \\ \alpha, & \beta, & \gamma \\ x, & y, & z \end{vmatrix} + (\alpha d_xU + \beta d_yU + \gamma d_zU) \begin{vmatrix} X, & Y, & Z \\ \alpha', & \beta', & \gamma' \\ x, & y, & z \end{vmatrix} = 0.$$

Hence for the curve in question the equation of the normal is

$$\{(k-1)\alpha'\theta^k - k\beta'\theta^{k-1} + \gamma'\}\,\{(\beta X - \alpha Y)\,\theta^k + (\alpha Z - \gamma X)\,\theta + (\gamma Y - \beta Z)\}$$
$$+\{(k-1)\,\alpha\theta^k - k\beta\theta^{k-1} + \gamma\}\,.\,\{(\beta' X - \alpha' Y)\,\theta^k + (\alpha' Z - \gamma' X)\,\theta + (\gamma' Y - \beta' Z)\} = 0,$$

or, expanding and reducing, this equation is

$$\begin{aligned}
&\theta^{2k}\,.\,(k-1)\ \ \{(\alpha\beta' + \alpha'\beta)\,X - 2\alpha\alpha'\,Y\}\\
+\ &\theta^{2k-1}\,.\ \ -k\,\{2\beta\beta' X - (\alpha\beta' + \alpha'\beta)\,Y\}\\
+\ &\theta^{k+1}\,.\,(k-1)\,\{2\,\alpha\alpha'\,Z - (\alpha\gamma' + \alpha'\gamma)\,X\}\\
+\ &\theta^{k}\,.\,\{(k+1)\,(\beta\gamma' + \beta'\gamma)\,X + (k-2)\,(\gamma\alpha' + \gamma'\alpha)\,Y - (2k-1)\,(\alpha\beta' + \alpha'\beta)\,Z\}\\
+\ &\theta^{k-1}\,.\ \ -k\,\{(\beta\gamma' + \beta'\gamma)\,Y - 2\beta\beta' Z\}\\
+\ &\theta\ \ .\ \ \{(\gamma\alpha' + \gamma'\alpha)\,Z - 2\gamma\gamma' X\}\\
+\ &\ \ \{2\gamma\gamma' Y - (\beta\gamma' + \beta'\gamma)\,Z\} = 0,
\end{aligned}$$

where k is a positive integer not less than 2; hence except in the case $k=2$, all the terms θ^{2k}, θ^{2k-1}, ... θ, θ^0, have different indices, and the coefficients $k-1$, k, &c. none of them vanish; if however $k=2$, then the terms θ^{2k-1}, θ^{k+1} coalesce into a single term, as do also the terms θ^{k-1} and θ; moreover the coefficient $k-2$ is $=0$.

The evolute is the envelope of the line represented by the foregoing equation, considering therein θ as an arbitrary parameter; viz. the equation is obtained by equating to zero the discriminant of the foregoing equation in θ. Hence in general the class of the evolute is $=2k$, and its order is $=2\,(2k-1)$; results which agree with the formulæ for n'', m'', since in the present case $m+n$, $=k+k$, $=2k$, $\alpha = 3n+\kappa$, $=3k+(k-2)$, $=4k-2$. And moreover there are not any inflexions, $\iota''=0$ as before.

The equation may however contain a factor in θ independent of (X, Y, Z), and throwing out this factor, say its order is $=s$, the expression for the class is $2k-s$, $=m+n-s$, and that for the order is $4k-2-2s$, $=\alpha-2s$. Moreover, in the original equation or in the equation thus reduced, it may happen that the equation will on writing therein $\Omega=0$ (Ω a linear function of X, Y, Z) acquire a factor of the order ω, independent of (X, Y, Z); the line $\Omega=0$ is in this case a stationary tangent, $=\omega-1$ inflexions; and the discriminant contains the factor $\Omega^{\omega-1}$, which may be thrown out; that is we have here $n''=2k-s$, $\iota''=\omega-1$, $m''=4k-2-2s-(\omega-1)$; agreeing with the relation $m''-2n''+2+\iota''=0$ which holds good in virtue of the evolute being a unicursal curve. It is in this manner that the values of m'', n'', ι'' are obtained in the several cases to be considered, viz.:

A_k Inflexion situate on IJ, which is not a tangent.
B_k Inflexion situate on IJ, which is a tangent.
C_k Cusp situate on IJ, which is not a tangent.
D_k Cusp situate on IJ, which is a tangent.
P_k Inflexion at J, IJ not a tangent.
Q_k Inflexion at J, IJ a tangent.
R_k Cusp at J, IJ not a tangent.
S_k Cusp at J, IJ a tangent.

The results are respectively as follows:

	A_k	B_k	C_k	D_k	P_k	Q_k	R_k	S_k
$m'' = \alpha\ -$	0	$3k-3$	$k-2$	$k+1$	k	$3k-2$	$2k-2$	$2k$
$n'' = m+n\ -$	0	$k-1$	0	1	1	k	$k-1$	k
$\iota'' = 0\ +$	0	$k-1$	$k-2$	$k-1$	$k-2$	$k-2$	0	0
$\kappa'' = -3m-3n+3\alpha\ -$	0	$5k-5$	$2k-4$	$2k+1$	$2k-1$	$5k-4$	$3k-3$	$3k$

	A_2C_2	B_2D_2	A_3	B_3	C_3	D_3	$\bar{P}_2\bar{R}_2$	Q_2S_2	P_3	Q_3	R_3	S_3
$m'' = \alpha\ -$	0	3	0	6	0	4	3	4	3	7	4	6
$n'' = m+n\ -$	0	1	0	2	0	1	1	2	1	3	2	3
$\iota'' = 0\ +$	0	1	0	2	0	2	1	0	1	1	0	0
$\kappa'' = -3m-3n+3\alpha\ -$	0	5	0	10	0	7	5	6	5	11	6	9

read for instance in B_k, $m'' = \alpha - (3k-3)$, $n'' = m+n-(k-1)$, $\iota'' = 0+(k-1)$, and $\kappa'' = -3m-3n+3\alpha+(5k-5)$; and so in other cases.

A_2C_2 (that is indifferently A_2 or C_2) is when there is on IJ an ordinary point, IJ not a tangent; and so B_2D_2 when there is on IJ an ordinary point, IJ a tangent. Similarly P_2R_2 when there is at J an ordinary point, IJ not a tangent; only instead thereof I have written $\bar{P}_2\bar{R}_2$ to indicate that (for a reason which will appear) the numbers are *not* deducible from those for P_2 or R_2 by writing therein $k=2$; and Q_2S_2 is when there is at J an ordinary point, IJ a tangent.

Case A_k. We have to take the line IJ passing through the inflexion; the condition for this is $\beta\gamma' - \beta'\gamma = 0$: there is no speciality, or we have $n'' = 2k$, $m'' = 4k-2$, $\iota'' = 0$; whence also $\kappa'' = 0$; the value of κ'' being in every case deduced from those of m'', n'', ι'' by the formula

$$3m'' + \iota'' = 3n'' + \kappa''.$$

Case B_k. I write $\gamma = \gamma' = 0$, the equation of the normal is

$$
\begin{aligned}
&\theta^{2k} \quad .\,(k-1)\left\{(\alpha\beta' + \alpha'\beta)X - 2\alpha\alpha' Y\right\} & & \theta^{k+1}\ (X,\ Y) \\
&+\, \theta^{2k-1}.\, -k\left\{2\beta\beta' X - (\alpha\beta' + \alpha'\beta)Y\right\} & & +\, \theta^{k}\ \ (X,\ Y) \\
&+\, \theta^{k+1} \quad .\,(k-1)\,2\alpha\alpha' Z & & +\, \theta^{2}\ \ \ Z \\
&+\, \theta^{k} \quad .\, -(2k-1)(\alpha\beta' + \alpha'\beta)Z & & +\, \theta\ \ \ \ Z \\
&+\, \theta^{k-1}.\, k. \quad 2\beta\beta' Z \qquad\qquad = 0, & & +\ \qquad Z = 0,
\end{aligned}
$$

where throwing out the factor θ^{k-1}, the form is as shown on the right hand. Writing $Z=0$, we have a factor θ^k, whence $\iota''=k-1$, and then $n''=k+1$, $m''=2k-(k-1)=k+1$, agreeing with the table. The process holds good for $k=2$.

Case C_k. I write $\beta=\beta'=0$; this brings as well the inflexion as the cusp upon the line IJ; but it has been seen (Case A_k) that there is not any reduction on account of this position of the inflexion, hence the whole effect will be due to the cusp. The equation is

$$\begin{array}{ll|l} & \theta^{2k} \;.\, (k-1)\{-2\alpha\alpha' Y\} & \theta^{2k} \;\; Y \\ + & \theta^{k+1} .\, (k-1)\{2\alpha\alpha' Z-(\alpha\gamma'+\alpha'\gamma)X\} & +\,\theta^{k+1}(Z,\,X) \\ + & \theta^{k} \;.\, (k-2)(\gamma\alpha'+\gamma'\alpha)Y & +\,\theta^{k}(k-2)\,Y \\ + & \theta \;\;.\;\; \{(\gamma\alpha'+\gamma'\alpha)Z-2\gamma\gamma' X\} & +\,\theta\,(Z,\,X) \\ + & \qquad 2\gamma\gamma' Y=0, & + \quad Y=0, \end{array}$$

so that here $n''=2k$. On writing $Y=0$, there is a factor $\left(1-\dfrac{\theta}{\infty}\right)^{k-1}$ thrown out (indicated by the reduction of the order from $2k$ to $k+1$), whence

$$\iota''=k-2,\quad m''=2(2k-1)-(k-2),\quad =3k.$$

The process holds good for $k=2$.

Case D_k. We may write $\alpha=\alpha'=0$; the equation is

$$\begin{array}{ll|l} & \theta^{2k-1} .\, -k\,.\,2\beta\beta' X & \theta^{2k-1}\; X \\ + & \theta^{k} \;.\; (k+1)(\beta\gamma'+\beta'\gamma)X & +\,\theta^{k} \;\; X \\ + & \theta^{k-1} .\, -k\,[(\beta\gamma'+\beta'\gamma)\,Y-2\beta\beta' Z] & +\,\theta^{k-1}(Y,\,Z) \\ + & \theta \;.\, -2\gamma\gamma' X & +\,\theta \;\; X \\ & \quad +2\gamma\gamma' Y-(\beta\gamma'+\beta'\gamma)Z=0, & + \quad (Y,\,Z)=0, \end{array}$$

so that $n''=2k-1$. Writing $X=0$, we have the factor $\left(1-\dfrac{\theta}{\infty}\right)^{k}$ (indicated by the reduction of order from $2k-1$ to $k-1$), whence $\iota''=k-1$, and then $m''=(4k-2)-2-(k-1)$, $=3k-3$, agreeing with the table. The process holds good for $k=2$.

Case P_k. We have $\beta'=\gamma'=0$; the equation is

$$\begin{array}{ll|l} & \theta^{2k} \;.\, (k-1)\{\alpha'\beta X-2\alpha\alpha' Y\} & \theta^{2k-1}(X,\,Y) \\ + & \theta^{2k-1} .\, -k \quad \{-\alpha'\beta Y\} & +\,\theta^{2k-2}\,Y \\ + & \theta^{k+1} .\, (k-1)\{-\alpha'\gamma X\} & +\,\theta^{k} \;\; X \\ + & \theta^{k} \;.\, (k-2)\gamma\alpha' Y-(2k-1)\,\alpha'\beta Z & +\,\theta^{k-1}(k-2)\,Y+Z \\ + & \theta \;\;.\;\; \gamma\alpha' Z=0, & + \quad Z=0, \end{array}$$

so that here $n''=2k-1$. Writing $Z=0$, we have the factor θ^{k-1}, whence $\iota''=k-2$, $m''=4k-4-(k-2)=3k-2$, agreeing with the table. If, however, $k=2$, then on writing $Z=0$ the equation (instead of the factor θ^{k-1}) acquires the factor $\theta^k\,(=\theta^2)$; so that here $n''=3$, $\iota''=1$, $m''=3$, agreeing with the column P_2R_2 of the table.

Case Q_k. We have $\gamma=0$, $\beta'=\gamma'=0$, viz. $\beta'=0$ in the formulæ of B_k. The equation is

$$\begin{array}{ll|l} \theta^{2k} & .\ (k-1)\,\alpha'\beta Y & \theta^k\ X \\ +\,\theta^{2k-1} & .\ -k\,(-\alpha'\beta Y) & +\,\theta^{k-1}Y \\ +\,\theta^{k+1} & .\ (k-1)\,2\alpha\alpha' Z & +\,\theta\ \ Z \\ +\,\theta^k & .\ -(2k-1)\,\alpha'\beta Z=0, & +\quad Z=0, \end{array}$$

so that $n''=k$. For $Z=0$ there is the factor θ^{k-1}, hence $\iota''=k-2$, $m''=2(k-1)-(k-2)$, $=k$. The process holds good for $k=2$.

Case R_k. We have $\beta=0$, $\alpha'=0$, $\beta'=0$, viz. $\alpha'=0$ in the formulæ of C_k. The equation is

$$\begin{array}{ll|l} \theta^{k+1} & .\ (k-1)\,(-\alpha\gamma' X) & \theta^{k+1}X \\ +\,\theta^k & .\ (k-2)\,(\alpha\gamma' Y) & +\,\theta^k\ \ Y(k-2) \\ +\,\theta & .\quad (\alpha\gamma' Z-2\gamma\gamma' X) & +\,\theta\ \ (Z,\ X) \\ + & \quad 2\gamma\gamma' Y=0, & +\quad Y=0, \end{array}$$

so that $n''=k+1$, $\iota''=0$, $m''=2k$. But observe that in the particular case $k=2$, the form is $\theta^3X+\theta(Z,\ X)+Y=0$, the term θ^kY disappearing on account of the factor $k-2$. Here on writing $X=0$, there is a factor $\left(1-\dfrac{\theta}{\infty}\right)^2$ (indicated by the reduction of order from 3 to 1), hence $\iota''=1$, $n''=3$, $m''=2.2-1=3$, agreeing with the column $\bar{P}_2\bar{R}_2$.

Case S_k. We have $\alpha=0$, $\alpha'=\beta'=0$, viz. $\beta'=0$ in the formulæ of D_k. The equation is

$$\begin{array}{ll|l} \theta^k & .\ (k+1)\,\beta\gamma' X & \theta^k\ X \\ +\,\theta^{k-1} & .\quad -k\beta\gamma' Y & +\,\theta^{k-1}\ Y \\ +\,\theta & .\quad -2\gamma\gamma' X & +\,\theta\ \ X \\ + & \quad 2\gamma\gamma' Y-\beta\gamma' Z=0, & +\quad Y+Z=0, \end{array}$$

so that $n''=k$, $\iota''=0$, $m''=2k-2$. The process applies to the case $k=2$.

As to the formula for $A_3, B_3, \ldots S_3$, there is nothing special in these; they are simply deduced from those for $A_k, B_k, \ldots S_k$ by writing therein $k=3$. And we have thus the foregoing series of formulæ, which will apply to the greater part of the cases which ordinarily arise. For instance suppose there is at I or J a triple point = cusp + 2 nodes; there is here an ordinary branch and a cuspidal (ordinary cuspidal) branch and according as IJ touches neither branch, the ordinary branch, or the cuspidal branch, the corrections to m'', n'', ι'', κ'' are $\bar{R}_2+R_3$, S_2+R_3, $\bar{R}_2+S_3$ respectively. Observe moreover that A_2C_2 is no speciality, B_2D_2 is the speciality $g=1$, $\bar{P}_2\bar{R}_2$ the speciality $f=1$.

There is a remarkable case in which the fundamental assumption of the (1, 1) correspondence of the evolute with the original curve ceases to be correct. In fact, in the case about to be considered of a parallel curve; the parallel to any given curve is *in general* a curve not breaking up into two distinct curves of the same order with such given curve, and when this is so (viz. when the parallel curve does not break up) each normal of the parallel curve is a normal at two distinct points thereof: the evolute of the parallel curve is thus the evolute of the given curve taken twice; and the parallel curve and its evolute have not a (1, 1) but (1, 2) correspondence. Hence, (m, n, δ, κ, τ, ι) the unaccented letters referring to the parallel curve, or say rather to a curve which has a (1, 2) correspondence with its evolute, and, as before, the twice accented letters to the evolute, it is *not true* that $m'' - 2n'' + \iota'' = m - 2n + \iota$; it will subsequently appear that the values of m'', n'' are correct, those of ι'', κ'' suffering a modification; viz. the formulæ are

$$\begin{aligned} m'' &= \alpha - 3f - 3g, \\ n'' &= m + n - f - g, \\ \iota'' &= f + g - \Theta, \\ \kappa'' &= -3m - 3n + 3\alpha - 5f - 5g - \Theta, \end{aligned}$$

where, for the present, I leave Θ undetermined.

Coming now to the parallel curve, let the numbers in regard to it be m', n', δ', κ', τ', ι'; α', D'. Supposing in the first instance that the given curve does not stand in any special relation to I, J, the formulæ are

$$\begin{aligned} m' &= 2m + 2n, & \alpha &= 6n + 2\alpha, \\ n' &= 2n, & 2D' &= -4m + 2 + 2\alpha, = -2m + 2n + \alpha. \\ \iota' &= 2\alpha - 6m, \\ \kappa' &= 2\alpha. \end{aligned}$$

Considering the parallel curve as the envelope of a circle of constant radius having its centre on the given curve, it appears (e.g. by consideration of the case of the ellipse) that when the radius of the circle is $=0$, there is not any depression in the order of the parallel curve, but that the parallel curve reduces itself to the given curve twice, together with the system of tangents from the points I, J to the given curve: the order of the parallel curve is thus $m' = 2m + 2n$.

To find the class, consider the tangents from a given point to the parallel curve; about the point as centre describe a circle, radius k; then the tangents in question are respectively parallel to, and correspond each to each with, the common tangents of the circle and the given curve, and the number of these is $= 2n$, that is $n' = 2n$.

Each inflexion of the given curve gives rise to two inflexions of the parallel curve; and the inflexions of the parallel curve arise in this way only: that is $\iota' = 2\iota$, $= 2\alpha - 6m$. And the Plückerian relations then give $\kappa' = 2\alpha$; a value which may be investigated independently.

Attending now to the singularities f and g; the values of n', ι' are unaltered: to obtain m' we as before consider the particular case where the radius of the variable circle is $=0$: the parallel curve here breaks up into the original curve, together with the focal tangents from the points I, J; viz. we have

$$m' = 2n + (n - 2f_1 - g) + (n - 2f_2 - g), \quad = 2m + 2n - 2f - 2g;$$

and knowing m', n', ι' we have κ'.

The points on IJ of the original curve are I, J counting as f_1 and f_2 respectively; or together as f points: the points of contact counting as $2g$: and besides $m - f - 2g$ points. As regards the parallel curve, we have the same points on IJ; but here I is $(n - g)$ tuple point, having in respect of each branch of the f_1-tuple point on the original curve a pair of branches touching each other, and in respect of each of the tangents from I to the given curve a single branch, together $2f_1 + (n - 2f_1 - g)$, $= n - g$ branches; and thus counting $n - g$ times: similarly J counts $n - g$ times. Hence also for the parallel curve $f_1' = f_2' = n - g$. In respect of each of the points g, we have a point where there are two branches touching each other and the line IJ; and thus counting 4 times, or together as $4g$: moreover, on account of the two branches at each of these points, $g' = 2g$. Lastly, each of the $m - f - 2g$ points is a node on the parallel curve; and as such counts twice; $m' = 2(n - g) + 4g + 2(m - f - 2g)$, $= 2m + 2n - 2f - 2g$ as above.

And we have thus the formulæ

$$\begin{aligned}
m' &= 2m + 2n - 2f - 2g, \\
n' &= 2n, \\
\iota' &= -6m + 2\alpha, \; = 2\iota, \\
\kappa' &= 2\alpha - 6f - 6g, \; = 6n + 2\kappa - 6f - 6g, \\
f' &= 2n - 2g, \\
g' &= \phantom{-2n - {}} 2g,
\end{aligned}$$

where observe that m' is $= 2n''$; that is, twice the class of the evolute (which relation however is not in all cases true for a curve with singularities); and further that $n' - f' - g'$ is $= 0$.

The case of a curve for which $n - f - g = 0$ is very interesting and remarkable. Recurring to the formulæ for the evolute, we have here $m'' = \kappa$, $n'' = m$, $\iota'' = n$, $\kappa'' = n - 3m + 3\kappa$. And for the parallel curve $m' = 2m$, $n' = 2n$, $\iota' = 2\iota$, $\kappa' = 2\kappa$, $f' = 2f$, $g' = 2g$; formulæ which lead to the assumption that the parallel curve here breaks up into two distinct curves, each such as the given curve.

Observe further that for a curve possessing the singularities f and g, but where $n - f - g$ is not $= 0$; then for the parallel curve we have as above $n' - f' - g' = 0$; or the parallel of the parallel curve should, according to the assumption, break up into two distinct curves such as the parallel curve; this is of course correct.

Consider the evolute of the parallel curve: since for the parallel curve $n' - f' - g' = 0$, the formulæ for the evolute thereof (viz. those containing the undetermined quantity Θ)

are $m'''=\kappa'$, $n'''=m'$, $\iota'''=n'-\Theta$, $\kappa'''=n'-3m'+3\kappa'-\Theta$, or substituting for m', n', κ' their values, and comparing with the formulæ in regard to the evolute, we have

$$\begin{aligned}
m''' &= 2\alpha - 6f - 6g, && = 2m'',\\
n''' &= 2m + 2n - 2f - 2g, && = 2n'',\\
\iota''' &= 2n - \Theta, && = 2\iota'' + 2(n-f-g) - \Theta,\\
\kappa''' &= -6m - 4n + 6\alpha - 12f - 12g - \Theta, && = 2\kappa'' + 2(n-f-g) - \Theta,
\end{aligned}$$

where m'', n'', ι'', κ'' refer to the evolute. Hence by assuming $\Theta = 2(n-f-g)$, the values of m''', n''', ι''', κ''' become $2m''$, $2n''$, $2\iota''$, $2\kappa''$, viz. the evolute of the parallel curve is the evolute of the original curve taken twice. Observe that in the foregoing value of Θ, the letters n, f, g refer not to the parallel curve, the evolute whereof is under consideration, but to the curve from which such parallel curve was derived; this value $\Theta = 2(n-f-g)$ is not a value of Θ applicable to be substituted in the evolute-formulæ for the case of a curve which has with its evolute a (1, 2) correspondence.

Instead of the foregoing case of the f, viz. f- and g-singularities, we may, as regards the parallel curve, consider the original curve as having any I- and J-singularities whatever. Suppose in this case (excluding always the line IJ and the tangents at I or J) the number of tangents from I to the curve is $=n-I$, and the number of tangents from J to the curve is $=n-J$, then when the radius of the variable curve is $=0$, the parallel curve becomes the original curve twice together with the $(n-I)+(n-J)$, $=2n-I-J$ tangents; so that the order is $m'=2m+2n-I-J$ (1); we have, as before, $n'=2n$ and $\iota'=2\iota$, and these values give κ', so that the equations are

$$\begin{aligned}
m' &= 2m + 2n - I - J,\\
n' &= 2n,\\
\iota' &= -6m + 2\alpha, && = 2\iota,\\
\kappa' &= 2\alpha - I - J, && = 6n + 2\kappa - 3I - 3J.
\end{aligned}$$

Suppose $2n-I-J=0$; this implies $n-I=0$, $n-J=0$ since neither $n-I$ nor $n-J$ can be negative; viz. that there are no I- or J-tangents; and conversely, when this is the case $2n-I-J=0$: and we have then m', n', ι', $\kappa' = 2m$, $2n$, 2ι, 2κ; viz. it is assumed, as before, that the parallel curve breaks up into two distinct curves such as the original curve; that is, the condition in order that the parallel curve should break up, is that the original curve has no focal tangents. Observe that the number of foci is $=(n-I)(n-J)$ which is $=0$ if only $n-I=0$ or $n-J=0$; but as regards real curves $I=J$, so that the equations $n-I=0$ and $n-J=0$ are one and the same equation, satisfied if $(n-I)(n-J)=0$; so that for a real curve without foci (real or imaginary) the parallel curve will break up. An instance given to me by Dr Salmon is the curve $x^{\frac{2}{3}}+y^{\frac{2}{3}}-c^{\frac{2}{3}}=0$ or $(x^2+y^2-c^2)^3+27c^2x^2y^2=0$, here $m=6$, $n=4$,

1 That the order of the evolute is not (in every case of a curve with singularities) one-half this, or $=m+n-\frac{1}{2}(I+J)$, is at once seen by remarking that there is no reason why $I+J$ should be even.

$\delta=4$, $\kappa=6$, $\iota=0$, $\tau=3$: the points I, J are each of them a cusp, the tangents being the line IJ; the number of tangents from a cusp is $n-3$, $=1$, but for the cusp I or J, this tangent is the line IJ itself, so that we have $I=J=4$.

Theory when the Absolute is a conic.

When the Absolute is a conic the formulæ for the evolute are essentially the same as those in the former case, but the formulæ for the parallel curve are modified essentially and in a very remarkable manner. I observe that corresponding to a passage of the given curve through I or J we have a contact with the Absolute, so that in the present case f will properly denote the number of contacts of the given curve with the Absolute, and attending to this singularity only, viz. considering a given curve (m, n, δ, κ, ι, τ; α, D) having f contacts with the Absolute, the formulæ for the evolute are

$$\begin{aligned} m'' &= \alpha - 3f, \\ n'' &= m + n - f, \\ \iota'' &= f, \\ \kappa'' &= -3m - 3n + 3\alpha - 5f. \end{aligned}$$

In the case $f=0$, these at once follow from the two equations $n''=m+n$, and $\iota''=0$. The normal is the line joining a point of the given curve with the pole of the tangent; or, what is the same thing, it is the line joining the point of the given curve with the corresponding point of the reciprocal curve: the degrees of the two curves are m, n, and the correspondence is a (1, 1) correspondence. Hence, by the general theorem previously referred to, it follows that we have $n''=m+n$, and $\iota''=0$. Compare herewith the demonstration of the theorem in the case where the Absolute is a point-pair.

The formulæ for the parallel curve are

$$\begin{aligned} m' &= 2m + 2n - 2f, \\ n' &= 2m + 2n - 2f, \\ \iota' &= 2\alpha - 6f, \\ \kappa' &= 2\alpha - 6f, \\ f' &= 2m + 2n - 2f, \end{aligned}$$

(so that $m'=n'$, $\iota'=\kappa'$). The intersections of the curve and Absolute are in this case the points f each twice, and besides $2m-2f$ points; similarly the common tangents are the tangents at f each twice and besides $2n-2f$ tangents. Now I remark that the parallel curve, when the radius of the variable circle is $=0$, reduces itself to the original curve twice, together with the $2n-2f$ common tangents, and the $2m-2f$ common points; the order is thus $=2m+(2n-2f)$, and the class $=2n+(2m-2f)$: and these are the values in the general case where the radius of the variable circle is not $=0$.

But in the remarkable case where the curve and its evolute have a (1, 2) correspondence, then I correct the formulæ by adding $-\Theta$ to the expressions for ι', κ' respectively. We have for the evolute of the parallel curve

$$\begin{aligned} m''' &= 2m'', \\ n''' &= 2n'', \\ \iota''' &= 2\iota'' + (2m + 2n - 4f) - \Theta, \\ \kappa''' &= 2\kappa'' + (2m + 2n - 4f) - \Theta, \end{aligned}$$

viz. assuming $\Theta = 2m + 2n - 4f$, this means that the evolute is the evolute of the original curve taken twice.

A very interesting case is when $m = n = f$: observe that neither $m - f$ nor $n - f$ can be negative, so that the assumed relation $m + n - 2f = 0$ would imply these two relations. We have here for the parallel curve $m' = 2m$, $n' = 2n$, $\iota' = 2\iota$, $\kappa' = 2\kappa$; the parallel curve in fact breaking up into two curves such as the given curve. And in this case the formulæ for the evolute assume the very simple form $m'' = \kappa$, $n'' = f$, $\iota'' = f$, $\kappa'' = -2f + 3\kappa$.

Whatever the original curve may be, we have for the parallel curve $m' = n' = f'$, so that the formulæ for the evolute of the parallel curve are of the foregoing form $m''' = \kappa'$, $n''' = f'$, $\iota''' = f' - \Theta$, $\kappa''' = -2f' + 3\kappa' - \Theta$, which agree with the above values of m''', n''', ι''', κ'''. In the particular case $m = n = f$, we have $\Theta = 0$, so that the evolute-formulæ, if originally written down without the terms in Θ, would still be $m''' = 2m''$, $n''' = 2n''$, $\iota''' = 2\iota''$, $\kappa''' = 2\kappa''$; viz. the evolute is here the original evolute taken twice; as already seen, the parallel curve consisted of two curves such as the original curve, and each of these has for its evolute the evolute of the original curve.

494.

EXAMPLE OF A SPECIAL DISCRIMINANT.

[From the *Quarterly Journal of Pure and Applied Mathematics*, vol. XI. (1871), pp. 211—213.]

If we have a function $(a, \ldots \between x, y, z)^n$, where the coefficients $(a, \ldots)$ are such that the curve $(a, \ldots \between x, y, z)^n = 0$ has a node, and *à fortiori* if this curve has any number of nodes or cusps, the discriminant of the function (that is, the discriminant of the general function $(* \between x, y, z)^n$, substituting in such discriminant for the coefficients their values for the particular function in question) vanishes *identically*. But the particular function has nevertheless a *special discriminant*, viz. this is a function of the coefficients which, equated to zero, gives the condition that the curve may have (besides the nodes or cusps which it originally possesses) one more node; and the determination of this special discriminant (which, observe, is not deducible from the expression of the discriminant of the general function $(* \between x, y, z)^n$) is an interesting problem. I have, elsewhere, shown that if the curve in question $(a, \ldots \between x, y, z)^n = 0$ has δ nodes and κ cusps, then the degree of the special discriminant in regard to the coefficients a, &c., of the function is $= 3(n-1)^2 - 7\delta - 11\kappa$: and I propose to verify this in the case of a quartic curve with two cusps.

Consider the curve

$$6nx^2y^2 + 12rz^2xy + (4gx + 4iy + cz)\, z^3 = 0,$$

where $x = 0$ is the tangent at a cusp; $y = 0$ the tangent at a cusp; and $z = 0$ the line joining the two cusps.

For the special discriminant we have

$$3nxy^2 + 3ryz^2 + gz^3 = 0,$$

$$3nx^2y + 3rxz^2 + iz^3 = 0,$$

$$z\,\{6rxy + (3gx + 3iy + 4cz)\, z\} = 0\,;$$

the last of which may be replaced by the equation of the curve.

Assume $x = \lambda z$, $y = \mu z$, the first two equations give

$$3(n\lambda\mu + r)\mu + g = 0,$$

$$3(n\lambda\mu + r)\lambda + i = 0,$$

whence also

$$6n\lambda^2\mu^2 + 6r\lambda\mu + g\lambda + i\mu = 0,$$

and the equation of the curve gives

$$6n\lambda^2\mu^2 + 12r\lambda\mu + 4g\lambda + 4i\mu + c = 0,$$

whence eliminating $g\lambda + i\mu$ we find

$$18n\lambda^2\mu^2 + 12r\lambda\mu - c = 0.$$

Moreover the first two equations give

$$9(n\lambda\mu - r)^2\lambda\mu - ig = 0,$$

or putting $\lambda\mu = \theta$ we have

$$18n\theta^2 + 12r\theta - c = 0,$$

$$9(n\theta + r)^2\theta - ig = 0,$$

from which θ is to be eliminated.

The equations are

$$18n\theta^2 + 12r\theta - c = 0,$$

$$9n^2\theta^3 + 18nr\theta^2 + 9r^2\theta - ig = 0,$$

and thence

$$\begin{aligned}
18n^2\theta^3 + 36nr\theta^2 + 18r^2\theta - 2ig &= 0,\\
18n^2\theta^3 + 12nr\theta^2 - cn\theta &= 0,\\
24nr\theta^2 + (18r^2 + cn)\theta - 2ig &= 0,\\
18nr\theta^2 + 12r^2\theta - cr &= 0,\\
(6r^2 + 3cn)\theta - 6ig + 4cr &= 0,
\end{aligned}$$

$$\theta = \frac{6ig - 4cr}{6r^2 + 3cn} = \tfrac{2}{3}\,\frac{3ig - 2cr}{2r^2 + cn};$$

or substituting in $18n\theta^2 + 12r\theta - c = 0$, this is

$$8n(3ig - 2cr)^2 + 8r(3ig - 2cr)(2r^2 + cn) - c(2r^2 + cn)^2 = 0.$$

Hence, developing, the special discriminant is

$$\begin{aligned}
\square = &- 1\, c^3n^2\\
&+ 12\, c^2nr^2\\
&- 72\, cginr\\
&- 36\, cr^4\\
&+ 72\, g^2i^2n\\
&+ 48\, gir^3,
\end{aligned}$$

which is as it should be of the degree 5, $= 3 \cdot 3^2 - 11 \cdot 2$.

495.

ON THE ENVELOPE OF A CERTAIN QUADRIC SURFACE.

[From the *Quarterly Journal of Pure and Applied Mathematics*, vol. XI. (1871), pp. 244—246.]

To find the envelope of the quadric surface

$$ax^2 + by^2 + cz^2 + dw^2 = 0,$$

where the coefficients vary subject to the conditions

$$\begin{cases} a\alpha^2 + b\beta^2 + c\gamma^2 + d\delta^2 = 0, \\ \dfrac{p^2}{a} + \dfrac{q^2}{b} + \dfrac{r^2}{c} + \dfrac{s^2}{d} = 0, \end{cases}$$

$(\alpha, \beta, \gamma, \delta)$ and (p, q, r, s) being respectively constant.

We have in the usual manner

$$x^2 + \lambda\alpha^2 + \mu\frac{p^2}{a^2} = 0,$$

$$y^2 + \lambda\beta^2 + \mu\frac{q^2}{b^2} = 0,$$

$$z^2 + \lambda\gamma^2 + \mu\frac{r^2}{c^2} = 0,$$

$$w^2 + \lambda\delta^2 + \mu\frac{s^2}{d^2} = 0$$

and thence $a^2 = \dfrac{-\mu p^2}{x^2+\lambda\alpha^2}$, &c., and substituting these values μ disappears and we have

$$p\sqrt{(x^2+\lambda\alpha^2)}+q\sqrt{(y^2+\lambda\beta^2)}+r\sqrt{(z^2+\lambda\gamma^2)}+s\sqrt{(w^2+\lambda\delta^2)}=0,$$

$$\frac{\alpha^2 p}{\sqrt{(x^2+\lambda\alpha^2)}}+\frac{\beta^2 q}{\sqrt{(y^2+\lambda\beta^2)}}+\frac{\gamma^2 r}{\sqrt{(z^2+\lambda\gamma^2)}}+\frac{\delta^2 s}{\sqrt{(w^2+\lambda\delta^2)}}=0,$$

from which λ is to be eliminated; the second equation is here the derived function of the first in regard to λ, so that rationalising the first equation, the result is, as will be shown, of the form $(*\between\lambda,\ 1)^4=0$, and the result is obtained by equating to zero the discriminant of the quartic function.

Denoting for shortness the first equation by

$$A+B+C+D=0,$$

the rationalised form is

$$(A^4+B^4+C^4+D^4-2A^2B^2-2A^2C^2-2A^2D^2-2B^2C^2-2B^2D^2-2C^2D^2)^2-64A^2B^2C^2D^2=0,$$

which is of the form

$$-(\mathfrak{A}+2\mathfrak{B}\lambda+\mathfrak{C}\lambda^2)^2+(\mathrm{a},\ \mathrm{b},\ \mathrm{c},\ \mathrm{d},\ \mathrm{e}\between 1,\ \lambda)^4=0,$$

where

$$\begin{aligned}
\mathfrak{A} &= p^4x^4 \ldots - 2p^2q^2x^2y^2\ldots,\\
\mathfrak{B} &= p^4\alpha^2x^2 \ldots - p^2q^2(\alpha^2y^2+\beta^2x^2)\ldots,\\
\mathfrak{C} &= p^4\alpha^4 \ldots - 2p^2q^2\alpha^2\beta^2 \ldots,\\[4pt]
\mathrm{a} &= 8\,.\,x^2y^2z^2w^2,\\
4\mathrm{b} &= 8\,.\,\alpha^2y^2z^2w^2+\ldots,\\
6\mathrm{c} &= 8\,.\,\alpha^2\beta^2z^2w^2+\ldots,\\
4\mathrm{d} &= 8\,.\,\alpha^2\beta^2\gamma^2w^2+\ldots,\\
\mathrm{e} &= 8\,.\,\alpha^2\beta^2\gamma^2\delta^2.
\end{aligned}$$

Writing I', J' for the two invariants we find without difficulty

$$\begin{aligned}
I' &= I-\tfrac{4}{3}P+\Delta^2,\\
J' &= J-Q+\tfrac{1}{3}\Delta P-\tfrac{8}{27}\Delta^3,
\end{aligned}$$

where

$$\begin{aligned}
I &= \mathrm{ae}-4\mathrm{bd}+3\mathrm{c}^2,\\
J &= \mathrm{ace}-\mathrm{ad}^2-\mathrm{b}^2\mathrm{e}-\mathrm{c}^3+2\mathrm{bcd},\\
\Delta &= \mathfrak{AC}-\mathfrak{B}^2,\\
P &= \mathrm{a}\mathfrak{C}^2-4\mathrm{b}\mathfrak{BC}+2\mathrm{c}(\mathfrak{AC}+2\mathfrak{B}^2)-4\mathrm{d}\mathfrak{AB}+\mathrm{e}\mathfrak{A}^2,\\
Q &= (\mathrm{ce}-\mathrm{d}^2)\mathfrak{A}^2+(\mathrm{ae}+2\mathrm{bd}-3\mathrm{c}^2)\,.\,\tfrac{1}{3}(\mathfrak{AC}+2\mathfrak{B}^2)+(\mathrm{ac}-\mathrm{b}^2)\mathfrak{C}^2\\
&\quad -2(\mathrm{ad}-\mathrm{bc})\mathfrak{BC} \qquad\qquad -2(\mathrm{be}-\mathrm{cd})\mathfrak{AB}.
\end{aligned}$$

The result thus is

$$(I - P + \tfrac{4}{3}\Delta^2)^3 - 27\,(J - Q + \tfrac{1}{3}\Delta P - \tfrac{8}{27}\Delta^3)^2 = 0,$$

or, what is the same thing, it is

$$\begin{aligned}(I - P)^3 - 27\,(J - Q)^2 - 9\Delta P\,(J - 2Q)&\\ +\ \Delta^2\,(4I^2 - 8IP + P^2)&\\ +\,8\Delta^3\,(J - 2Q)&\\ +\ \Delta^4 . \tfrac{16}{3} I = 0,&\end{aligned}$$

where the left-hand side is of the order 24 in (x, y, z, w). I apprehend that the order should be $=12$ only; for writing (x, y, z, w) in place of (x^2, y^2, z^2, w^2), the equations which connect (a, b, c, d) express that these quantities are the coordinates of a point on a plane cubic; and the problem is in fact that of finding the reciprocal of the plane cubic: this is a sextic cone, or restoring (x^2, y^2, z^2, w^2) instead of (x, y, z, w), we should have a surface of the order 12. I cannot explain how the reduction is effected.

496.

TABLES OF THE BINARY CUBIC FORMS FOR THE NEGATIVE DETERMINANTS, $\equiv 0$ (MOD. 4), FROM -4 to -400; AND $\equiv 1$ (MOD. 4), FROM -3 TO -99; AND FOR FIVE IRREGULAR NEGATIVE DETERMINANTS.

[From the *Quarterly Journal of Pure and Applied Mathematics*, vol. XI. (1871), pp. 246—261.]

THE theory of binary cubic forms for determinants, as well positive as negative, has been studied by M. Arndt in the memoir "Versuch einer Theorie der homogenen Functionen des dritten Grades mit zwei Variabeln," *Grunert's Archiv*, t. XVII. (1851, pp. 1—54); and in the later memoir, "Tabellarische Berechnung der reducirten binären cubischen Formen und Klassification derselben für alle negativen Determinanten $(-D)$ von $D=3$ bis $D\equiv 2000$," *ditto*, t. XXXI. (1858), pp. 335—445, he has given a very valuable Table of the forms for a Negative Determinant. It has appeared to me suitable to arrange this Table in the manner made use of for Quadratic Forms in my memoir "Tables des formes quadratiques binaires pour les déterminants négatifs $D\equiv -1$ jusqu'à $D\equiv -100$, pour les déterminants positifs non carrés depuis $D\equiv 2$ jusqu'à $D\equiv 99$, et pour les treize déterminants négatifs du premier millier," *Crelle*, t. LX. (1862), pp. 357—372, [335]; and confining myself to the limits of the last-mentioned tables I deduce from that of M. Arndt the three Tables which follow.

To explain the arrangement, I give in the first instance the following extract from M. Arndt's Table:

D.	*Reducirte Formen mit Charakteristik.*	*Klassen.*
3	(0, 1, 1, 0) (1, 0, − 1, − 1) (1, 1, 0, − 1) (2, 1, 2) (2, 1, 2) (2, 1, 2)	(0, 1, 1, 0), (1, 0, − 1, ± 1)
4	(0, 1, 0, − 1) (1, 0, − 1, 0) (2, 0, 2) (2, 0, 2)	(0, − 1, 0, 1)

D.	*Reducirte Formen mit Charakteristik.*	*Klassen.*
7	(0, 1, 1, − 1) (2, 1, 4)	(0, − 1, − 1, 1)
8	(0, 1, 0, − 2) (2, 0, 4)	(0, − 1, 0, 2)
11	(0, 1, 1, − 2) (2, 1, 6)	(0, − 1, − 1, 2)
12	(0, 1, 0, − 3) (2, 0, 6)	(0, − 1, 0, 3)
15 ⋮	(0, 1, 1, − 3) (2, 1, 8)	(0, − 1, − 1, 3)
44 ⋮	(0, 1, 0, − 11) (1, − 1, − 2, 0) (2, 0, 22) (6, 2, 8)	(0, − 1, 0, 11), (0, − 2, ± 1, 1)
112 ⋮	(0, 1, 0, − 28) (0, 2, 2, − 2) (1, 2, 0, − 4) (2, 0, 56) (8, 4, 16) (8, 4, 16) (1, − 1, − 3, − 1) (8, 4, 16)	(0, − 1, 0, 28), (0, 2, 2, − 2), (1, ± 1, − 3, ± 1)
144	(0, 1, 0, − 36) (0, 2, 2, − 3) (2, 0, 72) (8, 4, 20)	(0, − 1, 0, 36), (0, − 2, − 2, 3)
156	(0, 1, 0, − 39) (1, − 1, − 3, 1) (2, 0, 78) (8, 2, 20)	(0, − 1, 0, 39), (1, ∓ 1, − 3, ± 1)
216	(0, 1, 0, − 54) (1, − 2, − 3, 0) (2, 0, − 3, 0) (2, 0, 108) (14, 6, 18) (12, 0, 18)	(0, − 1, 0, 54), (0, ∓ 3, 0, ± 2), (0, ∓ 3, 2, ± 1)

The first column contains the value of the determinant, the second column contains the *reduced* forms, omitting the *contrary* and *opposite* forms; viz. for the cubic form (a, b, c, d), the contrary form (equal, that is, properly equivalent to the given form) is $(-a, -b, -c, -d)$; and the opposite form (improperly equivalent to the given form) is $(a, -b, c, -d)$ or $(-a, b, -c, d)$; this second column contains also the *characteristic* of each cubic form, viz. the cubic form (a, b, c, d) has for its characteristic the quadratic form

$$\{2(b^2 - ac),\ bc - ad,\ 2(c^2 - bd)\},$$

(so that the cubic form and its characteristic have the same determinant

$$-D = (bc - ad)^2 - 4(b^2 - ac)(c^2 - bd),\ \equiv 1 \text{ or } 0 \pmod{4}),$$

and a cubic form which corresponds to a reduced characteristic is itself a reduced form. The third column contains for each determinant the entire series of *unequal* cubic forms (that is of the forms whereof no two are properly equivalent to each other), the representatives of the *classes* for this determinant. M. Arndt has included in his table the non-primitive classes (for example Det. $= -112$, the form $(0, 2, 2, -2)$), for which the terms (a, b, c, d) have a common divisor μ, but as these are at once deducible from the classes which belong to the determinant $= -\dfrac{D}{\mu^4}$, it seems better to omit the non-primitive classes.

The two opposite forms included in a single expression by means of the sign $\pm$ have opposite characteristics which are for the most part unequal to each other, for instance

$$\begin{array}{lll} \text{Det.} -44\,;\ (0,\ -2,\ 1,\ 1) & \text{has the characteristic} & (6,\ -2,\ 8), \\ \qquad (0,\ 2,\ 1,\ -1) & \quad ,, \qquad ,, & (6,\ \ 2,\ 8), \end{array}$$

where (6, -2, 8), (6, 2, 8) are unequal forms, but this is not always the case, for instance

$$\begin{array}{lll} \text{Det.} -112\,;\ (1,\ -1,\ -3,\ -1) & \text{has the characteristic} & (8,\ -4,\ 16), \\ \qquad (1,\ 1,\ -3,\ 1) & \quad ,, \qquad ,, & (8,\ \ 4,\ 16), \end{array}$$

where (8, -4, 16) = (8, 4, 16), since each is an ambiguous form. Instead of the two unequal forms (1, -1, -3, -1), (1, 1, -3, 1) which correspond to the opposite (though equal) characteristics (8, -4, 16), (8, 4, 16), M. Arndt might have given the two forms (1, 2, 0, -4) and (1, -1, -3, -1) corresponding to the *same* characteristic (8, 4, 16); but then it would not have appeared at a glance that the two classes were opposite to each other; and I presume that it is for this reason that he has selected the two representative forms (1, -1, -3, -1) and (1, 1, -3, 1). It must not, however, be imagined that the opposite cubic forms which correspond to opposite characteristics, which are ambiguous (and therefore equal to each other), are always, as in the last preceding example, unequal: for example Det. -144, there is only the form (0, -2, -2, 3) given as corresponding to the ambiguous characteristic (8, 4, 20); the opposite form (0, 2, -2, -3) corresponding to the opposite but equal characteristic (8, -4, 20) is equal to (0, -2, -2, 3), and so does not give rise to a distinct opposite class. In the new tables, the sign $\pm$ is only employed in regard to opposite ambiguous characteristics; for instance, Det. -4×28 there are given (not included in a single expression by means of the sign $\pm$) the two forms (1, -1, -3, 1), (1, 1, -3, 1) corresponding to the characteristic $2\,(2,\ \pm 1,\ 4)$.

I remark that, in a few instances M. Arndt, in passing from the second to the third column, has modified the expression for a cubic form in such manner that the characteristic has ceased to be a reduced form; for instance, Det. -216, he has given in the third column the two forms (0, ∓ 3, 2, ± 1) belonging to the characteristic (18, ∓ 6, 14); it would have been better, it appears to me, to preserve the expression of the second column (1, -2, -3, 0), and adopt the two representative forms (1, ∓ 2, -3, 0); I have accordingly made this change.

I divide M. Arndt's table into two tables; the first of them corresponding to the determinants $\equiv 0$ (mod. 4), the second to the determinants $\equiv 1$ (mod. 4). In the first table I take for the characteristic the form

$$\{b^2 - ac,\ \tfrac{1}{2}(bc - ad),\ bd - c^2\},$$

which belongs to the determinant $-\frac{1}{4}D$, and I arrange the cubic classes according to their *order*; viz. we have the properly primitive order (*pp*) when the terms $(a, 3b, 3c, d)$ have no common divisor; and the improperly primitive order (*ip*) when the terms $(a, 3b, 3c, d)$ have no common divisor other than 3, or what is the same thing when a and d being each of them divisible by 3, the terms (a, b, c, d) have no common divisor. But, moreover, the characteristic $\{b^2 - ac, \frac{1}{2}(bc - ad), bd - c^2\}$, may be of the properly primitive order *pp*; or of the improperly primitive order *ip*; or it may be of a derived order $\mu(A', B', C')$, $= \mu . pp$ or $\mu . ip$, according as (A', B', C') considered as a form belonging to the determinant $B'^2 - A'C'$, $= -\frac{1}{4\mu^2}D$, is of the properly primitive or the improperly primitive order. And in these different cases, the cubic class is said to be of the order *pp* on *pp*, *pp* on *ip*, *pp* on $\mu . pp$, *pp* on $\mu . ip$, *ip* on *pp*, &c., as the case may be.

For the determinants $\equiv 1$ (mod. 4), I retain the characteristic

$$\{2(b^2 - ac),\ bc - ad,\ 2(c^2 - bd)\},$$

and this being so, the division into orders is the same as in the former case; only as the characteristic, when primitive, is of necessity improperly primitive, the orders *pp* on *pp* and *ip* on *pp* no longer exist.

To every characteristic I annex in the tables the symbol of its composition; viz. 1 denotes the principal form, c a form which by its duplication, d a form which by its triplication, &c., produces the principal form, σ denotes the most simple form of order *ip*, σc, σd, &c., the forms obtained by combining σ with the forms c, d, &c., of the order *pp*. Similarly to a characteristic $\mu(A', B', C')$ I annex the symbol of composition of the form (A', B', C'), $\left(\text{considered as belonging to the determinant } B'^2 - A'C', = -\frac{D}{4\mu^2}\right)$ multiplying this symbol by the number μ; for instance, $\mu . 1$ denotes that (A', B', C') is the principal form, and similarly in other cases.

I have given a third table for the determinants

$$-4 \times 243,\ -4 \times 307,\ -4 \times 339,\ -4 \times 459,\ -4 \times 675,$$

where -243, -307, -339, -459, -675 are those of the thirteen irregular negative determinants in the first thousand for which the number of classes is divisible by 3. The number -4×675, $= -2700$, is beyond the limits of M. Arndt's Table, but the calculation (at least for the order *pp* on *pp*) presents no difficulty.

I remark that, according to M. Arndt (*Grunert*, t. XVII. p. 19), the number of cubic forms corresponding to a given characteristic (A, B, C) is equal to the number of proper transformations of $(A, -B, C)$, Det. D, into $(\frac{1}{2}A^2, B^2 - \frac{1}{2}AC, \frac{1}{2}C^2)$, Det. DB^2, so that when there is no such transformation, there exists no cubic form corresponding to the characteristic (A, B, C). This includes, I believe, the theorem in a letter of mine to M. Hermite, *Quarterly Mathematical Journal*, t. I. (1857), p. 85, [162], viz. that for a *pp* form (A, B, C) of negative determinant, there is either no corresponding cubic form, or else a single corresponding cubic form, according as (A, B, C) does not, or does, produce by its triplication the principal form; but the particular theorem, in the cases to which it applies, is the more convenient one: it shows at once that for a regular negative determinant the number of cubic forms corresponding to a properly primitive characteristic (or, what is the same thing, number of cubic classes of the order (*pp* or *ip*) on *pp*) is 1 or 3, according as the number of quadratic classes is not, or is, divisible by 3.

The inspection of the tables gives rise to other remarks, but at present I abstain from pursuing the subject further; I will only notice that in some instances, for example Det. -224, the classes which correspond to characteristics of the principal genus are partly of the order *pp* on *pp* and partly of the order *ip* on *pp*.

Table I. *of the binary cubic forms, the determinants of which are the negative numbers* $\equiv 0$ (*mod.* 4) *from* -4 *to* -400.

Det. 4×	Classes	Order	on	Charact.	Compn.
1	0, −1, 0, 1	*pp*	*pp*	1, 0, 1	1
2	0, −1, 0, 2	*pp*	*pp*	1, 0, 2	1
3	0, −1, 0, 3	*ip*	*pp*	1, 0, 3	1
4	0, −1, 0, 4	*pp*	*pp*	1, 0, 4	1
	1, −1, −1, 1	*pp*	2*pp*	2 (1, 0, 1)	2 . 1
5	0, −1, 0, 5	*pp*	*pp*	1, 0, 5	1
6	0, −1, 0, 6	*ip*	*pp*	1, 0, 6	1
7	0, −1, 0, 7	*pp*	*pp*	1, 0, 7	1
	1, 0, −2, 2 1, 0, −2, −2	*pp*	2*ip*	2 (2, ± 1, 4)	2σ
8	0, −1, 0, 8	*pp*	*pp*	1, 0, 8	1
	0, −2, 0, 1	*pp*	2*pp*	2 (1, 0, 2)	2 . 1
9	0, −1, 0, 9	*ip*	*pp*	1, 0, 9	1
10	0, −1, 0, 10	*pp*	*pp*	1, 0, 10	1
11	0, −1, 0, 11	*pp*	*pp*	1, 0, 11	1
	0, −2, −1, 1			3, 1, 4	α
	0, −2, 1, 1			3, −1, 4	α^2
12	0, −1, 0, 12	*ip*	*pp*	1, 0, 12	1
13	0, −1, 0, 13	*pp*	*pp*	1, 0, 13	1
14	0, −1, 0, 14	*pp*	*pp*	1, 0, 14	1

Det. 4×	Classes	Order	on	Charact.	Compn.
15	0, − 1, 0, 15	*ip*	*pp*	1, 0, 15	1
	1, − 2, 0, 2	} *pp*	*ip*	4, ± 1, 4	*σc*
	1, 2, 0, − 2				
16	0, − 1, 0, 16	*pp*	*pp*	1, 0, 16	1
17	0, − 1, 0, 17	*pp*	*pp*	1, 0, 17	1
18	0, − 1, 0, 18	*ip*	*pp*	1, 0, 18	1
	1, 1, − 2, − 2	} *pp*	3*pp*	3 (1, 0, 2)	3 . 1
	1, − 1, − 2, 2				
19	0, − 1, 0, 19	} *pp*	*pp*	1, 0, 19	1
	0, 2, 1, − 2			4, 1, 5	*d*
	0, − 2, 1, 2			4, − 1, 5	d^2
20	0, − 1, 0, 20	*pp*	*pp*	1, 0, 20	1
	0, − 2, − 2, 1	*pp*	2*pp*	2 (2, 1, 3)	2*c*
21	0, − 1, 0, 21	*ip*	*pp*	1, 0, 21	1
22	0, − 1, 0, 22	*pp*	*pp*	1, 0, 22	1
23	0, − 1, 0, 23	} *pp*	*pp*	1, 0, 23	1
	1, − 1, − 2, 4			3, − 1, 8	d^2
	1, 1, − 2, − 4			3, 1, 8	*d*
24	0, − 1, 0, 24	*ip*	*pp*	1, 0, 24	1
	0, − 2, 0, 3	*ip*	2*pp*	2 (2, 0, 3)	2*c*
25	0, − 1, 0, 25	*pp*	*pp*	1, 0, 25	1
	1, − 2, − 1, 2	} *pp*	5*pp*	5 (1, 0, 1)	5 . 1
	1, 2, − 1, − 2				
26	0, − 1, 0, 26	} *pp*	*pp*	1, 0, 26	1
	1, 0, − 3, 2			3, − 1, 9	g^2
	1, 0, − 3, − 2			3, 1, 9	g^4
27	0, − 1, 0, 27	} *ip*	*pp*	1, 0, 27	1
	0, − 2, 1, 3			4, − 1, 7	d^2
	0, 2, 1, − 3			4, 1, 7	*d*
	0, − 3, 0, 1	*pp*	3*pp*	3 (1, 0, 3)	3 . 1
28	0, − 1, 0, 28	*pp*	*pp*	1, 0, 28	1
	1, 1, − 3, 1	} *pp*	2*ip*	2 (2, ± 1, 4)	2*σ*
	1, − 1, − 3, − 1				

Det. 4 ×	Classes	Order	on	Charact.	Compn.
29	0, − 1, 0, 29			1, 0, 29	1
	2, 1, − 2, − 2	*pp*	*pp*	5, 1, 6	g^2
	2, − 1, − 2, 2			5, − 1, 6	g^4
30	0, − 1, 0, 30	*ip*	*pp*	1, 0, 30	1
31	0, − 1, 0, 31			1, 0, 31	1
	2, 1, − 2, − 3	*pp*	*pp*	5, 2, 7	d
	2, − 1, − 2, 3			3, − 2, 7	d^2
32	0, − 1, 0, 32	*pp*	*pp*	1, 0, 32	1
33	0, − 1, 0, 33	*ip*	*pp*	1, 0, 33	1
34	0, − 1, 0, 34	*pp*	*pp*	1, 0, 34	1
35	0, − 1, 0, 35			1, 0, 35	1
	0, 2, 1, − 4	*pp*	*pp*	4, 1, 9	g^2
	0, − 2, 1, 4			4, − 1, 9	g^4
36	0, − 1, 0, 36	*ip*	*pp*	1, 0, 36	1
	0, − 2, − 2, 3	*ip*	2*pp*	2 (2, 1, 5)	2c
37	0, − 1, 0, 37	*pp*	*pp*	1, 0, 37	1
38	0, − 1, 0, 38			1, 0, 38	1
	2, − 2, − 1, 3	*ip*	*pp*	6, 2, 7	g^2
	2, 2, − 1, − 3			6, − 2, 7	g^4
39	0, − 1, 0, 39	*ip*	*pp*	1, 0, 39	1
	1, 1, − 3, − 1	*pp*	*ip*	4, − 1, 10	σe
	1, − 1, − 3, 1			4, 1, 10	σe^3
40	0, − 1, 0, 40	*pp*	*pp*	1, 0, 40	1
	0, − 2, 0, 5	*pp*	2*pp*	2 (2, 0, 5)	2 . 1
41	0, − 1, 0, 41	*pp*	*pp*	1, 0, 41	1
42	0, − 1, 0, 42	*ip*	*pp*	1, 0, 42	1
43	0, − 1, 0, 43			1, 0, 43	1
	0, − 2, 1, 5	*pp*	*pp*	4, − 1, 11	d^2
	0, 2, 1, − 5			4, 1, 11	d
44	0, − 1, 0, 44	*pp*	*pp*	1, 0, 44	1
	1, 2, − 1, − 4	*pp*	2*ip*	2 (2, ± 1, 6)	2σ
	1, − 2, − 1, 4				

Det. 4 ×	Classes	Order		Charact.	Compn.
			on		
45	0, − 1, 0, 45	ip	pp	1, 0, 45	1
	2, 0, − 3, 3	} pp	3pp	3 (2, ± 1, 3)	3c
	2, 0, − 3, − 3				
46	0, − 1, 0, 46	pp	pp	1, 0, 46	1
47	0, − 1, 0, 47	pp	pp	1, 0, 47	1
	1, − 2, − 2, 2	} pp	ip	6, 1, 8	σf
	1, 2, − 2, − 2			6, − 1, 8	σf^4
48	0, − 1, 0, 48	ip	pp	1, 0, 48	1
	1, − 1, − 3, 3	} pp	4pp	4 (1, 0, 3)	4.1
	1, 1, − 3, − 3				
49	0, − 1, 0, 49	pp	pp	1, 0, 49	1
50	0, − 1, 0, 50	} pp	pp	1, 0, 50	1
	2, 0, − 3, 2			6, − 2, 9	g^4
	2, 0, − 3, − 2			6, 2, 9	g^2
51	0, − 1, 0, 51	} ip	pp	1, 0, 51	1
	0, − 2, 1, 6			4, − 1, 13	g^4
	0, 2, 1, − 6			4, 1, 13	g^2
52	0, − 1, 0, 52	pp	pp	1, 0, 52	1
	0, − 2, − 2, 5	pp	2pp	2 (2, 1, 7)	2c
53	0, − 1, 0, 53	} pp	pp	1, 0, 53	1
	1, − 3, 0, 2			6, − 1, 9	g^2
	1, 3, 0, − 2			6, 1, 9	g^4
54	0, − 1, 0, 54	ip }	pp	1, 0, 54	1
	1, 2, − 3, 0	pp }		7, − 3, 9	g^4
	1, − 2, − 3, 0	pp }		7, 3, 9	g^2
	0, − 3, 0, 2	} pp	3pp	3 (2, 0, 3)	3c
	0, 3, 0, − 2				
55	0, − 1, 0, 55	pp	pp	1, 0, 55	1
	1, − 1, − 3, 5	} pp	ip	4, − 1, 14	σe^3
	1, 1, − 3, − 5			4, 1, 14	σe
56	0, − 1, 0, 56	pp	pp	1, 0, 56	1
	0, − 2, 0, 7	pp	2pp	2 (2, 0, 7)	$2e^2$
57	0, − 1, 0, 57	ip	pp	1, 0, 57	1
58	0, − 1, 0, 58	pp	pp	1, 0, 58	1

Det. 4×	Classes	Order	on	Charact.	Compn.
59	0, − 1, 0, 59			1, 0, 59	1
	0, − 2, 1, 7	*pp*	*pp*	4, − 1, 15	j^6
	0, 2, 1, − 7			4, 1, 15	j^3
60	0, − 1, 0, 60	*ip*	*pp*	1, 0, 60	1
	1, 0, − 4, 4	*pp*	2*ip*	2 (2, ± 1, 8)	2σ
	1, 0, − 4, − 4				
61	0, − 1, 0, 61			1, 0, 61	1
	1, − 2, − 1, 6	*pp*	*pp*	5, − 2, 13	g^2
	1, 2, − 1, − 6			5, 2, 13	g^4
62	0, − 1, 0, 62	*pp*	*pp*	1, 0, 62	1
63	0, − 1, 0, 63	*ip*	*pp*	1, 0, 63	1
	1, 0, − 4, 2	*pp*	*ip*	4, − 1, 16	σe
	1, 0, − 4, − 2			4, 1, 16	σe^3
64	0, − 1, 0, 64	*pp*	*pp*	1, 0, 64	1
	1, 0, − 4, 0	*pp*	4*pp*	4 (1, 0, 4)	4 . 1
65	0, − 1, 0, 65	*pp*	*pp*	1, 0, 65	1
66	0, − 1, 0, 66	*ip*	*pp*	1, 0, 66	1
67	0, − 1, 0, 67			1, 0, 67	1
	0, − 2, 1, 8	*pp*	*pp*	4, − 1, 17	d^2
	0, 2, 1, − 8			4, 1, 17	d
68	0, − 1, 0, 68	*pp*	*pp*	1, 0, 68	1
	0, − 2, − 2, 7	*pp*	2*pp*	2 (2, 1, 9)	$2e^2$
69	0, − 1, 0, 69	*ip*	*pp*	1, 0, 69	1
70	0, − 1, 0, 70	*pp*	*pp*	1, 0, 70	1
71	0, − 1, 0, 71	*pp*	*pp*	1, 0, 71	1
	1, − 3, 1, 3	*pp*	*ip*	8, − 3, 10	σh^3
	1, 3, 1, − 3			8, 3, 10	σh^4
72	0, − 1, 0, 72	*ip*	*pp*	1, 0, 72	1
	0, − 2, 0, 9	*ip*	2*pp*	2 (2, 0, 9)	$2c$
	1, − 2, − 2, 4	*pp*	6*pp*	6 (1, 0, 2)	6 . 1
	1, 2, − 2, − 4				
	2, − 3, 0, 3	*pp*	3*pp*	3 (3, ± 1, 3)	$3c$
	2, 3, 0, − 3				
73	0, − 1, 0, 73	*pp*	*pp*	1, 0, 73	1

Det. 4×	Classes	Order	on	Charact.	Compn.
74	0, – 1, 0, 74	pp	pp	1, 0, 74	1
75	0, – 1, 0, 75	ip	pp	1, 0, 75	1
	0, – 2, 1, 9			4, – 1, 19	g^2
	0, 2, 1, – 9			4, 1, 19	g^4
76	0, – 1, 0, 76	pp	pp	1, 0, 76	1
	0, – 4, 1, 1			5, – 2, 16	g^2
	0, 4, 1, – 1			5, 2, 16	g^4
77	0, – 1, 0, 77	pp	pp	1, 0, 77	1
78	0, – 1, 0, 78	ip	pp	1, 0, 78	1
79	0, – 1, 0, 79	pp	pp	1, 0, 79	1
	2, – 2, – 2, 3	pp	ip	8, – 1, 10	σf
	2, 2, – 2, – 3			8, 1, 10	σf^4
80	0, – 1, 0, 80	pp	pp	1, 0, 80	1
81	0, – 1, 0, 81	pp	pp	1, 0, 81	1
	0, – 3, 2, 2	ip		9, – 3, 10	g^2
	0, 3, 2, – 2	ip		9, 3, 10	g^4
82	0, – 1, 0, 82	pp	pp	1, 0, 82	1
83	0, – 1, 0, 83	pp	pp	1, 0, 83	1
	0, – 2, 1, 10			4, – 1, 21	j^6
	0, 2, 1, – 10			4, 1, 21	j^3
84	0, – 1, 0, 84	ip	pp	1, 0, 84	1
	0, – 2, – 2, 9	ip	2pp	2 (2, 1, 11)	2σ
85	0, – 1, 0, 85	pp	pp	1, 0, 85	1
86	0, – 1, 0, 86	pp	pp	1, 0, 86	1
87	0, – 1, 0, 87	ip	pp	1, 0, 87	1
	1, – 2, – 3, 2	pp		7, – 2, 13	g^4
	1, 2, – 3, – 2	pp		7, 2, 13	g^2
88	0, – 1, 0, 88	pp	pp	1, 0, 88	1
	0, – 2, 0, 11	pp	2pp	2 (2, 0, 11)	2σ
89	0, – 1, 0, 89	pp	pp	1, 0, 89	1
	1, – 1, – 4, 2			5, – 1, 18	m^8
	1, 1, – 4, – 2			5, 1, 18	m^4
90	0, – 1, 0, 90	ip	pp	1, 0, 90	1

Det. 4×	Classes	Order	on	Charact.	Compn.
91	0, − 1, 0, 91	pp	pp	1, 0, 91	1
	0, − 2, 1, 11			4, − 1, 23	g^4
	0, 2, 1, − 11			4, 1, 23	g^2
92	0, − 1, 0, 92	pp	pp	1, 0, 92	1
	2, − 3, 0, 4			9, − 4, 12	g^4
	2, 3, 0, − 4			9, 4, 12	g^2
93	0, − 1, 0, 93	ip	pp	1, 0, 93	1
94	0, − 1, 0, 94	ip	pp	1, 0, 94	1
95	0, − 1, 0, 95	pp	pp	1, 0, 95	1
	1, − 2, − 2, 6	pp	ip	6, − 1, 16	σi^7
	1, 2, − 2, − 6			6, 1, 16	σi
96	0, − 1, 0, 96	ip	pp	1, 0, 96	1
97	0, − 1, 0, 97	pp	pp	1, 0, 97	1
98	0, − 1, 0, 98	pp	pp	1, 0, 98	1
99	0, − 1, 0, 99	ip	pp	1, 0, 99	1
	0, − 2, 1, 12			4, − 1, 25	g^2
	0, 2, 1, − 12			4, 1, 25	g^4
100	0, − 1, 0, 100	pp	pp	1, 0, 100	1
	0, − 2, − 2, 11	pp	2pp	2 (2, 1, 13)	2.1
	1, − 1, − 4, 4	pp	5pp	5 (1, 0, 4)	5.1
	1, 1, − 4, − 4				
	1, − 3, − 1, 3	pp	10pp	10 (1, 0, 1)	10.1
	1, 3, − 1, − 3				

Table II. *of the binary cubic forms the determinants of which are the positive numbers* ≡ 1 (mod. 4) *from* − 3 to − 99.

Det. 4×	Classes	Order	on	Charact.	Compn.
3	0, 1, 1, 0	ip	ip	2, ± 1, 2	σ
	1, 0, − 1, 1	pp			
	1, 0, − 1, − 1	pp			
7	0, − 1, − 1, 1	pp	ip	2, 1, 4	σ
11	0, − 1, − 1, 2	pp	ip	2, 1, 6	σ
15	0, − 1, − 1, 3	ip	ip	2, 1, 8	σ
19	0, − 1, − 1, 4	pp	ip	2, 1, 10	σ

TABLES OF THE BINARY CUBIC FORMS

Det.		Order			
4 ×	Classes		on	Charact.	Compn.
23	0, − 1, − 1, 5			2, 1, 12	σ
	1, − 1, − 1, 2	pp	ip	4, − 1, 6	σd
	1, 1, − 1, − 2			4, 1, 6	σd^2
27	0, − 1, − 1, 6	ip	ip	2, 1, 14	σ
	1, 1, − 2, − 1				
	1, − 1, − 2, 1	pp	$3ip$	3 (2, ± 1, 2)	3σ
	2, − 1, − 1, 2				
31	0, − 1, − 1, 7			2, 1, 16	σ
	1, 0, − 2, 1	pp	ip	4, − 1, 8	σd
	1, 0, − 2, − 1			4, 1, 8	σd^2
35	0, − 1, − 1, 8	pp	ip	2, 1, 18	σ
39	0, − 1, − 1, 9	ip	ip	2, 1, 20	σ
43	0, − 1, − 1, 10	pp	ip	2, 1, 22	σ
47	0, − 1, − 1, 11	pp	ip	2, 1, 24	σ
51	0, − 1, − 1, 12	ip	ip	2, 1, 26	σ
55	0, − 1, − 1, 13	pp	ip	2, 1, 28	σ
59	0, − 1, − 1, 14			2, 1, 30	σ
	1, − 1, − 2, 1	pp	ip	6, 1, 10	σj
	1, 1, − 2, − 1			6, − 1, 10	σj^2
63	0, − 1, − 1, 15	ip	ip	2, 1, 32	σ
67	0, − 1, − 1, 16	pp	ip	2, 1, 34	σ
71	0, − 1, − 1, 17	pp	ip	2, 1, 36	σ
75	0, − 1, − 1, 18	ip	ip	2, 1, 38	σ
79	0, − 1, − 1, 19	pp	ip	2, 1, 40	σ
83	0, − 1, − 1, 20			2, 1, 42	σ
	1, − 1, − 2, 3	pp	ip	6, − 1, 14	σj^2
	1, 1, − 2, − 3			6, 1, 14	σj
87	0, − 1, − 1, 21	ip		2, 1, 44	σ
	1, − 2, 0, 3	pp	ip	8, − 3, 12	σg^2
	1, 2, 0, − 3	pp		8, 3, 12	σg^4
91	0, − 1, − 1, 22	pp	ip	2, 1, 46	σ
95	0, − 1, − 1, 23	pp	ip	2, 1, 48	σ
99	0, − 1, − 1, 24	ip	ip	2, 1, 50	σ
	1, 0, − 3, 3				
	1, 0, − 3, − 3	pp	$3ip$	3 (2, ± 1, 6)	3σ

Table III. *of the binary cubic forms the determinants of which are the negative numbers* – 972, – 1228, 1336, – 1836 *et* – 2700. (– 4 × 675 = – 2700 is beyond Arndt's Tables.)

Det. 4 ×	Classes	Order	on	Charact.	Compn.
243	0, – 1, 0, 243	*ip*	*pp*	1, 0, 243	1
	1, – 1, – 6, 0	*pp*		7, 3, 36	d
	1, 1, – 6, 0	*pp*		7, – 3, 36	d^2
	0, 2, 1, – 30	*ip*		4, 1, 61	d_1
	2, 3, – 2, – 5	*pp*		13, – 2, 19	dd_1
	0, 3, 2, – 8	*pp*		9, 3, 28	d^2d_1
	0, – 2, 1, 30	*ip*		4, – 1, 61	d_1^2
	0, – 3, 2, 8	*pp*		9, – 3, 28	dd_1^2
	2, – 3, – 2, 5	*pp*		13, 2, 19	$d^2d_1^2$
307	0, – 1, 0, 307	*pp*	*pp*	1, 0, 307	1
	1, 1, – 6, – 8			7, 1, 44	d
	1, – 1, – 6, 8			7, – 1, 44	d^2
	0, 2, 1, – 38			4, 1, 77	d_1
	1, – 3, – 2, 8			11, – 1, 28	dd_1
	4, 1, – 4, – 3			17, 4, 19	d^2d_1
	0, – 2, 1, 38			4, – 1, 77	d_1^2
	4, – 1, – 4, 3			17, – 4, 19	dd_1^2
	1, 3, – 2, 8			11, 1, 28	$d^2d_1^2$
339	0, – 1, 0, 339	*ip*	*pp*	1, 0, 339	1
	1, 0, – 7, – 4	*pp*		7, 2, 49	d
	1, 0, – 7, 4	*pp*		7, – 2, 49	d^2
	0, 2, 1, – 42	*ip*		4, 1, 85	d_1
	3, 0, – 5, – 4	*pp*		15, 6, 25	dd_1
	2, – 3, – 2, 8	*pp*		13, – 5, 28	d^2d_1
	0, – 2, 1, 42	*ip*		4, – 1, 85	d_1^2
	2, 3, – 2, – 8	*pp*		13, 5, 28	dd_1^2
	3, 0, – 5, 4	*pp*		15, – 6, 25	$d^2d_1^2$
459	0, – 1, 0, 459	*ip*	*pp*	1, 0, 459	1
	0, 3, 2, – 16	*pp*		9, 3, 52	d
	0, – 3, 2, 16	*pp*		9, – 3, 52	d^2
	0, 2, 1, – 57	*ip*		4, 1, 115	d_1
	1, 4, – 3, – 4	*pp*		19, – 4, 25	dd_1
	2, – 1, 6, 0	*pp*		13, 3, 36	d^2d_1
	0, – 2, 1, 57	*ip*		4, – 1, 115	d_1^2
	2, 1, – 6, 0	*pp*		13, – 3, 36	dd_1^2
	1, – 4, – 3, 4	*pp*		19, 4, 25	$d^2d_1^2$
	0, – 3, 0, 17	*pp*	3*pp*	3 (3, 0, 17)	$3g^3$

Det. 4×	Classes	Order	on	Charact.	Compn.
675	0, − 1, 0, 675	ip	pp	1, 0, 675	1
	0, 3, 2, − 24			9, 3, 76	d
	0, − 3, 2, 24			9, − 3, 76	d^2
	0, 2, 1, − 168			4, 1, 169	d_1
	0, 5, − 4, − 3			25, − 10, 31	dd_1
	3, − 1, − 6, 0			19, 3, 36	d^2d_1
	0, − 1, 1, 168			4, − 1, 169	d_1^2
	3, 1, − 6, 0			19, − 3, 36	dd_1^2
	0, − 5, − 4, 3			25, 10, 31	$d^2d_1^2$

N.B. For this last determinant -4×675, there may possibly be other cubic classes based on a non-primitive characteristic; I have not ascertained whether such forms do or do not exist.

497.

NOTE ON THE CALCULUS OF LOGIC.

[From the *Quarterly Journal of Pure and Applied Mathematics*, vol. XI. (1871), pp. 282, 283.]

IT appears to me that the theory of the Syllogism, as given in Boole's paper, "The Calculus of Logic," *Camb. and Dubl. Math. Jour.*, t. III. (1848), pp. 183—198, may be presented in a more concise and compendious form as follows:

We are concerned with complementary classes, X, X'; viz. these together make up the universe (of things under consideration), $X + X' = 1$; viz. X' is the class not-X, and X the class not-X'.

Any kind whatever of simple relation between two classes (if we attend also to the complementary classes) can be expressed as a relation of total exclusion, $XY = 0$, or as a relation of partial (it may be total) inclusion, YX not $= 0$; viz. the relation $XY = 0$ may be read in any of the forms

No X's are X's,
No Y's are X's,
All X's are not-Y's,
All Y's are not-X's,

and the relation XY not $= 0$ in either of the forms

Some X's are Y's,
Some Y's are X's.

I say the above are the *only* kinds of simple relations; it being understood that X' may be substituted for X, or Y' for Y; so that the example $X'Y = 0$ (all Y's are X's) is the same kind of relation as $XY = 0$; and $X'Y$ not $= 0$ (some Y's are not-X's) the same kind of relation as XY not $= 0$.

Now taking X or X' and Z or Z' for the extreme terms, and Y or Y' for the middle term, of a syllogism; the only combinations of premises are

(1) $XY = 0$, $ZY = 0$.

(2) $XY = 0$, ZY not $= 0$, therefore $X'Z$ not $= 0$.

(3) XY not $= 0$, ZY not $= 0$.

(4) $XY = 0$, $ZY' = 0$, therefore $XZ = 0$.

(5) $XY = 0$, ZY' not $= 0$.

(6) XY not $= 0$, ZY' not $= 0$.

And of these, there are (as shown by the third column) only two which give rise to a conclusion (or relation between the extreme terms). As regards the negative cases, this is at once seen to be so; thus $XY = 0$, $ZY = 0$ (no X's are Y's, no Z's are Y's) leads to no conclusion in regard to X, Z. As regards the positive cases, it is also at once seen that the conclusions do follow; but we may obtain the conclusions by symbolical reasoning, thus

(2) $Y = YX + YX'$, $= YX'$;

therefore $ZY = ZYX'$, not $= 0$; therefore ZX' not $= 0$.

(4) $XZ = XZY + XZY'$, where on the right-hand side each term (the first as containing XY, the second as containing ZX') is $= 0$; that is, $XZ = 0$; where the logical signification of each step is obvious.

498.

ON THE INVERSION OF A QUADRIC SURFACE.

[From the *Quarterly Journal of Pure and Applied Mathematics*, vol. XI. (1871), pp. 283—288.]

THE inversion intended to be considered is that by reciprocal radius vectors, viz. if x, y, z are rectangular coordinates, and $r^2 = x^2 + y^2 + z^2$, then x, y, z are to be changed into $\frac{x}{r^2}, \frac{y}{r^2}, \frac{z}{r^2}$. But it is convenient to introduce for homogeneity a fourth coordinate w, $=1$; and the change then is x, y, z into $\frac{xw^2}{r^2}, \frac{yw^2}{r^2}, \frac{zw^2}{r^2}$.

Starting from the quadric surface

$$(a, b, c, d, f, g, h, l, m, n \between x, y, z, w)^2 = 0,$$

or, what is the same thing,

$$\begin{aligned} &(a, b, c, f, g, h \between x, y, z)^2 \\ &+ 2w\,(lx + my + nz) \\ &+ dw^2 \qquad\qquad = 0, \end{aligned}$$

the equation of the inverse surface is

$$\begin{aligned} &w^2\,(a, b, c, f, g, h \between x, y, z)^2 \\ &+ 2w\,(lx + my + nz)\,r^2 \\ &+ dr^4 \qquad\qquad = 0, \end{aligned}$$

where $r^2 = x^2 + y^2 + z^2$. The inverse surface is thus a quartic having the nodal conic $w=0$, $x^2 + y^2 + z^2 = 0$ (circle at infinity); and having the node $x=0$, $y=0$, $z=0$ (the centre of inversion); or say it is a nodal bicircular quartic surface, or nodal anallagmatic.

For x, y, z write $x-\frac{1}{2}\frac{l}{d}w$, $y-\frac{1}{2}\frac{m}{d}w$, $z-\frac{1}{2}\frac{n}{d}w$, and put for shortness

$$\begin{aligned} &lx+my+nz=u, \quad l^2+m^2+n^2=\alpha,\\ &al+hm+gn=\mathrm{a}, \qquad (a,\ b,\ c,\ f,\ g,\ h \between l,\ m,\ n)^2=A,\\ &hl+bm+fn=\mathrm{b},\\ &gl+fm+cn=\mathrm{c}, \end{aligned}$$

then

$$r^2 \qquad \text{becomes } r^2-\frac{uw}{d}+\frac{1}{4}\frac{\alpha}{d^2}w^2,$$

$$lx+my+nz \qquad ,, \qquad u-\frac{1}{2}\frac{\alpha}{d}w,$$

$$(a,\ldots\between x,\ y,\ z)^2 \qquad ,, \qquad (a,\ldots\between x,\ y,\ z)^2-(\mathrm{a}x+\mathrm{b}y+\mathrm{c}z)\frac{w}{d}+\frac{1}{4}A\frac{w^2}{d^2}.$$

Hence the equation is

$$\begin{aligned} &d\left\{r^4-2r^2\frac{uw}{d}+w^2\left(\frac{1}{2}\frac{\alpha}{d^2}r^2+\frac{u^2}{d^2}\right)-\frac{1}{2}\frac{\alpha uw^3}{d^3}+\frac{1}{16}\frac{\alpha^2}{d^4}w^4\right\}\\ &+2\left(wr^2-\frac{uw^2}{d}+\frac{1}{4}\frac{\alpha}{d^2}w^3\right)\left(u-\frac{1}{2}\frac{\alpha}{d}w\right)\\ &+w^2\left\{(a,\ldots\between x,\ y,\ z)^2-(\mathrm{a}x+\mathrm{b}y+\mathrm{c}z)\frac{w}{d}+\frac{1}{4}A\frac{w^2}{d^2}\right\}=0; \end{aligned}$$

viz. arranging and reducing, this is

$$\begin{aligned} &dr^4\\ &+w^2\left\{-\frac{1}{2}\frac{\alpha}{d}r^2-\frac{u^2}{d}+(a,\ldots\between x,\ y,\ z)^2\right\}\\ &+w^3\left\{\ \frac{\alpha u}{d^2}\ -\frac{1}{d}(\mathrm{a}x+\mathrm{b}y+\mathrm{c}z)\right\}\\ &+w^4\left\{-\frac{3}{16}\frac{\alpha^2}{d^3}+\frac{1}{4}A\frac{1}{d^2}\right\} \qquad\qquad =0; \end{aligned}$$

and we may without loss of generality assume

$$\begin{aligned} &-\frac{mn}{d}+f=0, \text{ that is } && df-mn=0,\\ &-\frac{nl}{d}+g=0, \quad ,, && dg-nl=0,\\ &-\frac{lm}{d}+h=0, \quad ,, && dh-lm=0. \end{aligned}$$

The equation then is

$$\begin{aligned} & r^4 \\ & + w^2\left\{-\tfrac{1}{2}\frac{\alpha}{d^2}(x^2+y^2+z^2)+\left(\frac{a}{d}-\frac{l^2}{d^2}\right)x^2+\left(\frac{b}{d}-\frac{m^2}{d^2}\right)y^2+\left(\frac{c}{d}-\frac{n^2}{d^2}\right)z^2\right\} \\ & + w^3\left\{\ \frac{\alpha u}{d^3}-\frac{1}{d^2}(\mathrm{a}x+\mathrm{b}y+\mathrm{c}z)\right\} \\ & + w^4\left\{-\tfrac{3}{16}\frac{\alpha^2}{d^4}+\tfrac{1}{4}A\frac{1}{d^3}\right\}=0. \end{aligned}$$

Write

$$\begin{aligned} ad-l^2 &= a'd, \\ bd-m^2 &= b'd, \\ cd-n^2 &= c'd. \end{aligned}$$

We have

$$\mathrm{a}=al+hm+gn=gn=al+\frac{lm^2}{d}+\frac{ln^2}{d}=\frac{l}{d}(ad-l^2+\alpha),$$

that is

$$\mathrm{a}=la'+\frac{l\alpha}{d},$$

and similarly

$$\mathrm{b}=mb'+\frac{m\alpha}{d},$$

$$\mathrm{c}=nc'+\frac{n\alpha}{d}.$$

Hence also

$$A=l^2a'+m^2b'+n^2c'+\frac{\alpha^2}{d},$$

and the equation is

$$\begin{aligned} & r^4 \\ & + w^2\left\{\left(-\tfrac{1}{2}\frac{\alpha}{d^2}+\frac{a'}{d}\right)x^2+\left(-\tfrac{1}{2}\frac{\alpha}{d^2}+\frac{b'}{d}\right)y^2+\left(-\tfrac{1}{2}\frac{\alpha}{d^2}+\frac{c'}{d}\right)z^2\right\} \\ & + w^3\left\{\ -\frac{la'}{d^2}x-\frac{mb'}{d^2}y-\frac{nc'}{d^2}z\right\} \\ & + w^4\left\{\ \frac{1}{4d^3}(l^2a'+m^2b'+n^2c')+\tfrac{1}{16}\frac{\alpha^2}{d^4}\right\}=0. \end{aligned}$$

This is Kummer's form, say

$$r^4=4w^2\{\alpha_1x^2+\beta_1y^2+\gamma_1z^2+\delta_1w^2+2w(a_1x+b_1y+c_1z)\},$$

where

$$-4\alpha_1 = -\tfrac{1}{2}\frac{\alpha}{d^2} + \frac{a'}{d},$$

$$-4\beta_1 = -\tfrac{1}{2}\frac{\alpha}{d^2} + \frac{b'}{d},$$

$$-4\gamma_1 = -\tfrac{1}{2}\frac{\alpha}{d^2} + \frac{c'}{d},$$

$$-4\delta_1 = \frac{1}{4d^3}(l^2a' + m^2b' + n^2c') + \tfrac{1}{16}\frac{\alpha^2}{d^4},$$

$$-8a_1 = -\frac{la'}{d^2},$$

$$-8b_1 = -\frac{mb'}{d^2},$$

$$-8c_1 = -\frac{nc'}{d^2}.$$

Hence Kummer's equation

$$\delta_1 + \lambda^2 = \frac{a_1^2}{\lambda + \alpha_1} + \frac{b_1^2}{\lambda + \beta_1} + \frac{c_1^2}{\lambda + \gamma_1},$$

or say

$$64\delta_1 + 64\lambda^2 = \frac{256a_1^2}{4\lambda + 4\alpha_1} + \frac{256b_1^2}{4\lambda + 4\beta_1} + \frac{256c_1^2}{4\lambda + 4\gamma_1},$$

becomes

$$64\lambda^2 - \frac{4}{d^3}(l^2a' + m^2b' + n^2c') - \frac{\alpha^2}{d^4} = \frac{4l^2a'^2}{d^4\left(\tfrac{1}{2}\frac{\alpha}{d^2} - \frac{a'}{d} + 4\lambda\right)} + \frac{4m^2b'^2}{d^4\left(\tfrac{1}{2}\frac{\alpha}{d^2} - \frac{b'}{d} + 4\lambda\right)} + \frac{4n^2c'^2}{d^4\left(\tfrac{1}{2}\frac{\alpha}{d^2} - \frac{c'}{d} + 4\lambda\right)},$$

which is satisfied by $4\lambda = -\tfrac{1}{2}\frac{\alpha}{d^2}$. Writing therefore

$$4\lambda + \tfrac{1}{2}\frac{\alpha}{d^2} = -\frac{\theta}{d},$$

that is

$$8\lambda = -\frac{2\theta}{d} - \frac{\alpha}{d^2},$$

$$64\lambda^2 = \frac{4\theta^2}{d^2} + \frac{4\theta\alpha}{d^3} + \frac{\alpha^2}{d^4};$$

the equation is

$$\frac{4\theta^2}{d^3} + \frac{4\theta\alpha}{d^3} - \frac{4}{d^3}(l^2a' + m^2b' + n^2c') = \frac{4l^2a'^2}{d^4\left(-\frac{\theta}{d} - \frac{a'}{d}\right)} + \frac{4m^2b'^2}{d^4\left(-\frac{\theta}{d} - \frac{b'}{d}\right)} + \frac{4n^2c'^2}{d^4\left(-\frac{\theta}{d} - \frac{c'}{d}\right)},$$

viz. this is

$$l^2a' + m^2b' + n^2c' - \theta\alpha - \theta^2 d = \frac{l^2a'^2}{\theta + a'} + \frac{m^2b'^2}{\theta + b'} + \frac{n^2c'^2}{\theta + c'},$$

which is of course satisfied by $\theta = 0$. Moreover the derived equation

$$-\alpha - 2\theta d = -\frac{l^2a'^2}{(\theta+a')^2} - \frac{m^2b'^2}{(\theta+b')^2} - \frac{n^2c'^2}{(\theta+c')^2}$$

is also satisfied by $\theta = 0$, so that this is a double root. The equation in fact is

$$\{\theta^2 d + \theta\alpha - (l^2a' + m^2b' + n^2c')\}(\theta+a')(\theta+b')(\theta+c')$$
$$+\{l^2a'^2(\theta+b')(\theta+c') + m^2b'^2(\theta+c')(\theta+a') + n^2c'^2(\theta+a')(\theta+b')\} = 0,$$

or, expanding and dividing by θ^2, this is

$$\begin{aligned} &d(\theta+a')(\theta+b')(\theta+c') \\ &+\alpha\{\theta^2 + \theta(a'+b'+c') + b'c' + c'a' + a'b'\} \\ &-\quad (l^2a' + m^2b' + n^2c')(\theta + a' + b' + c') \\ &+\quad l^2a'^2 + m^2b'^2 + n^2c'^2 = 0, \end{aligned}$$

which gives the remaining three roots.

If $a' = b' = c'$ the equation is

$$(\theta + a' + \alpha)(\theta + a')^2 = 0.$$

I recall that we have

$$a,\ b,\ c,\ d,\ f = \frac{mn}{d},\quad g = \frac{nl}{d},\quad h = \frac{lm}{d},\quad l,\ m,\ n,$$

$$a' = a - \frac{l^2}{d},\quad b' = b - \frac{m^2}{d},\quad c' = c - \frac{n^2}{d},\quad \alpha = l^2 + m^2 + n^2,$$

so that the quadric surface is

$$d(a'x^2 + b'y^2 + c'z^2) + (lx + my + nz + dw)^2 = 0,$$

and that, $\alpha_1, \beta_1, \gamma_1, \delta_1, a_1, b_1, c_1$ denoting as before, the equation of the inverse surface (referred to a different origin) is

$$r^4 = 4w^2\{\alpha_1x^2 + \beta_1y^2 + \gamma_1z^2 + \delta_1w^2 + 2w(a_1x + b_1y + c_1z)\}.$$

499.

ON THE THEORY OF THE CURVE AND TORSE.

[From the *Quarterly Journal of Pure and Applied Mathematics*, vol. XI. (1871), pp. 294—317.]

THE fundamental relations in the theory of the Curve and Torse were first established in my "Mémoire sur les Courbes à double courbure et sur les Surfaces developpables," *Liouv.* t. X. (1845), [30] see also *Camb. and Dubl. Math. Jour.*, t. IV. (1850), [83], viz. I showed that the systems (m, r, β, h, n, y) and (r, n, m, x, α, g), (the notation is subsequently explained), each of them satisfied the Plückerian relations. An additional set of equations giving the values of $(\gamma, t, k, q, \gamma', t', k', q')$ was furnished by Dr Salmon's "Theory of Reciprocal Surfaces," *Trans. R. I. Acad.*, t. XXIII. (1857), see also the *Solid Geometry*. The theory as thus established is complete in itself, but it does not take account of certain singularities υ, ω, H, G; the singularity υ was first considered in my paper "On a special sextic Developable," *Quart. Math. Jour.* t. VII. (1865), [373], (there called θ), and I afterwards endeavoured to take account of the remaining singularities ω, H, G. I was in correspondence on the subject with Prof. Cremona, and the discovery of the complete forms of several of the formulæ is due to him.

There has recently appeared a very valuable memoir by M. Zeuthen, "Sur les singularités ordinaires d'une courbe gauche et d'une surface developpable," *Annali di Matem.* t. III. (1869); he excludes, however, from consideration the singularity ω, and does not throughout attend to H, G.

I propose in the present memoir to reproduce and develope the whole theory.

Explanations and Notation.

1. We have a singly infinite series of points, lines, and planes; viz. each line passes through two consecutive points and lies in two consecutive planes; each plane passes through three consecutive points and contains two consecutive lines; each point

is the intersection of three consecutive planes and of two consecutive lines. The points describe and the lines envelope a *curve*; the lines describe and the planes envelope a *torse*; the entire system of points, lines, and planes is thus the system of the curve and torse. The curve is the edge of regression or cuspidal curve of the torse; in regard to the curve the points are points (or ineunts), the lines tangents, and the planes osculating planes: in regard to the torse the points are points on the edge of regression, the lines generating lines, and the planes tangent planes.

2. Each line of the system is met by a certain number of non-consecutive lines, and the locus of the points of intersection (or say the locus of the intersection of two intersecting non-consecutive lines) is a nodal curve on the torse, or say simply it is the *nodal curve.* The plane containing the two intersecting non-consecutive lines envelopes a torse which is called the *nexal torse.*

3. There is occasion to consider

m, order of the system; this is the number o points of the system which lie in a given plane; or it is the order of the curve.

r, rank of the system; this is the number of lines of the system which meet a given line. It is thus the class of the curve; and the order of the torse.

n, class of the system; this is the number of planes of the system which pass through a given point. It is thus the class of the torse.

α, number of stationary planes; that is, planes each passing through four consecutive points of the system.

β, number of stationary points, that is, points each of them the intersection of four consecutive planes of the system.

g, number of *lines in two planes* (that is, lines each of them the intersection of two non-consecutive planes of the system) contained in a given plane; or say, number of apparent double planes of the torse.

G, number of double planes, or *tropes*, of the torse; viz. considering the torse as the envelope of a variable plane, if the plane in the course of its motion comes twice into the same position, we have then a double plane or trope.

h, number of *lines through two points* (that is, lines each through two non-consecutive points of the system) passing through a given point; or say, number of apparent double points of the curve.

H, number of double points, or *nodes* of the curve; viz. considering the curve as described by a variable point, if the point in the course of its motion comes twice into the same position we have then a double point or node of the curve.

x, number of *points in two lines* (that is, points each of them the intersection of two non-consecutive lines of the system) contained in a given plane; or what is the same thing, order of nodal curve.

y, number of *planes through two lines* (that is, planes each containing two non-consecutive lines of the system) passing through a given point; or what is the same thing, class of the nexal torse.

v, number of stationary lines of the system, that is, lines each containing three consecutive points of the system.

ω, number of double lines of the system, that is, lines each containing two pairs of consecutive points of the system.

t, number of *points on three lines* (that is, points each of them the common intersection of three non-consecutive lines of the system): these are also triple points on the curve.

γ, number of points of the system, through each of which passes a non-consecutive line of the system: these are intersections of the curve with the nodal curve, stationary points on the latter curve.

k, number of apparent double points of nodal curve.

q, class of nodal curve.

t', number of *planes through three lines* (that is, planes each of them through three non-consecutive lines of the system): these are also triple tangent planes of the torse.

γ', number of planes of the system each of them passing through a non-consecutive line of the system: these are common tangent planes of the torse and nexal torse, stationary planes of the latter torse.

k', number of apparent double planes of nexal torse.

q', order of nexal torse.

4. The formulæ thus contains in all the 21 quantities

$$m,\ r,\ n,\ \alpha,\ \beta,\ g,\ G,\ h,\ H,\ x,\ y,\ v,\ \omega\ \|\ t,\ \gamma,\ k,\ q\ |\ t',\ \gamma',\ k',\ q'.$$

My own Plückerian equations, or, say, the Plücker-Cayley equations, establish in all 6 relations between the first 13 quantities, and thus enable the expression of them in terms of any seven, say of

$$m,\ r,\ n,\ G,\ H,\ v,\ \omega,$$

and the Salmon-Cremona equations then lead to the expressions in terms of these, of the remaining eight quantities $t,\ \gamma,\ k,\ q,\ t',\ \gamma',\ k',\ q'$.

I also consider

D_m, the deficiency of the curve,

D_x, the deficiency of the nodal curve.

I will first consider the equations themselves, and the mere algebraical transformations thereof; and afterwards the geometrical theory.

The Plücker-Cayley Equations.

5. These are found (as will be further explained) by considering first the cone, vertex an arbitrary point, which passes through the curve; and secondly, the section of the torse by an arbitrary plane. We have in the two figures

Cone.	*Section.*
m, order,	n, class,
r, class,	r, order,
$h+H$, double lines,	$g+G$, double tangents,
β, stationary lines,	α, stationary tangents,
$y+\omega$, double planes,	$x+\omega$, double points,
$n+v$, stationary planes.	$m+v$, stationary points.

And hence the two sets of quantities respectively are connected by the Plückerian relations, viz. these are

$$r = m(m-1) - 2(h+H) - 3\beta,$$

$$n+v = 3m(m-2) - 6(h+H) - 8\beta,$$

$$\begin{aligned} y+\omega = \tfrac{1}{2}m(m-2)(m^2-9) & \\ -(m^2-m-6)\{2(h+H)+3\beta\} & \\ +2(h+H)(h+H-1) & \\ +6(h+H)\beta & \\ +\tfrac{9}{2}\beta(\beta-1), & \end{aligned}$$

$$m = r(r-1) - 2(y+\omega) - 3(n+v),$$

$$\beta = 3r(r-2) - 6(y+\omega) - 8(n+v),$$

$$\begin{aligned} h+H = \tfrac{1}{2}r(r-2)(r^2-9) & \\ -(r^2-r-6)\{2(y+\omega)+3(n+v)\} & \\ +2(y+\omega)(y+\omega-1) & \\ +6(y+\omega)(n+v) & \\ +\tfrac{9}{2}(n+v)(n+v-1), & \end{aligned}$$

$$n+v-\beta = 3(r-m),$$

$$y+\omega-h-H = \tfrac{1}{2}(r-m)(r+m-9),$$

$$\begin{aligned} \tfrac{1}{2}(r-1)(r-2)-(y+\omega)-(n+v) & \\ = \tfrac{1}{2}(m-1)(m-2) & \\ -(h+H)-\beta, & \end{aligned}$$

$$n = r(r-1) - 2(x+\omega) - 3(m+v),$$

$$\alpha = 3r(r-2) - 6(x+\omega) - 8(m+v),$$

$$\begin{aligned} g+G = \tfrac{1}{2}r(r-2)(r^2-9) & \\ -(r^2-r-6)\{2(x+\omega)+3(m+v)\} & \\ +2(x+\omega)(x+\omega-1) & \\ +6(x+\omega)(m+v) & \\ +\tfrac{9}{2}(m+v)(m+v-1), & \end{aligned}$$

$$r = n(n-1) - 2(g+G) - 3\alpha,$$

$$m+v = 3n(n-2) - 6(g+G) - 8\alpha,$$

$$\begin{aligned} x+\omega = \tfrac{1}{2}n(n-2)(n^2-9) & \\ -(n^2-n-6)\{2(g+G)+3\alpha\} & \\ +2(g+G)(g+G-1) & \\ +6(g+G)\alpha & \\ +\tfrac{9}{2}\alpha(\alpha-1), & \end{aligned}$$

$$\alpha-(m+v) = 3(n-r),$$

$$g+G-(x+\omega) = \tfrac{1}{2}(n-r)(n+r-9),$$

$$\begin{aligned} \tfrac{1}{2}(r-1)(r-2)-(x+\omega)-(m+v) & \\ = \tfrac{1}{2}(n-1)(n-2) & \\ -(g+G)-\alpha, & \end{aligned}$$

and combining the two systems

$$\beta = \alpha + 2(m-n),$$
$$y = x + (m-n),$$
$$h + H = g + G + \tfrac{1}{2}(m-n)(m+n-7),$$
$$\tfrac{1}{2}(m-1)(m-2) - (h+H) - \beta = \tfrac{1}{2}(n-1)(n-2) - (g+G) - \alpha,$$
$$\tfrac{1}{2}(r-1)(r-2) - (x+\omega) - (m+v) = \tfrac{1}{2}(r-1)(r-2) - (y+\omega) - (n+v).$$

6. Taking as data r, m, n, G, H, v, ω, we find very easily

$$h = \tfrac{1}{2}(m^2 - 10m - 3n + 8r - 3v - 2H),$$
$$g = \tfrac{1}{2}(n^2 - 10n - 3m + 8r - 3v - 2G),$$
$$x = \tfrac{1}{2}(r^2 - r - n - 3m - 3v - 2\omega),$$
$$y = \tfrac{1}{2}(y^2 - r - m - 3n - 3v - 2\omega),$$
$$\alpha = m - 3r + 3n + v,$$
$$\beta = n - 3r + 3m + v.$$

The Salmon-Cremona Equations.

7. These are

$$m(r-2) = 2n + 4\beta + \gamma + 4v + 4\omega + 4H,$$
$$x(r-2) = n(r-4) + 2\beta + 3\gamma + 3t + v(3r-14) + \omega(2r-10) + 12H,$$
$$x(r-2)(r-3) = n(x-2r+8) + 3mx + 4k - 3\alpha - 9\beta - 6\gamma$$
$$+ v(3x - 6r + 18) + \omega(2x - 4r + 8) - 12H,$$
$$q = x^2 - x - 2k - 3\gamma - 6t - 3v(r-6) - 2\omega(r-8) - 2G - 18H,$$
$$\{= r(n-3) - 3\alpha - 2G\},$$

and

$$n(r-2) = 2m + 4\alpha + \gamma' + 4v + 4\omega + 4G,$$
$$y(r-2) = m(r-4) + 2\alpha + 3\gamma' + 3t' + v(3r-14) + \omega(2r-10) + 12G,$$
$$y(r-2)(r-3) = m(y-2r+8) + 3ny + 4k' - 3\beta - 9\alpha - 6\gamma'$$
$$+ v(3y - 6r + 18) + \omega(2y - 4r + 8) - 12G,$$
$$q' = y^2 - y - 2k' - 3\gamma' - 6t' - 3v(r-6) - 2\omega(r-8) - 2H - 18G,$$
$$\{= r(m-3) - 3\beta - 2H\},$$

where the second values of q, q' respectively are reduced forms obtained by the aid of the foregoing Plückerian relations.

8. Expressing these in terms of r, m, n, G, H, v, ω, we obtain

$$\gamma = rm + 12r - 14m - 6n - 8v - 4\omega - 4H,$$
$$t = \tfrac{1}{6}[r^3 - 3r^2 - 58r - 3r(n + 3m + 3v + 2\omega) + 42n + 78m + 78v + 48\omega],$$
$$k = \tfrac{1}{8}[r^4 - 6r^3 + 11r^2 + 66r - (2r^2 - 10r)(n + 3m + 3v + 2\omega)$$
$$+ (n + 3m + 3v + 2\omega)^2 - 58n - 126m - 126v - 76\omega - 24H],$$
$$q = rn + 6r - 3m - 9n - 3v - 2G,$$

and

$$\gamma' = rn + 12r - 14n - 6m - 8\upsilon - 4\omega - 4G,$$
$$t' = \tfrac{1}{6}[r^3 - 3r^2 - 58r - 3r(m + 3n + 3\upsilon + 2\omega) + 42m + 78n + 78\upsilon + 48\omega],$$
$$k' = \tfrac{1}{8}[r^4 - 6r^3 + 11r^2 + 66r - (2r^2 - 10r)(m + 3n + 3\upsilon + 2\omega) + (m + 3n + 3\upsilon + 2\omega)^2 - 58m - 126n - 126\upsilon - 76\omega - 24G],$$
$$q' = rm + 6r - 3n - 9m - 3\upsilon - 2H.$$

9. We have thence

$$\begin{aligned}
\gamma' - \gamma &= -(r-8)(m-n) - 4(G-H),\\
t' - t &= (r-6)(m-n),\\
q' - q &= (r-6)(m-n) + 2(G-H),\\
k' - k &= \tfrac{1}{2}(m-n)(r^2 - 5r - 2m - 2n - 3\upsilon - 2\omega + 17) - 3(G-H)\\
&= \tfrac{1}{2}(m-n)(x + y - 4r + 17) - 3(G-H).
\end{aligned}$$

10. Instead of obtaining the above values of γ, t, k, q directly it is convenient to verify them by substitution in the equations from which they were obtained; viz. writing for shortness $n + 3m + 3\upsilon + 2\omega = P$, these may be written

$$-m(r-2) + 2n + 4\beta + 4\upsilon + 4\omega + 4H + \gamma = 0,$$
$$-2x(r-2) + 2r(P - 3m) - 8n + 4\beta - 28\upsilon - 20\omega + 6\gamma + 24H + 6t = 0,$$
$$-2x(r^2 - 5r + 6 - P) - 4(r-4)n - 18\beta - 12\gamma - 6\alpha + 36\upsilon + 16\omega - 24H - 4r(-n - 3m + P) + 8k = 0,$$
$$-4q + 2x(2x-2) - 8k - 8G - 4r(-n - 3m + P) + 72\upsilon + 64\omega - 12\gamma - 24t - 72H = 0,$$

which are to be satisfied by

$$\begin{aligned}
\gamma &= rm + 12r - 14m - 6n - 8\upsilon - 4\omega - 4H,\\
t &= \tfrac{1}{6}[r^3 - 3r^2 - 58r - 3rP + 42n + 78m + 78\upsilon + 48\omega],\\
k &= \tfrac{1}{8}[r^4 - 6r^3 + 11r^2 + 66r - (2r^2 - 10r)P + P^2 - 58n - 126m - 126\upsilon - 76\omega - 24H],\\
q &= rn + 6r - 3m - 9n - 3\upsilon - 2G,\\
x &= \tfrac{1}{2}(r^2 - r - P),\\
\alpha &= m - 3r + 3n + \upsilon,\\
\beta &= n - 3r + 3m + \upsilon,\\
P &= n + 3m + 3\upsilon + 2\omega.
\end{aligned}$$

11. We have, in fact, first

$$\left.\begin{array}{r}
-mr + 2m\\
+2n\\
+12m + 4n - 12r + 4\upsilon\\
+4\upsilon\\
+4\omega\\
+4H\\
+mr - 14m - 6n + 12r - 8\upsilon - 4\omega - 4H
\end{array}\right\} = 0,$$

secondly

$$\begin{array}{l}
-r^3+r^2 \qquad +rP \\
\quad +2r^2-2r \qquad -2P \\
\qquad\qquad +2rP \qquad -6mr \\
\qquad\qquad\qquad\qquad\qquad -8n \\
\qquad -12r \qquad\qquad\qquad\quad +4n+12m+4v \\
\qquad\qquad\qquad\qquad\qquad\qquad\qquad -28v \\
\qquad\qquad\qquad\qquad\qquad\qquad\qquad\quad -20\omega \\
\qquad +72r \qquad\qquad\quad +6mr-36n-84m-48v-24\omega-24H \\
\qquad\qquad\qquad\qquad\qquad\qquad\qquad\qquad\qquad +24H \\
+r^3-3r^2-58r-3rP \qquad\quad +42n+78m+78v+48\omega \\
\hline
\text{that is} \qquad\qquad -2P \qquad +2n+6m+6v \qquad = 0,
\end{array}$$

thirdly

$$\begin{array}{l}
-r^4+5r^3-6r^2 \\
\quad +r^3-5r^2+6r+P(2r^2-6r+6)-P^2 \\
\qquad\qquad\qquad\qquad -4rn+16n \\
\qquad +54r \qquad\qquad\qquad\quad -18n-54m-18v \\
\qquad -144r \qquad -12rm \qquad +72n+168m+96v+48\omega+48H \\
\qquad +18r \qquad\qquad\qquad\quad -18n-6m-6v \\
\qquad\qquad\qquad\qquad\qquad\qquad\qquad +36v \\
\qquad\qquad\qquad\qquad\qquad\qquad\qquad\quad +16\omega \\
\qquad\qquad\qquad\qquad\qquad\qquad\qquad\qquad -24H \\
\qquad\qquad +P(-4r)+12rm+4rn \\
+r^4-6r^3+11r^2+66r+P(-4r^2+10r)+P^2-58n-126m-126v-76\omega-24H \\
\hline
\text{that is} \qquad\qquad 6P \qquad\qquad -6n-18m-18v-12\omega-24H=0,
\end{array}$$

and lastly

$$\begin{array}{l}
\qquad -24r \qquad\qquad\qquad\quad -4rn \qquad +12m+36n+12v \qquad\qquad +8G \\
+r^4-2r^3-r^2+2r+P(-2r^2+2r+2)+P^2 \qquad +126m+58n+126v+76\omega+24H \\
-r^4+6r^3-11r^2-66r+P(2r^2-10r)-P^2 \qquad\qquad\qquad\qquad -8G \\
\qquad\qquad +P(-4r) \quad +4rn+12rm \qquad\qquad +72v+64\omega \\
\qquad -144r \qquad\qquad\qquad -12rm+168m+72n+96v+48\omega+48H \\
\quad -4r^3+12r^2+232r+P(12r) \qquad -312m-168n-312v-192\omega-72H \\
\hline
\text{that is} \qquad\qquad 2P \qquad\qquad -6m-2n-6v-4\omega \qquad = 0,
\end{array}$$

which completes the verification.

12. The deficiency of the curve m is given by the equation

$$2D_m = r - 2m + 2 + \beta,$$

or substituting for β its value $= n - 3r + 3m + v$, this is

$$2D_m = m + n - 2r + v + 2.$$

The deficiency of the nodal curve x is given by

$$2D_x = q - 2x + 2 + \gamma + v(r-6) + 2H,$$

which, substituting for q, x, and γ, the values

$$rn + 6r - 3m - 9n - 3v - 2G,$$
$$\tfrac{1}{2}(r^2 - r - n - 3m - 3v - 2\omega),$$
$$rm + 12r - 14m - 6n - 8v - 4\omega - 4H,$$

respectively, becomes

$$2D_x = (r-14)(m+n+v) - r^2 + 19r + 2 - 2\omega - 2G - 2H;$$

whence, also, writing herein $m + n + v = 2D_m + 2r - 2$, we have

$$2D_x - (r-14).\,2D_m = (r-5)(r-6) - 2\omega - 2G - 2H,$$

a relation between the two deficiencies.

Geometrical Theory of the foregoing Relations.

13. In considering the geometrical theory, we have to speak of the original curve, or curve of the system, and also of the nodal curve; it will be convenient to call them the curve m and the curve x respectively. I speak of the torse absolutely, to signify the torse of the system, as in what follows there is not the like occasion to speak of the nexal torse. I speak also of a plane α, meaning thereby any one of the stationary planes, the number of which is $= \alpha$; and so of a line α, meaning the line in the stationary plane α; and a point α, meaning the point of contact of such line with the curve m; or in the plural, the planes α, lines α, &c. And so in other cases; thus we have the stationary tangents v, and the points v, which are the points of contact hereof with the curve m. As regards a double tangent ω, we have here two points of contact; one of these separately would be a point ω; and we may speak of the points (or pair) 2ω, meaning thereby the two points of contact of the same tangent ω; or of the points 2ω, meaning the system of the 2ω points of contact of the tangents ω.

14. Observe that the expressions, the planes α, lines α, &c., have an absolute signification; there are other such expressions which have only a relative signification, in regard to the system considered in connexion with a given point, line, or plane, as the case may require. Thus the expression, *the lines* g, must be understood of the system in connexion with *a given plane*, to signify the lines in two planes contained in the *given plane*; the planes g, points g, would of course mean the planes or points of the system belonging to the lines g.

In particular the points m are the points of the system which lie in a given plane, the lines r are the lines which meet a given line; the planes n are the planes which pass through a given point. There will be occasion to speak, not only of the planes n, but also of the lines n and points n; these are of course the lines and points in the planes n.

15. It is to be remarked that, considering the torse as a surface, the nodal curve thereof is made up of the curve x and of the double lines ω (or its order is $= x + \omega$); the cuspidal curve is made up of the curve m, and of the stationary lines v (or its order is $= m + v$).

The Plücker-Cayley Equations.

16. The mode of obtaining these equations has already been indicated. We in fact consider the system in connexion with an arbitrary point, and with an arbitrary plane. The point is the vertex of a cone passing through the curve m, and this cone is of the order m, the class r, with $h + H$ double lines, β stationary lines, $y + \omega$ double tangent planes, and $n + v$ stationary tangent planes; viz. the order of the cone is equal to the number of lines in which this is intersected by a plane through the vertex; but each of these is determined as the line joining the vertex with an intersection of the plane by the curve m, and the number of them is thus $= m$. The class of the cone is equal to the number of the tangent planes which can be drawn through an arbitrary line through the vertex; but this is in fact the number of lines of the system which meet the arbitrary line, viz. it is $= r$. Again, any line drawn from the vertex to meet the curve twice, and also any line drawn to one of the points H, is a double line of the cone, that is, the whole number of double lines is $= h + H$. A line from the vertex to one of the points β is a stationary line of the cone; the number of these is $= \beta$. A plane through the vertex, and containing two tangents of the curve m, or containing a double tangent ω, is a double tangent plane of the cone, the number of these is thus $= y + \omega$. A plane through the vertex, which is also a plane of the system, is a stationary tangent plane; in fact, we have here on the curve m three consecutive points lying in a plane with the vertex, the tangent plane of the cone is the plane through the vertex, and the first and second of the points on the curve; but this is also the plane through the vertex, and the second and third points, or the plane is a stationary tangent plane. But the plane through the vertex and the tangent v is also a stationary tangent plane of the cone; and the number of stationary tangent planes is thus $= n + v$.

17. Similarly for the section by the arbitrary plane; this is a curve of the order r and class n with $x + \omega$ double points, $m + v$ stationary points, $g + G$ double tangents, and α stationary tangents. In fact, the order of the curve is equal to that of the torse; that is, to the number of lines which meet an arbitrary line, or $= r$. The class of the curve is equal to the number of tangents which pass through an arbitrary point of the plane; or, what is the same thing, the number of planes of the system which pass through this same arbitrary point, viz. it is $= n$. Each point of the plane which is the intersection of two lines of the system, and also each intersection of the plane by a line ω, is a double point of the curve; viz. the number of these is $= x + \omega$. Each intersection with the curve m, and also each intersection with the tangent v, is a stationary point of the curve; viz. the number is $= m + v$. Each line

in the plane, which is the intersection of two planes of the system, and also each intersection of the plane with a double tangent plane G, is a double tangent of the curve; viz. the number is $=g+G$. And, finally, each intersection of the plane with a stationary plane α is a stationary tangent of the curve; viz. the number of these is $=\alpha$.

We have thus the Plückerian relations as above between

$$m,\ r,\ h+H,\ \beta,\ y+\omega,\ n+v,$$

and between

$$n,\ r,\ g+G,\ \alpha,\ x+\omega,\ n+v.$$

Zeuthen's Tables.

18. The vertex of the cone and the plane of the section may occupy special positions. We have a table given by Zeuthen, but which I have completed so far as regards the double lines ω as follows:

Cone through the Curve.

	Vertex	Order	Class	Double lines	Stationary lines	Double tangent planes	Stationary tangent planes
1	Arbitrary	m	r	h $+H$	β	y $+\omega$	n $+v$
2	On a tangent	m	$r-1$	$h-1$ $+H$	β $+1$	$y-r+4$ $+\omega$	$n-2$ $+v$
3	On the curve	$m-1$	$r-2$	$h-m+2$ $+H$	β	$y-2r+8$ $+\omega$	$n-3$ $+v$
4	At a point H	$m-2$	$r-4$	$h-2m+6$ $+H-1$	β	$y-4r+20$ $+\omega$	$n-6$ $+v$
5	At a point β	$m-2$	$r-3$	$h-2m+6$ $+H$	$\beta-1$	$y-3r+13$ $+\omega$	$n-4$ $+v$
6	On stationary tangent v	m	$r-2$	$h-2$ $+H$	β $+2$	$y-2r+9$ $+\omega$	$n-3$ $+v-1$
7	At point of contact of ditto	$m-1$	$r-3$	$h-m+1$ $+H$	β $+1$	$y-3r+14$ $+\omega$	$n-4$ $+v-1$
8	On double tangent ω	m	$r-2$	$h-2$ $+H$	β $+2$	$y-2r+10$ $+\omega-1$	$n-4$ $+v$
9	At point of contact of ditto	$m-1$	$r-3$	$h-m+1$ $+H$	β $+1$	$y-3r+15$ $+\omega-1$	$n-5$ $+v$

Plane Section of the Torse.

	Plane	Class	Order	Double tangents	Stationary tangents	Double points	Stationary points
1	Arbitrary	n	r	g $+ G$	α	x $+ \omega$	m $+ \upsilon$
2	Through a tangent	n	$r-1$	$g-1$ $+ G$	α $+1$	$x-r+4$ $+ \omega$	$m-2$ $+ \upsilon$
3	A tangent plane	$n-1$	$r-2$	$g-n+2$ $+ G$	α	$x-2r+8$ $+ \omega$	$m-3$ $+ \upsilon$
4	A double tangent plane G	$n-2$	$r-4$	$g-2n+6$ $+ G-1$	α	$x-4r+20$ $+ \omega$	$m-6$ $+ \upsilon$
5	A stationary tangent plane α	$n-2$	$r-3$	$g-2n+6$ $+ G$	$\alpha-1$	$x-3r+13$ $+ \omega$	$m-4$ $+ \upsilon$
6	Through stationary tangent υ	n	$r-2$	$g-2$ $+ G$	α $+2$	$x-2r+9$ $+ \omega$	$m-3$ $+ \upsilon - 1$
7	Tangent plane at contact of ditto	$n-1$	$r-3$	$g-n+1$ $+ G$	α $+1$	$x-3r+14$ $+ \omega$	$m-4$ $+ \upsilon - 1$
8	Through double tangent ω	n	$r-2$	$g-2$ $+ G$	α $+2$	$x-2r+10$ $+ \omega - 1$	$m-4$ $+ \upsilon$
9	A tangent plane at one of contacts of ditto	$n-1$	$r-3$	$g-n+1$ $+ G$	α $+1$	$x-3r+15$ $+ \omega - 1$	$m-5$ $+ \upsilon$

19. To avoid confusion with the geometrical term line, I will speak (not of the lines, but) of the cases of these tables; the numbers in each case satisfy the foregoing Plückerian relations. To fix the ideas, I attend to the second table. We require to know in each case, say the numbers in the (n, r, α) columns; these being known, the other three numbers will be determined.

20. First for the n column; for the Cases 2, 6, 8, the plane is not a tangent plane; the number of tangent planes which pass through a fixed point in the plane is still $=n$. For Cases 3, 7, 9, the plane is a tangent plane, it therefore counts 1 among the tangent planes which pass through a fixed point thereof; and the number of the remaining tangent planes is $=n-1$. And so for Cases 4 and 5, the plane counts for 2 among the tangent planes which pass through a fixed point thereof, and the number of the remaining tangent planes is $=n-2$.

21. Next as to the r and α columns.

Case (2). The plane passes through a generating line, and therefore besides cuts the torse in a curve of the order $r-1$; this generating line is a stationary tangent cutting the curve $r-1$ in the point of contact with the curve m, counting 3 times (in all $\alpha+1$ stationary tangents), and in $r-4$ other points.

Case (3). The plane cuts the torse in the generating line counting twice, and in a curve of the order $r-2$; the generating line is in regard to this curve an ordinary tangent at the point of contact with the curve m (so that the number of stationary tangents remains $=\alpha$), and besides cuts the curve in $r-4$ points as in Case 2.

Case (4). The plane cuts the torse in two generating lines, each counting twice, and in a curve of the order $r-4$; each of the generating lines is in regard to this curve an ordinary tangent at the point of contact with the curve m (number of stationary tangents remains $=\alpha$), and besides cuts the curve in $r-6$ other points.

Case (5). The plane cuts the torse in a generating line counting 3 times, and in a curve of the order $r-3$; the generating line is in regard to this curve an ordinary tangent at the point of contact with the curve m, and besides cuts it in $r-5$ points. The plane being in the present case a plane α, its intersections with the remaining planes α, give the $\alpha-1$ stationary tangents.

Case (6). The plane meets the torse in a generating line counting twice, and in a curve of the order $r-2$; the generating line is in regard to the curve a singular tangent meeting it in the point of contact with the curve m, counting 4 times, and besides meeting it in $r-6$ points. The generating line in respect of this four-pointic intersection counts as a stationary tangent twice; and the number of stationary tangents is $=\alpha+2$.

Case (7). The plane meets the torse in the generating line counting 3 times, and in a curve of the order $r-3$; the generating line is in regard to this curve a stationary tangent at the point of contact with the curve m, counting 3 times, and besides meeting it in $r-6$ points as in Case 6. The whole number of stationary tangents is thus $=\alpha+1$.

Case (8). The plane meets the torse in a generating line counting twice and in a curve of the order $r-2$. The generating line is in respect of the curve a stationary tangent at each of the points of contact with the curve m, viz. each of these points counts 3 times, and there are besides $r-8$ intersections. Moreover, the generating line counting as 2 stationary tangents, the whole number of stationary tangents is $=\alpha+2$.

Case (9). The plane meets the torse in a generating line counting 3 times, and in a curve of the order $r-3$. The generating line is in respect of the curve an ordinary tangent at one of the points of contact with the curve m (viz. the point at which the plane is a tangent plane of the torse), so that we have here two intersections; and it is a stationary tangent at the other of the points of contact with the curve m (viz. the point at which the plane is not a tangent plane of the torse), so that there are here 3 intersections; there are therefore $r - 8$ other intersections. The generating line reckons once as a stationary tangent, and the whole number of stationary tangents is $= \alpha + 1$.

22. The $r - 6$ points in Cases 6 and 7 are the points of intersection of the stationary tangent υ by other tangents of the curve m; and so the $r - 8$ points in Cases 8 and 9 are the points of intersection of the double tangent ω by other tangents of the curve m; these numbers $r - 6$ and $r - 8$ will present themselves in the sequel.

We may as to the α-column sum up by saying, that in Cases 2, 7, 9 there is a generating line which reckons as a stationary tangent; in Case 6 a generating line, which, in respect of 4 consecutive intersections, reckons as two stationary tangents; and in Case 8 a generating line, which, in respect of two pairs of 3 consecutive intersections, reckons as two stationary tangents.

In the $(x + \omega)$ and $(m + \upsilon)$-columns, observe that in Cases 6, 7, 8, 9, we have in the first two ω, $\upsilon - 1$, and in the last two $\omega - 1$, υ; viz. these numbers refer to the intersections of the plane with the tangent ω, υ respectively; the actual numbers $\begin{pmatrix} x - 2r + 9 \\ \quad + \omega \end{pmatrix}$ and $\begin{pmatrix} x - 2r + 10 \\ \quad + \omega - 1 \end{pmatrix}$, &c. are equal for the Cases 6 and 8, and also for the Cases 7 and 9. So in the $g + G$ column in Case 4, we have $G - 1$ for G.

The Nodal Curve x; Intersections with the Curve m; and Singularities.

23. The intersections of the curve m with the nodal curve x, are points α, β, γ, H, υ or ω.

24. At a point α, four consecutive points of the curve m lie in a plane; the point may be considered as the intersection of two consecutive tangents, viz. of the line through two consecutive points with that through the next two consecutive points; and it is thus a point on the curve x. We may imagine the points A, A' starting from α in opposite directions along the curve m, and moving in such manner that the tangents at these two points respectively continually intersect; we have thus a portion of the curve x, proceeding apparently in one direction only from the point α; and being, as regards the portion in question, an intersection of two real sheets of the torse; that is, a crunodal curve. The curve x, however, really extends in the opposite direction from α, but it is as to this portion thereof an intersection of two imaginary sheets of the torse; that is, an acnodal curve. The nodal curve x thus meets the curve m in the several points α, the curve x, each time that it passes through such a point of intersection, changing its character from crunodal to acnodal. The two *half-sheets*([1]) of the torse cross each other in the crunodal portion, extending

---

[1] In a curve (plane or twisted), the portions extending each way from a cusp (and considered without reference to a termination) are called half-branches; and so in a surface which has a cuspidal curve, or in particular, in a torse, the portions extending each way from the cuspidal curve (and considered without reference to a termination) are called half-sheets. In the case of higher singularities of a like nature, we may speak of a partial branch or partial sheet (as the case may be).

in one direction from the point α of the nodal curve; and the acnodal portion extends in the opposite direction from the same point.

25. A point β, or stationary point (cusp) on the curve m, is the intersection of four consecutive osculating planes; and it is thus a point of intersection of two consecutive tangents (viz. of the line of intersection of two consecutive osculating planes, and of that of the next two consecutive osculating planes), consequently a point on the curve x. We may imagine the points B, B' starting from β, and moving along the two half-branches of the curve m, in such manner that the tangents at the two points respectively continually intersect. We have thus a portion of the curve x, proceeding apparently in one direction only from the point β (and having at this point a common tangent with the curve m), and being as regards this portion thereof an intersection of two real sheets of the torse; that is, a crunodal curve. The curve x, however, really extends in the opposite direction from the point β, but it is as to this portion thereof an intersection of two imaginary sheets of the torse; that is, an acnodal curve. The nodal curve thus meets the curve m in each of the points β, the curve x, each time that it passes through such a point, changing its character from crunodal to acnodal. There are at β three partial sheets of the torse; viz. if we imagine at this point the half-tangent b in the sense of the two half-branches of the curve m, and the half-tangent b' in the opposite sense, then we have through b' and one of the half-branches a partial sheet, and through b' and the other half-branch a partial sheet; these two partial sheets touching along b', and intersecting in the crunodal portion of the curve x: and a third partial sheet through b and the two half-branches of the curve m.

26. At a point γ on the cuspidal curve m, we have, traversing the curve and the two half-sheets which meet along it, another sheet of the torse, meeting the two half-sheets respectively in two half-branches, which are a portion of the nodal curve x, and which unite together (as at a cusp), in the point γ, which is thus a cusp or stationary point on the curve x.

27. A point H is the intersection of two branches of the cuspidal curve m. There are for each branch two half-sheets; and we have thus at the point H four (say) quarter-branches of the curve x, touching each other at the point (viz. the common tangent is the intersection of the osculating planes belonging to the two branches of the curve m respectively). I find in a special manner that in regard to the curve x a point H is equivalent to six double points, plus two stationary points; it thus causes a reduction $2\cdot 6 + 3\cdot 2 = 18$ in the class of the curve.

28. The nodal curve x has in each of the double tangent planes G an actual double point. In fact the plane is an osculating plane of the curve m at two points thereof; that is, it contains two consecutive tangents R, R', and two other consecutive tangents S, S'; hence RS is a point on the nodal curve x; and not only so, but this is an actual double point, the two tangents being R and S; for since R is met by S and S', there is a consecutive point on the line R; that is, R is a tangent; and similarly since S is met by R and R' there is a consecutive point on the line S; that is, S is also a tangent.

29. At a point υ the tangent has with the curve x a 3-pointic intersection (whence also the tangent is a stationary tangent υ in regard to the curve x), the curves m and x have also a 3-pointic intersection at υ.

30. At each of the two points ω the tangent has with the curve x a 3-pointic intersection (viz. the tangent is in regard to the curve x more than a double tangent ω, instead of two 2-pointic intersections, or ordinary contacts, there are two 3-pointic intersections; I am unable to perceive this directly, but accept it on other grounds). But as at each of the points ω the intersection of the tangent with the curve m is 2-pointic, the intersection of the curves x and m is only 2-pointic.

31. The curves m and x meet in the points α each once, β each 3 times, γ each twice, υ each 3 times, 2ω each twice, and H each 8 times: we have thus

$$\alpha + 3\beta + 2\gamma + 3\upsilon + 4\omega + 8H$$

for the number of actual intersections of the two curves; and the number of apparent intersections is therefore

$$= mx - \alpha - 3\beta - 2\gamma - 3\upsilon - 4\omega - 8H,$$

a result which is required in the sequel.

32. Consider the cone (vertex an arbitrary point) through the curve x, or say simply the cone x.

Each line n (*quà* ordinary line of the system) is met by $r-4$ other lines, the $r-4$ points being situate on the curve x; the line n at each of these points *touches* the cone x, and it therefore besides intersects it in $x-2r+8$ points.

33. A line υ meets the curve x in the point υ counting 3 times, and in $r-6$ points each a stationary point of the curve; it consequently meets the cone x, in the point υ counting 3 times, in each of the $r-6$ points counting twice, and besides in $x-2r+9$ points.

34. A line ω meets the curve x in the two points ω each counting 3 times, and in $r-8$ points each an actual double point of the curve; it consequently meets the cone x in the two points ω each counting 3 times, in the $r-8$ points each counting twice, and besides in $x-2r+10$ points.

35. Each of the $r-8$ points in which the double tangent ω meets another tangent of m is an actual double point of the curve x; and each of the $r-6$ points in which a stationary tangent υ meets another tangent of m is a stationary point or cusp on the curve x. We have thus as singularities of the curve x, the k apparent double points, G actual double points, $\omega(r-8)$ ditto, t triple points, H 4-branch cuspidal points, γ stationary or cuspidal points, $\upsilon(r-6)$ ditto. In regard to the effect upon the class of the curve x, each of the points t is equivalent to 3 double points and produces a reduction $=6$; each of the points H is equivalent to 6 double points $+2$ cusps, and produces a reduction $2\cdot6+3\cdot2=18$, as already mentioned. We have thus the relation

$$q = x(x-1) - 2k - 2G - 3\upsilon(r-6) - 2\omega(r-8) - 3\gamma - 6t - 18H,$$

which is one of the Salmon-Cremona equations.

36. The stationary points of the curve x are the points γ, the $\upsilon(r-6)$ points, and the points H, each counting as 2 stationary points; that is, we have

$$2D_x = q - 2x + 2 + \gamma + \upsilon(r-6) + 2H,$$

used above for finding the value of the deficiency D_x.

The remaining Salmon-Cremona Equations.

37. The formulæ for $m(r-2)$, $x(r-2)$ and $x(r-2)(r-3)$ correspond to those for $c(n-2)$, $b(n-2)$ and $b(n-2)(n-3)$ in the case of a general surface (*Solid Geometry*, 2nd Ed., Nos. 522 and 525 [4th Ed., Nos. 610, 613]), being obtained as follows; considering as before the cone, vertex an arbitrary point, which passes through the curve m, the total number of lines hereof which have with the torse a 3-pointic intersection at the curve m is $= m(r-2)$; and, similarly, considering the cone, vertex the same arbitrary point, which passes through the curve x, the total number of lines hereof which have with the torse a 3-pointic intersection at the curve x is $= x(r-2)$; viz. these numbers $m(r-2)$ and $x(r-2)$ are the numbers of the intersections of the two curves respectively with the surface of the order $(r-2)$, which is the second polar of the arbitrary point in regard to the torse. And so also in the cone, vertex the arbitrary point, which passes through the curve x, the total number of lines which touch the torse at a point not on the curve x is $= x(r-2)(r-3)$; viz. this is obtained by Salmon, the number of intersections of the curve x with a certain surface of the order $(r-2)(r-3)$ (Salmon, Nos. 269 and 273, writing r for n), but in a preferable manner by Cremona thus; the cone through the curve x meets the torse in the curve x counting twice and in a residual curve of the order $x(r-2)$; the lines in question are the tangents from the vertex to this curve, which is a curve meeting each line of the cone $r-2$ times; we may on the surface of the cone metrically, as in the plane, construct the polar of the vertex in regard to the curve of the order $x(r-2)$; viz. we have thus on the cone a curve meeting each generating line $(r-3)$ times; and this polar curve meets the curve of the order $x(r-2)$ in $x(r-2)(r-3)$ points (this would be clearly the case if only the curve of the order $x(r-2)$ were the complete intersection of the cone by a surface of the order $r-2$, for then the polar curve would be the intersection of the cone by the polar surface of the order $r-3$; but in the case in hand, where this is not so, some additional considerations would be required in order to sustain the result), and we have thus the number $x(r-2)(r-3)$ of the lines in question.

38. I consider the second polar surface, order $r-2$, which belongs to a given arbitrary point. Any point on the torse, such that the line joining it with the arbitrary point cuts the torse 3-pointically at the point on the torse, is a point on the second polar $r-2$. Such points are, as will be shewn, the points n, $n(r-4)$, β, γ, υ, 2ω, $\upsilon(r-6)$, $\omega(r-8)$, t, H.

39. We may consider through any such point and the arbitrary point either a particular section of the torse or any section whatever. If the line joining the two points has at the point in question a 3-pointic intersection with the section, then it has a 3-pointic intersection with the torse, viz. the point possesses the required

property. For the points n and $n(r-8)$, the plane may be taken to be the osculating plane of the curve m. This meets the torse in the line n twice, and in a curve of the order $r-2$ touching this line at the point n and meeting it in the $r-4$ points. Hence, considering the complete section made up of the line twice and the curve of the order $r-2$, the line from the arbitrary point to the point n or to any one of the $r-4$ points meets the section 3-pointically; viz. the lines to the points n and to the points $n(r-4)$ meet the torse 3-pointically.

40. For the points β; we may consider any section through the arbitrary point and β; in effect any section through the point β. A plane section through a point β has at this point an invisible triple point, that is a point not in appearance differing from an ordinary point of the curve, but which by considering a consecutive position of the plane of section is seen to be equivalent to a double point and two cusps; viz. the node is a point of intersection of the plane with the curve x, the cusps are the two intersections with the curve m, in the neighbourhood of the point β. The line from β to the arbitrary point has thus with the torse a 3-pointic intersection at β.

41. Similarly for the points γ; we take any section through the arbitrary point and γ; in effect any section through the point γ. The section through a point γ has at this point a triple point, at which an ordinary branch passes through a cusp, and thus equivalent to a cusp $+2$ nodes; in fact, for a consecutive position of the cutting plane, the section has actually a cusp and two nodes; the cusp at the intersection of the plane with the curve m, the nodes at the two intersections of the plane with the curve x in the neighbourhood of the cusp γ. The line through the point γ has thus a 3-pointic intersection with the torse.

42. For a point υ I consider the section through the tangent υ; this is made up of the tangent twice, and of a curve of the order $r-2$ having with the tangent a 4-pointic intersection at the point υ, and besides meeting it in $r-6$ points. Hence in the plane, a line through υ, or through one of the $r-6$ points has at such point a 3-pointic intersection with the curve. And thus the lines through the points υ and $\upsilon(r-6)$ respectively have a 3-pointic intersection with the torse.

43. Similarly for a point ω, I consider the section through a tangent ω; this is made up of the tangent twice, and of a curve of the order $r-2$ having with the tangent a 3-pointic intersection at each of the points ω, and besides meeting it in $r-8$ points. Hence in the plane a line through either of the points ω or through one of the $r-8$ points has with the curve a 3-pointic intersection, and thus the lines through the points 2ω and $\omega(r-8)$ respectively have a 3-pointic intersection with the torse.

44. For a point H I consider any section through the arbitrary point and H; in effect any section through H. There is at H a singularity $=6$ nodes $+2$ cusps. But a line through H in the plane of the section cuts the section in 4 points only; that is, the line from the arbitrary point to H has with the curve a 4-pointic intersection at H; and à fortiori it may be regarded as a line of 3-pointic intersection. The lines to the points H have thus a (4-pointic, that is, more than) 3-pointic intersection with the torse.

45. For a point t I consider any section through the arbitrary point and t; in effect any section through t; there is at t a triple point, and the line through t has thus a 3-pointic intersection at t. Hence a line through t has 3-pointic intersections with the torse.

46. The several points above referred to lie on the curve m or on the curve x, or on both of these curves; and each curve has at these points respectively a simple or multiple intersection with the polar surface $r-2$, as shown in the table.

	Intersection of second polar $r-2$ with curve m	curve x
Points n	2	0
,, $n(r-4)$	0	1
,, β	4	2
,, γ	1	3
,, v	4	4
,, 2ω	2	3
,, $v(r-6)$	0	3
,, $\omega(r-8)$	0	2
,, H	4	12
,, t	0	3

where the figures 1, 2, &c. denote a simple intersection, 2-pointic intersection, &c. of the curve and surface; 0 denotes of course that there is not any intersection, viz. that the curve does not pass through the point referred to.

47. Several of the foregoing numbers are obtained without difficulty; thus we see that the points n, $n(r-4)$ are ordinary points on the second polar $r-2$, the surface at each of the points n *touching* the curve m, but at each of the points $n(r-4)$ simply cutting the curve x. So also the points β, γ, v, 2ω, $v(r-6)$, $\omega(r-8)$ are ordinary points on the second polar; at a point β the two half-branches of the curve m touch the surface in a special manner so as to give a 4-pointic intersection; whereas the curve x simply touches the surface. At γ the curve m cuts the surface, but the two half-branches of x *touch* the surface. At v, each of the curves m, x has a 4-pointic intersection with the surface; at each of the two points ω, the curve m touches the surface, but the curve x has with it a 3-pointic intersection, and at $\omega(r-8)$ it simply touches the surface.

48. The point H is in the nature of a biplanar point on the polar surface; this appears, or is at least indicated, by the circumstance that the line to the arbitrary point has with the torse (not a 3-pointic but) a 4-pointic intersection; the two branches of the curve m each simply cut the two coincident sheets of the polar surface, giving 2×2, $=4$ intersections; but for the curve x, the four partial branches each touch the

two coincident sheets; for a single sheet the number of intersections would be 6, but for the two coincident sheets it is twice this, or $=12$. Finally, a point t is an ordinary point on the second polar, each of the three branches of x simply cuts the surface, or the number of intersections is $=3$.

49. The last table gives at once

$$m\,(r-2) = \qquad 2n + 4\beta + \quad \gamma + 4\upsilon + 2\,.\,2\omega + 4H,$$

$$x\,(r-2) = (r-4) + 2\beta + 3\gamma + 4\upsilon + 3\,.\,2\omega + 3\upsilon\,(r-6) + 2\omega\,(r-8) + 12H + 3t,$$

which are the true theoretical forms of the equations for $m\,(r-2)$ and $x\,(r-2)$, in which these were obtained by Cremona.

50. The $x\,(r-2)\,(r-3)$ points are those points in which the Cremona $x\,(r-2)$ curve is met 2-pointically by the line from the arbitrary point (I recall that taking the arbitrary point as the vertex of a cone through the curve x, this cone, say the cone x, meets the torse in the curve x twice, and in the $x\,(r-2)$ curve in question); viz. these points are either points of contact of tangents from the vertex to the $x\,(r-2)$ curve; or they are double points, or else cusps of the $x\,(r-2)$ curve; in which several cases respectively they count 1, 2 or 3 times, among the $x\,(r-2)\,(r-3)$ points.

51. The points of contact are the $n\,(x-2r+8)$ points of intersection of the lines n with the cone x. We have in fact a plane n through the vertex of the cone, and in this plane two consecutive lines of the system; hence at each of the $x-2r+8$ points the generating line of the cone meets the two consecutive lines of the system; that is, there is with the curve $x\,(r-2)$ a 2-pointic intersection, not arising out of any singularity of the curve, and consequently a contact of this curve with the generating line of the cone.

52. The actual double points of the curve $x\,(r-2)$ are first the $2k$ apparently coincident points of the curve x, and secondly the $\omega\,(x-2r+10)$ points on the lines ω. For first if we consider through the vertex a line meeting the curve x in two points, say A, B, this meets the torse in these points each twice and in $r-4$ other points. Now imagine a line from the vertex to the point P in the vicinity of A, this meets the torse in the point P twice and in $r-2$ points, which are points on the $x\,(r-2)$ curve; hence as P travels through A, 2 of the $r-2$ points come together at B, and again separate, that is B is an actual double point on the $x\,(r-2)$ curve; and similarly A is an actual double point on the curve; and we have thus the $2k$ double points. Secondly, since the line ω is a nodal line on the torse, a generating line of the cone, in the neighbourhood of and considered as travelling through one of the $x-2r+10$ points, meets the torse in two points which come to coincide and then again separate; that is each of the $x-2r+10$ points is an actual double point on the curve $x\,(r-2)$; and the whole number of these is $=\omega\,(x-2r+10)$.

53. The stationary points of the curve $x\,(r-2)$ are first the points on the curve m which apparently coincide with the curve x; viz. the number of these, as was seen, is $=mx-\alpha-3\beta-2\gamma-3\upsilon-4\omega-8H$; secondly, the $\upsilon\,(x-2r+9)$ points on the lines υ;

thirdly, the points H each counting as 4 cusps. For first consider a generating line meeting the curve x in B and the curve x in A; if we imagine on the curve x a point Q which approaches and ultimately coincides with B, the generating line through Q meets the torse in the neighbourhood of its cuspidal edge in two points which come ultimately to coincide with the point A, and we thus see that A is a stationary point on the $x(r-2)$ curve.

54. Secondly, observing that the line υ is a cuspidal line on the torse, and considering in like manner a generating line of the x cone, which approaches and comes ultimately to coincide with one of the $x-2r+9$ points, we see that this is a stationary point on the $x(r-2)$ curve. And thirdly, any line through a point H meets the torse in this point counting 4 times, and in $r-4$ other points. Hence considering the generating line of the x cone, which travelling along any one of the four partial branches of the x curve comes ultimately to coincide with H, 2 of the $r-2$ points on such generating line come to coincide at the point H; and we have thus the point H as a singular point on the $x(r-2)$ curve; viz. it reckons as a stationary point once in respect of each of the four partial branches of the curve x (it must be assumed that this is so, but a further proof is required), that is as 4 cusps on the $x(r-2)$ curve.

55. By what precedes we have

$$\begin{aligned}x(r-2)(r-3) = \;& n(x-2r+8)\\ &+2\{2k+\omega(x-2r+10)\}\\ &+3\{(mx-\alpha-3\beta-2\gamma-3\upsilon-4\omega-8H)+\upsilon(x-2r+9)+4H\},\end{aligned}$$

which is the true theoretical form in which the equation for $x(r-2)(r-3)$ was obtained by Cremona.

500.

ON A THEOREM RELATING TO EIGHT POINTS ON A CONIC.

[From the *Quarterly Journal of Pure and Applied Mathematics*, vol. XI. (1871), pp. 344—346.]

THE following is a known theorem:

"In any octagon inscribed in a conic, the two sets of alternate sides intersect in the 8 points of the octagon and in 8 other points lying in a conic."

In fact the two sets of sides are each of them a quartic curve, hence any quartic curve through 13 of the 8 + 8 points passes through the remaining 3 points: but the original conic together with a conic through 5 of the 8 new points form together such a quartic curve; and hence the remaining 3 of the new points (inasmuch as obviously they are not situate on the original conic) must be situate on the conic through the 5 new points, that is the 8 new points must lie on a conic.

We may without loss of generality take $(\alpha_1^2, \alpha_1, 1)$, $(\alpha_2^2, \alpha_2, 1)$, ... $(\alpha_8^2, \alpha_8, 1)$, as the coordinates (x, y, z) of the 8 points of the octagon; and obtain hereby an *à posteriori* verification of the theorem, by finding the equation of the conic through the 8 new points: the result contains cyclical expressions of an interesting form.

Calling the points of the octagon 1, 2, 3, 4, 5, 6, 7, 8, the 8 new points are

$$12.45,\quad 23.56,\quad 34.67,\quad 45.78,\quad 56.81,\quad 67.12,\quad 78.23,\quad 81.34,$$

viz. 12.45 is the intersection of the lines 12 and 45; and so on. The 8 points lie on a conic, the equation of which is to be found.

The equation of the line 12 is

$$x-(\alpha_1+\alpha_2)\,y+\alpha_1\alpha_2 z=0,$$

or as it is convenient to write it

$$x-(1+2)\,y+12\,.\,z=0,$$

viz. 1, 2, &c., are for shortness written in place of α_1, α_2, &c. respectively.

The coordinates of the 12.45 are consequently proportional to the terms of

$$\begin{matrix} 1, & -(1+2), & 12, \\ 1, & -(4+5), & 45, \end{matrix}$$

or say they are as

$$12(4+5)-45(1+2) : 12-45 : 1+2-(4+5).$$

The equation of the line (12.45)(23.56) which joins the points 12.45 and 23.56 thus is

$$\begin{vmatrix} x & , & y & , & z \\ 12(4+5)-45(1+2), & & 12+45, & & (1+2)-(4+5) \\ 23(5+6)-56(2+3), & & 23-56, & & (2+3)-(5+6) \end{vmatrix} = 0,$$

where the determinant vanishes identically if $2-5=0$ $(\alpha_2-\alpha_5=0)$; it in fact thereby becomes

$$\begin{vmatrix} x & , & y & , & z \\ 2^2(1-4), & & 2(1-4), & & (1-4) \\ 2^2(3-6), & & 3(3-6), & & (3-6) \end{vmatrix},$$

which is $=0$; the determinant divides therefore by $2-5$; the coefficient of x is easily found to be

$$=(2-5)(12-23+34-45+56-61),$$

and so for the other terms; and omitting the factor $2-5$ the equation is

$$\begin{aligned} & x\{12-23+34-45+56-61\} \\ & -y\{12(4+5)-23(5+6)+34(6+1)-45(1+2)+56(2+3)-61(3+4)\} \\ & +z\{1234-2345+3456-4561+5612-6123\}=0. \end{aligned}$$

There is now not much difficulty in forming the equation of the required conic; viz. this is

$$\begin{aligned} & (2-8)\{x-(6+7)y+67z\} \times \\ & \left\{\begin{aligned} & x[12-23+34-45+56-61] \\ & -y[12(4+5)-23(5+6)+34(6+1)-45(1+2)+56(2+3)-61(3+4)] \\ & +z[1234-2345+3456-4561+5612-6123] \end{aligned}\right\} \\ & +(6-8)\{x-(1+2)y+12z\} \times \\ & \left\{\begin{aligned} & x[23-34+45-56+67-72] \\ & -y[23(5+6)-34(6+7)+45(7+2)-56(2+3)+67(3+4)-72(4+5)] \\ & +z[2345-3456+4567-5672+6723-7234] \end{aligned}\right\} \\ & =0. \end{aligned}$$

In fact this equation written with an indeterminate coefficient λ, say, for shortness, thus

$$67\,[(12\,.\,45)(23\,.\,56)] = \lambda 12\,.\,[(23\,.\,56)(34\,.\,67)] = 0,$$

is the general equation of the conic through the 4 points $12\,.\,67$, $34\,.\,67$, $12\,.\,45$, and $23\,.\,56$; and by making the conic pass through 1 of the remaining 4 of the 8 points, I succeeded in finding the value $\lambda = \dfrac{6-8}{2-8}$, so that the conic in question passes through 5 of the 8 points, and is therefore by the theorem the conic through the 8 points. But as thus written down the equation contains the extraneous factor $2-6$, as appears at once by the observation that the left-hand side on writing therein $6 = 2\,(\alpha_6 = \alpha_2)$ becomes identically $=0$; the value in fact is

$$\begin{aligned}&-(2-8)\,[x-(2+7)\,y+27z]\,(23-34+45-52)\,[x-(1+2)\,y+12z]\\&+(2-8)\,[x-(1+2)\,y+12z]\,(23-34+45-52)\,[x-(2+7)\,y+27z]\end{aligned}$$

which is $=0$; there is consequently the factor $2-6$ to be rejected, and throwing this out the equation assumes a symmetrical form in regard to the 8 symbols 1, 2, 3, 4, 5, 6, 7, 8. The coefficient of x^2 is very easily found to be

$$= (2-6)\,(12-23+34-45+56-67+78-81),$$

and similarly that of z^2 to be

$$= (2-6)\,\{123456-234567+345678-456781+567812-678123+781234-812345\}:$$

those of the other terms are somewhat more difficult to calculate; but the final result, throwing out the factor $(2-6)$, and introducing an abbreviated notation

$$\Sigma 12 = (12-23+34-45+56-67+78-81),$$

and the like in other cases, is found to be

$$\begin{aligned}&x^2\,.\;\Sigma 12\\&+y^2\,.\,[\Sigma 12\,(4+5)\,(6+7)-\tfrac{1}{2}\,\Sigma 1256]\\&+z^2\,.\;\Sigma 123456\\&-yz\,.\;\Sigma 16\,(234+235+245+345)\\&+zx\,.\,[\Sigma 1234 \qquad\qquad +\tfrac{1}{2}\,\Sigma 1256]\\&-xy\,.\;\Sigma 12\,(4+5+6+7) \qquad\qquad = 0,\end{aligned}$$

where it is to be observed that $\Sigma 1256$ consists of 4 distinct terms each twice repeated: $\tfrac{1}{2}\Sigma 1256$ consists therefore of these 4 terms; and in the coefficient of y^2 they destroy 4 of the 32 terms of $\Sigma 12\,(4+5)\,(6+7)$ so that the coefficient of y^2 contains $32-4$, $=28$ terms. In the coefficient of zx there is no destruction, and this contains therefore $12+4$, $=16$ terms.

501.

REVIEW. *Tables de Logarithmes vulgaires à dix decimales construites d'après un nouveau mode* par S. Pineto, approuvées par l'Académie des Sciences de S. Pétersbourg. S. Pétersbourg, 1871.

[From the *Quarterly Journal of Pure and Applied Mathematics*, vol. XI. (1871), pp. 375—376.]

THE tables occupy 56 pages—the principal one being a table in 44 pages, the 22 left-hand pages containing the 10 figure logarithms of the numbers from 1,000,000 to 1,010,999, and the 22 right-hand pages the proportional parts ·01,·02, ... ·99 of the differences. A like table 100,000 to 999,999 would occupy 3600 pages. By means of an auxiliary table of 3 pages, and of a slight increase of the numerical calculation, *the table of* 44 *pages does the work of the table of* 3600 *pages*. To explain how this is: the auxiliary table gives for any number A the initial four digits of which are equal to or exceed 1011, a multiplier M, such that in the product MA the initial four digits are between 1000 and 1011; this multiplier M contains only 1, 2 or 3 figures, and when there are 3 figures, then in general either the middle figure is 0, or two of the figures are equal; the table gives also $\log \frac{1}{M}$ to 12 decimals; and there is a third column, as will be explained. Hence A being as above, the auxiliary table gives M, we form the product MA, obtain the logarithm thereof from the principal table, and adding thereto $\log \frac{1}{M}$, we have the required $\log A$. Conversely, when there is given a logarithm B the first five digits in the mantissa of which are not included between 00000 and 00474 (being the limits of the first five digits of the logarithms in the principal table), the auxiliary table by means of its third column gives M; adding $\log \frac{1}{M}$ to B, we have a logarithm included in the limits of the principal table, and seeking for the corresponding number, this is $=\frac{1}{M}$ number having B for its logarithm: that is, the required number is $=M$ times the number obtained as above. Of course as regards the principal Table, the proportional parts are employed in the usual manner; the tabulation of them to hundredths (instead of tenths) facilitates the interpolation; for better securing the accuracy of the last figure, directions are given in regard to the 11th and 12th figures. An example of the determination of a logarithm is as follows:

$$\begin{array}{r} \pi = \quad 3{\cdot}14159\ 26536 \\ M = 32 \\ \hline M\pi = 100{\cdot}53096\ 49152 \end{array}$$

$$\log\frac{1}{M} = 8{\cdot}49485\ 00216.80 - 10$$

$$\left.\begin{array}{lr} \log = 2{\cdot}00229 & 95705.75 \\ 64 & 2764.80 \\ 91 & 39.31 \\ 52 & 22 \end{array}\right\} D = 4320$$

$$\log\pi = 0{\cdot}49714\ 98726.88$$

(correct value of last two figures $= 94$).

The labour saved by the small bulk of the Tables goes far to balance that occasioned by the additional steps in the calculation.

502.

ON THE SURFACES DIVISIBLE INTO SQUARES BY THEIR CURVES OF CURVATURE.

[From the *Proceedings of the London Mathematical Society*, vol. IV. (1871—1873), pp. 8, 9. Read December 14, 1871.]

GEOMETRICALLY, the question is as follows:—Consider any two curves of curvature AB, CD of one set, and any two AC, BD of the other set, as shown by the continuous lines of the figure: drawing the consecutive curves as shown by dotted lines, the curve consecutive to AB at an arbitrary (infinitesimal) distance from AB, the other three curves may be drawn at such distances that the elements at A, B, and C shall be each of them a square; but this being so, the element at D will not be in general a square, and it is only for certain surfaces that it is so. But if (whatever the curves

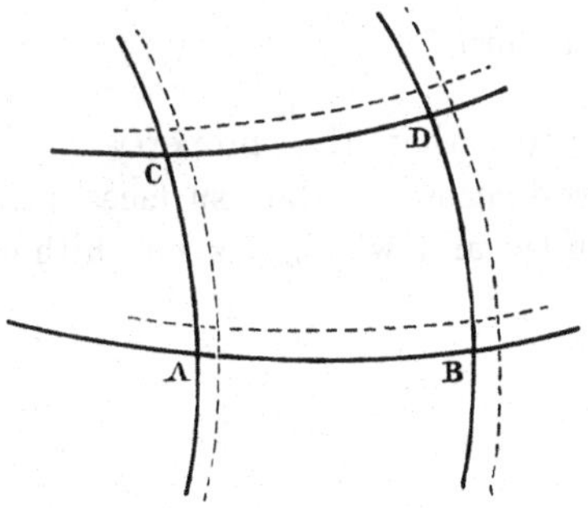

of curvature AB, CD, AD, BC may be) the element at D is a square, then it is clear that the whole surface can be, by means of its curves of curvature, divided into infinitesimal squares.

Analytically, if for a given surface the equations of its curves of curvature are expressed in the form $h=f(x, y, z)$, $k=\phi(x, y, z)$; then the coordinates x, y, z can be expressed each of them as a function of the parameters h, k, and we have for the element of distance between two consecutive points on the surface

$$dx^2 + dy^2 + dz^2 = A dh^2 + C dk^2,$$

where A, C are in general each of them a function of h and k. The condition for the divisibility into squares is that the quotient $A \div C$ shall be of the form function h $\div$ function k.

It was shown by M. Bertrand that, in a triple system of orthotomic isothermal surfaces, each surface possesses the property in question of divisibility into squares by means of its curves of curvature. But in such a triple system, each surface of the system is necessarily a quadric; so that the theorem comes to this, that a quadric surface is, by means of its curves of curvature, divisible into squares. The analytical verification is at once effected: taking the equation of the surface to be

$$\frac{x^2}{a} + \frac{y^2}{b} + \frac{z^2}{c} = 1,$$

then the expressions for the coordinates in terms of the parameters h, k of a curve of curvature are

$$x^2 = \frac{a(a+h)(a+k)}{(a-b)(a-c)},$$

$$y^2 = \frac{b(b+h)(b+k)}{(b-c)(b-a)},$$

$$z^2 = \frac{c(c+h)(c+k)}{(c-a)(c-b)},$$

and we have

$$4(dx^2 + dy^2 + dz^2) = (h-k)\left\{\frac{h\,dh^2}{(a+h)(b+h)(c+h)} - \frac{k\,dk^2}{(a+k)(b+k)(c+k)}\right\};$$

so that $A \div C$ is of the required form.

But there is nothing to show that the property is confined to quadric surfaces; and the question of the determination of the surfaces possessing the property appears to be one of considerable difficulty, and which has not hitherto been examined.

503.

ON THE SURFACES EACH THE LOCUS OF THE VERTEX OF A CONE WHICH PASSES THROUGH m GIVEN POINTS AND TOUCHES $6-m$ GIVEN LINES.

[From the *Proceedings of the London Mathematical Society*, vol. IV. (1871—1873), pp. 11—47. Read January 11, 1872.]

I CONSIDER the surfaces, each of them the locus of the vertex of a (quadri-)cone which passes through m given points and touches $6-m$ given lines; viz. calling the given points $a, b, c, \ldots$ and the given lines $\alpha, \beta, \gamma, \ldots$, the surfaces in question are:

	Order
$abcdef$	4
$abcde\alpha$	8
$abcd\alpha\beta$	16
$abc\alpha\beta\gamma$	24
$ab\alpha\beta\gamma\delta$	24
$a\alpha\beta\gamma\delta\epsilon$	14
$\alpha\beta\gamma\delta\epsilon\zeta$	8

I remark that the orders of these several surfaces are in effect determined by the investigations of M. Chasles in regard to the conics in space which satisfy seven conditions. The surface $abcdef$ was long ago considered by M. Chasles, and it is treated of in my "Memoir on Quartic Surfaces," [445], and in the same Memoir the surface $\alpha\beta\gamma\delta\epsilon\zeta$ is also referred to: these two surfaces, and also the surfaces $a\alpha\beta\gamma\delta\epsilon$ and $ab\alpha\beta\gamma\delta$ are considered by Dr Hierholzer[1] in his excellent paper "Ueber Kegelschnitte im Raume," *Math. Annalen*, t. II. (1870), pp. 563—586, and to him are due the equations given in the sequel for the surfaces $abcdef$ and $\alpha\beta\gamma\delta\epsilon\zeta$: the researches of the present Memoir are in fact a continuation and development of those in the Memoir last referred to.

[1] I was grieved to hear of Dr Hierholzer's death last autumn, at Carlsruhe, at the early age of 30.

Table of Singularities, and Explanations in regard thereto.

1. We have on the before-mentioned surfaces respectively certain simple or multiple points, right lines, and curves, as shown in the following Table:

	abcdef, 4	*abcdea*, 8	*abcdaβ*, 16	*abcaβγ*, 24	*abaβγδ*, 24	*aaβγδε*, 14	*aβγδεζ*, 8	
Points *a*	6 × (2)	5 × (4)	4 × (8)	3 × (8)	2 × (4)	1 × (2)	0	(0)
Lines *ab*	15 × (1), *C*	10 × (2), *C*	6 × (4), *C*	3 × (4), *C*	1 × (2), *C*	0	0	(1)
a	0	1 × (2), *C*	2 × (4), *C*	3 × (8), *C*	4 × (8), *C*	5 × (4), *C*	6 × (2), *C*	(2)
[*ab*, *a*, *β*, *γ*]	0	0	0	6 × (4), *P*	[3] 8 × (2+2), *L*	0	0	(3)
[*a*, *β*, *γ*, *δ*]	0	0	0	0	2 × (8), *P*	10 × (2), *L*	30 × (1), *L*	(4)
[*ab*, *cd*, *a*, *β*]	0	0	6 × (2), *P*	0	0	0	0	(5)
abc, *def*	10 × (1), *P*	0	0	0	0	0	0	(6)
abc, *de*, *a*	0	[1] 10 × (1), *P*	0	0	0	0	0	(7)
abc, *a*, *β*	0	0	[2] 4 × (2+2), *P*	3 × (4), *P*	0	0	0	(8)
Cubic *abcdef*	1 × (1), *C*	0	0	0	0	0	0	(9)
Quadriquadric *aβγ*, *δεζ*	0	0	0	0	0	0	(10) × (1), *L*	(10)
Excuboquartic *aβγ*, *δε*, *a*	0	0	0	0	0	[4] 10 × (1), *L*	0	(11)

[1] Tangent plane along line is plane *abc*.
[2] Tacnodal line, each sheet touched along line by plane *abc*.
[3] Tacnodal line, each sheet of surface touched along line by hyperboloid *aβγ*.
[4] Surface touched along line by hyperboloid *aβγ*.

2. In the Table, the upper margin refers to the surfaces, and the left-hand margin to the points, lines, and curves situate on these surfaces respectively; the body of the Table showing the number, and in ( ) the multiplicity, of these points, lines, and curves in regard to the several surfaces respectively. Thus, points a; for the surface $abcdef$, $6 \times (2)$, there are 6 such points, each of them a 2-conical (ordinary conical) point on the surface: so $abcde\alpha$, $5 \times (4)$, there are 5 such points, each a 4-conical point on the surface (viz. instead of the tangent plane there is a quartic cone); and so on. Similarly, lines ab (viz. these are the lines joining two points a, b); for the surface $abcdef$, $15 \times (1)$, there are 15 such lines, each a simple line on the surface; surface $abcde\alpha$, $10 \times (2)$, there are 10 such lines, each a double (ordinary nodal) line on the surface; and so on. We have in two places the multiplicity $(2+2)$, which refers to a tacnodal line, as presently explained. The corner letters C, P, L denote respectively proper cone, plane-pair, and line-pair, as afterwards explained.

3. The lines and curves referred to in the left-hand margin are:

(1) ab, line joining the points a and b.

(2) α, line α.

(3) $[ab, \alpha, \beta, \gamma]$, pair of lines meeting each of the four lines, or say the tractors of the four lines ab, α, β, γ. As regards the surface $ab\alpha\beta\gamma\delta$, the multiplicity is given as $(2+2)$, viz. the line is (not an ordinary nodal, but) a tacnodal line, each sheet touching along the whole line the hyperboloid $\alpha\beta\gamma$.

(4) $[\alpha, \beta, \gamma, \delta]$, tractors of the four lines α, β, γ, δ.

(5) $[ab, cd, \alpha, \beta]$ tractors of the four lines ab, cd, α, β.

(6) abc, def, line of intersection of the planes abc and def.

(7) abc, de, α, line in the plane abc joining the intersections of this plane by the lines de and α respectively.

(8) abc, α, β, line in the plane abc joining the intersections of this plane by the lines α and β respectively. As regards the surface $abcd\alpha\beta$, the multiplicity is given as $(2+2)$, viz. each line is (not an ordinary nodal, but) a tacnodal line, each sheet touching along the whole line the plane abc.

(9) Cubic $abcdef$, cubic curve through the six points a, b, c, d, e, f, common intersection of the cones each having its vertex at one of the points and passing through the other five.

(10) Quadriquadric $\alpha\beta\gamma$, $\delta\epsilon\zeta$, intersection of the quadric surfaces $\alpha\beta\gamma$ and $\delta\epsilon\zeta$, that is, the quadric surfaces through the lines α, β, γ and δ, ϵ, ζ respectively.

(11) Excuboquartic $\alpha\beta\gamma$, $\delta\epsilon$, a, quartic curve generated as follows: viz. taking any line whatever which meets the lines α, β, γ (or say any generating line of the quadric $\alpha\beta\gamma$), the plane through this line and the point a meets the lines δ, ϵ in two points respectively; and the line joining these meets the generating line in a point having for its locus the excuboquartic curve in question (theory further considered in the sequel).

Special forms of (Quadri-)Cones.

4. We have to consider the special forms of (quadri-)cones; these are: 1°. The sharp-cone, or plane-pair; that is, a pair of two planes, intersecting in a line called the axis, the vertex being in this case an indeterminate point on the axis. Observe that a plane-pair passes through a given point when either of its planes passes through such point; it touches a given line when its axis meets the given line. 2°. The flat-cone, or line-pair; viz. this is a pair of intersecting lines, their point of intersection being the vertex of the line-pair, and the plane of the two lines being the *diametral* of the line-pair. Observe that the line-pair passes through a given point when its diametral passes through such point; it touches a given line when either of its lines meets the given line. 3°. There is a third kind, the line-pair-plane; viz. the two planes of the plane-pair may come to coincide, retaining, however, a definite line of intersection, or axis: or again, the two lines of a line-pair may come to coincide, retaining a definite plane or diametral; that is, in either case we have a plane passing through a line; and which is to be considered indifferently as two coincident planes intersecting in the line, or as two coincident lines lying in the plane. But there is not, in the present Memoir, any occasion to consider this third kind of special cone.

The letters C, P, L in the Table denote that the cone is a (proper) cone, plane-pair, or line-pair, as the case may be.

Singular Lines and Curves on the Surfaces.

5. We may establish *à priori* the existence, and even to some extent the multiplicity, of the several lines and curves on the surfaces $abcdef, \ldots \alpha\beta\gamma\delta\epsilon\zeta$. Thus:

1°. Lines ab: take for the vertex of the cone a point at pleasure on the line ab; the cone passing through b will *ipso facto* pass through a; and the conditions are thus that the cone shall pass through b and satisfy four other conditions—in all, five conditions: and there is thus a cone with the point in question as vertex; that is, the line ab is situate on the surface. Moreover, for the surfaces $abcdef$, $abcde\alpha$, $abcd\alpha\beta$, $abc\alpha\beta\gamma$, $ab\alpha\beta\gamma\delta$ respectively, for a given position of the vertex on the line ab, the number of cones is 1, 2, 4, 4, 2 respectively: and these are the multiplicities of the line ab on the several surfaces respectively.

2°. Lines α: take for the vertex of the cone a point at pleasure on the line α; then the cone *ipso facto* touches the line α, and there are only five other conditions to be satisfied; that is, we have a cone with the vertex in question; or the line α is situate on the surface. Moreover, for the surfaces $abcde\alpha$, $abcd\alpha\beta$, $abc\alpha\beta\gamma$, $ab\alpha\beta\gamma\delta$, $a\alpha\beta\gamma\delta\epsilon$, $\alpha\beta\gamma\delta\epsilon\zeta$ respectively, the number of cones is 1, 2, 4, 4, 2, 1 respectively: and it may be seen that the multiplicities of the line α are the doubles of these numbers, or are $=2$, 4, 8, 8, 4, 2 for the several surfaces respectively.

3°. Lines $[ab, \alpha, \beta, \gamma]$: taking the vertex in one of these tractors, the cone cannot be a proper cone, but (if it exist) it must be either a line-pair having the tractor for one of its lines, or else a plane-pair having the tractor for its axis. The two cases are:

Surface $abc\alpha\beta\gamma$. Cone is a plane-pair, the two planes intersecting in the tractor, and passing, the one of them through the points a, b, the other through the point c. The vertex being an indeterminate point on the tractor, the tractor is situate on the surface.

Surface $ab\alpha\beta\gamma\delta$. Cone is a line-pair, one line being the tractor, the other a line drawn in the plane of the tractor and ab to meet δ, and which meets the tractor in an arbitrary point thereof: the tractor is thus a line on the surface.

4°. Lines $[\alpha, \beta, \gamma, \delta]$: taking the vertex in one of these tractors, then, as in the last case, the cone is either a line-pair having the tractor for one of its lines or a plane-pair having the tractor for its axis. The three cases are:

Surface $ab\alpha\beta\gamma\delta$. Cone is a plane-pair, the two planes intersecting in the tractor and passing through the points a, b respectively.

Surface $a\alpha\beta\gamma\delta\epsilon$. Cone is a line-pair, one line being the tractor, the other a line in the plane of the tractor and a, meeting the line ϵ and meeting the tractor in an indeterminate point.

Surface $\alpha\beta\gamma\delta\epsilon\zeta$. Cone is a line-pair, one line being the tractor, the other a line drawn from an indeterminate point of the tractor to meet the lines ϵ and ζ.

5°. Lines $[ab, cd, \alpha, \beta]$. Cone is a plane-pair, the two planes intersecting in the tractor, and passing through the points a, b and the points c, d respectively.

6°. Line abc, def. Cone is a plane-pair, consisting of the two planes abc and def.

7°. Line abc, de, α. Cone is a plane-pair, the two planes intersecting in the line; one plane being abc, the other a plane through the line de.

8°. Line abc, α, β. There are two cases:

Surface $abcd\alpha\beta$. Cone is a plane-pair, the two planes intersecting in the line; the one being abc, and the other passing through the point d.

Surface $abc\alpha\beta\gamma$. Cone is a line-pair; one line being abc, α, β, the other a line in the plane abc meeting the line δ, and meeting the line abc, α, β in an indeterminate point.

9°. Cubic $abcdef$. Each point of the cubic is the vertex of a proper cone passing through the cubic, and therefore through the six points; that is, the cubic is a line on the surface $abcdef$.

10°. Quadriquadric $\alpha\beta\gamma$, $\delta\epsilon\zeta$. Cone is a line-pair; viz. it is composed of the lines drawn from any point of the curve, one of them to meet the lines α, β, γ, and the other to meet the lines δ, ϵ, ζ.

11°. Excuboquartic $\alpha\beta\gamma$, $\delta\epsilon$, a. Cone is a line-pair; the two lines being, one of them a line at pleasure meeting α, β, γ, the other the line which, in the plane of the other line and the point a, meets the lines δ, ϵ.

Mode of obtaining the several Equations: Notations and Formulæ.

6. The equations of the several surfaces are obtained by taking as centre of projection an assumed position of the vertex, and projecting everything upon an arbitrary plane; the projections of the given points and lines are points and lines in the arbitrary plane, and the section of the cone by this plane is a conic; the equation of the surface is thus obtained as the condition that there shall be a conic passing through m given points and touching $6-m$ given lines.

7. We take as current coordinates (X, Y, Z, W), or when plane-coordinates are employed $(\xi, \eta, \zeta, \omega)$: the coordinates of the vertex are throughout represented by (x, y, z, w); but in explanations &c., these are also used as current coordinates. The plane of projection is taken to be $W=0$. The coordinates of the given points a, &c., are taken to be (x_a, y_a, z_a, w_a), &c. There is no confusion occasioned by so doing, and I retain the ordinary letters (a, b, c, f, g, h) for the six coordinates of a line, it being understood that these letters so used have no reference whatever to the given points a, b, &c.; viz. the coordinates of the given lines α, &c., are $(a_\alpha, b_\alpha, c_\alpha, f_\alpha, g_\alpha, h_\alpha)$, &c.; there is sometimes occasion to consider the coordinates of other lines ab, &c., but the notation will always be explained.

8. I write l, m, n, p, q, r for the coordinates of the line joining the vertex (x, y, z, w) with a point (x', y', z', w'); viz.

$$\begin{aligned} l &= yz' - y'z, & p &= xw' - x'w, \\ m &= zx' - z'x, & q &= yw' - y'w, \\ n &= xy' - x'y, & r &= zw' - z'w, \end{aligned}$$

$(l_a = yz_a - y_a z$, &c., this being explained when necessary); and also

$$\begin{aligned} P &= \quad . \quad hy - gz + aw, \\ Q &= -hx \quad . \quad + fz + bw, \\ R &= \quad gx - fy \quad . \quad + cw, \\ S &= -ax - by - cz \quad . \,, \end{aligned}$$

$(P_\alpha = h_\alpha y - g_\alpha z + a_\alpha w$, &c., this being explained when necessary).

This being so, then projecting from the vertex (x, y, z, w), say on the plane $W=0$, the x, y, z coordinates of the projection of a point a are as $p_a : q_a : r_a \, (p_a = xw_a - x_a w$, &c.); and the equation of the projection of a line α is

$$P_\alpha X + Q_\alpha Y + R_\alpha Z = 0,$$

$(P_\alpha = h_\alpha y - g_\alpha z + a_\alpha w$, &c.). We thus have, in the projection on the plane $W=0$, the m points and $6-m$ lines situate in and touched by the conic.

The following notations and formulæ are convenient:

9. $pabc=0$ is the equation of the plane through the points a, b, c; viz.

$$pabc=\begin{vmatrix} x, & y, & z, & w \\ x_a, & y_a, & z_a, & w_a \\ x_b, & y_b, & z_b, & w_b \\ x_c, & y_c, & z_c, & w_c \end{vmatrix}.$$

Of course $pbac=-pabc$, &c. Observe that here, and in the notations which follow, the letter p is used as referring to the coordinates (x, y, z, w), and that the index of $p\,(=1$ when no index is expressed) shows the degree in these coordinates.

10. $pa\alpha=0$ is the equation of the plane through the point a and the line α; viz. $pa\alpha$ is the foregoing determinant, if for a moment b, c are any two points on the line α; or, what is the same thing,

$$pa\alpha=P_a x+Q_a y+R_a z+S_a w,$$

where

$$\begin{aligned} P_a &= \quad . \qquad\quad hy_a-gz_a+aw_a, \\ Q_a &= -hx_a \quad . \quad +fz_a+bw_a, \\ R_a &= \ \ gx_a-fy_a \quad . \quad +cw_a, \\ S_a &= -ax_a-by_a-cz_a \quad . \ ; \end{aligned}$$

and (a, b, c, f, g, h) are the coordinates of the line α: observe that $pa\alpha=p\alpha a$.

11. $p^2\alpha\beta\gamma=0$ is the equation of the quadric surface through the lines α, β, γ; viz. we have

$$\begin{aligned} p^2\alpha\beta\gamma=(agh)\,x^2+(bhf)\,y^2+(cfg)\,z^2+(abc)\,w^2 & \\ +[(abg)-(cah)]\,xw & \\ +[(bch)-(abf)]\,yw & \\ +[(caf)-(bcg)]\,zw & \\ +[(bfg)+(chf)]\,yz & \\ +[(cgh)+(afg)]\,zx & \\ +[(ahf)+(bgh)]\,xy, & \end{aligned}$$

where

$$agh=\begin{vmatrix} a_\alpha, & g_\alpha, & h_\alpha \\ a_\beta, & g_\beta, & h_\beta \\ a_\gamma, & g_\gamma, & h_\gamma \end{vmatrix}\ \&\text{c.}$$

$(a_\alpha, b_\alpha, c_\alpha, f_\alpha, g_\alpha, h_\alpha)$, $(a_\beta, \ldots)$, $(a_\gamma, \ldots)$ being the coordinates of the given lines α, β, γ. Observe that $p^2\beta\alpha\gamma=-p^2\alpha\beta\gamma$, &c.

12. It is to be noticed that, writing

$$\begin{aligned} P &= \quad . \quad hy - gz + aw, \\ Q &= -hx \quad . \quad + fz + bw, \\ R &= \quad gx - fy \quad . \quad + cw, \\ S &= -ax - by - cz \quad . \ , \end{aligned}$$

viz. $P_\alpha = h_\alpha y - g_\alpha z + a_\alpha w$, &c., then that we have identically

$$\begin{vmatrix} \lambda, & \mu, & \nu, & \rho \\ P_\alpha, & Q_\alpha, & R_\alpha, & S_\alpha \\ P_\beta, & Q_\beta, & R_\beta, & S_\beta \\ P_\gamma, & Q_\gamma, & R_\gamma, & S_\gamma \end{vmatrix} = -(\lambda x + \mu y + \nu z + \rho w) \,.\, p^2 \alpha\beta\gamma,$$

and further that we have identically

$$-p^2\alpha\beta\gamma = L_{\alpha\beta} P_\gamma + M_{\alpha\beta} Q_\gamma + N_{\alpha\beta} R_\gamma + \Omega_{\alpha\beta} S_\gamma,$$

where

$$\begin{aligned} L &= (af' - a'f)\, x + (bf' - b'f)\, y + (cf' - c'f)\, z - (bc' - b'c)\, w, \\ M &= (ag' - a'g)\, x + (bg' - b'g)\, y + (cg' - c'g)\, z - (ca' - c'a)\, w, \\ N &= (ah' - a'h)\, x + (bh' - b'h)\, y + (ch' - c'h)\, z - (ab' - a'b)\, w, \\ \Omega &= (gh' - g'h)\, x + (hf' - h'f)\, y + (fg' - f'g)\, z + (af' - a'f + bg' - b'g + ch' - c'h)\, w; \end{aligned}$$

and $L_{\alpha\beta}$, &c. are the values of L, &c. on substituting therein $(a_\alpha, \ldots)$ and $(a_\beta, \ldots)$ for the unaccented and accented letters respectively.

13. Observe that we have

$$\begin{aligned} L + (a'f + b'g + c'h)\, x &= \quad . \quad - c'Q + b'R - f'S, \\ M + (\quad \text{,,} \quad)\, y &= \quad c'P \quad . \quad - a'R - g'S, \\ N + (\quad \text{,,} \quad)\, z &= -b'P + a'Q \quad . \quad - h'S, \\ \Omega + (\quad \text{,,} \quad)\, w &= \quad f'P + g'Q + h'R \quad . \ ; \end{aligned}$$

and similarly

$$\begin{aligned} -L + (af' + bg' + ch')\, x &= \quad . \quad - cQ' + bR' - fS', \\ -M + (\quad \text{,,} \quad)\, y &= \quad cP' \quad . \quad - aR' - gS', \\ -N + (\quad \text{,,} \quad)\, z &= -bP' + aQ' \quad . \quad - hS', \\ -\Omega + (\quad \text{,,} \quad)\, w &= \quad fP' + gQ' + hR' \quad . \ ; \end{aligned}$$

whence also

$$\begin{aligned} . \quad h'M - g'N + a'\Omega &= -(a'f + b'g + c'h)\, P', \\ -h'L \quad . \quad + f'N + b'\Omega &= -(\quad \text{,,} \quad)\, Q', \\ g'L - f'M \quad . \quad + c'\Omega &= -(\quad \text{,,} \quad)\, R', \\ -a'L - b'M - c'N \quad . \quad &= -(\quad \text{,,} \quad)\, S'; \end{aligned}$$

and

$$\begin{array}{rrrrl} . & hM & -gN & +a\Omega = & (af'+bg'+ch')P, \\ -hL & . & +fN & +b\Omega = & (\quad ,, \quad)Q, \\ gL & -fM & . & +c\Omega = & (\quad ,, \quad)R, \\ -aL & -bM & -cN & . \quad = & (\quad ,, \quad)S. \end{array}$$

14. $p^3a\,.\,a\beta\,.\,\gamma\delta=0$ is the equation of the cubic surface through the lines α, β, γ, δ and $a\alpha\beta$, $a\gamma\delta$ (viz. $a\alpha\beta$ is the line from a to meet α, β, and so $a\gamma\delta$ is the line from a to meet γ, δ). Observe that the conditions which determine this cubic surface thus are that the cubic shall pass through

a; the points of $a\alpha\beta$ on α and β respectively, 3 other points on α, 3 on β, and 1 on $a\alpha\beta$;

also the points of $a\gamma\delta$ on γ and δ respectively, 3 other points on γ, 3 on δ, and 1 on $a\gamma\delta$; in all, $1+9+9=19$ points;

viz. the conditions completely determine the surface.

15. We have

$$p^3a\,.\,a\beta\,.\,\gamma\delta=\begin{vmatrix} x, & y, & z, & w \\ x_a, & y_a, & z_a, & w_a \\ L_{\alpha\beta}, & M_{\alpha\beta}, & N_{\alpha\beta}, & \Omega_{\alpha\beta} \\ L_{\gamma\delta}, & M_{\gamma\delta}, & N_{\gamma\delta}, & \Omega_{\gamma\delta} \end{vmatrix},$$

viz. this determinant, equated to zero, gives the equation of the surface.

To prove this, take as before the unaccented letters (a, b, c, f, g, h) to refer to the line α, and the letters with one, two, and three accents to refer to the lines β, γ, δ respectively; write also L, M, N, Ω and L', M', N', Ω' for $L_{\alpha\beta}$, &c., and $L_{\gamma\delta}$, &c., respectively. Referring to the foregoing expressions for L, M, N, Ω, and observing that for a point on the line α, the values of P, Q, R, S are each $=0$, then for such a point we have $L+(a'f+b'g+c'h)x=0$, &c., that is, $L:M:N:\Omega=x:y:z:w$, and these values satisfy the equation of the surface, which is thus a surface passing through the line α; and similarly it passes through the lines β, γ, δ.

To show that the surface passes through the line $a\alpha\beta$, take the coordinates of the point a to be 0, 0, 0, 1; then the line $a\alpha\beta$ is given as the intersection of the planes $ax+by+cz=0$ and $a'x+b'y+c'z=0$, that is, $S=0$ and $S'=0$. And the equation of the surface, writing therein x_a, y_a, z_a, $w_a=0$, 0, 0, 1, becomes

$$\begin{vmatrix} x, & y, & z \\ L, & M, & N \\ L', & M', & N' \end{vmatrix}=0,$$

or, as this may be written,

$$\begin{vmatrix} S & , S' & , z \\ aL+bM+cN, & a'L+b'M+c'N, & N \\ aL'+bM'+cN', & a'L'+b'M'+c'N', & N' \end{vmatrix} = 0;$$

and, for a point on the line $a\alpha\beta$, this is

$$\begin{vmatrix} aL+bM+cN, & a'L+b'M+c'N \\ aL'+bM'+cN', & a'L'+b'M'+c'N' \end{vmatrix} = 0.$$

But in the equations $-a'L-b'M-c'N=-(a'f+b'g+c'h)S'$, and $-aL-bM-cN=(af'+bg'+ch')S$, writing $S=0$ and $S'=0$, we have $aL+bM+cN=0$ and $a'L+b'M+c'N=0$, and the equation is satisfied; that is, the surface passes through the line $a\alpha\beta$, and similarly it passes through the line $a\gamma\delta$.

Surface abcdef.

16. The equation may be written

$$pabe\,.\,pcde\,.\,pacf\,.\,pdbf-pabf\,.\,pcdf\,.\,pace\,.\,pdbe=0,$$

where $pabe=0$ is the equation of the plane through the points a, b, e; and the like for the other symbols. The form is one out of 45 like forms, depending on the partitionment

$$\left\{\begin{matrix} ab\,.\,cd \\ ac\,.\,db \\ ad\,.\,bc \end{matrix}\right\}(ef),$$

of the six letters.

17. *Investigation.* In the projection, the six points (p_a, q_a, r_a) are situate on a conic; the condition for this is

$$(p,\ q,\ r)^2=0,$$

where the left-hand side represents the determinant obtained by writing successively (p_a, q_a, r_a), &c., for (p, q, r). The equation in question may be written

$$abe\,.\,cde\,.\,acf\,.\,dbf-abf\,.\,cdf\,.\,ace\,.\,dbe=0;$$

where

$$abe=\begin{vmatrix} p_a, & q_a, & r_a \\ p_b, & q_b, & r_b \\ p_e, & q_e, & r_e \end{vmatrix}, \text{ \&c.};$$

and substituting for $p_a, \ldots,$ their values, we have $abe=w^2\,.\,pabe$, whence the foregoing result.

{**Surface** *abcdef.*}

18. *Singularities.* The form of the equation shows at once that [1]

(0)[2] The point a is a 2-conical point; in fact, for this point we have $pabe = 0$, $pacf = 0$, $pabf = 0$, $pace = 0$.

(1) The line ab a simple line; in fact, for any point of this line we have $pabe = 0$, $pabf = 0$.

(2) The line $abe\,.\,cdf$ a simple line; in fact, for any point of this line we have $pabe = 0$, $pcdf = 0$.

(9) To show analytically that the cubic curve $abcdef$ is a line on the surface, observe that the equation of the surface is satisfied if we have simultaneously (λ being arbitrary)

$$pabe\,.\,pacf - \lambda\,.\,pabf\,.\,pace = 0,$$
$$\lambda\,.\,pcde\,.\,pdbf - \quad pcdf\,.\,pdbe = 0.$$

The first of these equations is a cone, vertex a, which passes through the points b, e c, f, and which, if λ is properly determined, will pass through the point d; the second is a cone, vertex d, which passes through the points b, e, c, f, and which, if λ is properly determined, will pass through the point a; the two determinations of λ are

$$dabe\,.\,dacf - \lambda\,.\,dabf\,.\,dace = 0,$$
$$\lambda\,.\,acde\,.\,adbf - \quad acdf\,.\,adbe = 0;$$

giving the same value of λ; and the equations then represent cones, the first having a for its vertex, and passing through d, b, e, c, f; the second having d for its vertex, and passing through a, b, e, c, f; the two intersect in the line ad, and in the cubic curve $abcdef$, which is thus a curve on the surface.

Surface abcdeα.

19. The equation may be written

$$(pabe\,.\,pcde\,.\,p^2\alpha ac\,.\,db - pace\,.\,pdbe\,.\,p^2\alpha ab\,.\,cd)^2$$
$$+ 4pabe\,.\,pcde\,.\,pace\,.\,pdbe\,.\,pabc\,.\,pdbc\,.\,pa\alpha\,.\,pd\alpha = 0,$$

or, what is the same thing,

$$(pabe\,.\,pcde\,.\,p^2\alpha ac\,.\,db + pace\,.\,pdbe\,.\,p^2\alpha ab\,.\,cd)^2$$
$$+ 4pabe\,.\,pcde\,.\,pace\,.\,pdbe\,.\,pbad\,.\,pcad\,.\,pb\alpha\,.\,pc\alpha = 0,$$

(the equivalence of the two depending on the identity

$$- p^2\alpha ab\,.\,cd\,.\,p^2\alpha ac\,.\,db + pabc\,.\,pdbc\,.\,pa\alpha\,.\,pd\alpha - pbad\,.\,pcad\,.\,pb\alpha\,.\,pc\alpha = 0)$$

[1] Or course, as regards the present surface and the other surfaces for which the equation is given in an unsymmetrical form, the conclusion obtained in regard to any point or line of the surface applies to every point or line of the same kind. Thus ab being a simple line, we have also ad a simple line, although the equation, as written down, does not put this in evidence.

[2] The bracketed numbers refer to the lines of the Table.

{Surface *abcdeα*.}

where, as before, $pabe=0$ is the equation of the plane through the points a, b, e; $p^2\alpha acdb=0$ the equation of the quadric surface through the lines α, ac, db; and $pa\alpha=0$ the equation of the plane through the point a and the line α.

The above forms are 2 out of 30 like forms, as appears by the partitionment

$$\left\{\begin{matrix} ab, & cd \\ ac, & db \\ ad, & bc \end{matrix}\right\} e\alpha.$$

20. *Investigation.* In the projection, the equation of the conic through the five points may be written

$$\left|\begin{matrix} (X, & Y, & Z)^2 \\ (p, & q, & r)^2 \end{matrix}\right| = 0,$$

where the symbol denotes a determinant the last five lines of which are obtained by giving to (p, q, r) the suffixes a, b, c, d, e respectively. This is at once transformed into

$$abe.\ cde.ac\Delta.db\Delta - ace.\ dbe.ab\Delta.cd\Delta = 0,$$

or, what is the same thing,

$$pabe.pcde.ac\Delta.db\Delta - pace.pdbe.ab\Delta.cd\Delta = 0,$$

or say,

$$pabe.pcde(A''X+B''Y+C''Z)(A'''X+B'''Y+C'''Z)$$
$$-pace.pdbe(AX+BY+CZ)(A'X+B'Y+C'Z),$$

where $pabe$, &c. signify as before, and

$$AX+BY+CZ = \left|\begin{matrix} X, & Y, & Z \\ p_a, & q_a, & r_a \\ p_b, & q_b, & r_b \end{matrix}\right|$$

and so for $A'X+B'Y+C'Z$, &c., the suffixes for A', B', C' being (c, d), and those for A'', B'', C'' and A''', B''', C''' being (a, c) and (d, b) respectively.

21. Passing to the reciprocal equation, and making the conic touch the line α, we obtain the equation of the surface in the form

$$\left\{pabe.pcde\left|\begin{matrix} P_\alpha, & Q_\alpha, & R_\alpha \\ A'', & B'', & C'' \\ A''', & B''', & C''' \end{matrix}\right| - pace.pdbe\left|\begin{matrix} P_\alpha, & Q_\alpha, & R_\alpha \\ A, & B, & C \\ A', & B', & C' \end{matrix}\right|\right\}^2$$
$$+4pace.pdbe.pabe.pcde\left|\begin{matrix} P_\alpha, & Q_\alpha, & R_\alpha \\ A, & B, & C \\ A'', & B'', & C'' \end{matrix}\right|\left|\begin{matrix} P_\alpha, & Q_\alpha, & R_\alpha \\ A', & B', & C' \\ A''', & B''', & C''' \end{matrix}\right| = 0,$$

{**Surface** *abcdea.*}

(where $P_\alpha = h_\alpha y - g_\alpha z + a_\alpha w = 0$) or in the equivalent form wherein we have in the first term $+$ instead of $-$, and in the second term the determinants

$$\left|\begin{matrix} P_\alpha, & Q_\alpha, & R_\alpha \\ A, & B, & C \\ A''', & B''', & C''' \end{matrix}\right|, \quad \left|\begin{matrix} P_\alpha, & Q_\alpha, & R_\alpha \\ A', & B', & C' \\ A'', & B'', & C'' \end{matrix}\right|.$$

22. {The question, in fact, is to find the reciprocal of the form

$$\lambda (ax + by + cz)(a'x + b'y + c'z) - \mu (a''x + b''y + c''z)(a'''x + b'''y + c'''z) = 0;$$

taking ξ, η, ζ for the reciprocal variables, the coefficient of ξ^2 is

$$\{\lambda (bc' + b'c) - \mu (b''c''' + b'''c'')\}^2 - (2\lambda bb' - 2\mu b''b''')(2\lambda cc' - 2\mu c''c'''),$$

viz. this is

$$\lambda^2 (bc' - b'c)^2 + \mu^2 (b''c''' - b'''c'')^2 + 2\lambda\mu \{2bb'c''c''' + 2b''b'''cc' - (bc' + b'c)(b''c''' + b'''c'')\},$$

or, as it may be written,

$$\{\lambda (bc' - b'c) \pm \mu (b''c''' - b'''c'')\}^2 + 2\lambda\mu \left\{\begin{matrix} 2bb'c''c''' + 2b''b'''cc' \\ \mp (bc - b'c)(b''c''' - b'''c'') \\ -(bc' + b'c)(b''c''' + b'''c'') \end{matrix}\right\}.$$

Taking the upper signs, this is

$$\{\lambda (bc' - b'c) + \mu (b''c''' - b'''c'')\}^2 + 4\lambda\mu \left(\begin{matrix} bb'c''c''' + b''b'''cc' \\ - bc'b''c''' - b'cb'''c'' \end{matrix}\right);$$

viz. the term in $\lambda\mu$ is

$$= + 4\lambda\mu (bc''' - b'''c)(b'c'' - b''c').$$

Taking the lower signs, it is

$$\{\lambda (bc' - b'c) - \mu (b''c''' - b'''c'')\}^2 + 4\lambda\mu \left(\begin{matrix} bb'c''c''' + b''b'''cc' \\ - bc'b'''c'' - b'cb''c''' \end{matrix}\right);$$

viz. the term in $\lambda\mu$ is

$$4\lambda\mu (bc'' - b''c)(b'c''' - b'''c');$$

and it is thence easy to infer the forms of the other coefficients, and to obtain the reciprocal equation in the two equivalent forms

$$\left\{\lambda \left|\begin{matrix} \xi, & \eta, & \zeta \\ a, & b, & c \\ a', & b', & c' \end{matrix}\right| + \mu \left|\begin{matrix} \xi, & \eta, & \zeta \\ a'', & b'', & c'' \\ a''', & b''', & c''' \end{matrix}\right|\right\}^2 + 4\lambda\mu \left|\begin{matrix} \xi, & \eta, & \zeta \\ a, & b, & c \\ a''', & b''', & c''' \end{matrix}\right| \left|\begin{matrix} \xi, & \eta, & \zeta \\ a', & b', & c' \\ a'', & b'', & c'' \end{matrix}\right| = 0,$$

$$\left\{\lambda \left|\begin{matrix} \xi, & \eta, & \zeta \\ a, & b, & c \\ a', & b', & c' \end{matrix}\right| - \mu \left|\begin{matrix} \xi, & \eta, & \zeta \\ a'', & b'', & c'' \\ a''', & b''', & c''' \end{matrix}\right|\right\}^2 + 4\lambda\mu \left|\begin{matrix} \xi, & \eta, & \zeta \\ a, & b, & c \\ a'', & b'', & c'' \end{matrix}\right| \left|\begin{matrix} \xi, & \eta, & \zeta \\ a', & b', & c' \\ a''', & b''', & c''' \end{matrix}\right| = 0,$$

which are the required auxiliary formulæ.}

{**Surface** ***abcdea.***}

23. To reduce the foregoing result, we have

$$A,\ B,\ C = \left\| \begin{matrix} xw_a - wx_a, & yw_a - wy_a, & zw_a - wz_a \\ xw_b - wx_b, & yw_b - wy_b, & zw_b - wz_b \end{matrix} \right\|$$

proportional to the three determinants which contain w, of the set

$$\left\| \begin{matrix} x\ , & y\ , & z\ , & w \\ x_a, & y_a, & z_a, & w_a \\ x_b, & y_b, & z_b, & w_b \end{matrix} \right\|, \quad \text{viz. } A = w \left| \begin{matrix} y\ , & z\ , & w \\ y_a, & z_a, & w_a \\ y_b, & z_b, & w_b \end{matrix} \right|, \ \&\text{c.};$$

and similarly A', B', C' are proportional to the three determinants which contain w, of the set

$$\left\| \begin{matrix} x\ , & y\ , & z\ , & w \\ x_c, & y_c, & z_c, & w_c \\ x_d, & y_d, & z_d, & w_d \end{matrix} \right\|, \quad \text{viz. } A' = w \left| \begin{matrix} y\ , & z\ , & w \\ y_c, & z_c, & w_c \\ y_d, & z_d, & w_d \end{matrix} \right|, \ \&\text{c.}$$

Hence, omitting the factor w, and writing (a, b, c, f, g, h) and (a′, b′, c′, f′, g′, h′) for the coordinates of the lines ab and cd respectively, we have

$$\begin{aligned} A &= \quad \mathrm{h}y - \mathrm{g}z + \mathrm{a}w, & A' &= \quad \mathrm{h}'y - \mathrm{g}'z + \mathrm{a}'w, \\ B &= -\mathrm{h}x \quad + \mathrm{f}z + \mathrm{b}w, & B' &= -\mathrm{h}'x \quad + \mathrm{f}'z + \mathrm{b}'w, \\ C &= \mathrm{g}x - \mathrm{f}y \quad + \mathrm{c}w, & C' &= \mathrm{g}'x - \mathrm{f}'y \quad + \mathrm{c}'w; \end{aligned}$$

and thence

$$\begin{aligned} BC' - B'C &= \Omega x - Lw, \\ CA' - C'A &= \Omega y - Mw, \\ AB' - A'B &= \Omega z - Nw, \end{aligned}$$

where

$$\begin{aligned} L &= (\mathrm{af}' - \mathrm{a}'\mathrm{f})x + (\mathrm{bf}' - \mathrm{b}'\mathrm{f})y + (\mathrm{cf}' - \mathrm{c}'\mathrm{f})z - (\mathrm{bc}' - \mathrm{b}'\mathrm{c})w, \\ M &= (\mathrm{ag}' - \mathrm{a}'\mathrm{g})x + (\mathrm{bg}' - \mathrm{b}'\mathrm{g})y + (\mathrm{cg}' - \mathrm{c}'\mathrm{g})z - (\mathrm{ca}' - \mathrm{c}'\mathrm{a})w, \\ N &= (\mathrm{ah}' - \mathrm{a}'\mathrm{h})x + (\mathrm{bh}' - \mathrm{b}'\mathrm{h})y + (\mathrm{ch}' - \mathrm{c}'\mathrm{h})z - (\mathrm{ab}' - \mathrm{a}'\mathrm{b})w, \\ \Omega &= (\mathrm{gh}' - \mathrm{g}'\mathrm{h})x + (\mathrm{hf}' - \mathrm{h}'\mathrm{f})y + (\mathrm{fg}' - \mathrm{f}'\mathrm{g})z - (\mathrm{af}' - \mathrm{a}'\mathrm{f} + \mathrm{bg}' - \mathrm{b}'\mathrm{g} + \mathrm{ch}' - \mathrm{c}'\mathrm{h})w; \end{aligned}$$

and consequently

$$\left| \begin{matrix} P_a, & Q_a, & R_a \\ A\ , & B\ , & C \\ A', & B', & C' \end{matrix} \right| = \Omega\,(xP_a + yQ_a + zR_a) - w\,(LP_a + MQ_a + NR_a)$$

$$= -\,w\,(LP_a + MQ_a + NR_a + \Omega S_a);$$

or omitting the factor $-w$, say it is $= LP_a + MQ_a + NR_a + \Omega S_a$, viz. this is $= p^2 \alpha\, ab \,.\, cd$.

{**Surface** ***abcdea***.}

We have similarly

$$\begin{vmatrix} P_\alpha, & Q_\alpha, & R_\alpha \\ A'', & B'', & C'' \\ A''', & B''', & C''' \end{vmatrix}$$

taken to be $=p^2\alpha\, ac \,.\, db$.

24. We have in like manner the other two determinants

$$\begin{vmatrix} P_\alpha, & Q_\alpha, & R_\alpha \\ A\,, & B\,, & C \\ A'', & B'', & C'' \end{vmatrix} \text{ and } \begin{vmatrix} P_\alpha, & Q_\alpha, & R_\alpha \\ A', & B', & C' \\ A''', & B''', & C''' \end{vmatrix}$$

taken to be $=p^2\alpha\, ab \,.\, ac$ and $p^2\alpha\, cd \,.\, db$ respectively.

But we have

$$p^2\alpha\, ab \,.\, ac = pa\alpha \,.\, pabc,$$

(viz. geometrically the hyperboloid through the lines α, ab, ac breaks up into the plane $pa\alpha$ through the line α and point a, and the plane $pabc$ through the points a, b, c).

And similarly

$$p^2\alpha\, cd \,.\, db = -p^2\alpha\, dc \,.\, db = +p^2\alpha\, db \,.\, dc = pad \,.\, pdbc\,;$$

whence, substituting for the several determinants, we have the foregoing equation of the surface.

25. *Singularities.* The form of the equation shows that

(0) The point a is a 4-conical point: in fact, for this point we have $pabe=0$, $p^2\alpha\, ac \,.\, db=0$, $pace=0$, $p^2\alpha\, ab \,.\, cd=0$.

(1) The line ab is a double line: in fact, for any point of the line we have $pabe=0$, $p^2\alpha\, ab \,.\, cd=0$, $pabc=0$.

(2) The line α is a double line: in fact, for any point of the line we have $p^2\alpha\, ac \,.\, db=0$, $p^2\alpha\, ab \,.\, cd=0$, $pa\alpha=0$, $pd\alpha=0$.

(7) The line $abe \,.\, cd \,.\, \alpha$ is a simple line: in fact, for any point of the line we have $pabe=0$, $p^2\alpha\, ab \,.\, cd=0$. Observe that, on writing in the equation $pabe=0$ the equation becomes $(p^2\alpha\, ab \,.\, cd)^2=0$; so that the surface along the line in question touches the plane $pabe$.

Surface abcdαβ.

26. The equation of the surface is

$$\text{Norm } \{\sqrt{pa\alpha \,.\, pa\beta} \,.\, pbcd - \sqrt{pb\alpha \,.\, pb\beta} \,.\, pcda + \sqrt{pc\alpha \,.\, pc\beta} \,.\, pdab - \sqrt{pd\alpha \,.\, pd\beta} \,.\, pabc\} = 0,$$

where the norm is the product of 8 factors.

As before, $pa\alpha=0$ is the equation of the plane through the point a and the line α; and $pbcd=0$ the equation of the plane through the points b, c, d. The form is unique.

{Surface $abcd\alpha\beta$.}

27. *Investigation.* In the projection, the equation of the conic touching the projections of the lines α, β is

$$\sqrt{(P_\alpha X + Q_\alpha Y + R_\alpha Z)(P_\beta X + Q_\beta Y + R_\beta Z)} + AX + BY + CZ = 0,$$

where A, B, C are arbitrary coefficients. To make this pass through the projection of the point a, we must write $X : Y : Z = p_a : q_a : r_a$; viz. we thus have

$$\begin{aligned} P_\alpha X + Q_\alpha Y + R_\alpha Z = &\ w_a (x\ P_\alpha + y\ Q_\alpha + z\ R_\alpha) \\ &- w\ (x_a P_\alpha + y_a Q_\alpha + z_a R_\alpha), \\ = &- w\ (x_a P_\alpha + y_a Q_\alpha + z_a R_\alpha + w_a S_\alpha), \\ = &- w\,.\,pa\alpha; \end{aligned}$$

and similarly

$$P_\beta X + Q_\beta Y + R_\beta Z = - w\,.\,pa\beta.$$

We thus have

$$w\sqrt{pa\alpha\,.\,pa\beta} + Ap_a + Bq_a + Cr_a = 0.$$

Or, forming the like equations for the points b, c, d respectively and eliminating, the equation is

$$\begin{vmatrix} \sqrt{pa\alpha\,.\,pa\beta}, & p_a, & q_a, & r_a \\ \sqrt{pb\alpha\,.\,pb\beta}, & p_b, & q_b, & r_b \\ \sqrt{pc\alpha\,.\,pc\beta}, & p_c, & q_c, & r_c \\ \sqrt{pd\alpha\,.\,pd\beta}, & p_d, & q_d, & r_d \end{vmatrix} = 0;$$

which, substituting for (p_a, q_a, r_a), &c., their values, viz. $p_a = xw_a - x_a w$, &c., is readily converted into

$$\begin{vmatrix} & x\ , & y\ , & z\ , & w \\ \sqrt{pa\alpha\,.\,pa\beta}, & x_a, & y_a, & z_a, & w_a \\ \sqrt{pb\alpha\,.\,pb\beta}, & x_b, & y_b, & z_b, & w_b \\ \sqrt{pc\alpha\,.\,pc\beta}, & x_c, & y_c, & z_c, & w_c \\ \sqrt{pd\alpha\,.\,pd\beta}, & x_d, & y_d, & z_d, & w_d \end{vmatrix} = 0.$$

or, what is the same thing,

$$\sqrt{pa\alpha\,.\,pa\beta}\,.\,pbcd - \sqrt{pb\alpha\,.\,pb\beta}\,.\,pcda + \sqrt{pc\alpha\,.\,pc\beta}\,.\,pdab - \sqrt{pd\alpha\,.\,pd\beta}\,.\,pabc = 0;$$

viz. taking the norm, we have the form mentioned above.

28. *Singularities.* The equation shows that

(0) The point a is an 8-conical point; in fact, for the point in question $pa\alpha = 0$, $pa\beta = 0$, $pcda = 0$, $pdab = 0$, $pabc = 0$; each factor is of the form 0^1, and the norm is 0^8.

{Surface $abcd\alpha\beta$.}

(1) The line ab is a 4-tuple line. To show this, observe in the first instance, that we may obtain the 8 factors of the norm by giving to the radical $\sqrt{pa\alpha \,.\, pa\beta}$ the sign $+$, and to the other three radicals the signs $+$, $-$, at pleasure. For a point on the line in question, we have $pdab = 0$, $pabc = 0$; hence the norm is the product of the four equal factors

$$\sqrt{pa\alpha \,.\, pa\beta} \,.\, pbcd - \sqrt{pb\alpha \,.\, pb\beta} \,.\, pcda,$$

and the other four equal factors obtained by writing herein $+$ instead of $-$.

Now for a point on the line ab, we may write for x, y, z, w the values $ux_a + vx_b$, $uy_a + vy_b$, $uz_a + vz_b$, $uw_a + vw_b$, where u, v are arbitrary coefficients. We have

$$\begin{aligned}
pa\alpha &= u \,.\, aa\alpha + v \,.\, ba\alpha = v \,.\, ba\alpha = - v \,.\, ab\alpha,\\
pa\beta &= v \,.\, ba\beta = - v \,.\, ab\beta,\\
pb\alpha &= u \,.\, ab\alpha + v \,.\, bb\alpha = u \,.\, ab\alpha,\\
pb\beta &= u \,.\, ab\beta,\\
pbcd &= u \,.\, abcd + v \,.\, bbcd = u \,.\, abcd,\\
pcda &= u \,.\, acda + v \,.\, bcda = v \,.\, bcda = - v \,.\, abcd,
\end{aligned}$$

where $ab\alpha = 0$ is the condition that the points a, b and the line α may be in the same plane (or, what is the same thing, that the lines ab and α may intersect), viz. $ba\alpha$ is $= P_a x_b + Q_a y_b + R_a z_b + S_a w_b$. And similarly $abcd = 0$ is the condition that the four points a, b, c, d may be in a plane; viz. we have

$$abcd = \begin{vmatrix} x_a, & y_a, & z_a, & w_a \\ x_b, & y_b, & z_b, & w_b \\ x_c, & y_c, & z_c, & w_c \\ x_d, & y_d, & z_d, & w_d \end{vmatrix}.$$

Substituting, we have $\sqrt{pa\alpha \,.\, pa\beta} \,.\, pbcd$ and $\sqrt{pb\alpha \,.\, pb\beta} \,.\, pcda$, each equal (save as to sign) to $uv\sqrt{ab\alpha \,.\, ab\beta} \,.\, abcd$; that is, the four equal factors of one set will vanish. The vanishing factors are of the form 0^1, and the norm is 0^4, that is, the line in question, ab, is a 4-tuple line.

(2) The line α is a 4-tuple line; in fact, for any point of the line we have $pa\alpha = 0$, $pb\alpha = 0$, $pc\alpha = 0$, $pd\alpha = 0$; each factor of the norm is therefore evanescent, of the form $0^{\frac{1}{2}}$, and the norm itself is thus $= 0^4$.

29. (5) The line (ab, cd, α, β) is a double line. To show this, take $z = 0$, $w = 0$ as the equations of the line in question; then we have $h_\alpha = 0$, $h_\beta = 0$, $z_a w_b - z_b w_a = 0$; or say $w_a = \lambda z_a$, $w_b = \lambda z_b$: and $z_c w_d - z_d w_c = 0$; or say $w_c = \mu z_c$, $w_d = \mu z_d$ (λ and μ arbitrary coefficients). Putting for shortness

$$I = (g - \lambda a)\, x - (f + \lambda b)\, y, \quad J = (g - \mu a)\, x - (f + \mu b)\, y;$$

{Surface $abcda\beta$.}

viz. $I_\alpha = (g_\alpha - \lambda a_\alpha)x - (f_\alpha + \lambda b_\alpha)y$, &c., and writing $z=0$, $w=0$, we have $pa\alpha \,.\, pa\beta = z_a^2 I_\alpha I_\beta$, $pb\alpha \,.\, pb\beta = z_b^2 I_\alpha I_\beta$, $pc\alpha \,.\, pc\beta = z_c^2 J_\alpha J_\beta$, $pd\alpha \,.\, pd\beta = z_d^2 J_\alpha J_\beta$; and the factor of the norm (reverting to the expression thereof as a determinant) is

$$\begin{vmatrix} & x, & y & & \\ z_a\sqrt{I_\alpha I_\beta}, & x_a, & y_a, & z_a, & \lambda z_a \\ z_b\sqrt{I_\alpha I_\beta}, & x_b, & y_b, & z_b, & \lambda z_b \\ z_c\sqrt{J_\alpha J_\beta}, & x_c, & y_c, & z_c, & \mu z_c \\ z_d\sqrt{J_\alpha J_\beta}, & x_d, & y_d, & z_d, & \mu z_d \end{vmatrix}$$

which vanishes. In fact, resolving the determinant into a set of products of the form $\pm 2 \,.\, 13 \,.\, 45$, where the single symbol denotes a term of the top line, and the binary symbols refer to the second and third lines, and the fourth and fifth lines respectively (denoting minors composed with the terms in these pairs of lines respectively); then each product will contain a term 14, 15, or 45, and the minor so designated (to whichever of the two pairs of lines it belongs) is $=0$. The factor is thus evanescent, being, as it is easy to see, $=0^1$. There are two factors which vanish; viz. taking the first radical to be $+$, the second radical must be also $+$, but the third and fourth radicals may be either both $+$ or both $-$; the norm is thus $=0^2$, viz. the line (ab, cd, α, β) is a double line.

30. (8) The line abc, α, β is a double line. To prove this, take $w=0$ for the equation of the plane abc, and $(z=0, w=0)$ for those of the line in question; we have $h_\alpha=0$, $h_\beta=0$, $w_a=0$, $w_b=0$, $w_c=0$; and writing $I_\alpha = -g_\alpha x + f_\alpha y$, $I_\beta = -g_\beta x + f_\alpha y$, then for $z=0$, $w=0$, the factor expressed as a determinant is

$$\begin{vmatrix} & x, & y & . & . \\ z_a\sqrt{I_\alpha I_\beta}, & x_a, & y_a, & z_a, & . \\ z_b\sqrt{I_\alpha I_\beta}, & x_b, & y_b, & z_b, & . \\ z_c\sqrt{I_\alpha I_\beta}, & x_c, & y_c, & z_c, & . \\ \sqrt{pd\alpha \,.\, pd\beta}, & x_d, & y_d, & z_d, & w_d \end{vmatrix}$$

which is

$$= w_d\sqrt{I_\alpha I_\beta}\begin{vmatrix} . & x, & y & . \\ z_a, & x_a, & y_a, & z_a \\ z_b, & x_b, & y_b, & z_b \\ z_c, & x_c, & y_c, & z_c \end{vmatrix}$$

and consequently vanishes, the form being 0^1. There are two such factors, viz. the radical $\sqrt{pd\alpha \,.\, pd\beta}$ may be either $+$ or $-$, hence the norm is $=0^2$.

31. But it is to be further shown that the line is tacnodal, each sheet of the surface being touched along the line by the plane $w=0$: we have to show that the

factor operated upon by $\Delta = X\delta_x + Y\delta_y + Z\delta_z + W\delta_w$, reduces itself for $z=0$, $w=0$ to a multiple of W. Considering the factor in the form of a determinant, the result of the operation is

$$\begin{vmatrix} & X, & Y, & Z, & W \\ \sqrt{pa\alpha \,.\, pa\beta}, & x_a, & y_a, & z_a, & . \\ \sqrt{pb\alpha \,.\, pb\beta}, & x_b, & y_b, & z_b, & . \\ \sqrt{pc\alpha \,.\, pc\beta}, & x_c, & y_c, & z_c, & . \\ \sqrt{pd\alpha \,.\, pd\beta}, & x_d, & y_d, & z_d, & w_d \end{vmatrix} + \begin{vmatrix} & x, & y, & . & . \\ \Delta\sqrt{pa\alpha \,.\, pa\beta}, & x_a, & y_a, & z_a, & . \\ \Delta\sqrt{pb\alpha \,.\, pb\beta}, & x_b, & y_b, & z_b, & . \\ \Delta\sqrt{pc\alpha \,.\, pc\beta}, & x_c, & y_c, & z_c, & . \\ \Delta\sqrt{pd\alpha \,.\, pd\beta}, & x_d, & y_d, & z_d, & w_d \end{vmatrix};$$

the first term is

$$\begin{vmatrix} & X, & Y, & Z, & W \\ z_a\sqrt{I_\alpha I_\beta}, & x_a, & y_a, & z_a, & . \\ z_b\sqrt{I_\alpha I_\beta}, & x_b, & y_b, & z_b, & . \\ z_c\sqrt{I_\alpha I_\beta}, & x_c, & y_c, & z_c, & . \\ \sqrt{pd\alpha \,.\, pd\beta}, & x_d, & y_d, & z_d, & w_d \end{vmatrix},$$

where the first column may be replaced by

$$\begin{matrix} -Z\sqrt{I_\alpha I_\beta} \\ : \\ \sqrt{pd\alpha \,.\, pd\beta} - z_d\sqrt{I_\alpha I_\beta}, \end{matrix}$$

and the term in question thus becomes

$$\{w_d Z\sqrt{I_\alpha I_\beta} + W(-z_d\sqrt{I_\alpha I_\beta} + \sqrt{pd\alpha \,.\, pd\beta})\} \,.\, abc,$$

if for shortness

$$\begin{vmatrix} x_a, & y_a, & z_a \\ x_b, & y_b, & z_b \\ x_c, & y_c, & z_c \end{vmatrix} = abc.$$

As regards the second term, we have

$$\Delta\sqrt{pa\alpha \,.\, pa\beta} = \frac{pa\alpha \,.\, \Delta pa\beta + pa\beta \,.\, \Delta pa\alpha}{2\sqrt{pa\alpha \,.\, pa\beta}},$$

which is

$$= \frac{I_\alpha \Delta pa\beta + I_\beta \Delta pa\alpha}{2\sqrt{I_\alpha I_\beta}}.$$

But

$$pa\alpha = x\;(-g_\alpha z_a) + y\,(f_\alpha z_a) + z\,(g_\alpha x_a - f_\alpha y_a) + w\,(a_\alpha x_a - b_\alpha y_a - c_\alpha z_a),$$

$$= x_a\;(g_\alpha z - a_\alpha w) + y_a\,(-f_\alpha z - b_\alpha w) + z_a\,(-g_\alpha x + f_\alpha y - c_\alpha w);$$

and thence

$$\Delta pa\alpha = x_a\,(g_\alpha Z - a_\alpha W) + y_a\,(-f_\alpha Z - b_\alpha W) + z_a\,(-g_\alpha X + f_\alpha Y - c_\alpha W),$$

with the like formula for $\Delta p a \beta$; hence

$$\frac{I_\alpha \Delta p a \beta + I_\beta \Delta p a \alpha}{2\sqrt{I_\alpha I_\beta}} = Ax_a + By_a + Cz_a,$$

where

$$A = \frac{1}{2\sqrt{I_\alpha I_\beta}}\{I_\beta(\ g_\alpha Z - a_\alpha W) + I_\alpha(\ g_\beta Z - a_\beta W)\},$$

$$B = \frac{1}{2\sqrt{I_\alpha I_\beta}}\{I_\beta(-f_\alpha Z - b_\alpha W) + I_\alpha(-f_\beta Z - b_\beta W)\},$$

$$C = \frac{1}{2\sqrt{I_\alpha I_\beta}}\{I_\beta(-g_\alpha X + f_\alpha Y - c_\alpha W) + I_\alpha(-g_\beta X + f_\beta Y - c_\beta W)\}.$$

The term in question is thus

$$\left|\begin{array}{ccccc} & x, & y, & . & . \\ Ax_a + By_a + Cz_a, & x_a, & y_a, & z_a & . \\ Ax_b + By_b + Cz_b, & x_b, & y_b, & z_b, & . \\ Ax_c + By_c + Cz_c, & x_c, & y_c, & z_c, & . \\ \Delta\sqrt{pd\alpha \,.\, pd\beta}\ , & x_d, & y_d, & z_d, & w_d \end{array}\right|$$

viz. replacing the first column by

$$\begin{array}{l} -\ Ax - By \\ \quad\vdots \\ \Delta\sqrt{pd\alpha \,.\, pd\beta} - Ax_d - By_d - Cz_d\,; \end{array}$$

this is

$$= (Ax + By)\, w_d \,.\, abc\,;$$

and we have

$$Ax + By = \frac{1}{2\sqrt{I_\alpha I_\beta}}\begin{array}{l}[\ \ I_\beta(\ \ g_\alpha x - f_\alpha y) + I_\alpha(\ \ g_\beta x - f_\beta y)]\, Z \\ [+\, I_\beta(-a_\alpha x - b_\alpha y) + I_\alpha(-a_\beta x - b_\beta y)]\, W, \end{array}$$

$$= \frac{1}{2\sqrt{I_\alpha I_\beta}}(-2I_\alpha I_\beta Z - MW),$$

if for shortness

$$M = (-g_\beta x + f_\beta y)(a_\alpha x + b_\alpha y) + (-g_\alpha x + f_\alpha y)(a_\beta x + b_\beta y)\,;$$

viz. the whole term is

$$w_d\left\{-\sqrt{I_\alpha I_\beta}\, Z - \frac{\tfrac{1}{2}M}{\sqrt{I_\alpha I_\beta}}\, W\right\} abc.$$

Hence the first and second terms together are

$$= W\left\{-z_d\sqrt{I_\alpha I_\beta} + \sqrt{pd\alpha \,.\, pd\beta} - \frac{\tfrac{1}{2}M}{\sqrt{I_\alpha I_\beta}}\, w_d\right\} abc\,;$$

viz. this is a multiple of W, which was the theorem to be proved.

{**Surface** $abcd\alpha\beta$.}

Surface abcαβγ.

32. The equation is

$$\text{Norm}\begin{vmatrix} \sqrt{pa\alpha}, & \sqrt{pb\alpha}, & \sqrt{pc\alpha} \\ \sqrt{pa\beta}, & \sqrt{pb\beta}, & \sqrt{pc\beta} \\ \sqrt{pa\gamma}, & \sqrt{pb\gamma}, & \sqrt{pc\gamma} \end{vmatrix} = 0,$$

where the norm is a product of 16 factors, each of the order $\frac{3}{2}$. As before, $pa\alpha = 0$ is the equation of the plane through the point a and the line α; viz. $pa\alpha$ has the value already mentioned.

33. *Investigation.* In the projection, the equation of the conic touching the projections of the lines α, β, γ is

$$A\sqrt{P_\alpha X + Q_\alpha Y + R_\alpha Z} + B\sqrt{P_\beta X + Q_\beta Y + R_\beta Z} + C\sqrt{P_\gamma X + Q_\gamma Y + R_\gamma Z} = 0;$$

and to make this pass through the projection of the point a, we must write herein $X : Y : Z = p_a : q_a : r_a$. As before, we have

$$\begin{aligned} P_\alpha X + Q_\alpha Y + R_\alpha Z = & \; w_a(x\, P_\alpha + y\, Q_\alpha + z\, R_\alpha) \\ & - w\,(x_a P_\alpha + y_a Q_\alpha + z_a R_\alpha), \\ = & - w\,(x_a P_\alpha + y_a Q_\alpha + z_a R_\alpha + w_a S_\alpha), \\ = & - w\,.\,pa\alpha; \end{aligned}$$

and so for the other terms; the equation thus is

$$A\sqrt{pa\alpha} + B\sqrt{pa\beta} + C\sqrt{pa\gamma} = 0;$$

or forming the like equations in regard to the points b, c respectively, and eliminating we have a determinant $= 0$, and then, taking the norm, we obtain the above-written equation of the surface.

34. *Singularities.* The equation of the surface shows that

(0) The point a is 8-conical: in fact, for the point in question we have $pa\alpha = 0$, $pa\beta = 0$, $pa\gamma = 0$; each factor is $0^{\frac{1}{2}}$, and the norm is 0^8.

(1) The line ab is 4-tuple. To prove this, observe that the sixteen factors are obtained by attributing at pleasure the signs $+$, $-$ to the radicals $\sqrt{pb\beta}$, $\sqrt{pc\beta}$, $\sqrt{pb\gamma}$, $\sqrt{pc\gamma}$; hence there are four factors in which $\sqrt{pb\beta}$, $\sqrt{pb\gamma}$ have determinate signs, but in which we attribute to the radicals $\sqrt{pc\beta}$, $\sqrt{pc\gamma}$ the signs $+$ or $-$ at pleasure. It is to be shown that the four factors each vanish for a point on the line ab; that is, on writing therein for x, y, z, w the values $ux_a + vx_b$, $uy_a + vy_b$, &c. But we thus

{Surface $abc\alpha\beta\gamma$.}

have, as before, $pa\alpha = -v . ab\alpha$ and $pb\alpha = u . ab\alpha$, with the like formulæ with β and γ in place of α. The factor thus becomes

$$\sqrt{-uv}\begin{vmatrix} \sqrt{ab\alpha}, & \sqrt{ab\alpha}, & \sqrt{pc\alpha} \\ \sqrt{ab\beta}, & \sqrt{ab\beta}, & \sqrt{pc\beta} \\ \sqrt{ab\gamma}, & \sqrt{ab\gamma}, & \sqrt{pc\gamma} \end{vmatrix},$$

which vanishes, being $=0^1$; and the norm is thus $=0^4$, viz. the line is 4-tuple.

(2) The line α is 8-tuple: in fact, for a point on the line we have $pa\alpha = 0$, $pb\alpha = 0$, $pc\alpha = 0$, whence each factor vanishes, being $=0^{\frac{3}{2}}$, and the norm is therefore 0^8.

(3) The line $(ab, \alpha, \beta, \gamma)$ is 4-tuple: in fact, writing $z = 0$, $w = 0$ for the equations of the line, we have $h_\alpha = 0$, $h_\beta = 0$, $h_\gamma = 0$, and $z_a w_b - z_b w_a = 0$, or say $w_a = \lambda z_a$, $w_b = \lambda z_b$. Hence, writing

$$I = (g - \lambda a)\, x - (f + \lambda b)\, y,$$

viz. $I_\alpha = (g_\alpha - \lambda a_\alpha)\, x - (f_\alpha + \lambda b_\alpha)\, y$, &c., for $z = 0$, $w = 0$, we have $pa\alpha = z_a I_\alpha$, $pb\alpha = z_b I_\alpha$; and similarly $pa\beta = z_a I_\beta$, $pb\beta = z_b I_\beta$, and $pa\gamma = z_a I_\gamma$, $pb\gamma = z_b I_\gamma$. The factor thus is

$$\sqrt{z_a z_b}\begin{vmatrix} \sqrt{I_\alpha}, & \sqrt{I_\alpha}, & \sqrt{pc\alpha} \\ \sqrt{I_\beta}, & \sqrt{I_\beta}, & \sqrt{pc\beta} \\ \sqrt{I_\gamma}, & \sqrt{I_\gamma}, & \sqrt{pc\gamma} \end{vmatrix},$$

which vanishes, being $=0^1$; there are four such factors, or the norm is 0^4; whence the line is 4-tuple.

(8) The line $abc . \alpha . \beta$ is a 4-tuple line. To prove it, take as before $w = 0$ for the equation of the plane abc, and $(z = 0, w = 0)$ for the equations of the line in question. We have $h_\alpha = 0$, $h_\beta = 0$, $w_a = 0$, $w_b = 0$, $w_c = 0$; whence (if $z = 0$, $w = 0$), writing for shortness $I = gx - fy$ (viz. $I_\alpha = g_\alpha x - f_\alpha y$, $I_\beta = g_\beta x - f_\beta y$), we have $pa\alpha$, $pb\alpha$, $pc\alpha = I_\alpha z_a$, $I_\alpha z_b$, $I_\alpha z_c$, and similarly $pa\beta$, $pb\beta$, $pc\beta = I_\beta z_a$, $I_\beta z_b$, $I_\beta z_c$: the factor thus is

$$\begin{vmatrix} \sqrt{I_\alpha z_a}, & \sqrt{I_\alpha z_b}, & \sqrt{I_\alpha z_c} \\ \sqrt{I_\beta z_a}, & \sqrt{I_\beta z_b}, & \sqrt{I_\beta z_c} \\ \sqrt{pa\gamma}, & \sqrt{pb\gamma}, & \sqrt{pc\gamma} \end{vmatrix}$$

which vanishes, being $=0^1$: and there are four such factors, obtained by giving to the radicals the signs $+$, $-$ at pleasure: hence the norm is $=0^4$.

{**Surface** $abc\alpha\beta\gamma$.}

Surface abαβγδ.

35. The equation is

$$\text{Norm}\ \{\sqrt{pa\alpha\,.\,pb\alpha}\,.\,p^2\beta\gamma\delta - \sqrt{pa\beta\,.\,pb\beta}\,.\,p^2\gamma\delta\alpha + \sqrt{pa\gamma\,.\,pb\gamma}\,.\,p^2\delta\alpha\beta - \sqrt{pa\delta\,.\,pb\delta}\,.\,p^2\alpha\beta\gamma\} = 0,$$

where the norm is the product of 8 factors each of the order 3. As before, $pa\alpha = 0$ is the equation of the plane through the point a and the line α; viz. $pa\alpha$ has the value previously mentioned: and $p^2\beta\gamma\delta = 0$ is the equation of the quadric surface through the lines β, γ, δ.

36. *Investigation.* In the projection, taking ξ, η, ζ as current line-coordinates, the equation of the conic passing through the projections of the points a, b is

$$\sqrt{(p_a\xi + q_a\eta + r_a\zeta)(p_b\xi + q_b\eta + r_b\zeta)} + A\xi + B\eta + C\zeta = 0,$$

where A, B, C are arbitrary coefficients. To make this touch the projection of the line α, we must write $\xi : \eta : \zeta = P_\alpha : Q_\alpha : R_\alpha$; and then

$$\begin{aligned} p_a\xi + q_a\eta + r_a\zeta = {} & p_aP_\alpha + q_aQ_\alpha + r_aR_\alpha, \\ = {} & w_a(x\ P_\alpha + y\ Q_\alpha + z\ R_\alpha) \\ & - w\ (x_aP_\alpha + y_aQ_\alpha + z_aR_\alpha), \\ = {} & -w\ (x_aP_\alpha + y_aQ_\alpha + z_aR_\alpha + w_aS_\alpha), \\ = {} & -w\,.\,pa\alpha, \end{aligned}$$

and similarly

$$p_b\xi + q_b\eta + r_b\zeta = -w\,.\,pb\alpha.$$

Hence the equation is

$$w\sqrt{pa\alpha\,.\,pb\alpha} + AP_\alpha + BQ_\alpha + CR_\alpha = 0;$$

and forming the like equations for the lines β, γ, δ respectively, and eliminating, we have

$$\begin{vmatrix} \sqrt{pa\alpha\,.\,pb\alpha}, & P_\alpha, & Q_\alpha, & R_\alpha \\ \sqrt{pa\beta\,.\,pb\beta}, & P_\beta, & Q_\beta, & R_\beta \\ \sqrt{pa\gamma\,.\,pb\gamma}, & P_\gamma, & Q_\gamma, & R_\gamma \\ \sqrt{pa\delta\,.\,pb\delta}, & P_\delta, & Q_\delta, & R_\delta \end{vmatrix} = 0;$$

which, throwing out a factor w, becomes

$$\sqrt{pa\alpha\,.\,pb\alpha}\,.\,p^2\beta\gamma\delta - \sqrt{pa\beta\,.\,pb\beta}\,.\,p^2\gamma\delta\alpha + \sqrt{pa\gamma\,.\,pb\gamma}\,.\,p^2\delta\alpha\beta - \sqrt{pa\delta\,.\,pb\delta}\,.\,p^2\alpha\beta\gamma = 0;$$

or, taking the norm, we have the above written equation.

37. *Singularities.* The equation shows that

(0) The point a is a 4-conical point; in fact, for the point in question we have $pa\alpha = 0$, $pa\beta = 0$, $pa\gamma = 0$, $pa\delta = 0$; each factor is $= 0^{\frac{1}{2}}$, and the norm is $= 0^4$.

{Surface $ab\alpha\beta\gamma\delta$.}

(1) The line ab is a 2-tuple line. To prove this, we have for the coordinates of a point on the line in question $ux_a + vx_b$, $uy_a + vy_b$, &c.; the values of $pa\alpha$, $pb\alpha$ become as before $-v \,.\, ab\alpha$, $+u \,.\, ab\alpha$, and similarly for $pa\beta$, $pb\beta$, &c.; so that, omitting the constant factor $\sqrt{-uv}$, the value of the factor is

$$ab\alpha \,.\, p^2\beta\gamma\delta - ab\beta \,.\, p^2\gamma\delta\alpha + ab\gamma \,.\, p^2\delta\alpha\beta - ab\delta \,.\, p^2\alpha\beta\gamma.$$

Taking (a, b, c, f, g, h) for the coordinates of the line ab, we have

$$ab\alpha = \mathrm{a}f_\alpha + \mathrm{b}g_\alpha + \mathrm{c}h_\alpha + \mathrm{f}a_\alpha + \mathrm{g}b_\alpha + \mathrm{h}c_\alpha,$$

with the like expressions for $ab\beta$, &c.; and substituting for $p^2\beta\gamma\delta$, &c., their values, the factor is

	x^2	y^2	z^2	w^2	xw	yw	zw	yz	zx	xy
a	*fagh*			*fabc*	*fabg* − *fach*	*fbch*	− *fbcg*		*fcgh*	*fbgh*
b		*gbhf*		*gabc*	− *gach*	*gbch* − *gabf*	*gcaf*	*gchf*	*hafg*	*gahf*
c			*hcfg*	*habc*	*habg*	− *habf*	*hcaf* − *hbcg*	*hbfg*		
f		*abhf*	*acfg*			*abch*	− *abcg*	*abfg* + *achf*	*acgh*	*abgh*
g	*bagh*		*bcfg*		− *bach*		*bcaf*	*bchf*	*bcgh* + *bafg*	*bahf*
h	*cagh*	*cbhf*			*cabg*	− *cabf*		*cbfg*	*cafg*	*cahf* + *cbgh*

viz. the value of the factor is $\{\mathrm{a}(fagh) + \mathrm{g}(bagh) + \mathrm{h}(cagh)\}\, x^2 + \&c.$, where $fagh = f_\alpha a_\beta g_\gamma h_\delta$ is the determinant

$$\begin{vmatrix} f, & a, & g, & h \\ \vdots & & & \end{vmatrix},$$

the suffixes in the four lines being α, β, γ, δ respectively.

Collecting, this is

$$\begin{aligned}
&(\quad . \quad cbhfy - bcfgz + fabcw)(\quad . \quad \mathrm{h}y - \mathrm{g}z + \mathrm{a}w)\\
&(-caghx \quad . \quad + acfgz + gabcw)(-\mathrm{h}x \quad . \quad + \mathrm{f}z + \mathrm{b}w)\\
&(+baghx - abhfy \quad . \quad + habcw)(\ \mathrm{g}x - \mathrm{f}y \quad . \quad + \mathrm{c}w)\\
&(-afghx - bfghy - cfghz \quad . \quad)(\ \mathrm{a}x + \mathrm{b}y + \mathrm{c}z \quad . \quad)\\
&\quad + bcgh\,[w(\mathrm{a}x + \mathrm{b}y + \mathrm{c}z) - x(\quad . \quad \mathrm{h}y - \mathrm{g}z + \mathrm{a}w)]\\
&\quad + cahf\,[w(\mathrm{a}x + \mathrm{b}y + \mathrm{c}z) - y(-\mathrm{h}x \quad . \quad + \mathrm{f}z + \mathrm{b}w)]\\
&\quad + abfg\,[w(\mathrm{a}x + \mathrm{b}y + \mathrm{c}z) - z(\ \mathrm{g}x - \mathrm{f}y \quad . \quad + \mathrm{c}w)] = 0;
\end{aligned}$$

or, what is the same thing,

$$AP + BQ + CR + DS = 0,$$

{**Surface** $abcda\beta$.}

where

$$\begin{aligned}
A &= \{-\,bcghx + cbhfy - bcfgz + fabcw\},\\
B &= \{-\,caghx + cahfy + acfgz + gabcw\},\\
C &= \{+\,baghx - abhfy + abfgz + habcw\},\\
D &= \{-\,afghx - bfghy - cfghz + (bcgh + cahf + abfg)\,w\},\\
P &= (\quad . \quad hy - gz + aw),\\
Q &= (-hx \quad . \quad + fz + bw),\\
R &= (\;gx - fy \quad . \quad + cw),\\
S &= (\;ax + by + cz \quad . \quad) = 0,
\end{aligned}$$

the right-hand factors vanishing for the values $ux_a + vx_b$ of the coordinates.

38. It thus appears so far that the factor is $=0^1$; it is, in fact, $=0^2$, viz. we can show that, operating upon it with

$$\Delta = Xd_x + Yd_y + Zd_z + Wd_w,$$

the value (for any point of the line ab) is $=0$. We have

$$\Delta\sqrt{pa\alpha\,.\,pb\alpha}\,.\,p^2\beta\gamma\delta = \frac{pa\alpha\,.\,lb\alpha + pb\alpha\,.\,la\alpha}{2\sqrt{pa\alpha\,.\,pb\alpha}}\,p^2\beta\gamma\delta + \sqrt{pa\alpha\,.\,pb\alpha}\,.\,\Delta\,.\,p^2\beta\gamma\delta,$$

where $lb\alpha\,(=\Delta pb\alpha)$ is what $pb\alpha$ becomes on writing therein (X, Y, Z, W) in place of (x, y, z, w). Writing, as before, for x, y, z, w the values $ux_a + vx_b$, &c., we have $pa\alpha = -v\,.\,ab\alpha$, $pb\alpha = u\,.\,ab\alpha$; and putting for shortness

$$-v\,.\,lb\alpha + u\,.\,la\alpha = lk\alpha, \text{ \&c.},$$

the expression in question, divided by $\sqrt{-uv}$, is

$$\begin{aligned}
&= -2vu\,\{ab\alpha\,.\,\Delta p^2\beta\gamma\delta - \text{\&c.}\}\\
&\quad + \{lk\alpha\,.\;\; p^2\beta\gamma\delta - \text{\&c.}\},
\end{aligned}$$

where, denoting the determinants

$$\begin{vmatrix} X & Y & Z & W \\ ux_a - vx_b, & uy_a - vy_b, & uz_a - vz_b, & uw_a - vw_b \end{vmatrix}$$

by (a′, b′, c′, f′, g′, h′), we have

$$lk\alpha = \text{a}'\text{f}_\alpha + \text{b}'\text{f}_\alpha + \text{c}'\text{g}_\alpha + \text{f}'a_\alpha + \text{g}'b_\alpha + \text{h}'c_\alpha.$$

But $ab\alpha\,.\,\Delta p^2\beta\gamma\delta = \Delta ab\alpha\,.\,p^2\beta\gamma\delta$, since $ab\alpha$ is independent of (x, y, z, w); and the expression is

$$\begin{aligned}
&= -2vu\Delta\,(AP + BQ + CR + DS)\\
&\quad + AP' + BQ' + CR' + DS',
\end{aligned}$$

{Surface $abcda\beta$.}

where P', Q', R', S' denote $\mathrm{h}'y - \mathrm{g}'z + \mathrm{a}'w$, &c., and where, finally, x, y, z, w are to be replaced by $ux_a + vx_b$, &c. Since for these values P, Q, R, S vanish, the expression becomes

$$\begin{aligned} = &- 2vu\,(A\Delta P + B\Delta Q + C\Delta R + D\Delta S) \\ &+ AP' + BQ' + CR' + DS'; \end{aligned}$$

that is

$$= A\,(P' - 2uv\Delta P) + B\,(Q' - 2uv\Delta Q) + C\,(R' - 2uv\Delta R) + D\,(S' - 2uv\Delta S)$$

and we have, in fact, $P' - 2uv\Delta P = 0$, &c. For, writing for a moment

$$\begin{aligned} x,\ y,\ z,\ w &= ux_a + vx_b,\ uy_a + vy_b,\ uz_a + vz_b,\ uw_a + vw_b, \\ x',\ y',\ z',\ w' &= ux_a - vx_b,\ uy_a - vy_b,\ uz_a - vz_b,\ uw_a - vw_b; \end{aligned}$$

then, for instance,

$$S' = \mathrm{a}'x + \mathrm{b}'y + \mathrm{c}'z,$$

where

$$\mathrm{a}',\ \mathrm{b}',\ \mathrm{c}' = Yz' - Zy',\ Zx' - Xz',\ Xy' - Yx';$$

and thence

$$\begin{aligned} S' = &- \begin{vmatrix} X, & Y, & Z \\ x, & y, & z \\ x', & y', & z' \end{vmatrix} \\ = &\ \ 2uv\,(\mathrm{a}X + \mathrm{b}Y + \mathrm{c}Z) \\ = &\ \ 2uv\Delta S; \end{aligned}$$

and similarly for the other equations. The factor is thus $= 0^2$; there is only one such factor, and the line ab is double.

(2) The line α is an 8-tuple line: in fact, for a point on the line we have $pa\alpha = 0$, $pb\alpha = 0$, $p^2\gamma\delta\alpha = 0$, $p^2\delta\alpha\beta = 0$, $p^2\alpha\beta\gamma = 0$; and the factor vanishes, being $= 0^1$. Each of the factors is 0^1, and the norm is $= 0^8$.

39. (3) The line $[ab, \alpha, \beta, \gamma]$ is a double line. To prove this, observe first that for a point on this line we have $p^2\alpha\beta\gamma = 0$.

Taking as before $z = 0$, $w = 0$ for the equation of the line ab, α, β, γ, we have $h_\alpha = 0$, $h_\beta = 0$, $h_\gamma = 0$, and $z_a w_b - z_b w_a = 0$; or say $w_a = \lambda z_a$, $w_b = \lambda z_b$; whence, writing for shortness $I = -(g - \lambda a)\,x + (f + \lambda b)\,y$, viz. $I_\alpha = -(g_\alpha - \lambda a_\alpha)\,x + (f_\alpha + \lambda b_\alpha)\,y$, we have (when $z = 0$, $w = 0$) $pa\alpha = z_a I_\alpha$, $pb\alpha = z_b I_\alpha$, or omitting the factor $\sqrt{z_a z_b}$, $\sqrt{pa\alpha \,.\, pb\alpha} = I_\alpha$; and so for $\sqrt{pa\beta \,.\, pb\beta}$ and $\sqrt{pa\gamma \,.\, pb\gamma}$. The factor thus is

$$I_\alpha \,.\, p^2\beta\gamma\delta - I_\beta \,.\, p^2\gamma\delta\alpha + I_\gamma \,.\, p^2\delta\alpha\beta;$$

viz. writing $z = 0$, $w = 0$ in the expressions of $p^2\beta\gamma\delta$, &c., this may be written

$$\Sigma\,[(g - \lambda a)\,x - (f + \lambda b)\,y]\,\{(agh)\,x^2 + [(ahf) + (bgh)]\,xy + (bhf)\,y^2\},$$

where observe that Σ denotes a sum of three terms of the form

$$\alpha \,.\, \beta\gamma\delta - \beta \,.\, \gamma\delta\alpha + \gamma \,.\, \delta\alpha\beta.$$

{Surface $abcda\beta$.}

Adding thereto a fourth term $-\delta.\alpha\beta\gamma$, the value of the sum would be $=\alpha\beta\gamma\delta$, or the sum of the three terms is $=\alpha\beta\gamma\delta+\delta.\alpha\beta\gamma$, where the symbols represent determinants. But in each case the determinant $\alpha\beta\gamma$ is $=0$, as containing the column h_α, h_β, h_γ, the terms of which are each $=0$: thus $\Sigma g.agh$ is $=gagh-g_\delta.agh$, where in $gagh$ the suffixes are α, β, γ, δ, and in agh they are α, β, γ: that is, we have $\Sigma g.agh=gagh$. And the whole expression thus is

$$\begin{aligned}= \;& x^3\,(gagh-\lambda aagh)\\ &+x^2y\,(gahf-\lambda aahf+gbgh-\lambda abgh-fagh-\lambda bagh)\\ &+xy^2\,(gbhf-\lambda abhf \qquad\qquad -fahf-\lambda bahf-fbgh-\lambda bbgh)\\ &+y^3\,(\qquad\qquad\qquad\qquad\qquad\qquad -fhbf-\lambda bbhf),\end{aligned}$$

where $gahf$ denotes the determinant $\left|\begin{matrix} g, & a, & h, & f \\ \vdots \end{matrix}\right|$, with the suffixes α, β, γ, δ, in the four lines respectively, and so in other cases: the terms, such as $gagh$, which contain a twice-repeated letter, vanish of themselves; and in the coefficients of x^2y and xy^2, the terms which do not separately vanish destroy each other in pairs, $gahf-fagh=0$, &c.; whence the factor vanishes, being $=0^1$; there are two such factors (viz. the zero term $\sqrt{pa\delta.pb\delta}.p^2\alpha\beta\gamma$ may be taken with the sign + or − at pleasure), and the norm is thus $=0^2$.

40. But the line is tacnodal, each sheet of the surface touching along the line in question the hyperboloid $p^2\alpha\beta\gamma$. To prove this, write

$$\Delta=X\delta_x+Y\delta_y+Z\delta_z+W\delta_w;$$

we have for the hyperboloid, writing $z=0$, $w=0$,

$$\Delta p^2\alpha\beta\gamma=(afg.x+bfg.y)Z+(abg.x-abf.y)W;$$

and it is to be shown that

$$\Delta(\sqrt{pa\alpha.pb\alpha}.p^2\beta\gamma\delta-\sqrt{pa\beta.pb\beta}.p^2\gamma\delta\alpha+\sqrt{pa\gamma.pb\gamma}.p^2\delta\alpha\beta\mp\sqrt{pa\delta.pb\delta}.p^2\alpha\beta\gamma)$$

each contain the factor $\Delta p^2\alpha\beta\gamma$; or, what is the same thing, that

$$\Delta\Sigma\sqrt{pa\alpha.pb\alpha}.p^2\beta\gamma\delta$$

contains the factor in question, Σ denoting the sum of the first three terms of the original expression. The value is

$$=\Sigma\left(\frac{pa\alpha.Pb\alpha+pb\alpha.Pa\alpha}{2\sqrt{pa\alpha.pb\alpha}}p^2\beta\gamma\delta+\sqrt{pa\alpha.pb\alpha}.\Delta p^2\beta\gamma\delta\right);$$

where $Pa\alpha$, $=\Delta pa\alpha$, denotes what $pa\alpha$ becomes on writing therein X, Y, Z, W for x, y, z, w; and the like as to $Pb\alpha$. Substituting for $pa\alpha$ and $pb\alpha$ their values z_aI_α and z_bI_α, and multiplying by $\sqrt{z_az_b}$, the expression is

$$=\Sigma\{(z_aPb\alpha+z_bPa\alpha)p^2\beta\gamma\delta+2z_az_bI_\alpha\Delta p^2\beta\gamma\delta\},$$

{Surface $abcda\beta$.}

where we have

$$z_a P b\alpha + z_b P a\alpha$$
$$= z_b \{x_a (Zg_a - Wa_a) + y_a(-Zf_a - Wb_a) + z_a [X(-g_a + \lambda a_a) + Y(f_a + \lambda b_a) + (\lambda Z - W) c_a]\}$$
$$+ z_a \{x_b (Zg_a - Wa_a) + y_b(-Zf_a - Wb_a) + z_b [X(-g_a + \lambda a_a) + Y(f_a + \lambda b_a) + (\lambda Z - W) c_a]\},$$
$$= (z_b x_a + z_a x_b)(\ \ Zg_a - Wa_a)$$
$$+ (z_b y_a + z_a y_b)(-Zf_a - Wb_a)$$
$$+ 2z_a z_b \{X(-g_a + \lambda a_a) + Y(f_a + \lambda b_a) + (\lambda Z - W) c_a\}.$$

Also

$$z_a z_b I_a = z_a z_b \{(-g_a + \lambda a_a) x + (f_a + \lambda b_a) y\},$$
$$p^2\beta\gamma\delta = x^2 . agh + xy(ahf + bgh) + y^2 . hbf,$$
$$\Delta p^2\beta\gamma\delta = X . 2x . agh + \quad y(ahf + bgh)$$
$$+ Y . x(ahf + bgh) + 2y . hbf$$
$$+ Z . x(cgh + afg) + \ y(bfg + chf)$$
$$+ W . x(abg - cah) + \ y(bch - abf).$$

41. The whole expression is a linear function of X, Y, Z, W, and it is easy to see *à priori*, or to verify, that the coefficients of X, Y, each of them vanish. The coefficient of Z is

$$= \Sigma \{(z_b x_a + z_a x_b) g_a - (z_b y_a + z_a y_b) f_a + 2\lambda z_a z_b c_a\} p^2\beta\gamma\delta$$
$$+ \Sigma z_a z_b [(-g_a + \lambda a_a) x + (f_a + \lambda b_a) y] [x(cgh + afg) + y(bfg + chf)],$$

with a like expression for the coefficient of W.

The foregoing expression may be written

$$(z_b x_a + z_a x_b) \Sigma g [agh . x^2 + (ahf + bgh) xy + bhf . y^2]$$
$$-(z_b y_a + z_a y_b) \Sigma f [agh . x^2 + (ahf + bgh) xy + bhf . y^2]$$
$$+ 2\lambda z_a z_b \Sigma \{c [agh . x^2 + (ahf + bgh) xy + bhf . y^2]$$
$$+ (ax + by) [(cgh + afg) x + (bfg + chf) y]\}$$
$$+ 2z_a z_b \Sigma (-gx + fy) [(cgh + afg) x + (bfg + chf) y].$$

The first sum is

$$x^2 . gagh + xy(gahf + gbgh) + y^2 . gbhf,$$
$$= -xy\ . afgh - y^2 . bfgh,$$
$$= -h_\delta y (afg . x + bfg . y);$$

where afg, bfg denote determinants with the suffixes α, β, γ. Similarly the second sum is

$$= -h_\delta x (afg . x + bfg . y);$$

the third sum is

$$(a_\delta x + b_\delta y)(afg . x + bfg . y),$$

and the fourth sum is

$$(-g_\delta x + f_\delta y)(afg . x + bfg . y).$$

{Surface $abcda\beta$.}

The whole coefficient of Z thus contains the factor $(afg\,.\,x+bfg\,.\,y)$; and similarly it would appear that the whole coefficient of W contains the factor $(abg\,.\,x-abf\,.\,y)$, the other factor being the same in each case; viz. the two terms together are

$$\left\{\begin{array}{l}-(z_bx_a+z_ax_b)\,h_\delta y\\ +(z_by_a+z_ay_b)\,h^\delta x\\ +2\lambda z_az_b(a_\delta x+b_\delta y)\\ +2z_az_b(-g_\delta x+f_\delta y)\end{array}\right\}\{Z\,(afg\,.\,x+bfg\,.\,y)+W\,(abg\,.\,x-abf\,.\,y)\};$$

where the second factor is $\Delta p^2\alpha\beta\gamma$, which is the required result. See *post*, Nos. 59 *et seq.*

42. (4) The line $[\alpha, \beta, \gamma, \delta]$ is an 8-tuple line; in fact, for any point of the line in question we have $p^2\beta\gamma\delta=0$, $p^2\gamma\delta\alpha=0$, $p^2\delta\alpha\beta=0$, $p^2\alpha\beta\gamma=0$; whence each factor is 0^1, or the norm is 0^8.

I notice that the surface meets the quadric $p^2\alpha\beta\gamma$ in

lines α, β, γ	each	8	times	24
„ $(\alpha, \beta, \gamma, \delta)$	„		„	16
„ $(ab, \alpha, \beta, \gamma)$	„	4	„	8
			$24\times 2=$	48

Surface $a\alpha\beta\gamma\delta\epsilon$.

43. The equation is

$$(p^2\alpha\beta\epsilon\,.\,p^2\gamma\delta\epsilon\,.\,p^3a\alpha\gamma\,.\,\delta\beta+p^2\alpha\gamma\epsilon\,.\,p^2\delta\beta\epsilon\,.\,p^3a\alpha\beta\,.\,\gamma\delta)^2$$
$$-4p^2\alpha\beta\epsilon\,.\,p^2\gamma\delta\epsilon\,.\,p^2\alpha\gamma\epsilon\,.\,p^2\delta\beta\epsilon\,.\,p^2\alpha\beta\gamma\,.\,p^2\delta\beta\gamma\,.\,pa\alpha\,.\,p\delta a=0;$$

or, what is the same thing,

$$(p^2\alpha\beta\epsilon\,.\,p^2\gamma\delta\epsilon\,.\,p^3a\alpha\gamma\,.\,\delta\beta-p^2\alpha\gamma\epsilon\,.\,p^2\delta\beta\epsilon\,.\,p^3a\alpha\beta\,.\,\gamma\delta)^2$$
$$-4p^2\alpha\beta\epsilon\,.\,p^2\gamma\delta\epsilon\,.\,p^2\alpha\gamma\epsilon\,.\,p^2\delta\beta\epsilon\,.\,p^2\beta\alpha\delta\,.\,p^2\gamma\alpha\delta\,.\,p\beta a\,.\,p\gamma a=0;$$

the equivalence of the two depending on the identity

$$p^3a\alpha\beta\,.\,\gamma\delta\,.\,p^3a\alpha\gamma\,.\,\delta\beta$$
$$-p^2\alpha\beta\gamma\,.\,p^2\delta\beta\gamma\,.\,pa\alpha\,.\,p\delta a$$
$$+p^2\beta\alpha\delta\,.\,p^2\gamma\alpha\delta\,.\,p\beta a\,.\,p\gamma a=0;$$

where, as before, $p^2\alpha\beta\epsilon=0$ is the equation of the quadric through the lines α, β, ϵ, and $pa\alpha=0$ is the equation of the plane through the line α and the point a; viz. $p^2\alpha\beta\epsilon$, &c., and $pa\alpha$, &c., have the values already mentioned: $p^3a\alpha\beta\,.\,\gamma\delta=0$ as already mentioned is the cubic surface through the lines α, β, γ, δ and $a\alpha\beta$, $a\gamma\delta$.

{Surface $a\alpha\beta\gamma\delta\epsilon$.}

44. *Investigation.* In the projection, using line-coordinates, the equation of the conic touching the five lines may be written

$$\left|\begin{array}{l}(\xi,\ \eta,\ \zeta\ )^2\\ (P,\ Q,\ R)^2\end{array}\right| = 0;$$

where the symbol denotes a determinant the last five lines of which are obtained by giving to (P, Q, R) the suffixes α, β, γ, δ, ϵ respectively. This is at once transformed into

$$\alpha\beta\epsilon\,.\,\gamma\delta\epsilon\,.\,\alpha\gamma\Delta\,.\,\delta\beta\Delta - \alpha\gamma\epsilon\,.\,\delta\beta\epsilon\,.\,\alpha\beta\Delta\,.\,\gamma\delta\Delta = 0,$$

or, what is the same thing,

$$p^2\alpha\beta\epsilon\,.\,p^2\gamma\delta\epsilon\,.\,\alpha\gamma\Delta\,.\,\delta\beta\Delta - p^2\alpha\gamma\epsilon\,.\,p^2\delta\beta\epsilon\,.\,\alpha\beta\Delta\,.\,\gamma\delta\Delta = 0;$$

or say

$$p^2\alpha\beta\epsilon\,.\,p^2\gamma\delta\epsilon\,(A''\xi + B''\eta + C''\zeta)\,(A'''\xi + B'''\eta + C'''\zeta)$$
$$- p^2\alpha\gamma\epsilon\,.\,p^2\delta\beta\epsilon\,(A\xi + B\eta + C\zeta)\,(A'\xi + B'\eta + C'\zeta) = 0;$$

where $p^2\alpha\beta\epsilon$, &c., signify as before; and

$$A\xi + B\eta + C\zeta = \left|\begin{array}{ccc}\xi, & \eta, & \zeta\\ P_\alpha, & Q_\alpha, & R_\alpha\\ P_\beta, & Q_\beta, & R_\beta\end{array}\right|,$$

and so for $A'\xi + B'\eta + C'\zeta$, &c., the suffixes for A', B', C' being (γ, δ); and those for $A''\xi + B''\eta + C''\zeta$ and $A'''\xi + B'''\eta + C'''\zeta$ being (α, γ) and (δ, β) respectively.

45. Passing to the reciprocal equation, and making the conic pass through the point a, we obtain the equation of the surface in the form

$$\left\{p^2\alpha\gamma\epsilon\,.\,p^2\delta\beta\epsilon\left|\begin{array}{ccc}p_a, & q_a, & r_a\\ A, & B, & C\\ A', & B', & C'\end{array}\right| - p^2\alpha\beta\epsilon\,.\,p^2\gamma\delta\epsilon\left|\begin{array}{ccc}p_a, & q_a, & r_a\\ A'', & B'', & C''\\ A''', & B''', & C'''\end{array}\right|\right\}^2$$
$$+\ 4p^2\alpha\gamma\epsilon\,.\,p^2\delta\beta\epsilon\,.\,p^2\alpha\beta\epsilon\,.\,p^2\gamma\delta\epsilon\left|\begin{array}{ccc}p_a, & q_a, & r_a\\ A, & B, & C\\ A'', & B'', & C''\end{array}\right|\left|\begin{array}{ccc}p_a, & q_a, & r_a\\ A', & B', & C'\\ A''', & B''', & C'''\end{array}\right| = 0;$$

or in the equivalent form, where in the first term we have + instead of −, and in the second term the determinants are

$$\left|\begin{array}{ccc}p_a, & q_a, & r_a\\ A, & B, & C\\ A''', & B''', & C'''\end{array}\right|, \quad \left|\begin{array}{ccc}p_a, & q_a, & r_a\\ A', & B', & C'\\ A'', & B'', & C''\end{array}\right|.$$

46. To reduce this result, observe that we have

$$A,\ B,\ C = \left\|\begin{array}{ccc}hy - gz + aw, & -hx + fz + bw, & gx - fy + cw\\ h'y - g'z + a'w, & -h'x + f'z + b'w, & g'x - f'y + c'w\end{array}\right\|$$

where, for convenience, I retain the unaccented and accented letters $(a, \ldots)$, $(a', \ldots)$ instead of these letters with the suffixes α and β respectively. Writing as before

$$L = (af' - a'f)\,x + \ldots$$
$$M = (ag' - a'g)\,x + \ldots$$
$$N = (ah' - a'h)\,x + \ldots$$
$$\Omega = (gh' - g'h)\,x + \ldots$$

then

$$A = \Omega x - Lw,$$
$$B = \Omega y - Mw,$$
$$C = \Omega z - Nw,$$

and similarly

$$A' = \Omega'x - L'w,$$
$$B' = \Omega'y - M'w,$$
$$C' = \Omega'z - N'w\,;$$

where for L', M', N', Ω' we have $(a'', \ldots)$ and $(a''', \ldots)$. Hence

$$BC' - B'C = w \begin{vmatrix} y, & z, & w \\ M, & N, & \Omega \\ M', & N', & \Omega' \end{vmatrix}$$

with like expressions for $CA' - C'A$ and $AB' - A'B$; and substituting, we have

$$\begin{vmatrix} p_a, & q_a, & r_a \\ A, & B, & C \\ A', & B', & C' \end{vmatrix} = w \begin{vmatrix} p_a, & q_a, & r_a, \\ x, & y, & z, & w \\ L, & M, & N, & \Omega \\ L', & M', & N', & \Omega' \end{vmatrix}$$

or substituting for p_a, q_a, r_a their values $xw_a - wx_a$, $yw_a - wy_a$, $zw_a - wz_a$, this is

$$\begin{vmatrix} p_a, & q_a, & r_a \\ A, & B, & C \\ A', & B', & C' \end{vmatrix} = -w^2 \begin{vmatrix} x, & y, & z, & w \\ x_a, & y_a, & z_a, & w_a \\ L, & M, & N, & \Omega \\ L', & M', & N', & \Omega' \end{vmatrix}$$

whence, omitting the factors w^2, the equation is

$$\left\{ p^2\alpha\gamma\epsilon \, . \, p^2\delta\beta\epsilon \begin{vmatrix} x, & y, & z, & w \\ x_a, & y_a, & z_a, & w_a \\ L, & M, & N, & \Omega \\ L', & M', & N', & \Omega' \end{vmatrix} - p^2\alpha\beta\epsilon \, . \, p^2\gamma\delta\epsilon \begin{vmatrix} x, & y, & z, & w \\ x_a, & y_a, & z_a, & w_a \\ L'', & M'', & N'', & \Omega'' \\ L''', & M''', & N''', & \Omega''' \end{vmatrix} \right\}^2$$

$$+ \, 4p^2\alpha\gamma\epsilon \, . \, p^2\delta\beta\epsilon \, . \, p^2\alpha\beta\epsilon \, . \, p^2\gamma\delta\epsilon \begin{vmatrix} x, & y, & z, & w \\ x_a, & y_a, & z_a, & w_a \\ L, & M, & N, & \Omega \\ L'', & M'', & N'', & \Omega'' \end{vmatrix} \begin{vmatrix} x, & y, & z, & w \\ x_a, & y_a, & z_a, & w_a \\ L', & M', & N', & \Omega' \\ L''', & M''', & N''', & \Omega''' \end{vmatrix} = 0,$$

{Surface $a\alpha\beta\gamma\delta\epsilon$.}

where I recall that for $(L, \ldots)$, $(L', \ldots)$, $(L'', \ldots)$, $(L''', \ldots)$ the suffixes are (α, β), (γ, δ), (α, γ), and (δ, β) respectively. The values of the first two determinants thus are $p^3 a\alpha\beta \,.\, \gamma\delta$ and $p^3 a\alpha\gamma \,.\, \delta\beta$ respectively: that of the third is $p^3 a\alpha\beta \,.\, \alpha\gamma$; viz. this is $= p^2\alpha\beta\gamma \,.\, pa\alpha$; similarly, that of the fourth is $p^3 a\gamma\delta \,.\, \delta\beta$, which is $= -p^3 a\delta\gamma \,.\, \delta\beta = +p^3 a\delta\beta \,.\, \delta\gamma$; or finally this is $= p^2\delta\beta\gamma \,.\, pa\delta$. And we have thus the before-mentioned equation of the surface.

47. *Singularities.* The equation of the surface shows that

(0) The point a is a 2-conical point: in fact, we have for this point $p^3 a\alpha\beta \,.\, \gamma\delta = 0$, $p^3 a\alpha\gamma \,.\, \delta\beta = 0$, $pa\alpha = 0$, $pa\delta = 0$.

(2) The line α is a 4-tuple line: in fact, for any point on this line $p^2\alpha\beta\epsilon = 0$, $p^3 a\alpha\beta \,.\, \gamma\delta = 0$, $p^2\alpha\gamma\epsilon = 0$, $p^3 a\alpha\gamma \,.\, \delta\beta = 0$, $p^2\alpha\beta\gamma = 0$, $p^2 a\alpha = 0$.

(4) The line $(\alpha, \beta, \gamma, \epsilon)$ is a 2-tuple line: in fact, for any point on the line we have $p^2\alpha\beta\epsilon = 0$, $p^2\alpha\gamma\epsilon = 0$.

(10) The excuboquartic $\alpha\beta\epsilon \,.\, \gamma\delta \,.\, a$ is a simple curve: in fact, for any point of this curve we have $p^2\alpha\beta\epsilon = 0$, $p^3 a\alpha\beta \,.\, \gamma\delta = 0$, these two surfaces intersecting in the lines α, β and the curve. It is, moreover, obvious that the surface is touched along the curve by the hyperboloid $p^2\alpha\beta\epsilon$.

I notice that the surface meets the quadric $p^2\alpha\beta\gamma$ in

lines (α, β, γ)	each 4 times,		12
„ $(\alpha, \beta, \gamma, \delta)$	„	twice,	4
„ $(\alpha, \beta, \gamma, \epsilon)$	„	„	4
curve $a\alpha\beta\gamma \,.\, \delta\epsilon$	„	„	8
		$14 \times 2 =$	28

Surface $\alpha\beta\gamma\delta\epsilon\zeta$.

48. The equation of the surface may be written

$$p^2\alpha\beta\epsilon \,.\, p^2\gamma\delta\epsilon \,.\, p^2\alpha\gamma\zeta \,.\, p^2\delta\beta\zeta - p^2\alpha\beta\zeta \,.\, p^2\gamma\delta\zeta \,.\, p^2\alpha\gamma\epsilon \,.\, p^2\delta\beta\epsilon = 0,$$

where $p^2\alpha\beta\epsilon = 0$ is the equation of the quadric through the lines α, β, ϵ; viz. $p^2\alpha\beta\epsilon$ has the value already mentioned.

The form is one of 45 like forms depending on the partitionment

$$\left\{\begin{matrix} \alpha\beta \,.\, \gamma\delta \\ \alpha\gamma \,.\, \delta\beta \\ \alpha\delta \,.\, \beta\gamma \end{matrix}\right\} (\epsilon, \zeta)$$

of the six letters.

{Surface $\alpha\beta\gamma\delta\epsilon\zeta$.}

49. *Investigation.* The projections of the six lines are tangents to a conic: the condition for this is $(P, Q, R)^2=0$, where the left-hand side represents the determinant obtained by writing successively $(P_\alpha, Q_\alpha, R_\alpha)$, &c. for (P, Q, R). The equation may be written

$$\alpha\beta\epsilon\,.\,\gamma\delta\epsilon\,.\,\alpha\gamma\zeta\,.\,\delta\beta\zeta-\alpha\beta\zeta\,.\,\gamma\delta\zeta\,.\,\alpha\gamma\epsilon\,.\,\delta\beta\gamma=0,$$

where

$$\alpha\beta\epsilon=\begin{vmatrix} P_\alpha, & Q_\alpha, & R_\alpha \\ P_\beta, & Q_\beta, & R_\beta \\ P_\epsilon, & Q_\epsilon, & R_\epsilon \end{vmatrix}$$

and substituting for P_α, &c., their values, we have $\alpha\beta\epsilon=w\,.\,p^2\alpha\beta\epsilon$; whence the foregoing result.

50. *Singularities.* The equation shows that

(2) The line α is a 2-tuple line: in fact, for each point of the line we have $p^2\alpha\beta\epsilon=0,\ p^2\alpha\gamma\zeta=0,\ p^2\alpha\beta\zeta=0,\ p^2\alpha\gamma\epsilon=0.$

(4) The line $(\alpha, \beta, \epsilon, \zeta)$ is a simple line: in fact, for each point of the line we have $p^2\alpha\beta\epsilon=0,\ p^2\alpha\beta\zeta=0.$

(9) The quadriquadric $\alpha\beta\epsilon\,.\,\gamma\delta\zeta=0$ is a simple curve on the surface: in fact, for each point of the curve we have $p^2\alpha\beta\epsilon=0,\ p^2\gamma\delta\zeta=0.$

It may be remarked that the surface meets the hyperboloid $p^2\alpha\beta\epsilon$ in

lines $(\alpha, \beta, \epsilon)$	each twice,		6
„ $(\alpha, \beta, \epsilon, \gamma)$	„	once,	2
„ $(\alpha, \beta, \epsilon, \delta)$	„	„	2
„ $(\alpha, \beta, \epsilon, \zeta)$	„	„	2
curve $\alpha\beta\epsilon\,.\,\gamma\delta\zeta$	„	„	4
			$2\times 8=16$

51. It might be thought that there should be on the surface some curve $\alpha\beta\gamma\delta\epsilon\zeta$, such as the cubic *abcdef* on the surface *abcdef*; but I cannot find that this is so. The equation of the surface is satisfied if we have simultaneously (λ being arbitrary)

$$p^2\alpha\beta\epsilon\,.\,p^2\alpha\gamma\zeta-\lambda p^2\alpha\beta\zeta\,.\,p^2\alpha\gamma\epsilon=0,$$

$$\lambda p^2\gamma\delta\epsilon\,.\,p^2\delta\beta\zeta-\ p^2\gamma\delta\zeta\,.\,p^2\delta\beta\epsilon=0;$$

which equations represent quartic surfaces, the first of them having α for a double line, and passing through the lines $\beta, \gamma, \epsilon, \zeta$ $(13+4\times 5=33$ conditions, so that the equation of such a surface contains only an arbitrary parameter $\lambda)$; and the second having δ for a double line, and passing through the lines $\beta, \gamma, \epsilon, \zeta$. But I see no condition by which λ can be determined so as to have the same value in the two equations respectively. Of course, leaving it arbitrary, the two quartic surfaces intersect in the lines $\beta, \gamma, \epsilon, \zeta$ and in a curve of the order 12 depending on the arbitrary value of λ, which curve lies on the surface $\alpha\beta\gamma\delta\epsilon\zeta$.

{Surface $\alpha\beta\gamma\delta\epsilon\zeta$.}

The Excuboquartic $\alpha\beta\gamma,\ \delta\epsilon,\ a$.

52. The notion is, that we have a fixed point a, two fixed lines δ, ϵ, and a singly infinite series of lines, or say the generating lines of a skew surface: each generating line determines, with the point a, a plane; and if in this plane we draw, meeting the lines δ, ϵ, a line to meet the generating line in a point P, then the locus of this point P is the curve about to be considered.

53. In the case in question, the singly infinite series of lines is that of the lines which meet each of the lines α, β, γ, or say these are the generatrices of the hyperboloid $\alpha\beta\gamma$: the locus, or curve $\alpha\beta\gamma,\ \delta\epsilon,\ a$, is (as mentioned above) an excuboquartic. It is not necessary for the purpose of the memoir, but it is interesting to consider in conjunction therewith the excuboquartic arising in like manner from the directrices of the hyperboloid; it will appear that the two curves are the complete intersection of the quadric $\alpha\beta\gamma$ by a quartic surface. Observe that the two curves are given as follows: viz. considering for the quadric $\alpha\beta\gamma$ any tangent-plane through the point a, and drawing in this plane, to meet the lines δ and ϵ, a line, this meets the section of the quadric surface by the tangent-plane in two points, the locus of which is the aggregate of the two curves: viz. the section being a line-pair, the two points belong, one of them to a generatrix and the other to a directrix of the quadric surface.

54. It is convenient to take $x=0$, $y=0$ for the equations of the line δ; $z=0$, $w=0$ for those of the line ϵ: for then, for any plane $Ax+By+Cz+Dw=0$, the line in this plane and meeting the lines δ and ϵ, has for its equations $Ax+By=0$, $Cz+Dw=0$; or, what is the same thing, for the plane $P=0$ the equations of the line are $P_{xy}=0$, $P_{zw}=0$, where P_{xy}, P_{zw} denote the terms in x, y and in z, w respectively.

I take also x_0, y_0, z_0, w_0 for the coordinates of the point a, and $PS-QR=0$ for the equation of the quadric surface, P, Q, R, S being given linear functions of (x, y, z, w): we have then say $P-\theta R=0$, $Q-\theta S=0$ for the equations of any generatrix, and $P-\phi Q=0$, $R-\phi S=0$ for the equations of any directrix of the hyperboloid.

The equation of the plane through the point a and the generatrix $P-\theta R=0$, $Q-\theta S=0$, is clearly

$$(Q_0-\theta S_0)(P\ \ -\theta R\ \ )-(P_0-\theta R_0)(Q\ \ -\theta S\ \ )=0;$$

so that for the line in this plane, meeting the lines δ and ϵ, we have

$$(Q_0-\theta S_0)(P_{xy}-\theta R_{xy})-(P_0-\theta R_0)(Q_{xy}-\theta S_{xy})=0,$$
$$(Q_0-\theta S_0)(P_{zw}-\theta R_{zw})-(P_0-\theta R_0)(Q_{zw}-\theta S_{zw})=0;$$

and joining thereto the equations

$$\theta=\frac{P}{R}=\frac{Q}{S}=\frac{P_{xy}+P_{zw}}{R_{xy}+R_{zw}}=\frac{Q_{xy}+Q_{zw}}{R_{xy}+R_{zw}},$$

(equivalent in all to three equations,) the elimination of θ gives the required curve: the equations thus are

$$PS-QR=0,$$
$$(Q_0S-QS_0)(P_{xy}R-PR_{xy})-(P_0R-PR_0)(Q_{xy}S-QS_{xy})=0,$$

or, as the second equation may also be written,

$$(Q_0S - QS_0)(P_{xy}R_{zw} - P_{zw}R_{xy}) - (P_0R - PR_0)(Q_{xy}S_{zw} - Q_{zw}S_{xy}) = 0;$$

viz. the second equation represents a cubic surface having upon it the lines $(P=0, R=0)$ and $(Q=0, S=0)$: it therefore intersects the quadric $PS - QR = 0$ in these two lines, and besides in an excuboquartic curve, which is the required locus.

55. Representing the determinants

$$\begin{vmatrix} P, & Q, & R, & S \\ P_0, & Q_0, & R_0, & S_0 \end{vmatrix} \text{ by } (\text{a}', \text{b}', \text{c}', \text{f}', \text{g}', \text{h}'), \text{ viz. } \text{a}' = QR_0 - Q_0R, \ldots \quad \text{f}' = PS_0 - P_0S, \ldots;$$

$$\begin{vmatrix} P_{xy}, & Q_{xy}, & R_{xy}, & S_{xy} \\ P_{zw}, & Q_{zw}, & R_{zw}, & S_{zw} \end{vmatrix} \text{ by } (\text{a}, \text{b}, \text{c}, \text{f}, \text{g}, \text{h}), \text{ viz. } \text{a} = Q_{xy}R_{zw} - Q_{zw}R_{xy}, \ldots;$$

so that $(\text{a}', \ldots)$ are linear functions, $(\text{a}, \ldots)$ quadric functions, of the coordinates; the equation of the cubic surface is $\text{gb}' - \text{bg}' = 0$, viz. the excuboquartic arising from the generatrices is the partial intersection of the quadric $PS - QR = 0$ and the cubic $\text{gb}' - \text{g}'\text{b} = 0$; the two surfaces besides intersecting in the lines $(P=0, R=0)$ and $(Q=0, S=0)$.

It appears, in the same manner, that the excuboquartic arising from the directrices is the partial intersection of the quadric $PS - QR = 0$ and the cubic $\text{hc}' - \text{ch}' = 0$; the two surfaces besides intersecting in the lines $(P=0, Q=0)$ and $(R=0, S=0)$.

56. But the elimination may be performed in a different manner, as follows: from the first two equations in θ, multiplying by $P_{zw}, -P_{xy}$ and adding, and so with $Q_{zw}, -Q_{xy}$, &c., we obtain

$$\begin{aligned}
(Q_0 - \theta S_0)(\quad\;\; - \theta\text{b}) &- (P_0 - \theta R_0)(-\text{c} + \theta\text{f}) = 0,\\
(Q_0 - \theta S_0)(\;\text{c} + \theta\text{a}) &- (P_0 - \theta R_0)(\quad\;\; \theta\text{g}) = 0,\\
(Q_0 - \theta S_0)(-\text{b} \quad\;\;) &- (P_0 - \theta R_0)(\;\text{a} + \theta\text{h}) = 0,\\
(Q_0 - \theta S_0)(\;\text{f} - \theta\text{h}) &- (P_0 - \theta R_0)(\;\text{g} \quad\;\;) = 0.
\end{aligned}$$

We then have

$$\theta = \frac{-\text{c} + \theta\text{f}}{\text{a} + \theta\text{h}} = \frac{\text{c} + \theta\text{a}}{\text{f} - \theta\text{h}},$$

or, what is the same thing,

$$\text{h}\theta^2 + (\text{a} - \text{f})\,\theta + \text{c} = 0.$$

Using this equation, written in the form $(\text{a} + \theta\text{h})\,\theta = -\text{c} + \theta\text{f}$, to transform the first or third of the four equations in θ, we obtain

$$-\text{a}P_0 - \text{b}Q_0 - \text{c}R_0 + \theta(-\text{h}P_0 \quad . \quad + \text{f}R_0 + \text{b}S_0) = 0;$$

and using the same equation, written in the form $(\text{f} - \theta\text{h})\,\theta = \text{c} + \theta\text{a}$, to transform the second or fourth equation, we obtain

$$\text{g}P_0 - \text{f}Q_0 \;\; + \text{c}S_0 + \theta(\qquad \text{h}Q_0 - \text{g}R_0 + \text{a}S_0) = 0;$$

and hence, eliminating θ, we obtain

$$(\mathrm{h}Q_0 - \mathrm{g}R_0 + \mathrm{a}S_0)(-\mathrm{a}P_0 - \mathrm{b}Q_0 - \mathrm{c}R_0) - (-\mathrm{h}P_0 + \mathrm{f}R_0 + \mathrm{b}S_0)(\mathrm{g}P_0 - \mathrm{f}Q_0 . + \mathrm{c}S_0) = 0,$$

which, as being of the second order in (a, ...), represents a quartic surface. The equation remains unaltered by the interchange of Q, R, and the consequent interchanges among (a, b, c, f, g, h): hence the quartic surface contains not only the excuboquartic arising from the generatrices, but also that arising from the directrices; and these two curves are the complete intersection of the quartic by the quadric $PS - QR = 0$.

57. I obtain this same result also as follows. Consider a point (P_1, Q_1, R_1, S_1) on the quadric surface; $P_1S_1 - Q_1R_1 = 0$; the tangent plane at the point is

$$PS_1 - QR_1 - RQ_1 + SP_1 = 0;$$

and if this passes through the point a, then

$$P_0S_1 - Q_0R_1 - R_0Q_1 + S_0P_1 = 0.$$

The line which in the tangent-plane meets the lines δ, ϵ is given, as before, by the equations

$$P_{xy}S_1 - Q_{xy}R_1 - R_{xy}Q_1 + S_{xy}P_1 = 0,$$
$$P_{zw}S_1 - Q_{zw}R_1 - R_{zw}Q_1 + S_{zw}P_1 = 0.$$

Remembering the significations of (a, ...), the last three equations give

$$\begin{aligned} S_1 : R_1 : -Q_1 : -P_1 = &\quad . \qquad \mathrm{h}Q_0 - \mathrm{g}R_0 + \mathrm{a}S_0 \\ &: -\mathrm{h}P_0 \quad . \quad + \mathrm{f}R_0 + \mathrm{b}S_0 \\ &: \mathrm{g}P_0 - \mathrm{f}Q_0 \quad . \quad + \mathrm{c}S_0 \\ &: -\mathrm{a}P_0 - \mathrm{b}Q_0 - \mathrm{c}R_0 \quad . \ ; \end{aligned}$$

and substituting these values in $S_1P_1 - Q_1R_1 = 0$, we have the above equation of the quadric surface.

58. Or again, changing the notation, I take the equation of the quadric surface to be

$$(a, b, c, d, f, g, h, l, m, n \between x, y, z, w)^2 = 0.$$

A tangent-plane hereof is

$$\xi x + \eta y + \zeta z + \omega w = 0,$$

where ξ, η, ζ, ω are any quantities satisfying the relation

$$(A, B, C, D, F, G, H, L, M, N \between \xi, \eta, \zeta, \omega)^2 = 0,$$

the capitals denoting the inverse coefficients.

Supposing that the tangent-plane passes through a fixed point a, coordinates $(\alpha, \beta, \gamma, \delta)$, we have

$$\alpha\xi + \beta\eta + \gamma\zeta + \delta\omega = 0;$$

and if the equations of the lines δ, ϵ are as before $(x=0,\ y=0)$ and $(z=0,\ w=0)$; then for the line in the tangent-plane meeting the lines δ, ϵ, we have

$$\xi x+\eta y=0,\quad \zeta z+\omega w=0.$$

These last equations may be represented by

$$\xi=ly,\quad \eta=-lx,\quad \zeta=mw,\quad \omega=-mz;$$

and, substituting these values, we have

$$(A,\ldots\between ly,\ -lx,\ mw,\ -mz)^2=0,$$
$$(\alpha,\ldots\between ly,\ -lx,\ mw,\ -mz)^1=0,$$

that is

$$(Ay^2-2Hxy+Bx^2,\ -Fxw+Gyw-Lyz+Mxz,\ Cw^2-2Nwz+Dz^2\between l,\ m)^2=0,$$

and

$$(\alpha y-\beta x,\ \gamma w-\delta z\between l,\ m)=0.$$

Whence, eliminating l, m, we have the quartic equation

$$(Ay^2-2Hxy+Bx^2,\ -Fxw+Gyw-Lyz+Mxz,\ Cw^2-2Nzw+Dz^2\between \gamma w-\delta z,\ \beta x-\alpha y)^2=0.$$

Further Investigation as to the Surface $ab\alpha\beta\gamma\delta$.

59. The theorem that in the surface $ab\alpha\beta\gamma\delta$, the equation of which is

$$\text{Norm}\,\{\sqrt{pa\alpha\,.\,pb\alpha}\,.\,p^2\beta\gamma\delta-\sqrt{pa\beta\,.\,pb\beta}\,.\,p^2\gamma\delta\alpha+\sqrt{pa\gamma\,.\,pb\gamma}\,.\,p^2\delta\alpha\beta-\sqrt{pa\delta\,.\,pb\delta}\,.\,p^2\alpha\beta\gamma\}=0;$$

the lines $(ab,\ \alpha,\ \beta,\ \gamma)$ are tacnodal, each sheet touching along the line the quadric $p^2\alpha\beta\gamma$, may be proved in a different manner by investigating the intersection of the surface with the quadric $p^2\alpha\beta\gamma$.

For this purpose take the equation of the quadric to be $yz-xw=0$; the equations of the lines α, β, γ will be

$$\begin{pmatrix}z-\lambda_\alpha w=0\\ x-\lambda_\alpha y=0\end{pmatrix},\quad \begin{pmatrix}z-\lambda_\beta w=0\\ x-\lambda_\beta y=0\end{pmatrix},\quad \begin{pmatrix}z-\lambda_\gamma w=0\\ x-\lambda_\gamma y=0\end{pmatrix};$$

and we may write $(a,\ b,\ c,\ f,\ g,\ h)$ for the coordinates of the line δ. The equation of the surface will be

$$\text{Norm}\left\{\Sigma\left[\pm\sqrt{pa\alpha\,.\,pb\alpha}\,(\lambda_\beta-\lambda_\gamma)\begin{pmatrix}(a-f)\,xz-(\lambda_\beta+\lambda_\gamma)\,yz+\lambda_\beta\lambda_\gamma yw\\ +(b-g)\,\lambda_\beta\lambda_\gamma\,(yz-xw)\\ +c\,(z-\lambda_\beta w)\,(z-\lambda_\gamma w)\\ +h\,(x-\lambda_\beta y)\,(x-\lambda_\gamma z)\end{pmatrix}\right]\right.$$
$$\left.-\sqrt{pa\delta\,.\,pb\delta}\,(\lambda_\beta-\lambda_\gamma)\,(\lambda_\gamma-\lambda_\alpha)\,(\lambda_\alpha-\lambda_\beta)\,(yz-xw)\right\};$$

where Σ denotes the sum of the three terms obtained by the cyclical interchange of α, β, γ; and

$$p a \alpha = (z_a - \lambda w_a)(x - \lambda y) - (x_a - \lambda y_a)(z - \lambda w),$$

$$p b \alpha = (z_b - \lambda w_b)(x - \lambda y) - (x_b - \lambda y_b)(z - \lambda w);$$

λ here standing for λ_α; and similarly for $pa\beta$, &c.

60. To obtain the intersection with $xw - yz = 0$, writing $w = \frac{yz}{x}$, then

$$p a \alpha = [z_a - \lambda w_a - \frac{z}{x}(x_a - \lambda y_a)](x - \lambda y), \quad (\lambda = \lambda_\alpha),$$

$$p b \alpha = [z_b - \lambda w_b - \frac{z}{x}(x_b - \lambda y_b)](x - \lambda y);$$

or say

$$\sqrt{p a \alpha \,.\, p b \alpha} = \sqrt{M_\alpha}\,(x - \lambda_\alpha y);$$

also the expression in { } becomes

$$= \{(a - f)\frac{z}{x} + c\frac{z^2}{x^2} + h\}\,(x - \lambda_\beta y)(x - \lambda_\gamma y);$$

so that the norm in question is

$$\text{Norm } \Sigma \sqrt{M_\alpha}\,(\lambda_\beta - \lambda_\gamma)\,\{(a - f)\frac{z}{x} + c\frac{z^2}{x^2} + h\}\,(x - \lambda_\alpha y)(x - \lambda_\beta y)(x - \lambda_\gamma y);$$

or say

$$\text{Norm } \Sigma \sqrt{M_\alpha}\,(\lambda_\beta - \lambda_\gamma)\,\{hx^2 + (a - f)zx + cz^2\}(x - \lambda_\alpha y)(x - \lambda_\beta y)(x - \lambda_\gamma y);$$

where M_α is now considered to stand for

$$\{(z_a x - z x_a) - \lambda(w_a x - y_a z)\}\,\{(z_b x - z x_b) - \lambda(w_b x - y_b z)\}.$$

Observing that the norm was originally the product of 8 factors, this breaks up into

$$\{hx^2 + (a - f)zx + cz^2\}^8\,\{(x - \lambda_\alpha y)(x - \lambda_\beta y)(x - \lambda_\gamma y)\}^8 = 0,$$

and

$$\text{Norm}^2 \sqrt{M_\alpha}\,(\lambda_\beta - \lambda_\gamma) = 0,$$

where the new norm is the product of 4 factors.

61. Writing for greater convenience λ, μ, ν in place of λ_α, λ_β, λ_γ, and observing that M_α is a quadric function of λ_α, that is of λ, the last-mentioned norm is

$$\text{Norm } \sqrt{A + B\lambda + C\lambda^2}\,(\mu - \nu),$$

which is easily seen to be

$$= (4AC - B^2)(\mu - \nu)^2(\nu - \lambda)^2(\lambda - \mu)^2;$$

or writing for a moment

$$(A + B\lambda + C\lambda^2) = (P - Q\lambda)(P' - Q'\lambda),$$

whence

$$A = PP', \quad B = -(PQ' + P'Q), \quad C = QQ';$$

then

$$4AC - B^2 = -(PQ' - P'Q)^2;$$

and we have

$$\begin{aligned} P,\ Q &= z_a x - z x_a, \quad w_a x - y_a z, \\ P',\ Q' &= z_b x - z x_b, \quad w_b x - y_b z, \end{aligned}$$

whence

$$\begin{aligned} PQ' - P'Q = &\ (z_a w_b - z_b w_a)\, x^2 \\ &+ [y_a z_b - y_b z_a - (x_a w_b - x_b w_a)]\, xz \\ &+ (x_a y_b - x_b y_a)\, z^2; \end{aligned}$$

viz. if (a, b, c, f, g, h) are the coordinates of the line ab, this is

$$= \text{h}x^2 + (\text{a} - \text{f})\, xz + \text{c}z^2.$$

Hence, omitting the constant factor $(\mu - \nu)^4 (\nu - \lambda)^4 (\lambda - \mu)^4$ {that is $(\lambda_\beta - \lambda_\gamma)^4 (\lambda_\gamma - \lambda_\alpha)^4 (\lambda_\alpha - \lambda_\beta)^4$}, the foregoing equation $\text{norm}^2 = 0$ becomes

$$[\text{h}x^2 + (\text{a} - \text{f})\, xz + \text{c}z^2]^4 = 0,$$

and the intersections of the quadric with the surface are obtained by combining the equation $xw - yz = 0$ with the several equations

$$\begin{aligned} &\{\text{h}x^2 + (\text{a} - \text{f})\, zx + \text{c}z^2\}^8 = 0, \\ &\{(x - \lambda_\alpha y)(x - \lambda_\beta y)(x - \lambda_\gamma y)\}^8 = 0, \\ &\{\text{h}x^2 + (\text{a} - \text{f})\, zx + \text{c}z^2\}^4 = 0; \end{aligned}$$

viz. these are

lines (α, β, γ, δ) each	8 times	16	
line ($x = 0$, $z = 0$)	16 „		16
lines α, β, γ each	8 „	24	
line ($x = 0$, $y = 0$)	24 „		24
lines [ab, α, β, γ] each	4 „	8	
line ($x = 0$, $z = 0$)	8 „		8
	$(16 + 24 + 8) \times 2 =$	48	+ 48

But it is clear that the lines ($x = 0$, $y = 0$) and ($x = 0$, $z = 0$) are introduced by the process of elimination, and are no part of the intersection. The complete intersection consists of the lines (α, β, γ, δ) each 8 times, the lines (α, β, γ) each 8 times, and the lines [ab, α, β, γ] each 4 times. But the last-mentioned lines being only double lines on the surface, this means that the two sheets each touch the quadric surface, or that the lines are tacnodal.

504.

ON THE MECHANICAL DESCRIPTION OF CERTAIN SEXTIC CURVES.

[From the *Proceedings of the London Mathematical Society*, vol. IV. (1871—1873), pp. 105—111. Read April 11, 1872.]

THE curves in question might be taken to be those described by a point C rigidly connected with points A and B, each of which describes a circle: but the construction is considered under a somewhat more general form. I consider a quadrilateral, the sides of which are a, b, c, d, and the inclinations of these to a fixed line $\alpha, \beta, \gamma, \delta$. This being so, if a, b, c, d, and one of the angles, say δ, are constant, then we have between the three variable angles the relations

$$a\cos\alpha + b\cos\beta + c\cos\gamma + d\cos\delta = 0,$$

$$a\sin\alpha + b\sin\beta + c\sin\gamma + d\sin\delta = 0,$$

giving rise to a single relation between any two of the variable angles; and we consider a curve such that the coordinates x, y of any point thereof are given linear functions of the sines and cosines of the three variable angles, or, what is the same thing, of the sines and cosines of any two of these angles. We thus unite together what would otherwise be distinct cases; for everything is symmetrical in regard to the sides a, b, c and the corresponding variable angles α, β, γ, irrespectively of the order of succession of these sides: and we can thus, in the discussion of the curve, employ any two at pleasure, say α, β, of the variable angles, without determining whether the sides a, b are contiguous or opposite.

Eliminating, then, the variable angle γ, we obtain between α, β a relation which, if we write therein $\tan\frac{1}{2}\alpha = u$, $\tan\frac{1}{2}\beta = v$, takes the form $(*\between u, 1)^2 (v, 1)^2 = 0$; viz. either of the variables u, v is expressible rationally in terms of the other of them and of the root of a quartic function thereof; say v is a rational function of u and $\sqrt{U}$. And hence a curve for which the coordinates x, y are rational functions of u, v, is a curve having a deficiency $D=1$, or, what is the same thing, having a number of dps. less by unity than the maximum number $\{=\frac{1}{2}(n-1)(n-2)$, if n be the order of the curve$\}$.

It will further appear that the relation $(*\between u,\ 1)^2(v,\ 1)^2=0$ is satisfied by the values $u=v=i$ and $u=v=-i$ (if, as usual, $i=\sqrt{-1}$).

In the curve in question, the coordinates (x, y) are given linear functions of the sines and cosines of α, β; and if we make the curve meet an arbitrary line $Ax+By+C=0$, we obtain between the sines and cosines of α, β a linear relation which, substituting therein the expressions in terms of u, v, takes the form

$$(*\between u,\ 1)^2.(1+v^2)+(*\between v,\ 1)^2.(1+u^2)=0,$$

viz. this is a relation of the form $(\dagger\between u,\ 1)^2(v,\ 1)^2=0$, such that it is satisfied by the four sets of values $u=\pm i$, $v=\pm i$, and therefore in particular by the values $u=v=i$ and $u=v=-i$.

Hence, considering the intersections of the curve by the arbitrary line, the values of (u, v) are given by the two equations $(*\between u,\ 1)^2(v,\ 1)^2=0$, $(\dagger\between u,\ 1)^2(v,\ 1)^2=0$; these, regarding for a moment u, v as ordinary rectangular coordinates, represent each of them a quartic curve having two dps. at infinity on the axes $u=0$, $v=0$ respectively: each of these points reckons therefore as 4 intersections, and the number of the remaining intersections therefore is $4.4-2.4$, $=8$. But, by what precedes, the two quartic curves have also in common the points $u=v=i$ and $u=v=-i$; and rejecting these, there remain $8-2$, $=6$ intersections.

The conclusion is, that the curve is a sextic curve of deficiency 1, that is, having 9 dps. The reasoning may be presented under a slightly different form as follows: regarding u, v as coordinates, we have the curve $(*\between u,\ 1)^2(v,\ 1)^2=0$, a binodal quartic curve, and having therefore the deficiency 1; the curve passes, as above-mentioned, through the points $u=v=i$ and $u=v=-i$. The required curve is obtained as a transformation of the quartic curve by formulæ of the form $x : y : z(=1)=P : Q : R$, where P, Q, and R $\{=(1+u^2)(1+v^2)\}$ are quartic functions of the coordinates u, v, such that $P=0$, $Q=0$, $R=0$ are each of them a quartic curve passing twice through each of the nodes and once through each of the before-mentioned points, $(u=v=i)$ and $(u=v=-i)$, of the binodal quartic curve. Hence the curve in question is a curve of the order $4.4-2.4-2.1$, $=6$, and having the same deficiency as the binodal quartic, that is, the deficiency is $=1$.

I observe that the sextic curve does not, in general, pass through the circular points at infinity, but it intersects the line at infinity in three distinct pairs of points; one of these, or all three of them, (but not two pairs only,) may coincide with the circular points at infinity, the circular points at infinity being, in the latter case, triple points, or the curve being tricircular: this will appear presently.

To obtain the foregoing equation $(*\between u,\ 1)^2(v,\ 1)^2=0$, the elimination of γ gives

$$(a\cos\alpha+b\cos\beta+d\cos\delta)^2+(a\sin\alpha+b\sin\beta+d\sin\delta)^2=c^2,$$

that is

$$a^2+b^2-c^2+d^2+2ab\cos(\alpha-\beta)+2ad\cos(\alpha-\delta)+2bd\cos(\beta-\delta)=0\,;$$

or, substituting herein the values

$$\cos\alpha=\frac{1-u^2}{1+u^2},\quad \sin\alpha=\frac{2u}{1+u^2},\quad \cos\beta=\frac{1-v^2}{1+v^2},\quad \sin\beta=\frac{2v}{1+v^2},$$

this is easily found to be $0=$

	1	$2u$	u^2
1	$(a+b)^2-2(a+b)d\cos\delta+d^2-c^2$	$2ad\sin\delta$	$(a-b)^2+2(a-b)d\cos\delta+d^2-c^2$
$2v$	$2bd\sin\delta$	$2ab$	$2bd\sin\delta$
v^2	$(a-b)^2-2(a-b)d\cos\delta+d^2-c^2$	$2ad\sin\delta$	$(a+b)^2+2(a+b)d\cos\delta+d^2-c^2$

or writing, for greater convenience, $\delta=0$, that is, taking α, β, γ to be the inclinations of the sides a, b, c to the fixed side d, this is $0=$

	1	$2u$	u^2
1	$(a+b-d)^2-c^2$	0	$(a-b+d)^2-c^2$
$2v$	0	$2ab$	0
v^2	$(a-b-d)^2-c^2$	0	$(a+b+d)^2-c^2$

,

or say this is

$$(A+Bu^2)+8abuv+v^2(C+Du^2)=0,$$

where, writing for shortness

$$\begin{aligned} a+d-b-c&=\lambda, & -a+b+c+d&=\lambda', \\ b+d-c-a&=\mu, & -b+c+d+a&=\mu', \\ c+d-a-b&=\nu, & -c+d+a+b&=\nu', \\ a+b+c+d&=\sigma, & -d+a+b+c&=\rho'; \end{aligned}$$

then

$$A=-\nu\rho',\quad B=\lambda\mu',\quad C=\lambda'\mu,\quad D=\nu'\sigma.$$

We at once verify that the equation is satisfied by $u=v=\pm i$, viz. this will be so if only

$$A-B-C+D-8ab=0,$$

and we have

$$A + D = 2(a+b)^2 + 2d^2 - 2c^2,$$
$$B + C = 2(a-b)^2 + 2d^2 - 2c^2;$$

whence the relation in question.

Writing $u = i$, we have

$$\{v(C-D) + 4abi\}^2 = -16a^2b^2 - (A-B)(C-D)$$
$$= -\tfrac{1}{4}(A - B + C - D)^2,$$

whence

$$v(C-D) = \pm \tfrac{1}{2}i\{\mp 8ab + A - B + C - D\}$$
$$= i(C-D) \text{ or } -i(A-B);$$

so that, corresponding to $u = i$, we have the values $v = i$ and $v = -i\dfrac{A-B}{C-D}$; and similarly, corresponding to $u = -i$, the values $v = -i$ and $v = i\dfrac{A-B}{C-D}$.

And in like manner, corresponding to $v = i$, we have the values $u = i$ and $u = -i\dfrac{A-C}{B-D}$; and to $v = -i$, the values $u = -i$ and $u = i\dfrac{A-C}{B-D}$.

It is easy to show that, if $u = i + \epsilon$, $v = i + \zeta$ are the values consecutive to $u = v = i$, then $a\epsilon + b\zeta = 0$: in fact, substituting the foregoing values in the relation between u, v, and writing for $8ab$ its value, $= A - B - C + D$, we have

$$A + B(-1 + 2i\epsilon) + (A - B - C + D)\{-1 + i(\epsilon + \zeta)\} + C(-1 + 2i\zeta) + D\{1 - 2i(\epsilon + \zeta)\} = 0,$$

which is

$$= \epsilon(A + B - C - D) + \zeta(A - B + C - D) = 0;$$

or finally $a\epsilon + b\zeta = 0$. And similarly, if $u = -i + \epsilon$, $v = -i + \zeta$ are the values consecutive to $u = v = -i$, then we have the same relation $a\epsilon + b\zeta = 0$.

The points at infinity on the sextic curve are those for which $1 + u^2$ or $1 + v^2$, or each of these, is $= 0$; viz. the values of u, v for the six points are

$$u = i + \epsilon,\quad -i - \epsilon,\quad i\quad,\quad -i\quad,\quad -i\left(\frac{A-C}{B-D}\right),\quad i\left(\frac{A-C}{B-D}\right);$$

$$v = i + \zeta,\quad -i - \zeta,\quad -i\frac{A-B}{C-D},\quad i\frac{A-B}{C-D},\quad i\quad,\quad -i,$$

where, instead of $u = v = i$ and $u = v = -i$, I have written down the consecutive values of u, v, and as before ϵ, ζ are infinitesimals such that $a\epsilon + b\zeta = 0$.

Suppose that the coordinates x, y of a point on the sextic curve are

$$x = \mathrm{a}\cos\alpha + \mathrm{c}\sin\alpha + \mathrm{b}\cos\beta + \mathrm{d}\sin\beta,$$
$$y = \mathrm{a}'\cos\alpha + \mathrm{c}'\sin\alpha + \mathrm{b}'\cos\beta + \mathrm{d}'\sin\beta;$$

then, if $u = \pm(i+\epsilon)$, $\cos\alpha = \frac{-i}{\epsilon}$, $\sin\alpha = \frac{\pm 1}{\epsilon}$, and similarly if $v = \pm(i+\zeta)$, then $\cos\beta = -\frac{i}{\zeta}$, $\sin\beta = \frac{\pm 1}{\zeta}$. Hence the points at infinity of the sextic curve are as follows:

1°. $u = \pm(i+\epsilon)$, v not $= \pm(i+\zeta)$,

$$x = \frac{-\mathrm{a}i \pm \mathrm{c}}{\epsilon},\quad y = \frac{-\mathrm{a}'i \pm \mathrm{c}'}{\epsilon},$$ first pair of points;

2°. $v = \pm(i+\zeta)$, u not $= \pm(i+\epsilon)$,

$$x = \frac{-\mathrm{b}i \pm \mathrm{d}}{\zeta},\quad y = \frac{-\mathrm{b}'i \pm \mathrm{d}'}{\zeta},$$ second pair of points;

3°. $u = \pm(i+\epsilon)$, $v = \pm(i+\zeta)$,

$$x = \frac{-\mathrm{a}i \pm \mathrm{c}}{\epsilon} + \frac{-\mathrm{b}i \pm \mathrm{d}}{\zeta},\quad y = -\frac{\mathrm{a}'i \pm \mathrm{c}'}{\epsilon} + \frac{-\mathrm{b}'i \pm \mathrm{d}}{\zeta};$$

$a\epsilon + b\zeta = 0$, as above, third pair of points;

which six points are in general distinct from each other, and from the circular points at infinity.

The foregoing values of x, y may be said to be "circular *quoad* α," if $\mathrm{a} = \mathrm{c}'$, $\mathrm{a}' = -\mathrm{c}$; and similarly to be "circular *quoad* β," if $\mathrm{b} = \mathrm{d}'$, $\mathrm{b}' = -\mathrm{d}$.

And we see at once that if the values are circular *quoad* α, then the first pair of points coincide with the circular points at infinity; and that, in like manner, if the values are circular *quoad* β, then the second pair of points coincide with the circular points at infinity; but if the values are circular *quoad* α and β respectively, then each of the three pairs of points coincides with the circular points at infinity: so that these are then triple points on the curve; or the curve is tricircular, having besides the two triple points, 3 dps.

The relation between u, v gives

$$\{v(C+Du^2) + 4abu\}^2 = 16a^2b^2 \,.\, u^2 - (A+Bu^2)(C+Du^2),$$

and it thus appears that if any one of the functions A, B, C, D is $=0$, the function under the radical sign is a mere linear function of u^2, say it is $L + Mu^2$; introducing a new parameter θ such that $u = \sqrt{\left(\frac{L}{M}\right)}\frac{2\theta}{1+\theta^2}$, we have $\sqrt{L+Mu^2} = \sqrt{L}\frac{1+\theta^2}{1-\theta^2}$, and consequently u, v are each of them a rational function of θ. Hence, when any one of the relations in question is satisfied, or say, when $a+d = b+c$, $b+d = c+a$, or $c+d = a+b$, the curve becomes unicursal: there is no diminution of the order, and the curve is consequently a unicursal sextic, or sextic with 10 dps.

It would at first sight appear that the curve might become unicursal in a different manner; viz. it would be unicursal if

$$16a^2b^2u^2 - (A + Bu^2)(C + Du^2)$$

was a perfect square; but this is only the case when one of the four sides a, b, c, d is $= 0$. The condition in fact is

$$AD + BC - 16a^2b^2 = 2\sqrt{ABCD};$$

that is

$$\lambda\lambda'\mu\mu' - \nu\nu'\rho'\sigma - 16a^2b^2 = 2\sqrt{-\lambda\lambda'\mu\mu'\rho'\sigma};$$

where, putting for shortness $M = d^2 - a^2 - b^2 - c^2$, we have

$$\begin{aligned} \lambda\lambda' &= M - 2bc + 2ca + 2ab, \\ \mu\mu' &= M + 2bc - 2ca + 2ab, \\ \nu\nu' &= M + 2bc + 2ca - 2ab, \\ -\rho'\sigma &= M - 2bc - 2ca - 2ab; \end{aligned}$$

and thence

$$\begin{aligned} \lambda\lambda'\mu\mu' &= M^2 - 4b^2c^2 - 4c^2a^2 + 4a^2b^2 + 4abM + 8c^2ab, \\ -\nu\nu'\rho'\sigma &= M^2 - 4b^2c^2 - 4c^2a^2 + 4a^2b^2 - 4abM - 8c^2ab, \end{aligned}$$

and the equation thus becomes

$$M^2 - 4b^2c^2 - 4c^2a^2 - 4a^2b^2 = \sqrt{(M^2 - 4b^2c^2 - 4c^2a^2 + 4a^2b^2)^2 - 16a^2b^2(M + 2c^2)^2},$$

viz. putting for a moment $X = M^2 - 4c^2(a^2 + b^2)$, this is

$$(X - 4a^2b^2)^2 = (X + 4a^2b^2)^2 - 16a^2b^2(M + 2c^2)^2,$$

that is

$$16a^2b^2\{X - (M + 2c^2)^2\} = 0;$$

or, substituting for X its value, the equation is

$$64a^2b^2c^2(M + a^2 + b^2 + c^2) = 0,$$

that is $a^2b^2c^2d^2 = 0$.

We may have simultaneously 1°, $a = d$, $b = c$; 2°, $b = d$, $a = c$; 3°, $c = d$, $a = b$; the three cases are really equivalent, but the results present themselves in different forms.

1°. Here $A = 0$, $B = 4a(a - b)$, $C = 0$, $D = 4a(a + b)$; the relation between u, v contains the factor u, and throwing this out, and also the constant factor $4a$, it is

$$u[(a - b) + (a + b)v^2] + 2bv = 0,$$

viz. u is given as a rational function of v.

2°. Here $A=0$, $B=0$, $C=4b(b-a)$, $D=4b(b+a)$; the equation contains the factor v, and throwing out this and also the constant factor $4b$, the equation is

$$v[(b-a)+(b+a)u^2]+2au=0,$$

viz. v is given as a rational function of u.

3°. Here $A=4a(a-c)$, $B=0$, $C=0$, $D=4a(a+c)$; or, dividing by $4a$, the equation is

$$(a-c)+2a\,uv+(a+c)\,u^2v^2=0;$$

viz. this is

$$(uv+1)[(a+c)\,uv+a-c]=0,$$

which may be reduced to

$$(a+c)\,uv+a-c=0,$$

giving u or v each a rational function of the other.

I do not discuss the theory in detail, but only remark that in each case there is a conic thrown off, and that in place of the sextic we have a unicursal (or trinodal) quartic curve.

505.

ON THE SURFACES DIVISIBLE INTO SQUARES BY THEIR CURVES OF CURVATURE.

[From the *Proceedings of the London Mathematical Society*, vol. IV. (1871—1873), pp. 120, 121. Read June 13, 1872.]

PROFESSOR CAYLEY gave an account of an investigation recently communicated by him to the Academy of Sciences at Paris. The fundamental theorem is that, if the coordinates x, y, z of a point on a surface are expressed as functions of two parameters p, q (such expressions, of course replacing the equation of the surface); and if these parameters are such that $p=\text{const.}$, $q=\text{const.}$ are the equations of the two sets of curves of curvature respectively; then (writing for shortness

$$\frac{dx}{dp}=x_1,\quad \frac{dx}{dq}=x_2,\quad \frac{d^2x}{dp^2}=x_3,\quad \frac{d^2x}{dpdq}=x_4,\quad \frac{d^2x}{dq^2}=x_5,$$

and the like for y, z), the coordinates x, y, z, considered always as functions of p, q, satisfy the equations

$$x_1x_2+y_1y_2+z_1z_2=0,$$

$$\begin{vmatrix} x_1, & y_1, & z_1 \\ x_2, & y_2, & z_2 \\ x_4, & y_4, & z_4 \end{vmatrix}=0.$$

The last equation is equivalent to

$$x_4+Ax_1+Bx_2=0,$$

$$y_4+Ay_1+By_2=0,$$

$$z_4+Az_1+Bz_2=0;$$

and if in the notation of Gauss we write

$$x_1^2 + y_1^2 + z_1^2 = E,$$

$$x_2^2 + y_2^2 + z_2^2 = G,$$

then adding the equations multiplied by x_1, y_1, z_1 respectively, and also adding the equations multiplied by x_2, y_2, z_2 respectively, we find

$$A = -\tfrac{1}{2}\frac{1}{E}\frac{dE}{dq}, \quad B = -\tfrac{1}{2}\frac{1}{G}\frac{dG}{dq}$$

and the equations thus become

$$2x_4 - \frac{1}{E}\frac{dE}{dq}x_1 - \frac{1}{G}\frac{dG}{dq}x_2 = 0,$$

$$\&c. \quad \&c. \quad \&c.,$$

which, in fact, agree with the equations (10 bis) in Lamé's "Leçons sur les coordonnées curvilignes," Paris (1859), p. 89. The surface will be divisible into squares if only $E : G$ is the quotient of a function of p by a function of q, or say if

$$E = \Theta P, \quad G = \Theta Q,$$

where Θ is any function of (p, q), but P and Q are functions of p and q respectively; we then have

$$\frac{1}{E}\frac{dE}{dq} = \frac{1}{\Theta}\frac{d\Theta}{dq}, \quad \frac{1}{G}\frac{dG}{dp} = \frac{1}{\Theta}\frac{d\Theta}{dp},$$

and the equations for x, y, z are

$$2x_4 - \frac{1}{\Theta}\frac{d\Theta}{dq}x_1 - \frac{1}{\Theta}\frac{d\Theta}{dp}x_2 = 0,$$

$$\&c. \quad \&c. \quad \&c.,$$

viz. x, y, z being functions of p, q such that $x_1x_2 + y_1y_2 + z_1z_2 = 0$, and which besides satisfy these equations, or say which each of them satisfy the equation

$$2u_4 - \frac{1}{\Theta}\frac{d\Theta}{dq}u_1 - \frac{1}{\Theta}\frac{d\Theta}{dp}u_2 = 0,$$

then the values of x, y, z in terms of (p, q) determine a surface which has the property in question.

506.

ON THE MECHANICAL DESCRIPTION OF A CUBIC CURVE.

[From the *Proceedings of the London Mathematical Society*, vol. IV. (1871—1873), pp. 175—178. Read November 14, 1872.]

IF the coordinates x, y of a point on a curve are rational functions of $\sin\phi$, $\cos\phi$, $\sqrt{1-k^2\sin^2\phi}$, the curve has the deficiency 1, and conversely in any curve of deficiency 1 the coordinates x, y can be thus expressed in terms of the parameter ϕ. Hence writing $\sin\theta = k\sin\phi$, the coordinates will be rational functions of $\sin\phi$, $\cos\phi$, $\cos\theta$, or say of $\sin\phi$, $\cos\phi$, $\sin\theta$, $\cos\theta$; and for the mechanical representation of the relation $k\sin\phi=\sin\theta$, we require only a rod OA rotating about the fixed point O, and connected with it by a pin at A, a rod AB, the other extremity of which, B, moves in a fixed line Ox. The curve most readily obtained by such an arrangement is that described by a point C rigidly connected with the rod AB; this is however a quartic curve (with two dps., since its deficiency is $=1$). I first considered the cubic curve

$$xy-1 = \sqrt{(1-x^2)(1-k^2x^2)},$$

or say

$$xy-1 = -\sqrt{(1-x^2)(1-k^2x^2)};$$

writing herein $x=\sin\phi$, and as before $k\sin\phi=\sin\theta$, we have then $y\sin\phi = 1-\cos\theta\cos\phi$; which values may be written

$$x=\sin\phi,$$
$$y=\frac{1-\cos(\theta+\phi)}{\sin\phi}-\sin\theta.$$

I found, however, that this was *not* the cubic curve most easily constructed; and I ultimately devised a mechanical arrangement consisting of

1. Rod OH, and connected with it by a pin at H, rod HI [1].

[1] There was a mechanical convenience in this, but observe that producing OH to meet IP in I', the single straight rod OHI' might have been made use of.

2 Square ACD, and connected with it by a pin at D, rod DG.

3. Square ECF; the two squares being connected by a pin at C.

4. Rod IJ.

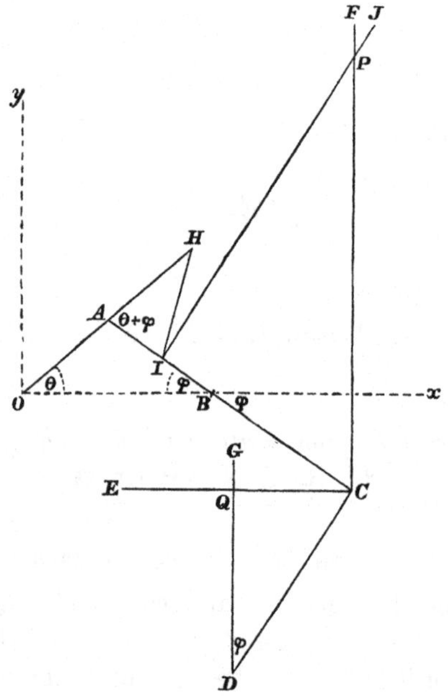

The rod OH rotates about a pin at O; taking $HA = HI$, there is a pin at A connecting a fixed point of this rod with the extremity A of the square ACD: the fixed point B of this square moves along the line Ox. There is a pin at I connecting the extremities of the rods HI, IJ; and this slides along the leg AC of the square ACD, the rod IJ being always at right angles thereto: finally the legs of the square ECF are always parallel to Ox, Oy, and the rod DG at right angles to EC. I have omitted from the description the parallel-motion rods or other arrangements necessary for giving these fixed directions to the rod IJ, the square ECF, and the rod DG. It will be seen that the angles AOB, ABO are variable angles connected by an equation of the form above referred to; and that the lines IJ, CF determine by their intersection the point P; and the lines CE, DG determine by their intersection the point Q; the curve about to be considered is that determined by the relative motion of P in regard to Q; or say the curve the coordinates of a point of which are

$$x = QC, \quad y = CP.$$

I write

$$\angle AOB = \theta, \quad \angle ABO = \phi,$$

$$OA = a, \quad AB = b, \quad AC = c, \quad CD = d,$$

$$AH = HI = \tfrac{1}{2}h.$$

We then have $a \sin \theta = b \sin \phi$; and moreover the length AI being $= h \cos (\theta + \phi)$, and therefore $IC = c - h \cos (\theta + \phi)$, we have

$$y = CP = \frac{c - h \cos (\theta + \phi)}{\sin \phi},$$

$$x = QC = d \sin \phi ;$$

whence also

$$xy = d \{c - h \cos (\theta + \phi)\} ;$$

or we have

$$xy = d \left\{c - h \sqrt{\left(1 - \frac{x^2}{d^2}\right)} \sqrt{\left(1 - \frac{b^2}{a^2} \frac{x^2}{d^2}\right)} + h \frac{b}{a} \frac{x^2}{d^2}\right\},$$

that is

$$x \left(y - \frac{bh}{ad} x\right) - cd = - dh \sqrt{\left(1 - \frac{x^2}{d^2}\right)} \sqrt{\left(1 - \frac{b^2}{a^2} \frac{x^2}{d^2}\right)} :$$

or rationalising and reducing, this is

$$x^2 y^2 - \frac{2bh}{ad} x^3 y - 2cdxy + \left\{2 \frac{b}{a} ch + h^2 \left(1 + \frac{b^2}{a^2}\right)\right\} x^2 + d^2 (c^2 - h^2) = 0,$$

a quartic curve with two dps.

In the particular case $a = b$, the relation between θ, ϕ is simply $\theta = \phi$; the curve should become unicursal.

Writing in the equation $\frac{b}{a} = 1$, the equation takes the form

$$\left\{x \left(y - \frac{2h}{d} x\right) - d (c - h)\right\} \{xy - d (c + h)\} = 0 ;$$

the second factor is extraneous, and the curve is the hyperbola

$$x \left(y - \frac{2h}{d} x\right) - d (c - h) = 0,$$

as at once appears from the foregoing irrational form of the equation.

In the particular case $h = c$, the equation contains the factor x, and omitting this it becomes

$$x \left(y^2 - \frac{2bc}{ad} x^2\right) - 2cdy + c \left(1 + \frac{b}{a}\right)^2 x = 0 ;$$

viz. we have here a cubic curve with three real asymptotes meeting in a point which is also the centre of the curve.

If simultaneously $a = b$ and $h = c$, then the equation is

$$x \left(y - \frac{2c}{d} x\right) (xy - 2cd) = 0,$$

the actual locus being in this case the line $y - \frac{2c}{d} x = 0$.

Writing $h=c$, and for greater convenience $h=c=d=1$; also to fix the ideas supposing $b<a$, and writing $\frac{b}{a}=k$, $=\sin\lambda$, then we have

$$\sin\theta = k\sin\phi,$$

$$x = \sin\phi,$$

$$y=\frac{1-\cos(\theta+\phi)}{\sin\phi};$$

that is

$$xy = 1-\sqrt{1-x^2}\sqrt{1-k^2x^2}+kx^2,$$

giving the rationalised equation

$$x(y^2-2kx^2)-2y+4x=0;$$

the angle ϕ may be anything whatever, but θ varies between the limits $\pm\lambda$, the simultaneous values of these angles and of the coordinates being

$\phi=0$	$\theta=0$	$x=0$	$y=0$
$\phi=90^\circ$	$\theta=\lambda$	$x=1$	$y=1+\sin\lambda$
$\phi=180^\circ$	$\theta=0$	$x=0$	$y=\pm\infty$
$\phi=270^\circ$	$\theta=-\lambda$	$x=-1$	$y=-(1+\sin\lambda)$
$\phi=360^\circ$	$\theta=0$	$x=0$	$y=0$;

and it thus appears that the mechanism gives the continuous branch which belongs to the asymptote $x=0$ of the cubic curve; the other two branches belong to $x=\sin\phi$, $y=\frac{1+\cos(\theta+\phi)}{\sin\phi}$, which would require a slight alteration in the arrangement of the mechanism.

I remark that if AH, HI had been unequal, then writing $\angle HIA=\chi$, this would be connected with $\theta+\phi$ by an equation of the form

$$\sin(\theta+\phi)=m\sin\chi,$$

and the coordinates x, y would be rational functions of the sines and cosines of θ, ϕ, χ; the deficiency is in this case >1.

507.

ON THE MECHANICAL DESCRIPTION OF CERTAIN QUARTIC CURVES BY A MODIFIED OVAL CHUCK.

[From the *Proceedings of the London Mathematical Society*, vol. IV. (1871—1873), pp. 186—190. Read December 12, 1872.]

THE geometrical principle of the oval chuck is the well-known one that if a plane move in such manner that two lines Ox, Oy, fixed in the plane and moveable with it, pass through two fixed points A, B, respectively, then any fixed point P traces out on the plane an ellipse. The point A is on the (geometrical) axis of the mandril; there is connected with the head a guide-ring moving horizontally; the point B is the centre of the guide-ring, this being a ring connected with the head, moveable horizontally at right angles to the axis in such wise that the distance AB of the two centres is adjustable to any given value; the fixed point P is the tool, which practically is held on the level of the axis, that is, at a point in the line AB. The guide-ring remains fixed during the motion of the lathe.

It occurred to me that a chuck applicable to ornamental turning might be constructed by giving to the guide-ring a reciprocating motion synchronous with the rotation of the mandril; viz. for this purpose it is only necessary to affix to the axis of the mandril an eccentric, working in a frame attached to the guide-ring so as to move the centre B of the guide-ring backwards and forwards along the line AB: the curve is thus that described by the fixed point P upon a plane moving in such manner that the lines Ox, Oy pass always through the points A, B respectively; the former of these being a fixed point, the latter of them a point moving according to determinate law backwards and forwards along a fixed line through A.

The plan is carried out in a drawing apparatus which I have had constructed in wood, the axis being here vertical instead of horizontal, and the details of course different from what they would be for a lathe.

The apparatus consists of a piece of inch-board, about 10 inches long by 7 inches broad, pierced with a circular hole of 1 inch diameter for a vertical axis: the edges of the board serve as guides for the frame *L*, which carries the guide-ring, and resting on the board we have the frame *M*, itself guided by the frame *L*: the two frames move independently of each other, but they can be clamped together; the axis has

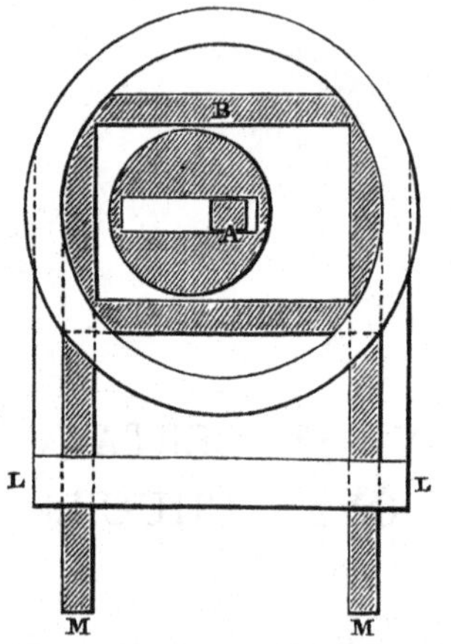

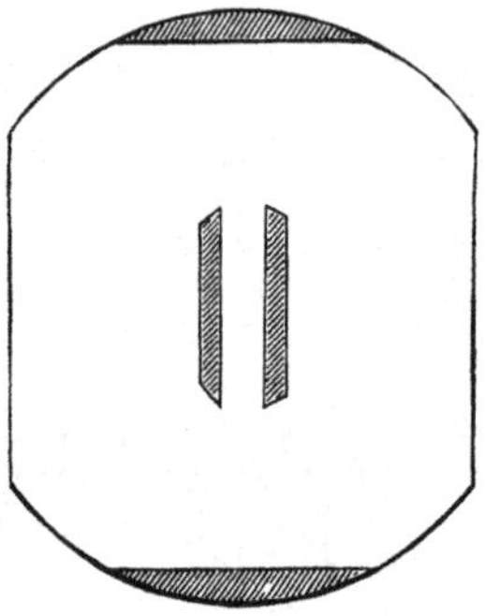

upon it a square nut, the sides of which work in the slot of an eccentric, the throw of this being adjustable by means of a screw passing through the nut and axis, and there is above the eccentric a square nut shown in the figure. This is capable of rotation round the axis, so that two of its sides may be placed either parallel with or inclined to the sides of the slot; but I fix it with two sides parallel to those of the slot by means of a screw run into the axis. The upper surfaces of the last-mentioned nut and of the guide-ring are flush with each other; and we then have a table or bed having, on its under-surface, guides which work on the outer edges of the guide-ring and on two edges of the nut. It will be observed that the bed may be placed in two different positions, viz. the guides may work on either pair of edges of the nut, those which are parallel to the sides of the slot, or those which are at right angles to it.

Supposing the bed placed as above upon the guide-ring and nut, then if the frames *L* and *M* are disconnected, and the former of them is fixed, the frame *M* will, on rotation of the axis, be carried backwards and forwards by the eccentric, but this will in no wise affect the motion of the bed; the arrangement is then equivalent to the oval chuck, and a pencil fixed above the bed in any given position will trace out upon it an ellipse. If, however, the frame *L*, instead of being fixed, is clamped to the frame *M*, then the two frames, and therefore the guide-ring, are carried backwards and forwards by the eccentric, and the curve traced out by the pencil is no longer an ellipse; it is, as I proceed to show, a special form of trinodal quartic; viz. there is a tacnode (= two nodes) at infinity, and a third node, which may be a crunode, cusp, or acnode. In the last-mentioned case, the acnode or conjugate point is, as usual, not exhibited by the mechanical description, and the curve has no visible singularity.

Let the coordinates of the fixed point P, referred to axes through A, the first of them perpendicular to, and the second coincident with, AB, be b, c; let the distance AB be $=a$; and let θ denote the angle BAO: then, if x, y are the coordinates of P referred to the origin O and axes Ox, Oy, we have

$$x + a\cos\theta = b\sin\theta + c\cos\theta,$$
$$y = -b\cos\theta + c\sin\theta,$$

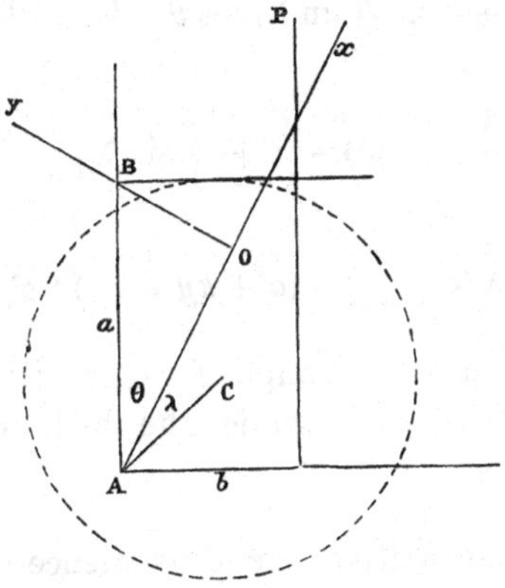

which, if a be constant, gives a quadric equation, or the curve is an ellipse; and, in particular, if $b=0$, that is if the point P is on the line AB, then we have

$$x = (c-a)\cos\theta,\quad y = c\sin\theta,$$

or the curve is

$$\frac{x^2}{(c-a)^2} + \frac{y^2}{c^2} = 1.$$

But if a is a given function of θ, then the equation is still found by eliminating θ between the two equations for x and y. In particular, if the distance AB is given as the perpendicular upon the tangent of a circle, as shown in the figure, then if k be the radius AC of this circle, and λ the inclination of AC to Ax (k and λ being taken to be constants), we have

$$a = k\cos(\theta+\lambda),$$

and the equations are

$$x = b\sin\theta + \{c - k\cos(\theta+\lambda)\}\cos\theta,$$
$$y = -b\cos\theta + c\sin\theta.$$

The elimination is nearly the same as if b were $=0$; viz. we may determine γ, α in such wise that

$$b\sin\theta + c\cos\theta = \gamma\cos(\theta+\alpha),\quad = \gamma\cos\phi \text{ suppose},$$
$$-b\cos\theta + c\sin\theta = \gamma\sin(\theta+\alpha),\quad = \gamma\sin\phi;$$

and then

$$x = \gamma \cos \phi - k \cos (\phi + \lambda - \alpha) \cos (\phi - \alpha),$$
$$y = \gamma \sin \phi.$$

I will, for greater simplicity, at once write $b = 0$: the equations thus are

$$x = (c - k \cos \theta \cos \lambda + k \sin \theta \sin \lambda) \cos \theta,$$
$$y = c \sin \theta;$$

or say the first is

$$x = (c + k \sin \theta \sin \lambda) \cos \theta - k \cos^2 \theta \cos \lambda;$$

whence we have

$$x + \frac{k \cos \lambda}{c^2} (c^2 - y^2) = \left(c + \frac{ky}{c} \sin \lambda\right) \frac{1}{c} \sqrt{c^2 - y^2},$$

that is

$$c^2 x + k \cos \lambda (c^2 - y^2) = (c^2 + ky \sin \lambda) \sqrt{c^2 - y^2},$$

an equation which will assume a more simple form if either $\lambda = 0$ or $\lambda = 90°$; that is, if in the apparatus the nut-sides which guide the bed are either at right angles, or parallel, to the sides of the slot.

Taking the general case, and writing for convenience $\cos \theta = \xi$, $\sin \theta = \eta$, the curve is given by equations of the form

$$x = (\xi, \eta, 1)^2,$$
$$y = (\xi, \eta, 1)^2,$$
$$\xi^2 + \eta^2 = 1;$$

viz. the elimination of ξ, η from these equations leads to the equation of the curve. The points of the curve have thus a (1, 1) correspondence with those of the circle $\xi^2 + \eta^2 = 1$; or, the circle being unicursal, the curve is also unicursal. Moreover, considering the intersections of the curve with an arbitrary line $ax + by + c = 0$, the points of intersection correspond to the points of intersection of the circle by the quadric $a(\xi, \eta, 1)^2 + b(\xi, \eta, 1)^2 + c = 0$; viz. there are four points of intersection, or the curve is a quartic, and hence it is a binodal quartic. But it is a binodal quartic of a special form: to show this more clearly, I introduce for homogeneity the coordinates z, ζ, so that the foregoing equations become

$$x : y : z = (\xi, \eta, \zeta)^2 : (\xi, \eta, \zeta)^2 : \zeta^2, \text{ where } \xi^2 + \eta^2 - \zeta^2 = 0;$$

the curve corresponding to these equations is, as just seen, a binodal quartic. But in the case in hand the form is the more special one,

$$x : y : z = (\xi, \eta, \zeta)^2 : \xi\zeta \quad : \zeta^2, \text{ where } \xi^2 + \eta^2 - \zeta^2 = 0.$$

The intersections by the arbitrary line $ax + by + cz = 0$ are the points corresponding to the intersections of the circle $\xi^2 + \eta^2 - \zeta^2 = 0$ by the quadric $a(\xi, \eta, \zeta)^2 + b\xi\zeta + c\zeta^2 = 0$, giving four intersections. But the intersections by the line $by+cz=0$ (that is, by any line through the point $y=0$, $z=0$) are obtained from the equation $b\xi\zeta+c\zeta^2=0$; viz. this breaks up into $b\xi+c\zeta=0$, $\zeta=0$, and the last factor combined with the equation of the circle gives $\zeta=0$, $\xi^2+\eta^2=0$, the two circular points at infinity, corresponding each to the point $y=0$, $z=0$: the other factor gives points corresponding to two variable points on the curve; that is, a line through the point $y=0$, $z=0$ meets the curve in this point twice and in two other points. Again, making $b=0$, or taking the line to be the line at infinity $z=0$, the equations then are $\zeta^2=0$, $\xi^2+\eta^2=0$; viz. we then have the circular points at infinity each twice, corresponding to the point $y=0$, $z=0$ four times, and no other point; that is, the line $z=0$ meets the curve in the point $y=0$, $z=0$ four times. We thus see that the curve has at $y=0$, $z=0$, that is at infinity on the line $y=0$, a tacnode (counting as two nodes), the tangent at this point being the line at infinity $z=0$. The curve being trinodal has of course one other node.

508.

ON GEODESIC LINES, IN PARTICULAR THOSE OF A QUADRIC SURFACE.

[From the *Proceedings of the London Mathematical Society*, vol. IV. (1871—1873), pp. 191—211. Read December 12, 1872.]

THE present Memoir contains an investigation of the differential equation (of the second order) of the geodesic lines on a surface, the coordinates of a point on the surface being regarded as given functions of two parameters p, q, and researches in connection therewith; a deduction of Jacobi's differential equation of the first order in the case of a quadric surface, the parameters p, q being those which determine the two sets of curves of curvature; formulæ where the parameters are those which determine the two right lines through the surface; and a discussion of the forms of the geodesic lines in the two cases of an ellipsoid and a skew hyperboloid respectively.

Preliminary Formulæ.

1. I call to mind the fundamental formulæ in the Memoir by Gauss, "Disquisitiones generales circa superficies curvas," *Comm. Gott. recent.* t. VI., 1827, (reprinted as an Appendix in Liouville's edition of Monge,) together with some that I have added to them. The coordinates x, y, z of a point on a surface are regarded as given functions of two parameters p, q, these expressions of x, y, z in effect determining the equation of the surface, and we have

$$dx + \tfrac{1}{2}d^2x = a\,dp + a'dq + \tfrac{1}{2}(\alpha\,dp^2 + 2\alpha'\,dp\,dq + \alpha''\,dq^2),$$
$$dy + \tfrac{1}{2}d^2y = b\,dp + b'dq + \tfrac{1}{2}(\beta\,dp^2 + 2\beta'dp\,dq + \beta''dq^2),$$
$$dz + \tfrac{1}{2}d^2z = c\,dp + c'dq + \tfrac{1}{2}(\gamma\,dp^2 + 2\gamma'dp\,dq + \gamma''dq^2),$$
$$A,\ B,\ C = bc' - b'c,\ ca' - c'a,\ ab' - a'b\,;$$

whence differential equation of surface is

$$Adx + Bdy + Cdz = 0.$$

Also

$$E,\ F,\ G = a^2 + b^2 + c^2,\quad aa' + bb' + cc',\quad a'^2 + b'^2 + c'^2;$$

so that element of length on the surface is given by

$$dx^2 + dy^2 + dz^2 = Edp^2 + 2Fdp\,dq + Gdq^2;$$

or, as I write it,

$$= (E,\ F,\ G \between dp,\ dq)^2;$$

and moreover

$$V^2 = A^2 + B^2 + C^2 = EG - F^2.$$

The equation $(E,\ F,\ G \between dp,\ dq)^2 = 0$ determines at each point on the surface two directions (necessarily imaginary) which are called the "circular" directions. Passing on the surface from point to point along the circular directions, we obtain two series of curves (always imaginary) which are the "circular" curves; the equation $(E,\ F,\ G \between dp,\ dq)^2 = 0$ is the differential equation of these curves; and if we have $E = 0$, $G = 0$, then this becomes $dp\,dq = 0$; viz. we have in this case $p = \text{const.}$ and $q = \text{const.}$ as the equations of the two sets of circular curves respectively. It is clear *à priori*, and will be shown analytically in the sequel, that the circular curves are geodesic lines.

I write also

$$E',\ F',\ G' = A\alpha + B\beta + C\gamma,\quad A\alpha' + B\beta' + C\gamma',\quad A\alpha'' + B\beta'' + C\gamma'',$$

or, what is the same thing, E', F', G' represent the determinants

$$\begin{vmatrix} a, & b, & c \\ a', & b', & c' \\ \alpha, & \beta, & \gamma \end{vmatrix},\quad \begin{vmatrix} a, & b, & c \\ a', & b', & c' \\ \alpha', & \beta', & \gamma' \end{vmatrix},\quad \begin{vmatrix} a, & b, & c \\ a', & b', & c' \\ \alpha'', & \beta'', & \gamma'' \end{vmatrix},\ \text{respectively,}$$

(these last symbols do not occur in Gauss). [They are the D, D', D'' of Gauss.]

2. The radius of curvature of normal section corresponding to direction $dp : dq$ is given by

$$\frac{\rho}{V} = \frac{(E,\ F,\ G \between dp,\ dq)^2}{(E',\ F',\ G' \between dp,\ dq)^2};$$

whence it appears that the directions of the inflexional or chief tangents (the *Haupt-tangenten*) are determined by the equation

$$(E',\ F',\ G' \between dp,\ dq)^2 = 0.$$

The directions in question are imaginary on a surface such as the ellipsoid where the curvatures are in the same direction, but on a concavo-convex surface they are real; and in particular on the hyperboloid they coincide with the directions of the generating lines. We may on any surface pass from point to point along the chief directions; we have thus on the surface two sets of curves which are the chief curves;

the differential equation of these is $(E', F', G'\between dp, dq)^2=0$; and in particular if $E'=0$, $G'=0$, then this becomes $dp\,dq=0$, or we have $p=$ const., $q=$ const. for the equations of the two sets of chief curves respectively. On the hyperboloid the chief curves are the two sets of generating lines. The chief curves are not in general geodesic lines, but on the hyperboloid, *quâ* straight lines, they are, it is clear, geodesic lines.

3. The directions of the curves of curvature, or the principal tangents, and the corresponding values of the radius of curvature are determined by

$$\frac{\rho}{V}=\frac{Edp+Fdq}{E'dp+F'dq}=\frac{Fdp+Gdq}{F'dp+G'dq};$$

or, what is the same thing, these directions are determined by the equation

$$\begin{vmatrix} dq^2, & -dq\,dp, & dp^2 \\ E\;, & F\;, & G \\ E', & F'\;, & G' \end{vmatrix}=0.$$

The same equations may be written

$$-dq : dp=\frac{\rho E'-VE}{\rho F'-VF}=\frac{\rho F'-VF}{\rho G'-VG};$$

that is the principal radii of curvature are determined by the equation

$$\rho^2(E'G'-F'^2)-\rho V(EG'+E'G-2FF')+V^2(EG-F^2)=0,$$

(last term is $=V^4$, but it is better to retain the original form): and then, ρ being either root, the last preceding equations give the direction of the curve of curvature corresponding to the given value of the radius of curvature.

If $p=$ const., $q=$ const. are the equations of the two systems of curves of curvature respectively, then the quadric equation in (dp, dq) must become $dp\,dq=0$; this will be so if $F=0$, $F'=0$; and we thus have these equations, viz. written at full length they are

$$d_px\,d_qx+d_py\,d_qy+d_pz\,d_qz=0,$$

$$\begin{vmatrix} d_px\;, & d_py\;, & d_pz \\ d_qx\;, & d_qy\;, & d_qz \\ d_pd_qx, & d_pd_qy, & d_pd_qz \end{vmatrix}=0,$$

as the conditions in order that $p=$ const., $q=$ const. may be the two systems of curves of curvature. The former of these equations merely expresses that the two sets of curves always intersect at right angles.

General Theory of the Geodesic Lines on a Surface.

4. I now proceed to investigate the theory of geodesic lines on a surface, the surface being determined as above by means of given expressions of the coordinates x, y, z in terms of the parameters p, q.

The differential equation obtained by Gauss for the geodesic lines is in a form not symmetrical in regard to the two variables; viz. his equation is

$$\frac{dE}{dq}dp^2 + 2\frac{dF}{dp}dp\,dq + \frac{dG}{dp}dp^2 = 2ds\,d\frac{Edp + Fdq}{ds},$$

where, as above,

$$ds^2 = (E,\ F,\ G\between dp,\ dq)^2.$$

If we herein consider p, q as functions of a parameter θ, and write for shortness

$$d_\theta p,\ d_\theta q,\ d_\theta^2 p,\ \&\text{c.} = p',\ q',\ p'',\ \&\text{c.},$$

also

$$\Omega = (E,\ F,\ G\between p',\ q')^2,$$

and

$$d_p E = E_1,\quad d_q E = E_2,\ \&\text{c.},$$

then the equation is

$$(E_1,\ F_1,\ G_1\between p',\ q')^2 - 2\sqrt{\Omega}\left(\frac{Ep' + Fq'}{\sqrt{\Omega}}\right)' = 0.$$

We have

$$\left(\frac{Ep' + Fq'}{\sqrt{\Omega}}\right)' = \frac{1}{\Omega\sqrt{\Omega}}(M + N),$$

where N is the part containing p'', q'', which I will first calculate; viz. we have

$$\begin{aligned} N &= \Omega(Ep'' + Fq'') - (Ep' + Fq')\tfrac{1}{2}\Omega', \\ &= \Omega(Ep'' + Fq'') - (Ep' + Fq')\{(Ep' + Fq')p'' + (Fp' + Gq')q''\}, \\ &= p''\{E\Omega - (Ep' + Fq')^2\} + q''\{F\Omega - (Ep' + Fq')(Fp' + Gq')\}; \end{aligned}$$

or substituting for Ω its value, this is

$$= p''(EG - F^2)q'^2 - q''(EG - F^2)p'q', \quad = -q'(EG - F^2)(p'q'' - p''q');$$

wherefore

$$\left(\frac{Ep' + Fq'}{\sqrt{\Omega}}\right)' = \frac{1}{\Omega\sqrt{\Omega}}\{M - q'(EG - F^2)(p'q'' - p''q')\},$$

and the equation becomes

$$\Omega(E_1 p'^2 + 2F_1 p'q' + G_1 q'^2) - 2M + 2q'(EG - F^2)(p'q'' - p''q') = 0;$$

whence we foresee that the whole equation must divide by q'.

5. We have

$$\begin{aligned} M = &\ (p'dE + q'dF)\,\Omega - (Ep' + Fq')\,(\tfrac{1}{2}p'^2dE + p'q'dF + \tfrac{1}{2}q'^2dG), \\ = &\ dE\,\{p'\Omega - \tfrac{1}{2}p'^2(Ep' + Fq')\} \\ & + dF\,\{q'\Omega - p'q'(Ep' + Fq')\} \\ & + dG\,\{\quad - \tfrac{1}{2}q'^2(Ep' + Fq')\}, \end{aligned}$$

or say

$$\begin{aligned} 2M = &\ dE\,[\ p'^2(p'E + q'F) + 2p'q'(p'F + q'G)] \\ & + dF\,[\qquad\qquad 2q'^2(p'F + q'G)] \\ & + dG\,[-q'^2(p'E + q'F)\qquad\qquad], \\ = &\ (E_1p' + E_2q')\,[\ p'^2(Ep' + Fq') + 2p'q'\,(Fp' + Gq')] \\ & + (F_1p' + F_2q')\,[\qquad\qquad 2q'^2(Fp' + Gq')] \\ & + (G_1p' + G_2q')\,[-q'^2(Ep' + Fq')\qquad\qquad]. \end{aligned}$$

The term in p'^4 is $EE_1p'^4$, which is also the term in p'^4 of $\Omega\,(E_1p'^2 + 2F_1p'q' + G_1q'^2)$; whence $\Omega\,(E_1p'^2 + 2F_1p'q' + G_1q'^2) - 2M$ divides by q'.

Proceeding to the reduction:

Term in E_1 is

$$E_1\,.\,p'^2\,\Omega - p'^3(Ep' + Fq') - 2p'^2q'\,(Fp' + Gq'),\ = E_1\,.-p'^2q'\,(Fp' + Gq');$$

term in F_1 is

$$F_1\,.\,2p'q'\Omega - 2p'q'^2(Fp' + Gq'),\ = F_1\,.\,2p'^2q'\,(Ep' + Fq');$$

term in G_1 is

$$G_1\,.\,q'^2\Omega + p'q'^2(Ep' + Fq'),\ = G_1\,\{2p'q'^2(Ep' + Fq') + q'^3(Fp' + Gq')\}.$$

6. The remaining terms in E_2, F_2, G_2 require no reduction, and the result is

$$\begin{aligned} & E_1\,\{-p'^2(Fp' + Gq')\} - E_2\,\{\ p'^2(Ep' + Fq') + 2p'q'\,(Fp' + Gq')\} \\ & + F_1\,\{2\,p'^2(Ep' + Fq')\} - F_2\,\{2q'^2(Fp' + Gq')\} \\ & + G_1\,\{2p'q'\,(Ep' + Fq') + q'^2(Fp' + Gq')\}\ - G_2\,\{-q'^2(Ep' + Fq')\} \\ & \quad + 2\,(EG - F^2)\,(p'q'' - p''q') = 0, \end{aligned}$$

or, what is the same thing,

$$\begin{aligned} & (Ep' + Fq')\,\{(2F_1 - E_2)\,p'^2 + 2G_1p'q' + G_2q'^2\} \\ & - (Fp' + Gq')\,\{E_1p'^2 + 2E_2p'q' + (2F_2 - G_1)\,q'^2\} \\ & \quad + 2\,(EG - F^2)\,(p'q'' - p''q') = 0, \end{aligned}$$

which is the required differential equation of the second order: the independent variable has been taken to be the arbitrary quantity θ; but taking $\theta = p$, or q, say $\theta = p$, we have $p' = 1$, $p'' = 0$, and the equation then contains (besides p and q) only q'' and q', that is, d_p^2q and d_pq, and is therefore a differential relation between p and q.

7. Instead of starting, as above, from the equation given by Gauss, we may use the geometrical property that at each point of a geodesic line the osculating plane passes through the normal of the surface.

Considering, as above, p, q as functions of a variable θ, then, θ becoming $\theta+d\theta$, the new values of p, q are

$$p+p'd\theta+\tfrac{1}{2}p''d\theta^2,\quad q+q'd\theta+\tfrac{1}{2}q''d\theta^2;$$

and that of x is

$$x+a\,(p'd\theta+\tfrac{1}{2}p''d\theta^2)+a'\,(q'd\theta+\tfrac{1}{2}q''d\theta^2)+\tfrac{1}{2}\,(\alpha p'^2+2\alpha'p'q'+\alpha''q'^2)\,d\theta^2;$$

or calling this $x+x'd\theta+\frac{1}{2}x''d\theta^2$, we have

$$x'=ap'+a'q',\quad x''=ap''+a'q''+\alpha p'^2+2\alpha'p'q'+\alpha''q'^2,$$

and so

$$y'=bp'+b'q',\quad y''=bp''+b'q''+\beta p'^2+2\beta'p'q'+\beta''q'^2,$$

$$z'=cp'+c'q',\quad z''=cp''+c'q''+\gamma p'^2+2\gamma'p'q'+\gamma''q'^2.$$

The condition in question is expressed by the equation

$$\begin{vmatrix} A, & B, & C \\ x', & y', & z' \\ x'', & y'', & z'' \end{vmatrix}=0,$$

that is

$$\begin{vmatrix} A, & B, & C \\ ap'+a'q', & bp'+b'q', & cp'+c'q' \\ ap''+a'q'', & bp''+b'q'', & cp''+c'q'' \end{vmatrix}$$

$$+\begin{vmatrix} A, & B, & C \\ ap'+a'q', & bp'+b'q', & cp'+c'q' \\ \alpha p'^2+2\alpha'p'q'+\alpha''q'^2, & \beta p'^2+2\beta'p'q'+\beta''q'^2, & \gamma p'^2+\gamma'p'q'+\gamma''q'^2 \end{vmatrix}=0.$$

8. The first determinant is the sum of three terms such as $A\,(bc'-b'c)\,(p'q''-p''q')$; viz. this is $A^2\,(p'q''-p''q')$, or the determinant is

$$(A^2+B^2+C^2)\,(p'q''-p''q'),\ =(EG-F^2)\,(p'q''-p''q').$$

The second determinant is the sum of three terms such as

$$(\alpha p'^2+2\alpha'p'q'+\alpha''q'^2)\,[B\,(cp'+c'q')-C\,(bp'+b'q')],$$

where the factor in [ ] is

$$p'\,[c\,(ca'-c'a)-b\,(ab'-a'b)]+q'\,[c'\,(ca'-c'a)-b'\,(ab'-a'b)],$$

which is $=p'(a'E-aF)+q'(a'F-aG)$. We have thus the second determinant, and the equation becomes

$$\begin{aligned}&(EG-F^2)(p'q''-p''q')\\&\quad+(\alpha p'^2+2\alpha' p'q'+\alpha'' q'^2)\{p'(a'E-aF)+q'(a'F-aG)\}\\&\quad+(\beta p'^2+2\beta' p'q'+\beta'' q'^2)\{p'(b'E-bF)+q'(b'F-bG)\}\\&\quad+(\gamma p'^2+2\gamma' p'q'+\gamma'' q'^2)\{p'(c'E-cF)+q'(c'F-cG)\}=0,\end{aligned}$$

an equation of the same form as that previously obtained, and which can be without difficulty identified therewith.

The Circular Curves are Geodesics.

9. I proceed to show that the circular curves are geodesics; viz. that an integral of the geodesic equation is

$$(E,\ F,\ G\between p',\ q')^2=0.$$

Starting from this equation, we have

$$2\{(Ep'+Fq')p''+(Fp'+Gq')q''\}+(E_1p'+E_2q')p'^2+2(F_1p'+F_2q')p'q'+(G_1p'+G_2q')q'^2=0.$$

Now the equation, writing therein $Ep'+Fq'=\lambda q'$, gives $Fp'+Gq'=-\lambda p'$: these equations may be written

$$\begin{aligned}Ep'+(F-\lambda)q'&=0,\\(F+\lambda)p'+G\qquad q'&=0,\end{aligned}$$

the value of λ being therefore $-\lambda^2=EG-F^2$. The result just obtained thus becomes

$$\begin{aligned}&2\lambda(p''q'-p'q'')\\&\quad+[(E_1p'+E_2q')p'+(F_1p'+F_2q')q'].-\frac{1}{\lambda}(Fp'+Gq')\\&\quad+[(F_1p'+F_2q')p'+(G_1p'+G_2q')q'].\ \ \frac{1}{\lambda}(Ep'+Fq'),\end{aligned}$$

that is

$$\begin{aligned}&2(EG-F^2)(p'q''-p''q')\\&\quad-(Fp'+Gq')[E_1p'^2+(E_2+F_1)p'q'+F_2q'^2]\\&\quad+(Ep'+Fq')[F_1p'^2+(F_2+G_1)p'q'+G_2q'^2]=0;\end{aligned}$$

or what is the same thing, adding hereto the zero value

$$\Delta(E,\ F,\ G\between p',\ q')^2,\ =(Fp'+Gq')\Delta q'+(Ep'+Fq')\Delta p',$$

where Δ is arbitrary, the equation is

$$\begin{aligned}&2(EG-F^2)(p'q''-p''q')\\&\quad+(Ep'+Fq')[F_1p'^2+(F_2+G_1)p'q'+G_2q'^2+\Delta p']\\&\quad-(Fp'+Gq')[E_1p'^2+(E_2+F_1)p'q'+F_2q'^2-\Delta q']=0;\end{aligned}$$

viz. taking $\Delta=(F_1-E_2)p'+(G_1-F_2)q'$, this agrees with the geodesic equation.

The foregoing integral, $(E,\ F,\ G\between p',\ q')^2=0$, is, I believe, a particular, not a singular, solution of the geodesic equation.

The Chief Lines are not in general Geodesics.

10. That the chief lines are not in general geodesics appears most readily as follows:

To find the condition in order that $p=\text{const.}$ may be a geodesic, we write in the geodesic equation $p'=0$, $p''=0$: the equation thus becomes

$$Fq' \,.\, G_2q'^2 - Gq'(2F_2 - G_1)q'^2 = 0;$$

that is we have

$$FG_2 - 2F_2G + GG_1 = 0$$

as the condition in order that $p=\text{const.}$ may be a geodesic: the condition that it may be a chief curve is $G'=0$, which is a different condition.

We have of course, in like manner,

$$2EF_1 - E_1F - EE_2 = 0$$

as the condition in order that $q=\text{const.}$ may be a geodesic; and $E'=0$ as the condition that this may be a chief curve. If $p=\text{const.}$, $q=\text{const.}$ are each of them at once a geodesic and a chief curve, then the four equations must all be satisfied, viz. we must have

$$FG_2 - 2F_2G + GG_1 = 0,\quad 2EF_1 - E_1F - EE_2 = 0,$$
$$G' = 0,\qquad E' = 0.$$

Special Form of the Geodesic Equation.

11. In the case where the curves $p=\text{const.}$, $q=\text{const.}$ intersect at right angles (and in particular when these are the curves of curvature), we have $F=0$; whence also $F_1=0$, $F_2=0$; and the geodesic equation assumes the more simple form,

$$\begin{aligned}&Ep'\;(-E_2p'^2 + 2G_1p'q' + G_2q'^2)\\ &- Gq'\;(\;\;E_1p'^2 + 2E_2p'q' - G_1q'^2)\\ &+ 2EG(p'q'' - p''q') = 0.\end{aligned}$$

[11[a]. In the case of a surface of revolution we have

$$x = p\cos q,\quad y = p\sin q,\quad z = p;$$

E is of the form $1+P'^2$, $P'=d_pP$, where P is a function of p only, and we have

$$ds^2 = (1+P'^2)\,dp^2 + p^2dq^2,$$

that is

$$\begin{aligned}&E = 1 + P'^2,\quad E_1 = 2P'P'',\quad E_2 = 0,\\ &F = 0,\\ &G = p^2\quad,\quad G_1 = 2p\quad,\quad G_2 = 0;\end{aligned}$$

hence the differential equation is

$$(1+P'^2)\{p^2(p'q''-p''q')+2pp'^2q'\}-p^2p'^2q'P'P''+p^3q'^3=0;$$

this has an integral

$$\frac{(1+P'^2)p'^2}{p^4q'^2}+\frac{1}{p^2}=\frac{1}{C^2},$$

or say

$$C^2s'^2=p^4q'^2$$

where

$$s'^2=(1+P'^2)p'^2+p^2q'^2.$$

Writing here ρ, ψ for p, q, where ρ is the distance of the point from the axis, and ψ is the longitude reckoned from an arbitrary meridian, then the equation is

$$Cds=\rho^2d\psi,$$

which is the equation given by Legendre, *Théorie des fonctions elliptiques*, t. I. p. 361. This may also be written

$$\frac{C}{\rho}=\cos\gamma$$

if γ be the inclination of the geodesic line to the parallel of latitude.]

Geodesics on a Quadric Surface.

12. In the case of a quadric surface $\frac{x^2}{a}+\frac{y^2}{b}+\frac{z^2}{c}=1$, writing for shortness $\alpha, \beta, \gamma = b-c, c-a, a-b$, we may express the coordinates x, y, z in terms of two parameters p, q as follows:

$$-\beta\gamma\, x^2=a\,(a+p)\,(a+q),$$
$$-\gamma\alpha\, y^2=b\,(b+p)\,(b+q),$$
$$-\alpha\beta\, z^2=c\,(c+p)\,(c+q),$$

where, in fact, $p=$ const., $q=$ const. are the equations of the two sets of curves of curvature respectively. Writing moreover

$$P=\frac{p}{(a+p)(b+p)(c+p)},\quad Q=\frac{q}{(a+q)(b+q)(c+q)},$$

we have

$$ds^2=\tfrac{1}{4}(p-q)(Pdp^2-Qdq^2),$$

that is

$$E,\ F,\ G=\tfrac{1}{4}(p-q)P,\ 0,\ \tfrac{1}{4}(q-p)Q;$$

and the geodesic equation becomes

$$\begin{aligned}&Pp'\,\{Pp'^2-2Qp'q'+(Q+\overline{q-p}\,Q')\,q'^2\}\\ &+Qq'\,\{(P+\overline{p-q}\,P')\,p'^2-2Pp'q'+Qq'^2\}\\ &-\quad 2\,(p-q)\,PQ\,(p'q''-p''q')=0,\end{aligned}$$

where P', Q' stand for d_pP and d_qQ respectively; viz. this is

$$\begin{aligned}&p'^3 \,.\; P^2\\ &+p'^2q' \,.\, -PQ+(p-q)\,P'Q\\ &+p'q'^2 \,.\, -PQ+(q-p)\,PQ'\\ &+\;q'^3 \,.\; Q^2\\ &-2\,(p-q)\,PQ\,(p'q''-p''q')=0.\end{aligned}$$

13. This has a first integral,

$$p'\sqrt{\left(\frac{P}{\theta+p}\right)}+q'\sqrt{\left(\frac{Q}{\theta+q}\right)}=0,$$

where θ is the constant of integration; or say this is

$$\theta\,(p'^2P-q'^2Q)+p'^2qP-q'^2pQ=0\,;$$

viz. differentiating logarithmically, this gives

$$\frac{2p'p''P-2q'q''Q+p'^3P'-q'^3Q'}{p'^2P-q'^2Q}=\frac{2p'p''qP-2q'q''pQ+p'^2\,(qp'P'+q'P)-q'^2\,(pq'Q'+p'Q)}{p'^2qP-q'^2pQ},$$

which, multiplying out the denominators, is in fact the foregoing geodesic equation. To verify, consider first the part involving p'', q'': this is

$$(2p'p''P-2q'q''Q)\,(p'^2qP-q'^2pQ)-(2p'p''qP-2q'q''pQ)\,(p'^2P-q'^2Q),$$

which is

$$=2p'p''P\,.\,q'^2Q\,(q-p)-2q'q''Q\,.\,p'^2P\,(q-p),$$

that is

$$=2\,(q-p)\,PQp'q'\,(p''q'-p'q''),$$

or say

$$=2\,(p-q)\,PQp'q'\,(p'q''-p''q').$$

We have next the part

$$(p'^3P'-q'^3Q')\,(p'^2qP-q'^2pQ)-\{p'^2\,(qp'P'+q'P)+q'^2\,(pq'Q'+p''Q)\}\,(p'^2P-q'^2Q),$$

which is readily found to be

$$=-p'q'\,\{p'^3P^2+p'^2q'\,(-PQ+\overline{p-q}P'Q)+p'q'^2\,(-PQ+\overline{q-p}PQ')+q'^3Q^2\}$$

and the equation is thus verified.

14. We have consequently

$$dp\sqrt{\left(\frac{p}{(a+p)\,(b+p)\,(c+p)\,(\theta+p)}\right)}+dq\sqrt{\left(\frac{q}{(a+q)\,(b+q)\,(c+q)\,(\theta+q)}\right)}=0,$$

involving the arbitrary constant θ as the differential equation of the first order of the geodesic lines on the quadric surface $\frac{x^2}{a}+\frac{y^2}{b}+\frac{z^2}{c}=1$: the geodesic lines in question all touch the curve of curvature determined by the parameter θ, that is the curve which is the intersection of the surface by the confocal surface

$$\frac{x^2}{a+\theta}+\frac{y^2}{b+\theta}+\frac{z^2}{c+\theta}=1.$$

15. In the particular case $\theta=\infty$, the equation becomes

$$Pdp^2-Qdq^2=0,$$

that is

$$\frac{pdp^2}{(a+p)(b+p)(c+p)}-\frac{qdq^2}{(a+q)(b+q)(c+q)}=0,$$

which is the differential equation of the circular curves on the surface.

16. The signification of the case $\theta=0$ is not at first sight so obvious. Supposing that θ is first indefinitely small, and writing the equation in the form

$$\frac{x^2}{a}+\frac{y^2}{b}+\frac{z^2}{c}-1-\theta\left(\frac{x^2}{a^2}+\frac{y^2}{b^2}+\frac{z^2}{c^2}\right)+\&\text{c.}=0,$$

we have the series of geodesics touching the (imaginary) curve of curvature, the intersection of the surface by the imaginary cone $\frac{x^2}{a^2}+\frac{y^2}{b^2}+\frac{z^2}{c^2}=0$. *These are, in fact, the right lines on the surface:* I apprehend that the intersection in question is not a proper envelope, but is the locus of nodes of the geodesics, viz. each geodesic is to be considered as a pair of lines belonging to the two sets: I do not, however, quite understand this.

17. I say that the geodesics in question are the right lines on the surface; viz. writing in the differential equation $\theta=0$, it is to be shown that the differential equation of the right lines is

$$\frac{dp}{\sqrt{(a+p)(b+p)(c+p)}}+\frac{dq}{\sqrt{(a+q)(b+q)(c+q)}}=0,$$

or what is the same thing, that the integral of this equation represents the right lines on the surface.

Writing the equation of the surface in the form

$$\frac{x^2}{a}+\frac{y^2}{b}=1-\frac{z^2}{c},$$

we have at once

$$\frac{x}{\sqrt{a}}+\frac{iy}{\sqrt{b}}=\sigma\left(1+\frac{z}{\sqrt{c}}\right),$$

$$\frac{x}{\sqrt{a}}-\frac{iy}{\sqrt{b}}=\frac{1}{\sigma}\left(1-\frac{z}{\sqrt{c}}\right),$$

(σ an arbitrary parameter) as the equations of a right line on the surface; viz. considering x, y, z as denoting the foregoing functions of p and q, these two equations are forms of a single relation between p, q, σ, which relation expresses that the point (p, q) is situate on the right line determined by the parameter σ. We may from this integral equation deduce without difficulty the foregoing differential equation; viz. we have

$$\frac{dx}{\sqrt{a}} + \frac{idy}{\sqrt{b}} = \sigma \frac{dz}{\sqrt{c}},$$

$$\frac{dx}{\sqrt{a}} - \frac{idy}{\sqrt{b}} = -\frac{1}{\sigma}\frac{dz}{\sqrt{c}},$$

or multiplying these equations,

$$\frac{dx^2}{a} + \frac{dy^2}{b} + \frac{dz^2}{c} = 0;$$

and substituting herein for dx, dy, dz their values in terms of dp and dq, we find the required equation

$$\frac{dp^2}{(a+p)(b+p)(c+p)} - \frac{dq^2}{(a+q)(b+q)(c+q)} = 0.$$

18. I return to the integral equation involving σ: we have to rationalise this equation, that is, obtain from it an equation containing x^2, y^2, z^2, and then substituting for these their values in terms of p, q, we have the required relation between p, q, σ. We at once obtain

$$\left\{2\left(\frac{x^2}{a} - \frac{y^2}{b}\right) - \left(\sigma^2 + \frac{1}{\sigma^2}\right)\left(1 + \frac{z^2}{c}\right)\right\}^2 - 4\left(\sigma^2 - \frac{1}{\sigma^2}\right)^2 \frac{z^2}{c} = 0;$$

or if for greater convenience we introduce in place of σ a new parameter ϕ, determined by the equation $\sigma^2 + \frac{1}{\sigma^2} = \frac{2\phi}{\gamma}$, the equation is

$$\left\{\gamma\left(\frac{x^2}{a} - \frac{y^2}{b}\right) - \phi\left(1 + \frac{z^2}{c}\right)\right\}^2 - 4(\phi^2 - \gamma^2)\frac{z^2}{c} = 0.$$

Writing for shortness $p + q = X$, $pq = Y$, we have

$$-\beta\gamma\frac{x^2}{a} = a^2 + aX + Y, \quad -\gamma\alpha\frac{y^2}{b} = b^2 + bX + Y, \quad -\alpha\beta\frac{z^2}{c} = c^2 + cX + Y;$$

and substituting these values, the equation becomes

$$\{\beta(b^2 + bX + Y) - \alpha(a^2 + aX + Y) - \phi\gamma(\alpha\beta - c^2 - cX - Y)\}^2 + 4\alpha\beta(\phi^2 - \gamma^2)(c^2 + cX + Y) = 0,$$

or, what is the same thing,

$$\{\beta b^2 - \alpha a^2 - \phi\gamma(\alpha\beta - c^2) + X(\beta b - \alpha a + \phi\gamma c) + Y(\beta - \alpha + \phi\gamma)\}^2 + 4\alpha\beta(\phi^2 - \gamma^2)(c^2 + cX + Y) = 0.$$

19. This is an equation, quadric as regards p, and also as regards q, viz. it is of the form

$$\begin{aligned}&(\mathrm{a} + 2\mathrm{h}p + \mathrm{g}p^2)\\ &+ 2q\,(\mathrm{h} + 2\mathrm{b}p + \mathrm{f}p^2)\\ &+ \;q^2\,(\mathrm{g} + 2\mathrm{f}p + \mathrm{c}p^2) = 0,\end{aligned}$$

and it leads to a differential equation

$$\frac{dp}{\sqrt{P}}+\frac{dq}{\sqrt{Q}}=0,$$

where

$$P=(\mathrm{h}+2\mathrm{b}p+\mathrm{f}p^2)^2-(\mathrm{a}+2\mathrm{h}p+\mathrm{g}p^2)(\mathrm{g}+2\mathrm{f}p+\mathrm{c}p^2),$$
$$Q=(\mathrm{h}+2\mathrm{b}q+\mathrm{f}q^2)^2-(\mathrm{a}+2\mathrm{h}q+\mathrm{g}q^2)(\mathrm{g}+2\mathrm{f}q+\mathrm{c}q^2);$$

and upon effecting the calculation, it is found that we have

$$P=-8\alpha^2\beta^2(\phi^2-\gamma^2)(a+b-2c-\phi)(a+p)(b+p)(c+p),$$
$$Q=-8\alpha^2\beta^2(\phi^2-\gamma^2)(a+b-2c-\phi)(a+q)(b+q)(c+q),$$

viz. P, Q are the same multiples of $(a+p)(b+p)(c+p)$, $(a+q)(b+q)(c+q)$ respectively; so that, omitting the common factor, or taking P, Q to represent the last-mentioned functions respectively, we have

$$\frac{dp}{\sqrt{P}}+\frac{dq}{\sqrt{Q}}=0;$$

and since the parameter ϕ has disappeared, we see that the original equation involving ϕ is the general integral of this differential equation; viz. that the differential equation belongs to the right lines on the surface.

20. The form of the integral equation may be simplified by introducing instead of ϕ a new parameter K, connected with it by the equation

$$K=\frac{\beta b-\alpha a+\phi c}{\beta-\alpha+\phi},\ \text{ or }\ \phi=\frac{(\beta-\alpha)K-\beta b+\alpha a}{c-K},$$

viz. we thence deduce

$$\begin{aligned}\beta-\alpha+\phi&=-2\alpha\beta && (\div),\\ \beta b-\alpha a+\phi c&=-2\alpha\beta K && (\div),\\ \beta b^2-\alpha a^2-\phi(\alpha\beta-c^2)&=\ \ 2\alpha\beta(ac+bc-ab+2cK) && (\div),\\ \phi+\gamma&=\ \ 2\beta(K-b) && (\div),\\ \phi-\gamma&=-2\alpha(K-a) && (\div),\end{aligned}$$

where the sign $(\div)$ is used to signify that the functions preceding it have to be divided by a denominator which in fact is $=c-K$. The equation thus becomes

$$(ac+bc-ab-2cK-KX-Y)^2-4(K-a)(K-b)(c^2+cX+Y)=0;$$

and if we moreover write

$$\nu,\ \mu,\ \lambda=abc,\ bc+ca+ab,\ a+b+c,$$

and instead of K introduce the parameter C, $=\lambda-K$, the equation becomes

$$\{-\mu-2c^2+2cC-(\lambda-C)X-Y\}^2+(b+c-C)(c+a-C)(c^2+cX+Y)=0;$$

or expanding and reducing, this is

$$\begin{aligned}&\{Y+(\lambda-C)X\}^2\\&\quad+Y(-2\mu+4\lambda C-4C^2)\\&\quad+X(2\mu\lambda-4\nu-2\mu C)\\&\quad+\mu^2-4\nu C=0,\end{aligned}$$

or say

$$\begin{aligned}&\mu^2-4\nu C\\&+(2\mu\lambda-4\nu-2\mu C)(p+q)\\&+(-2\mu+4\lambda C-4C^2)\,pq\\&+(\lambda-C)^2(p^2+q^2+2pq)\\&+2(\lambda-C)\,pq\,(p+q)\\&+\qquad p^2q^2\qquad=0,\end{aligned}$$

viz. this, containing the constant C, is the general integral of the differential equation

$$\frac{dp}{\sqrt{P}}+\frac{dq}{\sqrt{Q}}=0,$$

where

$$\begin{aligned}P&=(a+p)(b+p)(c+p),\quad=\nu+\mu p+\lambda p^2+p^3,\\Q&=(a+q)(b+q)(c+q),\quad=\nu+\mu q+\lambda q^2+q^3.\end{aligned}$$

21. The constant C is connected with the parameter σ, which originally served to determine the right line, by the equation

$$\sigma+\frac{1}{\sigma}=\frac{2}{\gamma}\,\frac{(\beta-\alpha)(\lambda-C)-\beta b+\alpha a}{c-(\lambda-C)},$$

or, what is the same thing,

$$\sigma+\frac{1}{\sigma}=\frac{2}{b-c}\,\frac{2c^2-a^2-b^2-C(2c-a-b)}{C-a-b}.$$

Reverting to the equation between p, q, ϕ, I remark that if ϕ be therein considered as variable, we have the differential equation

$$\sqrt{Q}\,dp+\sqrt{P}\,dq+\sqrt{\Phi}\,d\phi=0,$$

where P, Q have the foregoing values

$$P=-8\alpha^2\beta^2(\phi^2-\gamma^2)(a+b-c-\phi)(a+p)(b+p)(c+p),\quad Q=\&c.;$$

and where, if the integral equation be written in the form

$$L+2M\phi+N\phi^2=0,$$

then we have $\Phi=M^2-NL$, viz. we thus find

$$\Phi=16\alpha^2\beta^2(a+p)(b+p)(c+p)(a+q)(b+q)(c+q).$$

22. Changing the notation, and writing

$$P = (a+p)(b+p)(c+p),$$

$$Q = (a+q)(b+q)(c+q),$$

$$\Phi = (\phi^2 - \gamma^2)(a+b-2c-\phi),$$

the equation is

$$\frac{dp}{\sqrt{P}} + \frac{dq}{\sqrt{Q}} + \frac{\sqrt{2}\,d\phi}{\sqrt{\Phi}} = 0;$$

or if, instead of ϕ, we introduce the original parameter σ, then, observing that

$$\frac{2d\sigma}{\sigma} = \frac{d\phi}{\sqrt{\phi^2 - \gamma^2}},$$

we at once find

$$\frac{dp}{\sqrt{P}} + \frac{dq}{\sqrt{Q}} + \frac{4d\sigma}{\sqrt{\Sigma}} = 0,$$

where

$$\Sigma = \gamma(1+\sigma^4) - 2(a+b-2c)\sigma^2,$$

or, what is the same thing,

$$\Sigma = a(\sigma^2-1)^2 - b(\sigma^2+1)^2 + c\,.\,4\sigma^2;$$

viz. passing from a point (p, q) on the line σ to a consecutive point $(p+dp,\ q+dq)$ on the line $\sigma + d\sigma$, the above is the relation between the variations dp, dq, $d\sigma$. If τ be the parameter of the other line through the same point, then we have in like manner, say

$$\frac{dp}{\sqrt{P}} - \frac{dq}{\sqrt{Q}} + \frac{4d\tau}{\sqrt{\mathrm{T}}} = 0,$$

(viz. one of the radicals $\sqrt{P}$, $\sqrt{Q}$ must present itself with a reversed sign): and we thus have dp, dq each expressed in terms of $d\sigma$, $d\tau$; viz. we have the increments dp, dq when a point passes from (σ, τ) to $(\sigma + d\sigma,\ \tau + d\tau)$. These results will be presently obtained in a more simple manner.

Formulæ where the position of a Point on the Surface is determined by means of the two Lines through the Point.

23. We may determine the position of a point by means of the parameters σ, τ of the two lines through the point. The equations of these are

$$\frac{x}{\sqrt{a}} + \frac{iy}{\sqrt{b}} = \sigma\left(1 + \frac{z}{\sqrt{c}}\right), \quad \frac{x}{\sqrt{a}} + \frac{iy}{\sqrt{b}} = \tau\left(1 - \frac{z}{\sqrt{c}}\right),$$

$$\frac{x}{\sqrt{a}} - \frac{iy}{\sqrt{b}} = \frac{1}{\sigma}\left(1 - \frac{z}{\sqrt{c}}\right), \quad \frac{x}{\sqrt{a}} - \frac{iy}{\sqrt{b}} = \frac{1}{\tau}\left(1 + \frac{z}{\sqrt{c}}\right),$$

and from these equations we deduce

$$\frac{x}{\sqrt{a}}=\frac{\sigma\tau+1}{\tau+\sigma},\quad \frac{iy}{\sqrt{b}}=\frac{\sigma\tau-1}{\tau+\sigma},\quad \frac{z}{\sqrt{c}}=\frac{\tau-\sigma}{\tau+\sigma}.$$

We have thence

$$\frac{dx}{\sqrt{a}}=(\tau^2-1)\,d\sigma+(\sigma^2-1)\,d\tau\quad(\div),$$

$$\frac{idy}{\sqrt{b}}=(\tau^2+1)\,d\sigma+(\sigma^2+1)\,d\tau\quad(\div),$$

$$\frac{dz}{\sqrt{c}}=\quad-2\tau d\sigma+\quad 2\sigma d\tau\quad(\div),$$

where denom. $=(\tau+\sigma)^2$: regarding σ, τ as the parameters in place of p, q, these show the values of the first differential coefficients a, a'; b, b'; c, c'. We deduce

$$A=-2i\sqrt{bc}\,(\sigma\tau+1)\ \div,\quad B=-2i\sqrt{ca}\,(\sigma\tau-1)\ \div,\quad C=-2i\sqrt{ba}\,(\tau-\sigma)\ \div,$$

where denom. $=(\tau+\sigma)^3$. We have, moreover,

$$\frac{d^2x}{\sqrt{a}}=-2(\tau^2-1)\,d\sigma^2+2(\sigma\tau+1)\,2d\sigma d\tau-2(\sigma^2-1)\,d\tau^2\quad(\div),$$

$$\frac{2d^2y}{\sqrt{b}}=-2(\tau^2+1)\,d\sigma^2+2(\sigma\tau-1)\,2d\sigma d\tau-2(\sigma^2+1)\,d\tau^2\quad(\div),$$

$$\frac{d^2z}{\sqrt{c}}=+\quad 4\tau\,d\sigma^2+2(\tau-\sigma)\,2d\sigma d\tau+\quad 4\sigma d\tau^2\quad(\div),$$

where denom. $=(\tau+\sigma)^3$: giving $\alpha, \alpha', \alpha''$; β, β', β''; $\gamma, \gamma', \gamma''$. We deduce as the numerators of $E'(=A\alpha+B\beta+C\gamma)$ and $G'(=A\alpha''+B\beta''+C\gamma'')$,

$$4i\sqrt{abc}\,\{(\sigma\tau+1)(\tau^2-1)-(\sigma\tau-1)(\tau^2+1)-2\tau(\tau-\sigma)\},\ =0,$$

and

$$4i\sqrt{abc}\,\{(\sigma\tau+1)(\sigma^2-1)-(\sigma\tau-1)(\sigma^2+1)+2\sigma(\tau-\sigma)\},\ =0;$$

that is, $E'=0$ and $G'=0$; or the differential equation of the chief lines is $d\sigma d\tau=0$, which is right. The value of $F'(=A\alpha'+B\beta'+C\gamma')$ is hardly required, but it is readily found to

$$=4i\sqrt{abc}\,\{-(\sigma\tau+1)^2+(\sigma\tau-1)^2-(\tau-\sigma)\}^2\div(\tau+\sigma)^6,$$

or since the term in $\{\ \}$ is $=-(\tau+\sigma)^2$, we have

$$F'=\frac{-4i\sqrt{abc}}{(\tau+\sigma)^4}.$$

24. The values of E, F, G $(ds^2=Ed\sigma^2+2Fd\sigma d\tau+Gd\tau^2)$ are

$$E=a(\tau^2-1)^2\qquad -b(\tau^2+1)^2\qquad +c\,.\,4\tau^2\quad(\div),$$

$$F=a(\tau^2-1)(\sigma^2-1)-b(\tau^2+1)(\sigma^2+1)-c\,.\,4\tau\sigma\quad(\div),$$

$$G=a(\sigma^2-1)^2\qquad -b(\sigma^2+1)^2\qquad +c\,.\,4\sigma^2\quad(\div),$$

where denom. $=(\tau+\sigma)^4$.

We have, it is clear, ($E_1 = d_\sigma E$, $E_2 = d_\tau E$, &c.)

$$E_1 = -\frac{4}{\sigma+\tau}E, \quad G_2 = -\frac{4}{\sigma+\tau}G.$$

Hence the condition $FG_2 - 2F_2G + GG_1 = 0$, in order that $\sigma = \text{const.}$ may be a geodesic, reduces itself to

$$-\frac{4}{\sigma+\tau}F - 2F_2 + G_1 = 0;$$

and similarly, the condition $-2EF_1 + E_1F + EE_2 = 0$, in order that $\tau = \text{const.}$ may be a geodesic, reduces itself to

$$-\frac{4}{\sigma+\tau}F - 2F_1 + E_2 = 0.$$

We have at once

$$\begin{aligned}
E_2 &= 4a(\tau^2-1)(\tau\sigma+1) \quad - 4b(\tau^2+1)(\tau\sigma-1) \quad + 4c\,.\,2\tau(\sigma-\tau) \quad (\div),\\
G_1 &= 4a(\sigma^2-1)(\tau\sigma+1) \quad - 4b(\sigma^2+1)(\tau\sigma-1) \quad - 4c\,.\,2\sigma(\sigma-\tau) \quad (\div),\\
F_1 &= 2a(\tau^2-1)(\tau\sigma-\sigma^2+2) - 2b(\tau^2+1)(\tau\sigma-\sigma^2-2) - 2c\,.\,2\tau(\tau-3\sigma) \quad (\div),\\
F_2 &= 2a(\sigma^2-1)(\tau\sigma-\tau^2+2) - 2b(\sigma^2+1)(\tau\sigma-\tau^2-2) - 2c\,.\,2\sigma(\sigma-3\tau) \quad (\div),
\end{aligned}$$

where denom. $= (\sigma+\tau)^5$; and substituting these values, the conditions are verified: we thus again see *à posteriori* that the right lines $\sigma = \text{const.}$ and $\tau = \text{const.}$ are geodesics.

25. The last-mentioned values of E, G are $E = \mathrm{T} \div (\tau+\sigma)^4$, $G = \Sigma \div (\tau+\sigma)^4$; and writing for a moment

$$A = a(\tau^2-1)(\sigma^2-1) - b(\tau^2+1)(\sigma^2+1) - c\,.\,4\sigma\tau,$$

we have $F = A \div (\tau+\sigma)^4$, the value of ds^2 is thus

$$= \mathrm{T}d\sigma^2 + 2A\,d\sigma\,d\tau + \Sigma d\tau^2 \div (\tau+\sigma)^4,$$

which should be

$$= \tfrac{1}{4}(p-q)\left(p\frac{dp^2}{P} - q\frac{dq^2}{Q}\right),$$

where, as before,

$$P,\ Q = (a+p)(b+p)(c+p), \quad (a+q)(b+q)(c+q),$$

respectively. We have already found

$$\frac{dp}{\sqrt{P}} + \frac{dq}{\sqrt{Q}} + \frac{4d\sigma}{\sqrt{\Sigma}} = 0,$$

$$\frac{dp}{\sqrt{P}} - \frac{dq}{\sqrt{Q}} + \frac{4d\tau}{\sqrt{\mathrm{T}}} = 0;$$

or, what is the same thing,

$$\frac{dp}{\sqrt{P}} = -2\left(\frac{d\sigma}{\sqrt{\Sigma}} + \frac{d\tau}{\sqrt{\mathrm{T}}}\right),$$

$$\frac{dq}{\sqrt{Q}} = -2\left(\frac{d\sigma}{\sqrt{\Sigma}} - \frac{d\tau}{\sqrt{\mathrm{T}}}\right);$$

and we ought therefore to have identically

$$(p-q)\left\{p\left(\frac{d\sigma}{\sqrt{\Sigma}}+\frac{d\tau}{\sqrt{\mathrm{T}}}\right)^2-q\left(\frac{d\sigma}{\sqrt{\Sigma}}-\frac{d\tau}{\sqrt{\mathrm{T}}}\right)^2\right\}=\mathrm{T}d\sigma^2+2Ad\sigma d\tau+\Sigma d\tau^2\div(\tau+\sigma)^4;$$

that is

$$\begin{aligned}(p-q)^2&=\quad \mathrm{T}\Sigma\div(\tau+\sigma)^4,\\ p^2-q^2&=A\sqrt{\mathrm{T}\Sigma}\div(\tau+\sigma)^4;\end{aligned}$$

or, what is the same thing,

$$\begin{aligned}(p-q)^2&=\quad \mathrm{T}\Sigma\div(\tau+\sigma)^4,\\ p+q\;&=\quad A\div(\tau+\sigma)^2,\end{aligned}$$

which are easily verified.

26. In fact, the equation

$$\frac{x^2}{a+u}+\frac{y^2}{b+u}+\frac{z^2}{c+u}=1$$

gives for u a quadric equation, the roots of which are $u=p$, $u=q$; that is, we have

$$\begin{aligned}p+q&=\quad x^2+y^2+z^2-a-b-c,\\ pq\;&=-(b+c)x^2-(c+a)y^2-(a+b)z^2+bc+ca+ab;\end{aligned}$$

and substituting herein for x^2, y^2, z^2 their values in terms of σ, τ, we find

$$\begin{aligned}p+q&=a(\sigma^2-1)(\tau^2-1)-b(\sigma^2+1)(\tau^2+1)-4c\tau\sigma\quad\div(\tau+\sigma)^2,\\ pq\;&=bc(\sigma\tau+1)^2-ca(\sigma\tau-1)^2+ab(\sigma-\tau)^2\qquad\div(\tau+\sigma)^4,\end{aligned}$$

the first of which is, in fact, $p+q=A\div(\tau+\sigma)^2$. And from the two equations, forming the combination $(p+q)^2-4pq$, we at once obtain the other equation

$$(p-q)^2=\Sigma\mathrm{T}\div(\tau+\sigma)^4.$$

27. The most ready way of obtaining the relations between the differentials of p, q, σ, τ, is from the foregoing expressions of $p+q$, pq. Writing for a moment $p+q=A\div(\sigma+\tau)^2$, $pq=B\div(\sigma+\tau)^2$, we have $p^2(\sigma+\tau)^2-Ap+B=0$, $q^2(\sigma+\tau)^2-Aq+B=0$; viz. the first of these equations is

$$p^2(\sigma+\tau)^2-p\left[a(\sigma^2-1)(\tau^2-1)-b(\sigma^2+1)(\tau^2+1)-4c\sigma\tau\right]$$
$$+bc(\sigma\tau+1)^2-ca(\sigma\tau-1)^2+ab(\sigma-\tau)^2=0,$$

which is quadric in p, σ, τ. The negative discriminants in regard to these variables respectively are $\Sigma\mathrm{T}$, $4P\mathrm{T}$, $4P\Sigma$ respectively, and we have thus the equation

$$\frac{dp}{\sqrt{P}}+2\left(\frac{d\sigma}{\sqrt{\Sigma}}+\frac{d\tau}{\sqrt{\mathrm{T}}}\right)=0;$$

and the like equation for dq.

28. In the first integral of the geodesic lines, introducing σ, τ instead of p, q, the equation becomes

$$\sqrt{\left(\frac{p}{p+\theta}\right)}\left(\frac{d\sigma}{\sqrt{\Sigma}}+\frac{d\tau}{\sqrt{\mathrm{T}}}\right)-\sqrt{\left(\frac{q}{q+\theta}\right)}\left(\frac{d\sigma}{\sqrt{\Sigma}}-\frac{d\tau}{\sqrt{\mathrm{T}}}\right)=0;$$

or, what is the same thing,

$$p(q+\theta)\left(\frac{d\sigma}{\sqrt{\Sigma}}+\frac{d\tau}{\sqrt{\mathrm{T}}}\right)^2-q(p+\theta)\left(\frac{d\sigma}{\sqrt{\Sigma}}-\frac{d\tau}{\sqrt{\mathrm{T}}}\right)^2=0;$$

that is,

$$(p-q)\theta\left(\frac{d\sigma^2}{\Sigma}+\frac{d\tau^2}{\mathrm{T}}\right)+2\left[2pq+\theta(p+q)\right]\frac{d\sigma d\tau}{\sqrt{\Sigma\mathrm{T}}}=0;$$

or substituting herein for $p-q$, pq, $p+q$ the values $\sqrt{\Sigma\mathrm{T}}$, B, A, each divided by $(\sigma+\tau)^2$, this is

$$\theta(\mathrm{T}d\sigma^2+\Sigma d\tau^2)+2(2B+\theta A)\,d\sigma d\tau=0,$$

or say

$$\theta(\mathrm{T}d\sigma^2+2Ad\sigma d\tau+\Sigma d\tau^2)+4Bd\sigma d\tau=0;$$

viz. writing herein $\theta=0$, the equation is $d\sigma d\tau=0$, giving the right lines on the surface; and writing $\theta=\infty$, it is $\mathrm{T}d\sigma^2+2Ad\sigma d\tau+\Sigma d\tau^2=0$, giving the circular lines.

29. The equation $ds^2=\mathrm{T}d\sigma^2+2Ad\sigma d\tau+\Sigma d\tau^2\div(\tau+\sigma)^4$ shows that the right lines σ, $\sigma+d\sigma$, τ, $\tau+d\tau$ form on the surface an indefinitely small parallelogram, the sides whereof are $\sqrt{\mathrm{T}}\,d\sigma\div(\tau+\sigma)^2$ and $\sqrt{\Sigma}\,d\tau\div(\tau+\sigma)^2$, viz. the ratio of the coefficients of $d\sigma$, $d\tau$ is of the form function $\sigma\div$ function τ; and it thus appears that it is possible to draw on the surface the two sets of right lines, the lines of each set being at such intervals that the surface is divided into parallelograms, the sides of which have to each other any given ratio (the angles being variable); viz. if this ratio be as $m:1$, then, to determine the relation between σ, τ, we must have $\sqrt{\mathrm{T}}\,d\sigma=\pm m\sqrt{\Sigma}\,d\tau$, or what is the same thing, $\frac{d\sigma}{\sqrt{\Sigma}}=\pm m\frac{d\tau}{\sqrt{\mathrm{T}}}$. In particular, if $m=1$, the parallelograms will be rhombs; and we must then have

$$\frac{d\sigma}{\sqrt{\Sigma}}=\pm\frac{d\tau}{\sqrt{\mathrm{T}}};$$

viz. this being in terms of σ, τ, the differential equation of the curves of curvature, it appears that the two sets of lines may be taken so as to divide the surface into indefinitely small rhombs, such that, drawing the diagonals of these, we have the two sets of curves of curvature.

The Ellipsoid and the Skew Hyperboloid.

30. I have thus far considered a quadric surface in general, the various theorems being applicable as well to the ellipsoid and the hyperboloid of two sheets as to the skew hyperboloid, the right lines being of course imaginary for the first-mentioned surfaces; but I will now consider the ellipsoid and the skew hyperboloid separately.

31. First the ellipsoid. We have here a, b, c all positive, and I assume as usual $a>b>c$. The principal sections are all ellipses, viz. $\frac{x^2}{a}+\frac{y^2}{b}=1$ is the major-

mean, or say the minor section, $\frac{y^2}{b}+\frac{z^2}{c}=1$ the minor-mean, or say the major section, and $\frac{x^2}{a}+\frac{z^2}{c}=1$ the mean, or umbilicar section. The elliptic coordinates p, q enter into the equations symmetrically, but we distinguish them by taking p to extend from $-c$ to $-b$, and q to extend from $-b$ to $-a$. Thus $p=$ const. denotes the curves of

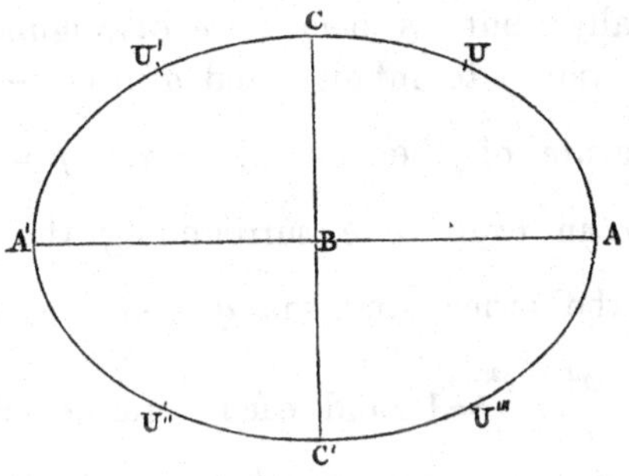

curvature of the one kind; viz. $p=-c$ denotes the major-mean section $\frac{x^2}{a}+\frac{y^2}{b}=1$, $p=-b$ the portions UU' and $U''U'''$ of the umbilicar section; and $q=$ const. denotes the curves of curvature of the other kind, viz. $q=-b$ the remaining portions $U'U''$ and $U'''U$ of the umbilicar section, $q=-a$ the minor-mean section $\frac{y^2}{b}+\frac{z^2}{c}=1$; say $p=$ const. the major-mean curves, and $q=$ const. the minor-mean curves.

32. Hence, in order that the equation

$$dp\sqrt{\left(\frac{p}{(a+p)(b+p)(c+p)(\theta+p)}\right)} \pm dq\sqrt{\left(\frac{q}{(a+q)(b+q)(c+q)(\theta+q)}\right)}=0$$

of the geodesic lines may be real {observing that we have $a+p$, $b+p=+$, $c+p$, $p=-$, and $a+q=+$, $b+q$, $c+q$, $q=-$, consequently $p\div(a+p)(b+p)(c+p)=+$, but $q\div(a+q)(b+q)(c+q)=-$}, we must have $\theta+p$, $\theta+q$ of opposite signs, that is $\theta+p=+$ and $\theta+q=-$; or θ included between the limits a, c. Or, what is the same thing, $-\theta$ is included between the limits $-c$, $-b$, say $-\theta$ has a p-value; or else between the limits $-b$, $-a$, say $-\theta$ has a q-value. This is conveniently shown

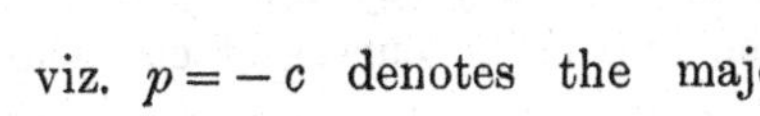

in the annexed diagram of the values of $-p$, $-q$, $-\theta$. Hence on the ellipsoid we have two kinds of geodesic lines, each of them touching a real curve of curvature; viz. those which touch a major-mean curve and those which touch a minor-mean curve: the transition case, answering to the value $\theta=b$, is that of the geodesic lines which pass through an umbilicus. I have considered the theory more in detail in my memoir "On the Geodesic Lines of an Ellipsoid," *Mem. R. Ast. Soc.*, t. XXX., pp. 31—53, 1872, [478].

33. Next, for the skew hyperboloid, we have a and $b = +$, $c = -$, and I assume for convenience $a > b$. Attending to the signs, we still have therefore $a > b > c$. The principal sections are one of them an ellipse, and the other two hyperbolas, viz. the minor section is the ellipse $\frac{x^2}{a} + \frac{y^2}{b} = 1$, the major section is the hyperbola $\frac{y^2}{b} + \frac{z^2}{c} = 1$, and the mean section is the hyperbola $\frac{x^2}{a} + \frac{z^2}{c} = 1$: there are no umbilici. The elliptic coordinates enter symmetrically; but, as before, we distinguish them, viz. we take p to extend from $-c$ (a positive value) to infinity, and q from $-b$ to $-a$. Thus $p = \text{const.}$ denotes the curves of curvature of the one kind, viz. $p = -c$ the ellipse $\frac{x^2}{a} + \frac{y^2}{b} = 1$, and every other value of p an oval curve surrounding the hyperboloid; and $q = \text{const.}$ the curves of curvature of the other kind, viz. $q = -c$ the major hyperbola $\frac{y^2}{b} + \frac{z^2}{c} = 1$, $q = -b$ the mean hyperbola $\frac{x^2}{a} + \frac{z^2}{c} = 1$, and each intermediate value gives a curve of curvature of a hyperbolic form: we may say that $p = \text{const.}$ determines the oval curves of curvature, and $q = \text{const.}$ the hyperbolic curves of curvature.

34. In the equation of the geodesic lines we have $a+p$, $b+p$, $c+p$, p all positive; but $a+q = +$, $b+q$, $c+q$, q each $= -$; hence $p \div (a+p)(b+p)(c+p) = +$, but $q \div (a+q)(b+q)(c+q) = -$; therefore $\theta + p$ and $\theta + q$ must be of opposite signs, or we must have $\theta + p = +$ and $\theta + q = -$; or what is the same thing, θ may have any value from $-p$ to $-q$, or say $-\theta$ any value from p to q; that is, the value of $-\theta$ may be positive and greater than $-c$, positive and less than $-c$, negative and less than $-b$, negative and between $-b$ and $-a$; viz. in the first case $-\theta$ has a p-value,

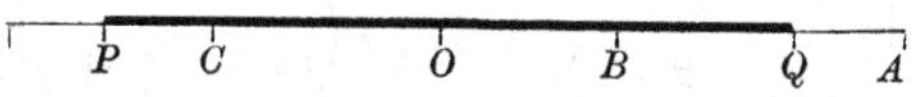

and in the fourth case it has a q-value, but in the second and third cases it has neither a p- nor a q-value. This is better seen from the diagram. It follows that we have, on the hyperboloid, geodesic lines of four different kinds: those which touch a real curve of curvature, oval or hyperbolic, and those which touch no real curve of curvature, but for which $-\theta$ has a positive value from 0 to $-c$, or a negative value from 0 to $-b$. And there are the transitional cases $-\theta = -c$, where the geodesic touches the ellipse $\frac{x^2}{a} + \frac{y^2}{b} = 1$; $\theta = 0$, where the geodesic becomes a right line; and $-\theta = -b$, where the geodesic touches the mean hyperbola $\frac{x^2}{a} + \frac{z^2}{c} = 1$.

35. To explain this more in detail, consider the geodesics which start from a point M of the hyperboloid. To fix the ideas, consider the axis of z as vertical, and take the point M in the positive octant of the hyperboloid; and let $M1$ represent the direction of the oval curve of curvature, $M9$ that of the hyperbolic curve of curvature, $M5$ that of one of the right lines.

The geodesic of initial direction $M1$ touches at M the oval curve of curvature $M1$, and lies wholly above this curve; it makes an infinity of convolutions round the upper part of the hyperboloid, cutting all the oval curves of curvature for which p has a (positive) value greater than p_1 (if p_1 is the value of p corresponding to the oval curve through M), and ascending to infinity: or considering the curve as described in the opposite sense, it descends from infinity to touch the oval curve through M, after which it again ascends to infinity.

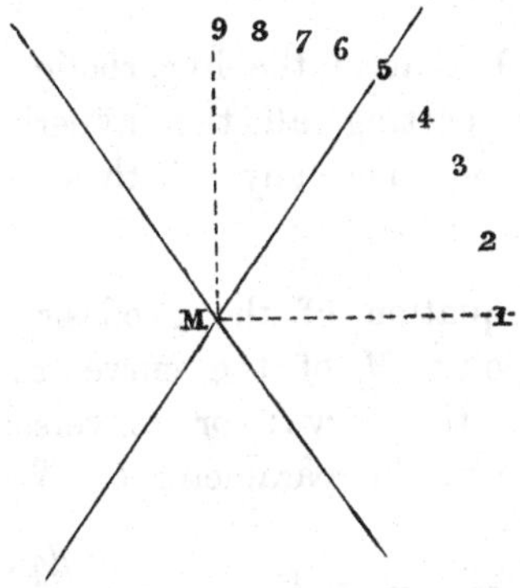

Next, if the initial direction is $M2$, we have a geodesic of the same kind, only descending below M to touch a certain oval curve having for its parameter $p_2 (p_2 > -c < p_1)$.

We come next to a critical direction $M3$, for which the geodesic descends below M to touch the oval curve of parameter $p_3 = -c$, that is, the ellipse $\frac{x^2}{a} + \frac{y^2}{b} = 1$. But it is to be observed that, whatever the initial point M may be, the geodesic makes below M an infinity of convolutions round the hyperboloid, so that it does not in fact ever actually touch the ellipse, but has this ellipse for an asymptote. That this is so appears from the consideration that the ellipse, *quâ* plane curve of curvature, is a geodesic; so that, starting from a point of the ellipse in the direction of the ellipse, the geodesic coincides with the ellipse, or, besides the ellipse itself, there is not any geodesic which touches the ellipse.

Next, if the initial direction be $M4$, the geodesic does not here touch any oval curve; it descends through M below the ellipse $\frac{x^2}{a} + \frac{y^2}{b} = 1$, lying in the upper and lower portions of the hyperboloid, and making round it an infinity of convolutions.

36. We come, then, to the initial direction $M5$, which is that of the right line; the geodesic here coincides with the right line.

In the cases which follow, the geodesic lies in the upper and lower portions of the hyperboloid, cutting all the oval curves of curvature.

Initial direction $M6$: the geodesic does not touch any hyperbolic curve of curvature, but makes round the hyperboloid an infinity of convolutions.

Initial direction *M*7: the geodesic touches at opposite infinities the mean hyperbola $\frac{x^2}{a}+\frac{z^2}{c}=1$, it lies wholly in front of the plane $y=0$ of this hyperbola.

Initial direction *M*8: the geodesic touches a hyperbolic curve of curvature parameter q_8 where q_8 (negative) is between $-b$ and q_9 the parameter of the hyperbolic curve of curvature through *M*; viz. it cuts all the hyperbolic curves the parameters of which are between $-b$ and q_8, but does not cut the remaining curves the parameters of which extend from q_8 to $-a$.

Lastly, initial direction is *M*9, that of the hyperbolic curve of curvature through *M*; the geodesic touches this curve, cutting all the hyperbolic curves the parameters of which are between $-b$ and q_9, but not any of those the parameters of which are between q_9 and $-a$.

37. If in the differential equation of the geodesic line we consider p, q as the elliptic coordinates of a given point *M* of the curve, the equation for a given value of θ determines the direction of the curve; or conversely, if the direction be given, the equation determines the value of the parameter θ. Writing

$$\frac{dp\,\sqrt{p}}{\sqrt{(a+p)(b+p)(c+p)}}=P,\quad \frac{dq\,\sqrt{q}}{\sqrt{(a+q)(b+q)(c+q)}}=Q,$$

then P, Q are proportional to the rectangular coordinates of a consecutive point *M'*, measured from *M* in the directions of the hyperbolic and oval curves of curvature respectively; and the differential equation of the geodesic lines gives

$$\frac{P}{\sqrt{p+\theta}}\pm\frac{Q}{\sqrt{q+\theta}}=0;$$

viz. if ϕ be the inclination of the geodesic to the hyperbolic curve of curvature, then $Q=P\tan\phi$, or we have $\frac{1}{p+\theta}=\frac{\tan^2\phi}{q+\theta}$, that is, $p\tan^2\phi-q=\theta(1-\tan^2\phi)$; hence, if for the right line $\phi=\lambda$, then $p\tan^2\lambda-q=0$; and therefore $\theta=\frac{p(\tan^2\phi-\tan^2\lambda)}{1-\tan^2\phi}$; viz. $\phi=0$, $\theta=-q$, $=\infty$, $\theta=-p$, as it should be.

509.

PLAN OF A CURVE-TRACING APPARATUS.

[From the *Proceedings of the London Mathematical Society*, vol. IV. (1871—1873), pp. 345—347. Read May 8, 1873.]

I HAVE devised a curve-tracing apparatus on the following plan:

Imagine two planes Π, Π' moving in the same horizontal plane, and above the two planes respectively the two points P, P' moving in the same or a parallel plane.

To fix the ideas, suppose that the two planes each move according to a law (that is, let the motion of each of them depend on a single variable parameter; for instance, they may each of them rotate about an axis); but let the motion of the two points be free.

Suppose, next, that the planes are connected together in such manner that the motion of one of them determines the motion of the other (e.g. by a train of wheel-work); and that the two points are also connected together in such manner that the motion of one of them determines the motion of the other (e.g. by a pentagraph; or by a slotted rod, the slot of which works on an axle, so as to allow the rod to move lengthways as well as rotate).

Suppose, finally, that one of the points, say P', is attached to a point of the plane Π'; then the plane Π being set in motion, this determines the motion of Π', consequently of P', consequently of P; and the moving point P, or say the pencil P, will describe on the moving plane Π a curve, the nature of which will of course depend on the nature of the motions of Π, Π', and on that of the connection between these planes and of the connections between the points P and P'.

I propose to describe the apparatus as nearly as I have actually constructed it. (See sketch-plan Fig. 1, and side-elevation Fig. 2.)

The framework of the instrument consists of two longitudinal bars (B) each about three feet long, one inch thick, and three inches broad, supported edgewise at a distance of about eighteen inches on the cross-pieces C, C; and half-way between them, supported by the same cross-pieces, is an axis carrying at each extremity two mitre wheels. The bars B support three cross-pieces D, D, D, and between these are the moveable cross-pieces E, E carrying the axes A, A with the attached mitre wheels, and circular disks X, X. Each of these wheels separately may be placed (as in the figure) out of gear with the two vertical wheels, or it can (by moving the

Fig. 1.

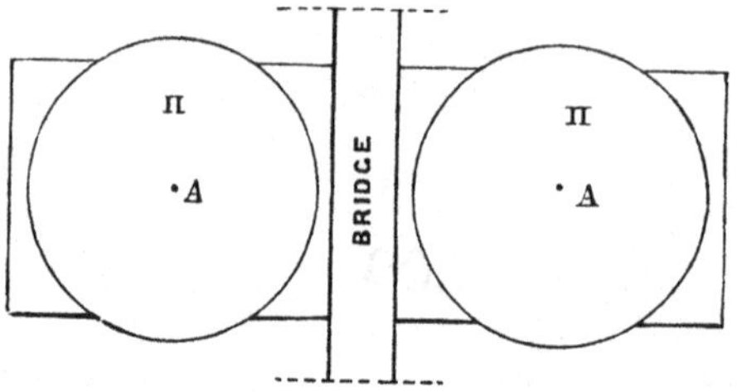

cross-piece E) be brought into gear with either of these wheels. Each axis A passes through a circular disk H, capable of rotating about it, so that it may be fixed in any position, and serving as the bearing for the plane X.

The disks X, X may be regarded as being themselves the planes Π, Π (say these planes are rigidly attached to X, X respectively); or we may, in any other manner, move the plane Π by means of the disk X; for instance, X may carry a spur-wheel gearing into a spur-wheel on the under surface of Π, and thereby communicating a

Fig. 2.

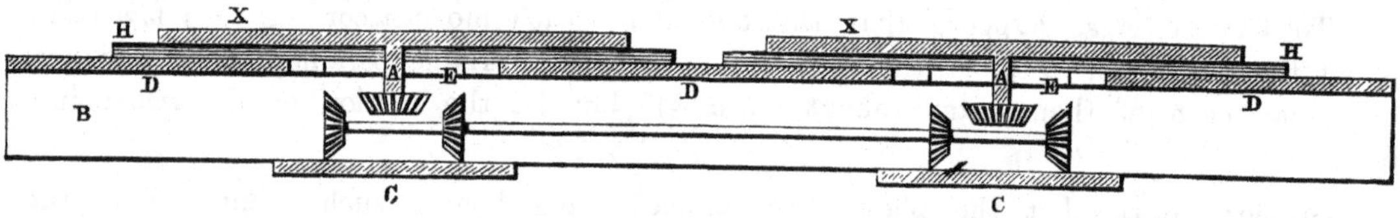

rotation of different velocity to the plane Π; or the connection may be as in the ordinary oval chuck. In any such case, the disk H (which, for this reason, was made to project beyond X) serves as a support for the plane Π, and the apparatus connected therewith; and observe that the angular position of such apparatus, and therefore of the path of any point of Π, may be varied at plèasure by moving the disk H through any angle.

Detached altogether from the rest of the instrument, or what is better, supported on longitudinal pieces carried by the cross-pieces C, C we have a bridge (see fig. 1) capable of adjustment as regards height, and serving as a support for the pentagraph-apparatus, or other connection of the one plane Π with the pencil which works upon the other plane Π.

It is hardly necessary to remark that in the simple form of the instrument where the disks X, X are themselves the planes Π, Π, then putting the mitre wheel of one plane X in gear with either of the corresponding vertical wheels, and making the plane rotate, the other plane X will rotate with the same angular velocity, in the same or the opposite direction, or it will remain at rest, according as its mitre wheel is in gear with one or the other of the corresponding vertical wheels, or is out of gear with each of them.

510.

ON BICURSAL CURVES.

[From the *Proceedings of the London Mathematical Society*, vol. IV. (1871—1873), pp. 347—352. Read May 8, 1873.]

A CURVE of deficiency 1 may be termed bicursal: there is some distinction according as the order is even or odd, and to fix the ideas I take it to be even.

A bicursal curve of the order n contains

$$\tfrac{1}{2}n(n+3) - \{\tfrac{1}{2}(n-1)(n-2) - 1\}, \ = 3n \text{ constants};$$

hence, if the order is $= 2n$, the number of constants is $= 6n$; such a curve is normally represented by a system of equations

$$(x,\ y,\ z) = (1,\ \theta)^n + (1,\ \theta)^{n-2}\sqrt{\Theta},$$

where Θ is a quartic function, which may be taken to be of the form $(1-\theta^2)(1-k^2\theta^2)$, or otherwise to depend on a single constant; viz. (x, y, z) are proportional to n-thic functions involving such a radical: since in the values of (x, y, z) one constant divides out, the number of constants is $3\{(n+1)+(n-1)\} - 1 + 1, = 6n$, as it should be.

But the curve of the order $2n$ may be abnormally or improperly represented by a system of equations

$$(x,\ y,\ z) = (1,\ \theta)^{n+k} + (1,\ \theta)^{n+k-2}\sqrt{\Theta},$$

viz. these equations, instead of representing a curve of the order $2n+2k$, will represent a curve of the order $2n$, provided only there exist $2k$ common values of θ for which each of the three functions vanish. The passage to a normal representation is effected by finding θ' a function of θ, $\sqrt{\Theta}$ (viz. an irrational function of θ) such that the foregoing equations become

$$(x,\ y,\ z) = (1,\ \theta')^n + (1,\ \theta')^{n-2}\sqrt{\Theta'};$$

it is shown that such a transformation is possible, and a mode of effecting it, derived from a theorem of Hermite's in relation to the Jacobian H, Θ functions, is given in Clebsch's Memoir "Ueber diejenigen Curven, deren Coordinaten sich als elliptische Functionen eines Parameters darstellen lassen," *Crelle*, t. LXIV. (1865), pp. 210—270. The demonstration is a very interesting one, and I reproduce it at the end of this paper. I remark, in passing, that the analogous reduction in the case of unicursal curves is self-evident; the equations

$$(x,\ y,\ z) = (1,\ \theta)^{n+k}$$

will represent a curve, not of the order $n+k$, but of the order n, provided there exist k common values of θ for which the three functions vanish; in fact, the three functions have then a common factor of the order k, and omitting this, the form is $(x,\ y,\ z) = (1,\ \theta)^n$.

Returning to the curves of deficiency 1, we see that a curve of the order $2m+2n$ contains $6(m+n)$ constants, and is normally represented by a system of equations

$$(x,\ y,\ z) = (1,\ \theta)^{m+n} + (1,\ \theta)^{m+n-2}\sqrt{\Theta}.$$

Such a curve may be otherwise represented: we may derive it by a rational transformation from the curve $(D=1)$ of the order 4 (binodal quartic), the equation of which is

$$(1,\ u)^2(1,\ v)^2 = 0;$$

viz. the coordinates are here connected by a quadriquadric equation; and we then express x, y, z in terms of these by a system of equations

$$(x,\ y,\ z) = (1,\ u)^m(1,\ v)^n.$$

It is, however, to be observed that the form of these functions is not determinate: each of them may be altered by adding to it a term $\{(1,\ u)^{m-2}(1,\ v)^{n-2}\}\{(1,\ u)^2(1,\ v)^2\}$, where the second factor is that belonging to the quadriquadric transformation, and the first factor is arbitrary. Using the arbitrary function to simplify the form, the real number of constants is reduced to $(m+1)(n+1)-(m-1)(n-1)$, $=2(m+n)$; or the three functions contain together $6(m+n)$ constants, one of which divides out. The quadriquadric equation, dividing out one constant, contains eight constants; but reducing by linear transformations on u, v respectively, the number of constants is $8-6$, $=2$. Hence, in the system of equations, the whole number of constants is $6(m+n)-1+2$, $=6(m+n)+1$; viz. this is greater by unity than the number of constants in the curve $(D=1)$ of the order $2(m+n)$. The explanation of the excess is that the same curve of the order $2(m+n)$ may be derived from the different quartic curves $(1,\ u)^2(1,\ v)^2=0$; this will be further examined.

The transition from the one form to the other is not immediately obvious; in fact, if from the quadriquadric equation $(1,\ u)^2(1,\ v)^2=0$ (say this is $A+2Bv+Cv^2=0$, where A, B, C are quadric functions of u) we determine v; this gives $Cv=-B+\sqrt{B^2-AC}$,

$= -B + \sqrt{\Omega}$ suppose; and then substituting in the equations $(x, y, z) = (1, u)^m (1, v)^n$, we find

$$(x, y, z) = (1, u)^m (C, -B + \sqrt{\Omega})^n;$$

viz. we have (x, y, z) proportional to functions of u, involving the quartic radical $\sqrt{\Omega}$; but these functions are of the order (not $m + n$, but) $m + 2n$.

In particular, if $n = m$, then, instead of functions of the order $2n$, we have functions of the order $3n$. The reduction in this last case to the form where the order is $2n$ can be effected without difficulty, but in the general case where m and n are unequal, I do not know how it is to be effected except by the general process explained in Clebsch's Memoir.

We may, in fact, by linear transformations on u, v, reduce the quadriquadric relation to

$$\begin{array}{ccc} 1 & & +u^2 \\ & +2buv & \\ +v^2 & & +cu^2v^2 = 0, \end{array}$$

that is

$$1 + u^2 + 2buv + v^2 + cu^2v^2 = 0;$$

or putting herein $u + v = p$, $uv = q$, the relation is

$$1 + p^2 - 2q + 2bq + cq^2 = 0,$$

that is

$$\begin{aligned} p^2 &= -1 + (\ \ 2 - 2b)\, q - cq^2, \\ p^2 - 4q &= -1 + (-2 - 2b)\, q - cq^2; \end{aligned}$$

viz. extracting the square roots,

$$u + v = \sqrt{Q}, \quad u - v = \sqrt{Q'},$$

if for shortness

$$\begin{aligned} Q &= -1 + (\ \ 2 - 2b)\, q - cq^2, \\ Q' &= -1 + (-2 - 2b)\, q - cq^2; \end{aligned}$$

we may then rationalise one of the radicals, for instance, Q; viz. writing

$$-c\{-1 + (2 - 2b)\, q - cq^2\} = \{cq - (1 - b)\}^2 + c - (1 - b)^2,$$

then, if

$$cq - (1 - b) = \sqrt{c - (1 - b)^2} \,.\, \tfrac{1}{2}\left(\theta - \frac{1}{\theta}\right),$$

this becomes

$$\begin{aligned} -cQ &= \left\{c - (1 - b)^2\right\} \left\{\tfrac{1}{4}\left(\theta - \frac{1}{\theta}\right)^2 + 1\right\} \\ &= \left\{c - (1 - b)^2\right\} . \tfrac{1}{4}\left(\theta + \frac{1}{\theta}\right)^2, \end{aligned}$$

that is

$$\sqrt{Q} = \tfrac{1}{2}\sqrt{\left(\frac{c-(1-b)^2}{-c}\right)}\left(\theta + \frac{1}{\theta}\right);$$

and the corresponding value of $\sqrt{Q'}$ is

$$\sqrt{Q'} = \sqrt{\left\{\tfrac{1}{4}\frac{c-(1-b)^2}{-c}\left(\theta + \frac{1}{\theta}\right)^2 - 4q\right\}},$$

where q stands for its value

$$\frac{1}{c}\left\{1 - b + \sqrt{c-(1-b)^2}\,.\,\tfrac{1}{2}\left(\theta - \frac{1}{\theta}\right)\right\}.$$

We may write these in the form

$$1 : \tfrac{1}{2}\sqrt{Q} : \tfrac{1}{2}\sqrt{Q'} = M\theta : 1 + \theta^2 : \sqrt{\Theta},$$

where M is a constant, and Θ is a quartic function of θ, such that $(1+\theta^2)^2 - \Theta$ is a quadric function only of θ.

The equations

$$(x,\ y,\ z) = (1,\ u)^m (1,\ v)^n$$

thus assume the form

$$(x,\ y,\ z) = (M\theta,\ 1 + \theta^2 + \sqrt{\Theta})^m (M\theta,\ 1 + \theta^2 - \sqrt{\Theta})^n;$$

and on the right-hand side the term of the highest order in θ is

$$(1 + \theta^2 + \sqrt{\Theta})^m (1 + \theta^2 - \sqrt{\Theta})^n,$$

viz. if $n = \text{ or } > m$, then this is

$$\{(1+\theta^2)^2 - \Theta\}^m (1 + \theta^2 - \sqrt{\Theta})^{n-m}.$$

This is

$$= (1,\ \theta)^{2m} (1 + \theta^2 - \sqrt{\Theta})^{n-m},$$

which is of the order $2m + 2(n-m)$, $= 2n$ (which, in virtue of $n = \text{ or } > m$, is $= \text{ or } > m + n$). In particular, if $n = m$, then the highest order is $= 2n$; or the curve of the order $2n$, as represented by the equations

$$(x,\ y,\ z) = (1,\ u)^n (1,\ v)^n,$$

where $(u,\ v)$ are connected by a quadriquadric equation, is also represented by the equations

$$(x,\ y,\ z) = (1,\ \theta)^n + (1,\ \theta)^{n-2}\sqrt{\Theta},$$

which is the required transformation of the original equations.

It is to be noticed that the foregoing form,

$$\begin{array}{lll} 1 & & + u^2 \\ & + 2buv & \\ + v^2 & & + cu^2v^2 = 0, \end{array}$$

is the most special form to which the quadriquadric relation can be reduced by the linear transformations of u, v; in fact, by mere division, the equation is made to have the constant term 1; the number of the coefficients of transformation is then $3+3, =6$; and to reduce the relation to the foregoing form, we have the six conditions, coeff. $u=0$, coeff. $v=0$, coeff. $vu^2=0$, coeff. $v^2u=0$, coeff. $u^2=1$, coeff. $v^2=1$; but in this form, expressing v in terms of u, the radical is $\sqrt{\Omega}$, where

$$\Omega=(1+u^2)(1+cu^2)-b^2u^2,$$

which is not more general than if we had

$$\Omega=(1+u^2)(1+cu^2),$$

viz. there is a superfluous constant b. And we thus see how it is that the system of equations

$$(x,\ y,\ z)=(1,\ u)^n(1,\ v)^n,$$

(u, v) connected by a quadriquadric relation, contains, not $6n$, but $6n+1$ constants, one of these being superfluous.

Clebsch's transformation, above referred to, is as follows:

Starting with the equations

$$(x,\ y,\ z)=(1,\ \theta)^{n+k}+(1,\ \theta)^{n+k-2}\sqrt{\Theta},$$

if the function Θ is not originally of the standard form, we may, by a linear substitution, reduce it to this form, viz. we may write

$$\Theta=\theta(1-\theta)(1-k^2\theta);$$

and then writing $\theta=\text{sn}^2 u$, $(\sin^2 \text{am}\, u)$, we have

$$\begin{aligned}\sqrt{\Theta}&=\text{sn}\, u\, \text{cn}\, u\, \text{dn}\, u\\ &=\text{sn}\, u\, \text{sn}'\, u;\end{aligned}$$

so that the formulæ become

$$(x,\ y,\ z)=(1,\ \text{sn}^2 u)^{n+k}+(1,\ \text{sn}^2 u)^{n+k-2}\, \text{sn}\, u\, \text{sn}'\, u.$$

Hermite's theorem, used in the demonstration, is that any such function of $\text{sn}\, u$ is expressible in the form

$$C\frac{H(u-\alpha_1)\, H(u-\alpha_2)\ldots H(u-\alpha_{2n+2k})}{\Theta^{2n+2k}(u)}$$

(H, Θ denoting here the two Jacobian functions), where

$$\alpha_1+\alpha_2\ldots+\alpha_{2n+2k}=0.$$

{Observe, in passing, that the equation

$$0=(1,\ \text{sn}^2 u)^{n+k}+(1,\ \text{sn}^2 u)^{n+k-2}\, \text{sn}\, u\, \text{sn}'\, u$$

gives, when rationalised, an equation in $\text{sn}^2 u$ of the order $2n+2k$; the roots of this equation are $\text{sn}^2\alpha_1$, $\text{sn}^2\alpha_2 \ldots \text{sn}^2\alpha_{2n+2k}$. Considering the functions $(1, \text{sn}^2 u)^{n+k}$ and $(1, \text{sn}^2 u)^{n+k-2}$ as indeterminate, the coefficients can be found so that all but one of the roots of the equation in $\text{sn}^2 u$ shall have any given values whatever, $\text{sn}^2\alpha_1$, $\text{sn}^2\alpha_2, \ldots \text{sn}^2\alpha_{2n+2k-1}$; the theorem then shows that the remaining root is $\text{sn}^2\alpha_{2n+2k}$, where

$$-\alpha_{2n+2k} = \alpha_1 + \alpha_2 \ldots + \alpha_{2n+2k-1},$$

which is, in fact, Abel's theorem.}

Now, supposing that the three functions of θ all vanish for $2k$ common values of θ, each of the functions of u will contain the same $2k$ H functions, say these are $H(u-\alpha_{2n+1}) \ldots H(u-\alpha_{2n+2k})$. Omitting these and also the denominator factor $\Theta^{2k}(u)$, we have the set of equations

$$(x,\ y,\ z) = C\frac{H(u-\alpha_1)\,H(u-\alpha_2)\ldots H(u-\alpha_{2n})}{\Theta^{2n}(u)},$$

where, however,

$$\alpha_1+\alpha_2\ldots+\alpha_{2n} \neq 0,$$

{the values α_1, $\alpha_2 \ldots \alpha_{2n}$ are of course different for the three coordinates x, a, z respectively}; viz. we have

$$\begin{aligned}\alpha_1+\alpha_2\ldots+\alpha_{2n} &= -(\alpha_{2n+1}+\ldots+\alpha_{2n+2k})\\ &= 2ns \text{ suppose.}\end{aligned}$$

Writing then

$$u=u'+s,\quad \alpha_1=\alpha_1'+s,\quad \alpha_2=\alpha_2'+s,\ \ldots\ \alpha'_{2n}=\alpha'_{2n}+s,$$

and consequently

$$\alpha_1'+\alpha_2'+\ldots+\alpha'_{2n}=0,$$

we have

$$(x,\ y,\ z)=C\frac{H(u'-\alpha_1')\ldots H(u'-\alpha'_{2n})}{\Theta^{2n}(u'+s)};$$

or, changing the common denominator,

$$(x,\ y,\ z)=C\frac{H(u'-\alpha_1')\ldots H(u'-\alpha'_{2n})}{\Theta^{2n}(u')},$$

where

$$\alpha_1'+\alpha_2'\ldots+\alpha'_{2n}=0;$$

or, what is the same thing,

$$(x,\ y,\ z)=(1,\ \text{sn}^2 u')^n+(1,\ \text{sn}^2 u')^{n-2}\,\text{sn}\,u'\,\text{sn}'.u';$$

viz. writing $\text{sn}\,u'=\theta'$, and $\Theta'=\theta'(1-\theta')(1-k^2\theta')$, this is

$$(x,\ y,\ z)=(1,\ \theta')^n+(1,\ \theta')^{n-2}\sqrt{\Theta'},$$

a normal representation of the curve of the order $2n$.

The relation between the parameters θ, θ' is given by $\theta' = \text{sn}^2(u-s)$, $\text{sn}^2 u = \theta$, that is, we have

$$\sqrt{\theta'} = \frac{\sqrt{\theta}\,\text{cn}\,s\,\text{dn}\,s - \text{sn}\,s\,\sqrt{(1-\theta)(1-k^2\theta)}}{1-k^2\,\text{sn}^2 s\,.\,\theta};$$

or, writing $\text{sn}^2 s = \sigma$, this is

$$\theta' = \frac{\sqrt{\theta}\,\sqrt{(1-\sigma)(1-k^2\sigma)} - \sqrt{\sigma}\,\sqrt{(1-\theta)(1-k^2\theta)}}{1-k^2\sigma\theta};$$

and, conversely, we have

$$\theta = \frac{\sqrt{\theta'}\,\sqrt{(1-\sigma)(1-k^2\sigma)} + \sqrt{\sigma}\,\sqrt{(1-\theta')(1-k^2\theta')}}{1-k^2\sigma\theta'},$$

and the theorem, in fact, shows that, substituting this value of θ in the functions of θ which serve to express x, y, z, these become *proportional* to the functions of θ'; viz. they become equal to these functions, each multiplied by an irrational function $A' + B'\sqrt{\Theta'}$, (A', B' rational functions of θ').

511.

ADDITION TO THE MEMOIR ON GEODESIC LINES, IN PARTICULAR THOSE OF A QUADRIC SURFACE (509). [1]

[From the *Proceedings of the London Mathematical Society*, vol. IV. (1871—1873), pp. 368—380. Read June 12, 1873.]

38. In the Memoir above referred to, speaking of the geodesic lines on the skew hyperboloid, I say (No. 35), "The geodesic of initial direction $M1$ touches at M the oval curve of curvature $M1$, and lies wholly above this curve; *it makes an infinity of convolutions round the upper part of the hyperboloid,* cutting all the oval curves of curvature for which p has a positive value greater than p_1 (if p_1 is the value of p corresponding to the oval curve through M), and ascending to infinity." The statement as to the infinity of convolutions is incorrect; I was led to it by the assumption that the geodesic could not touch any hyperbolic curve of curvature. The fact is, that it touches at infinity (has for asymptotes) in general two hyperbolic curves of curvature; viz. the geodesic descending from infinity in the direction of a hyperbolic curve of curvature, so as to touch the oval curve through M, again ascends to infinity in the direction of a hyperbolic curve of curvature (the same as the first-mentioned one, or a different curve), making in its whole course say k convolutions, where m is a positive finite number; if $k < 1$, there is no complete convolution, and when $k = 1$ or any integer number, then the two hyperbolic curves are one and the same curve; k is infinite only in the special case afterwards referred to in the same No. 35, where the oval curve of curvature is the ellipse which is a principal section of the hyperboloid, and does not even attain to the value 1 except for an oval curve exceedingly close to this ellipse. The error was on consideration obvious enough, though I was in fact led to perceive it by the numerical calculations about to be referred to, which gave me geodesics not making a complete convolution.

[1] The articles are numbered consecutively with those of the Memoir, (509).

39. I have effected, for a particular skew hyperboloid and oval curve of curvature thereof, the numerical calculations for laying down the geodesic lines which touch this curve of curvature. Taking in general the equation of the hyperboloid to be

$$\frac{x^2}{a} + \frac{y^2}{b} + \frac{z^2}{c} = 1,$$

and the θ-curve of curvature to be the intersection by the confocal surface

$$\frac{x^2}{a-\theta} + \frac{y^2}{b-\theta} + \frac{z^2}{c-\theta} = 1,$$

then the selected values for the hyperboloid, and oval curve of curvature touched by the geodesics, are

$$\begin{aligned} a &= 900, \\ b &= 400, \\ c &= -c' = -1600, \\ \theta &= -\theta' = -1650; \end{aligned}$$

so that a, b, c', θ' are the positive values 900, 400, 1600, and 1650 respectively. I recall that $p=c'$ to $p=\infty$ gives the oval curves of curvature, viz. $p=c'$, the elliptic principal section; $p=\theta'$, the given oval curve: and that we are in the sequel concerned only with the oval curves above this, for which p extends from θ' to ∞. Moreover, $q=-b$ to $-a$ gives the hyperbolic curves of curvature, viz. $q=-b$ the xz-principal section; and $q=-a$ the yz-principal section of the hyperboloid. We have, in fact, to deal with the integrals

$$\Pi(p) = \int_{\theta'} dp \sqrt{\frac{p}{(p+a)(p+b)(p-c')(p-\theta')}},$$

and

$$\Psi(q) = \int_{-b} dq \sqrt{\frac{q}{(q+a)(q+b)(q-c')(q-\theta')}};$$

or if $p=\theta'+u$, u extending from 0 towards ∞, then

$$\Pi(p) = \int_0^u du \sqrt{\frac{u+\theta'}{(u+a+\theta')(u+b+\theta')(u+\theta'-c')\,u}},$$

and so if $q=-b-v$, v extending from 0 towards $(a-b)$, which is its limit,

$$\Psi(q) = \int_0^v dv \sqrt{\frac{v+b}{(a-b-v)\,v\,(v+b+c')(v+b+\theta')}};$$

the relation for any particular geodesic of the series being

$$\pm\Pi(p) \pm \Psi(q) = C.$$

40. To avoid discontinuity as to sign, it is convenient to take the integral $\Psi(q)$ in a particular manner. The hyperboloid is by the xz- and yz-principal planes divided into four quadrants; or since we attend only to the upper half of the hyperboloid, say this upper half is thus divided into four quadrants, x to y, y to x', x' to y', and y' to x; or call them the first, second, third, and fourth quadrants. But we may consider the quadrants as forming an infinite succession, first, second, third, fourth, fifth,

and so on; or we may take them in the reverse order, -1, -2, -3, &c. For a hyperbolic curve $q=-b-v$ in the first quadrant the integral is to be taken $v=0$ to $v=v$; for a curve in the second quadrant $v=0$ to $a-b$, and thence positively $a-b$ to v; for a curve in the third quadrant $v=0$ to $a-b$, thence $a-b$ to 0, and thence 0 to v; and so on: and so for a point in the quadrant -1, the integral is from 0 to v, taken negatively, &c.; that is, as the hyperbolic curve travels from the xz-position in the positive direction, the integral $\Psi(q)$ continually increases from zero; and if the curve travels from the xz-position in the negative direction, then the integral $\Psi(q)$ continually decreases from zero; that is, it increases negatively. It is to be remarked that the integral $v=0$ to $a-b$ is finite, say it is $=K'$; and of course it is only thus far that the integral requires to be calculated, the subsequent values differing from the preceding ones only by multiples of this complete integral.

The integral $\Pi(p)$ requires no explanation; it is taken from $u=0$, giving a certain oval curve, up to any positive value of u, giving the oval curves above this one; and, in particular, taking the integral to $u=\infty$ (or, what is the same thing, to $p=\infty$) its value is finite, $=K$, suppose.

41. Consider the geodesic which touches the given oval curve at a point for which $\Psi(q)$ has a given value Q; at this point $p=\theta'$, or $\Pi(p)=0$; so that, taking for the equation of the geodesic $\Psi(q)\pm\Pi(p)=C$, we have $Q=C$, and consequently

$$\Psi(q)=Q\mp\Pi(p).$$

Taking the positive sign, then as p increases from θ', $\Psi(q)$ increases, or the describing point of the geodesic moves upwards from the point of contact in the direction of positive rotation; and taking the negative sign, then $\Psi(q)$ decreases, or the describing point of the geodesic moves upwards from the point of contact in the direction of negative rotation; and, in particular, p becoming infinite, then the first-mentioned branch touches at infinity the hyperbolic curve, for which q is such that $\Psi(q)=Q+K$, and the second branch that for which q is such that $\Psi(q)=Q-K$.

42. The graphical process is as follows: we describe, on the hyperboloid, a series of hyperbolic curves of curvature, numbering them according to the values of $\Psi(q)$; viz. considering the hyperbolic branches which form the xz, yz, $x'z$ and $y'z$ sections respectively, these are 0, K', $2K'$, $3K'$, and on going round a second time they would be $4K'$, $5K'$, $6K'$, $7K'$, and so on respectively. We similarly describe, say on the upper half of the hyperboloid, the oval curves of curvature, numbering them according to the values of $\Pi(p)$, viz. beginning with that for which $p=\theta'$, which is 0, we go successively up to the oval curve at infinity, which is K.

In the example, and drawing belonging thereto[1], where, for convenience, the values of the integrals have been multiplied by 100,000, we have, as will appear, $K=12490$, $K'=34726$: the two sets of curves are drawn at intervals of 2000; viz. we have the hyperbolic curves 0, 2000, 4000, ... 34000, 34726; and the oval curves 0, 2000, 4000, 6000, 8000, 10000; the distance of the successive oval curves increases very rapidly, since the curve at infinity would be K, $=12490$, and the curve $K=10000$ is the last which comes into the limits of the figure.

[1] This drawing was exhibited at the meeting of the Society.

43. The two sets of curves of curvature being thus drawn at equal intervals of $\Pi(p)$ and $\Psi(q)$ respectively, dividing the hyperboloid into quadrilateral spaces (which of course should theoretically be indefinitely small), the diagonals of these quadrilateral spaces are the elements of the geodesic lines; and by a series of such elements we have a particular geodesic line. The general character comes out in the drawing very distinctly; viz. the geodesic is a hyperbola-like curve descending from infinity to touch the oval curve, and again ascending to infinity; by reason of the small value of K in comparison of K', there is nothing like a complete convolution, but the whole curve is included within a quadrant of the hyperboloid.

44. I remark that the calculations were performed roughly. I made no attempt to estimate or allow for errors arising from the intervals being too great; and there are very probably accidental errors of calculation. But starting with the value 10411 of $\Pi(p)$ for $p=10000$, I found, with some care, superior and inferior limits of the remainder of the integral, $p=10000$ to $p=\infty$; and the process is, I think, an interesting one. Consider in general the integral

$$\begin{aligned} I = & \int_{\theta'}^{\infty} dp \sqrt{\frac{p}{(p+a)(p+b)(p-c')(p-\theta')}} \\ = & I_1 + \int_{p_1}^{\infty} dp \sqrt{\frac{p}{(p+a)(p+b)(p-c')(p-\theta')}}, \end{aligned}$$

where I_1 is the integral calculated up to a somewhat large value $p=p_1$.

Writing

$$\begin{aligned} \alpha &= \tfrac{1}{2}(a+b), \\ \beta &= \tfrac{1}{2}(c'+\theta'), \\ \gamma &= \tfrac{1}{2}(a+b)-m, \\ \delta &= \tfrac{1}{2}(c'+\theta')+n, \end{aligned}$$

where m and n are as yet undetermined, we have

$$\begin{aligned} (p+a)(p+b) &= (p+\alpha)^2 - \tfrac{1}{4}(a-b)^2, \quad > (p+\alpha)^2, \\ (p-c')(p-\theta') &= (p-\beta)^2 - \tfrac{1}{4}(\theta'-c')^2, \quad > (p-\beta)^2; \end{aligned}$$

and the integral is thus

$$> I_1 + \int_{p_1}^{\infty} dp \frac{\sqrt{p}}{(p+\alpha)(p-\beta)}.$$

But we may determine m and n, so that for $p=$ or $>p_1$,

$$\begin{aligned} (p+a)(p+b) &< (p+\gamma)^2, \\ (p-c')(p-\theta') &< (p-\delta)^2; \end{aligned}$$

and the integral is then

$$< I_1 + \int_{p_1}^{\infty} dp \frac{\sqrt{p}}{(p+\gamma)(p-\delta)}.$$

The determination thus depends on the last-mentioned integrals, the values of which are at once obtainable by writing therein $\sqrt{p}=x$; viz. we have

$$\int dp \frac{\sqrt{p}}{(p+\alpha)(p-\beta)} = 2\int \frac{x^2dx}{(x^2+\alpha)(x^2-\beta)} = C + \frac{2}{\alpha+\beta}\left\{\tfrac{1}{2}\beta\, h\,.\,l\frac{x-\sqrt{\beta}}{x+\sqrt{\beta}}\right\} + \frac{2\sqrt{\alpha}}{\alpha+\beta}\tan^{-1}\frac{x}{\sqrt{\alpha}};$$

and hence, substituting in the formula, for $h.l.x$ its value $\dfrac{\log x}{\log e}$, the superior limit is

$$I_1 + \left\{\frac{\sqrt{\delta}}{\gamma+\delta}\,\frac{1}{\log e}\log\frac{\sqrt{p_1}+\sqrt{\delta}}{\sqrt{p_1}-\sqrt{\delta}} + \frac{2\sqrt{\gamma}}{\gamma+\delta}\cot^{-1}\frac{\sqrt{p_1}}{\sqrt{\gamma_1}}\right\},$$

and the inferior limit is

$$I_1 + \left\{\frac{\sqrt{\beta}}{\alpha+\beta}\,\frac{1}{\log e}\log\frac{\sqrt{p_1}+\sqrt{\beta}}{\sqrt{p_1}-\sqrt{\beta}} + \frac{2\sqrt{\alpha}}{\alpha+\beta}\cot^{-1}\frac{\sqrt{p_1}}{\sqrt{\alpha_1}}\right\}.$$

45. The numerical values are $p_1 = 10{,}000$; $a, b, c', \theta' = 900, 400, 1600, 1650$; and thence determining by trial values of m and n,

$$\alpha = 650, \qquad \gamma = 650 - 4 = 646,$$
$$\beta = 1625, \qquad \delta = 1625 + 160 = 1785,$$

I obtained for the logarithmic and circular terms of the two limits respectively

	Superior	Inferior
Logarithmic	·015668	·015144
Circular	·005202	·005593
	·020870	·020737

The value of I_1 was $10411 \div 100{,}000 = {\cdot}104110$, and the two limits thus are ·124980 and ·124850; or restoring the factor 100,000, they are 12498 and 12485; the mean of these, say 12490, was taken for the value $\Pi(p)$, $p = \infty$; that is $K = 12490$.

46. As regards the calculation of the integrals $\Pi(p)$ and $\Psi(q)$, introducing the numerical values, and multiplying by the before-mentioned factor 100,000, we have $(q = -400 - v)$,

$$\Psi(q) = 100{,}000\int_0^v dv\sqrt{\frac{v+400}{(500-v)\,v\,(v+2000)(v+2050)}},$$

which for any small value of v is

$$= 100{,}000\sqrt{\frac{400}{500\,.\,2000\,.\,2050}}\left(\int\frac{dv}{\sqrt{v}},\ = 2\sqrt{v}\right);$$

viz. this is

$$= 883{\cdot}45\ (\log = 2{\cdot}9461830)\sqrt{v},$$

which was used for the values $v = 1, 2, \ldots 10$, that is, to $q = -410$; after which the calculation was continued by quadratures giving to v the values 10, 20, 30,... up to $v = 490$, or $q = -890$. For the remainder of the integral, writing $500 - v = w$ (that is, $q = -900 + w$), we have

$$\Psi(q) - \Psi(-890) = 100{,}000\int dw\sqrt{\frac{900-w}{w\,(500-w)(2500-w)(2550-w)}},$$

$$= 100{,}000\sqrt{\frac{900}{500\,.\,2500\,.\,2550}}\left\{\int\frac{dw}{\sqrt{w}},\ = 2(\sqrt{10}-\sqrt{w})\right\}$$

$$= 1062{\cdot}7\ (\log = 3{\cdot}0264261)(\sqrt{10}-\sqrt{w}),$$

which was used for the values $w=9, 8, 7, \ldots 1, 0$, to complete the calculation up to $q=-900$.

47. We have in like manner, $p=1650+u$,

$$\Pi(p)=100000\int_0^u du\sqrt{\frac{u+1650}{(u+2550)(u+2050)(u+50)u}},$$

which for small values of u is

$$=100000\sqrt{\frac{1650}{2550\,.\,2050\,.\,50}}\left(\int\frac{du}{\sqrt{u}},\ =2\sqrt{u}\right);$$

viz. this is

$$502{\cdot}5\ (\log=2{\cdot}7011399)\sqrt{u},$$

used for $u=1, 2, \ldots 10$, that is to $p=1660$. The calculation was afterwards continued by quadratures, by giving to u a succession of values, at intervals at first of 10, and afterwards of 20, 50, 100, 200, and 500, up to $p=10{,}000$, giving for the integral the value 10411; and thence, as appearing above, the value for $p=\infty$ was found to be $=12490$.

48. After the calculation of the values of $\Pi(p)$ and $\Psi(q)$, it was easy by interpolation to revert these tables, so as to obtain a table which, for Π or Ψ as argument, gives the values of p and q. The arguments are taken at intervals of 500; up to 10000 as regards p, since the original table was only calculated thus far; and up to 34726 as regards q. I had thus calculated the annexed Table III., when it occurred to me that there was a convenience in taking the arguments to be submultiples of the complete integral 34726; say we divide this into 90 parts, or, as it were, graduate the quadrant of the hyperboloid by means of hyperbolic curves of curvature adapted for the geodesics in question. Taking every fifth part, or in fact dividing the quadrant into 18 parts, we have the Table IV.

49. It will be remembered that the foregoing results apply only to the geodesics which touch the oval curve of curvature $p=+1650$; for the geodesics touching any other oval curve of curvature, the values of the integrals, and the mutual distances of the curves of curvature used for tracing the geodesics, would be completely altered. But it is possible to derive some general conclusions as to the geodesics that touch a given oval curve of curvature.

Observe that the integral K' (= 34726 in the case considered) measures the quadrant of the hyperboloid; viz. $\Psi(q)=0$, $\Psi(q)=K'$ determine two hyperbolic curves of curvature (principal sections), the mutual distance whereof is a quadrant. Each geodesic touches the given oval curve of curvature, and it touches at infinity the two hyperbolic curves $\Psi(q)=Q\pm K$ ($K=12490$ in the case considered); viz. the distance of these in regard to the circuit of four quadrants, or say the amplitude of the geodesic, is measured by the ratio $\dfrac{K}{2K'}$.

50. Now it is easy to see that as the oval curve of curvature approaches the principal elliptic section, that is, as θ' approaches c' (or writing $\theta'=c'+m$, as m

diminishes towards zero), the integral K' alters its value only slowly, increasing towards a certain constant limit; but, contrariwise, K increases without limit, its value for any small value of m being of the form $A - B\log m$, $=\infty$ in the limit; wherefore, as m diminishes, the value of $\frac{K}{2K'}$, the amplitude of the geodesic, continually increases. If this is $=1$, the geodesic touching at infinity a certain hyperbolic curve of curvature, in descending to touch the oval curve, makes round the hyperboloid a half-convolution, and then again ascends through another half-convolution to touch at infinity the same hyperbolic curve of curvature; viz. it makes in all one entire convolution, or say in descending it makes a half-convolution. But if $K \div 2K' = 2$, then the curve makes in descending a complete convolution; and so, if $K \div 2K' = 2s$, then the geodesic makes in descending s convolutions; and, as already mentioned, ultimately when $m=0$ the geodesic makes an infinity of convolutions; that is, it never actually reaches the elliptic principal section, but has this line for an asymptote.

51. To sustain the foregoing statements, I write $\theta'\,(=c'+m) = 1600+m$, and I consider the integral

$$K'_m = 100000\int_0^\infty du\sqrt{\frac{u+1600+m}{(u+2500+m)(u+2000+m)(u+m)u}},$$

say for a moment this is

$$K'_m = \int_0^\infty U_m\,du.$$

Supposing m to be small, we divide the integral into two parts, say from 0 to α [where α, $=$ for example 50 or 100, is large in comparison with m, but small in comparison with the numbers (c', &c.), 1600, &c.], and from α to ∞. In the second part, the expression under the integral sign and the value of the integral varies slowly with m, and we may, as an approximation, write $m=0$. We have thus

$$K_m = \int_0^\alpha U_m\,du + \int_\alpha^\infty U_0\,du,$$

and the first part hereof is

$$=100000\sqrt{\left(\frac{1600}{2500\,.\,2000}\right)}\int_0^\alpha \frac{du}{\sqrt{u(u+m)}};$$

viz. the integral is here

$$h.l.\,\{u+\tfrac{1}{2}m+\sqrt{u(u+m)}\} = h.l.\,\frac{\alpha+\frac{1}{2}m+\sqrt{\alpha(\alpha+m)}}{\frac{1}{2}m} = h.l.\,\frac{4\alpha}{m}\ \text{approximately};$$

or say this is

$$=\frac{1}{\log e}\log\frac{4\alpha}{m}.$$

The first term is thus

$$=100000\sqrt{\frac{1600}{2500\,.\,2000}}\,\frac{1}{\log e}\log\frac{4\alpha}{m},$$

which is

$$=4119\log\frac{4\alpha}{m},$$

or we have

$$K'_m = 4119 \log \frac{4\alpha}{m} + \int_{\alpha}^{\infty} U_0 \, du.$$

We ought to have the same value of the integral, whatever, within proper limits, the assumed value of α may be. Taking, for instance, $\alpha = 50$ and $\alpha = 100$, we ought to have

$$K'_m = 4119 \log \frac{400}{m} + \int_{100}^{\infty} U_0 \, du$$

$$= 4119 \log \frac{200}{m} + \int_{50}^{\infty} U_0 \, du ;$$

that is,

$$4119 \log 2 = \int_{50}^{100} U_0 \, du.$$

In verification, I calculated the second side by quadratures; viz. for the values 50, 60, 70, 80, 90, 100, the values of U_0 are 35·532, 29·570, 25·311, 22·373, 19·632, 17·645; whence, adding the half sum of the extreme terms to the sum of the mean terms, and multiplying by 10, the value of the integral is $=1234{\cdot}74$. The value of the left-hand side is $=1239{\cdot}94$, which is a sufficient agreement.

52. Returning to the formula for K'_m, this may be written

$$K'_m = \left(4119 \log 400 + \int_{100}^{\infty} U_0 \, du\right) - 4119 \log m.$$

I did not calculate the value of the integral in this formula, but determined the term in ( ) in such wise that the formula should be correct for the foregoing value $m = 50$; viz. the term thus is

$$= 12490 + 4119 \log 50 = 12490 + 6998, \ = 19488, \text{ or say } 19500;$$

we thus have

$$K_m = 19500 - 4119 \log m ;$$

and we may roughly assume that, for any small value of m, K'_m has the same value as for $m = 50$; viz. we may write

$$K'_m = 34726, \text{ or say } = 35000.$$

We thus see how to give to m such a value that the quantity $\frac{K}{2K'}$, which is the number of convolutions of the geodesic, may have any given value; and, in particular, we see how exceedingly small m must be for any moderately large number of convolutions; for instance, $m = \frac{1}{100{,}000000}$ or $\log m = -8$, $K = 19500 + 32952$, $=$ say 52500, or the number is $= \frac{525}{700}$, about five-sevenths of a convolution.

CORRECTION. Instead of speaking, as above, of a geodesic as touching at infinity a hyperbolic curve of curvature, the accurate expression is that the geodesic at infinity is parallel to a certain hyperbolic curve of curvature. The geodesic has, in fact, for asymptote the right line on the surface parallel at infinity to such curve of curvature. Added Dec. 1873.

TABLE I.

p	$\Pi(p)$
1650	000
1	502
2	711
3	870
4	1005
5	1124
6	1231
7	1329
8	1421
9	1507
1660	1589
70	2188
80	2605
90	2933
1700	3205
10	3438
20	3642
30	3824
40	3989
50	4130
60	4287
70	4414
80	4532
90	4642
1800	4745
20	4945
40	5106
60	5261
80	5403
1900	5533
20	5654
40	5766
60	5871
80	5970
2000	6063
50	6275
100	6462
150	6629
200	6779
250	6916
300	7041
350	7156
400	7262
450	7362

TABLE II.

q	$\Psi(q)$
− 400	000
1	883
2	1249
3	1530
4	1767
5	1975
6	2164
7	2337
8	2499
9	2650
410	2794
20	4016
30	4952
40	5752
50	6467
60	7124
70	7739
80	8321
90	8877
500	9413
10	9931
20	10436
30	10923
40	11406
50	11880
60	12347
70	12809
80	13266
90	13720
600	14171
10	14619
20	15067
30	15513
40	15960
50	16408
60	16857
70	17308
80	17762
90	18220
700	18682
10	19149
20	19623
30	20101
− 40	20588

TABLE I. (*continued*).

p	$\Pi(p)$
2500	7454
600	7625
700	7777
800	7913
900	8037
3000	8151
200	8352
400	8526
600	8679
800	8815
4000	8936
500	9216
5000	9423
500	9593
6000	9737
500	9861
7000	9970
500	10066
8000	10152
500	10229
9000	10299
500	10363
10000	10411
⋮	
∞	12490

TABLE II. (*continued*).

q	$\Psi(q)$
− 750	21086
60	21595
70	22117
80	22653
90	23206
800	23778
10	24374
20	24998
30	25655
40	26355
50	27105
60	27928
70	28861
80	29936
890	31365
1	31538
2	31720
3	31914
4	32123
5	32350
6	32600
7	32886
8	33224
9	33663
− 900	34726

TABLE III.

Π or Ψ	p	Diff.	q	Diff.
0	1650·0	1	− 400·0	·3
500	1651·0	3	400·3	1·0
1000	1654·0	5	401·3	1·6
1500	1659·0	7·8	402·9	2·2
2000	1666·8	10·7	405·1	2·9
2500	1677·5	14·9	408·0	3·7
3000	1692·4	20·6	411·7	4·0
3500	1713·0	27·8	415·7	4·3
4000	1740·8	36·6	420·0	5·2
4500	1777·4	49·4	425·2	5·4
5000	1826·8	68·1	430·6	6·2
5500	1894·9	91·5	436·8	6·6
6000	1986·4	124·6	− 443·4	7·1

TABLE III. (*continued*).

Π or Ψ	p	Diff.	q	Diff.
6500	2111	124·6	− 450·5	7·1
7000	2283	172	458·1	7·6
7500	2527	244	466·1	8·0
8000	2870	353	474·5	8·4
8500	3285	415	483·2	8·7
9000	4114	829	492·3	9·1
9500	5226	1112	501·7	9·4
10000	7156	1930	511·3	9·6
500	.		521·1	9·8
11000	.		531·5	10·1
500	.		542·0	10·5
12000	.		552·5	10·8
500	∞		563·3	10·8
13000			574·2	10·9
500			585·1	10·9
14000			596·2	11·1
500			607·3	11·1
15000			618·5	11·2
500			629·7	11·2
16000			641·0	11·3
500			652·1	11·1
17000			663·2	11·1
500			674·2	11·0
18000			685·2	11·0
500			696·1	10·9
19000			706·8	10·7
500			717·4	10·6
20000			728·0	10·6
500			738·4	10·4
21000			748·3	9·9
500			758·1	9·8
22000			767·7	9·6
500			777·4	9·4
23000			786·4	9·0
500			795·1	8·7
24000			803·7	8·6
500			812·0	8·3
25000			820·0	8·0
500			827·6	7·6
26000			834·9	7·3
500			− 841·9	7·0
				6·7

TABLE III. (*continued*).

Π or Ψ	p	Diff.	q	Diff.
				6·7
27000			– 848·6	6·2
500			854·8	6·0
28000			860·8	5·3
500			866·1	5·2
29000			871·3	4·7
500			876·0	4·6
30000			880·6	3·4
500			884·0	3·5
31000			887·5	3·3
500			890·8	3·6
32000			893·4	2·2
500			895·6	1·7
33000			897·5	1·1
500			898·6	0·7
34000			899·3	0·5
500			899·8	0·2
34726			– 900	

TABLE IV.

	Π or Ψ	p	Diff.	q	Diff.
0	0	1650	15·7	– 400	4·8
1	1929	1665·7	66·4	404·8	13·9
2	3858	1732·1	212·1	418 7	21·8
3	5788	1944·2	713·5	440·5	29·1
4	7717	2657·7	3040·0	469·6	34·9
5	9646	5697·7		504·5	39·1
6	11575			543·6	41·6
7	13504			585·2	42·8
8	15434			628·0	43·2
9	17363			671·2	41·8
10	19292			713·0	39·7
11	21221			752·7	36·3
12	23150			789·0	32·2
13	25079			821·2	27·4
14	27008			848·6	22·1
15	28938			870·7	15·8
16	30868			886·5	10·2
17	32797			896·7	3·3
18	34726			– 900	

512.

ON A CORRESPONDENCE OF POINTS IN RELATION TO TWO TETRAHEDRA.

[From the *Proceedings of the London Mathematical Society*, vol. IV. (1871—1873), pp. 396—404. Read June 12, 1873.]

THE following question has been considered by R. Sturm in an interesting paper, "Das Problem der Projectivität und seine Anwendung auf die Flächen zweiten Grades," *Math. Ann.*, t. I. (1870), pp. 533—574: Given *in plano* two groups of the same number (5, 6, or 7) of points, to find points P, P' homographically related to these two groups respectively; viz. the lines from P to the points of the first group and those from P' to the points of the second group are to be homographic pencils. In the present paper I require only a particular form of these results; viz. in each group two of the points are the circular points at infinity; or, disregarding these, we have two groups of 3, 4, or 5 points such that the points of the first group at P, and those of the second group at P', subtend equal angles. I give for this particular case an independent analytical investigation; but I will first state the results included in the more general ones obtained by Sturm.

If the points A, B, C at P and the points A', B', C' at P' subtend equal angles, then to any given position of the one point corresponds a single position of the other point; viz. the two points have a (1, 1) correspondence; the nature of this being, that to any line in the one figure corresponds in the other figure a quintic curve, having 6 dps.; viz. the three points, the two circular points at infinity I, J, and one other fixed point of that figure {say for the first figure this fixed point is (ABC)}.

If the points A, B, C, D at P and the points A', B', C', D' at P' subtend equal angles, then the locus of each point is a cubic curve; viz. the locus of P passes through A, B, C, D, I, J and the four fixed points (ABC), (ABD), (ACD), (BCD); and the like for the locus of P'.

Finally (although this is a theorem which I do not require), if the points A, B, C, D, E at P and the points A', B', C', D', E' at P' subtend equal angles, then there are three positions of each point.

The problem I propose to consider is: Given the tetrahedra $ABCD$ and $A'B'C'D'$, it is required in the planes ABC and $A'B'C'$ respectively to find the points P, P' such that A, B, C, D at P, and A', B', C', D' at P', subtend equal angles. I was led to this by the more general problem, which I do not at present discuss: Given the two tetrahedra, it is required to find the loci of the points P, P' such that A, B, C, D at P, and A', B', C', D' at P', subtend equal angles.

Here, drawing from D, D' the perpendiculars $DK, D'K'$ on the planes ABC and $A'B'C'$ respectively, we have A, B, C, K at P, and A', B', C', K' at P', subtending equal angles, and such that the distances PK and $P'K'$ are proportional to the heights of the tetrahedra (for the triangles PDK and $P'D'K'$ are obviously similar). The required points P, P' are each the intersection of two loci, viz.:

1. P is such that A, B, C, K at P, and A', B', C', K' at P', subtend equal angles; locus is a cubic through $A, B, C, K, I, J, (ABC), (ABK), (ACK), (BCK)$.
2. P is such that A, B, K at P, and A', B', K' at P', subtend equal angles, and that PK and $P'K'$ are in a given ratio; locus is a certain octic curve Ω;

and the required positions of P are obtained as the intersections of the two loci.

I proceed to the analytical investigation.

Preliminary Formulæ.

1. Consider a triangle ABC, and let the position of a point P be determined by means of its coordinates x, y, z, which are equal to the perpendicular distances of P from the sides, each divided by the perpendicular distance of the opposite vertex (as usual, x, y, z are positive for a point within the triangle); or what is the same thing, $x, y, z = PBC, PCA, PAB$, divided each by ABC, whence identically $x + y + z = 1$.

Suppose for a moment the rectangular coordinates of A, B, C are (α_1, β_1), (α_2, β_2), (α_3, β_3) respectively; and that those of P are X, Y. Also let the sides BC, CA, AB be $= a, b, c$ respectively.

We have

$$X = \alpha_1 x + \alpha_2 y + \alpha_3 z,$$
$$Y = \beta_1 x + \beta_2 y + \beta_3 z,$$
$$1 = \quad x + \quad y + \quad z;$$

and if we consider a second point P', the coordinates of which are x', y', z' and X', Y', we have the like relations between these quantities. Calling δ the distance of the two points P, P', we may write

$$\begin{aligned}\delta^2 = &\ \{\alpha_1 (x-x') + \alpha_2 (y-y') + \alpha_3 (z-z')\}^2 \\ &+ \{\beta_1 (x-x') + \beta_2 (y-y') + \beta_3 (z-z')\}^2 \\ &- \{(\alpha_1^2+\beta_1^2)(x-x') + (\alpha_2^2+\beta_2^2)(y-y') + (\alpha_3^2+\beta_3^2)(z-z')\} \\ &\quad \times \{(x-x') + (y-y') + (z-z')\},\end{aligned}$$

the last term being in fact $=0$; viz. this is

$$\begin{aligned}\delta^2 = &- \{(\alpha_2-\alpha_3)^2 + (\beta_2-\beta_3)^2\}(y-y')(z-z') \\ &- \{(\alpha_3-\alpha_1)^2 + (\beta_3-\beta_1)^2\}(z-z')(x-x') \\ &- \{(\alpha_1-\alpha_2)^2 + (\beta_1-\beta_2)^2\}(x-x')(y-y');\end{aligned}$$

or what is the same thing, the expression for the distance δ^2 of the two points P, P' is

$$\delta^2 = -a^2 (y-y')(z-z') - b^2 (z-z')(x-x') - c^2 (x-x')(y-y'),$$

which expression may be modified by means of the identical equations

$$1 = x+y+z, \quad 1 = x'+y'+z';$$

viz. writing

$$yz'-y'z,\ zx'-z'x,\ xy'-x'y = \xi,\ \eta,\ \zeta,$$

we have

$$\begin{aligned}x-x' &= x(x'+y'+z') - x'(x+y+z) = \zeta - \eta, \\ y-y' &= \xi - \zeta, \\ z-z' &= \eta - \xi;\end{aligned}$$

and consequently

$$\begin{aligned}\delta^2 = &\ a^2 (-\xi^2 - \eta\zeta + \zeta\xi + \xi\eta) \\ &+ b^2 (-\eta^2 + \eta\zeta - \zeta\xi + \xi\eta) \\ &+ c^2 (-\zeta^2 + \eta\zeta + \zeta\xi - \xi\eta).\end{aligned}$$

2. Treating x', y', z' as constants and x, y, z as current coordinates, the formula for δ^2 is of course the equation of a circle, centre x', y', z' and radius δ. It thus also appears that the general equation of a circle is

$$-a^2yz - b^2zx - c^2xy + (Lx+My+Nz)(x+y+z) = 0;$$

viz. writing $-a^2yz - b^2zx - c^2xy = U$, and $x+y+z = \Omega$, this is

$$U + (Lx+My+Nz)\,\Omega = 0,$$

where $U=0$ is the circle circumscribed about the triangle ABC, and $\Omega=0$ is the line infinity. Of course the general equation of a circle passing through the points

(B, C) is $U + Lx\,\Omega = 0$, and similarly those of circles through (C, A) and through (A, B) are $U + My\,\Omega = 0$ and $U + Nz\,\Omega = 0$ respectively. But we require the interpretation of the coefficients L, M, N which enter into these equations.

3. Considering the triangle ABC, if through B, C we have a circle, this is by the side BC divided into two segments, and I consider that lying on the same side with A as the positive segment, and define the angle of the circle to be the angle in this positive segment. It is clear that if we have *within* the triangle a point P, and, through this point and (B, C), (C, A), (A, B) respectively, three circles, then if α, β, γ be the angles of these circles, we have $\alpha + \beta + \gamma = 2\pi$; and conversely, if the circles through (B, C), (C, A), (A, B) are such that their angles α, β, γ satisfy the relation $\alpha + \beta + \gamma = 2\pi$, then the three circles meet in a point. But it is further to be noticed, that if, producing the sides of the triangle so as to divide the plane into seven spaces, the triangle, three trilaterals, and three bilaterals, we take the point P within one of the bilaterals, we still have $\alpha + \beta + \gamma = 2\pi$; but taking it within one of the trilaterals, we have $\alpha + \beta + \gamma = \pi$. And the converse theorem is, that if the three circles (B, C), (C, A), (A, B) are such that $\alpha + \beta + \gamma = \pi$ or 2π, then the circles meet in a point; viz. if the sum is 2π, then this point lies in the triangle or one of the bilaterals; but if the sum is $= \pi$, then this point lies in a trilateral.

4. I seek for the equation of a circle through the points B, C, and containing the angle L. The equation in rectangular coordinates is easily seen to be

$$(X - \alpha_2)(X - \alpha_3) + (Y - \beta_2)(Y - \beta_3) - \cot L\,\{(\beta_2 - \beta_3) X - (\alpha_2 - \alpha_3) Y + \alpha_2\beta_3 - \alpha_3\beta_2\} = 0.$$

In fact this is the equation of a circle through (B, C); and taking for a moment the origin at B, and axis of X to coincide with BC, or writing $\alpha_2, \beta_2 = 0, 0$; $\alpha_3, \beta_3 = a, 0$, the equation is

$$X(X - a) + Y^2 - aY \cot L = 0,$$

viz. the equation of the tangent at B is $-aX - aY \cot L = 0$, that is, $Y = -X \tan L$, or the angle in the positive segment is $= L$.

If for a moment λ, μ, ν are the inclinations of the sides of the triangle ABC to the axis of X, then A, B, C being the angles, we may write

$$\begin{aligned} \mu - \nu &= \pi - A, \\ \nu - \lambda &= \pi - B, \\ \lambda - \mu &= -\pi - C, \end{aligned}$$

and

$$\begin{aligned} X - \alpha_2 &= (\alpha_1 x + \alpha_2 y + \alpha_3 z) - \alpha_2 (x + y + z) = c \cos\nu \,.\, x - a \cos\lambda \,.\, z, \\ X - \alpha_3 &= (\alpha_1 x + \alpha_2 y + \alpha_3 z) - \alpha_3 (x + y + z) = -b \cos\mu \,.\, x + a \cos\lambda \,.\, y, \\ Y - \beta_2 &= \beta_1 x + \beta_2 y + \beta_3 z - \beta_2 (x + y + z) = c \sin\nu \,.\, x - a \sin\lambda \,.\, z, \\ Y - \beta_3 &= \beta_1 x + \beta_2 y + \beta_3 z - \beta_3 (x + y + z) = -b \sin\nu \,.\, x + a \sin\lambda \,.\, z; \end{aligned}$$

whence

$$\begin{aligned} &(X - \alpha_2)(X - \alpha_3) + (Y - \beta_2)(Y - \beta_3) \\ &= -a^2 yz - b^2 zx - c^2 xy + bc \cos A \,.\, x^2 + (b^2 - ab \cos C)\, zx + (c^2 - ac \cos B)\, xy; \end{aligned}$$

viz. this is

$$= - a^2yz - b^2zx - c^2xy + bc\cos A \,.\, x(x+y+z).$$

Moreover, if $\Delta =$ twice the area of the triangle, then

$$(\beta_2 - \beta_3) X - (\alpha_2 - \alpha_3) Y + \alpha_2\beta_3 - \alpha_3\beta_2 = \Delta x (x+y+z) = bc \sin A \,.\, x(x+y+z);$$

so that the equation becomes

$$- a^2yz - b^2zx - c^2xy + bc\sin A\,(\cot A - \cot L)\, x(x+y+z) = 0,$$

or, what is the same thing,

$$- a^2yz - b^2zx - c^2xy + \Delta\,(\cot A - \cot L)\, x(x+y+z) = 0,$$

or, if we please,

$$- a^2yz - b^2zx - c^2xy + \Delta\,(\cot A - \cot L)\, x = 0.$$

Writing as before,

$$- a^2yz - b^2zx - c^2xy = U, \qquad x+y+z = \Omega,$$

the equation is

$$U + \Delta\,(\cot A - \cot L)\,\Omega x = 0;$$

or forming the like equations of two other similar circles, we have the circles (B, C), (C, A), (A, B) containing the angles L, M, N respectively; and the equations are

$$U + \Delta\,(\cot A - \cot L)\,\Omega x = 0,$$
$$U + \Delta\,(\cot B - \cot M)\,\Omega y = 0,$$
$$U + \Delta\,(\cot C - \cot N)\,\Omega z = 0.$$

Correspondence, A, B, C at P, and A', B', C' at P', subtending equal angles.

5. Consider now the two figures A', B', C', subtending at P' the same angles L, M, N which A, B, C subtend at P; then we have

$$\frac{U}{\Omega\Delta x} + \cot A - \cot L = 0, \quad \frac{U'}{\Omega'\Delta' x'} + \cot A' - \cot L = 0,$$
$$\frac{U}{\Omega\Delta y} + \cot B - \cot M = 0, \quad \frac{U'}{\Omega'\Delta' y'} + \cot B' - \cot M = 0,$$
$$\frac{U}{\Omega\Delta z} + \cot C - \cot N = 0, \quad \frac{U'}{\Omega'\Delta' z'} + \cot C' - \cot N = 0;$$

and thence

$$\frac{U}{\Omega\Delta x} + \cot A = \frac{U'}{\Omega'\Delta' x'} + \cot A',$$
$$\frac{U}{\Omega\Delta y} + \cot B = \frac{U'}{\Omega'\Delta' y'} + \cot B',$$
$$\frac{U}{\Omega\Delta z} + \cot C = \frac{U'}{\Omega'\Delta' z'} + \cot C';$$

and consequently

$$\frac{1}{x'} : \frac{1}{y'} : \frac{1}{z'} = \frac{U}{\Omega\Delta x} + \cot A - \cot A'$$

$$: \frac{U}{\Omega\Delta y} + \cot B - \cot B'$$

$$: \frac{U}{\Omega\Delta z} + \cot C - \cot C',$$

or, what is the same thing,

$$x' : y' : z' = \frac{x}{U + \Delta(\cot A - \cot A')\,\Omega x}$$

$$: \frac{y}{U + \Delta(\cot B - \cot B')\,\Omega y}$$

$$: \frac{z}{U + \Delta(\cot C - \cot C')\,\Omega z},$$

where observe that the equations

$$U + \Delta(\cot A - \cot A')\,\Omega x = 0,$$
$$U + \Delta(\cot B - \cot B')\,\Omega y = 0,$$
$$U + \Delta(\cot C - \cot C')\,\Omega z = 0,$$

represent circles (B, C), (C, A), (A, B) containing the angles A', B', C'; and since $A' + B' + C' = \pi$, these meet in a point O. We may for convenience write

$$x' : y' : z' = \frac{BC}{BCO} : \frac{CA}{CAO} : \frac{AB}{ABO},$$

where $BC = 0$ denotes $(x = 0)$ the line BC; $BCO = 0$ the circle through B, C, O. And of course, in like manner,

$$x : y : z = \frac{B'C'}{B'C'O'} : \frac{C'A'}{C'A'O'} : \frac{A'B'}{A'B'O'};$$

so that the points P, P' have a rational, or (1, 1), correspondence.

Writing

$$\begin{aligned} x' : y' : z' &= BC\,.\,CAO\,.\,ABO : CA\,.\,ABO\,.\,BCO : AB\,.\,BCO\,.\,CAO \\ &= \quad X \quad : \quad Y \quad : \quad Z \end{aligned}$$

suppose, X, Y, Z are quintic functions of x, y, z, and the curve in the first figure corresponding to the line $\alpha x' + \beta y' + \gamma z' = 0$ of the second figure is

$$\alpha X + \beta Y + \gamma Z = 0;$$

viz. this is a quintic curve having dps. at each of the points A, B, C, O, I, J. In fact, if for BCO we write $BCOIJ$, and so for the other two circles respectively, we have in an algorithm which will be at once understood $X = BC\,.\,CAOIJ\,.\,ABOIJ$, $= (ABCOIJ)^2$, and similarly $Y = Z$, $= (ABCOIJ)^2$, or the curve is $(ABCOIJ)^2 = 0$.

Correspondence, A, B, C, D at P and A′, B′, C′, D′ at P′ subtending equal angles.

6. Consider now *in plano* the points A, B, C, D which at P, and the points A', B', C', D' which at P', subtend equal angles. Let a, b, c, f, g, h denote the perpendicular distances of P from the lines BC, CA, AB, AD, BD, CD respectively; and the like as to a', b', c', f', g', h'. Observe that, neglecting constant factors, a, b, c are what were before represented by x, y, z; we may consider the coordinates of P in regard to the triangles ABC, BCD, CAD, ABD to be (a, b, c), (a, h, g), (b, f, h), (c, g, f) respectively. We have in regard to ABC the point O as before, and in regard to BCD, CAD, ABD the points O_1, O_2, O_3 respectively. Then A, B, C at P and A', B', C' at P' subtending equal angles, we may write

$$a' : b' : c' = \frac{a}{BCO} : \frac{b}{CAO} : \frac{c}{ABO};$$

viz. $BCO = 0$ is here the circle through B, C, O, and the like for CAO and ABO, the expressions being multiplied into the proper constant factors to take account of the constant factors whereby a, b, c and a', b', c' differ from x, y, z and x', y', z' respectively.

We have in like manner

$$a' : h' : g' = \frac{a}{BCO_1} : \frac{h}{CDO_1} : \frac{g}{BDO_1},$$

$$b' : f' : h' = \frac{b}{CAO_2} : \frac{f}{ADO_2} : \frac{h}{CDO_2},$$

$$c' : g' : f' = \frac{c}{ABO_3} : \frac{g}{BDO_3} : \frac{f}{ADO_3}.$$

From the ratios of (f', g', h'), (b', c', f'), (c', a', g'), (a', b', h') respectively we deduce

$$CDO_1 . ADO_2 . BDO_3 - BDO_1 . CDO_2 . ADO_3 = 0,$$

$$CAO . ABO_3 . ADO_2 - ABO . ADO_3 . CAO_2 = 0,$$

$$ABO . BCO_1 . BDO_3 - BCO . BDO_1 . ABO_3 = 0,$$

$$BCO . CAO_2 . CDO_1 - CAO . CDO_2 . BCO_1 = 0,$$

each of which equations represents a sextic curve; and admitting that it can be shown that these pass through O, O_1, O_2, O_3 respectively, the forms are

$$D^3 A\, B\, C\, O\, O_1 O_2 O_3 I^3 J^3 = 0,$$

$$D\, A^3 B\, C\, O\, O_1 O_2 O_3 I^3 J^3 = 0,$$

$$D\, A\, B^3 C\, O\, O_1 O_2 O_3 I^3 J^3 = 0,$$

$$D\, A\, B\, C^3 O\, O_1 O_2 O_3 I^3 J^3 = 0.$$

7. Now the locus of P is evidently a curve, and this can only happen by reason that the four left-hand functions contain a common factor, and the form of them suggests that this common factor is $ABCDOO_1O_2O_3IJ$, the four extraneous factors being $D^2I^2J^2$, $A^2I^2J^2$, $B^2I^2J^2$, $C^2I^2J^2$; viz. $ABCDOO_1O_2O_3IJ=0$ is a cubic curve passing through the ten points; and $D^2I^2J^2=0$ a cubic curve through each of the points D, I, J twice; viz. it is the triad of lines IJ, DI, DJ; and the like as to the other extraneous factors $A^2I^2J^2$, $B^2I^2J^2$, and $C^2I^2J^2$. I have not worked out the analysis to verify this *à posteriori*; but, the conclusion agreeing with Sturm, I accept it without further investigation, viz. the result is that A, B, C, D at P and A', B', C', D' at P' subtending equal angles, the locus of P is a cubic curve $ABCDOO_1O_2O_3O_4IJ=0$ through the ten points thus represented; and of course the locus of P' is in like manner a cubic curve $A'B'C'D'O'O_1'O_2'O_3'O_4'IJ=0$ through the ten points thus represented.

Correspondence, A, B, C, D, E at P and A', B', C', D', E' at P' subtending equal angles.

8. We may go a step further, and consider A, B, C, D, E at P and A', B', C', D', E' at P' subtending equal angles. Attending only to the points A, B, C, D and A', B', C', D', the locus of P is a cubic curve

$$ABCDOO_1O_2O_3O_4IJ=0;$$

and similarly attending to the points A, B, C, E and A', B', C', E', the locus of P is a cubic curve

$$ABCEOQ_1Q_2Q_3IJ=0.$$

(Observe that O, as depending only on A, B, C, is the same point as before; but that Q_1, Q_2, Q_3, as depending on E instead of D, are not the same as O_1, O_2, O_3.) The two cubic curves have in common the points A, B, C, I, J, O, and they consequently intersect in three other points; that is, there are three positions of the point P, and of course three corresponding positions of P'.

Correspondence, A, B, C at P and A', B', C' at P' subtending equal angles, and AP, $A'P'$ in a given ratio.

9. Consider, as before, A, B, C at P and A', B', C' at P' subtending equal angles, and the points P, P' being moreover such that the distances AP, $A'P'$ are in a given ratio. I write for shortness

$$x' : y' : z' = \frac{x}{L} : \frac{y}{M} : \frac{z}{N},$$

where L, M, N denote

$$U+\Delta(\cot A-\cot A')\,\Omega x,\quad U+\Delta(\cot B-\cot B')\,\Omega y,\quad U+\Delta(\cot C-\cot C')\,\Omega z,$$

respectively. We have

$$\begin{aligned}(AP)^2 &= -a^2yz-b^2z(x-1)-c^2(x-1)y,\\ &= -a^2yz-b^2zx-c^2xy+(b^2z+c^2y)(x+y+z),\\ &= \quad c^2y^2+b^2z^2+(b^2+c^2-a^2)\,yz;\end{aligned}$$

or, what is the same thing,

$$(A\,P\,)^2 = c^2y^2 + b^2z^2 + 2bc\,\cos A\,.\,y\,z\,;$$

and similarly

$$(A'P')^2 = c'^2y'^2 + b'^2z'^2 + 2b'c'\cos A'\,.\,y'z'.$$

The required relation therefore is

$$\frac{c'^2y'^2 + b'^2z'^2 + 2b'c'\cos A'\,.\,y'z'}{(x'+y'+z')^2} = \theta^2\,.\,\frac{c^2y^2 + b^2z^2 + 2bc\cos A\,.\,yz}{(x+y+z)^2}\,;$$

viz. substituting for x', y', z' their values, this is

$$L^2(x+y+z)^2(b'^2z^2M^2 + c'^2y^2N^2 + 2b'c'yzMN\cos A)$$
$$= \theta^2(b^2z^2 + c^2y^2 + 2bcyz\cos A)(xMN + yNL + zLM)^2,$$

which is an equation of the 12th order. I say that the points A, B, C, O, I, J are each quadruple. In fact, according to the foregoing algorithm, we may write

$$x+y+z = IJ,\quad zM = AB\,.\,CAOIJ,\ \&\text{c.},$$
$$xL = yM = zN = A^2BCOIJ,$$
$$y = z = A,\quad xMN = BC\,.\,CAOIJ\,.\,ABOIJ,\ \&\text{c.},$$
$$xMN = yNL = zLM = (ABCOIJ)^2;$$

and the equation is

$$(BCOIJ)^2(IJ)^2(A^2BCOIJ)^2 = \theta^2\,.\,A^2(ABCOIJ)^4,$$

that is

$$(IJ)^2(ABCOIJ)^4 = \theta^2\,.\,A^2(ABCOIJ)^4;$$

so that the points are each quadruple.

The two Tetrahedra; A, B, C, D at P in ABC and A′, B′, C′, D′ at P′ in A′B′C′ subtending equal angles.

10. I consider now the before-mentioned problem of the two tetrahedra; viz. on the two bases ABC and $A'B'C'$ respectively, letting fall the perpendiculars DK and $D'K'$, then first A, B, C, K at P and A', B', C', K' at P' subtend equal angles; the locus of P is a cubic curve $ABCKOO_1O_2O_3IJ=0$ through these ten points. ($O = ABC$ is derived from the points A, B, C; and in like manner $O_1 = BCK$, $O_2 = CAK$, $O_3 = ABK$.)

Next, B, C, K at P and B', C', K' at P' subtend equal angles, and moreover the distances KP and $K'P'$ are in a given ratio; the locus of P is a 12-thic curve

$$(BCKO_1IJ)^4 = 0,$$

having each of these six points as a quadruple point. Hence among the 36 intersections of the two curves we have the points B, C, K, O_1, I, J each 4 times, and there remain $36-24$, $=12$ intersections.

The conclusion is that A, B, C, D at a point P of ABC, and A', B', C', D' at a point P' of $A'B'C'$, subtending equal angles, there are 12 positions of P, and of course 12 corresponding positions of P'.

513.

ON A BICYCLIC CHUCK.

[From the *Philosophical Magazine*, vol. XLIII. (1872), pp. 365—367.]

THE apparatus, although I have called it a chuck, is constructed not for turning, but for drawing; viz. it rotates horizontally on a table (being moved, not from the inside by the axle of the lathe, but from the outside by a handle-frame), carrying a drawing-board which works under a fixed pencil supported by a bridge. Two points of the drawing-board describe circles; and the curve traced out on the drawing-board is consequently that described by a fixed point upon a moving plane two points of which describe circles; or, what is really the same thing, it is the curve described on a fixed plane by a point rigidly connected with two points each of which describes a circle. The apparatus is at once convertible into an oval chuck of nearly the ordinary construction; viz. it may be arranged so that the curve described on the drawing-board shall be an ellipse.

Bottom plane is a rectangular board (1) (see figure) about 30 inches by 24 inches, having in the middle a sliding-piece (2) carrying a block (3).

Second plane contains two circular segments (4) fixed to the bottom plane, serving as an axle for the moving piece (5) next referred to, and allowing the block (3) to move between them. And in the same plane we have a moving piece (5) in the form of a rectangle with a circle cut out thereof, rotating about the segments (4), and having upon it a groove in which works a sliding-piece (6) carrying a block (7); there is in this block a circular hole, *D*. The second plane includes also two sides (8) of a handle-frame, which two sides slide along two of the sides of the piece (5).

Third plane consists of a rectangular piece (9) rotating about an axle fixed to the block (3), and having a sliding-piece (10) in which is a circular hole, *C*. The third plane includes also the before-mentioned block (7), having upon it the hole *D*;

and it includes also the remaining two sides (11) of the handle-frame, and, let into the same so as to be flush therewith on the upper surface, two slips (12) completing, in this plane, the handle-frame.

We have thus on a level the sides (11), (12) of the handle-frame and the holes C, D, where C rotates about the point B, which is the centre of the block (3); and D rotates about the point A, which is the centre of the segments (4), each hole being capable of describing a complete circle; and the distances AB, BC, CD, and DA are (within limits) adjustable to any given values: the distance of the holes C, D is made equal to that of the two pegs next referred to.

Connected herewith by means of cylindrical pegs working in the holes C, D respectively, we have a carrying-frame; viz. the fourth plane contains two sides (13) of this carrying-frame, and two moveable bars (14), attached to the remaining two sides (15) of the carrying-frame, and having on their lower surfaces the pegs which work in the holes C and D respectively—each bar being free to rotate about one extremity, and being clampable at the other extremity so as to allow the two pegs to be adjusted at a given distance from each other. And then in the fifth plane we have the remaining two sides (15) of the carrying-frame.

Rigidly connected with the carrying-frame we have the drawing-board; or, to make the whole more complete, this should be adjustable to any given position in regard to the carrying-frame by giving it two sliding motions crosswise, and a rotating motion, in the manner of an eccentric chuck.

To convert the apparatus into an oval chuck, we remove altogether the carrying-frame; and in the third plane we fix to the sides (8) of the handle-frame two bars at right angles to these sides, by means of pegs on the lower surfaces of these bars fitting tightly into holes on the sides (8) (which holes and the ends of the bars are shown in the figure), in such wise that these bars include between them the piece (9), which is thereby kept in a direction at right angles to the sides (8), and thus slides between the two bars. There are thus in the handle-frame two lines at right angles to each other, which pass through the fixed points A and B respectively; so that, now connecting the drawing-board directly with the handle-frame, the apparatus has become an oval chuck, viz. the curve traced out on the drawing-board will be an ellipse. The drawing-board should be adjustable to any given position in regard to the handle-frame, in like manner as it was to any given position in regard to the carrying-frame; it is easy to arrange as to this.

It is hardly necessary to remark that the pencil should have two sliding motions crosswise, so as to allow it to be adjusted to any given position; and a small up-and-down motion, so that it may be loaded to press with the proper force upon the drawing-board.

The variety of forms, even with a fixed adjustment of the chuck, only the position of the pencil being altered, is very considerable: among them we have bent ovals and pear-shapes, passing through cuspidal forms into bent figures-of-eight.

Plan

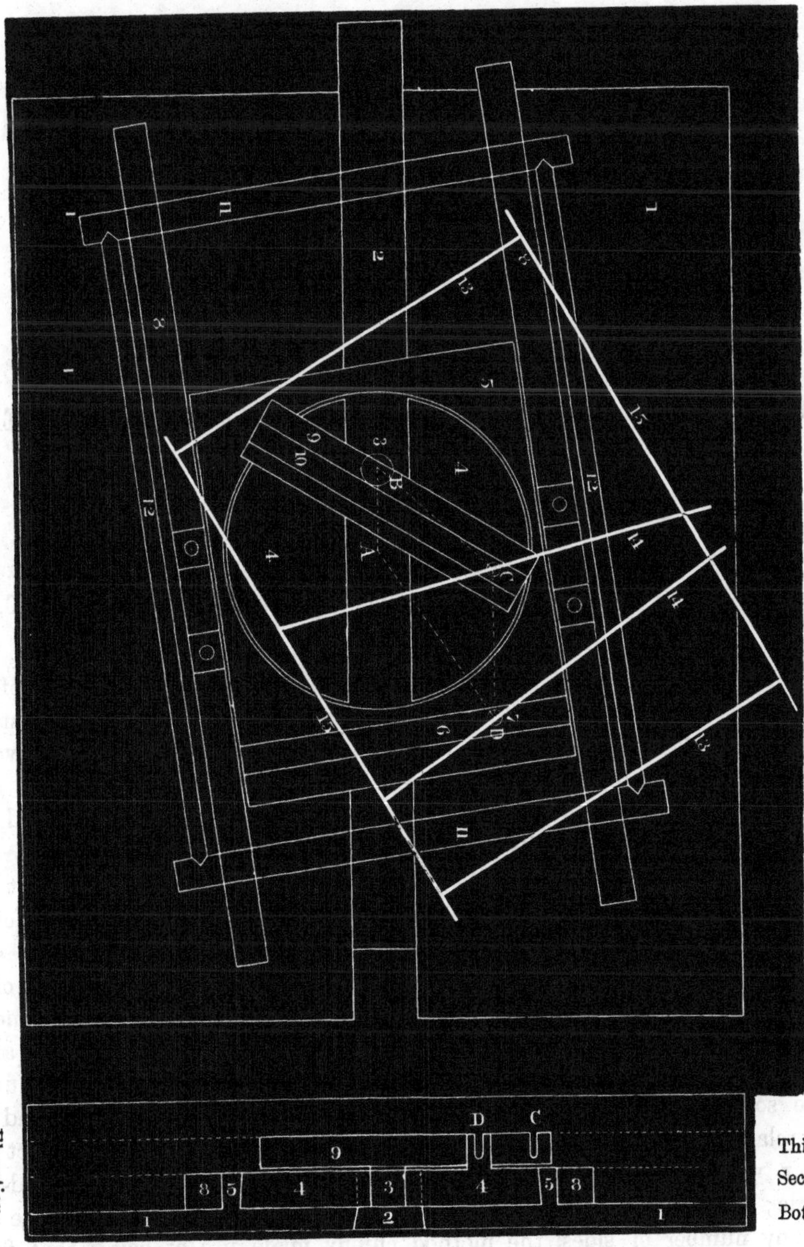

Elevation

Third plane.
Second plane.
Bottom.

514.

ON THE PROBLEM OF THE IN-AND-CIRCUMSCRIBED TRIANGLE.

[From the *Philosophical Transactions of the Royal Society of London*, vol. CLXI. (for the year 1871), pp. 369—412. Received December 30, 1870,—Read February 9, 1871.]

THE problem of the In-and-Circumscribed Triangle is a particular case of that of the In-and-Circumscribed Polygon: the last-mentioned problem may be thus stated—to find a polygon such that the angles are situate in and the sides touch a given curve or curves. And we may in the first instance inquire as to the number of such polygons. In the case where the curves containing the angles and touched by the sides respectively are all of them distinct curves, the number of polygons is obtained very easily and has a simple expression: it is equal to twice the product of the *orders* of the curves containing the several angles respectively into the product of the *classes* of the curves touched by the several sides respectively; or, say, it is equal to twice the product of the orders of the angle-curves into the product of the classes of the side-curves. But when several of the curves become one and the same curve, and in particular when the angles are all of them situate in and the sides all touch one and the same curve, it is a much more difficult problem to find the number of polygons. The solution of this problem when the polygon is a triangle, and for all the different relations of identity between the different curves, is the object of the present memoir, which is accordingly entitled "On the Problem of the In-and-Circumscribed Triangle;" the methods and principles, however, are applicable to the case of a polygon of any number of sides, the method chiefly made use of being that furnished by the theory of correspondence, as will be explained. The results (for the triangle) are given in the following Table; for the explanation of which I remark that the triangle is taken to be $aBcDeF$; viz. a, c, e are the angles, B, D, F the sides; that is, B, D, F are the sides ac, ce, ea respectively, and a, c, e are the angles FB, BD, DF

respectively. And I use the same letters a, c, e, B, D, F to denote the curves containing the angles and touched by the sides respectively; viz. the angle a is situate in the curve a, the side B touches the curve B, and so for the other angles and sides respectively. An equation such as $a = c$ or $a = B$ denotes that the curves a, c or, as the case may be, the curves a, B are one and the same curve: it is in general convenient to use a new letter for denoting these identical curves; viz. I write, for instance, $a = c = x$ or $a = B = x$, to denote that the curves a, c or, as the case may be, the curves a, B are one and the same curve x; the new letters thus introduced are x, y, z, there being in regard to them no distinction of small letters and capitals. The expression "no identities" denotes that the curves are all distinct. But I use also the letters a, c, e, b, d, f, x, y, z, and A, C, E, B, D, F, X, Y, Z quantitatively, to denote the orders and classes of the curves a, c, e, B, D, F, x, y, z respectively; thus, in the Table, for the case 1 "no identities" the number of triangles is given as $= 2aceBDF$, which agrees with the before-mentioned result for the polygon: for the case 2 the several separate identities $a = c$, $a = e$, $c = e$ are of course equivalent to each other; and selecting one of them, $a = c = x$, the number of triangles is given as $= 2x(x-1)\,eBDF$. There is a convenience in thus writing down the several forms $a = c$, $a = e$, $c = e$ of the identity or identities which constitute the 52 distinct cases of the Table; and I have accordingly done so throughout the Table, the expression for the number of triangles being however in each case given under one form only. It only remains to mention that for the curve x the Greek letter ξ denotes what may be termed the "stativity" of the curve, viz. this is = number of cusps + 3 times the class, or, what is the same thing, = number of inflections + 3 times the order; the curve being determined by its order x, class X, and ξ; and similarly for η and ζ.

Observe that, in the column "Specification," each line is to be read separately from the others, and, where the word "or" occurs, the two parts of the line are to be read separately; thus case 5, the six forms are $a = B$, $a = F$, $c = D$, $c = B$, $e = F$, $e = D$: the letter x (or, as the case may be, x, y, or x, y, z) accompanies the first of the given forms; in the present instance $a = B = x$, and it is to this first form that the number of triangles, here $2(Xx - X - x)\,ceDF$, applies.

I remark that what is primarily determined is the number of positions of a particular angle of the triangle, and that in some cases, on account of the symmetry of the figure, the number of triangles is a submultiple of this number; viz. the number of positions of the angle is to be divided by 2 or 6; this is expressly shown, by means of a separate column, in the Table.

No. of Case	Specification	No. of forms	Totals	No. of triangles	Divided by
1	No identities	1	1	$2aceBDF$	
2	$a=c=x$ $c=e$ $e=a$	3		$2x(x-1)\,eBDF$	
3	$D=F=x$ $F=B$ $B=D$	3		$2X(X-1)\,Bace$	
4	$a=D=x$ $c=F$ $e=B$	3		$2XxceBF$	
5	$a=B=x$, or $a=F$ $c=D$ „ $c=B$ $e=F$ „ $e=D$	6	15	$2(Xx-X-x)\,ceDF$	
6	$a=c=e=x$	1		$\{2x(x-1)(x-2)+X\}\,BDF$	
7	$B=D=F=x$	1		$\{2X(X-1)(X-2)+x\}\,ace$	
8	$a=c=B=x$ $c=e=D$ $e=a=F$	3		$2x(x-3)(X-2)\,eDF$	

9	$D=F=e=x$ $F=B=a$ $B=D=c$	3		$2X(X-3)(x-2)\,acB$	
10	$a=c=D=x$, or $a=c=F$ $c=e=F$,, $c=e=B$ $e=a=B$,, $e=a=D$	6		$2(x-1)(Xx-X-x)\,eBF$	
11	$D=F=a=x$, or $D=F=c$ $F=B=c$,, $F=B=e$ $B=D=e$,, $B=D=a$	6	20	$2(X-1)(Xx-X-x)\,ceB$	
12	$c=e=x$, $a=D=y$ $e=a$ $c=F$ $a=c$ $e=B$	3		$2x(x-1)\,yYBF$	
13	$F=B=x$, $a=D=y$ $B=D$ $c=F$ $D=F$ $e=B$	3		$2X(X-1)\,Yyce$	
14	$c=e=x$, $a=B=y$, or $c=e$, $a=F$ $e=a$ $c=D$,, $e=a$, $c=B$ $a=c$ $e=F$,, $a=c$, $e=D$	6		$2x(x-1)(Yy-Y-y)\,DF$	
15	$F=B=x$, $D=e=y$, or $F=B$, $D=c$ $B=D$ $F=a$,, $B=D$, $F=e$ $D=F$ $B=c$,, $D=F$, $B=a$	6		$2X(X-1)(Yy-Y-y)\,ac$	
	Carried over	(18)	(36)		

TABLE (*continued*).

No. of Case	Specification	No. of forms	Totals	No. of triangles	Divided by
	Brought over	(18)	(36)		
16	$c=e=x,\ D=F=y,$ or $c=e,\ D=B$ $e=a\quad F=B$,, $e=a,\ F=D$ $a=c\quad B=D$,, $a=c,\ B=F$	6		$2x(x-1)\,Y(Y-1)\,aB$	
17	$c=e=x,\ B=F=y$ $e=a\quad D=B$ $a=c\quad F=D$	3		$2x(x-1)\,Y(Y-1)\,aD$	2
18	$a=D=x,\ c=B=y,$ or $a=D,\ e=F$ $c=F\quad e=D$,, $c=F,\ a=B$ $e=B\quad a=F$,, $e=B,\ c=D$	6		$2xX(Yy-Y-y)\,eF$	
19	$c=F=x,\ e=B=y$ $e=B\quad a=D$ $a=D\quad c=F$	3		$2xyXYaD$	
20	$c=D=x,\ e=F=y,$ or $c=B,\ e=D$ $e=F\quad a=B$,, $e=D,\ a=F$ $a=B\quad c=D$,, $a=F,\ c=B$	6		$2\{xyXY-xy(X+Y)-XY(x+y)+2xy+2XY\}\,aB$	
21	$c=B=x,\ e=F=y$ $e=D\quad a=B$ $a=F\quad c=D$	3	45	$2\{xyXY-xy(X+Y)-XY(x+y)+2xY+2yX\}\,aD$	
22	$a=D=x,\ c=F=y,\ e=B=z$	1		$2xyzXYZ$	
23	$a=B=x,\ c=D=y,\ e=F=z$ $a=F\quad c=B\quad e=D$	2		$2\{xyzXYZ-xyz(YZ+ZX+XY)-XYZ(yz+zx+xy)$ $+2xyz(X+Y+Z)+2XYZ(x+y+z)-4xyz-4XYZ\}$	

24	$a=D=x,\ c=B=y,\ e=F=z$ $c=F \quad e=D \quad a=B$ $e=B \quad a=F \quad c=D$	3		$2xX\{yzYZ-yz(Y+Z)-YZ(y+z)+2yZ+2zY\}$ [Originally printed $2yz+2YZ\}$ but the correction was made in a footnote to Case 24.]	
25	$a=c=x,\ D=F=y,\ e=B=z$ $c=e \quad F=B \quad a=D$ $e=a \quad B=D \quad c=F$	3		$2x(x-1)\,Y(Y-1)\,zZ$	2
26	$a=c=x,\ B=D=y,\ e=F=z$ or $a=c,\ B=F,\ e=D$ $c=e \quad D=F \quad a=B$,, $c=e,\ D=B,\ a=F$ $e=a \quad F=B \quad c=D$,, $e=a,\ F=D,\ c=B$	6	15	$2x(x-1)\,Y(Y-1)(zZ-z-Z)$	
27	$a=c=e=x,\ B=F=y$ $a=c=e \quad D=B$ $a=c=e \quad F=D$	3		$\{2x(x-1)(x-2)+X\}\,Y(Y-1)\,D$	2
28	$B=D=F=x,\ c=e=y$ $B=D=F \quad e=a$ $B=D=F \quad a=c$	3		$\{2X(X-1)(X-2)+x\}\,y(y-1)\,a$	2
29	$a=c=B=x,\ D=F=y$ $c=e=D \quad F=B$ $e=a=F \quad B=D$	3		$2x(x-3)(X-2)\,Y(Y-1)\,e$	2
30	$e=D=F=x,\ a=c=y$ $a=F=B \quad c=e$ $c=B=D \quad e=a$	3		$2X(X-3)(x-2)\,y(y-1)\,B$	2
	Carried over	(12)	(96)		

TABLE (*continued*).

No. of Case	Specification	No. of forms	Totals	No. of triangles	Divided by
	Brought over	(12)	(96)		
31	$c=e=D=x,\ a=B=y,$ or $c=e=D,\ a=F$			$2x(x-3)(X-2)(yY-y-Y)F$	
	$e=a=F \quad c=D$,, $e=a=F,\ c=B$				
	$a=c=B \quad e=F$,, $a=c=B,\ e=D$	6			
32	$F=B=a=x,\ D=e=y,$ or $F=B=a,\ D=c$			$2X(X-3)(x-2)(yY-y-Y)c$	
	$B=D=c \quad F=a$,, $B=D=c,\ F=e$				
	$D=F=e \quad B=c$,, $D=F=e,\ B=a$	6			
33	$B=F=y,\ a=e=D=x,$ or $B=F,\ a=c=D$			$2(x-1)(xX-x-X)Y(Y-1)c$	
	$D=B \quad c=a=F$,, $D=B,\ c=e=F$				
	$F=D \quad e=c=B$,, $F=D,\ e=a=B$	6			
34	$c=e=y,\ B=D=a=x,$ [1] or $c=e,\ D=F=a$			$2(X-1)(xX-x-X)y(y-1)F$	
	$e=a \quad D=F=a$,, $e=a,\ F=B=c$				
	$a=c \quad F=B=c$,, $a=c,\ B=D=e$	6			
35	$a=D=y,\ c=e=B=x,$ or $a=D,\ c=e=F$			$2(x-1)(xX-x-X)yYF$	
	$c=F \quad e=a=D$,, $c=F,\ e=a=B$				
	$e=B \quad a=c=F$,, $e=B,\ a=c=D$	6			
36	$a=D=y,\ B=F=e=x,$ or $a=D,\ B=F=c$			$2(X-1)(xX-x-X)yYc$	
	$c=F \quad D=B=a$,, $c=F,\ D=B=e$				
	$e=B \quad F=D=c$,, $e=B,\ F=D=a$	6			
37	$a=e=D=x,\ c=B=y,$ or $a=e=B,\ c=D$			$2(x-1)\{xyXY-xy(X+Y)-XY(x+y)+2xy+2XY\}F$	
	$c=a=F \quad e=D$,, $c=a=D,\ e=F$				
	$e=c=B \quad a=F$,, $e=c=F,\ a=B$	6			

[1] Originally printed $B=D=e=x$. The correction was made *post*, Case 34.

38	$B=D=a=x,\ F=e=y,$ or $B=D=e,\ F=a$ $D=F=c\quad B=a\quad$,, $D=F=a,\ B=c$ $F=B=e\quad D=c\quad$,, $F=B=c,\ D=e$	6	60	$2(X-1)\{xyXY-xy(X+Y)-XY(x+y)+2xy+2XY\}c$	
39	$a=c=e=B=x$,, $=D$,, $=F$	3		$\{X^2+X(2x^3-10x^2+12x-1)-4x^3+20x^2-16x-3\xi\}\,DF$	
40	$B=D=F=e=x$,, $=a$,, $=c$	3		$\{x^2+x(2X^3-10X^2+12X-1)-4X^3+20X^2-16X-3\xi\}ac$	
41	$c=e=D=F=x,$ or $c=e=D=B$ $e=a=B=B\quad$,, $e=a=F=D$ $a=c=F=D\quad$,, $a=c=B=F$	6		$2(x-3)(X-3)(xX-x-X)\,aB$	
42	$a=c=D=F=x$ $c=e=F=B$ $e=a=B=D$	3		$\{X^2(2x^2-6x+4)+X(-6x^2+18x-4)+4x^2-4x-4\xi\}eB$	2
43	$a=c=e=x,\ B=D=F=y$	1	15	$2x(x-1)(x-2)\,Y(Y-1)(Y-2)+yx(x-1)(x-2)$ $+XY(Y-1)(Y-2)$	6
44	$e=D=F=x,\ a=c=B=y$ $a=F=B\quad c=e=D$ $c=B=D\quad e=a=F$	3		$2(x-2)\,X(X-3)(Y-2)\,y(y-3)$	2
	Carried over	(4)	(171)		

TABLE (*continued*).

No. of Case	Specification	No. of forms	Totals	No. of triangles	Divided by
	Brought over	(4)	(171)		
45	$a=D=B=x$, $c=e=F=y$, or $a=D=F$, $c=e=B$ $c=F=D$ $e=a=B$,, $c=F=B$, $e=a=D$ $e=B=F$ $a=c=D$,, $e=B=D$, $a=c=F$	6	10	$2(X-1)(y-1)\{XYxy-XY(x+y)$ $-xy(X+Y)+2xy+2XY\}$	
46	$a=c=y$, $B=D=F=e=x$ $c=e$ $B=D=F=a$ $e=a$ $B=D=F=c$	3		$y(y-1)\{x^2+x(2X^3-10X^2+12X-1)$ $-4X^3+20X^2-16X-3\xi\}$	2
47	$D=F=y$, $a=c=e=B=x$ $F=B$ $a=c=e=D$ $B=D$ $a=c=e=F$	3		$Y(Y-1)\{X^2+X(2x^3-10x^2+12x-1)$ $-4x^3+20x^2-16x-3\xi\}$	2
48	$a=c=D=F=x$, $e=B=y$ $c=e=F=B$ $a=D$ $e=a=B=D$ $c=F$	3		$\{X^2(2x^2-6x+4)+X(-6x^2+18x-4)+4x^2-4x-4\xi\}yY$	2
49	$a=B=y$, $c=e=D=F=x$, or $a=F$, $c=e=B=D$ $c=D$ $e=a=F=B$,, $c=B$, $e=a=D=F$ $e=F$ $a=c=B=D$,, $e=D$, $a=c=F=B$	6	15	$2(x-3)(X-3)\{xyXY-(x+y)XY$ $-(X+Y)xy+2xy+2XY\}$	
50	$c=e=B=D=F=x$ $e=a=B=D=F$ $a=c=B=D=F$	3		a into $x^3(\quad\quad\quad\quad+1)$ $+x^2(\quad 2X^3-14X^2+\ 28X-11)$ $+x\ (-10X^3+70X^2-116X-\ 8)$ $+12X^3-76X^2+\ 64X$ $+\xi\ (-\ 6x\ -\ 4X+\ 42)$	2

51	$F = B = a = c = e = x$ $B = D = a = c = e$ $D = F = a = c = e$	3	6	D into $X^3(\quad\quad\quad\quad + 1)$ $+X^2(\quad 2x^3 - 14x^2 + 28x - 11)$ $+X\,(-10x^3 + 70x^2 - 116x - 8)$ $+12x^3 - 76x^2 + 64x$ $+\xi(-6X - 4x + 42)$	2
52	$a = c = e = B = D = F = x$	1	1	$X^4(\quad\quad\quad\quad + 1)$ $+X^3(\quad 2x^3 - 18x^2 + 52x - 46)$ $+X^2(\quad -18x^3 + 162x^2 - 420x + 221)$ $+X\,(\quad 52x^3 - 420x^2 + 704x + 172)$ $+1x^4 - 46x^3 + 221x^2 + 172x$ $+\xi\left\{\begin{array}{l} X^2(\quad\quad - 9) \\ +X\,(\quad - 12x + 135) \\ -9x^2 + 135x - 600 \end{array}\right\}$	6
			203		

The foregoing results are chiefly obtained by means of the theory of correspondence; viz. if instead of the triangle *aBcDeF* we consider the unclosed trilateral *aBcDeFg*, where the points *a* and *g* are situate on one and the same curve, say the curve $a=g$, then the points *a* and *g* have a certain correspondence, say a (χ, χ') correspondence with each other; and when *a*, *g* are a "united point" of the correspondence, the trilateral in question becomes an in-and-circumscribed triangle *aBcDeF*; that is, the number of triangles is equal to that of the united points of the correspondence, subject however (in many of the cases) to a reduction on account of special solutions. It may be remarked that by the theory of correspondence the number of the united points is, in several of the cases, but not in all of them, $=\chi+\chi'$. But in some instances I employ a functional method, by assuming that the identical curves are each of them the aggregate of the two curves x, x': we here obtain for the number ϕx of the triangles belonging to the curve x a functional equation $\phi(x+x')-\phi x-\phi x'=$ given function; viz. the expression on the right-hand side depends on the solution of the preceding cases, wherein the number of identities between the several curves is less than in the case under consideration; and taking it to be known, the functional equation gives $\phi x=$ particular solution + linear function of (x, X, ξ). The particular solution is always easily obtainable, and the constants of the linear function can be determined by means of particular forms of the curve x.

Article Nos. 1 to 6. *The Principle of Correspondence as applied to the present Problem.*

1. Consider the unclosed trilateral *aBcDeFg*, where the points *a* and *g* are on one and the same curve, $a=g$. Starting from an arbitrary point *a* on the curve *a*, we have *aBc* any one of the tangents from *a* to the curve *B*, touching this curve, say at the point *B*, and intersecting the curve *c* in a point *c*; viz. *c* is any one of the intersections of *aBc* with the curve *c*; we have then similarly *cDe* any one of

Fig. 1.

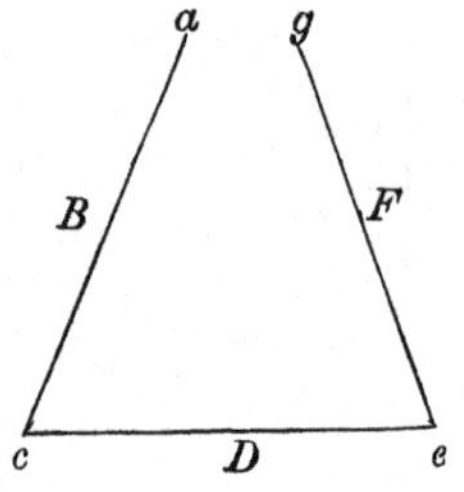

the tangents from *c* to the curve *D*, touching it, say at *D*, and intersecting the curve *e* in a point *e*; viz. the point *e* is any one of the intersections in question; and then in like manner we have *eFg* any one of the tangents from *e* to the curve *F*, touching it, say at *F*, and intersecting the curve *g* $(=a)$ in a point *g*; viz. *g* is any one of the intersections in question. Suppose that to a given position of *a* there correspond χ positions of *g*; it is easy to find the value of χ; viz. if (as above tacitly supposed) the curves a, B, c, D, e, F are all of them distinct curves, then the number of the tangents aBc is $=B$; there are on each of them c points c; through each of these we have D tangents cDe; on each of these e points e; through each of these F tangents cFg; and on each of these a points g; that is, $\chi = BcDeFa$. But if some of the curves become one and the same curve—if, for instance, $a = B = c$,—the line aBc is here a tangent from a point a on the curve, we exclude the tangent at the point a, and the number of the remaining tangents is $=(A-2)$; each tangent meets the curve in the point a counting once, the point B counting twice, and in $(a-3)$ other points; that is, the number of the points c is $=(A-2)(a-3)$, and so in other cases; the calculation is always immediate, and the only difference is that, instead of a factor a or A, we have such factor in its original form or diminished by 1, 2, or 3, as the case may be. Similarly starting from g, considered as a given point on the curve g $(=a)$, we find χ' the number of the corresponding points a; thus in the case where the curves are all distinct curves, we have $\chi' = FeDcBa$ $(=\chi)$; and so in other cases we find the value of χ'. The points (a, g) have thus a (χ, χ') correspondence, where the values of χ, χ' are found as above.

2. There will be occasion to consider the case where in the triangle $aBcDeF$ (or say the triangle $aBcDeFa$) the point a is not subjected to any condition whatever, but is a free point. There is in this case a "locus of a," which is at once constructed as follows: viz. starting with an arbitrary tangent aBc of the curve B, touching it at B and intersecting the curve c in a point c; through c we draw to the curve D the tangent cDe, touching it at D and intersecting the curve e in a point e; and finally from e to the curve F the tangent eFa, touching it at F and intersecting the original arbitrary tangent aBc in a point a, which is a point on the locus in question. We can, it is clear, at once determine how many points of the locus lie on an arbitrary tangent of the curve B (or of the curve F).

3. The general form of the equation of correspondence is

$$p(\mathrm{a} - \alpha - \alpha') + q(\mathrm{b} - \beta - \beta') + \ldots = k\Delta \ [1];$$

viz. if on a curve for which twice the deficiency is $=\Delta$ we have a point P corresponding to certain other points P', Q',... in such wise that P, P' have an (α, α') correspondence, P, Q' a (β, β') correspondence, &c.; and if (a) be the number of the united points (P, P'), (b) the number of the united points (P, Q'), &c.; and if moreover for a given position of P on the curve the points P', Q',... are obtained as the intersections of the curve with a curve Θ (depending on the point P) which meets the curve k times at P, p times at each of the points P', q times at each of the

[1] To avoid confusion with the notation of the present memoir, I abstain in the text from the use of D as denoting the deficiency, and there is a convenience in the use of a single symbol for twice the deficiency; but writing for the moment D to denote the deficiency, I remark, in passing, that perhaps the true theoretical form of the equation is

$$k(0 - D - D) + p(\mathrm{a} - \alpha - \alpha') + q(\mathrm{b} - \beta - \beta') + \ldots = 0;$$

viz. the point P is here considered as having with itself a (D, D) correspondence, the number of the united points therein being $=0$.

points Q', &c.; then the relation between the several quantities is as stated above: see my "Second Memoir on the Curves which satisfy given conditions," *Philosophical Transactions*, vol. 159 (1868), pp. 145—172, [407]. I omit for the present purpose the term "Supp.," treating it as included in the other terms.

4. In the present case we consider, as already mentioned, the unclosed trilateral $aBcDeFg$, where the angles a, g are on one and the same curve a $(=g)$ (the curve in the general theorem); and the curve Θ is the system of lines eFg which by their intersection with the curve a determine the points g. Considering these as the points (P, P') of the general theorem we have $p=1$: I change the notation, and instead of $\mathrm{a}-\alpha-\alpha'$ write $\mathrm{g}-\chi-\chi'$; viz. I take (g) for the number of the united points (a, g), and suppose that the points (a, g) have a (χ, χ') correspondence. The most simple case is when the curve a is distinct from each of the curves e, F; here all the intersections of the line-system eFg with the curve a are points g, that is we have *only* the correspondence (a, g); and since the line-system eFg does not pass through the point a, we have simply

$$\mathrm{g}-\chi-\chi'=0.$$

5. But suppose that the curves a, e, F are one and the same curve, say that $a=e=F$; understanding by the point F the point of contact of a line eFg with the curve a, then the intersections of the line-system eFg with the curve a are the points g each once, the points F each twice, and the points e each as many times as there are lines eFg through the point e, say each M times. (In the present case, where the curves e, F are identical, we have $M=F-2$ or $F-3$ according as the curve D is or is not distinct from the curve F; in the cases afterwards referred to, the values may be F or $F-1$; that is, we have always $M=F$, $F-1$, $F-2$, $F-3$, as the case may be.) We have to consider the several correspondences (a, g), (a, F), (a, e); k is as before $=0$; and the form of the theorem is

$$(\mathrm{g}-\chi-\chi')+2(\mathrm{f}-\phi-\phi')+M(\mathrm{e}-\epsilon-\epsilon')=0,$$

where the symbols denote as follows, viz.

(a, g) have a	(χ, χ')	correspondence, and No. of united points	$=\mathrm{g}$,
(a, F) ,,	(ϕ, ϕ')	,, ,, ,,	$=\mathrm{f}$,
(a, e) ,,	(ϵ, ϵ')	,, ,, ,,	$=\mathrm{e}$,

so that the determination of g here depends upon that of $\mathrm{f}-\phi-\phi'$ and $\mathrm{e}-\epsilon-\epsilon'$.

6. The curve a might however have been identical with only one of the curves e, F; viz. if $a=F$, but e is a distinct curve, then the equation will contain the term $2(\mathrm{f}-\phi-\phi')$, but not the term $M(\mathrm{e}-\epsilon-\epsilon')$; and so if $a=e$, but F is a distinct curve, then the equation will not contain $2(\mathrm{f}-\phi-\phi')$, but will contain $M(\mathrm{e}-\epsilon-\epsilon')$: it is to be noticed that in this last case we have $M=F$ or $M=F-1$, according as the curve D is not, or is, one and the same curve with F. The determination of (g) here depends upon that of $\mathrm{f}-\phi-\phi'$ or $\mathrm{e}-\epsilon-\epsilon'$, as the case may be. These sub-

sidiary values $f-\phi-\phi'$ and $e-\epsilon-\epsilon'$ are obtained by means of a more simple application of the principle of correspondence, as will appear in the sequel[1], but for the moment I do not pursue the question.

Article Nos. 7 to 14. *Locus of a free angle* (a).

7. I consider the case where a is a distinct curve $\neq e$, $\neq F$, and where, as was seen, the equation is simply

$$g-\chi-\chi'=0.$$

I suppose further that a is distinct from all the other curves, or say, *simpliciter*, that a is a distinct curve. The values of χ, χ' will here each of them contain the factor a, say we have $\chi=a\omega$, $\chi'=a\omega'$; and therefore the equation gives $g=a(\omega+\omega')$. It is obvious that ω, ω' are the values assumed by χ, χ' respectively in the particular case where the curve a is an arbitrary line $(a=1)$; and $\omega+\omega'$ is the number of the united points on this line.

8. Suppose now that in the triangle $aBcDeFa$ the point a is a free point, we have, as above-mentioned, a locus of a, and the united points on the arbitrary line are the intersections of the line with this locus; that is, the locus meets the arbitrary line in $\omega+\omega'$ points; or, what is the same thing, the order of the locus is $=\omega+\omega'$.

9. I stop for a moment to remark that in the particular case where the curve B is a point $(B=1)$, then in the construction of the locus of a the arbitrary tangent aBc is an arbitrary line through B, and the construction gives on this line ω positions of the point a. But drawing from B a tangent to the curve F, and thus constructing in order the points F, e, D, c, a, the construction shows that B is an ω'-tuple point on the locus; and (by what precedes) an arbitrary line through B meets the locus in ω other points; that is, in the particular case where the curve B is a point, the order of the locus of a is $=\omega+\omega'$, which agrees with the foregoing result.

10. The construction for the locus of a may be presented in the following form: viz. drawing to the curve D a tangent cDe, meeting the curves c, e in the points c, e respectively; then if from any point c we draw to the curve B a tangent cBa, and from any point e to the curve F a tangent eFa, the tangents cBa, eFa intersect in a point on the required locus. Hence if in any particular case (that is for any particular position of the tangent cDe) the lines cBa, eFa become one and the same line, the point a will be an indeterminate point on this line; that is, the line in question will be part of the locus of a.

11. The case cannot in general arise so long as the curves B, F are distinct from each other; but when these are one and the same curve, say when $B=F$, it will arise, and that in two distinct ways. To show how this is, suppose, to fix the ideas, that the curves c, D, e are distinct from each other and from the curve $B=F$. Then the first mode is that shown in the annexed "first-mode figure," viz. we have

[1] See *post*, Nos. 24 *et seq.*

here a tangent at D passing through a point ce of the intersection of the curves c, e, and from this point a tangent drawn to the curve $B=F$. For the position in question of the tangent of D, the points c, e coincide with each other, and we have thus the coincident tangents cBa and eFa to the identical curves $B=F$. It is further

Fig. 2. First-mode figure.

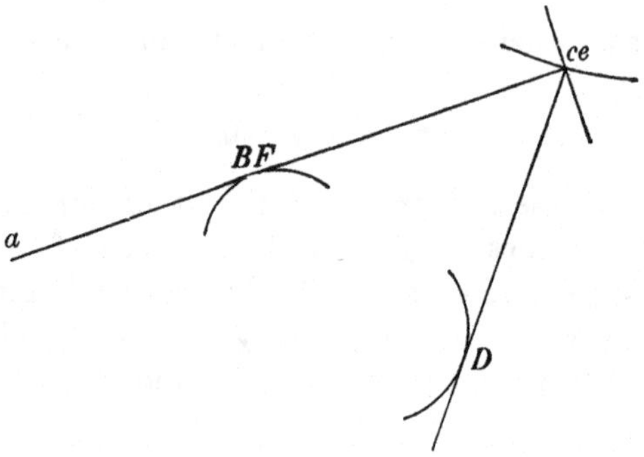

to be remarked that the number of the points of intersection is $=ce$; from each of these there are B tangents to the curve $B=F$ (in all $ce.B$ tangents), and each of these counts once in respect of each of the D tangents to the curve D, that is, it counts D times. We have thus, as part of the locus of a, $ce.B$ lines each D times, or, say, first-mode reduction $=ce.B.D$.

12. The second mode is that shown in the annexed "second-mode figure." The tangent from D is here a common tangent of the curves D, and $B=F$. This meets the curve c in c points, and the curve e in e points; and attending to any pair of points c, e, these give the tangents cBa, eFa, coinciding with the common tangent in

Fig. 3. Second-mode figure.

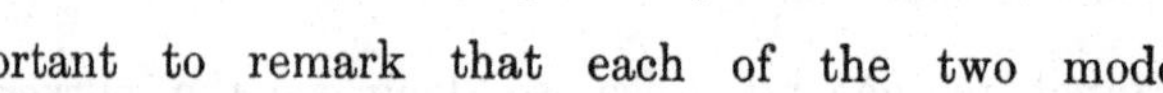

question, and forming part of the locus of a. The number of the common tangents is $=BD$; but each of these counts once in respect of each combination of the points c, e, that is in all ce times. And we have thus as part of the locus BD lines each $c.e$ times, or, say, second-mode reduction $=BD.c.e$. This is (as it happens) the same number as for the first mode; but to distinguish the different origins I have written as above $ce.B.D$ and $BD.c.e$ respectively.

13. It is important to remark that each of the two modes arises whatever relations of identity subsist between the curves c, e, D, and $B=F$, but with considerable modification of form. Thus if the curves c, e are identical $(c=e)$ but distinct from D, then in the first-mode figure ce may be a node or a cusp of the curve $c=e$, or it may be a point of contact of a common tangent of the curves D, and $c=e$. As regards the node, remark that if we consider a tangent of D meeting the curve $c=e$ in the neighbourhood of the node, then of the two points of intersection each

in succession may be taken for the point c, and the other of them will be the point e; so that the node counts twice. It requires more consideration to perceive, but it will be readily accepted that the cusp counts three times. Hence if for the curve $c=e$ the number of nodes be $=\delta$ and that of cusps $=\kappa$, the value of the first-mode reduction is $=(2\delta+3\kappa+C)\,BD$, or, what is the same thing, it is $=(c^2-c)\,BD$.

As regards the second-mode figure, the only difference is that c, e will be here any pair of intersections (each pair twice) of the tangent with the curve $c=e$; the value is thus $=(c^2-c)\,BD$.

It would be by no means uninteresting to enumerate the different cases, and indeed there might be a propriety in doing so here; but I have (instead of this) considered the several cases, when and as they arise in connexion with any of the cases of the in-and-circumscribed triangle.

14. Observe that the general result is, that in the case $B=F$ of the identity of the curves B and F, but not otherwise, the locus of a includes as part of itself a system of lines; or, say, that it is made up of these lines, and of a residual curve of the order $\omega+\omega'-$Red., which is the proper locus.

Article Nos. 15 to 17. *Application of the foregoing Theory as to the locus of* (a).

15. Reverting now to the case where the angle a is not a free angle but is situate on a given curve a, then if the curve a is distinct from the curves e, F, the number of positions of a is, as was seen, $g=\chi+\chi'$. But the points in question are the intersections of the curve a with the locus of a considered as a free angle; and hence in the case $B=F$, but not otherwise, they are made up of the intersections of the curve a with the system of lines, and of its intersections with the proper locus of a. But the intersections with the system of lines are improper solutions of the problem (or, to use a locution which may be convenient, they are "heterotypic" solutions): the true solutions are the intersections with the proper locus of a; and the number of these is not $\chi+\chi'$, $=a\,(\omega+\omega')$, but it is $=a\,(\omega+\omega'-\text{Red.})$; say it is $=\chi+\chi'-$Red., where the symbol "Red." is now used to signify a times the number of lines, or reduction in the expression $\omega+\omega'-$Red. of the order of the proper locus of a.

16. It is however to be noticed that if the curve a, being as is assumed distinct from the curves e, and $F=B$, is identical with one or both of the remaining curves c, D, the foregoing expression $\chi+\chi'-$Red. *may* include positions which are not true solutions of the problem, viz. the curve a may pass through special points on the proper locus of a, giving intersections which are a new kind of heterotypic solutions[1].

[1] More generally, if the curve a be a curve identical with any of the other curves, then if treating in the first instance the angle a as free we find in any manner the locus of a, the required positions of the angle a are the intersections of this locus and of the curve a; but these intersections will in general include intersections which give heterotypic solutions. The determination of these is a matter of some delicacy, and I have in general treated the problems in such manner that the question does not arise; but as an example see *post*, Case 43.

17. But this cannot happen if the curve a is distinct also from the curves c, D; or, say, simply when a is a distinct curve. The conclusion is, that in the case where a is a distinct curve we have

$$g = \chi + \chi' - \text{Red.},$$

where the term "Red." vanishes except in the case of the identity $B = F$ of the curves B, F; and that when this identity subsists it is $= a$ times the reduction in the order of the locus of a considered as a free angle; viz. this consists of a first-mode and a second-mode reduction as above explained.

Article Nos. 18 to 21. *Remarks in regard to the Solutions for the* 52 *Cases.*

18. Before going further I remark that the principle of correspondence applies to corresponding and united tangents in like manner as to corresponding and united points, and that all the investigations in regard to the in-and-circumscribed triangle might thus be presented in the reciprocal form, where, instead of points and lines, we have lines and points respectively. But there is no occasion to employ any such reciprocal process; the result to which it would lead is the reciprocal of a result given by the original process, and as such it can always be obtained by reciprocation of the original result, without any performance of the reciprocal process.

19. It is hardly necessary to remark that although reciprocal results would, by the employment of the two processes respectively, be obtained in a precisely similar manner, yet that this is not so when only one of the reciprocal processes is made use of; so that, using one process only, it may be and in general is easier and more convenient to obtain directly one than the other of two reciprocal results; for instance, to consider the case $B = D = F$ rather than $a = c = e$, or *vice versâ*; and that it is sufficient to do this, and having obtained the one result, directly to deduce from it the other by reciprocity; but that it may nevertheless be interesting to obtain each of the two results directly.

20. It is moreover obvious that although the several forms of the same case, for instance Case 2, $a = c$, $a = e$, or $c = e$, are absolutely equivalent to each other, yet that, when as above we select a vertex a, and seek for the number of the united points (a, g), the process of obtaining the result will be altogether different according to the different form which we employ. For instance, in the case just referred to, if the form is taken to be $a = c$ or $c = e$, then the equation $g = \chi + \chi'$ is applicable to it; but not so if the form is taken to be $a = e$. It would be by no means uninteresting in every case to consider the several forms successively and get out the result from each of them; I shall not, however, do this, but only consider two or more forms of the same case when for comparison, illustration, verification, or otherwise it appears proper so to do. The translation of a result, for instance, of a form $a = e$ or $c = e$ into that for the form $a = c = x$ is so easy and obvious, that it is not even necessary formally to make it.

21. I do not at present further consider the general theory, but proceed to consider in order the 52 cases, interpolating in regard to the general theory such further discussion or explanation as may appear necessary. In the several instances in which the equation $g=\chi+\chi'$ is applicable, it is sufficient to write down the values of χ, χ', the mode of obtaining these being already explained.

The 52 Cases for the in-and-circumscribed triangles.

Case 1. No identities.

$$\chi = BcDeFa,\quad \chi' = FeDcBa\,(=\chi),$$
$$g = 2aceBDF.$$

Case 2. $a=c=x$.

$$\chi = B(x-1)\,DeFx,\quad \chi' = FeDxB(x-1)\,(=\chi),$$
$$g = 2x(x-1)\,eBDF.$$

Second process, for form $a=e=x$. The equation of correspondence is here

$$g-\chi-\chi'+F(e-\epsilon-\epsilon')=0;$$

but the points e being given as all the intersections of the curve $a\,(=e)$ by the line-system cDe which does not pass through a, we have $e-\epsilon-\epsilon'=0$; so that $g=\chi+\chi'$; and then

$$\chi = BcDxF(x-1),\quad \chi' = F(x-1)\,DcBx,$$

giving the former result[1].

Case 3. $D=F=x$. Reciprocation from 2; or else, *second process*,

$$\chi = BcXe(X-1)\,a,\quad \chi' = Xe(X-1)\,cBa,$$
$$g = 2X(X-1)\,Bace.$$

Third process: form $F=B=x$. We have here $g=\chi+\chi'-\text{Red}$.

$$\chi = XcDeXa,\quad \chi' = XeDcXa\,(=\chi),$$
$$\chi+\chi' = 2X^2Dace;$$

and the reductions are those of the first and second mode, as explained *ante*, Nos. 11, 12, viz. each of these is $=XDace$, and together they are $=2XDace$; whence the foregoing result.

Case 4. $a=D=x$.

$$\chi = BcXeFx,\quad \chi' = FeXBx\,(=\chi),$$
$$g = 2Xx\,ceBF.$$

[1] Of course, the result is obtained in the form belonging to the new form of specification, viz. here it is $=2x(x-1)\,cBDF$; and so in other instances; but it is unnecessary to refer to this change.

Observe this is what the result for Case 1 becomes on writing therein $a = D = x$, viz. the opposite curves a, D may become one and the same curve without any alteration in the form of the result.

Case 5. $a = B = x$.

$$\chi = (X-2)\,cDeFx, \quad \chi' = FeDcX\,(x-2),$$

where

$$(X-2)\,x + X\,(x-2) = 2\,(Xx - X - x);$$

therefore

$$g = 2\,(Xx - X - x)\,ceDF.$$

Case 6. $a = c = e = x$: perhaps most easily by reciprocation of Case 7; or

Second process, functionally by taking the curve $a = c = e$ to be the aggregate curve $x + x'$. The triangle $aBcDeF$ is here in succession each of the eight triangles:

$x\ B\ x\ D\ x\ F$	$x'\ B\ x'\ D\ x'\ F$
x' ,, x ,, x ,,	x ,, x' ,, x' ,,
x ,, x' ,, x ,,	x' ,, x ,, x' ,,
x' ,, x ,, x ,,	x ,, x' ,, x' ,,

where the two top triangles give ϕx and $\phi x'$ respectively; the remaining triangles all belong to Case 2, and those of the first column give each $2\,(x^2 - x)\,x'BDF$, and those of the second column each $2\,(x'^2 - x')\,xBDF$. We have thus

$$\phi\,(x + x') - \phi x - \phi x' = \{6\,(x^2x' + xx'^2) - 12xx'\}\,BDF.$$

Hence obtaining a particular solution and adding the constants, we have

$$\phi x = (2x^3 - 6x^2 + \alpha x + \beta X + \gamma\xi)\,BDF;$$

it is easy to see that α, β, γ are independent of the curves B, D, F; and taking each of these to be a point, and the curve $a = c = e$ to be a conic, then it is known that $\phi x = 2$; we have therefore $2 = 16 - 24 + 2\alpha + 2\beta + 6\gamma$, that is $\alpha + \beta + 3\gamma = 5$.

The case where the curve $a = c = e$ is a line gives $0 = 2 - 6 + \alpha + 3\gamma$, that is $\alpha + 3\gamma = 4$; but it is not easy to find another condition; assuming however $\gamma = 0$, we have $\alpha = 4$, $\beta = 1$, and thence

$$\phi x = (2x^3 - 6x^2 + 4x + X)\,BDF,$$

or say

$$g = \{2x\,(x-1)\,(x-2) + X\}\,BDF:$$

this is a good easy example of the functional process, the use of which begins to exhibit itself; and I have therefore given it, notwithstanding the difficulty as to the complete determination of the constants.

Third process. The equation of correspondence is

$$g - \chi - \chi' + F(e - \epsilon - \epsilon') = 0,$$

but for the correspondence (a, e) we have

$$e - \epsilon - \epsilon' + D(c - \gamma - \gamma') = 0,$$

and for the correspondence (a, c) we have

$$c - \gamma - \gamma' = B\Delta,$$

whence

$$g = \chi + \chi' + BDF \,.\, \Delta;$$

and then

$$\chi = B(x-1)\,D(x-1)\,F(x-1), \quad \chi' = F(x-1)\,D(x-1)\,B(x-1)\,(=\chi);$$

that is

$$\chi + \chi' = BDF \,.\, 2(x-1)^3.$$

Moreover

$$\Delta = X - 2x + 2 + \kappa$$

(if κ be the number of cusps of the curve $a = c = e$), and the resulting value is

$$g = \{2(x-1)^3 + X - 2x + 2 + \kappa\}\,BDF;$$

that is

$$= \{2x(x-1)(x-2) + X + \kappa\}\,BDF,$$

where, however, the term κBDF is to be rejected. I cannot quite explain this; I should rather have expected a rejection $= 2\kappa BDF$, introducing the term $-\kappa$. For consider a tangent from the curve D from a cusp of the curve $a = c = e$: there are D such tangents; each gives in the neighbourhood of the cusp two points, say c, e; and from these we draw B tangents cBa to the curve B, and F tangents eFa to the curve F; we have thus in respect of the given tangent of D, BF positions of a, or in all BDF positions of a which will ultimately coincide with the cusp; that is, BDF infinitesimal triangles of which the angles a, c, e coincide together at the cusp; and for all the cusps together κBDF such triangles: this would be what is wanted; the difficulty is that as (of the two intersections at the cusp) each in succession might be taken for c, and the other of them for e, it would seem that the foregoing number κBDF should be multiplied by 2.

Case 7. $B = D = F = x$. Here $g = \chi + \chi' -$ Red. and

$$\chi = Xc(X-1)\,e(X-1)\,a, \quad \chi' = Xe(X-1)\,c(X-1)\,a\,(=\chi);$$

that is,

$$\chi + \chi' = 2X(X-1)^2\,ace.$$

The reductions of the two modes are as above, with only the variation that in the present case D is the same curve with the two curves $B = F$. That of the first mode is $= X(X-1)\,ace$, and that of the second mode is $(2\tau + 3\iota)\,ace$, which is $=\{X(X-1) - x\}\,ace$; together they are $= \{2X(X-1) - x\}\,ace$, or subtracting, we have

$$g = \{2X(X-1)(X-2) + x\}\,ace.$$

Case 8. $a=c=B=x$.

$$\chi=(X-2)(x-3)\,DeFx,\quad \chi'=FeDx\,(X-2)(x-3)\,(=\chi),$$
$$\mathrm{g}=2x\,(x-3)(X-2)\,eDF.$$

Case 9. $D=F=e=x$. By reciprocation of 8.

$$\text{No.}=2X\,(X-3)(x-2)\,acB.$$

Case 10. $a=c=D=x$.

$$\chi=B\,(x-1)(X-2)\,eFx,\quad \chi'=FeX\,(x-2)\,B\,(x-1),$$
$$\mathrm{g}=2\,(x-1)(Xx-X-x)\,eBF.$$

Case 11. $D=F=a=x$. By reciprocation of 10.

$$\text{No.}=2\,(X-1)(Xx-X-x)\,ceB.$$

Second process: form $a=B=D=x$.

$$\chi=(X-2)\,c\,(X-1)\,eFx,\quad \chi'=FeXc\,(X-1)(x-2),$$

giving the former result.

Case 12. $c=e=x,\ a=D=y$.

$$\chi=BxY\,(x-1)\,Fy,\quad \chi'=FxY\,(x-1)\,By\,(=\chi),$$
$$\mathrm{g}=2x\,(x-1)\,y\,YBF.$$

Case 13. $F=B=x,\ a=D=y$. By reciprocation of 12.

$$\text{No.}=2X\,(X-1)\,Yyce.$$

Case 14. $c=e=x,\ a=B=y$.

$$\chi=(Y-2)\,xD\,(x-1)\,Fy,\quad \chi'=FxD\,(x-1)\,Y\,(y-2),$$
$$\mathrm{g}=2x\,(x-1)\,(Yy-Y-y)\,DF.$$

Case 15. $F=B=x,\ D=e=y$. By reciprocation of 14.

$$\text{No.}=2X\,(X-1)(Yy-Y-y)\,ac.$$

Case 16. $c=e=x,\ D=F=y$.

$$\chi=BxY\,(x-1)(Y-1)\,a,\quad \chi'=Yx\,(Y-1)(x-1)\,Ba\,(=\chi),$$
$$\mathrm{g}=2x\,(x-1)\,Y\,(Y-1)\,aB.$$

Case 17. $c=e=x,\ B=F=y$.

$$\chi=D\,(x-1)\,Ya\,(Y-1)\,x,\quad \chi'=Ya\,(Y-1)\,xD\,(x-1)\,(=\chi),$$
$$\mathrm{g}=2x\,(x-1)\,Y\,(Y-1)\,aD.$$

But we have here aD as an axis of symmetry, so that each triangle is counted twice, or the number of distinct triangles is $=\frac{1}{2}\mathrm{g}$.

Case 18. $a = D = x,\ c = B = y.$

$$\chi = Y(y-2)XeFx, \quad \chi' = FeXy(Y-2)x\,(=\chi),$$
$$g = 2xX(Yy - Y - y)\,eF.$$

Case 19. $c = F = x,\ e = B = y.$

$$\chi = YxDyXa, \quad \chi' = XyDxYa\,(=\chi),$$
$$g = 2xyXYaD.$$

Case 20. $c = D = x,\ e = F = y.$

$$\chi = Bx(X-2)y(Y-2)a, \quad \chi' = Y(y-2)X(x-2)Ba,$$
$$\begin{aligned} g &= \{xy(X-2)(Y-2) + XY(x-2)(y-2)\}\,aB \\ &= 2\{xyXY - xy(X+Y) - XY(x+y) + 2xy + 2XY\}\,aB. \end{aligned}$$

Case 21. $c = B = x,\ e = F = y.$

$$\chi = X(x-2)Dy(Y-2)a, \quad \chi' = Y(y-2)Dx(X-2)a,$$
$$\begin{aligned} g &= \{X(Y-2)y(x-2) + Y(X-2)x(y-2)\}\,aD \\ &= 2\{xyXY - xy(X+Y) - XY(x+y) + 2xY + 2yX\}\,aD. \end{aligned}$$

Case 22. $a = D = x,\ c = F = y,\ e = B = z.$

$$\chi = ZyXzYx, \quad \chi' = YzXyZx\,(=\chi),$$
$$g = 2xyzXYZ.$$

Case 23. $a = B = x,\ c = D = y,\ e = F = z.$

$$\chi = (X-2)y(Y-2)z(Z-2)x, \quad \chi' = Z(z-2)Y(y-2)X(x-2),$$
$$\begin{aligned} g &= xyz(X-2)(Y-2)(Z-2) + XYZ(x-2)(y-2)(z-2) \\ &= 2\{xyzXYZ - xyz(YZ + ZX + XY) - XYZ(yz + zx + xy) \\ &\qquad + 2xyz(X+Y+Z) + 2XYZ(x+y+z) - 4xyz - 4XYZ\}. \end{aligned}$$

Case 24. $a = D = x,\ c = B = y,\ e = F = z.$

$$\chi = Y(y-2)Xz(Z-2)x, \quad \chi' = Z(z-2)Xy(Y-2)x,$$
$$\begin{aligned} g &= xX\{Y(Z-2)z(y-2) + Z(Y-2)y(z-2)\} \\ &= 2xX\{yzYZ - yz(Y+Z) - YZ(y+z) + 2yZ + 2zY\}. \end{aligned}$$

Case 25. $a = c = x,\ D = F = y,\ e = B = z.$

$$\chi = Z(x-1)Yz(Y-1)x, \quad \chi' = Yz(Y-1)xZ(x-1)\,(=\chi),$$
$$g = 2x(x-1)Y(Y-1)zZ.$$

But we have here eB as an axis of symmetry, so that each triangle is counted twice, or the number of distinct triangles is $=\tfrac{1}{2}g$.

Case 26. $a=c=x,\ B=D=y,\ e=F=z.$

$$\chi = Y(x-1)(Y-1)\,z\,(Z-2)\,x,\quad \chi' = Z(z-2)\,Yx\,(Y-1)(x-1),$$
$$\begin{aligned} g &= x(x-1)\,Y(Y-1)\,\{z(Z-2)+Z(z-2)\} \\ &= 2x(x-1)\,Y(Y-1)(zZ-z-Z). \end{aligned}$$

Case 27. $a=c=e=x,\ B=F=y.$ By reciprocation of 28.

$$\text{No.} = \{2x(x-1)(x-2)+X\}\,Y(Y-1)\,D,$$

where each triangle is counted twice, so that the number is really one half of this.

Case 28. $B=D=F=x,\ c=e=y.$

Here

$$\begin{aligned} g &= \chi+\chi' - \text{Red.} \\ \chi &= Xy(X-1)(y-1)(X-1)\,a,\quad \chi' = Xy(X-1)(y-1)(X-1)\,a\,(=\chi), \\ \chi+\chi' &= ay(y-1)\,.\,2X(X-1)^2. \end{aligned}$$

The reductions are those of the first and second mode as explained above, with the variation that the curves c and e are here identical, $c=e$, and that the curve D is identical with the curves $B=F$.

First-mode reduction is

$$a(C+2\delta+3\kappa)\,B(B-1)$$

(where δ, κ refer to the curve $c=e$), which is

$$=a\,c\,(c-1)\,B(B-1);$$

that is, the reduction is $=a\,y\,(y-1)\,X(X-1)$.

Second-mode reduction is

$$a(2\tau+3\iota)\,c\,(c-1)$$

(where τ, ι refer to the curve $B=D=F$), which is

$$=a\,\{B(B-1)-b\}\,c\,(c-1);$$

that is, the reduction is $=a\,y\,(y-1)\,\{X(X-1)-x\}$.

Hence the two together are $=a\,y\,(y-1)\,\{2X(X-1)-x\}$; and subtracting from $\chi+\chi'$ we have

$$g = a\,y\,(y-1)\,.\,\{2X(X-1)(X-2)+x\};$$

but on account of the symmetry each triangle is reckoned twice, and the number of triangles is $=\tfrac{1}{2}g$.

Case 29. $a=c=B=x,\ D=F=y.$

$$\begin{aligned} \chi &= (X-2)(x-3)\,Ye(Y-1)\,x,\quad \chi' = Ye(Y-1)\,x\,(X-2)(X-3)\,(=\chi), \\ g &= 2x(x-3)(X-2)\,Y(Y-1)\,e. \end{aligned}$$

Second process. Taking the form

$$C = D = e = x, \quad B = F = y;$$

here

$$\text{No.} = \chi + \chi' - \text{Red.},$$
$$\chi = Yx\,(X-2)\,(x-3)\;Ya, \;= \gamma',$$

and

$$\chi + \chi' = 2Y^2\,x\,(x-3)\,(X-2)\,a.$$

There is a first-mode reduction,

$$aY\,\{2\tau + 2\delta\,(X-4) + 3\kappa\,(X-3)\},$$

viz. this is

$$\begin{aligned} aY\{ \quad & X^2 - X + 8x - 3\xi \\ & + (X-4)\,(x^2 - x + 8X - 3\xi) \\ & + (X-3)\,(\qquad -9X + 3\xi)\}, \end{aligned}$$

which is

$$= aY\,\{X\,(x^2 - x - 6) - 4x^2 + 12x\};$$

and a second-mode reduction

$$= aYX\,(x-2)\,(x-3).$$

Hence the two together are

$$= aY\,\{X\,(2x^2 - 6x) - 4x^2 + 12x\}$$
$$= 2Yx\,(x-3)\,(X-2)\,a,$$

whence the result is

$$= 2\,(Y^2 - Y)\,x\,(x-3)\,(X-2)\,a,$$

which agrees with that obtained above.

On account of the symmetry we must divide by 2.

Case 30. $e = D = F = x,\; a = c = y$. By reciprocation of 29.

$$\text{No.} = 2X\,(X-3)\,(x-2)\,y\,(y-1)\,B.$$

On account of the symmetry we must divide by 2.

Case 31. $c = e = D = x,\; a = B = y$.

$$\chi = (Y-2)\,x\,(X-2)\,(x-3)\;Fy, \quad \chi' = Fx\,(X-2)\,(x-3)\;Y\,(y-2),$$
$$\begin{aligned} g &= x\,(x-3)\,(X-2)\,F\,\{(Y-2)\,y + Y\,(y-2)\} \\ &= 2x\,(x-3)\,(X-2)\,(yY - y - Y)\,F. \end{aligned}$$

Case 32. $F = B = a = x,\; D = e = y$. By reciprocation of 31.

$$\text{No.} = 2X\,(X-3)\,(x-2)\,(yY - y - Y)\,c.$$

Case 33. $B = F = y,\ a = e = D = x.$ By reciprocation of 34.

$$\text{No.} = 2(x-1)(xX - x - X)\,Y(Y-1)\,c.$$

Case 34. $c = e = y,\ B = D = a = x.$

$$\chi = (X-2)\,y\,(X-1)(y-1)\,Fx,\quad \chi' = FyX\,(y-1)(X-1)(x-2),$$
$$\begin{aligned} g &= y\,(y-1)(X-1)\,\{(X-2)\,x + X\,(X-2)\}\,F \\ &= 2\,(X-1)(xX - x - X)\,y\,(y-1)\,F. \end{aligned}$$

Case 35. $a = D = y,\ c = e = B = x.$

$$\chi = X\,(x-2)\,Y\,(x-1)\,Fy,\quad \chi' = FxY\,(x-1)(X-2)\,y,$$
$$\begin{aligned} g &= yY(x-1)\,\{X\,(x-2) + (X-2)\,x\}\,F \\ &= 2\,(x-1)(xX - x - X)\,yYF. \end{aligned}$$

Case 36. $a = D = y,\ B = F = e = x.$ By reciprocation of 35.

$$\text{No.} = 2\,(X-1)(Xx - x - X)\,yYc.$$

Case 37. $a = e = D = x,\ c = B = y.$ By reciprocation of 38.

$$\text{No.} = 2\,(x-1)\,\{xyXY - xy\,(X+Y) - XY\,(x+y) + 2xy + 2XY\}\,F.$$

Case 38. $B = D = a = x,\ F = e = y.$

$$\chi = (X-2)\,c\,(X-1)\,y\,(Y-2)\,x,\quad \chi' = Y(y-2)\,Xc\,(X-1)(x-2),$$
$$\begin{aligned} g &= (X-1)\,c\,\{xy\,(X-2)(Y-2) + XY\,(x-2)(y-2)\} \\ &= 2\,(X-1)\,\{xyXY - xy\,(X+Y) - XY\,(x+y) + 2xy + 2XY\}\,c. \end{aligned}$$

Case 39. $a = c = e = B = x.$

Functional process; the curve is assumed to be the aggregate of two curves, say $a = c = e = B = x + x'$. Forming the enumeration

		Case
$x\,X\,x\,DxF$	$x'X'x'Dx'F$	39
$x'X\,x\,.\,x\,.$	&c.	10
$x\,X'x\,.\,x\,.$	.	6
$x'X'x\,.\,x\,.$	.	14
$x\,X\,x'.\,x\,.$	.	10
$x'X\,x'.\,x\,.$	.	12
$x\,X'x'.\,x\,.$	.	14
$x'X'x'.\,x\,.$	.	8

(where the second column is derived from the first by a mere interchange of the accented and unaccented letters), I annex to each line the number of the case to which it belongs; thus $x'XxDxF$ is $B=c=e=x$, which is Case 10, and so in the other instances. Observing that cases 10 and 14 occur each twice, we have thus

$$\begin{aligned}\phi(x+x')-\phi x-\phi x' = DF &\text{ multiplied into}\\ &4(x-1)(Xx-X-x)x' + \ldots (10)\times 2\\ &+\{2x(x-1)(x-2)+X\}X' + \ldots (6)\\ &+4x(x-1)(X'x'-X'-x') + \ldots (14)\times 2\\ &+2x(x-1)x'X' + \ldots (12)\\ &+2x(x-3)(X-2)x' + \ldots (8)\end{aligned}$$

where the (. .)'s refer to the like functions with the two sets of letters interchanged. Developing and collecting, this is

$$\begin{aligned}\phi(x+x')-\phi x-\phi x' = DF &\text{ multiplied into}\\ &2XX'\\ &+2X(3x^2x'+3xx'^2+x'^3-10xx'-5x'^2+6x')\\ &+2X'(x^3+3x^2x'+3xx'^2-5x^2-10xx'+6x)\\ &-12(x^2x'+xx'^2)+40xx',\end{aligned}$$

and thence

$$\begin{aligned}\phi x = \quad DF &\text{ multiplied into}\\ &X^2\\ &+X(2x^3-10x^2+12x)-LX\\ &-4x^3+20x^2 \qquad -lx-\lambda\xi,\end{aligned}$$

where the constants L, l, λ have to be determined. Now for a cubic curve the number of triangles vanishes; that is, we have $\phi x=0$ in each of the three cases,

$$\begin{aligned}x=3,\quad &X=6,\quad \xi=18,\\ ,,\quad &X=4,\quad \xi=12,\\ ,,\quad &X=3,\quad \xi=10,\end{aligned}$$

and we thus obtain the three equations

$$\begin{aligned}0&=108-6L-3l-18\lambda,\\ 0&=\ 88-4L-3l-12\lambda,\\ 0&=\ 81-3L-3l-10\lambda,\end{aligned}$$

giving $L=1$, $l=16$, $\lambda=3$. Whence, finally,

$$\phi x=\{X^2+X(2x^3-10x^2+12x-1)-4x^3+20x^2-16x-3\xi\}DF.$$

Second process, by correspondence. We have

$$\begin{aligned}\mathrm{g}-\chi-\chi'+F(\mathrm{e}-\epsilon-\epsilon')&=0,\\ \mathrm{e}-\epsilon-\epsilon'+D(\mathrm{c}-\gamma-\gamma')&=0,\end{aligned}$$

and thence

$$\mathrm{g}-\chi-\chi'=DF(\mathrm{c}-\gamma-\gamma').$$

Moreover

$$\chi=(X-2)(x-3)\,D\,(x-1)\,F\,(x-1),$$
$$\chi'=F(x-1)\,D\,(x-1)\,(X-2)\,(x-1),\ =\chi,$$
$$\chi+\chi'=DF(X-2)\,2\,(x-3)(x-1)^2,$$

and

$$\mathrm{c}-\gamma-\gamma'=2\tau+(X-3)\,\kappa-2\,(X-2)\,(x-3),$$

as is easily obtained, but see also *post*, No. 29; hence

$$\mathrm{g}=DF \text{ multiplied into}$$
$$(X-2)\,.\,2\,(x-3)\,(x-1)^2$$
$$+\,(X-2)\,.-2\,(x-3)$$
$$+\,2\tau+(X-3)\,\kappa\,;$$

but I reject the term $DF\,.\,(X-3)\,\kappa$ as in fact giving a heterotypic solution; I do not go into the explanation of this. And then substituting for 2τ its value, we have

$$\mathrm{g}=DF \text{ multiplied into}$$
$$(X-2)\,.\,2x\,(x-1)\,(x-2)$$
$$+\,X^2-X+8x-3\xi,$$

where the second factor is

$$=X^2+X\,(2x^3-10x^2+12x-1)-4x^3+20x^2-16x-3\xi,$$

which is the foregoing result.

Case 40. $B=D=F=e=x$. By reciprocation of 39.

$$\text{No.}=\{x^2+x\,(2X^3-10X^2+12X-1)-4X^3+20X^2-16X-3\xi\}\,ac.$$

Case 41. $c=e=D=F=x$.

$$\chi=Bx\,(X-2)\,(x-3)\,(X-3)\,a,$$
$$\chi'=X\,(x-2)\,(X-3)\,(x-3)\,Ba,$$
$$\mathrm{g}=\quad(x-3)\,(X-3)\,aB\,\{x\,(X-2)+X\,(x-2)\},$$
$$=2\,(x-3)\,(X-3)\,(xX-x-X)\,aB.$$

Case 42. $a=c=D=F=x$.

Functional Process; the curve is supposed to be the aggregate of two curves, say $a=c=D=F=x+x'$.

The enumeration is

		Case
$x\ Bx\ X\ e\ X$	$x'Bx'X'eX'$,	(42)
$x'\ .\ x\ X\ .\ X$	&c.	(11)
$x\ .\ x'\ X\ .\ X$		(11)
$x'\ .\ x'\ X\ .\ X$		(17)
$x\ .\ x\ X'\ .\ X$		(10)
$x'\ .\ x\ X'\ .\ X$		(19)
$x\ .\ x'\ X'\ .\ X$		(21)
$x'\ .\ x'\ X'\ .\ X$		(10)

whence

$$\begin{aligned}\phi(x+x') - \phi x - \phi x' = eB \text{ multiplied into}&\\ 4(X-1)(Xx - X - x)x' &\quad + .. \ (11)\times 2\\ +2x(x-1)X'(X'-1) &\quad + .. \ (17)\\ +4(x-1)(Xx - x - x)X' &\quad + .. \ (10)\times 2\\ +2xXx'X' &\quad + .. \ (19)\\ +2xx'XX' - 2(x+x')XX' - 2(X+X')xx' + 4(Xx'+X'x) &\quad + .. \ (21)\end{aligned}$$

where the (. .)'s refer to the like functions with the two sets of letters interchanged. Developing and collecting, we have

$$\begin{aligned}\phi(x+x') - \phi x - \phi x' = eB &\text{ multiplied into}\\ &X^2\ (4xx' + 2x'^2 - 6x')\\ &+XX'(4x^2 + 8xx' + 4x'^2 - 12x - 12x' + 8)\\ &+X'^2\ (2x^2 + 4xx' - 6x)\\ &+X\ (-12xx' - 6x'^2 + 18x')\\ &+X'\ (-6x^2 - 12xx' + 18x)\\ &+\ 8xx',\end{aligned}$$

and consequently

$$\begin{aligned}\phi x = \quad &eB \text{ multiplied into}\\ &X^2(2x^2 - 6x + 4)\\ &+X\ (-6x^2 + 18x + L)\\ &+\ 4x^2 + lx \qquad + \lambda\xi,\end{aligned}$$

where the constants L, l, λ have to be determined. The number of triangles vanishes when the curve is a line or a conic, that is $\phi x = 0$ for $x=1$, $X=0$, $\xi=0$, and for $x=X=2$, $\xi=6$; we thus have

$$0 = 4 + l,$$

$$0 = 40 + 2L + 2l + 6\lambda.$$

Moreover, the data being sibireciprocal, the result must be so likewise; we must therefore have $L = l$. We thus obtain $L = l = \lambda = -4$; so that finally

$$\phi x = \{X^2(2x^2 - 6x + 4) + X(-6x^2 + 18x - 4) + 4x^2 - 4x - 4\xi\}\, eB.$$

Second process, by correspondence: form $a = c = D = F = x$. We have

$$\mathrm{c} - \chi - \chi' + 2(\mathrm{f} - \phi - \phi') = 0;$$

also from the special consideration that the points D, F are given as the intersections of the curve x, by the first polar of the point e, which first polar does not pass through a, we have

$$(\mathrm{f} - \phi - \phi') + e(\mathrm{d} - \delta - \delta') = 0,$$

and by the consideration that c, D are given as intersections, c a double intersection, of the curve with the first polar of the point c, which first polar does not pass through a,

$$\mathrm{d} - \delta - \delta' + 2(\mathrm{c} - \gamma - \gamma') = 0,$$

whence

$$\mathrm{g} - \chi - \chi' = -4e(\mathrm{c} - \gamma - \gamma')$$

and

$$\mathrm{c} - \gamma - \gamma' = B\Delta,$$

so that this is

$$\begin{aligned}\mathrm{g} - \chi - \chi' &= -4Be\Delta\\ &= -4Be(-2X - 2x + 2 + \xi).\end{aligned}$$

Also

$$\begin{aligned}\chi &= B(x-1)(X-2)\,e(X-1)(x-2),\\ \chi' &= (X-2)\,e(X-1)(x-2)\,B(x-1), \ = \chi,\end{aligned}$$

so that

$\mathrm{g} = Be$ multiplied into

$$2(X-1)(X-2)(x-1)(x-2) - 4(-2X - 2x + 2 + \xi),$$

viz. this is

$$Be\{X^2(2x^2 - 6x + 4) + X(-6x^2 + 18x - 4) + 4x^2 - 4x - 4\xi\}.$$

Third process: form $c = e = F = B = x$.

$$\begin{aligned}\mathrm{g} &= \chi + \chi' - \text{Red.},\\ \chi &= X(x-2)\,D(x-1)(X-2)\,a,\\ \chi' &= X(x-2)\,D(x-1)(X-2)\,a, \ = \chi,\\ \chi + \chi' &= aD\,.\,2X(X-2)(x-1)(x-2).\end{aligned}$$

The first-mode reduction is heie

$$aD\,[(X-2)\,X + (X-4)\,2\delta + (X-3)\,3\kappa + \kappa];$$

where the last term $aD\kappa$ arises from the tangents cBa and eFa, each coinciding with a cuspidal tangent, as shown in the figure.

Fig. 4.

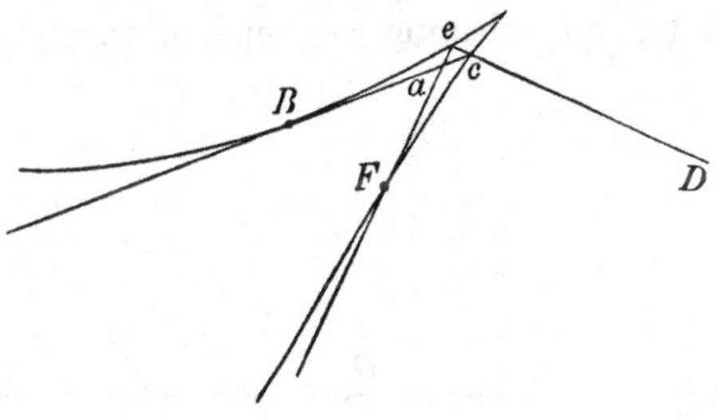

The second-mode reduction is

$$= aD \,.\, X(x-2)(x-3),$$

so that the two reductions together are

$$= aD\{(X-2)X + (X-4)\,2\delta + (X-3)\,3\kappa + \kappa + X(x-2)(x-3)\},$$

viz. this is

$$= aD\{(X-2)X + (X-4)(2\delta + 3\kappa) + 4\kappa + X(x-2)(x-3)\};$$

or substituting for $2\delta + 3\kappa$ and κ the values $x^2 - x - X$ and $-3X + \xi$ respectively, and reducing, it is

$$aD\{X(2x^2 - 6x - 4) - 4x^2 + 4x + 4\xi\}.$$

Hence subtracting from $\chi + \chi'$, written in the form

$$aD\{X^2(2x^2 - 6x + 4) + X(-4x^2 + 12x - 8)\},$$

the result is

$$= aD\{X^2(2x^2 - 6x + 4) + X(-6x^2 + 18x - 4) + 4x^2 - 4x - 4\xi\}.$$

On account of the symmetry we must divide by 2.

Case 43. $a = c = e = x$, $B = D = F = y$.

Suppose for a moment that the angle a is a free point; the locus of a is a curve the order of which is obtained from Case 28, by writing $c = e = x$, $B = D = F = y$; the locus in question meets a curve order a in $\{2Y(Y-1)(Y-2) + y\}\,x(x-1)\,a$ points; wherefore the order of the locus is

$$= \{2Y(Y-1)(Y-2) + y\}\,x\,(x-1),$$

and this locus meets the curve $a = c = e = x$ in a number of points

$$= \{2Y(Y-1)(Y-2) + y\}\,x^2(x-1),$$

viz. this is the number of positions of the angle a; but several of these belong to special forms of the triangle $aBcDeF$, giving heterotypic solutions, which are to be rejected; the required number is thus

$$\{2Y(Y-1)(Y-2) + y\}\,x^2(x-1) - \text{Red.}$$

The reduction is due first and secondly to triangles wherein the angle a coincides with an angle c or e, and thirdly to triangles wherein the angles a, c, e all coincide.

1°. Take for the side cDe a double tangent of the curve $B=D=F$, this meets the curve $a=c=e$ in x points, and selecting any one of them for e and any other for c,

Fig. 5.

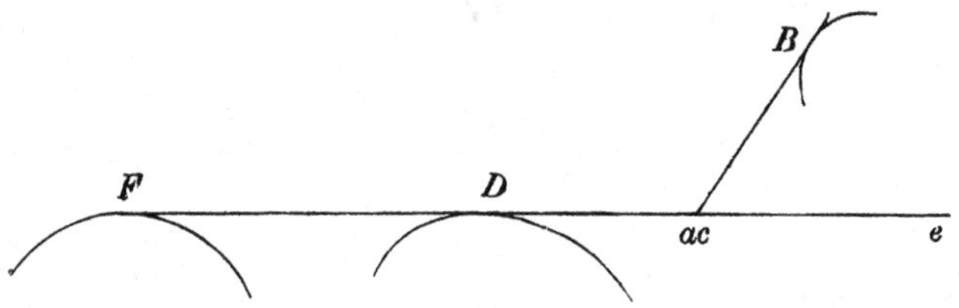

we have from the last-mentioned point $Y-2$ tangents to the curve $B=D=F$; and in respect of each of these a position of a coincident with c. The reduction on this account is $2\tau x(x-1)(Y-2)$; but since we may in the figure interchange c and e, B and F, we have the same number belonging to the coincidence of the angles a, e, or together the reduction is $=4\tau x(x-1)(Y-2)$.

Fig. 6.

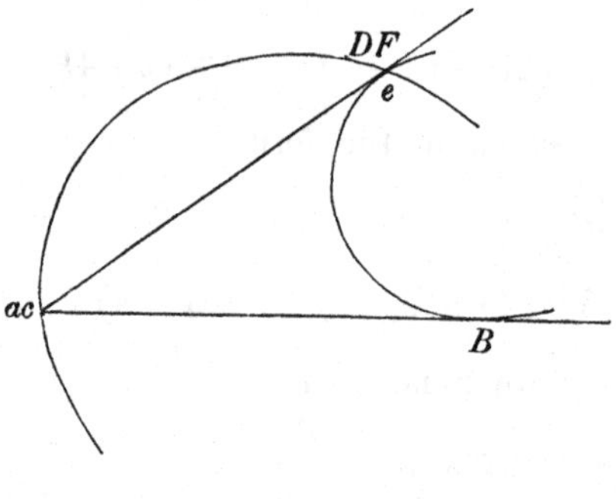

But instead of a double tangent we may have cDe a stationary tangent; we have thus reductions $3\iota\, x(x-1)(Y-2)$ and $3\iota\, x(x-1)(Y-2)$, together $6\iota\, x(x-1)(Y-2)$; and for the double and stationary tangents together we have

$$(4\tau+6\iota)\,x(x-1)(Y-2),$$

$$=2\{Y(Y-1)-y\}\,x(x-1)(Y-2),$$

that is

$$=2x(x-1)\,Y(Y-1)(Y-2)-2x(x-1)\,y(Y-2).$$

2°. The side cDe may be taken to be a tangent to the curve $B=D=F$ at any one of its intersections with the curve $a=c=e$. Taking then the point e at the intersection in question, and the point c at any other of the intersections of the tangent with the curve $a=c=e$, and from c drawing any other tangent to the curve $B=D=F$, there is in respect of each of these tangents a position of a at c; and the reduction on this account is $= xy(x-1)(Y-1)$. But interchanging in the figure the letters c, e, B, F, there is an equal reduction belonging to the coincidence of a, e; and the whole reduction in this manner is $= 2x(x-1)y(Y-1)$.

Fig. 7.

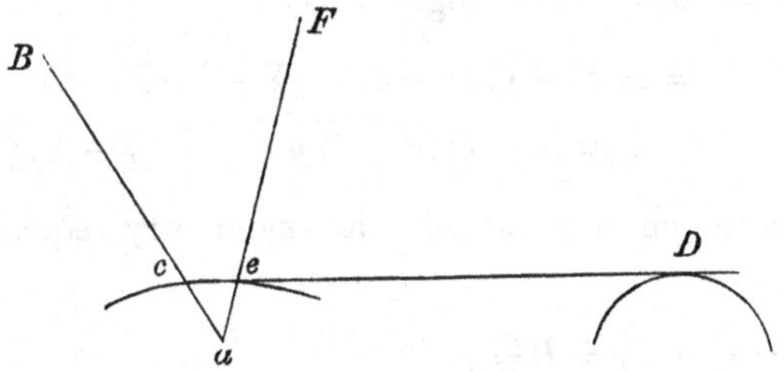

3°. If the side cDe intersects the curve $a=c=e$ in two coincident points, then taking these in either order for the points c, e, and from the two points respectively drawing two other tangents to the curve $D=B=F$, we have a triangle wherein the angles a, c, e all coincide. The side cDe may be a proper tangent to the curve $a=c=e$, or it may pass through a node or a cusp of this curve, viz. it is either a common tangent of the curves $B=D=F$ and $a=c=e$ (as in the figure, except that for greater distinctness the points c and e are there drawn nearly instead of actually coincident), or it may be a tangent to the curve $B=D=F$ from a node or a cusp of the curve $a=c=e$; we have thus the numbers

Common tangent	$XY(Y-1)(Y-2)$,
Tangent from node	$2\delta Y(Y-1)(Y-2)$,
Tangent from cusp	$2\kappa Y(Y-1)(Y-2)$;

but (as we are counting intersections with the curve $a=c=e$) the second of these, as being at a node of this curve, is to be taken 2 times; and the third, as being at a cusp, 3 times; and the three together are thus

$$(X+4\delta+6\kappa\quad)Y(Y-1)(Y-2),$$
$$=\{2x(x-1)-X\}\,Y(Y-1)(Y-2).$$

The reductions 1°, 2°, 3° altogether are

$$\begin{aligned}&2x(x-1)\,Y(Y-1)(Y-2)\\&-2x(x-1)\,y(Y-2)\\&+2x(x-1)\,y(Y-1)\\&+2x(x-1)\,Y(Y-1)(Y-2)\\&-XY(Y-1)(Y-2),\end{aligned}$$

which is

$$\begin{aligned}&=4x(x-1)\,Y(Y-1)(Y-2)\\&\quad+2x(x-1)\,y\\&\quad-XY(Y-1)(Y-2);\end{aligned}$$

and subtracting from the before-mentioned number

$$2x^2(x-1)Y(Y-1)(Y-2)$$
$$+x^2(x-1)y,$$

the required number of positions of the angle a is

$$=2x(x-1)(x-2)Y(Y-1)(Y-2)$$
$$+yx(x-1)(x-2)y+XY(Y-1)(Y-2).$$

The number of triangles is on account of the symmetry equal to one-sixth of this number.

Case 44. $e=D=F=x,\ a=c=B=y.$

$$\chi=(Y-2)(y-3)X(x-2)(X-3)y,$$
$$\chi'=X(x-2)(X-3)y(Y-2)(y-3)\ (=\chi),$$
$$g=2(x-2)X(X-3)(Y-2)y(y-3):$$

there is a division by 2 on account of the symmetry.

Case 45. $a=D=B=x,\ c=e=F=y.$

$$\chi=(X-2)y(X-1)(y-1)(Y-2)x,$$
$$\chi'=Y(y-2)X(y-1)(X-1)(x-2),$$
$$g=(X-1)(y-1)\{xy(X-2)(Y-2)+XY(x-2)(y-2)\}$$
$$=2(X-1)(y-1)\{XYxy-XY(x+y)-xy(X+Y)+2xy+2XY\}.$$

Case 46. $a=c=y,\ B=D=F=e=x.$ By reciprocation of 47,

$$\text{No.}=y(y-1)\{x^2+x(2X^3-10X^2+12X-1)-4X^3+20X^2-16X-3\xi\}:$$

there is a division by 2 on account of the symmetry.

Case 47. $D=F=y,\ a=c=e=B=x.$

The functional process is exactly the same as for No. 39 ($a=c=e=B=x$), with only $Y(Y-1)$ written instead of DF; hence

$$\text{No.}=Y(Y-1)\{X^2+X(2x^3-10x^2+12x-1)-4x^3+20x^2-16x-3\xi\}:$$

there is a division by 2 on account of the symmetry.

Case 48. $a=c=D=F=x,\ e=B=y.$

The functional process, writing $a=c=D=F=x+x'$, would be precisely the same as for Case 42, with only the factor yY written instead of eB; and we have thus the like result, viz.

$$\text{No.}=\{X^2(2x^2-6x+4)+X(-6x^2+18x-4)+4x^2-4x-4\xi)\,yY,$$

which on account of the symmetry must be divided by 2.

Case 49. $a = B = y,\ c = e = D = F = x.$

$$\begin{aligned}
\chi &= (Y-2)\,x\,(X-2)(x-3)(X-3)\,y,\\
\chi' &= X\,(x-2)(X-3)(x-3)\,Y\,(y-2),\\
g &= (x-3)(X-3)\,\{xy\,(X-2)(Y-2) + XY\,(x-2)(y-2)\}\\
&= 2\,(x-3)(X-3)\,\{xyXY - (x+y)\,XY - (X+Y)\,xy + 2xy + 2XY\}.
\end{aligned}$$

Case 50. $c = e = B = D = F = x.$

Functional process; by taking the curve $c = e = B = D = F$ as the aggregate of two curves, say $= x + x'$. The cases are

		Case
$aX\,x\,X\,x\,X$	$aX'x'X'x'X'$	50
$.\,X'x\,X\,x\ .$	&c.	41
$.\,X\,x'X\,x\ .$	.	40
$.\,X'x'X\,x\ .$	.	32
$.\,X\,x\,X'x\ .$	.	42
$.\,X'x\,X'x\ .$	.	33
$.\,X\,x'X'x\ .$	.	38
$.\,X'x'X'x\ .$	.	32
$.\,X\,x\,X\,x'\ .$	.	40
$.\,X'x\,X\,x'\ .$	.	36
$.\,X\,x'X\,x'\ .$	.	28
$.\,X'x'X\,x'\ .$	.	33
$.\,X\,x\,X'x'\ .$	.	38
$.\,X'x\,X'x'\ .$	.	36
$.\,X\,x'X'x'\ .$	.	29
$.\,X'x'X'x'X$	.	41

and we thus have

$\phi\,(x+x') - \phi x - \phi x' = a$ multiplied into

$$\begin{aligned}
= \quad & 4\,(x-3)(X-3)(xX - x - X)\,X' + \,.\,. && 2\,(41)\\
& + 2x'\,[x^2 + x\,(2X^3 - 10X^2 + 12X - 1) - 4X^3 + 20X^2 - 16X - 3\xi] \ + \,.\,. && 2\,(40)\\
& + 4X\,(X-3)(x-2)(x'X' - X' - x') \ + \,.\,. && 2\,(32)\\
& + [X^2\,(2x^2 - 6x + 4) + X\,(-6x^2 + 18x - 4) + 4x^2 - 4x - 4\xi]\,X' \ + \,.\,. && (42)\\
& + 4\,(x-1)(xX - x - X)(X'^2 - X') \ + \,.\,. && 2\,(33)\\
& + 4\,(X-1)\,[XX'xx' - XX'\,(x+x') - xx'\,(X+X') + 2XX' + 2xx'] + \,.\,. && 2\,(38)\\
& + 4xX\,(X'-1)(X'x' - X' - x') \ + \,. && 2\,(36)\\
& + (x^2 - x)(2X'^3 - 6X'^2 + 4X' + x') \ + \,.\,. && (28)\\
& + 2x\,(x-3)(X-2)(X'^2 - X') \ + \,.\,. && (29)
\end{aligned}$$

where as before the (. .)'s refer to the like functions with the two sets of letters interchanged. Developing and collecting, this is

$$\begin{aligned}
\phi(x+x') - \phi x - \phi x' = a \text{ multiplied into}&\\
&3x^2x' + 3xx'^2\\
&+ x^2 .\ 6X^2X' + 6XX'^2 + 2X'^3\\
&\quad - 28XX' - 14X'^2\\
&\quad + 28X'\\
&+ xx' .\ 4X^3 + 12X^2X' + 12XX'^2 + 4X'^3\\
&\quad - 28X^2 - 56XX' - 28X'^2\\
&\quad + 56X + 56X'\\
&\quad - 22\\
&+ x'^2 .\ 2X^3 + 6X^2X' + 6XX'^2\\
&\quad - 14X^2 - 28XX'\\
&\quad + 28X\\
&+ x .\ - 30X^2X' - 30XX'^2 - 10X'^3\\
&\quad + 140XX' + 70X'^2\\
&\quad - 116X' - 6\xi'\\
&+ x' .\ - 10X^3 - 30X^2X' - 30XX'^2\\
&\quad + 70X^2 + 140XX'\\
&\quad - 116X - 6\xi\\
&\quad + 36X^2X' + 36XX'^2\\
&\quad - 152XX'\\
&\quad - 4(X\xi' + X'\xi):
\end{aligned}$$

whence

$$\begin{aligned}
\phi x = a \text{ multiplied into}&\\
&x^3 (\qquad\qquad\qquad\qquad\qquad + 1)\\
&+ x^2 (\quad 2X^3 - 14X^2 + \ 28X - 11)\\
&+ x\ (- 10X^3 + 70X^2 - 116X + l)\\
&+ \qquad 12X^3 - 76X^2 \qquad\quad + LX\\
&+ \xi\ (- \ 6x - 4X \qquad\qquad + \lambda),
\end{aligned}$$

where the constants l, L, λ have to be determined. We should have $\phi x = 0$ for a cubic curve; viz. $x=3$: $X=6$, $\xi=18$; $X=4$, $\xi=12$; or $X=3$, $\xi=10$. Writing first $x=3$, the equation is

$$8X^2 - 96X - 72 - \xi(18+4X) + 3l + XL + \xi\lambda = 0,$$

giving in the three cases respectively

$$3l+6L+18\lambda=1116,$$
$$3l+4L+12\lambda=736,$$
$$3l+3L+10\lambda=588;$$

and we have then $l=-8$, $L=64$, $\lambda=42$, so that the required number is

$$\begin{aligned} =\ & x^3(\qquad\qquad\qquad\qquad +1)\\ +\ & x^2(\ \ 2X^3-14X^2+\ 28X-11)\\ +\ & x\,(-10X^3+70X^2-116X-\ 8)\\ +\ & \quad\ \ 12X^3-76X^2+\ 64X\\ +\ & \xi\,(-\ \ 6x-4X\ \ +42\qquad\ ). \end{aligned}$$

As a verification, observe that for a conic, $x=X=2$, $\xi=6$, this is $=0$.

Second process, by correspondence: form $c=e=B=D=F=x$.

We have

$$\begin{aligned} g\ &=\chi+\chi'-\text{Red.},\\ \chi\ &=X(x-2)(X-3)(x-3)(X-3)\,a,\\ \chi'&=X(x-2)(X-3)(x-3)(X-3)\,a,\ =\chi,\\ \chi+\chi'&=a\ \text{into}\\ &2(x-2)(x-3)X(X-3)^2. \end{aligned}$$

Fig. 8.

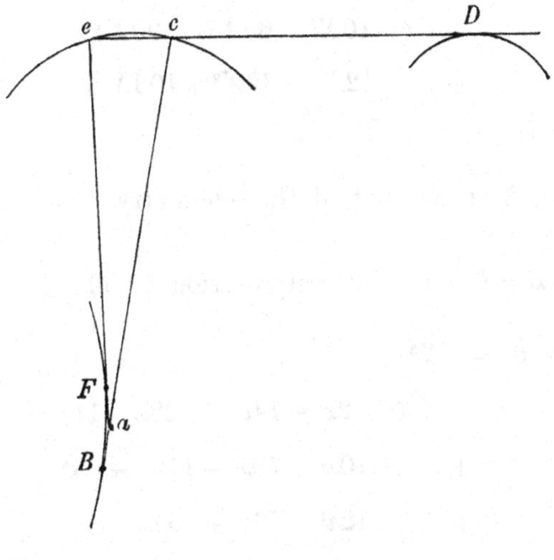

There is a first-mode reduction, which is

$$=a\,\{2\delta(X-4)(X-5)+3\kappa(X-3)(X-4)+\kappa(X-3)+2\tau(X-3)\},$$

where the term $a \,.\, 2\tau\,(X-3)$ arises, as shown in the figure, and a second-mode reduction, which is

$$=a\,\{2\tau\,(x-4)(x-5)+3\iota\,(x-3)(x-4)\}\,;$$

and the two together are $=a$ into

$$\begin{array}{llllll}
 & (X-4)(X-5) & (x^2-x+8X & & & -3\xi) \\
+ & (X-3)(X-4) & (& 9X & & +3\xi) \\
+ & (X-3) & \left(\begin{array}{l} -3X+\xi \\ +X^2-X+8x-3\xi \end{array}\right) & & & \\
+ & (x-4)(x-5) & (X^2-X+8x & & & -3\xi) \\
+ & (x-3)(x-4) & (& & -9x & +3\xi)\,;
\end{array}$$

that is, $=a$ into

$$\begin{array}{l}
-x^3 \\
+x^2\,.\,2X^2-10X+11 \\
+x\,.-10X^2+26X+8 \\
+\ 4X^2+44X \\
+\xi\,(6x+4X-42)\,;
\end{array}$$

and subtracting this from the foregoing value of $\chi+\chi'$, which is $=a$ into

$$\begin{array}{l}
x^2\,(\quad 2X^3-12X^2+\ 18X) \\
+x\,(-10X^3+60X^2-\ 90X) \\
+\quad 12X^3-72X^2+108X,
\end{array}$$

the result is as before.

There is a division by 2 on account of the symmetry.

Case 51. $a=c=e=B=D=x$. By reciprocation of 50,

$$\begin{array}{rl}
\text{No. is} = & X^3\,(\qquad\qquad\qquad\qquad +\ 1) \\
 & +X^2\,(\quad 2x^3-14x^2+\ 28x-11) \\
 & +X\,(-10x^3+70x^2-116x-\ 8) \\
 & +\quad 12x^3-76x^2+\ 64x \\
 & +\xi\,(-6X-4x+42).
\end{array}$$

There is a division by 2 on account of the symmetry.

Case 52. $a=c=e=B=D=F=x$.

Functional process, by taking the curve to be the aggregate of two curves, say $=x+x'$. The enumeration of the cases is conveniently made in a somewhat different manner from that heretofore employed, viz. we may write

x or x'	x' or x	Case	times
‖	‖		
all	none	(52)	1
a	residue	(50)	3
B	"	(51)	3
a, c	"	(46)	3
B, D	"	(47)	3
a, D	"	(48)	3
a, B	"	(49)	6
a, c, e	B, D, F	(43)	1
a, B, F	c, e, D	(44)	3
a, B, D	c, e, F	(45)	6
			32;

and the functional equation then is

$$\phi(x+x')-\phi x-\phi x'$$

$$= 3x'\left\{\begin{array}{l} x^3(\qquad\qquad\qquad\qquad\qquad + 1) \\ x^2(\ \ 2X^3-14X^2+\ 28X-11) \\ x\ (-10X^3+70X^2-116X-\ 8) \\ \quad +12X^3-76X^2+\ 64X \\ \quad +\xi(-6x-4X \qquad\qquad +42) \end{array}\right\}+\ldots \qquad (50)\times 3$$

$$+3X'\left\{\begin{array}{l} X^3(\qquad\qquad\qquad\qquad\qquad + 1) \\ X^2(\ \ 2x^3-14x^2+\ 28x-11) \\ X\ (-10x^3+70x^2-116x-\ 8) \\ \quad +12x^3-76x^2+\ 64x \\ \quad +\xi(-6X-4x \qquad\qquad +42) \end{array}\right\} \qquad (51)\times 3$$

$$+3(x'^2-x')\left\{\begin{array}{l} x^2 \\ +x(2X^3-10X^2+12X-1) \\ -\ \ 4X^3+20X^2-16X-3\xi \end{array}\right\}+\ldots \qquad (46)\times 3$$

$$+3(X'^2-X')\left\{\begin{array}{l} X^2 \\ +X\ (2x^3-10x^2+12x-1) \\ \quad -4x^3+20x^2-16x-3\xi \end{array}\right\} + .. \qquad (47)\times 3$$

$$+3x'X'\left\{\begin{array}{l} X^2(\ \ 2x^2-\ \ 6x+4) \\ +X\ (-6x^2+18x-4) \\ \quad +4x^2-\ \ 4x-4\xi \end{array}\right\} + .. \qquad (48)\times 3$$

$$+12(x'-3)(X'-3)\{xx'XX'-xx'(X+X')-XX'(x+x')+2xx'+2XX'\} + .. \quad (49)\times 6$$

$$+\{2x'(x'-1)(x'-2)X(X-1)(X-2)+xx'(x'-1)(x'-2)+X'X(X-1)(X-2)\} + .. \quad (43)$$

$$+6(x'-2)X'(X'-3)(X-2)(x-3) \qquad + .. \quad (44)\times 3$$

$$+12(X'-1)(x-1)\{xx'XX'-xx'(X+X')-XX'(x+x')+2xx'+2XX'\} + .. \quad (45)\times 6$$

where as before the (. .)'s refer to the like functions with the two sets of letters interchanged. Developing and collecting, this is found to be

$$\begin{array}{ll} = & 4X^3X'+6X^2X'^2+4XX'^3 \\ & +X^3\left\{\begin{array}{l} 6x^2x'+6xx'^2+2x'^3 \\ -36xx'-18x'^2 \\ +52x' \end{array}\right\} \\ & +(X^2X'+XX'^2)\left\{\begin{array}{l} \quad 6x^3+\ 18x^2x'+18xx'^2+6x'^3 \\ -\ \ 54x^2-108xx'\ -54x'^2 \\ +156x\ +156x' \\ -138 \end{array}\right\} \\ & +X'^3\left\{\begin{array}{l} 2x^3+6x^2x'+6xx'^2 \\ -18x^2-36xx' \\ +52x \end{array}\right\} \\ & +\ \&c.\ \&c. \end{array}$$

I abstain from writing down the remaining terms, as they can at once be obtained backwards from the value of ϕx; they were in fact found directly, and the integration of the functional equation then gives

$$\begin{array}{ll} \phi x = & X^4(\qquad\qquad\qquad\qquad\quad +\ \ 1) \\ & +X^3(\qquad 2x^3-\ \ 18x^2+\ \ 52x-\ \ 46) \\ & +X^2(\ -18x^3+162x^2-420x+221) \\ & +X\ (\qquad 52x^3-420x^2+704x+l\quad) \\ & +\qquad x^4-46x^3+221x^2+\quad lx \\ & +\xi\left\{\begin{array}{l} X^2(\qquad\qquad -\ \ 9) \\ +X\ (\ -\ \ 12x+135) \\ \quad -9x^2+135x+\ \ \lambda \end{array}\right\} \end{array}$$

where the constants l, λ have to be determined; I have in the first instance written $l(X+x)+\lambda\xi$, instead of $LX+lx+\lambda\xi$, thus introducing two constants only, since it is clear from the symmetry in regard to x, X that we must have $l=L$. We must have $\phi x=0$, when the curve is a conic or cubic. Writing $x=2$, we have

$$\phi x = X^4+2X^3-115X^2+144X+532+\xi(-9X^2+111X+234)+l(2+X)+\xi\lambda,$$

and then for the conic, $X=2$, $\xi=6$.

Writing $x=3$, we have

$$\phi x = X^4+2X^3-67X^2-264X+828+\xi(-9X^2+99X+324)+l(3+X)+\lambda\xi,$$

and then for the three cases of the cubic $X=6$, $\xi=18$; $X=4$, $\xi=12$; and $X=3$, $\xi=10$. We have thus the four equations

$$\begin{aligned} 2912+4l+\ \ 6\lambda&=0,\\ 9252+9l+18\lambda&=0,\\ 5796+7l+12\lambda&=0,\\ 4968+6l+10\lambda&=0,\end{aligned}$$

all satisfied by $l=+172$, $\lambda=-600$. Hence finally

$$\begin{aligned}\phi x = \quad & X^4(\qquad\qquad\qquad\qquad +1)\\ &+X^3(\quad 2x^3-\ 18x^2+\ 52x-\ 46)\\ &+X^2(-18x^3+162x^2-420x+221)\\ &+X\ (\quad 52x^3-420x^2+704x+172)\\ &+\quad x^4-46x^3+221x^2+172x\\ &+\xi\left\{\begin{array}{l} X^2(\qquad\qquad -\ 9)\\ +X\ (\quad -\ 12x+135)\\ \quad -9x^2+135x-600\end{array}\right\}\end{aligned}$$

but on account of the symmetry the number of triangles is = one-sixth of this expression.

Article Nos. 22 to 36. *The Case 52, as belonging to a different series of Problems.*

22. In the foregoing Case 52, where all the curves are one and the same curve, we have the unclosed trilateral *aBcDeFg*, and we seek for the number of the united points (a, g). But we may consider this as belonging to a series of questions, viz. we may seek for the number of the united points (a, B), (a, c), (a, D), (a, e), (a, F), (the last four of these giving by reciprocity the numbers of the united points (B, D), (B, e), (B, F), (B, g)), and finally the number of the united points (a, g). It is very instructive to consider this series of questions, and the more so that in those which precede (a, F) there are only special solutions having reference to the singular points and tangents of the curve, and that the solutions thus explain themselves.

23. Thus the first case is that of the united points (a, B), viz. we have here a point a on the curve, and from it we draw to the curve a tangent aB touching it at B; the points a and B are to coincide together. Observe that from a point in general a of the curve we have $X-2$ tangents (X the class as heretofore), viz. we disregard altogether the tangent *at* the point, counting as 2 of the X tangents from a point not on the curve, and attend exclusively to the $X-2$ tangents *from* the point. Now if the point a is an inflection, or if it is a cusp, there are only $X-3$ tangents, or, to speak more accurately, one of the $X-2$ tangents has come to coincide with the tangent *at* the point; such tangent is a tangent of three-pointic intersection, viz. we have the point a and the point B (counting, as a point of contact, twice) all three coinciding; that is, we have a position of the united point (a, B); and the number of these united points is $=\iota+\kappa$.

24. It is important to notice that neither a point of contact of a double tangent, nor a double point, is a united point. In the case of the point of contact of a double tangent, one of the tangents from the point coincides with the double tangent; but the point B is here the other point of contact of this tangent, so that the points a, B are not coincident. In the case of a double point, regarding the assumed position of a at the double point as belonging to one of the two branches, then of the $X-2$ tangents there are two, each coinciding with the tangent to the other branch; hence, attending to either of these, the point B belongs to the other branch, and thus, though a and B are each of them at the double point, the two do not constitute a united point. (In illustration remark that for a unicursal curve, the position of a answers to a value $=\lambda$, and that of B to a value $=\mu$ of the parameter θ, viz. λ, μ are the two values of θ at the double point; contrariwise in the foregoing case of a cusp, where there is a single value $\lambda=\mu$. Hence the whole number of the united points (a, B) is $=\iota+\kappa$, and this is in fact the value given, as will presently appear, by the theory of correspondence.)

I recall that I use Δ, $=2D$, to denote twice the deficiency of the curve, viz. that we have $\Delta = X-2x+2+\kappa$, $=-2x-2X+2+\xi$.

25. The several cases are

United points.	
(a, B)	$\mathrm{b}-\beta-\beta' = 2\Delta$,
(a, c)	$\mathrm{c}-\gamma-\gamma' + 2(\mathrm{b}-\beta-\beta') = (X-2)\Delta$,
(B, D)	$\mathrm{c}_0-\gamma_0-\gamma_0'$ by reciprocity,
(a, D)	$\mathrm{d}-\delta-\delta' + 2(\mathrm{c}_0-\gamma_0-\gamma_0') + (X-3)(\mathrm{b}-\beta-\beta') = 0$,
(a, e)	$\mathrm{e}-\epsilon-\epsilon' + 2(\mathrm{d}-\delta-\delta') + (X-3)(\mathrm{c}-\gamma-\gamma') = 0$,
(B, F)	$\mathrm{e}_0-\epsilon_0-\epsilon_0'$ by reciprocity,
(a, F)	$\mathrm{f}-\phi-\phi' + 2(\mathrm{e}_0-\epsilon_0-\epsilon_0') + (X-3)(\mathrm{d}-\delta-\delta') = 0$,
(a, g)	$\mathrm{g}-\chi-\chi' + 2(\mathrm{f}-\phi-\phi') + (X-3)(\mathrm{e}-\epsilon-\epsilon') = 0$,
(B, H)	$\mathrm{g}_0-\chi_0-\chi_0'$ by reciprocity,

and so on.

26. The mode of obtaining these equations appears *ante*, Nos. 5 and 6, but for greater clearness I will explain it in regard to a pair of the equations, say those for (a, e), (a, D). Regarding a as given, we draw from a the tangents aBc, touching at B and besides intersecting at c (viz. the number of tangents is $=X-2$, and the number of the points c is $=(X-2)(x-3)$); from each of the positions of c we draw to the curve the $(X-3)$ tangents cDe touching at D and intersecting at e; the whole number of these tangents is $=(X-2)(x-3)(X-3)$; and this is also the number of the points D, but the number of the points e is $=(X-2)(x-3)(X-3)(x-3)$. Now this system of the $(X-2)(x-3)(X-3)$ tangents is the curve Θ of the general theory (*ante*, Nos. 3, 4), viz. the curve Θ (which does not pass through a) intersects the given curve in the three classes of points c, D, e, the number of intersections at a point e being $=1$, at a point D being $=2$, and at a point c being $=X-3$. And we have thus the equation

$$\mathrm{e}-\epsilon-\epsilon'+2(\mathrm{d}-\delta-\delta')+(X-3)(\mathrm{c}-\gamma-\gamma')=0,$$

where e, d, c are the numbers of united points and (ϵ, ϵ'), (δ, δ'), (γ, γ') the correspondences in the three cases respectively.

27. Observe that we cannot, starting from a, obtain in this manner the equation for the number of the united points (a, D); for we introduce per force the points e, and thus obtain the foregoing equation for (a, e). But starting from D, the tangent at this point besides intersects the curve in $(x-2)$ points, each of which is a position of c; and from each of these drawing a tangent cBa to the curve, we have the curve Θ consisting of these $(x-2)(X-3)$ tangents, not passing through D, but intersecting the given curve in the three classes of points c, B, a, viz. the number of intersections at each point c is $=X-3$, at each point B it is $=2$, and at each point a it is $=1$; and we have thus the equation

$$(\mathrm{d}-\delta-\delta')+2(\mathrm{c}_0-\gamma_0-\gamma_0')+(X-3)(\mathrm{b}-\beta-\beta')=0,$$

where the numbers $(\mathrm{d}, \delta, \delta')$, $(\mathrm{c}_0, \gamma_0, \gamma_0')$, $(\mathrm{b}, \beta, \beta')$ refer to the correspondences (D, a), (D, B), and (D, c) (or what is the same thing (a, B)) respectively.

28. Correspondence (a, B).

We have

$$\beta=X-2,\quad \beta'=x-2,$$

and thence

$$\begin{aligned}\mathrm{b}&=x+X-4+2\Delta\\&=-3x-3X+2\xi,\end{aligned}$$

which is the solution: the value obtained above was $\mathrm{b}=\iota+\kappa$, and we in fact have identically

$$\iota+\kappa=-3x-3X+2\xi.$$

It was in this manner that I originally applied the principle of correspondence to investigating the number of inflections of a curve, regarding, however, the term κ as a special solution; it is better to put the cusp and inflection on the same footing as above.

29. Correspondence (a, c).

Since $\mathrm{b} - \beta - \beta' = 2\Delta$, we have here

$$\mathrm{c} - \gamma - \gamma' = (X-6)\,\Delta,$$

and

$$\gamma = \gamma' = (X-2)(x-3),$$

whence

$$\begin{aligned}\mathrm{c} &= 2(X-2)(x-3) + (X-6)(-2x-2X+2+\xi)\\ &= -2X^2 + 8X + 8x + (X-6)\,\xi;\end{aligned}$$

this is in fact $= 2\tau + (X-3)\,\kappa$, viz. we have

$$\begin{aligned}2\tau &= X^2 - X + 8x - 3\xi\\ (X-3)\,\kappa &= (X-3)(-3X+\xi) = -3X^2 + 9X + (X-3)\,\xi,\end{aligned}$$

and therefore

$$2\tau + (X-3)\,\kappa = \text{as above},$$

viz. the united points (a, c) are the 2τ points of contact of the double tangents, and the κ cusps each $(X-3)$ times in respect of the $(X-3)$ tangents from it to the curve. This is the way in which I originally applied the principle to finding the number of double tangents of a curve.

30. Correspondence (B, D). By reciprocation

$$\begin{aligned}\mathrm{c}_0 - \gamma_0 - \gamma_0' &= (x-6)\,\Delta,\\ \mathrm{c}_0 &= -2x^2 + 8x + 8X + (x-6)\,\xi\\ &= 2\delta + (x-3)\,\iota.\end{aligned}$$

31. It may be remarked, as regards the cases which follow, that although the result in terms of $(\delta, \kappa, \iota, \tau)$ when once known can be explained and verified easily enough, there is great risk of oversight if we endeavour to find it in the first instance; while on the other hand the transformation from the form in terms of (x, X, ξ), as given by the principle of correspondence, to the required form in terms of $(\delta, \kappa, \iota, \tau)$ is by no means easy. I in fact first obtained the expression in (x, X, ξ), and then, knowing in some measure the form of the other expression, was able to find it by the actual transformation of the expression in (x, X, ξ).

32. Correspondence (a, D).

From the values of $\mathrm{c}_0 - \gamma_0 - \gamma_0'$ and $\mathrm{b} - \beta - \beta'$ we have

$$\mathrm{d} - \delta - \delta' = -(2X + 2x - 18)\,\Delta,$$

and then

$$\delta = (X-2)(x-3)(X-3), \quad \delta' = (x-2)(X-3)(x-3),$$

whence

$$\begin{aligned}\mathrm{d} &= (x-3)(X-3)(X+x-4)\\ &\quad + (-2X-2x+18)(-2X-2x+2+\xi)\end{aligned}$$

which is

$$\begin{aligned}
&= X^2 (\qquad x + 1)\\
&+ X\ (x^2 - \ 2x - 19)\\
&+ \qquad x^2 - 19x\\
&+ \xi(-2X - 2x + 18).
\end{aligned}$$

And then, by means of the equations

$$\begin{aligned}
(x-4)\,2\tau &= (x-4)(X^2 - X + 8x - 3\xi),\\
(X-4)\,2\delta &= (X-4)(x^2 - x + 8X - 3\xi),\\
(x-3)\,\iota &= (x-3)(\quad -3x \qquad + \xi),\\
(X-3)\,\kappa &= (X-3)(\quad -3X \qquad + \xi),
\end{aligned}$$

we verify that

$$\mathrm{d} = (x-4)\,2\tau + (X-4)\,2\delta + (x-3)\,\iota + (X-3)\,\kappa.$$

33. Correspondence (a, e).

From the values of $\mathrm{d} - \delta - \delta'$, $\mathrm{c} - \gamma - \gamma'$ we have

$$\mathrm{e} - \epsilon - \epsilon' = (-X^2 + 13X + 4x - 54)\,\Delta,$$

and then

$$\epsilon \mp \epsilon' = (X-2)(x-3)(X-3)(x-3);$$

that is

$$\begin{aligned}
\mathrm{e} = {} & 2\,(x-3)^2 (X-2)(X-3)\\
& + (-X^2 + 13X + 4x - 54)(-2X - 2x + 2 + \xi),
\end{aligned}$$

which is

$$\begin{aligned}
&= X^3 (\qquad\qquad\qquad 2)\\
&+ X^2 (\quad 2x^2 - 10x \quad - 10)\\
&+ X\ (-10x^2 + 26x \quad + 44)\\
&+ \qquad 4x^2 + 44x\\
&+ \xi\ \ (-X^2 + 13X + 4x - 54),
\end{aligned}$$

and then

$$\begin{aligned}
(x-4)(x-5)\,2\tau &= (x-4)(x-5)(X^2 - X + 8x - 3\xi),\\
\{(X-4)(X-5) + x - 3\}\,2\delta &= \{(X-4)(X-5) + x - 3\}(x^2 - x + 8X - 3\xi),\\
\{3\,(x-3)(x-4) + x - 3\}\,\iota &= (x-3)(3x-11)(-3x + \xi),\\
2\,(X-3)(X-4)\,\kappa &= 2\,(X-3)(X-4)(-3X + \xi);
\end{aligned}$$

and summing these values and comparing,

$$\begin{aligned}
\mathrm{c} = {} & (x-4)(x-5)\,2\tau + 2\,(X-3)(X-4)\,\kappa\\
& + [(X-4)(X-5) + x - 3]\,2\delta + [3\,(x-3)(x-4) + x - 3]\,\iota.
\end{aligned}$$

The united points (a, e) are in fact, 1°, each of the $x-4$ intersections of a double tangent with the curve, in respect of the two contacts and of the remaining $x-5$ intersections; 2°, each double point in respect of the two branches and of the pairs of tangents from it to the curve; 3°, each of the $x-3$ intersections of each of the tangents at a double point with the curve; 4°, each of the $x-3$ intersections of a tangent at an inflection (stationary tangent) with the curve, in respect of the $(x-4)$ remaining intersections; 5°, each inflection in respect of the $x-3$ intersections of the

Fig. 9.

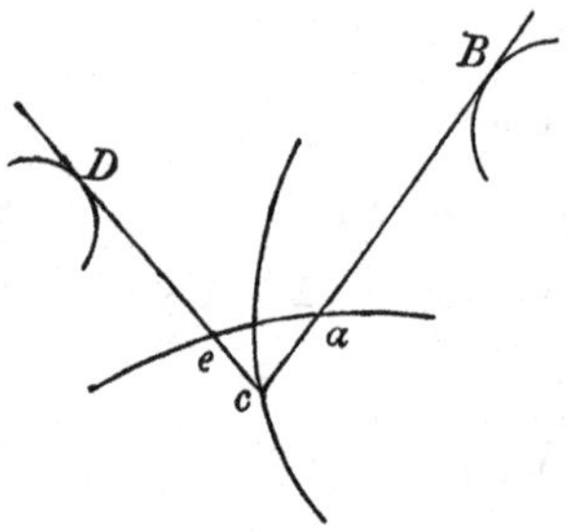

tangent with the curve; and 6°, each cusp in respect of the pairs of tangents from it to the curve. Thus (2°), the double point in respect of the branch which contains c, and of the two tangents from it to the curve, is a position of the united point (a, e), as appearing in the figure.

34. Correspondence (B, F). By reciprocation of (a, e)

$$\mathrm{e}_0 - \epsilon_0 - \epsilon_0' = (-x^2 + 13x + 4X - 54)\,\Delta,$$

$$\begin{aligned}\mathrm{e}_0 = {} & (X-4)(X-5)\,2\delta + 2(x-3)(x-4)\,\iota \\ & + [(x-4)(x-5) + X - 3]\,2\tau + [3(X-3)(X-4) + (X-3)]\,\kappa.\end{aligned}$$

35. Correspondence (a, F). By means of the values of $\mathrm{e}_0 - \epsilon - \epsilon'$ and $\mathrm{d} - \delta - \delta'$, we have

$$\mathrm{f} - \phi - \phi' = (2X^2 + 2Xx + 2x^2 - 32X - 32x + 162)\,\Delta,$$

and then

$$\begin{aligned}\phi &= (X-2)(x-3)(X-3)(x-3)(X-3), \\ \phi' &= (x-2)(X-3)(x-3)(X-3)(x-3),\end{aligned}$$

whence

$$\begin{aligned}\mathrm{f} = {} & (X + x - 4)(x-3)^2(X-3)^2 \\ & + (2X^2 + 2Xx + 2x^2 - 32X - 32x + 162)(-2X - 2x + 2 + \xi)\end{aligned}$$

which is

$$\begin{aligned} = {} & X^3(x^2 - 6x + 5) \\ & + X^2(x^3 - 16x^2 + 61x - 22) \\ & + X(-6x^3 + 61x^2 - 120x - 91) \\ & + 5x^3 - 22x^2 - 91x \\ & + \xi\left\{\begin{array}{l} X^2(2) \\ + X(2x - 32) \\ + 2x^2 - 32x + 132 \end{array}\right\}\end{aligned}$$

This result includes proper solutions of the problem of finding the number of the triangles *aBcDeF*, which are such that the side *ea* touches the curve at *a*; and also heterotypic solutions having reference to the singular points of the curve; but I have not determined the number of solutions of each kind.

36. Correspondence (a, g): from the values of $\mathrm{f} - \phi - \phi'$ and $\mathrm{e} - \epsilon - \epsilon'$, we have

$$\mathrm{g} - \chi - \chi' = (X^3 - 20X^2 - 8Xx - 4x^2 + 125X + 44x - 486)\,\Delta,$$

and then

$$\chi = \chi' = (X-2)(x-3)(X-3)(x-3)(X-3)(x-3),$$

wherefore

$$\begin{aligned}\mathrm{g} &= 2(X-2)(X-3)^2(x-3)^3\\ &\quad + (X^3 - 20X^2 - 8Xx + 125X + 44x - 486)(-2X - 2x + 2 + \xi),\end{aligned}$$

viz. this is

$$\begin{aligned}\mathrm{g} = \;& X^4(\qquad\qquad\qquad\qquad - \;\;2)\\ &+ X^3(\;\;\;2x^3 - \;18x^2 + \;\;52x - \;12)\\ &+ X^2(-16x^3 + 144x^2 - 376x + 142)\\ &+ X\;(\;\;42x^3 - 362x^2 + 780x + \;88)\\ &\qquad - 36x^3 + 236x^2 + \;88x\\ &+ \xi\left\{\begin{array}{l} X^3(\qquad\qquad\;\;1)\\ + X^2(\qquad\;\; - \;20)\\ + X\;(\;\; - \;\;8x + 125)\\ \qquad\quad + 44x - 486.\end{array}\right\}\end{aligned}$$

Comparing with the expression of ϕx, Case 52, we have

$$\begin{aligned}\mathrm{g} - \phi x = \;& X^4(\qquad\qquad\qquad\qquad\quad - \;3)\\ &+ X^3(\qquad\qquad\qquad\qquad\quad + 34)\\ &+ X^2(\qquad\;\; 2x^3 - 18x^2 + 44x - 79)\\ &+ X\;(\quad\; - 10x^3 + 58x^2 + 76x - 84)\\ &\qquad - x^4 + 10x^3 + 15x^2 - 84x\\ &+ \xi\left\{\begin{array}{l} X^3(\qquad\qquad\qquad\;\;1)\\ + X^2(\qquad\qquad\; - 11)\\ + X\;(\qquad\;\; 4x - \;10)\\ \qquad + 9x^2 - 91x + 114,\end{array}\right\}\end{aligned}$$

which difference must be the number of heterotypic solutions having relation to the singularities of the curve; but I have not further considered this.

515.

SUR LES COURBES APLATIES.

[From the *Comptes Rendus de l'Académie des Sciences de Paris*, tom. LXXIV. (*Janvier—Juin*, 1872), pp. 708—712.]

En lisant la thèse de M. S. Maillard, *Recherches des caractéristiques des systèmes élémentaires des courbes planes du troisième ordre* (Paris, 1871), j'ai été conduit à quelques réflexions sur la théorie générale des courbes aplaties de M. Chasles (1).

Je considère une courbe représentée par une équation de l'ordre n, $f(x, y, k) = 0$, laquelle pour $k = 0$ se réduit à la forme $P^\alpha Q^\beta \ldots = 0$. Pour k un infiniment petit, ou disons pour $k = 0^1$, cette courbe sera ce que je nomme la pénultième de $P^\alpha Q^\beta \ldots = 0$; la courbe $P^\alpha Q^\beta \ldots = 0$ elle-même sera la courbe finale; et les courbes $P = 0$, $Q = 0, \ldots$, les facteurs. Or en menant par un point donné quelconque les tangentes à la courbe pénultième, ces tangentes approchent continuellement aux droites que voici: 1° les tangentes aux courbes $P = 0$, $Q = 0, \ldots$, respectivement; 2° les droites par les points singuliers de ces mêmes courbes respectivement; 3° les droites par les intersections de deux quelconques de ces mêmes courbes $P = 0$, $Q = 0, \ldots$, respectivement; 4° les droites par certains points situés sur l'une quelconque des mêmes courbes $P = 0$, $Q = 0, \ldots$. En ne faisant aucune supposition particulière par rapport à la courbe pénultième, cette courbe sera une courbe sans points singuliers, et ainsi de la classe $n^2 - n$: le nombre des droites 1°, 2°, 3°, 4° (en faisant attention à la multiplicité de quelques-unes de ces droites) sera donc égal à $n^2 - n$. Les droites 3° sont comptées chacune un certain nombre de fois; en supposant que pour un point d'intersection $P = 0$, $Q = 0$ quelconque ce nombre soit θ, nous dirons qu'il y a à ce point un nombre θ de *sommets fixes*. Les droites 4° sont comptées en général chacune une seule fois; les points par lesquels passent ces droites (points sur l'une quelconque des courbes $P = 0$, $Q = 0, \ldots$) seront

1 *Comptes Rendus*, t. LXIV. p. 799—805 et 1079—1081; séances des 22 avril et 27 mai 1867.

nommés *sommets libres.* Cela étant, on peut considérer la courbe pénultième comme équivalente à la courbe finale $P^\alpha Q^\beta \ldots = 0$ *plus* les sommets: il s'agit, pour un cas donné quelconque, de trouver le nombre et la distribution de ces sommets.

Je considère d'abord le cas le plus simple, celui d'une conique aplatie, pénultième de $x^2 = 0$; l'équation d'une telle conique est

$$(a, b, c, f, g, h \between x, y, z)^2 = 0,$$

où, en prenant $a = 1$, tous les autres coefficients seront des infiniment petits, pas en général du même ordre. Les tangentes menées à la courbe par un point donné (α, β, γ) seront déterminées par l'équation

$$(bc - f^2,\ ca - g^2,\ ab - h^2,\ gh - af,\ hf - bg,\ fg - ch) \times (\gamma y - \beta z,\ \alpha z - \gamma x,\ \beta x - \alpha y)^2 = 0\,;$$

ou disons

$$(bc - f^2,\ c\ - g^2,\ \ b - h^2,\ gh - \ f,\ hf - bg,\ fg - ch) \times (\gamma y - \beta z,\ \alpha z - \gamma x,\ \beta x - \alpha y)^2 = 0.$$

En considérant pour un moment tous les coefficients comme étant des infiniment petits du même ordre, $= 0^1$, cette équation se réduit à

$$(0,\ c,\ b,\ -f,\ 0,\ 0 \between \gamma y - \beta z,\ \alpha z - \gamma x,\ \beta x - \alpha y)^2 = 0,$$

ou, ce qui est la même chose,

$$(c,\ -f,\ b \between \alpha z - \gamma x,\ \beta x - \alpha y)^2 = 0\,;$$

et ces tangentes coupent la droite $x = 0$ dans les deux points donnés par l'équation $(c, -f, b \between \alpha z, -\alpha y)^2 = 0$, c'est-à-dire $by^2 + 2fyz + cz^2 = 0$, points indépendants de la position du point donné (α, β, γ); ces points sont en effet *les intersections de la pénultième par la droite* $x = 0$.

Mais il y a là une restriction qu'on évite au moyen d'une supposition plus générale, savoir: en prenant g, h du premier, b, c, f du second ordre, ou disons $g, h = 0^1$, $b, c, f = 0^2$, l'équation des tangentes devient

$$(0,\ c - g^2,\ b - h^2,\ gh - f,\ 0,\ 0 \between \gamma y - \beta z,\ \alpha z - \gamma x,\ \beta x - \alpha y)^2 = 0,$$

ou

$$(c - g^2,\ gh - f,\ b - h^2 \between \alpha z - \gamma x,\ \beta x - \alpha y)^2 = 0.$$

Or, en écrivant $x = 0$, cette équation devient

$$(c - g^2,\ gh - af,\ b - h^2 \between \alpha z,\ -\alpha y)^2 = 0,$$

c'est-à-dire

$$by^2 + 2fyz + cz^2 - (hy + gz)^2 = 0\,;$$

nous avons ainsi, sur la droite $x = 0$, deux points indépendants de la position du point donné (α, β, γ), et qui ne sont plus les intersections de la conique par cette droite (autrement dit, ces points ne sont pas situés sur la conique); ces points sont en effet deux points quelconques sur cette droite. Il y a ainsi pour la conique aplatie pénultième de $x^2 = 0$ deux sommets situés à volonté sur la droite $x = 0$ (et qui ainsi ne sont pas situés sur la conique pénultième).

Je passe à un cas nouveau, celui de la courbe quartique pénultième de $x^2y^2=0$; mais pour simplifier l'analyse, au lieu d'un point quelconque (α, β, γ) je prends successivement les points $(y=0, z=0)$ et $(x=0, y=0)$. On conçoit, en effet, que s'il y a p sommets libres sur la droite $x=0$, q sommets libres sur la droite $y=0$, et r sommets fixes au point $(x=0, y=0)$, alors les droites par le point donné $(y=0, z=0)$ seront les droites par les p points, *plus* la droite $y=0$, $q+r$ fois; et de même les droites par le point donné $(x=0, z=0)$ seront les droites par les q points, *plus* la droite $x=0$, $p+r$ fois: de manière que le procédé donnera les nombres cherchés p, q, r.

J'écris l'équation de la pénultième sous les deux formes

$$\begin{aligned}&x^4 . a\\ &+4x^3(h, j \between y, z)\\ &+6x^2(1, p, m \between y, z)^2\\ &+4x\ (k, q, r, g \between y, z)^3\\ &+\quad (b, f, l, i, c \between y, z)^4=0,\end{aligned}$$

$$\begin{aligned}&y^4 . b\\ &+4y^3(k, f \between x, z)\\ &+6y^2(1, q, l \between x, z)^2\\ &+4y\ (k, p, r, l \between x, z)^3\\ &+\quad (a, j, m, g, c \between x, z)^4=0,\end{aligned}$$

où le coefficient de x^2y^2 est $=6$, et tous les autres coefficients sont des infiniment petits, pas nécessairement du même ordre. Je représente ces deux équations par

$$(A, B, y^2+C, D, E \between x, 1)^4=0, \quad (A', B', x^2+C', D', E' \between y, 1)^4=0$$

respectivement.

Cela étant, on obtient l'équation des tangentes par le point $(y=0, z=0)$ en égalant à zéro le discriminant de la fonction quartique de x; et de même pour les tangentes par le point $(x=0, z=0)$: les deux équations seront

$$\begin{aligned}0=\ &(y^2+C)^4 . 81AE\\ &+(y^2+C)^3(-54AD^2-54B^2E)\\ &+(y^2+C)^2(-18A^2E^2-180ABDE+36B^2D^2)\\ &+\dots\\ 0=\ &(x^2+C')^4 . 81A'E'\\ &+\dots\end{aligned}$$

En prenant pour le moment tous les coefficients $=0^1$, chaque équation contiendra un seul terme de l'ordre le plus bas 0^2, et en négligeant les autres termes, les équations deviendront simplement

$$y^8 . AE=0, \qquad x^8 . A'E'=0;$$

il y a ainsi sur la droite $x=0$ quatre sommets libres donnés par l'équation $E=0$; et de même sur la droite $y=0$, quatre sommets libres donnés par l'équation $E'=0$; donc quatre sommets fixes au point $x=0$, $y=0$. Les sommets libres sur les droites $x=0$ et $y=0$ sont *les intersections de la quartique par ces deux droites respectivement.*

Mais, au contraire, prenons $b, f, l, i, c = 0^2$, les autres coefficients étant $= 0^1$. On a d'abord $A, B, D = 0^1$, $E = 0^2$; la première équation se réduit à

$$27Ay^6(3Ey^2 - 2D^2) = 0,$$

ce qui donne, sur la droite $x = 0$, six sommets libres déterminés par l'équation

$$3Ey^2 - 2D^2 = 0.$$

On a depuis $A' = 0^2$, $B', D', E' = 0^1$; la seconde équation est donc

$$27E'x^6(3A'x^2 - 2B'^2) = 0;$$

mais ici

$$E' = (a, j, m, g, c \between x, z)^4, = x(ax^3 + 4jx^2z + 6mxz^2 + 4gz^3),$$

à cause de $c = 0^2$; et, de plus,

$$3A'x^2 - 2B'^2 = 3bx^2 - 2(kx + fz)^2, = (3b - 2k^2)x^2,$$

à cause de $f = 0^2$; donc l'équation se réduit à

$$x^9(ax^3 + 4jx^2z + 6mxz^2 + 4gz^3) = 0,$$

et il y a sur la droite $y = 0$, trois sommets libres déterminés par l'équation

$$ax^3 + 4jx^2z + 6mxz^2 + 4gz^3 = 0.$$

Remarquons que la droite $y = 0$ rencontre la quartique dans les quatre points donnés par l'équation $E' = 0$, c'est-à-dire un point infiniment près de $(x = 0, y = 0)$ et trois autres points, lesquels sont précisément les trois sommets libres sur la droite $y = 0$. Il y a de plus trois sommets fixes au point $(x = 0, y = 0)$.

Conclusion. Il y a ainsi une courbe quartique pénultième de $x^2y^2 = 0$, avec neuf sommets libres, trois sur l'une des deux droites (disons la droite $y = 0$) et qui sont trois des intersections de la quartique par cette même droite (la quatrième intersection étant infiniment près du point $x = 0$, $y = 0$), six situés à volonté sur l'autre droite $x = 0$, et trois sommets fixes à l'intersection des deux droites.

On peut se figurer une telle courbe quartique: elle peut consister en trois ovales aplaties *plus* une trigonoïde (savoir, figure fermée avec trois angles saillants et trois angles réentrants) rétrécie; l'une des ovales coïncide à peu près avec la droite $y = 0$, les deux autres à peu près avec la droite $x = 0$; la trigonoïde entoure le point $x = 0$, $y = 0$, de manière que les angles réentrants, très-approchés de ce point, soient les trois sommets fixes: mais il n'est pas facile d'en faire un dessin.

Je considère le système des courbes quartiques, qui satisfont chacune aux $(14 - 1 =) 13$ conditions que voici: toucher deux droites données 1, 2 en des points donnés A, B; passer par deux points donnés C, D; toucher sept droites données 3, 4, ..., 9. Prenons $y = 0$ pour la droite AB, et $x = 0$ pour la droite CD: il y aura dans le système une courbe quartique pénultième de $x^2y^2 = 0$, laquelle compte sept fois au moins; cette courbe pénultième est censée toucher les droites 1, 2 dans les points donnés A, B, et l'une quelconque des sept droites à son intersection avec la droite $y = 0\,(AB)$; les autres six droites à leurs intersections avec la droite $x = 0\,(CD)$. Cette courbe pénultième entre donc dans la théorie des caractéristiques d'un tel système de courbes quartiques.

516.

SUR UNE SURFACE QUARTIQUE APLATIE.

[From the *Comptes Rendus de l'Académie des Sciences de Paris,* tom. LXXIV. (*Janvier—Juin,* 1872), pp. 1393—1395.]

IL y a évidemment pour les surfaces une théorie analogue à celle des courbes aplaties: la pénultième d'une surface $P^{\alpha}Q^{\beta}\ldots=0$ est, pour ainsi dire, composée des surfaces $P=0$, $Q=0$, &c., *plus* des lignes courbes ou *arêtes,* lesquelles correspondent aux sommets d'une courbe aplatie[1]; par exemple une surface quadrique peut se réduire à $P^2=0$, un plan deux fois, *plus* une conique qui est l'arête de la surface aplatie. Pour les surfaces quartiques, un exemple assez intéressant se rencontre dans le beau Mémoire de M. Casey, "On cyclides and spheroquartics," (*Phil. Trans.*, vol. CLXI. pp. 585—721, 1871). L'auteur, d'après M. Darboux, nomme *cyclide* la surface quartique générale qui a pour ligne double le cercle à l'infini (*surface quartique anallagmatique* de M. Moutard), et *spheroquartic* la courbe d'intersection d'une sphère par une surface quadrique quelconque; et il est conduit à considérer la sphéroquartique comme cas particulier de la cyclide. J'aime mieux dire qu'il y a une cyclide aplatie ayant pour arête une courbe sphéroquartique.

Voici comment on y arrive: la cyclide est l'enveloppe des sphères dont chacune a son centre sur une surface quadrique nommée *focale,* et coupe orthogonalement une sphère fixe, nommée *sphère d'inversion,* disons la sphère S. Cela étant, en envisageant la focale comme une surface réglée, chaque droite sur la surface donne lieu à une infinité de sphères, qui passent toutes par un même cercle. En supposant que la droite coupe la sphère S aux points O, O', ce cercle est ce que j'appelle l'*anticircle* des points O, O', savoir, le plan du cercle est perpendiculaire à la corde OO' au point central M, et le rayon en est égal à $i\,OM\,(=i\,O'M)$, de manière que le cercle est réel ou imaginaire, selon que les points O, O' sont imaginaires ou réels: toute

[1] Voir *Comptes Rendus,* t. LXXIV. p. 708, [515].

sphère ayant son centre sur la droite OO', et coupant orthogonalement la sphère S, passe par le cercle dont il s'agit, disons le cercle L. On voit sans peine que chaque point du cercle L est situé sur la cyclide. Il y a donc sur la cyclide une série infinie single de cercles L qui correspondent un à une aux directrices de la surface focale; il y a de même une série infinie single de cercles L' qui correspondent un à une aux génératrices de la surface focale. La cyclide est le lieu des cercles de l'une ou l'autre série; chaque cercle de la première série coupe en deux points opposés chaque cercle de l'autre série, mais deux cercles de la même série ne se rencontrent pas, &c.

Or, en supposant avec M. Casey que la surface focale se réduise à un cône, les deux séries de cercles se réduisent à une seule série de cercles L, dont chacun est situé sur la sphère, centre le sommet du cône, qui coupe orthogonalement la sphère S, disons la sphère T. On a sur la sphère T la série des cercles L, lesquels ont pour enveloppe une courbe sphérique, la sphéroquartique de M. Casey. Les points des différents cercles L ne remplissent pas la surface sphérique entière, mais seulement une partie de cette surface, limitée par la courbe sphéroquartique. Cela étant, on pourrait dire que la surface cyclide se réduit à la sphère T deux fois, mais il vaut mieux la considérer comme une cyclide aplatie ayant pour arête la courbe sphéroquartique.

La sphéroquartique, considérée comme courbe sur une sphère T, est donnée (comme le remarque M. Cásey) par une construction tout à fait analogue à celle pour la cyclide comme surface dans l'espace, savoir (en considérant toujours les courbes sphériques sur une même sphère), la sphéroquartique est l'enveloppe des cercles qui ont leurs centres sur une sphéro-conique et qui coupent orthogonalement un cercle fixe. Le cône, sommet le centre de la sphère, qui passe par la sphéroquartique, est de l'ordre 4, avec deux droites doubles (la classe est donc $=8$); j'ajoute qu'il touche quatre fois la sphère-cône $x^2+y^2+z^2=0$, ayant le même sommet (1).

M. Casey dit que le cône quartique a 16 droites focales: cela a besoin d'explication. Le cône quartique et le sphère-cône ont en commun $8\times 2=16$ plans tangents, y compris les plans tangents selon les 4 droites de contact, chacun deux fois; hormis ceux-ci, il y a donc 8 plans tangents communs. L'intersection de deux quelconques de ces 8 plans est droite focale du cône quartique: donc $\frac{1}{2}(8\times 7)$, $=28$ droites focales. Mais je trouve que les 8 plans tangents forment deux systèmes de 4 plans chacun: les 4 points de l'un de ces systèmes coupent les 4 plans de l'autre système dans 16 droites, lesquelles sont les droites focales de M. Casey; il y a de plus $6+6$ droites, dont chacune est l'intersection de deux plans du même système. Je n'ai pas cherché les distinctions qui doivent exister entre ces différents systèmes de droites focales.

1 En général, en considérant une courbe quelconque sur une surface S, et un point O quelconque, les deux cônes, sommet O, dont l'un passe par la courbe et l'autre est circonscrit à la surface, se touchent partout où ils se rencontrent: autrement dit, ils n'ont que des droites d'intersection doubles ou de contact.

517.

SUR LES SURFACES DIVISIBLES EN CARRÉS PAR LEURS COURBES DE COURBURE ET SUR LA THÉORIE DE DUPIN.

[From the *Comptes Rendus de l'Académie des Sciences de Paris*, tom. LXXIV. (*Janvier—Juin*, 1872), pp. 1445—1449.]

Soient Θ une fonction arbitraire de h, k; x, y, z des fonctions de h, k telles que

$$2\Theta \frac{d^2x}{dh\,dk} - \frac{d\Theta}{dh}\frac{dx}{dk} - \frac{d\Theta}{dk}\frac{dx}{dh} = 0,$$

$$2\Theta \frac{d^2y}{dh\,dk} - \frac{d\Theta}{dh}\frac{dy}{dk} - \frac{d\Theta}{dk}\frac{dy}{dh} = 0,$$

$$2\Theta \frac{d^2z}{dh\,dk} - \frac{d\Theta}{dh}\frac{dz}{dk} - \frac{d\Theta}{dk}\frac{dz}{dh} = 0,$$

et que, de plus,

$$\frac{dx}{dh}\frac{dx}{dk} + \frac{dy}{dh}\frac{dy}{dk} + \frac{dz}{dh}\frac{dz}{dk} = 0;$$

en éliminant h, k, on a, entre x, y, z, l'équation $V=0$ d'une surface. Je dis que les équations $h=\text{const.}$, $k=\text{const.}$ déterminent les deux systèmes des courbes de courbure de cette surface, et, de plus, que cette surface est divisible en carrés par ses courbes de courbure.

En effet, les équations donnent

$$\Theta \frac{d}{dk}\left[\left(\frac{dx}{dh}\right)^2 + \left(\frac{dy}{dh}\right)^2 + \left(\frac{dz}{dh}\right)^2\right] - \frac{d\Theta}{dk}\left[\left(\frac{dx}{dh}\right)^2 + \left(\frac{dy}{dh}\right)^2 + \left(\frac{dz}{dh}\right)^2\right] = 0,$$

ce qui implique

$$\left(\frac{dx}{dh}\right)^2 + \left(\frac{dy}{dh}\right)^2 + \left(\frac{dz}{dh}\right)^2 = \Theta H,$$

où H est fonction de h seulement; et l'on trouve de même

$$\left(\frac{dx}{dk}\right)^2 + \left(\frac{dy}{dk}\right)^2 + \left(\frac{dz}{dk}\right)^2 = \Theta K,$$

où K est fonction de k seulement; donc en écrivant, comme à l'ordinaire,

$$dx^2 + dy^2 + dz^2 = E\,dh^2 + 2F\,dh\,dk + G\,dk^2,$$

cette expression se réduit à

$$dx^2 + dy^2 + dz^2 = \Theta\,(H\,dh^2 + K\,dk^2),$$

ce qui fait voir que la surface est divisible en carrés par les courbes

$$h = \text{const.}, \qquad k = \text{const.}$$

Les équations donnent aussi

$$\begin{vmatrix} \dfrac{dx}{dh}, & \dfrac{dy}{dh}, & \dfrac{dz}{dh} \\ \dfrac{dx}{dk}, & \dfrac{dy}{dk}, & \dfrac{dz}{dk} \\ \dfrac{d^2x}{dh\,dk}, & \dfrac{d^2y}{dh\,dk}, & \dfrac{d^2z}{dh\,dk} \end{vmatrix} = 0;$$

et, cela étant, l'équation différentielle des courbes de courbure se réduit, comme je vais le montrer, à $dh\,dk = 0$; on a donc $h = \text{const.}$, $k = \text{const.}$, pour les équations des courbes de courbure de la surface.

Pour cela, en considérant x, y, z comme des fonctions données de h, k, j'écris, comme à l'ordinaire,

$$\frac{dx}{dh} = a, \quad \frac{dx}{dk} = a', \quad \frac{d^2x}{dh^2} = \alpha, \quad \frac{d^2x}{dh\,dk} = \alpha', \quad \frac{d^2x}{dk^2} = \alpha'',$$

et de même b, b', β, β', β'', et c, c', γ, γ', γ'' pour les coefficients différentiels de y et z respectivement. J'écris aussi

$$\begin{array}{lll} A = bc' - b'c, & B = ca' - c'a, & C = ab' - a'b, \\ E = a^2 + a'^2 + a''^2, & F = aa' + bb' + cc', & G = a'^2 + b'^2 + c'^2. \end{array}$$

L'équation différentielle des courbes de courbure est

$$\begin{vmatrix} dx, & dy, & dz \\ A, & B, & C \\ dA, & dB, & dC \end{vmatrix} = 0.$$

Le premier terme de ce déterminant est $dx\,(B\,dC - C\,dB)$, savoir:

$$\begin{aligned} (a\,dh + a'dk)\{\; & B\,[(a\beta' - b\alpha' + b'\alpha - a'\beta)\,dh + (a\beta'' - b\alpha'' + b'\alpha' - a'\beta')\,dk] \\ & - C\,[(c\alpha' - a\gamma' + a'\gamma - c'\alpha)\,dh + (c\alpha'' - a\gamma'' + a'\gamma' - c'\alpha')\,dk]\}, \end{aligned}$$

ce qui se réduit tout de suite à

$$(adh + a'dk)\{\ [a(A\alpha' + B\beta' + C\gamma') - a'(A\alpha + B\beta + C\gamma)]\,dh \\ - [a(A\alpha'' + B\beta'' + C\gamma'') - a'(A\alpha' + B\beta' + C\gamma')]\,dk\}\,;$$

en formant les expressions analogues du second et du troisième terme, et en prenant la somme, l'équation devient

$$[E(A\alpha' + B\beta' + C\gamma') - F(A\alpha + B\beta + C\gamma)]\,dh^2 \\ + [E(A\alpha'' + B\beta'' + C\gamma'') - G(A\alpha + B\beta + C\gamma)]\,dh\,dk \\ + [F(A\alpha'' + B\beta'' + C\gamma'') - G(A\alpha' + B\beta' + C\gamma')]\,dk^2 = 0,$$

ou, ce qui est la même chose,

$$\begin{vmatrix} dk^2, & -dh\,dk, & dh^2 \\ E, & F, & G \\ A\alpha + B\beta + C\gamma, & A\alpha' + B\beta' + C\gamma', & A\alpha'' + B\beta'' + C\gamma'' \end{vmatrix} = 0\,;$$

celle-ci est l'équation différentielle des courbes de courbure d'une surface quand les coordonnées x, y, z d'un point de la surface sont données comme fonctions de deux paramètres h, k.

En supposant $F = 0$, l'équation se réduit à

$$(A\alpha' + B\beta' + C\gamma')(E\,dh^2 - G\,dk^2) \\ + [(A\alpha'' + B\beta'' + C\gamma'')E - (A\alpha + B\beta + C\gamma)G]\,dh\,dk = 0\,;$$

et en supposant de plus $A\alpha' + B\beta' + C\gamma' = 0$, l'équation se réduit simplement à $dh dk = 0$; mais cette équation $A\alpha' + B\beta' + C\gamma' = 0$, savoir

$$\begin{vmatrix} a, & b, & c \\ a', & b', & c' \\ \alpha', & \beta', & \gamma' \end{vmatrix} = 0,$$

ou

$$\begin{vmatrix} \frac{dx}{dh}, & \frac{dy}{dh}, & \frac{dz}{dh} \\ \frac{dx}{dk}, & \frac{dy}{dk}, & \frac{dz}{dk} \\ \frac{d^2x}{dh\,dk}, & \frac{d^2y}{dh\,dk}, & \frac{d^2z}{dh\,dk} \end{vmatrix} = 0,$$

et aussi $F = 0$, subsistent dans le cas actuel; et nous avons ainsi $dk\,dh = 0$ pour équation différentielle des courbes de courbure.

On vérifie sans peine les équations fondamentales, en prenant $\Theta = h - k$,

$$-(c-a)(a-b)x^2 = a(a+h)(a+k),$$
$$-(a-b)(b-c)y^2 = b(b+h)(b+k),$$
$$-(b-c)(c-a)z^2 = c(c+h)(c+k);$$

ce qui donne les courbes de courbure de l'ellipsoïde $\frac{x^2}{a}+\frac{y^2}{b}+\frac{z^2}{c}=1$; l'ellipsoïde étant, comme on sait, une surface divisible en carrés par des courbes de courbure; mais je n'ai pas encore cherché d'autres solutions.

Je remarque que l'équation pour x peut s'écrire sous la forme

$$\frac{d}{dh}\left(\frac{1}{\Theta}\frac{dx}{dk}\right)+\frac{d}{dk}\left(\frac{1}{\Theta}\frac{dx}{dh}\right)=0;$$

donc, en posant

$$-\frac{d}{dh}\left(\frac{1}{\Theta}\frac{dx}{dk}\right)=\frac{d}{dk}\left(\frac{1}{\Theta}\frac{dx}{dh}\right)=\frac{d^2\Omega}{dh\,dk},$$

on trouve

$$\frac{dx}{dh}=\Theta\frac{d\Omega}{dh}, \quad \frac{dx}{dk}=-\Theta\frac{d\Omega}{dk},$$

ce qui donne

$$\frac{d}{dk}\left(\Theta\frac{d\Omega}{dh}\right)+\frac{d}{dh}\left(\Theta\frac{d\Omega}{dk}\right)=0,$$

équation pour Ω de la même forme que celle pour x.

On déduit une démonstration très-simple du théorème de Dupin. En considérant comme auparavant (x, y, z) comme des fonctions données de (h, k), le point (x, y, z) sera situé sur une surface, et les conditions pour que les courbes de courbure soient $h = \text{const.}$, $k = \text{const.}$ seront

$$\frac{dx}{dh}\frac{dx}{dk}+\frac{dy}{dh}\frac{dy}{dk}+\frac{dz}{dh}\frac{dz}{dk}=0,$$

$$\begin{vmatrix} \frac{dx}{dh}, & \frac{dy}{dh}, & \frac{dz}{dh} \\ \frac{dx}{dk}, & \frac{dy}{dk}, & \frac{dz}{dk} \\ \frac{d^2x}{dh\,dk}, & \frac{d^2y}{dh\,dk}, & \frac{d^2z}{dh\,dk} \end{vmatrix}=0.$$

Cela étant, en introduisant un troisième paramètre l, soient h, k, l des fonctions données de (x, y, z), ou réciproquement (x, y, z) des fonctions données de (h, k, l). On a ici les trois systèmes de surfaces $h = \text{const.}$, $k = \text{const.}$, $l = \text{const.}$, et les conditions pour que ces surfaces se coupent orthogonalement peuvent s'écrire sous la forme

$$\frac{dx}{dk}\frac{dx}{dl} + \frac{dy}{dk}\frac{dy}{dl} + \frac{dz}{dk}\frac{dz}{dl} = 0,$$

$$\frac{dx}{dl}\frac{dx}{dh} + \frac{dy}{dl}\frac{dy}{dh} + \frac{dz}{dl}\frac{dz}{dh} = 0,$$

$$\frac{dx}{dh}\frac{dx}{dk} + \frac{dy}{dh}\frac{dy}{dk} + \frac{dz}{dh}\frac{dz}{dk} = 0.$$

On a donc

$$\frac{dx}{dl} : \frac{dy}{dl} : \frac{dz}{dl} = \frac{dy}{dh}\frac{dz}{dk} - \frac{dz}{dh}\frac{dy}{dk} : \frac{dz}{dh}\frac{dx}{dk} - \frac{dx}{dh}\frac{dy}{dk} : \frac{dx}{dh}\frac{dy}{dk} - \frac{dy}{dh} : \frac{dz}{dh}.$$

Pour abréger, j'écris

$$\frac{dx}{dh}\quad\frac{dx}{dk} + \frac{dy}{dh}\quad\frac{dy}{dk} + \frac{dz}{dh}\quad\frac{dz}{dk} = [h\,.\,k], \ldots,$$

et de même

$$\frac{dx}{dh}\frac{d^2x}{dk\,dl} + \frac{dy}{dh}\frac{d^2y}{dk\,dl} + \frac{dz}{dh}\frac{d^2z}{dk\,dl} = [h\,.\,kl], \ldots.$$

Les conditions données sont ainsi

$$[k\,.\,l] = 0, \quad [l\,.\,h] = 0, \quad [h\,.\,k] = 0\,;$$

en différentiant ces équations par rapport à h, k, l respectivement, on obtient

$$[k\,.\,lh] + [l\,.\,hk] = 0,$$
$$[l\,.\,kh] + [h\,.\,kl] = 0,$$
$$[h\,.\,kl] + [k\,.\,lh] = 0\,;$$

donc

$$[h\,.\,kl] = 0, \quad [k\,.\,lh] = 0, \quad [l\,.\,kh] = 0.$$

Mais l'équation $[h\,.\,k] = 0$ et l'équation $[l\,.\,hk] = 0$, en substituant dans celle-ci les valeurs de $\frac{dx}{dl}$, $\frac{dy}{dl}$, $\frac{dz}{dl}$, sont précisément les conditions pour que la surface $l = \text{const.}$ soit coupée par les autres surfaces selon ses courbes de courbure: donc le théorème.

518.

SUR LA CONDITION POUR QU'UNE FAMILLE DE SURFACES DONNÉES PUISSE FAIRE PARTIE D'UN SYSTÈME ORTHOGONAL.

[From the *Comptes Rendus de l'Académie des Sciences de Paris*, tom. LXXV. (*Juillet—Décembre*, 1872), pp. 177—185, 246—250, 324—330, 381—385, 1800—1803.]

[Pp. 177—185.]

1. SOIT $\rho = f(x, y, z)$ l'équation d'une famille de surfaces qui fait partie d'un système orthogonal. On sait que ρ satisfait à une équation à différences partielles du troisième ordre, et en suivant la route tracée par M. Levy, dans son excellent "Mémoire sur les coordonnées curvilignes orthogonales" (*Journal de l'École Polytechnique*, t. XXVI., pp. 157—200, 1870), je suis parvenu à trouver cette équation.

2. Je remarque que le théorème fondamental de M. Levy est, en effet, assez évident. Considérons une surface de la famille ρ: soit P un point quelconque de cette surface, et PT, PT_1, PT_2 la normale et les tangentes aux deux courbes de courbure par le point P. Passons, suivant la normale au point P' de la surface consécutive $\rho + d\rho$, et soient $P'T'$, $P'T_1'$, $P'T_2'$ la normale et les tangentes aux deux courbes de courbure par le point P'. Or, si les surfaces ρ forment partie d'un système orthogonal, évidemment PP' sera élément d'une courbe de courbure d'une surface ρ_1 et aussi d'une surface ρ_2 des deux autres familles du système orthogonal, et PT_1, $P'T_1'$ seront les normales à deux points consécutifs de cette courbe de courbure de la surface ρ_1: et de même PT_2 et $P'T_2'$ seront les normales à deux points consécutifs de cette courbe de courbure de la surface ρ_2. Donc PT_1 et $P'T_1'$ se rencontrent; et de même PT_2 et $P'T_2'$ se rencontrent. En se souvenant que PT_1, PT_2 sont perpendiculaires l'une à l'autre, et de même $P'T_1'$, $P'T_2'$, on voit sans peine que les deux conditions se réduisent à une seule. Réciproquement, si PT_1, $P'T_1'$ se rencontrent (ou, ce qui est la même chose, PT_2 et $P'T_2'$), la famille ρ fera partie d'un système orthogonal; ce qui est le théorème de M. Levy.

3. Soient (X, Y, Z) les fonctions dérivées de ρ du premier ordre; $(\mathrm{a}, \mathrm{b}, \mathrm{c}, \mathrm{f}, \mathrm{g}, \mathrm{h})$ celles du second ordre; $(a, b, c, f, g, h, i, j, k, l)$ celles du troisième ordre, savoir:

$$(X, Y, Z) = (\partial_x, \partial_y, \partial_z)\rho,$$

$$(\mathrm{a}, \mathrm{b}, \mathrm{c}, \mathrm{f}, \mathrm{g}, \mathrm{h}) = (\partial_x^2, \partial_y^2, \partial_z^2, \partial_y\partial_z, \partial_z\partial_x, \partial_x\partial_y)\rho,$$

$$(a, b, c, f, g, h, i, j, k, l) = (\partial_x^3, \partial_y^3, \partial_z^3, \partial_y^2\partial_z, \partial_z^2\partial_x, \partial_x^2\partial_y, \partial_y\partial_z^2, \partial_z\partial_x^2, \partial_x\partial_y^2, \partial_x\partial_y\partial_z)\rho;$$

soient de plus

$$A = 2(Z\mathrm{h} - Y\mathrm{g}),\quad F = X(\mathrm{c} - \mathrm{h}) + Y\mathrm{h} - Z\mathrm{g},$$

$$B = 2(X\mathrm{f} - Z\mathrm{h}),\quad G = Y(\mathrm{a} - \mathrm{c}) + Z\mathrm{f} - X\mathrm{h},$$

$$C = 2(Y\mathrm{g} - X\mathrm{f}),\quad H = Z(\mathrm{h} - \mathrm{a}) + X\mathrm{g} - Y\mathrm{f},$$

valeurs qui satisfont aux équations

$$A + B + C = 0 \text{ et } (A, B, C, F, G, H \between X, Y, Z)^2 = 0.$$

Alors les tangentes PT_1, PT_2 sont données par les équations

$$(A, B, C, F, G, H \between x, y, z)^2 = 0,$$

$$Xx + Yy + Zz = 0,$$

et en partant de ces équations, mais en supposant que pour le point P les valeurs de X, Y soient $X = 0$, $Y = 0$, M. Levy obtient comme condition de l'intersection dont il s'agit

$$\left(\frac{dX}{dx} - \frac{dY}{dy}\right)\frac{d^2}{dxdy}\frac{1}{Z} + \frac{dY}{dx}\left(\frac{d^2}{dy^2} - \frac{d^2}{dx^2}\right)\frac{1}{Z} = 0,$$

ou, ce qui est la même chose

$$2\mathrm{fg}(\mathrm{a} - \mathrm{b}) + 2\mathrm{h}(\mathrm{f}^2 - \mathrm{g}^2) - Z[(f - j)\mathrm{h} + l(\mathrm{a} - \mathrm{b})] = 0;$$

savoir: cette équation est ce que devient l'équation cherchée du troisième ordre en y écrivant $X = 0$, $Y = 0$.

4. Je passe à la recherche de l'équation générale; pour cela (X, Y, Z) dénotant comme auparavant, nous pouvons considérer ces quantités comme les coordonnées (mesurées du point P comme origine) d'un point sur la normale PT; soient de même X_1, Y_1, Z_1 les coordonnées d'un point sur la tangente PT_1 et X_2, Y_2, Z_2 les coordonnées d'un point sur la tangente PT_2. Il s'agit seulement des valeurs relatives de ces coordonnées; et celles de X_1, Y_1, Z_1 et X_2, Y_2, Z_2 sont les valeurs de (x, y, z), données par les équations

$$(A, B, C, F, G, H \between x, y, z)^2 = 0,$$

$$Xx + Yy + Zz = 0.$$

Ces équations impliquent $X_1X_2 + Y_1Y_2 + Z_1Z_2 = 0$, et en se rappelant une équation déjà mentionnée, on a le système

$$\begin{aligned}
&(A, \ldots \between X\ , Y\ , Z\ )^2 = 0,\\
&(A, \ldots \between X_1, Y_1, Z_1)^2 = 0,\\
&(A, \ldots \between X_2, Y_2, Z_2)^2 = 0,\\
&X_1X_2 + Y_1Y_2 + Z_1Z_2 = 0,\\
&XX_1 + YY_1 + ZZ_1 = 0,\\
&XX_2 + YY_2 + ZZ_2 = 0.
\end{aligned}$$

L'origine étant quelconque, prenons (x, y, z) pour coordonnées de P, et $x + \delta x$, $y + \delta y$, $z + \delta z$ pour coordonnées de P'; nous avons $\delta x : \delta y : \delta z = X : Y : Z$; et comme il ne s'agit que des valeurs relatives, nous pouvons omettre un facteur infinitésimal commun, et écrire simplement $\delta x, \delta y, \delta z = X, Y, Z$. De même, en supposant qu'une fonction quelconque u de (x, y, z) devient $u + \delta u$, en passant du point P au point P', la valeur de δu sera $X\dfrac{du}{dx} + Y\dfrac{du}{dy} + Z\dfrac{du}{dz}$, ou, ce qui est la même chose, nous aurons $\delta = X\dfrac{d}{dx} + Y\dfrac{d}{dy} + Z\dfrac{d}{dz}$. Dans tout ce qui suit, δ aura cette signification.

5. Cela étant, si pour un moment nous prenons ξ, η, ζ pour coordonnées courantes, et θ pour un paramètre arbitraire, les équations de PT seront

$$\xi = x + \theta X_1, \quad \eta = y + \theta Y_1, \quad \zeta = z + \theta Z_1,$$

et si cette droite rencontre PT_1, alors en prenant ξ, η, ζ pour les coordonnées du point d'intersection, nous aurons $0 = \delta x + X_1\delta\theta + \theta\delta X_1, \ldots$: ou en éliminant $\delta\theta$ et θ,

$$0 = \begin{vmatrix} \delta x, & X_1, & \delta X_1 \\ \delta y, & Y_1, & \delta Y_1 \\ \delta z, & Z_1, & \delta Z_1 \end{vmatrix},$$

ou, ce qui est la même chose,

$$0 = \begin{vmatrix} X, & X_1, & \delta X_1 \\ Y, & Y_1, & \delta Y_1 \\ Z, & Z_1, & \delta Z_1 \end{vmatrix}.$$

Mais nous avons $X_2 : Y_2 : Z_2 = YZ_1 - ZY_1 : ZX_1 - XZ_1 : XY_1 - YX_1$: donc cette équation devient $X_2\delta X_1 + Y_2\delta Y_1 + Z_2\delta Z_1 = 0$. Or nous avons $\delta(X_1X_2 + Y_1Y_2 + Z_1Z_2) = 0$; l'équation trouvée peut donc s'écrire sous la forme plus symétrique

$$X_2\delta X_1 + Y_2\delta Y_1 + Z_2\delta Z_1 - (X_1\delta X_2 + Y_1\delta Y_2 + Z_1\delta Z_2) = 0,$$

équation qui exprime la condition pour l'intersection des tangentes PT_1, $P'T_1'$ (ou PT_2, $P'T_2'$).

6. Dans la démonstration précédente, je me suis servi du théorème de Dupin; mais il convient de remarquer qu'en partant du système orthogonal, et dénotant par X, Y, Z; X_1, Y_1, Z_1; X_2, Y_2, Z_2 les dérivées de ρ, ρ_1, ρ_2, respectivement, il serait possible de déduire cette même équation des seules équations

$$XX_1 + YY_1 + ZZ_1 = 0,$$
$$XX_2 + YY_2 + ZZ_2 = 0,$$
$$X_1X_2 + Y_1Y_2 + Z_1Z_2 = 0.$$

En effet, l'équation fut démontrée de cette manière par R. L. Ellis, dans une démonstration du théorème de Dupin, publiée dans l'ouvrage de Gregory (*Examples of the processes of the differential and integral calculus;* Cambridge, 1841). Les premières deux équations donnent $X : Y : Z = Y_1Z_2 - Y_2Z_1 : Z_1X_2 - Z_2X_1 : X_1Y_2 - X_2Y_1$; on a donc l'expression

$$(Y_1Z_2 - Y_2Z_1)\,dx + (Z_1X_2 - Z_2X_1)\,dy + (X_1Y_2 - X_2Y_1)\,dz,$$

intégrable par un facteur; ce qui donne

$$(Y_1Z_2 - Y_2Z_1)\left\{\frac{d}{dy}(X_1Y_2 - X_2Y_1) - \frac{d}{dz}(Z_1X_2 - Z_2X_1)\right\} \ldots = 0.$$

Le terme en { } est égal à

$$\left(X_2\frac{dX_1}{dx} + Y_2\frac{dX_1}{dy} + Z_2\frac{dX_1}{dz}\right) - \left(X_1\frac{dX_2}{dx} + Y_1\frac{dX_2}{dy} + Z_1\frac{dX_2}{dz}\right)$$
$$- X_2\left(\frac{dX_1}{dx} + \frac{dY_1}{dy} + \frac{dZ_1}{dz}\right) + X_1\left(\frac{dX_2}{dx} + \frac{dY_2}{dy} + \frac{dZ_2}{dz}\right),$$

et la somme qui correspond à la deuxième ligne de cette expression s'évanouit identiquement; la première ligne peut s'écrire sous la forme $\delta_2X_1 - \delta_1X_2$; donc, en rétablissant X, Y, Z au lieu de $Y_1Z_2 - Y_2Z_1, \ldots$, la condition devient simplement

$$X(\delta_2X_1 - \delta_1X_2) + Y(\delta_2Y_1 - \delta_1Y_2) + Z(\delta_2Z_1 - \delta_1Z_2) = 0.$$

Mais nous avons

$$\delta_2X_1 = X_2\frac{dX_1}{dx} + Y_2\frac{dX_1}{dy} + Z_2\frac{dX_1}{dz}, \quad = X_2\frac{dX_1}{dx} + Y_2\frac{dY_1}{dx} + Z_2\frac{dZ_1}{dx},$$

$$\delta_1X_2 = X_1\frac{dX_2}{dx} + Y_1\frac{dX_2}{dy} + Z_1\frac{dX_2}{dz}, \quad = X_1\frac{dX_2}{dx} + Y_1\frac{dY_2}{dx} + Z_1\frac{dZ_2}{dx},$$

et ainsi $\delta_2X_1 + \delta_1X_2 = \frac{d}{dx}(X_1X_2 + Y_1Y_2 + Z_1Z_2) = 0$; c'est-à-dire $\delta_1X_2 = -\delta_2X_1$, et de même $\delta_1Y_2 = -\delta_2Y_1$, $\delta_1Z_2 = -\delta_2Z_1$, et l'équation trouvée se réduit à

$$X\delta_2X_1 + Y\delta_2Y_1 + Z\delta_2Z_1 = 0, \quad \text{ou} \quad X\delta_1X_2 + Y\delta_1Y_2 + Z\delta_1Z_2 = 0;$$

on a de même

$$X_1\delta X_2 + Y_1\delta Y_2 + Z_1\delta Z_2 = 0, \quad \text{ou} \quad X_1\delta_2X + Y_1\delta_2Y + Z_1\delta_2Z = 0,$$

et

$$X_2\delta_1X + Y_2\delta_1Y + Z_2\delta_1Z = 0, \quad \text{ou} \quad X_2\delta X_1 + Y_2\delta Y_1 + Z_2\delta Z_1 = 0,$$

et ainsi l'équation dont il s'agit

$$X_2\delta X_1 + Y_2\delta Y_1 + Z_2\delta Z_1 - (X_1\delta X_2 + Y_1\delta Y_2 + Z_1\delta Z_2) = 0.$$

On ne savait pas auparavant la signification géométrique de cette équation.

7. Dans la question actuelle, partant de cette équation, je rappelle que les valeurs de X, Y, Z, X_1, Y_1, Z_1 sont celles de (x, y, z) données par les équations

$$(A, B, C, F, G, H \between x, y, z)^2 = 0, \quad Xx + Yy + Zz = 0.$$

En supposant que ces équations donnent

$$X_1 : Y_1 : Z_1 = U + U' : V + V' : W + W',$$
$$X_2 : Y_2 : Z_2 = U - U' : V - V' : W - W',$$

la condition devient

$$U\delta U' + V\delta V' + W\delta W' - (U'\delta U + V'\delta V + W'\delta W) = 0.$$

8. Pour effectuer la réduction de cette formule, nous avons besoin de plusieurs formules subsidiaires. J'écris

$$(BC - F^2, CA - G^2, AB - H^2, GH - AF, HF - BG, FG - CH \between X, Y, Z)^2$$
$$= (\mathfrak{A}, \mathfrak{B}, \mathfrak{C}, \mathfrak{F}, \mathfrak{G}, \mathfrak{H} \between X, Y, Z)^2 = -\phi,$$

et je dénote par (a), (b), (c), (f), (g), (h) les coefficients de $\lambda^2, \ldots$ dans la fonction

$$(A, B, C, F, G, H \between \nu Y - \mu Z, \lambda Z - \nu X, \mu X - \lambda Y)^2,$$

savoir, j'écris

$$\begin{aligned}
(\mathrm{a}) &= BZ^2 + CY^2 - 2FYZ,\\
(\mathrm{b}) &= CX^2 + AZ^2 - 2GZX,\\
(\mathrm{c}) &= AY^2 + BX^2 - 2HXY,\\
(\mathrm{f}) &= -AYZ - FX^2 + GXY + HXZ,\\
(\mathrm{g}) &= -BZX + FXY - GY^2 + HYZ,\\
(\mathrm{h}) &= -CXY + FYZ + GYZ - HZ^2,
\end{aligned}$$

où je remarque qu'en vertu des valeurs de $A, \ldots$ nous avons

$$(\mathrm{a}) + (\mathrm{b}) + (\mathrm{c}) = 0.$$

Cela étant, nous avons les identités

$$[(\mathrm{a}), (\mathrm{h}), (\mathrm{g})]\,(X, Y, Z) = 0,$$
$$[(\mathrm{h}), (\mathrm{b}), (\mathrm{f})]\,(X, Y, Z) = 0,$$
$$[(\mathrm{g}), (\mathrm{f}), (\mathrm{c})]\,(X, Y, Z) = 0,$$

$$[(\mathrm{b})(\mathrm{c}) - (\mathrm{f})^2, (\mathrm{c})(\mathrm{a}) - (\mathrm{g})^2, (\mathrm{a})(\mathrm{b}) - (\mathrm{h})^2, (\mathrm{g})(\mathrm{h}) - (\mathrm{a})(\mathrm{f}), (\mathrm{h})(\mathrm{f}) - (\mathrm{b})(\mathrm{g}), (\mathrm{f})(\mathrm{g}) - (\mathrm{c})(\mathrm{h})]$$
$$= -(X^2, Y^2, Z^2, YZ, ZX, XY)\,\phi,$$

savoir, $(b)(c)-(f)^2=-X^2\phi, \ldots$ De plus

$$\begin{aligned}
(A,\ H,\ G)\,[(a),\ (h),\ (g)] &= -X(\mathfrak{A}X+\mathfrak{H}Y+\mathfrak{G}Z)-\phi,\\
(H,\ B,\ F)\,[(a),\ (h),\ (g)] &= -Y(\mathfrak{A}X+\mathfrak{H}Y+\mathfrak{G}Z),\\
(G,\ F,\ C)\,[(a),\ (h),\ (g)] &= -Z(\mathfrak{A}X+\mathfrak{H}Y+\mathfrak{G}Z),\\
(A,\ H,\ G)\,[(h),\ (b),\ (f)] &= -X(\mathfrak{H}X+\mathfrak{B}Y+\mathfrak{F}Z),\\
(H,\ B,\ F)\,[(h),\ (b),\ (f)] &= -Y(\mathfrak{H}X+\mathfrak{B}Y+\mathfrak{F}Z)-\phi,\\
(G,\ F,\ C)\,[(h),\ (b),\ (f)] &= -Z(\mathfrak{H}X+\mathfrak{B}Y+\mathfrak{F}Z),\\
(A,\ H,\ G)\,[(g),\ (f),\ (c)] &= -X(\mathfrak{G}X+\mathfrak{F}Y+\mathfrak{C}Z),\\
(H,\ B,\ F)\,[(g),\ (f),\ (c)] &= -Y(\mathfrak{G}X+\mathfrak{F}Y+\mathfrak{C}Z),\\
(G,\ F,\ C)\,[(g),\ (f),\ (c)] &= -Z(\mathfrak{G}X+\mathfrak{F}Y+\mathfrak{C}Z)-\phi\,;
\end{aligned}$$

aussi

$$A(a)+B(b)+C(c)+2F(f)+2G(g)+2H(h)+2\phi=0.$$

Multipliant cette dernière équation par l'un quelconque des coefficients (a), ..., et réduisant, on obtient six équations; mais je forme seulement celle qui se dérive de (g), savoir, nous avons

$$\begin{aligned}
&(g)\,[A(a)+B(b)+C(c)+2F(f)+2G(g)+2H(h)]\\
&\qquad\qquad +2\phi(-BZX+FXY-GY^2+HYZ)=0.
\end{aligned}$$

Ici la seconde ligne est égale à

$$2B[(f)(h)-(b)(g)]-2F[(f)(g)-(c)(h)]+2G[(c)(a)-(g)^2]-2H[(g)(h)-(a)(f)],$$

et l'équation est

$$A(a)(g)+B[2(h)(f)-(b)(g)]+C(c)(g)+2F(c)(h)+2G(c)(a)+2H(a)(f)=0.$$

Des équations $(g)(h)-(a)(f)=-YZ\phi$, $(h)(f)-(b)(g)=-ZX\phi$, multipliant par $-X$, $-Y$ et ajoutant, nous obtenons $-(h)[(g)X+(f)Y]+(a)(f)X+(b)(g)Y=2XYZ$, c'est-à-dire

$$(a)(f)X+(b)(g)Y+(c)(f)Z=2XYZ.$$

9. Je reviens à la question principale. A moins de se servir de quantités arbitraires qui rendraient les formules plus complexes, il n'y a pas d'expression symétrique pour les valeurs de $X_1 : Y_1 : Z_1$ et $X_2 : Y_2 : Z_2$: et ainsi j'écris

$$\begin{aligned}
X_1 : Y_1 : Z_1 &= (a) : (h)+Z\sqrt{\phi} : (g)-Y\sqrt{\phi},\\
X_2 : Y_2 : Z_2 &= (a) : (h)-Z\sqrt{\phi} : (g)+Y\sqrt{\phi},
\end{aligned}$$

et la condition devient

$$[(h)\,\delta Z-Z\delta(h)-(g)\,\delta Y+Y\delta(g)]\sqrt{\phi}+[(h)Z-(g)Y]\,\delta\sqrt{\phi}=0,$$

ou, puisque $\delta\sqrt{\phi}=\dfrac{1}{2\sqrt{\phi}}\,\delta\phi$, ceci est

$$2[(h)\,\delta Z-Z\delta(h)-(g)\,\delta Y+Y\delta(g)]\,\phi+[(h)Z-(g)Y]\,\delta\phi=0,$$

équation qui contient, comme nous le verrons, le facteur (a); et, en omettant ce facteur, l'équation deviendra symétrique.

J'écris

$$\delta(\mathrm{g}) = \Delta(\mathrm{g}) + \delta'(\mathrm{g}), \quad \delta(\mathrm{h}) = \Delta(\mathrm{h}) + \delta'(\mathrm{h}), \quad \delta\phi = \Delta\phi + \delta'\phi,$$

en dénotant par Δ les parties qui dépendent de δX, δY, δZ, et par δ' celles qui dépendent de $\delta A, \ldots$. La fonction à droite est ainsi la somme des deux parties

$$\Omega_1 = 2\,[(\mathrm{h})\,\delta Z - Z\Delta(\mathrm{h}) - (\mathrm{g})\,\delta Y + Y\Delta(\mathrm{g})]\,\phi + [(\mathrm{h})\,Z - (\mathrm{g})\,Y]\,\Delta\phi,$$

$$\Omega_2 = 2\,[\quad - Z\delta'(\mathrm{h}) \qquad + Y\delta'(\mathrm{g})]\,\phi + [(\mathrm{h})\,Z - (\mathrm{g})\,Y]\,\delta'\phi,$$

où cette seconde partie Ω_2 est la seule qui contient les dérivées de ρ du troisième ordre.

10. Je réduis l'expression de Ω_1. Nous avons

$$\Delta(\mathrm{h}) = (-CY + FZ)\,\delta X + (-CX + GZ \qquad)\,\delta Y + (FX + GY - 2HZ)\,\delta Z,$$

$$\Delta(\mathrm{g}) = (-BZ + FY)\,\delta X + (FX - 2GY + HZ)\,\delta Y + (-BX + HY \qquad)\,\delta Z,$$

et de là

$$\tfrac{1}{2}\Omega_1 = \phi\,\{[(C-B)\,YZ + F(Y^2 - Z^2)]\,\delta X + [-AXZ + G(Y^2 + Z^2)]\,\delta Y + [AXZ + H(Y^2 + Z^2)]\,\delta Z\}$$
$$+ \tfrac{1}{2}\,[(\mathrm{h})\,Z - (\mathrm{g})\,Y]\,\Delta\phi,$$

où la dernière ligne est égale à

$$[(\mathrm{g})\,Y - (\mathrm{h})\,Z]\,[(\mathfrak{A}X + \mathfrak{H}Y + \mathfrak{G}Z)\,\delta Z + (\mathfrak{H}X + \mathfrak{B}Y + \mathfrak{F}Z)\,\delta Z + (\mathfrak{G}X + \mathfrak{F}Y + \mathfrak{C}Z)\,\delta Z].$$

Ici le coefficient de δX est égal à

$$(C - B)\,[(\mathrm{a})\,(\mathrm{f}) - (\mathrm{g})\,(\mathrm{h})] + F\,(\mathrm{g})^2 - (\mathrm{a})\,(\mathrm{c}) - (\mathrm{h})^2 + (\mathrm{a})\,(\mathrm{b})]$$
$$+ [(\mathrm{g})\,Y - (\mathrm{h})\,Z]\,(\mathfrak{A}X + \mathfrak{H}Y + \mathfrak{G}Z),$$

où la seconde ligne est égale à

$$-(\mathrm{g})\,(H,\ B,\ F)\,[(\mathrm{a}),\ (\mathrm{b}),\ (\mathrm{g})] + (\mathrm{h})\,(G,\ F,\ C)\,[(\mathrm{a}),\ (\mathrm{h}),\ (\mathrm{g})],$$

et ainsi l'expression entière se réduit à

$$(\mathrm{a})\,\{-(B - C)\,\Theta + F[(\mathrm{b}) - (\mathrm{c})] - H\,(\mathrm{g}) + G\,(\mathrm{h})\}$$

c'est-à-dire le coefficient de δX contient le facteur (a).

Le coefficient de δY est

$$[-AXZ - F(Y^2 + Z^2)]\,\phi + [(\mathrm{g})\,Y - (\mathrm{h})\,Z]\,(\mathfrak{H}X + \mathfrak{B}Y + \mathfrak{F}Z),$$

où la seconde partie est

$$(\mathrm{g})\,\{-\phi - (H,\ B,\ F)\,[(\mathrm{h}),\ (\mathrm{b}),\ (\mathrm{f})]\} + (\mathrm{h})\,(G,\ F,\ C)\,[(\mathrm{h}),\ (\mathrm{b}),\ (\mathrm{f})]\,;$$

on a donc les termes

$$-\phi\,[(\mathrm{g}) + AZX + G(Y^2 + Z^2)],$$

c'est-à-dire

$$-\phi\,[(A - B)\,ZX + FXY + GZ^2 + HYZ],$$

et l'expression entière est ainsi égale à

$$(A-B)[(\mathrm{h})(\mathrm{f})-(\mathrm{b})(\mathrm{g})]+F[(\mathrm{f})(\mathrm{g})-(\mathrm{c})(\mathrm{h})]+G[(\mathrm{a})(\mathrm{b})-(\mathrm{h})^2]$$
$$+H[(\mathrm{g})(\mathrm{h})-(\mathrm{a})(\mathrm{f})]-(\mathrm{g})(H,\ B,\ F)[(\mathrm{h}),\ (\mathrm{b}),\ (\mathrm{f})]$$
$$+(\mathrm{h})(G,\ F,\ C)[(\mathrm{h}),\ (\mathrm{b}),\ (\mathrm{f})],$$

c'est-à-dire à

$$A[(\mathrm{h})(\mathrm{f})]-(\mathrm{b})(\mathrm{g})-B(\mathrm{h})(\mathrm{f})+C(\mathrm{h})(\mathrm{f})+F[(\mathrm{b})-(\mathrm{c})](\mathrm{h})+G(\mathrm{a})(\mathrm{b})-H(\mathrm{a})(\mathrm{f}),$$

ou enfin à

$$A-(\mathrm{b})(\mathrm{g})+B-2(\mathrm{h})(\mathrm{f})+F[(\mathrm{b})(\mathrm{c})](\mathrm{h})+G(\mathrm{a})(\mathrm{b})+H-(\mathrm{a})(\mathrm{f}).$$

J'ajoute la quantité nulle

$$A(\mathrm{a})(\mathrm{g})+B[2(\mathrm{h})(\mathrm{f})-(\mathrm{b})(\mathrm{g})]+C(\mathrm{c})(\mathrm{g})$$
$$+F2(\mathrm{c})(\mathrm{h})+G2(\mathrm{c})(\mathrm{a})\qquad\qquad+H2(\mathrm{a})(\mathrm{f}),$$

et, en réduisant au moyen de $(\mathrm{a})+(\mathrm{b})+(\mathrm{c})=0$ et $A+B+C=0$, le coefficient de δY devient

$$=(\mathrm{a})\{-(C-A)(\mathrm{g})+H(\mathrm{f})+G[(\mathrm{c})-(\mathrm{a})]-F(\mathrm{h})\},$$

et, de même, le coefficient de δZ est

$$=(\mathrm{a})\{-(A-B)(\mathrm{h})-G(\mathrm{f})+F(\mathrm{g})+H[(\mathrm{a})-(\mathrm{b})]\},$$

ce qui achève le calcul de Ω_1.

[Pp. 246—250.]

11. Pour trouver Ω_2, nous avons

$$\delta'(\mathrm{g})=-ZX\delta B+YX\delta F-Y^2\delta G+YZ\delta H,$$
$$\delta'(\mathrm{h})=-XY\delta C+XZ\delta F+YZ\delta G-Z^2\delta H,$$
$$\delta'\phi\ =-[(\mathrm{a})\delta A+(\mathrm{b})\delta B-(\mathrm{c})\delta C+2(\mathrm{f})\delta F+2(\mathrm{g})\delta G+2(\mathrm{h})\delta H],$$

et de là

$$\Omega_2=2\phi[-XYZ(\delta B-\delta C)+X(Y^2-Z^2)\delta F-Y(Y^2+Z^2)\delta G+Z(Y^2+Z^2)\delta H]$$
$$+[(\mathrm{g})Y-(\mathrm{h})Z][(\mathrm{a})\delta A+(\mathrm{b})\delta B+(\mathrm{c})\delta C+2(\mathrm{f})\delta F+2(\mathrm{g})\delta G+2(\mathrm{h})\delta H],$$

ce qui se réduit tout de suite à

$$-[X(\mathrm{a})(\mathrm{f})+Y(\mathrm{b})(\mathrm{g})+Z(\mathrm{c})(\mathrm{h})](\delta B-\delta C)$$
$$+[(\mathrm{g})Y-(\mathrm{h})Z][(\mathrm{a})\delta A+(\mathrm{b})\delta B+(\mathrm{c})\delta C]$$
$$+2[Y(\mathrm{c})(\mathrm{h})-Z(\mathrm{b})(\mathrm{g})]\delta F$$
$$+2(\mathrm{a})[Y(\mathrm{c})-Z(\mathrm{f})\qquad]\delta G$$
$$+2(\mathrm{g})[Y(\mathrm{f})-Z(\mathrm{b})\qquad]\delta H.$$

Les premières deux lignes se réduisent facilement à

$$(a)[-X(f)+Z(h)](\delta B-\delta C)+(a)[(g)Y-(h)Z](\delta A-\delta C),$$

et la troisième ligne à $2(\mathrm{a})[Z(\mathrm{g})-X(\mathrm{c})]\,\delta F$. Donc l'expression entière contient le facteur (a), et nous aurons

$$\begin{aligned}\Omega_2 : (\mathrm{a}) = -\ & [X(\mathrm{f})+Z(\mathrm{h})](\delta B-\delta C)\\ +\ & [Y(\mathrm{g})-Z(\mathrm{h})](\delta A-\delta C)\\ +\,2 & [Z(\mathrm{g})-X(\mathrm{c})]\,\delta F\\ +\,2 & [Y(\mathrm{c})-Z(\mathrm{f})]\,\delta G\\ +\,2 & [Y(\mathrm{f})-Z(\mathrm{b})]\,\delta H,\end{aligned}$$

expression qui se réduit sans peine à la forme symétrique sous laquelle je la présente dans l'équation finale.

12. Cette équation est $\Omega_1+\Omega_2=0$; savoir, en omettant le facteur (a), nous avons

$$\begin{aligned}2\,[&\{F[(\mathrm{b})-(\mathrm{c})]-(B-C)(\mathrm{f})-H(\mathrm{g})+G(\mathrm{h})\}\,\delta X\\ &+\{G[(\mathrm{c})-(\mathrm{a})]+H(\mathrm{f})-(C-A)(\mathrm{g})-F(\mathrm{h})\}\,\delta Y\\ &+\{H[(\mathrm{a})-(\mathrm{b})]-G(\mathrm{h})+F(\mathrm{g})-(A-B)(\mathrm{h})\}\,\delta Z]\\ &-X(\mathrm{f})(\delta B-\delta C)\\ &-Y(\mathrm{g})(\delta C-\delta A)\\ &-Z(\mathrm{h})(\delta A-\delta B)\\ &+\{X[(\mathrm{b})-(\mathrm{c})]-Y(\mathrm{h})+Z(\mathrm{g})\}\,\delta F\\ &+\{X(\mathrm{h})+Y[(\mathrm{c})-(\mathrm{a})]-Z(\mathrm{f})\}\,\delta G\\ &+\{-X(\mathrm{g})+Y(\mathrm{f})+Z[(\mathrm{a})-(\mathrm{b})]\}\,\delta H=0.\end{aligned}$$

On se rappelle que δ signifie

$$X\frac{d}{dx}+Y\frac{d}{dy}+Z\frac{d}{dz}.$$

13. Pour déduire de là le résultat de M. Levy, j'écris d'abord $X=0$, $Y=0$; nous avons alors

$$[(\mathrm{a}),\ (\mathrm{b}),\ (\mathrm{c}),\ (\mathrm{f}),\ (\mathrm{g}),\ (\mathrm{h})]=(BZ^2,\ AZ^2,\ 0,\ 0,\ 0,\ -HZ^2),$$

et l'équation devient

$$2\,[(AF-GH)\,\delta X+(-BG+FH)\,\delta Y]+HZ(\delta A-\delta B)-Z(A-B)\,\delta\Pi=0;$$

mais ici

$$(A,\ B,\ C,\ F,\ G,\ H)=[2Z\mathrm{h},\ -2Z\mathrm{h},\ 2\mathrm{h},\ 0,\ -Z\mathrm{g},\ Z\mathrm{f},\ -Z(\mathrm{a}-\mathrm{b})],$$

et l'équation devient

$$2\,\{[\mathrm{f}(\mathrm{a}-\mathrm{b})-2\mathrm{gh}]\,\delta X+[\mathrm{g}(\mathrm{a}-\mathrm{b})+2\mathrm{fh}]\,\delta Y\}-(\mathrm{a}-\mathrm{b})(\delta A-\delta B)-4\mathrm{h}\,\delta H=0.$$

Mais nous avons $\delta X = \mathrm{g}Z$, $\delta Y = \mathrm{f}Z$, $\delta Z = \mathrm{c}Z$, et, de plus,

$$\begin{aligned}\delta A &= 2\delta Z\mathrm{h} - 2\mathrm{g}\delta Y = 2lZ^2 + 2(\mathrm{ch} - \mathrm{fg})Z,\\ \delta B &= 2\mathrm{f}\delta X - 2\delta Z\mathrm{h} = -2lZ^2 - 2(\mathrm{ch} - \mathrm{fg})Z,\\ \delta H &= -\delta Z(\mathrm{a} - \mathrm{b}) + \mathrm{g}\delta X - \mathrm{f}\delta Y = (f - j)Z^2 + (-\mathrm{ac} + \mathrm{bc} - \mathrm{f}^2 + \mathrm{g}^2)Z;\end{aligned}$$

l'équation est donc

$$4\,\mathrm{fg}(\mathrm{a} - \mathrm{b}) + 4(\mathrm{f}^2 - \mathrm{g}^2)\mathrm{h} - (\mathrm{a} - \mathrm{b})[4lZ + 4(\mathrm{ch} - \mathrm{fg})] - 4\mathrm{h}[-\mathrm{c}(\mathrm{a} - \mathrm{b}) - (\mathrm{f}^2 - \mathrm{g}^2) + (f - j)Z] = 0,$$

ou enfin

$$2\,\mathrm{fg}(\mathrm{a} - \mathrm{b}) + 2\mathrm{h}(\mathrm{f}^2 - \mathrm{g}^2) - Z[(f - j)\mathrm{h} + l(\mathrm{a} - \mathrm{b})] = 0,$$

ce qui s'accorde avec le résultat cité.

14. En changeant la signification de X, Y, Z, écrivons $\rho = X + Y + Z$, où X, Y, Z dénotent à présent des fonctions de x, y, z respectivement; en dénotant par X', Y', Z' les fonctions dérivées de celles-ci, les fonctions premièrement représentées par X, Y, Z seront X', Y', Z'. Je cherche, au moyen de l'équation générale, la condition pour que la famille $\rho = X + Y + Z$ puisse faire partie d'un système orthogonal.

Dénotons par X', X'', X''' les dérivées de X, et de même celles de Y et Z, et écrivons, pour abréger, $\alpha, \beta, \gamma = Y'' - Z'', Z'' - X'', X'' - Y''$, nous avons

$$(\mathrm{a}, \mathrm{b}, \mathrm{c}, \mathrm{f}, \mathrm{g}, \mathrm{h}) = (X'', Y'', Z'', 0, 0, 0),$$

et de là

$$(A, B, C, F, G, H) = (0, 0, 0, -\alpha X', -\beta Y', -\gamma Z'),$$

et, de plus,

$$\begin{aligned}[(\mathrm{a}), (\mathrm{b}), (\mathrm{c}), (\mathrm{f}), (\mathrm{g}), (\mathrm{h})] = [&2\alpha X'Y'Z',\ 2\beta X'Y'Z',\ 2\gamma X'Y'Z',\\ &X'(\ \ \alpha X'^2 - \beta Y'^2 - \gamma Z'^2),\\ &Y'(-\alpha X'^2 + \beta Y'^2 - \gamma Z'^2),\\ &Z'(-\alpha X'^2 - \beta Y'^2 + \gamma Z'^2)].\end{aligned}$$

Nous avons aussi

$$\begin{aligned}(\delta X', \delta Y', \delta Z') &= (X'X'', Y'Y'', Z'Z''),\\ (\delta A, \delta B, \delta C) &= (0, 0, 0),\\ (\delta F, \delta G, \delta H) &= [X'(-\alpha X'' + Z'Z''' - Y'Y'''),\\ &\quad Y'(-\beta Y'' + X'X''' - Z'Z'''),\\ &\quad Z'(-\gamma Z'' + Y'Y''' - X'X''')].\end{aligned}$$

15. Donc, dans l'équation générale, la première ligne est

$$\begin{aligned}2[-\alpha X'.2X'Y'Z'(\beta - \gamma) + \gamma\, Y'Z'(-\alpha X'^2 + \beta Y'^2 - \gamma Z'^2)\\ - \beta Y'Z'(-\alpha X'^2 - \beta Y'^2 + \gamma Z'^2)]\,X'X'',\end{aligned}$$

c'est-à-dire

$$\begin{aligned}2X'Y'Z'.X''[-2\alpha(\beta - \gamma)X'^2 + \gamma(-\alpha X'^2 + \beta Y'^2 - \gamma Z'^2)\\ - \beta(-\alpha X'^2 - \beta Y'^2 + \gamma Z'^2)]\end{aligned}$$

ou, ce qui est la même chose,

$$2X'Y'Z'.\alpha X''[(\gamma-\beta)X'^2-\beta Y'^2+\gamma Z'^2],$$

et la somme des premières trois lignes sera aussi $= 2X'Y'Z'$ multiplié par

$$\begin{aligned}&\alpha X''[(\gamma-\beta)X'^2-\beta Y'^2+\gamma Z'^2]\\&+\beta Y''[\alpha X'^2+(\alpha-\gamma)Y'^2-\gamma Z'^2]\\&+\gamma Z''[-\alpha X'^2+\beta Y'^2+(\beta-\alpha)Z'^2],\end{aligned}$$

savoir dans ce second facteur le coefficient de $\alpha X'^2$ est $X''(\gamma-\beta)+\beta Y''-\gamma Z''$, $=-2\beta\gamma$, et de même les coefficients de $\beta Y'^2$ et $\gamma Z'^2$ sont $-2\gamma\alpha$, $-2\alpha\beta$ respectivement, donc le terme entier, ou première partie de l'équation est

$$4X'Y'Z'(X'^2+Y'^2+Z'^2)(-\alpha\beta\gamma).$$

Les termes en δA, δB, δC s'évanouissent, et il ne reste que les termes en δF, δG, δH qui forment la seconde partie de l'équation. Le premier de ceux-ci est

$$\begin{aligned}&[X'.2(\beta-\gamma)X'Y'Z'-Y'Z'(-\alpha X'^2-\beta Y'^2+\gamma Z'^2)\\&\qquad+Y'Z'(-\alpha X'^2+\beta Y'^2-\gamma Z'^2)]\times X'(-X''\alpha+Z'Z'''-Y'Y'''),\end{aligned}$$

c'est-à-dire

$$2X'Y'Z'[(\beta-\gamma)X'^2+\beta Y'^2-\gamma Z'^2](-X''\alpha+Z'Z'''-Y'Y''').$$

On a donc $2X'Y'Z'$ multiplié par

$$\begin{aligned}&[(\beta-\gamma)X'^2+\beta Y'^2-\gamma Z'^2](-X''\alpha+Z'Z'''-Y'Y''')\\&+[-\alpha X'^2+(\gamma-\alpha)Y'^2+\gamma Z'^2](-Y''\beta+X'X'''-Z'Z''')\\&+[\alpha X'^2-\beta Y'^2+(\alpha-\beta)Z'^2](-Z''\gamma+Y'Y'''-X'X'''),\end{aligned}$$

où dans le second facteur nous avons d'abord le terme $-2\alpha\beta\gamma\times(X'^2+Y'^2+Z'^2)$ et puis le terme $-2(\alpha X'X'''+\beta Y'Y'''+\gamma Z'Z''')\times(X'^2+Y'^2+Z'^2)$.

La seconde partie est donc

$$4X'Y'Z'(X'^2+Y'^2+Z'^2)[-\alpha\beta\gamma-(\alpha X'X'''+\beta Y'Y'''+\gamma Z'Z''')]$$

et en réunissant les deux parties et en omettant le facteur $-4X'Y'Z'\times(X'^2+Y'^2+Z'^2)$, l'équation devient

$$2\alpha\beta\gamma+\alpha X'X'''+\beta Y'Y'''+\gamma Z'Z'''=0,$$

savoir:

$$2(Y''-Z'')(Z''-X'')(X''-Y'')+(Y''-Z'')X'X'''+(Z''-X'')Y'Y'''+(X''-Y'')Z'Z'''=0,$$

équation trouvée par M. Bouquet dans sa "Note sur les surfaces orthogonales" (*Journal de M. Liouville*, t. XI., pp. 446—450, 1846), et reproduite par M. Serret dans son "Mémoire sur les surfaces orthogonales" (*Journal de M. Liouville*, t. XII., pp. 241—254, 1847).

[Pp. 324—330.]

En considérant une famille orthogonale (savoir: une famille de surfaces qui fait partie d'un système orthogonal), on peut se proposer la question: *Étant donnée une surface de la famille, trouver de la manière la plus générale la famille.* J'essaye de résoudre cette question en développant les trois coordonnées selon les puissances d'un paramètre; et, quoique je n'aie encore calculé que les trois premiers termes des trois développements, les résultats me paraissent assez intéressants pour les soumettre aux géomètres.

On peut, pour la surface donnée, considérer les coordonnées x, y, z d'un point quelconque de la surface comme des fonctions déterminées de deux paramètres p, q. Si, de plus, ces paramètres sont tels, que les équations des deux systèmes de courbes de courbure soient $p = \text{const.}$, $q = \text{const.}$ respectivement, alors (en écrivant pour abréger

$$\frac{dx}{dp} = x_1, \quad \frac{dx}{dq} = x_2, \quad \frac{d^2x}{dp^2} = x_3, \quad \frac{d^2x}{dp\,dq} = x_4, \quad \frac{d^2x}{dq^2} = x_5,$$

et de même pour y et z) ces coordonnées x, y, z, considérées toujours comme des fonctions de p, q, seront telles, que

$$x_1x_2 + y_1y_2 + z_1z_2 = 0, \qquad \begin{vmatrix} x_1 & y_1 & z_1 \\ x_2 & y_2 & z_2 \\ x_4 & y_4 & z_4 \end{vmatrix} = 0.$$

J'écris ici et dans la suite $X, Y, Z = y_1z_2 - y_2z_1,\ z_1x_2 - z_2x_1,\ x_1y_2 - x_2y_1$. On a donc identiquement

$$Xx_1 + Yy_1 + Zz_1 = 0,$$
$$Xx_2 + Yy_2 + Zz_2 = 0,$$

et les deux équations mentionnées sont

$$x_1x_2 + y_1y_2 + z_1z_2 = 0,$$
$$Xx_4 + Yy_4 + Zz_4 = 0.$$

Je m'arrête pour remarquer que la dernière équation, dans sa forme originale, peut être remplacée par trois équations de la forme $x_4 + Ax_1 + Bx_2 = 0$, et qu'en ajoutant les trois équations multipliées par x_1, y_1, z_1 respectivement, et aussi multipliées par x_2, y_2, z_2 respectivement, on obtient les valeurs de A, B, exprimées en termes de

$$E = x_1^2 + y_1^2 + z_1^2 \text{ et } G = x_2^2 + y_2^2 + z_2^2$$

(E, G de Gauss), et que l'on trouve de là

$$2\frac{d^2x}{dp\,dq} - \frac{1}{E}\frac{dE}{dq}\frac{dx}{dp} - \frac{1}{G}\frac{dG}{dp}\frac{dx}{dq} = 0,$$

avec les équations semblables en y et z. Ces équations sont, en effet, les équations (10 *bis*) de Lamé, "Mémoire sur les coordonnées curvilignes" (*Liouville*, t. V. 1840, p. 322).

Je suppose que les surfaces de la famille dépendent du paramètre r, lequel pour la surface donnée se réduit à $r=0$. Par le point (p, q) de la surface donnée on peut mener une trajectoire orthogonale aux différentes surfaces de la famille; les coordonnées ξ, η, ζ d'un point quelconque sur cette courbe seront des fonctions de p, q, r, lesquelles, pour $r=0$, se réduisent à x, y, z respectivement; et j'écris

$$\xi = x + ar + dr^2 + \ldots,$$

$$\eta = y + br + er^2 + \ldots,$$

$$\zeta = z + cr + fr^2 + \ldots,$$

où $a, b, c, d, e, f, \ldots$ sont des fonctions inconnues de p et q.

Pour exprimer que la courbe coupe orthogonalement les différentes surfaces de la famille, écrivons pour abréger

$$\eta_1\zeta_2 - \eta_2\zeta_1 = X + Ar + Dr^2 + \ldots, \quad X = y_1z_2 - y_2z_1,$$

$$\zeta_1\xi_2 - \zeta_2\xi_1 = Y + Br + Er^2 + \ldots, \quad A = y_1c_2 - y_2c_1 + b_1z_2 - b_2z_1,$$

$$\xi_1\eta_2 - \xi_2\eta_1 = Z + Cr + Fr^2 + \ldots, \quad \ldots\ldots$$

$\left(\text{où } \xi_1 = \dfrac{d\xi}{dp} \ldots, \text{ comme pour } x, y, z\right)$. La condition cherchée est

$$\frac{X + Ar + Dr^2 + \ldots}{a + 2dr + \ldots} = \frac{Y + Br + Er^2 + \ldots}{b + 2er + \ldots} = \frac{Z + Cr + Fr^2 + \ldots}{c + 2fr + \ldots},$$

laquelle doit être satisfaite pour une valeur quelconque de r; on a donc

$$(1) \qquad \frac{X}{a} = \frac{Y}{b} = \frac{Z}{c},$$

$$(2) \qquad \frac{A}{a} - \frac{2dX}{a^2} = \frac{B}{b} - \frac{2eY}{b^2} = \frac{C}{c} - \frac{2fZ}{c^2};$$

savoir, les équations (1) contiennent (a, b, c), les équations (2) contiennent de plus (d, e, f), et ainsi de suite.

Pour qu'il y ait un système orthogonal, il faut et il suffit que l'on ait

$$\xi_1\xi_2 + \eta_1\eta_2 + \zeta_1\zeta_2 = 0,$$

pour toute valeur de r; on aura donc

$$[0] \qquad x_1x_2 + y_1y_2 + z_1z_2 = 0,$$

$$[1] \qquad x_1a_2 + x_2a_1 + y_1b_2 + y_2b_1 + z_1c_2 + z_2c_1 = 0,$$

$$[2] \qquad x_1d_2 + x_2d_1 + y_1e_2 + y_2e_1 + z_1f_2 + z_2f_1 + a_1a_2 + b_1b_2 + c_1c_2 = 0,$$

savoir, l'équation [0] est satisfaite d'elle-même; l'équation [1] contient (a, b, c), l'équation [2] contient de plus (d, e, f), et ainsi de suite.

Il paraît donc qu'il y a les trois équations (1), [1] pour déterminer (a, b, c); les trois équations (2), [2] pour déterminer (d, e, f), et ainsi de suite. Mais les choses ne se comportent pas ainsi. On satisfait à (1), [1] par des valeurs de (a, b, c) qui contiennent une fonction arbitraire λ, fonction qui est ensuite déterminée au moyen d'une équation à différences partielles du second ordre, obtenue au moyen des équations (2), [2]; on satisfait alors à (2), [2] par des valeurs de (d, e, f) qui contiennent une fonction arbitraire θ; je présume que cette fonction serait ensuite déterminée au moyen des équations (3), [3], et ainsi de suite; mais je n'ai pas encore fait les calculs ultérieurs.

Par rapport à λ, en remplaçant cette fonction par $\rho = \lambda\sqrt{X^2+Y^2+Z^2}$, l'équation pour ρ est

$$2\frac{d^2\rho}{dp\,dq} - \frac{1}{E}\frac{dE}{dq}\frac{d\rho}{dp} - \frac{1}{G}\frac{dG}{dp}\frac{d\rho}{dq} = 0,$$

savoir, c'est la même équation que pour x, y, z: ainsi l'on y satisfait en prenant ρ égal à une fonction linéaire (avec terme constant) quelconque de x, y, z.

Pour obtenir ces conclusions, partant des équations (1), [1], les équations (1) donnent

$$a,\ b,\ c = \lambda X,\ \lambda Y,\ \lambda Z,$$

où λ est une fonction de p, q: ces valeurs satisfont d'elles-mêmes à l'équation [1]. La vérification se fait sans peine; j'écris pour abréger x_1x_2 pour dénoter $x_1x_2 + y_1y_2 + z_1z_2$, et ainsi dans les cas semblables: l'équation à vérifier est donc

$$x_1(\lambda X)_2 + x_2(\lambda X)_1 = 0,$$

c'est-à-dire

$$\lambda(x_1X_2 + x_2X_1) + \lambda_2x_1X + \lambda_1x_2X = 0,$$

où nous avons

$$x_1X = 0, \quad x_2X = 0;$$

reste à trouver le coefficient $x_1X_2 + x_2X_1$. Nous avons

$$X = y_1z_2 - y_2z_1,$$

et de là

$$X_1 = y_1z_4 - y_4z_1 + y_3z_2 - y_2z_3,$$

$$X_2 = y_1z_5 - y_5z_1 + y_4z_2 - y_2z_4,$$

et de là, en faisant la somme des trois termes de x_1X_2 et x_2X_1 respectivement, on trouve

$$x_1X_2 = -\begin{vmatrix} x_1, & y_1, & z_1 \\ x_2, & y_2, & z_2 \\ x_4, & y_4, & z_4 \end{vmatrix} = x_2X_1,$$

savoir: $x_1X_2 + x_2X_1$ est égal à -2 multiplié par ce déterminant, $= -2Xx_4$, c'est-à-dire $x_1X_2 + x_2X_1 = 0$. Donc la fonction λ est jusqu'ici indéterminée.

Passons aux équations (2), [2]. Substituant dans (2) les valeurs de (a, b, c), ces équations deviennent

$$\frac{A}{\lambda X} - \frac{2d}{\lambda^2 X} = \frac{B}{\lambda Y} - \frac{2e}{\lambda^2 Y} = \frac{C}{\lambda Z} - \frac{2f}{\lambda^2 Z}.$$

On y satisfait en écrivant

$$2d,\ 2e,\ 2f = \lambda(\theta X + A),\quad \lambda(\theta Y + B),\quad \lambda(\theta Z + C),$$

où θ est fonction de (p, q); en substituant ces valeurs dans l'équation [2], la fonction θ disparaît d'elle-même; mais on obtient pour λ une équation linéaire entre λ, λ_1, λ_2 et λ_4, laquelle est ainsi une équation à différences partielles du second ordre, et, cela étant, on a pour d, e, f les expressions mentionnées, qui contiennent la fonction θ, fonction qui n'est pas déterminée par les équations (2), [2].

L'équation [2], sous la forme abrégée, est

$$x_1 d_2 + x_2 d_1 + a_1 a_2 = 0,$$

c'est-à-dire

$$x_1 [\lambda(\theta X + A)]_2 + x_2 [\lambda(\theta X + A)]_1 + 2a_1 a_2 = 0,$$

ou, ce qui est la même chose,

$$\lambda [x_1 (\theta X + A)_2 + x_2 (\theta X + A)_1] + \lambda_2 x_1 (\theta X + A) + \lambda_1 x_2 (\theta X + A) + 2a_1 a_2 = 0.$$

Les termes en θ sont

$$\lambda [x_1 (\theta X_2 + \theta_2 X) + x_2 (\theta X_1 + \theta_1 X)] + \lambda_2 \theta x_1 X + \lambda_1 \theta x_2 X.$$

qui s'évanouissent d'eux-mêmes; l'équation se réduit donc à

$$\lambda (A_2 x_1 + A_1 x_2) + \lambda_2 A x_1 + \lambda_1 A x_2 + 2a_1 a_2 = 0,$$

ou, en substituant la valeur de $a_1 a_2$,

$$\lambda (A_2 x_1 + A_1 x_2) + \lambda_2 A x_1 + \lambda_1 A x_2 + 2(\lambda X)_1 (\lambda X)_2 = 0;$$

on a

$$(\lambda X)_1 (\lambda X)_2 = (\lambda X_1 + \lambda_1 X)(\lambda X_2 + \lambda_2 X),\ = \lambda_2 X_1 X_2 + \lambda\lambda_2 X X_1 + \lambda\lambda_1 X X_2 + \lambda_1 \lambda_2 X^2,$$

et l'on trouve sans peine $A x_1 = -a_1 X$, $A x_2 = -a_2 X$, et de là

$$A x_1 = -(\lambda X)_1 X,\ = -\lambda_1 X^2 - \lambda X X_1,$$
$$A x_2 = -(\lambda X)_2 X,\ = -\lambda_2 X^2 - \lambda X X_2.$$

Substituant ces valeurs, l'équation entière contiendra le facteur λ, et en l'écartant, elle devient

$$A_2 x_1 + A_1 x_2 + \lambda_2 X X_1 + \lambda_1 X X_2 + 2\lambda X_1 X_2 = 0.$$

Pour abréger encore la notation, au lieu de x_1^2, $(=x_1^2+y_1^2+z_1^2)$, j'écris simplement 11, et ainsi dans les cas semblables: savoir, je me sers des abréviations

$$11 = x_1^2 \;+ y_1^2 \;+ z_1^2,$$
$$12 = x_1x_2 + y_1y_2 + z_1z_2 (=0),$$
$$\vdots$$

et je remarque que l'équation $12=0$, en prenant les dérivées par rapport à p, q respectivement, donne $15+24=0$, $23+14=0$, équations qui servent pour éliminer des formules les expressions 15 et 23. Si pour un moment nous dénotons ainsi par 124 le déterminant $\begin{vmatrix} x_1 & y_1 & z_1 \\ x_2 & y_2 & z_2 \\ x_4 & y_4 & z_4 \end{vmatrix}$, alors, en multipliant par les déterminants analogues 123 et 125 respectivement, l'équation $124=0$ donne

$$\begin{vmatrix} 11 & . & 14 \\ . & 22 & 24 \\ 51 & 52 & 54 \end{vmatrix} = 0, \quad \begin{vmatrix} 11 & . & 14 \\ . & 22 & 24 \\ 31 & 32 & 34 \end{vmatrix} = 0,$$

dont chacune est une équation à trois termes entre les quantités 11, 22, ...

Nous avons

$$A = y_1(\lambda Z)_2 - y_2(\lambda Z)_1 + z_2(\lambda Y)_1 - z_1(\lambda Y)_2,$$
$$= \lambda(y_1Z_2 - y_2Z_1 + z_2Y_1 - z_1Y_2) + \lambda_1(z_2Y - y_2Z) + \lambda_2(y_1Z - z_1Y);$$

or nous avons

$$Y,\ Z = z_1x_2 - z_2x_1,\ x_1y_2 - x_2y_1,$$

et en formant de là les valeurs de Y_1, Y_2, Z_1, Z_2 on obtient sans peine

$$A = \lambda\,[x_1(15-24) + x_2(23-14) - x_3 . 22 + 2x_4 . 12 - x_5 . 11]$$
$$+ \lambda_1(x_2 . 12 - x_1 . 22) + \lambda_2(x_1 . 12 - x_2 . 11),$$

ou, ce qui est la même chose,

$$A = \lambda\,[-2x_1 . 24 - 2x_2 . 14 - x_3 . 22 - x_5 . 11] - \lambda_1x_1 . 22 - \lambda_2x_2 . 11.$$

Écrivons pour un moment

$$A = \lambda P + \lambda_1P' + \lambda_2P'';$$

nous avons

$$A_1 = P_1\lambda + (P+P_1')\lambda_1 + \quad P_1''\lambda_2 \quad + P'\lambda_3 + P''\lambda_4,$$
$$A_2 = P_2\lambda + \quad P_2'\ \lambda_1 + (P+P_2'')\lambda_2 + P'\lambda_4 + P''\lambda_5,$$

et de là

$$A_1x_2 + A_2x_1 = \lambda\,(P_1x_2 + P_2x_1) + \lambda_1[(P+P_1')x_2 + P_2'x_1]$$
$$+ \lambda_2[P_1''x_2 + (P+P_2'')x_1] + \lambda_3P'x_2$$
$$+ \lambda_4(P''x_2 + P'x_1) + \lambda_5P''x_1.$$

[Pp. 381—385.]

Les expressions de P_1, P_2,... contiennent les dérivées du troisième ordre

$$\frac{d^3x}{dp^3}=x_6,\quad \frac{d^3x}{dp^2dq}=x_7,\quad \frac{d^3x}{dpdq^2}=x_8,\quad \frac{d^3x}{dq^3}=x_9,\ldots$$

En formant les dérivées des équations $23+14=0$, $15+24=0$ par rapport à p et q respectivement, on obtient

$$\begin{aligned}17+26+\ 2\,.\,34\ &=0,\\ 18+27+35+44&=0,\\ 19+28+\ 2\,.\,45\ &=0.\end{aligned}$$

On obtient alors

$$\begin{aligned}P_1=&-2x_1(44+17)+2x_2(34+17)\\ &-4x_3\,.\,24-2x_4\,.\,14-2x_5\,.\,13-x_6\,.\,22-x_8\,.\,11,\end{aligned}$$

et de là la somme P_1x_2 est

$$=-2\,.\,22\,(34+17)-4\,.\,23\,.\,24-2\,.\,24\,.\,14-2\,.\,25\,.\,13-26\,.\,22-28\,.\,11,$$

ou, ce qui est la même chose,

$$P_1x_2=-22\,.\,17-11\,.\,28+2\,.\,14\,.\,24-2\,.\,25\,.\,13.$$

On a de même

$$\begin{aligned}P_2=&-2\,.\,x_1(45+28)-2\,.\,x_2(44+18)\\ &-2x_3\,.\,25-2x_4\,.\,24-4x_5\,.\,14-x_7\,.\,22-x_9\,.\,11,\end{aligned}$$

et de là la somme P_2x_1 est

$$=-2\,.\,11\,(45+28)-2\,.\,13\,.\,25-2\,.\,14\,.\,24-4\,.\,15\,.\,14-17\,.\,22-19\,.\,11,$$

ou, ce qui est la même chose,

$$P_2x_1\qquad=-22\,.\,17-11\,.\,28+2\,.\,14\,.\,24-2\,.\,25\,.\,13\ (=P_1x_2);$$

on a donc

$$P_1x_2+P_2x_1=-2\,.\,22\,.\,17-2\,.\,11\,.\,28+4\,.\,14\,.\,24-4\,.\,25\,.\,13.$$

On obtient sans peine les autres sommes

$$\begin{aligned}&P'x_2=0,\quad P''x_2+P'x_1=-2\,.\,11\,.\,22,\quad P''x_1=0,\\ &Px_1=-11\,.\,24-22\,.\,13,\quad Px_2=-11\,.\,25-22\,.\,14,\end{aligned}$$

et l'on a ainsi

$$\begin{aligned}A_1x_2+A_2x_1=\ &\lambda\ (-2\,.\,11\,.\,28-2\,.\,22\,.\,17+4\,.\,14\,.\,24-4\,.\,25\,.\,13)\\ &+\lambda_1(-3\,.\,11\,.\,25-2\,.\,22\,.\,14-22\,.\,23)\\ &+\lambda_2(-2\,.\,11\,.\,24-11\,.\,15-3\,.\,22\,.\,13)\\ &+\lambda_4(-2\,.\,11\,.\,22).\end{aligned}$$

L'équation en λ est

$$A_1x_2 + A_2x_1 + 2\lambda X_1X_2 + \lambda_1 XX_2 + \lambda_2 XX_1 = 0,$$

et l'on obtient sans peine

$$X_1X_2 = 11 . 45 + 22 . 34 + 3 . 14 . 24 + 25 . 13,$$
$$XX_2 = 11 . 25 + 22 . 14,$$
$$XX_1 = 11 . 24 + 22 . 13.$$

Donc enfin l'équation en λ est

$$\lambda[11(-28+45) + 22(-17+34) + 3 . 14 . 24 - 25 . 13] - \lambda_1 . 11 . 25 - \lambda_2 . 22 . 13 - \lambda_4 . 11 . 22 = 0.$$

Cette équation est vérifiée par la valeur $R = \frac{1}{V}(V = \sqrt{X^2 + Y^2 + Z^2})$; en effet, en dénotant pour un moment le premier coefficient par Λ, l'équation à vérifier est

$$\Lambda V^2 + 11 . 25 . XX_1 + 22 . 13 . XX_2 + (X^2 . X_1X_2 + X^2 . XX_4 - 3 . XX_1 . XX_2) = 0,$$

c'est-à-dire

$$11 . 22\Lambda + 11 . 25(22 . 13 + 11 . 24) + 22 . 13(22 . 14 + 11 . 25)$$
$$+ 11 . 22(11 . 45 + 22 . 34 + 3 . 14 . 24 + 24 . 13 + XX_4)$$
$$- 3(22 . 13 + 11 . 24)(22 . 14 + 11 . 25) = 0,$$

et l'on remarque qu'il n'y a ici que les termes $-2(11^2 . 24 . 25 + 22^2 . 13 . 24)$ qui ne contiennent pas le facteur $11 . 22$.

Savoir, l'équation est de la forme

$$11 . 22\,\Omega - 2(11^2 . 24 . 25 + 22^2 . 13 . 14) = 0;$$

mais, des équations mentionnées $123 . 124 = 0$ et $125 . 124 = 0$, on obtient

$$22^2 . 13 . 14 = 11 . 22(22 . 34 + 14 . 24),$$
$$11^2 . 24 . 25 = 11 . 22(11 . 45 + 14 . 24).$$

Donc l'équation entière contient le facteur $11 . 22$ et, en l'écartant, elle devient

$$\Omega - 2(11 . 45 + 22 . 34 + 2 . 14 . 24) = 0.$$

On a

$$\Omega = 11 . (-28 + 2 . 45) + 22(-17 + 2 . 34) - 25 . 13 + 3 . 14 . 24 + XX_4;$$

l'équation est donc

$$-11 . 28 - 22 . 17 - 25 . 13 - 14 . 25 + XX_4 = 0;$$

et l'on vérifie sans peine que la valeur de XX_4 est actuellement

$$XX_4 = 11 . 28 + 22 . 17 + 25 . 13 + 14 . 24.$$

Donc, en écrivant $\lambda = \frac{\rho}{V}$, l'équation en ρ ne contiendra que les termes en ρ_1, ρ_2, ρ_4. En effet, l'équation devient

$$-11.25\frac{\rho_1}{V} - 22.13\frac{\rho_2}{V} - V^2\left(\frac{\rho_4}{V} - \frac{\rho_1}{V^3}XX_2 - \frac{\rho_2}{V^3}XX_1\right),$$

où, comme auparavant, XX_1 dénote $XX_1 + YY_1 + ZZ_1$, et de même XX_2 dénote $XX_2 + YY_2 + ZZ_2$. Nous avons déjà trouvé

$$XX_1 = 22.13 + 11.24, \quad XX_2 = 22.14 + 11.25;$$

l'équation devient ainsi

$$11.22\,\rho_4 - 14.22\,\rho_1 - 24.11\,\rho_2 = 0.$$

Savoir, cette équation est

$$2\frac{d^2\rho}{dp\,dq} - \frac{1}{E}\frac{dE}{dq}\frac{d\rho}{dp} - \frac{1}{G}\frac{dG}{dp}\frac{d\rho}{dq} = 0.$$

Pour compléter la solution, il convient d'exprimer A, B, C en termes de ρ. Nous avons

$$A = \lambda(-2x_1.24 - 2x_2.14 - x_3.22 - x_5.11) - \lambda_1 x_1.22 - \lambda_2 x_2.11.$$

Substituant la valeur $\lambda = \frac{\rho}{V}$, le coefficient de ρ est

$$\frac{1}{V}(-2x_2.24 - 2x_2.14 - x_3.22 - x_5.11) + \frac{1}{V^3}(x_1.22XX_1 + x_2.11.XX_2)$$

$$= \frac{1}{V^3}[11.22(-2x_1.24 - 2x_2.14 - x_3.22 - x_5.11)$$

$$+ x_1.22(13.22 + 24.11) + x_2.11(14.22 + 25.11)],$$

ou, ce qui est la même chose

$$\frac{1}{V^3}[x_1.22(22.13 + 11.15) + x_2.11(11.25 + 22.23) - 11.22(x_3.22 + x_5.11)].$$

Le terme entre [ ] est fonction linéaire de x_3, y_3, z_3, x_5, y_5, z_5, et en réunissant les termes qui contiennent ces quantités respectivement, on le réduit sans peine à la forme

$$X[11(Xx_5 + Yy_5 + Zz_5) + 22(Xx_3 + Yy_3 + Zz_3)],$$

ou, ce qui est la même chose

$$X(11.125 + 22.123).$$

Nous avons donc

$$A = -\frac{\rho X}{V^3}(11.125 + 22.123) - \frac{1}{V}(x_1\rho_1.22 + x_2\rho_2.11).$$

Nous avons

$$11 = E, \quad 22 = G, \quad V = \sqrt{EG};$$

donc, en écrivant, pour abréger,

$$-\frac{\rho}{EG\sqrt{EG}}(11\,.\,125 + 22\,.\,123) = \theta',$$

la valeur est

$$A = \theta' X - \frac{1}{\sqrt{EG}}\left(G\frac{dx}{dp}\frac{d\rho}{dp} + E\frac{dx}{dq}\frac{d\rho}{dq}\right),$$

avec des expressions semblables pour B et C. Dans les expressions $2d = \frac{\rho}{\sqrt{EG}}(\theta X + A)\ldots,$ la fonction θ' se combine avec la fonction arbitraire θ, de manière qu'il serait permis de remplacer $\theta + \theta'$ par un seul symbole θ, mais je retiens $\theta + \theta'$.

Donc, enfin, les expressions de ξ, η, ζ deviennent

$$\xi = x + \frac{\rho X}{\sqrt{EG}} r + \tfrac{1}{2}\left[\frac{(\theta+\theta')\,\rho X}{\sqrt{EG}} - \frac{\rho}{EG}\left(G\frac{dx}{dp}\frac{d\rho}{dp} + E\frac{dx}{dq}\frac{d\rho}{dq}\right)\right] r^2 + \ldots,$$

$$\eta = y + \frac{\rho Y}{\sqrt{EG}} r + \tfrac{1}{2}\left[\frac{(\theta+\theta')\,\rho Y}{\sqrt{EG}} - \frac{\rho}{EG}\left(G\frac{dy}{dp}\frac{d\rho}{dp} + E\frac{dy}{dq}\frac{d\rho}{dq}\right)\right] r^2 + \ldots,$$

$$\zeta = z + \frac{\rho Z}{\sqrt{EG}} r + \tfrac{1}{2}\left[\frac{(\theta+\theta')\,\rho Z}{\sqrt{EG}} - \frac{\rho}{EG}\left(G\frac{dz}{dp}\frac{d\rho}{dp} + E\frac{dz}{dq}\frac{d\rho}{dq}\right)\right] r^2 + \ldots.$$

Je remarque que l'on satisfait à toutes les conditions en prenant $\rho = \text{const.}$ (ou, ce qui est la même chose, $\rho = 1$), $\theta + \theta' = 0$: cela donne

$$\xi,\ \eta,\ \zeta = x + \frac{rX}{\sqrt{EG}},\quad y = \frac{rY}{\sqrt{EG}},\quad z = \frac{rZ}{\sqrt{EG}};$$

savoir, la famille est ici celle des surfaces parallèles à la surface donnée.

[Pp. 1800—1803.]

J'ai trouvé que l'équation différentielle du troisième ordre, sous la forme [ci-dessus trouvée], contient le facteur étranger $X^2 + Y^2 + Z^2$, et que l'équation débarrassée de ce facteur devient beaucoup plus simple. La réduction et aussi la nouvelle méthode dont je me suis servi pour obtenir l'équation réduite sont toutes les deux assez pénibles; mais cette nouvelle méthode a l'avantage d'établir un résultat intermédiaire qui a quelque valeur. J'ai changé un peu la notation; aussi je commence en l'expliquant.

Je prends $U = 0$ l'équation d'une surface; X, Y, Z les coefficients différentiels du premier ordre; a, b, c, f, g, h les coefficients du second ordre;

$$(A,\ B,\ C,\ F,\ G,\ H \between dx,\ dy,\ dz)^2 = 0$$

l'équation différentielle des courbes de courbure; savoir:

$$\begin{aligned} A &= 2(hZ - gY), \\ B &= 2(fX - hZ), \\ C &= 2(gY - fX), \\ F &= hY - gZ - (b-c)X, \\ G &= fZ - hX - (c-a)Y, \\ H &= gX - fY - (a-b)Z; \end{aligned}$$

et

$$[(A), (B), (C), (F), (G), (H)](dx, dy, dz)^2$$
$$= (A, B, C, F, G, H \between Ydz - Zdy, Zdx - Xdz, Xdy - Ydx)^2;$$

savoir:

$$\begin{aligned} (A) &= BZ^2 + CY^2 - 2FYZ, \\ (B) &= CX^2 + AZ^2 - 2GZX, \\ (C) &= AY^2 + BX^2 - 2HXY, \\ (F) &= -aYZ - fX^2 + gXY + hXZ, \\ (G) &= -bZX + fYX - gY^2 + hYZ, \\ (H) &= -cXY + fZX + gZY - hZ^2, \end{aligned}$$

ce qui explique la signification des symboles (A), (B), ...; aussi

$$V^2 = X^2 + Y^2 + Z^2,$$

et

$$(\bar{a}, \bar{b}, \bar{c}, \bar{f}, \bar{g}, \bar{h}) = (bc - f^2, ca - g^2, ab - h^2, gh - af, hf - bg, fg - ch).$$

Cela étant, à chaque point P de la surface $U = 0$, je prends sur la normale une distance infinitésimale $PP' = \rho$, où ρ est une fonction quelconque des coordonnées x, y, z du point P; on obtient ainsi une surface, lieu des points P', laquelle se nomme la surface voisine, et les points P et P' sont des points correspondants sur les deux surfaces.

Pour que les deux surfaces forment partie d'un système orthogonal, je trouve que la distance ρ, considérée comme fonction des coordonnées (x, y, z), doit satisfaire à cette équation différentielle du second ordre

$$[(A), (B), (C), (F), (G), (H)](d_x, d_y, d_z)^2 \rho = 0.$$

Or, si la surface donnée $U = 0$ et la surface voisine sont des surfaces consécutives d'une famille $r - f(x, y, z) = 0$; savoir, si l'équation de la surface donnée est $r - f(x, y, z) = 0$, et celle de la surface voisine $r + dr - f(x, y, z) = 0$, on a

$$\rho = \frac{dr}{V}, \quad V = \sqrt{X^2 + Y^2 + Z^2},$$

où dr est une constante; l'équation devient ainsi

$$[(A), \ldots](d_x, d_y, d_z)^2 \frac{1}{V} = 0,$$

où, à présent, X, Y, Z, a, b, c, f, g, h dénotent les coefficients différentiels de $f(x, y, z)$, ou (ce qui est la même chose) du paramètre r, considéré comme fonction des coordonnées x, y, z. Cette équation est donc une équation du troisième ordre, à laquelle doit satisfaire la fonction ρ; multipliée par V^5, elle est en effet l'équation [dont il s'agit], laquelle, comme j'ai déjà dit, contient le facteur V^2: donc pour écarter le dénominateur, il suffira de multiplier par V^3.

J'écris $\delta = Xd_x + Yd_y + Zd_z$,

et de là

$$\delta X,\ \delta Y,\ \delta Z = aX + hY + gZ,\quad hX + bY + fZ,\quad gX + fY + cZ,$$

respectivement. On trouve

$$d_x^2 \frac{1}{V} = -\frac{1}{V^3}(a^2 + h^2 + g^2 + \delta a) + \frac{3}{V^5}(\delta X)^2,$$

$$d_y d_x \frac{1}{V} = -\frac{1}{V^3}(gh + bf + cf + \delta f) + \frac{3}{V^5}\delta Y \delta Z,$$

ou, en écrivant $\omega = a + b + c$, $\bar{\omega} = \bar{a} + \bar{b} + \bar{c}$, ces valeurs deviennent

$$d_x^2 \frac{1}{V} = -\frac{1}{V^3}(a\omega - \bar{\omega} + \bar{a} + \delta a) + \frac{3}{V^5}(\delta X)^2,$$

$$d_y d_x \frac{1}{V} = -\frac{1}{V^3}(f\omega \qquad + \bar{f} + \delta f) + \frac{3}{V^5}\delta Y \delta Z.$$

et, en substituant ces valeurs, les termes en ω, $\bar{\omega}$ disparaissent d'eux-mêmes, et l'équation, multipliée seulement par $-V^3$, se réduit à

$$[(A), \ldots](\bar{a}, \ldots) + [(A), \ldots](\delta a, \ldots) - \frac{3}{V^2}[(A), \ldots](\delta X,\ \delta Y,\ \delta Z)^2 = 0,$$

où le premier terme est

$$(A)\bar{a} + (B)\bar{b} + (C)\bar{c} + 2(F)\bar{f} + 2(G)\bar{g} + 2(H)\bar{h},$$

et de même pour le second terme.

Or je trouve que l'on a identiquement

$$[(A), \ldots](\delta X,\ \delta Y,\ \delta Z)^2 = -V^2(A, \ldots)(\delta X,\ \delta Y,\ \delta Z)^2,$$

de manière que l'équation est

$$[(A), \ldots](\bar{a}, \ldots) + [(A), \ldots](\delta a, \ldots) + 3(A, \ldots)(\delta X,\ \delta Y,\ \delta Z)^2 = 0,$$

et, de plus, que l'on a identiquement

$$[(A), \ldots](\bar{a}, \ldots) = -(A, \ldots)(\delta X,\ \delta Y,\ \delta Z)^2 = 2\begin{vmatrix} \delta X, & \delta Y, & \delta Z \\ X, & Y, & Z \\ \bar{\delta} X, & \bar{\delta} Y, & \bar{\delta} Z \end{vmatrix},$$

où $\bar{\delta}X$, $\bar{\delta}Y$, $\bar{\delta}Z$ dénotent respectivement $\bar{a}X+\bar{h}Y+\bar{g}Z$, $\bar{h}X+\bar{b}Y+\bar{f}Z$, $\bar{g}X+\bar{f}Y+\bar{c}Z$; l'équation se réduit donc à

$$[(A), \ldots](\delta a, \ldots)+\Omega=0,$$

où Ω peut être exprimé à volonté sous l'une quelconque des trois formes

$$=+2\,[(A), \ldots](\bar{a}, \ldots),$$

$$=+2\,(A, \ldots \between \delta X,\ \delta Y,\ \delta Z)^2,$$

$$=-4\begin{vmatrix} \delta X, & \delta Y, & \delta Z \\ X, & Y, & Z \\ \bar{\delta}X, & \bar{\delta}Y, & \bar{\delta}Z \end{vmatrix}.$$

Prenant la première forme, l'équation est

$$[(A), \ldots](\delta a, \ldots)-2\,[(A), \ldots](\bar{a}, \ldots)=0,$$

ou, ce qui est la même chose,

$$\begin{aligned}(A)\,\delta a+(B)\,\delta b+(C)\,\delta c+2\,(F)\,\delta f+2\,(G)\,\delta g+2\,(H)\,\delta h\\ -2\,[(A)\,\bar{a}+(B)\,\bar{b}+(C)\,\bar{c}+2\,(F)\bar{f}+2\,(G)\bar{g}+2\,(H)\,\bar{h}]=0,\end{aligned}$$

où les coefficients sont des fonctions données de X, Y, Z, a, b, c, f, g, h, les coefficients différentiels de r du premier et du second ordre, et δ dénote $Xd_x+Yd_y+Zd_z$.

519.

ON CURVATURE AND ORTHOGONAL SURFACES.

[From the *Philosophical Transactions of the Royal Society of London*, vol. CLXIII. (for the year 1873), pp. 229—251. Received December 27, 1872,—Read February 13, 1873.]

THE principal object of the present Memoir is the establishment of the partial differential equation of the third order satisfied by the parameter of a family of surfaces belonging to a triple orthogonal system. It was first remarked by Bouquet that a given family of surfaces does not in general belong to an orthogonal system, but that (in order to its doing so) a condition must be satisfied; it was afterwards shown by Serret that the condition is that the parameter, considered as a function of the coordinates, must satisfy a partial differential equation of the third order: this equation was not obtained by him or the other French geometers engaged on the subject, although methods of obtaining it, essentially equivalent but differing in form, were given by Darboux and Levy; the last-named writer even found a particular form of the equation, viz. what the general equation becomes on writing therein $X=0$, $Y=0$ (X, Y, Z the first derived functions, or quantities proportional to the cosine-inclinations of the normal). Using Levy's method, I obtained the general equation, and communicated it to the French Academy, [518]. My result was, however, of a very complicated form, owing, as I afterwards discovered, to its being encumbered with the extraneous factor $X^2+Y^2+Z^2$; I succeeded, by some difficult reductions, in getting rid of this factor, and so obtaining the equation in the form given in the present memoir, viz.

$$((A),\ (B),\ (C),\ (F),\ (G),\ (H)\between \delta a,\ \delta b,\ \delta c,\ 2\delta f,\ 2\delta g,\ 2\delta h)$$
$$-\,2\,((A),\ (B),\ (C),\ (F),\ (G),\ (H)\between \bar{a},\ \ \bar{b},\ \ \bar{c},\ 2\bar{f}\ ,\ 2\bar{g}\ ,\ 2\bar{h}\ )=0:$$

but the method was an inconvenient one, and I was led to reconsider the question. The present investigation, although the analytical transformations are very long, is in theory extremely simple: I consider a given surface, and at each point thereof take along the normal an infinitesimal length ρ (not a constant, but an arbitrary function of the coordinates), the extremities of these distances forming a new surface, say the vicinal surface; and the points on the same normal being considered as corresponding points, say this is the conormal correspondence of vicinal surfaces. In order that the two surfaces may belong to an orthogonal system, it is necessary and sufficient that at each point of the given surface the principal tangents (tangents to the curves of curvature) shall correspond to the principal tangents at the corresponding point of the vicinal surface; and the condition for this is that ρ shall satisfy a partial differential equation of the second order,

$$((A),\ (B),\ (C),\ (F),\ (G),\ (H)\between d_x,\ d_y,\ d_z)^2\,\rho=0,$$

where the coefficients depend on the first and second differential coefficients of U, if $U=0$ is the equation of the given surface. Now, considering the given surface as belonging to a family, or writing its equation in the form $r-r(x,\ y,\ z)=0$ (the last r a functional symbol), the condition in order that the vicinal surface shall belong to this family, or say that it shall coincide with the surface $r+\delta r-r(x,\ y,\ z)=0$, is $\rho=\frac{\delta r}{V}$, where $V=\sqrt{X^2+Y^2+Z^2}$, if $X,\ Y,\ Z$ are the first differential coefficients of $r(x,\ y,\ z)$, that is, of the parameter r considered as a function of the coordinates; we have thus the equation

$$((A),\ (B),\ (C),\ (F),\ (G),\ (H)\between d_x,\ d_y,\ d_z)^2\,\frac{1}{V}=0,$$

viz. the coefficients being functions of the first and second differential coefficients of r, and V being a function of the first differential coefficients of r, this is in fact a relation involving the first, second, and third differential coefficients of r, or it is the partial differential equation to be satisfied by the parameter r considered as a function of the coordinates. After all reductions, this equation assumes the form previously mentioned.

Article Nos. 1 to 21. *On the Curvature of Surfaces.*

1. Curvature is a metrical theory having reference to the circle at infinity; each point in space may be regarded as the vertex of a cone passing through this circle, say the circular cone; a line and plane through the vertex are at right angles to each other when they are polar line and polar plane in regard to the cone; and so two lines or two planes are at right angles when they are harmonics in regard to the cone, that is, when each line lies in the polar plane, or each plane passes through the polar line of the other. A plane through the vertex meets the cone in two lines, which are the "circular lines" in the plane and through the point; a line through the vertex has through it two tangent planes, which might be called the "circular planes" of the point and through the line; but the term is hardly required. Lines in the plane and through the point, at right angles to each other, are also harmonics (polar lines) in regard to the two circular lines.

2. Consider now a surface, and any point thereof; we have at this point a tangent plane and a normal. The tangent plane meets the surface in a curve having

at the point a node, and the tangents to the two branches of the curve (being of course lines in the tangent plane) are the "chief tangents" of the surface at the point.

3. The chief tangents are the intersections of the tangent plane by a quadric cone, which may be called the chief cone; but it is important to observe that this cone is not independent of the particular form under which the equation of the surface is presented. To explain this, suppose that the rational equation of the surface is $U=0$; taking ξ, η, ζ as current coordinates measured from the point as origin, the equation of the chief cone is $(\xi\partial_x+\eta\partial_y+\zeta\partial_z)^2\,U=0$, where x, y, z denote the coordinates of the point. But it is in the sequel necessary to present the equation of the surface in a different manner; say we have an equation between the coordinates (x, y, z) and a parameter r (r being therefore in general an irrational function of x, y, z), which, when $r=r_1$, reduces itself to $U=0$: we have then $r=r_1$ as the equation of the surface; and the corresponding equation of the chief cone is $(\xi\partial_x+\eta\partial_y+\zeta\partial_z)^2\,r=0$; this is not the same as the cone $(\xi\partial_x+\eta\partial_y+\zeta\partial_z)^2\,U=0$, although of course it intersects the tangent plane in the same two lines, viz. the chief lines; and so in general there is a distinct chief cone corresponding to each form of the equation of the surface. But adopting a definite form of equation, we have a definite chief cone intersecting the tangent plane in the chief tangents.

4. Observe that the equations $U=0$, $r=r_1$, although each relating to one and the same surface, serve to represent this surface, and that in different ways, as belonging to a family of surfaces, viz. one of these is the family $U=$ const., and the other the family $r=$ const. In order to represent a given surface as belonging to a certain family, we need the irrational form of equation; thus r denoting the irrational function of x, y, z determined by the equation $\dfrac{x^2}{a+r}+\dfrac{y^2}{b+r}+\dfrac{z^2}{c+r}=1$, we have $r=0$ as the equation of the ellipsoid $\dfrac{x^2}{a}+\dfrac{y^2}{b}+\dfrac{z^2}{c}=1$, considered as belonging to a family of confocal quadrics.

5. Although at first sight presenting some difficulty, it is convenient to use the same letter r to denote the parameter considered as a function of the coordinates, and the special value of the parameter; thus in general the equation of a surface may be written $r(x, y, z)-r=0$ (in which form the first r may be regarded as a functional symbol), or simply $r-r=0$, viz. the first r here denotes the given function of (x, y, z), and the second r the particular value of the parameter.

6. By what precedes we have through the point and in the tangent plane two circular lines, the intersections of the tangent plane by the circular cone having the point for its vertex.

We have also through the point and in the tangent plane two other lines, termed the principal tangents, viz. the definition of these is that they are the double (or sibiconjugate) lines of the involution formed by the circular lines and the chief tangents, or, what is the same thing, they are the bisectors (and as such at right angles to each other) of the angles formed by the chief tangents.

7. The principal tangents may also be considered as the intersections of the tangent plane by a quadric cone, called the principal cone; this being a cone constructed by means of the circular cone and the chief cone, and thus depending on the particular chief cone, that is, on the form of the equation of the surface. The definition is that the principal cone is the locus of a line (through the point), such that the line itself, the perpendicular (or harmonic in regard to the circular cone) of the polar plane of the line in regard to the chief cone, and the normal of the surface are *in plano*.

8. Analytically, taking, as before, (x, y, z) for the coordinates of the point, and u, v, w as current coordinates measured from the point as origin, then the equation of the circular cone is $u^2 + v^2 + w^2 = 0$; and taking $Xu + Yv + Zw = 0$ for the equation of the tangent plane, and $(a, b, c, f, g, h \between u, v, w)^2 = 0$ for that of the chief cone, then, if the line be $u : v : w = \xi : \eta : \zeta$, we have

$$(a, .. \between \xi, \eta, \zeta \between u, v, w) = 0$$

for the equation of the polar plane, and thence

$$u : v : w = a\xi + h\eta + g\zeta : h\xi + b\eta + f\zeta : g\xi + f\eta + c\zeta$$

for those of the perpendicular, or harmonic in regard to the circular cone; also for the normal $u, v, w = X : Y : Z$; whence, if the three lines are *in plano*, we have

$$\begin{vmatrix} \xi & , & \eta & , & \zeta \\ a\xi + h\eta + g\zeta, & & h\xi + b\eta + f\zeta, & & g\xi + f\eta + c\zeta \\ X & , & Y & , & Z \end{vmatrix} = 0$$

as the equation of the principal cone. This is in the sequel written, for shortness, as

$$\begin{vmatrix} \xi, & \eta, & \zeta \\ \delta\xi, & \delta\eta, & \delta\zeta \\ X, & Y, & Z \end{vmatrix} = 0.$$

9. Consider any point P', not in general on the surface, in the neighbourhood of the point on the surface, say P; then the point P' has in regard to the surface a polar plane, which plane, however, is dependent on the particular form of equation—viz. x', y', z' being the coordinates of P', and U' the same function of these that U is of x, y, z, then the form $U = 0$ of the equation of the surface gives for P' the polar plane $(ud_{x'} + vd_{y'} + wd_{z'}) U' = 0$; and we may through P' draw hereto a perpendicular (or harmonic in regard to the circular cone), say this is the normal line of P'. Then for points P' in the neighbourhood of P, when these are such that their normal lines meet the normal at P, the locus of P' is the before-mentioned principal cone. The analytical investigation presents no difficulty.

10. Taking P' on the surface, the normal line of P' becomes the normal at a consecutive point P' of the surface (being now a line independent of the particular form of equation), and this normal meets the normal at P; that is, we have the

principal cone meeting the tangent plane in two lines, the principal tangents, such that at a consecutive point P' on either of these the normal meets the normal at P; viz. we have the principal tangents at the tangents of the two curves of curvature through the point P.

The plane through the normal and a principal tangent is termed a principal plane; we have thus at the point of the surface two principal planes, forming with the tangent plane an orthogonal triad of planes.

11. I proceed to further develop the theory, commencing with the following lemma:

Lemma. Given the line $Xu + Yv + Zw = 0$, and conic

$$(a, b, c, f, g, h \between u, v, w)^2 = 0,$$

then, to determine the coordinates (u_1, v_1, w_1), (u_2, v_2, w_2) of the points of intersection of the line and conic, we have

$$\begin{aligned}&(a, .. \between Y\zeta - Z\eta,\ Z\xi - X\zeta,\ X\eta - Y\xi)^2\\ &= (\xi u_1 + \eta v_1 + \zeta w_1)(\xi u_2 + \eta v_2 + \zeta w_2),\end{aligned}$$

or, what is the same thing, we have

$$(a, ... \between Y\zeta - Z\eta,\ Z\xi - X\zeta,\ X\eta - Y\xi)^2 = 0$$

as the equation, in line coordinates, of the two points of intersection. The proof is obvious.

12. Making the equations refer to a plane and a cone, and writing throughout ξ, η, ζ as current point coordinates, the theorem is:

Given the plane $X\xi + Y\eta + Z\zeta = 0$, and cone

$$(a, b, c, f, g, h \between \xi, \eta, \zeta)^2 = 0;$$

then, to determine the lines of intersection of the plane and cone, we have

$$(a, .. \between Y\zeta - Z\eta,\ Z\xi - X\zeta,\ X\eta - Y\xi)^2 = 0$$

as the equation of the pair of planes at right angles to the two lines respectively.

13. Denoting the coefficients by (a), (b), &c., that is, writing

$$\begin{aligned}&(a, .. \between Y\zeta - Z\eta,\ Z\xi - X\zeta,\ X\eta - Y\xi)^2\\ &= ((a), (b), (c), (f), (g), (h) \between \xi, \eta, \zeta)^2,\end{aligned}$$

the values of these are

$$\begin{aligned}(a) &= bZ^2 + cY^2 - 2fYZ,\\ (b) &= cX^2 + aZ^2 - 2gZX,\\ (c) &= aY^2 + bX^2 - 2hXY,\\ (f) &= -aYZ - fX^2 + gXY + hXZ,\\ (g) &= -bZX + fYX - gY^2 + hYZ,\\ (h) &= -cXY + fZX + gZY - hZ^2.\end{aligned}$$

We have the following identities:

$$(a)\,X+(h)\ \ Y+(g\,)\,Z=0,$$
$$(h)\,X+(b)\ \ Y+(f\,)\,Z=0,$$
$$(g)\,X+(f\,)\,Y+(c)\ \ Z=0,$$
$$\big((b)(c)-(f)^2,\ldots,\ (g)(h)-(a)(f),\ldots\big)=-(X^2,\ Y^2,\ Z^2,\ YZ,\ ZX,\ XY)\,\phi,$$

that is, $(b)(c)-(f)^2=-X^2\phi$ &c., where

$$\phi=(bc-f^2,\ldots gh-af,\ldots\between X,\ Y,\ Z)^2.$$

Writing also

$$aX+hY+gZ,\ hX+bY+fZ,\ gX+fY+cZ=\delta X,\ \delta Y,\ \delta Z,$$

and $X^2+Y^2+Z^2=V^2$; also $a+b+c=\omega$, then

$$(a)\ =(b+c)\ V^2-\omega X^2\ +X\delta X-Y\delta Y-Z\delta Z,$$
$$(b)\ =(c+a)\ V^2-\omega Y^2\ -X\delta X+Y\delta Y-Z\delta Z,$$
$$(c)\ =(a+b)\ V^2-\omega Z^2\ -X\delta X-Y\delta Y+Z\delta Z,$$
$$(f)=-fV^2\qquad -\omega YZ\ +Y\delta Z\ +Z\delta Y,$$
$$(g)\ =-gV^2\qquad -\omega ZX\ +Z\delta X\ +X\delta Z,$$
$$(h)\ =-hV^2\qquad -\omega XY+X\delta Y+Y\delta X.$$

14. I give also the following lemma:

Lemma. The condition in order that the plane $X\xi+Y\eta+Z\zeta=0$ may meet the cones

$$(A\,,\ B\,,\ C\,,\ F\,,\ G\,,\ H\between\xi,\ \eta,\ \zeta)^2=0,$$
$$(A',\ B',\ C',\ F',\ G',\ H'\between\xi,\ \eta,\ \zeta)^2=0$$

in two pairs of lines harmonically related to each other, is

$$(BC'+B'C-2FF',\ldots,\ GH'+G'H-AF'-A'F,\ldots\between X,\ Y,\ Z)^2=0.$$

Writing here

$$(A,\ldots\between Y\zeta-Z\eta,\ Z\xi-X\zeta,\ X\eta-Y\xi)^2$$
$$=\big((A),\ (B),\ (C),\ (F),\ (G),\ (H)\between\xi,\ \eta,\ \zeta)^2,$$

that is, $(A)=BZ^2+CY^2-2FYZ$, &c., the condition may be written

$$(A)\,A'+(B)\,B'+(C)\,C'+2\,(F)\,F'+2\,(G)\,G'+2\,(H)\,H'=0,$$

or say

$$\big((A),\ldots\between A',\ldots)=0\,;$$

and we may, it is clear, interchange the accented and unaccented letters respectively.

15. I take $r - r = 0$ for the equation of a surface, X, Y, Z for the first derived functions of r, (a, b, c, f, g, h) for the second derived functions. The equation of the tangent plane at the point (x, y, z), taking ξ, η, ζ as current coordinates measured from this point, is

$$X\xi + Y\eta + Z\zeta = 0;$$

the equation of the chief cone in regard to this form of the equation of the surface is

$$(a, b, c, f, g, h \between \xi, \eta, \zeta)^2 = 0,$$

and the equation of the circular cone is $\xi^2 + \eta^2 + \zeta^2 = 0$, or, what is the same thing,

$$(1, 1, 1, 0, 0, 0 \between \xi, \eta, \zeta)^2 = 0.$$

Imagine a quadric cone

$$(A, B, C, F, G, H \between \xi, \eta, \zeta)^2 = 0,$$

such that it meets the tangent plane in the sibiconjugate lines of the involution formed by the intersections of the tangent plane by the chief cone and the circular cone respectively; that is, in a pair of lines harmonically related to the intersections with the chief cone, and also to the intersections with the circular cone; the conditions are

$$((A), \ldots \between a, \ldots) = 0,$$

and

$$(A) + (B) + (C) = 0,$$

viz. if only these two conditions are satisfied the cone will intersect the tangent plane in the two principal tangents.

16. The principal cone, writing, for shortness,

$$a\xi + h\eta + g\zeta,\ h\xi + b\eta + f\zeta,\ g\xi + f\eta + c\zeta = \delta\xi,\ \delta\eta,\ \delta\zeta,$$

was before taken to be the cone

$$\begin{vmatrix} \xi, & \eta, & \zeta \\ \delta\xi, & \delta\eta, & \delta\zeta \\ X, & Y, & Z \end{vmatrix} = 0.$$

Representing this equation by

$$\tfrac{1}{2}(A, B, C, F, G, H \between \xi, \eta, \zeta)^2 = 0,$$

the expressions of the coefficients are

$$\begin{aligned} A &= 2hZ - 2gY, \\ B &= 2fX - 2hZ, \\ C &= 2gY - 2fX, \\ F &= hY - gZ - (b - c)X, \\ G &= fZ - hX - (c - a)Y, \\ H &= gX - fY - (a - b)Z. \end{aligned}$$

These values give

$$AX + HY + GZ = Z\delta Y - Y\delta Z,$$
$$HX + BY + FZ = X\delta Z - Z\delta X,$$
$$GX + FY + CZ = Y\delta X - X\delta Y;$$

whence also

$$(A, \ldots \between X, Y, Z)^2 = 0,$$

as is, in fact, at once obvious from the determinant-form; and also

$$A + B + C = 0.$$

17. Writing for shortness

$$(\bar{a}, \bar{b}, \bar{c}, \bar{f}, \bar{g}, \bar{h}) = (bc - f^2, ca - g^2, ab - h^2, gh - af, hf - bg, fg - ch),$$

we find

$$Aa + Hh + Gg = \omega(hZ - gY) + \bar{h}Z - \bar{g}Y,$$
$$Hh + Bb + Ff = \omega(fX - hZ) + \bar{f}X - \bar{h}Z,$$
$$Gg + Ff + Cc = \omega(gY - fX) + \bar{g}Y - \bar{f}X;$$

whence

$$(A, \ldots \between a, \ldots) = 0.$$

18. By what precedes, we have

$$((A), \ldots \between \xi, \eta, \zeta)^2 = 0$$

for the equation of the two principal planes, where the coefficients (A), (B), &c. are functions of A, B, &c. and of X, Y, Z, as mentioned above. These coefficients satisfy of course the several relations similar to those satisfied by (a), (b), &c., and other relations dependent on the expressions of A, B, &c. in terms of a, b, &c. and X, Y, Z.

19. Proceeding to consider the coefficients (A), (B), &c., we have then

$$(A) + (B) + (C) = (A + B + C)V^2 - (A, \ldots \between X, Y, Z)^2,$$

that is

$$(A) + (B) + (C) = 0.$$

Observing the relation $A + B + C = 0$, the equations analogous to

$(a) = (b + c)V^2 - (a + b + c)X^2 + \&c.$, are $(A) = -AV^2 + X\delta'X - Y\delta'Y - Z\delta'Z$, &c.

if for a moment we write $\delta'X$, $\delta'Y$, $\delta'Z$ to denote the functions

$$AX + HY + GZ,\ HX + BY + FZ,\ GX + FY + CZ.$$

But, from the above values, $X\delta'X + Y\delta'Y + Z\delta'Z = 0$, or the equation is $(A) = -AV^2 + 2X\delta'X$, that is $= -AV^2 + 2X(Z\delta Y - Y\delta Z)$. The equation for (F) is $(F) = -FV^2 + Y\delta'Z + Z\delta'Y$, where $Y\delta'Z + Z\delta'Y$ is $= Y(Y\delta X - X\delta Y) + Z(X\delta Z - Z\delta X)$, viz. this is

$$= (Y^2 - Z^2)\delta X - XY\delta Y + XZ\delta Z.$$

We have thus the system of equations

$$\begin{array}{llll}
(A) = -AV^2 & . & +2XZ\delta Y & -2XY\delta Z, \\
(B) = -BV^2 & -2YZ\delta X & . & +2XY\delta Z, \\
(C) = -CV^2 & +2YZ\delta X & -2XZ\delta Y & . \\
(F) = -FV^2 & +(Y^2-Z^2)\delta X & -XY\delta Y & +XZ\delta Z, \\
(G) = -GV^2 & +XY\delta X & +(Z^2-X^2)\delta Y & -YZ\delta Z, \\
(H) = -HV^2 & -XZ\delta X & +YZ\delta Y & +(X^2-Y^2)\delta Z.
\end{array}$$

20. We hence find

$$\begin{aligned}
(A)\,a+(H)\,h+(G)\,g &= -(Aa+Hh+Gg)\,V^2+(Z\delta Y-Y\delta Z)\,\delta X+XP, \\
(H)\,h+(B)\,b+(F)f &= -(Hh+Bb+Ff)\,V^2+(X\delta Z-Z\delta X)\,\delta Y+YQ, \\
(G)\,g+(F)f+(C)\,c &= -(Gg+Ff+Cc)\,V^2+(Y\delta X-X\delta Y)\,\delta Z+ZR,
\end{aligned}$$

if for shortness

$$\begin{aligned}
P &= (gY-hZ)\,\delta X+(aZ-gX)\,\delta Y+(hX-aY)\,\delta Z, \\
Q &= (fY-bZ)\,\delta X+(hZ-fX)\,\delta Y+(bX-hY)\,\delta Z, \\
R &= (cY-fZ)\,\delta X+(gZ-cX)\,\delta Y+(fX-gY)\,\delta Z.
\end{aligned}$$

Forming the sum $PX+QY+RZ$, the coefficient of δX is found to be

$$= -Z(hX+bY+fZ)+Y(gX+fY+cZ), \;= -Z\delta Y+Y\delta Z;$$

hence the whole is

$$= \delta X\,(Y\delta Z-Z\delta Y)+\delta Y\,(Z\delta X-X\delta Z)+\delta Z\,(X\delta Y-Y\delta X), \text{ which is } =0, \text{ that is,}$$

$$PX+QY+RZ=0.$$

21. Hence, adding, we find

$$((A),\ldots \between a,\ldots)=0;$$

viz. in this and the before-mentioned equation

$$(A)+(B)+(C)=0$$

we have the *à posteriori* verification that the cone $(A,\ldots\between \xi, \eta, \zeta)^2=0$ cuts the tangent plane in the double lines of the involution.

In what precedes I have given only those relations between the several sets of quantities a, $\bar{a}$, (a), A, (A), &c. which have been required for establishing the results last obtained; but there are various other relations required in the sequel, and which will be obtained as they are wanted.

The Conormal Correspondence of Vicinal Surfaces.
Art. Nos. 22 to 35.

22. We consider a surface $U=0$ (or $r=r$), and at each point P thereof measure along the normal an infinitesimal length ρ, dependent on the position of the point P (that is, ρ is a function of x, y, z). We have thus a point P', the coordinates of which are

$$x',\ y',\ z' = x+\rho\alpha,\ y+\rho\beta,\ z+\rho\gamma,$$

where α, β, γ are the cosine-inclinations of the normal, that is,

$$\alpha,\ \beta,\ \gamma = \frac{X}{V},\ \frac{Y}{V},\ \frac{Z}{V},\ \text{if } V=\sqrt{X^2+Y^2+Z^2};$$

the locus of P' is of course a surface, say the vicinal surface, and we require to find the direction of the normal at P', or, what is the same thing, the differential equation $X'dx'+Y'dy'+Z'dz'$ of the surface. We have

$$\begin{aligned}
dx' &= (1+d_x\rho\alpha)\,dx + d_y\rho\alpha\,.\,dy + d_z\rho\alpha\,.\,dz,\\
dy' &= d_x\rho\beta\,.\,dx + (1+d_y\rho\beta)\,dy + d_z\rho\beta\,.\,dz,\\
dz' &= d_x\rho\gamma\,.\,dx + d_y\rho\gamma\,.\,dy + (1+d_z\rho\gamma)\,dz,\\
0 &= X\,dx + Y\,dy + Z\,dz;
\end{aligned}$$

whence, eliminating dx, dy, dz, we have between dx', dy', dz' a linear equation, the coefficients of which may be taken to be X', Y', Z'. Taking these only as far as the first power of ρ, we have

$$X' = X(1+d_y\rho\beta+d_z\rho\gamma) - Yd_x\rho\beta - Zd_x\rho\gamma,$$

or, what is the same thing,

$$X' = X(1+d_x\rho\alpha+d_y\rho\beta+d_z\rho\gamma) - Xd_x\rho\alpha - Yd_x\rho\beta - Zd_x\rho\gamma,$$

with the like expressions for Y' and Z'. I proceed to reduce these. The formula for X' is

$$\begin{aligned}
X' = X\{1&+\rho(d_x\alpha+d_y\beta+d_z\gamma)+\alpha d_x\rho+\beta d_y\rho+\gamma d_z\rho\}\\
&-\rho(Xd_x\alpha+Yd_x\beta+Zd_x\gamma)-(\alpha X+\beta Y+\gamma Z)\,d_x\rho.
\end{aligned}$$

23. I write, for shortness, $\delta = Xd_x+Yd_y+Zd_z$, whence δX, δY, $\delta Z = aX+hY+gZ$, $hX+bY+fZ$, $gX+fY+cZ$, agreeing with the former significations of δX, δY, δZ; also Vd_xV, Vd_yV, $Vd_zV = \delta X$, δY, δZ, and $V\delta V = X\delta X+Y\delta Y+Z\delta Z$. It is now easy to form the values of

$$d_x\alpha,\ d_x\beta,\ d_x\gamma,\ \text{viz. these are } \frac{a}{V}-\frac{X\delta X}{V^3},\ \frac{h}{V}-\frac{Y\delta X}{V^3},\ \frac{g}{V}-\frac{Z\delta X}{V^3},$$

$$d_y\alpha,\ d_y\beta,\ d_y\gamma,\quad \frac{h}{V}-\frac{X\delta Y}{V^3},\ \frac{b}{V}-\frac{Y\delta Y}{V^3},\ \frac{f}{V}-\frac{Z\delta Y}{V^3},$$

$$d_z\alpha,\ d_z\beta,\ d_z\gamma,\quad \frac{g}{V}-\frac{X\delta Z}{V^3},\ \frac{f}{V}-\frac{Y\delta Z}{V^3},\ \frac{c}{V}-\frac{Z\delta Z}{V^3};$$

and hence

$$d_x\alpha + d_y\beta + d_z\gamma = \frac{a+b+c}{V} - \frac{\delta V}{V^2},$$

$$Xd_x\alpha + Yd_x\beta + Zd_x\gamma = \frac{\delta X}{V} - \frac{V^2}{V^3}\delta X, \ = 0,$$

$$\alpha d_x\rho + \beta d_y\rho + \gamma d_z\rho = \frac{1}{V}\delta\rho,$$

$$\alpha X + \beta Y + \gamma Z \qquad = V;$$

and we have

$$X' = X\left\{1 + \rho\left(\frac{a+b+c}{V} - \frac{\delta V}{V^2}\right) + \frac{1}{V}\delta\rho\right\} - Vd_x\rho,$$

with the like values of Y' and Z'. But we are only concerned with the ratios $X' : Y' : Z'$; whence, dividing the foregoing values by the coefficient in { }, and taking the second terms only to the first order in ρ, we have simply

$$X',\ Y',\ Z' = X - Vd_x\rho,\ Y - Vd_y\rho,\ Z - Vd_z\rho.$$

24. We may investigate the condition in order that the surface x', y', z' may be the consecutive surface $r + dr = r(x, y, z)$. This will be the case of

$$r + dr = r\left(x + \rho\frac{X}{V},\ \ y + \rho\frac{Y}{V},\ \ z + \rho\frac{Z}{V}\right),$$

that is, $r + dr = r + \rho V$, or $\rho = \frac{dr}{V}$. This value of ρ gives $d_x\rho = -\frac{dr}{V^2}d_xV = -\frac{\rho}{V^2}\delta X$, and similarly $d_y\rho = -\frac{\rho}{V^2}\delta Y$, $d_z\rho = -\frac{\rho}{V^2}\delta Z$; whence

$$X',\ Y',\ Z' = X + \frac{\rho}{V}\delta X,\ Y + \frac{\rho}{V}\delta Y,\ Z + \frac{\rho}{V}\delta Z,$$

which is as it should be, viz. these are what X, Y, Z become on substituting therein for x, y, z the values $x + \rho\alpha,\ y + \rho\beta,\ z + \rho\gamma$.

25. I return to the case where ρ is arbitrary, and I investigate the values of $a, b, \ldots$ for the point P' on the vicinal surface; say these are a', b', &c., then we have $a' = d_{x'}X'$ &c. The relation between the differentials may be written

$$\begin{aligned}
dx &= (1 - d_x\rho\alpha)\,dx' - d_y\rho\alpha\ dy' - d_z\rho\alpha\ dz',\\
dy &= -d_x\rho\beta\ dx' + (1 - d_y\rho\beta)\,dy' - d_z\rho\beta\ dz',\\
dz &= -d_x\rho\gamma\ dx' - d_y\rho\gamma\ dy' + (1 - d_z\rho\gamma)\,dz',
\end{aligned}$$

and we thence have $d_{x'} = (1 - d_x\rho\alpha)\, d_x - d_x\rho\beta d_y - d_x\rho\gamma d_z$ &c.; hence

$$\begin{aligned} a' &= \{(1 - d_x\rho\alpha)\, d_x - d_x\rho\beta d_y - d_x\rho\gamma d_z\}\,(X - Vd_x\rho) \\ &= (1 - d_x\rho\alpha)\, a - d_x\rho\beta \,.\, h - d_x\rho\gamma \,.\, g - d_x\,(Vd_x\rho) \\ &= a - \rho\,(ad_x\alpha + hd_x\beta + gd_x\gamma) \\ &\quad - (a\alpha + h\beta + g\gamma)\, d_x\rho \\ &\quad - \frac{1}{V}\,\delta X d_x\rho - Vd_x^2\rho\,; \end{aligned}$$

and similarly, $f' = d_{y'}Z'$ (or $d_{z'}Y'$), that is

$$\begin{aligned} f' &= f - \rho\,(gd_y\alpha + fd_y\beta + cd_y\gamma) \\ &\quad - (g\alpha + f\beta + c\gamma)\, d_y\rho \\ &\quad - \frac{1}{V}\,\delta Y d_z\rho - Vd_yd_z\rho. \end{aligned}$$

26. Completing the reduction, we find

$$\begin{aligned} a' &= a - \rho\left(\frac{a\omega - \bar{b} - \bar{c}}{V} - \frac{(\delta X)^2}{V^3}\right) - \frac{2}{V}\,\delta X d_x\rho - Vd_x^2\rho, \\ b' &= b - \rho\left(\frac{b\omega - \bar{c} - \bar{a}}{V} - \frac{(\delta Y)^2}{V^3}\right) - \frac{2}{V}\,\delta Y d_y\rho - Vd_y^2\rho, \\ c' &= c - \rho\left(\frac{c\omega - \bar{a} - \bar{b}}{V} - \frac{(\delta Z)^2}{V^3}\right) - \frac{2}{V}\,\delta Z d_z\rho - Vd_z^2\rho, \\ f' &= f - \rho\left(\frac{\omega f + \bar{f}}{V} - \frac{\delta Y\delta Z}{V^3}\right) - \frac{1}{V}\,(\delta Y d_z\rho + \delta Z d_x\rho\ ) - Vd_yd_z\rho, \\ g' &= g - \rho\left(\frac{\omega g + \bar{g}}{V} - \frac{\delta Z\delta X}{V^3}\right) - \frac{1}{V}\,(\delta Z d_x\rho + \delta X d_y\rho) - Vd_zd_x\rho, \\ h' &= h - \rho\left(\frac{\omega h + \bar{h}}{V} - \frac{\delta X\delta Y}{V^3}\right) - \frac{1}{V}\,(\delta X d_y\rho + \delta Y d_z\rho) - Vd_xd_y\rho\,; \end{aligned}$$

say these expressions are $a' = a + \Delta a$, &c.

27. Taking ξ, η, ζ for the coordinates, referred to P as origin, of a point on the given surface near to P, and ξ', η', ζ' for the coordinates, referred to P' as origin, of the corresponding point on the vicinal surface, the relation between ξ', η', ζ' and ξ, η, ζ is the same as that between dx', dy', dz' and dx, dy, dz; viz. we have

$$\begin{aligned} \xi &= (1 - d_x\rho\alpha)\,\xi' - d_y\rho\alpha \,.\, \eta' \quad - d_z\rho\alpha \,.\, \zeta', \\ \eta &= - d_x\rho\beta \,.\, \xi' \quad + (1 - d_y\rho\beta)\,\eta' - d_z\rho\beta \,.\, \zeta', \\ \zeta &= - d_x\rho\gamma \,.\, \xi' \quad - d_y\rho\gamma \,.\, \eta' \quad + (1 - d_z\rho\gamma)\,\zeta' : \end{aligned}$$

or conversely

$$\begin{aligned}
\xi' &= (1 + d_x\rho\alpha)\,\xi + d_y\rho\alpha\,.\,\eta + d_z\rho\alpha\,.\,\zeta,\\
\eta' &= d_x\rho\beta\,.\,\xi + (1 + d_y\rho\beta)\,\eta + d_z\rho\beta\,.\,\zeta,\\
\zeta' &= d_x\rho\gamma\,.\,\xi + d_y\rho\gamma\,.\,\eta + (1 + d_z\rho\gamma)\,\zeta,
\end{aligned}$$

say $\xi', \eta', \zeta' = \xi + \Delta\xi, \eta + \Delta\eta, \zeta + \Delta\zeta$; hence

$$\begin{aligned}
X'\xi' + Y'\eta' + Z'\zeta' &= (X - Vd_x\rho)(\xi + \Delta\xi) + \&c.\\
&= X\xi + Y\eta + Z\zeta\\
&\quad + X\Delta\xi + Y\Delta\eta + Z\Delta\zeta\\
&\quad - V(\xi d_x\rho + \eta d_y\rho + \zeta d_z\rho),
\end{aligned}$$

where second line is

$$\begin{aligned}
&(X\alpha + Y\beta + Z\gamma)(\xi d_x\rho + \eta d_y\rho + \zeta d_z\rho)\\
&+ \rho\,\{(Xd_x\alpha + Yd_x\beta + Zd_x\gamma)\,\xi + (Xd_y\alpha + Yd_y\beta + Zd_y\gamma)\,\eta + (Xd_z\alpha + Yd_z\beta + Zd_z\gamma)\,\zeta\}.
\end{aligned}$$

But

$$\begin{aligned}
Xd_x\alpha + Yd_x\beta + Zd_x\gamma &= \frac{\delta X}{V} - \frac{V^2}{V^3}\delta X = 0,\\
Xd_y\alpha + Yd_y\beta + Zd_y\gamma &= 0,\\
Xd_z\alpha + Yd_z\beta + Zd_z\gamma &= 0,
\end{aligned}$$

or second line is $= V(\xi d_x\rho + \eta d_y\rho + \zeta d_z\rho)$; and we have therefore

$$X'\xi' + Y'\eta' + Z'\zeta' = X\xi + Y\eta + Z\zeta.$$

We require

$$(A', B', C', F', G', H' \between \xi', \eta', \zeta')^2;$$

viz., to the first order in ρ, this is

$$\begin{aligned}
&= (A', \ldots \between \xi, \eta, \zeta)^2\\
&\quad + 2(A, \ldots \between \Delta\xi, \Delta\eta, \Delta\zeta \between \xi, \eta, \zeta).
\end{aligned}$$

28. Here second line is

$$2\{(A\xi + H\eta + G\zeta)\,\Delta\xi + (H\xi + B\eta + F\zeta)\,\Delta\eta + (G\xi + F\eta + C\zeta)\,\Delta\zeta\}:$$

but

$$A\xi + H\eta + G\zeta = Z\delta\eta - Y\delta\zeta + \begin{vmatrix} a, & h, & g \\ X, & Y, & Z \\ \xi, & \eta, & \zeta \end{vmatrix},$$

$$H\xi + B\eta + F\zeta = X\delta\zeta - Z\delta\xi + \begin{vmatrix} h, & b, & f \\ X, & Y, & Z \\ \xi, & \eta, & \zeta \end{vmatrix},$$

$$G\xi + F\eta + C\zeta = Y\delta\xi + X\delta\eta + \begin{vmatrix} g, & f, & c \\ X, & Y, & Z \\ \xi, & \eta, & \zeta \end{vmatrix},$$

whence term in { } is

$$\begin{vmatrix} \Delta\xi, & \Delta\eta, & \Delta\zeta \\ \delta\xi, & \delta\eta, & \delta\zeta \\ X, & Y, & Z \end{vmatrix} + \begin{vmatrix} a\Delta\xi+h\Delta\eta+g\Delta\zeta, & h\Delta\xi+b\Delta\eta+f\Delta\zeta, & g\Delta\xi+f\Delta\eta+c\Delta\zeta \\ X, & Y, & Z \\ \xi, & \eta, & \zeta \end{vmatrix},$$

which might be written

$$\begin{vmatrix} \Delta\xi, & \Delta\eta, & \Delta\zeta \\ \delta\xi, & \delta\eta, & \delta\zeta \\ X, & Y, & Z \end{vmatrix} - \begin{vmatrix} \delta\Delta\xi, & \delta\Delta\eta, & \delta\Delta\zeta \\ \xi, & \eta, & \zeta \\ X, & Y, & Z \end{vmatrix}$$

but it is perhaps more convenient to retain the second term in its original form.

29. As regards the first line, we have

$$\begin{aligned} A' &= 2h'Z' - 2g'Y' \\ &= 2(h+\Delta h)(Z - Vd_z\rho) - 2(g+\Delta g)(Y - Vd_y\rho) \\ &= A + 2(Z\Delta h - Y\Delta g) - 2V(hd_z\rho - gd_y\rho), \end{aligned}$$

with similar expressions for the other coefficients. Attending only to the terms of the first order, we thus obtain

$$\begin{aligned} A' &= A + 2(Z\Delta h - Y\Delta g) - 2V(hd_z - gd_y)\rho, \\ B' &= B + 2(X\Delta f - Z\Delta h) - 2V(fd_x - hd_z)\rho, \\ C' &= C + 2(Y\Delta g - X\Delta f) - 2V(gd_y - fd_x)\rho, \\ F' &= F + Y\Delta h - Z\Delta g - X(\Delta b - \Delta c) - V\left(hd_y - gd_z - (b-c)\,d_x\right)\rho, \\ G' &= G + Z\Delta f - X\Delta h - Y(\Delta c - \Delta a) - V\left(fd_z - hd_x - (c-a)\,d_y\right)\rho, \\ H' &= H + X\Delta g - Y\Delta f - Z(\Delta a - \Delta b) - V\left(gd_x - fd_y - (a-b)\,d_z\right)\rho, \end{aligned}$$

say these are $A' = A + \theta A$, &c., where θ is a functional symbol; we thus have

$$(A', \ldots \between \xi', \eta', \zeta')^2 = (A, \ldots \between \xi, \eta, \zeta)^2 + (\theta A, \ldots \between \xi, \eta, \zeta)^2 + 2(A, \ldots \between \xi, \eta, \zeta \between \Delta\xi, \Delta\eta, \Delta\zeta),$$

which, for shortness, I represent by

$$= (A, \ldots \between \xi, \eta, \zeta)^2 + (A'', \ldots \between \xi, \eta, \zeta)^2;$$

and I proceed to complete the calculation of the coefficients A'', B'', &c.

30. We have

$$\begin{aligned} A'' = \theta A &+ \text{coeff. } \xi^2 \text{ in} \\ & 2\left[(A\xi + H\eta + G\zeta)\Delta\xi + (H\xi + B\eta + F\zeta)\Delta\eta + (G\xi + F\eta + C\zeta)\Delta\zeta\right] \\ = \theta A &+ 2(Ad_x\rho\alpha + Hd_x\rho\beta + Gd_x\rho\gamma), \end{aligned}$$

that is,

$$\begin{aligned} A'' = \theta A &+ 2\frac{1}{V}(AX + HY + GZ)\,d_x\rho \\ &+ 2\rho\,(Ad_x\alpha + Hd_x\beta + Gd_x\gamma), \end{aligned}$$

where coeff. 2ρ is

$$=\frac{Aa+Hh+Gg}{V}-\frac{(AX+HY+GZ)\,\delta X}{V^3}$$

$$=\frac{1}{V}\left\{\omega\,(hZ-gY)+\bar{h}Z-\bar{g}Y\right\}-\frac{\delta X}{V^3}(Z\delta Y-Y\delta Z).$$

31. And similarly,

$$\begin{aligned}F''&=\theta F+(H\alpha+B\beta+F\gamma)\,d_z\rho+(G\alpha+F\beta+C\gamma)\,d_y\rho\\&\quad+\rho\,\{(Hd_z\alpha+Bd_z\beta+Fd_z\gamma)+(Gd_y\alpha+Fd_y\beta+Cd_y\gamma)\}\\&=\theta F+\frac{1}{V}\{(HX+BY+FZ)\,d_z\rho+(GX+FY+CZ)\,d_y\rho\}\\&\quad+\rho\left\{\frac{Hg+Bf+Fc}{V}-\frac{(HX+BY+FZ)\,\delta Z}{V^3}\right.\\&\quad\left.+\frac{Gh+Fb+Cf}{V}-\frac{(GX+FY+CZ)\,\delta Y}{V^3}\right\},\end{aligned}$$

$$\begin{aligned}Gh+Fb+Cf&=\ \ \omega\,(hY-bX)+\bar{h}Y-\bar{b}X+\bar{\omega}X+\bar{a}X+\bar{h}Y+\bar{g}Z,\\Hg+Bf+Fc&=-\omega\,(gZ-cX)-\bar{g}Y+\bar{c}X-\bar{\omega}X-\bar{a}X-\bar{h}Y-\bar{g}Z.\end{aligned}$$

Sum is $\omega\,\{hY-gZ-(b-c)X\}+\bar{h}Y-\bar{g}Z-(\bar{b}-\bar{c})\,X$, which is $=\omega F+\bar{h}Y-\bar{g}Z-(\bar{b}-\bar{c})\,X$: hence

$$\begin{aligned}F''&=\theta F+(X\delta Z-Z\delta X)\left(\frac{1}{V}d_z\rho-\frac{\rho\delta Z}{V^3}\right)+(Y\delta X-X\delta Y)\left(\frac{1}{V}d_y\rho-\frac{\rho\delta Y}{V^3}\right)\\&\quad+\frac{\rho}{V}\{\omega F+\bar{h}Y-\bar{g}Z-(\bar{b}-\bar{c})\,X\}.\end{aligned}$$

32. We may write

$$\begin{aligned}A''&=\theta A+2\left(\frac{1}{V}d_x\rho-\frac{\rho\delta X}{V^3}\right)(Z\delta Y-Y\delta Z)+\frac{\rho}{V}\{\omega A+\bar{A}\},\\B''&=\theta B+2\left(\frac{1}{V}d_y\rho-\frac{\rho\delta Y}{V^3}\right)(X\delta Z-Z\delta X)+\frac{\rho}{V}\{\omega B+\bar{B}\},\\C''&=\theta C+2\left(\frac{1}{V}d_z\rho-\frac{\rho\delta Z}{V^3}\right)(Y\delta X-X\delta Y)+\frac{\rho}{V}\{\omega C+\bar{C}\},\\F''&=\theta F+\ \left(\frac{1}{V}d_z\rho-\frac{\rho\delta Z}{V^3}\right)(X\delta Z-Z\delta X)+\left(\frac{1}{V}d_y\rho-\frac{\rho\delta Y}{V^3}\right)(Y\delta X-X\delta Y)+\frac{\rho}{V}\{\omega F+\bar{F}\},\\G''&=\theta G+\ \left(\frac{1}{V}d_x\rho-\frac{\rho\delta X}{V^3}\right)(Y\delta X-X\delta Y)+\left(\frac{1}{V}d_z\rho-\frac{\rho\delta Z}{V^3}\right)(Z\delta Y-Y\delta Z)+\frac{\rho}{V}\{\omega G+\bar{G}\},\\H''&=\theta H+\ \left(\frac{1}{V}d_y\rho-\frac{\rho\delta Y}{V^3}\right)(Z\delta Y-Y\delta Z)+\left(\frac{1}{V}d_x\rho-\frac{\rho\delta X}{V^3}\right)(X\delta Z-Z\delta X)+\frac{\rho}{V}\{\omega H+\bar{H}\},\end{aligned}$$

in which equations $\bar{A}$, $\bar{B}$, &c. are the like functions of $\bar{a}$, $\bar{b}$, &c. that A, B, &c. are of a, b, &c.; viz. $\bar{A}=2\bar{h}Z-2\bar{g}Y$, &c.

The value of θA is

$$\theta A = 2Z\left(\left\{-\frac{\rho}{V}(h\omega+\bar{h})+\frac{\rho}{V^3}\delta X\delta Y\right\}-\frac{1}{V}(\delta Y d_x\rho+\delta X d_y\rho)-Vd_xd_y\rho\right)$$

$$-2Y\left(\left\{-\frac{\rho}{V}(g\omega+\bar{g})+\frac{\rho}{V^3}\delta Z\delta X\right\}-\frac{1}{V}(\delta Z d_x\rho+\delta X d_z\rho)-Vd_xd_z\rho\right)$$

$$-2V(hd_z-gd_y)\rho,$$

which is

$$=-\frac{\rho}{V}(\omega A+\bar{A}) \qquad +\frac{2\rho}{V^3}\delta X(Z\delta Y-Y\delta Z)$$

$$-\frac{2\delta X}{V}(Zd_y-Yd_z)\rho \qquad -2V(hd_z-gd_y)\rho$$

$$-\frac{2}{V}(Z\delta Y-Y\delta Z)d_x\rho-2V(Zd_y-Yd_z)d_x\rho.$$

Hence the value of A'' is equal to the last-mentioned expression, together with the following terms:—

$$+\frac{\rho}{V}(\omega A+A)-\frac{2\rho}{V^3}\delta X(Z\delta Y-Y\delta Z)+\frac{2}{V}(Z\delta Y-Y\delta Z)d_x\rho,$$

which destroy certain of the foregoing ones; viz. we have

$$A''=\left(2Vg-\frac{2Z\delta X}{V}\right)d_y\rho-2\left(Vh-\frac{2Y\delta X}{V}\right)d_z\rho-2V(Zd_y-Yd_z)d_x\rho.$$

33. Similarly, the value of θF is

$$\theta F=Y\left(-\frac{\rho}{V}(h\omega+\bar{h})+\frac{\rho}{V^3}\delta X\delta Y-\frac{1}{V}(\delta Y d_x\rho+\delta X d_y\rho)-Vd_xd_y\rho\right)$$

$$-Z\left(-\frac{\rho}{V}(g\omega+\bar{g})+\frac{\rho}{V^3}\delta Z\delta X-\frac{1}{V}(\delta Z d_x\rho+\delta X d_z\rho)-Vd_xd_z\rho\right)$$

$$-X\left(-\frac{\rho}{V}\{(b-c)\omega+\bar{b}-\bar{c}\}+\frac{\rho\delta Y^2}{V^3}-\frac{\rho\delta Z^2}{V^3}-\frac{2}{V}\delta Y d_y\rho+\frac{2}{V}\delta Z d_z\rho-Vd_y^2\rho+Vd_z^2\rho\right)$$

$$-V\left(hd_y-gd_z-(b-c)d_x\right)\rho,$$

which is

$$=\frac{\rho}{V}(-F\omega-\bar{F})+\frac{\rho\delta Z}{V^3}(X\delta Z-Z\delta X)+\frac{\rho\delta Y}{V^3}(Y\delta X-X\delta Y)$$

$$+\left\{-\frac{1}{V}(Y\delta Y-Z\delta Z)+V(b-c)\right\}d_x\rho$$

$$+\left\{-\frac{Y}{V}\delta X+\frac{2X}{V}\delta Y \quad -Vh\right\}d_y\rho$$

$$+\left\{\frac{Z}{V}\delta X+\frac{2X}{V}\delta Y \quad +Vg\right\}d_z\rho$$

$$+(-VYd_xd_y+VZd_xd_z+VXd_y^2-VXd_z^2)\rho.$$

Hence F'' is equal to the foregoing expression, together with the following terms:—

$$+\frac{\rho}{V}(F\omega+\overline{F})-\frac{\rho\delta Z}{V^3}(X\delta Z-Z\delta X)-\frac{\rho\delta Y}{V^3}(Y\delta X-X\delta Y)$$

$$+\frac{1}{V}(Y\delta X-X\delta Y)\,d_y\rho+\frac{1}{V}(X\delta Z-Z\delta X)\,d_z\rho,$$

which destroy certain of the foregoing terms; viz. we thus have

$$F''=\left\{-\frac{1}{V}(Y\delta Y-Z\delta Z)+V(b-c)\right\}d_x\rho+\left\{\frac{X}{V}\delta Y-Vh\right\}d_y\rho+\left\{-\frac{X}{V}\delta Z-Vg\right\}d_z\rho$$
$$+V(-Yd_xd_y+Zd_zd_x+Xd_y{}^2-Xd_z{}^2)\,\rho.$$

34. We thus have

$$A''=\quad\cdot\quad 2\left(Vg-\frac{Z\delta X}{V}\right)d_y\rho-2\left(Vh-\frac{Y\delta X}{V}\right)d_z\rho+2V(-Zd_yd_x+Yd_zd_x)\,\rho,$$

$$B''=-2\left(Vf-\frac{Z\delta Y}{V}\right)d_x\rho\quad\cdot\quad+2\left(Vh-\frac{X\delta Y}{V}\right)d_z\rho+2V(-Xd_zd_y+Zd_xd_y)\,\rho,$$

$$C''=+2\left(Vf-\frac{Y\delta Z}{V}\right)d_x\rho-2\left(Vg-\frac{X\delta Z}{V}\right)d_y\rho\quad\cdot\quad+2V(-Yd_xd_z+Xd_yd_z)\,\rho,$$

$$F''=\left\{V(b-c)-\frac{1}{V}(Y\delta Y-Z\delta Z)\right\}d_x\rho-\left(Vh-\frac{X\delta Y}{V}\right)d_y\rho+\left(Vg-\frac{X\delta Z}{V}\right)d_z\rho$$
$$+V(-Yd_xd_y+Zd_xd_z+Xd_y{}^2-Xd_z{}^2)\,\rho,$$

$$G''=\left(Vh-\frac{Y\delta X}{V}\right)d_x\rho+\left\{V(c-a)-\frac{1}{V}(Z\delta Z-X\delta X)\right\}d_y\rho-\left(Vf-\frac{Y\delta Z}{V}\right)d_z\rho$$
$$+V(-Zd_yd_z+Xd_yd_x+Yd_z{}^2-Yd_x{}^2)\,\rho,$$

$$H''=-\left(Vg-\frac{Z\delta X}{V}\right)d_x\rho-\left(Vf-\frac{Z\delta Y}{V}\right)d_y\rho+\left\{V(a-b)-\frac{1}{V}(X\delta X-Y\delta Y)\right\}d_z\rho$$
$$+V(-Xd_zd_x+Yd_zd_y+Zd_x{}^2-Zd_y{}^2)\,\rho.$$

35. It will be recollected that we have $X'\xi'+Y'\eta'+Z'\zeta'=X\xi+Y\eta+Z\zeta$; by what precedes it appears that for the given surface the principal tangents are determined by the equations

$$(A,\ldots\between\xi,\ \eta,\ \zeta)^2=0,$$
$$X\xi+Y\eta+Z\zeta=0,$$

and that the lines which (in the tangent plane of the given surface) correspond to the principal tangents of the corresponding point of the vicinal surface are determined by the equations

$$(A,\ldots\between\xi,\ \eta,\ \zeta)^2+(A'',\ldots\between\xi,\ \eta,\ \zeta)^2=0,$$
$$X\xi+Y\eta+Z\zeta=0.$$

Condition that the two surfaces may belong to an Orthogonal System.
Art. Nos. 36 to 41.

36. The condition in order that the two surfaces may belong to an orthogonal system is that the principal tangents shall correspond, or, what is the same thing, the lines which (in the tangent plane of the given surface) correspond to the principal tangents of the vicinal surface must be the principal tangents of the given surface. When this is the case, the plane and cone $X\xi+Y\eta+Z\zeta=0$, $(A'',..\between\xi,\eta,\zeta)^2=0$ intersect in the principal tangents, and this is therefore the required condition.

The plane $X\xi+Y\eta+Z\zeta=0$ meets the cone $(A'',..\between\xi,\eta,\zeta)^2=0$ in the principal tangents, that is, in a pair of lines harmonically related to the circular lines and also to the chief tangents. Forming then the coefficients (A''), (B''), (C''), (F''), (G''), (H'') from A'', &c. in the same way as (A) &c. are formed from A, &c., that is, writing $(A'')=B''Z^2+C''Y^2-2F''YZ$, &c., the conditions are

$$(A'')+(B'')+(C'')=0,$$
$$((A''),\ldots\between a,\ldots)=0,$$

or, what is the same thing,

$$(A'',\ldots\between(a),\ldots)=0.$$

The former of these, as about to be shown, is satisfied identically; we have therefore the second of them, say $(A'',..\between(a),..)=0$ as the required condition.

37. We have

$$(A'')+(B'')+(C'')=(A''+B''+C'')V^2-(A'',..\between X,Y,Z)^2,$$

$$A''+B''+C''=\frac{2}{V}\{(Z\delta Y-Y\delta Z)\,d_x\rho+(X\delta Z-Z\delta X)\,d_y\rho+(Y\delta X-X\delta Y)\,d_z\rho\}.$$

Forming next the expressions of $A''X+H''Y+G''Z$ &c., and, for convenience, writing down separately the terms which involve the second differential coefficients of ρ, we have

$$A''X+H''Y+G''Z=$$
$$d_x\rho\,.\,V(hZ-gY)+d_y\rho\,[V\delta Z-Z\delta V+V(gX-aZ)]+d_z\rho\,[-(V\delta Y-Y\delta V)-V(hX-aY)],$$
$$H''X+B''Y+F''Z=$$
$$d_x\rho\,[-(V\delta Z-Z\delta V)-V(fY-bZ)]+d_y\rho\,.\,V(fX-hZ)+d_z\rho\,[(V\delta X-X\delta V)+V(hY-bX)],$$
$$G''X+F''Y+C''Z=$$
$$d_x\rho\,[V\delta Y-Y\delta V+V(fZ-cY)]+d_y\rho\,[-(V\delta X-X\delta V)-V(gZ-cX)]+d_z\rho\,.\,V(gY-fX),$$

where δV stands for $\frac{1}{V}(X\delta X+Y\delta Y+Z\delta Z)$, and where the three expressions contain also the following terms respectively:

$$\{\quad .\quad -YZd_y^2+YZd_z^2+(Y^2-Z^2)\,d_yd_z+XYd_zd_x-XZd_xd_y\}\,\rho,$$
$$\{ZXd_x^2\quad .\quad -ZXd_z^2-XYd_yd_z+(Z^2-X^2)\,d_zd_x+YZd_xd_y\}\,\rho,$$
$$\{-XYd_x^2+XYd_y^2\quad .\quad +XZd_yd_z-YZd_zd_x+(X^2-Y^2)\,d_xd_y\}\,\rho.$$

Multiplying by X, Y, Z, and adding, the terms which contain the second differential coefficients disappear, and we obtain

$$(A'', \ldots \between X, Y, Z)^2 = 2V[(Z\delta Y - Y\delta Z) d_x\rho + (X\delta Z - Z\delta X) d_y\rho + (Y\delta X - X\delta Y) d_z\rho];$$

so that, attending to the above value of $A'' + B'' + C''$, we have the required equation

$$(A'') + (B'') + (C'') = 0.$$

38. Proceeding now to form the value of $(A'', \ldots \between (a), \ldots)$, that is,

$$A''(a) + B''(b) + C''(c) + 2F''(f) + 2G''(g) + 2H''(h),$$

it will be shown that the terms involving the first differential coefficients of ρ vanish of themselves; as regards those containing the second differential coefficients, forming the auxiliary equations

$$\begin{aligned}
(A) &= 2(h)Z - 2(g)Y, \\
(B) &= 2(f)X - 2(h)Z, \\
(C) &= 2(g)Y - 2(f)X, \\
(F) &= (h)Y - (g)Z - ((b) - (c))X, \\
(G) &= (f)Z - (h)X - ((c) - (a))Y, \\
(H) &= (g)X - (f)Y - ((a) - (b))Z,
\end{aligned}$$

we find without difficulty that the terms in question (being, in fact, the complete value of the expression) are

$$= V((A), \ldots \between d_x, d_y, d_z)^2 \rho.$$

39. As regards the terms involving the first differential coefficients, observe that the whole coefficient of $d_x\rho$ is

$$\begin{aligned}
&-2(b)\left(Vf - \frac{Z\delta Y}{V}\right) \\
&+2(c)\left(Vg - \frac{Y\delta Z}{V}\right) \\
&+2(f)\left(V(b-c) - \frac{1}{V}(Y\delta Y - Z\delta Z)\right) \\
&+2(g)\left(Vh - \frac{Y\delta X}{V}\right) \\
&-2(h)\left(Vg - \frac{Z\delta X}{V}\right),
\end{aligned}$$

which is

$$\begin{aligned}
&= 2V\{(g)h + (f)b + (c)g - ((h)g + (b)f + (f)c)\} \\
&+ \frac{2}{V}\{Z((h)\delta X + (b)\delta Y + (f)\delta Z) - Y((g)\delta X + (f)\delta Y + (c)\delta Z)\}.
\end{aligned}$$

40. The reduction depends on the following auxiliary formulæ:

$$\begin{array}{l|l|l} a(a)+h(h)+g(g)=V\bar{\delta}V-X\bar{\delta}X, & a(h)+h(b)+g(f)=\quad -X\bar{\delta}Y, & a(g)+h(f)+g(c)=\quad -X\bar{\delta}Z, \\ h\;,,\;+b\;,,\;+f\;,,\;=\quad -Y\bar{\delta}X, & h\;,,\;+b\;,,\;+f\;,,\;=V\bar{\delta}V-Y\bar{\delta}Y, & h\;,,\;+b\;,,\;+f\;,,\;=\quad -Y\bar{\delta}Z, \\ g\;,,\;+f\;,,\;+c\;,,\;=\quad -Z\bar{\delta}X, & g\;,,\;+f\;,,\;+c\;,,\;=\quad -Z\bar{\delta}Y, & g\;,,\;+f\;,,\;+c\;,,\;=V\bar{\delta}V-Z\bar{\delta}Z, \end{array}$$

where, for shortness, I have written $\bar{\delta}X$, $\bar{\delta}Y$, $\bar{\delta}Z$ to stand for $\bar{a}X+\bar{h}Y+\bar{g}Z$, $\bar{h}X+\bar{h}Y+\bar{f}Z$, $\bar{g}X+\bar{f}Y+\bar{c}Z$ respectively, and $V\bar{\delta}V$ for $X\bar{\delta}X+Y\bar{\delta}Y+Z\bar{\delta}Z$, $(=\bar{a},\ldots\between X,\ Y,\ Z)^2$.

From these we immediately have

$$\begin{aligned} (a)\,\delta X+(h)\,\delta Y+(g)\,\delta Z &= V(X\bar{\delta}V-V\bar{\delta}X), \\ (h)\,\delta X+(b)\,\delta Y+(f)\,\delta Z &= V(Y\bar{\delta}V-V\bar{\delta}Y), \\ (g)\,\delta X+(f)\,\delta Y+(c)\,\delta Z &= V(Z\bar{\delta}V-V\bar{\delta}Z). \end{aligned}$$

Hence, in the coefficient of $d_x\rho$, the first line is

$$=2V(-Y\bar{\delta}Z+Z\bar{\delta}Y),$$

and the second line is

$$=\frac{2}{V}\{VZ(Y\bar{\delta}V-V\bar{\delta}Y)-VY(Z\bar{\delta}V-V\bar{\delta}Z)\},\ =2V(Y\bar{\delta}Z-Z\bar{\delta}Y);$$

so that the sum, or whole coefficient of $d_x\rho$, is $=0$. Similarly, the coefficients of $d_y\rho$ and $d_z\rho$ are each $=0$.

41. We have thus arrived at the equation

$$((A),\ldots\between d_x,\ d_y,\ d_z)^2\,\rho=0$$

as the condition to be satisfied by the normal distance ρ in order that the given surface and the vicinal surface may belong to an orthogonal system, viz. this is a partial differential equation of the second order, its coefficients being given functions of X, Y, Z, a, b, c, f, g, h, the first and second differential coefficients of r (where $r=r(x,\ y,\ z)$ is the equation of the given surface).

The equation, it is clear, may also be written in the two forms

$$(A,\ldots\between Zd_y-Yd_z,\ Xd_z-Zd_x,\ Yd_x-Xd_y)^2\,\rho=0,$$

and

$$\begin{vmatrix} P, & Q, & R \\ aP+hQ+gR, & hP+bQ+fR, & gP+fQ+cR \\ X, & Y, & Z \end{vmatrix}\rho=0,$$

if, for shortness, P, Q, R are written to denote Zd_y-Yd_z, Xd_z-Zd_x, Yd_x-Xd_y respectively, it being understood that in each of these forms the d_x, d_y, d_z operate on the ρ only.

Condition that a family of surfaces may belong to an Orthogonal System.
Art. Nos. 42 to 49.

42. We pass at once to the condition in order that the family of surfaces

$$r - r(x, y, z) = 0$$

may belong to an orthogonal system, viz. when the vicinal surface belongs to the family, we have ρ proportional to $\frac{1}{V}\left(=\frac{1}{\sqrt{X^2+Y^2+Z^2}}\right)$, and the condition is

$$((A), \ldots \between d_x, d_y, d_z)^2 \frac{1}{V} = 0,$$

where r is a function of (x, y, z), the first and second differential coefficients of which are $X, Y, Z, a, b, c, f, g, h$; and the equation is thus a partial differential equation of the third order satisfied by r. The form is by no means an inconvenient one, but it admits of further reduction.

43. We have $d_x \frac{1}{V}$, $d_y \frac{1}{V}$, $d_z \frac{1}{V}$ equal to $-\frac{1}{V^3}\delta X$, $-\frac{1}{V^3}\delta Y$, $-\frac{1}{V^3}\delta Z$ respectively, and thence

$$d_x{}^2 \frac{1}{V} = -\frac{1}{V^3}(a^2 + h^2 + g^2 + \delta a) + \frac{3}{V^5}(\delta X)^2,$$

$$d_y d_z \frac{1}{V} = -\frac{1}{V^3}(gh + bf + cf + \delta f) + \frac{3}{V^5}\delta Y \delta Z,$$

or, as these may be written,

$$d_x{}^2 \frac{1}{V} = -\frac{1}{V^3}(a\omega - \bar{\omega} + \bar{a} + \delta a) + \frac{3}{V^5}(\delta X)^2,$$

$$d_y d_z \frac{1}{V} = -\frac{1}{V^3}(f\omega \qquad + \bar{f} + \delta f) + \frac{3}{V^5}\delta Y \delta Z,$$

with the like values for $d_y{}^2 \frac{1}{V}$, &c. Substituting, the equation contains a term multiplied by ω, viz. this is

$$-\frac{1}{V^3}\omega\big((A), \ldots \between a, \ldots),$$

which vanishes; and a term multiplied by $\bar{\omega}$, viz. this is

$$\frac{1}{V^3}\bar{\omega}\big((A) + (B) + (C)\big),$$

which also vanishes. Writing down the remaining terms, and multiplying the whole by $-V^3$, the equation becomes

$$\big((A), \ldots \between \bar{a}, \ldots) + \big((A), \ldots \between \delta a, \ldots) - \frac{3}{V^2}\big((A), \ldots \between \delta X, \delta Y, \delta Z)^2 = 0.$$

44. The last term admits of reduction; from the equations

$$(A) = -AV^2 + 2XZ\delta Y - 2XY\delta Z, \text{ \&c., we find}$$

$$\begin{aligned}
(A)\,\delta X + (H)\,\delta Y + (G)\,\delta Z &= -\,V^2 (A\delta X + H\delta Y + G\delta Z) + V\delta V(Z\delta Y - Y\delta Z),\\
(H)\,\delta X + (B)\,\delta Y + (F)\,\delta Z &= -\,V^2 (H\delta X + B\delta Y + F\delta Z) + V\delta V(X\delta Z - Z\delta X),\\
(G)\,\delta X + (F)\,\delta Y + (C)\,\delta Z &= -\,V^2 (G\delta X + F\delta Y + C\delta Z) + V\delta V(Y\delta X - X\delta Y),
\end{aligned}$$

and hence

$$((A),\,..\between \delta X,\ \delta Y,\ \delta Z)^2 = -\,V^2 (A,\,...\between \delta X,\ \delta Y,\ \delta Z)^2;$$

wherefore the equation becomes

$$((A),\,..\between \bar{a},\,..) + ((A),\,..\between \delta a\,..) + 3\,(A,\,..\between \delta X,\ \delta Y,\ \delta Z)^2 = 0.$$

45. It will be shown that we have identically

$$((A),\,...\between \bar{a},\,...) = -\,(A,\,..\between \delta X,\ \delta Y,\ \delta Z)^2 = 2\begin{vmatrix} \delta X, & \delta Y, & \delta Z \\ X, & Y, & Z \\ \bar\delta X, & \bar\delta Y, & \bar\delta Z \end{vmatrix}.$$

The partial differential equation thus assumes the form

$$((A),\,..\between \delta a,\,...) + \Omega = 0,$$

where Ω may be expressed indifferently in the three forms,

$$\begin{aligned}
&= +\,2\,(A,\,..\between \bar{a},\,..),\\
&= +\,2\,(A,\,..\between \delta X,\ \delta Y,\ \delta Z)^2,\\
&= -\,4\begin{vmatrix} \delta X, & \delta Y, & \delta Z \\ X, & Y, & Z \\ \bar\delta X, & \bar\delta Y, & \bar\delta Z \end{vmatrix}.
\end{aligned}$$

46. Taking the first of these, the partial differential equation is

$$((A),\,...\between \delta a,\,..) - 2\,((A),\,..\between \bar{a},\,...) = 0\,;$$

or, written at full length, it is

$$\begin{aligned}
&(A)\,\delta a + (B)\,\delta b + (C)\,\delta c + 2\,(F)\,\delta f + 2\,(G)\,\delta g + 2\,(H)\,\delta h\\
&-2\,\{(A)\ \bar{a} + (B)\ \bar{b} + (C)\ \bar{c} + 2\,(F)\ \bar{f} + 2\,(G)\ \bar{g} + 2\,(H)\,\bar{h}\} = 0,
\end{aligned}$$

where the coefficients are given functions of X, Y, Z, a, b, c, f, g, h, the first and second differential coefficients of r; and δ is written to denote $Xd_x + Yd_y + Zd_z$.

47. It remains to prove the above-mentioned identities.

To reduce the term $(A, .. \between \delta X, \delta Y, \delta Z)^2$, we have

$$\begin{aligned} &A\delta X + H\delta Y + G\delta Z \\ &\quad = A(aX + hY + gZ) + H(hX + bY + fZ) + G(gX + fY + cZ) \\ &\quad = X\{\ \omega(hZ - gY) + \bar{h}Z - \bar{g}Y\} \\ &\qquad + Y\{-\omega(fY - bZ) - (\bar{f}Y - \bar{b}Z) - \omega Z - \bar{\delta}Z\} \\ &\qquad + Z\{\ \omega(fZ - cY) + fZ - \bar{c}Y + \bar{\omega}Y + \bar{\delta}Z\} \\ &\quad = \omega(Z\delta Y - Y\delta Z) + (Z\bar{\delta}Y - Y\bar{\delta}Z) + (Z\bar{\delta}Y - Y\bar{\delta}Z), \end{aligned}$$

that is,

$$A\delta X + H\delta Y + G\delta Z = \omega(Z\delta Y - Y\delta Z) + 2(Z\bar{\delta}Y - Y\bar{\delta}Z);$$

and similarly

$$H\delta X + B\delta Y + F\delta Z = \omega(X\delta Z - Z\delta X) + 2(X\bar{\delta}Z - Z\bar{\delta}X),$$

$$G\delta X + F\delta Y + C\delta Z = \omega(Y\delta X - X\delta Y) + 2(Y\bar{\delta}X - X\bar{\delta}Y),$$

whence

$$(A, .. \between \delta X, \delta Y, \delta Z)^2 = -2 \begin{vmatrix} \delta X, & \delta Y, & \delta Z \\ X, & Y, & Z \\ \bar{\delta}X, & \bar{\delta}Y, & \bar{\delta}Z \end{vmatrix}.$$

48. Now, from the equations $AX + HY + GZ = Z\delta Y - Y\delta Z$, &c. we have for the value of twice the foregoing determinant

$$\begin{aligned} 2\,\text{det.} = 2\{&(\bar{a}X + \bar{h}Y + \bar{g}Z)(AX + HY + GZ) \\ &+ (\bar{h}X + \bar{b}Y + \bar{f}Z)(HX + BY + FZ) \\ &+ (\bar{g}X + \bar{f}Y + \bar{c}Z)(GX + FY + CZ)\}; \end{aligned}$$

and subtracting herefrom the function $((A), .. \between \bar{a}, ..)$, which is

$$\begin{aligned} = \quad & (BZ^2 + CY^2 - 2FYZ)\bar{a} \\ & + (CX^2 + AZ^2 - 2GZX)\bar{b} \\ & + (AY^2 + BX^2 - 2HYZ)\bar{c} \\ & + 2(-AYZ - FX^2 + GXY + HXZ)\bar{f} \\ & + 2(-BZX + FXY - GY^2 + HYZ)\bar{g} \\ & + 2(-CXY + FXZ + GYZ - HZ^2)\bar{h}, \end{aligned}$$

the difference is found to be

$$\begin{aligned} = \quad & \bar{a}\{(A, .. \between X, Y, Z)^2 + AV^2\} \\ & + \bar{b}\{(A, .. \between X, Y, Z)^2 + BV^2\} \\ & + \bar{c}\{(A, .. \between X, Y, Z)^2 + CV^2\} \\ & + 2\bar{f}\{(A + B + C)YZ + FV^2\} \\ & + 2\bar{g}\{(A + B + C)ZX + GV^2\} \\ & - 2\bar{h}\{(A + B + C)XY + HV^2\}, \end{aligned}$$

which, on account of $(A,..\between X,\ Y,\ Z)^2=0$, and $A+B+C=0$, reduces itself to

$$(A,..\between \bar{a},\ldots).\ V^2.$$

49. We have

$$\begin{aligned} A\bar{a}+H\bar{h}+G\bar{g} = &\ \ \bar{a}\,(2hZ-2gY) \\ &+\bar{h}\,(\ gX-\ fY-(a-b)\,Z) \\ &+g\,(\ fZ-\ hX-(c-a)\,Y) \\ = &\ \ X\,(g\bar{h}-h\bar{g}) \\ &+Y\,(a\bar{g}-g\bar{a}-(g\bar{a}+f\bar{h}+c\bar{g})) \\ &+Z\,(h\bar{a}-a\bar{h}+(h\bar{a}+b\bar{h}+f\bar{g})); \end{aligned}$$

or, observing that in the coefficients of Y and Z the second terms each vanish, this is

$$A\bar{a}+H\bar{h}+G\bar{g}=X\,(\bar{h}g-\bar{g}h)+Y\,(\bar{g}a-\bar{a}g)+Z\,(\bar{a}h-\bar{h}a);$$

and similarly

$$H\bar{h}+B\bar{b}+F\bar{f}=X\,(\bar{b}f-\bar{f}b)+Y\,(\bar{f}h-\bar{h}f)+Z\,(\bar{h}b-\bar{b}h),$$

$$G\bar{g}+H\bar{f}+C\bar{c}=X\,(\bar{f}c-\bar{c}f)+Y\,(\bar{c}g-\bar{g}c)+Z\,(\bar{g}f-\bar{f}g).$$

Adding these equations, the coefficient of X is the difference of two expressions each of which vanishes; and the like as regards the coefficients of Y and Z; that is, we have

$$(A,..\between \bar{a},..)=0;$$

and consequently

$$2\begin{vmatrix} \delta X, & \delta Y, & \delta Z \\ X, & Y, & Z \\ \bar{\delta} X, & \bar{\delta} Y, & \bar{\delta} Z \end{vmatrix} = ((A),..\between \bar{a},\ldots) = -(A,\ldots\between \delta X,\ \delta Y,\ \delta Z)^2,$$

the required relation.

520.

ON THE CENTRO-SURFACE OF AN ELLIPSOID.

[From the *Transactions of the Cambridge Philosophical Society*, vol. XII. Part I. (1873), pp. 319—365. Read March 7, 1870.]

THE Centro-surface of any given surface is the locus of the centres of curvature of the given surface, or say it is the locus of the intersections of consecutive normals, (the normals which intersect the normal at any particular point of the surface being those at the consecutive points along the two curves of curvature respectively which pass through the point on the surface). The terms, *normal*, *centre of curvature*, *curve of curvature*, may be understood in their ordinary sense, or in the generalised sense referring to the case where the Absolute (instead of being the imaginary circle at infinity) is any quadric surface whatever; viz. the normal at any point of a surface is here the line joining that point with the pole of the tangent plane in respect of the quadric surface called the Absolute: and of course the centre of curvature and curve of curvature refer to the normal as just defined.

The question of the centro-surface of a quadric surface has been considered in the two points of view, viz. 1°, when the terms "normal," &c. are used in the ordinary sense, and the equation of the quadric surface (assumed to be an ellipsoid) is taken to be $\frac{X^2}{a^2}+\frac{Y^2}{b^2}+\frac{Z^2}{c^2}=1$; 2°, when the Absolute is the surface $X^2+Y^2+Z^2+W^2=0$, and the equation of the quadric surface is taken to be $\alpha X^2+\beta Y^2+\gamma Z^2+\delta W^2=0$: in the first of them by Salmon, *Quart. Math. Jour.* t. II. pp. 217—222 (1858), and in the second by Clebsch, *Crelle*, t. LXII. pp. 64—107 (1863): see also Salmon's *Solid Geometry*, 2nd Ed. 1865, pp. 143, 402, &c. In the present Memoir, as shown by the title, the quadric surface is taken to be an Ellipsoid; and the question is considered exclusively from the first point of view: the theory is further developed in various respects, and in particular as regards the nodal curve upon the centro-surface: the distinction of real and imaginary is of course attended to. The new results suitably modified would be applicable to the theory treated from the second point of view; but I do not on the present occasion attempt so to present them.

The Ellipsoid; Parameters ξ, η, &c. Art. Nos. 1—6.

1. The position of a point (X, Y, Z) on the ellipsoid

$$\frac{X^2}{a^2}+\frac{Y^2}{b^2}+\frac{Z^2}{c^2}=1$$

may be determined by means of the parameters, or elliptic coordinates, ξ, η; viz. these are such that we have

$$\frac{X^2}{a^2+\xi}+\frac{Y^2}{b^2+\xi}+\frac{Z^2}{c^2+\xi}=1,$$

$$\frac{X^2}{a^2+\eta}+\frac{Y^2}{b^2+\eta}+\frac{Z^2}{c^2+\eta}=1;$$

or, what is the same thing, ξ, η are the roots of the quadric equation

$$\frac{X^2}{a^2+v}+\frac{Y^2}{b^2+v}+\frac{Z^2}{c^2+v}=1.$$

(In its actual form this is a cubic equation, but there is a root $v=0$, which is to be thrown out, and the quadric equation is thus

$$\begin{aligned}&v^2\\ &+v\,(a^2+b^2+c^2-X^2-Y^2-Z^2)\\ &+\{b^2c^2+c^2a^2+a^2b^2-(b^2+c^2)\,X^2-(c^2+a^2)\,Y^2-(a^2+b^2)\,Z^2\}=0,\end{aligned}$$

or putting

$$\begin{aligned}P&=a^2+b^2+c^2,\\ Q&=b^2c^2+c^2a^2+a^2b^2,\\ R&=a^2b^2c^2,\end{aligned}$$

the equation is

$$v^2+v\,(P-X^2-Y^2-Z^2)+Q-(b^2+c^2)\,X^2-(c^2+a^2)\,Y^2-(a^2+b^2)\,Z^2=0.)$$

2. It is convenient to write throughout

$$\begin{aligned}b^2-c^2&=\alpha,\\ c^2-a^2&=\beta,\\ a^2-b^2&=\gamma,\end{aligned}$$

(whence $\alpha+\beta+\gamma=0$).

As usual, a is taken to be the greatest and c the least of the semi-axes; we have thus α, γ each of them positive, and β negative, $= -\beta'$ where β' is a positive quantity $= \alpha + \gamma$. A distinction arises in the sequel between the two cases $a^2 + c^2 > 2b^2$ and $a^2 + c^2 < 2b^2$, but the two cases are not essentially different, and it is convenient to assume $a^2 + c^2 > 2b^2$, that is, $a^2 - b^2 > b^2 - c^2$ or $\gamma > \alpha$, say $\gamma - \alpha$ positive. The limiting case $a^2 + c^2 = 2b^2$ or $\gamma = \alpha$ requires special consideration.

3. We have

$$-\beta\gamma\, X^2 = a^2 (a^2 + \xi)(a^2 + \eta),$$
$$-\gamma\alpha\, Y^2 = b^2 (b^2 + \xi)(b^2 + \eta),$$
$$-\alpha\beta\, Z^2 = c^2 (c^2 + \xi)(c^2 + \eta).$$

It is in fact easy to verify that these values satisfy as well the equation of the ellipsoid as the assumed equations defining the elliptic coordinates ξ, η. We may also obtain the relations

$$X^2 + Y^2 + Z^2 = a^2 + b^2 + c^2 + \xi + \eta,$$
$$a^2X^2 + b^2Y^2 + c^2Z^2 = a^4 + b^4 + c^4 + b^2c^2 + c^2a^2 + a^2b^2 + (a^2 + b^2 + c^2)(\xi + \eta) + \xi\eta.$$

These, however, are obtained more readily from the equation in v, viz. the roots thereof being ξ, η, we have

$$-\xi - \eta = a^2 + b^2 + c^2 - X^2 - Y^2 - Z^2,$$
$$\xi\eta = b^2c^2 + c^2a^2 + a^2b^2 - (b^2 + c^2) X^2 - (c^2 + a^2) Y^2 - (a^2 + b^2) Z^2,$$

which lead at once to the relations in question.

4. Considering ξ as constant, the locus of the point (X, Y, Z) is the intersection of the ellipsoid with the confocal ellipsoid

$$\frac{X^2}{a^2 + \xi} + \frac{Y^2}{b^2 + \xi} + \frac{Z^2}{c^2 + \xi} = 1;$$

viz. this is one of the curves of curvature through the point; and similarly considering η as constant, the locus of the point is the intersection of the ellipsoid with the confocal ellipsoid

$$\frac{X^2}{a^2 + \eta} + \frac{Y^2}{b^2 + \eta} + \frac{Z^2}{c^2 + \eta} = 1;$$

viz. this is the other of the curves of curvature through the point.

5. If instead of ξ and η we write h and k, we may consider h as extending between the values $-a^2$, $-b^2$, and k as extending between the values $-b^2$, $-c^2$.

$h =$ const. will thus give the series of curves of curvature one of which is the section by the plane $X = 0$, or ellipse semi-axes b, c; say this is the *minor-mean* series. In particular $h = -a^2$ gives the ellipse just referred to; and $h = -b^2$, or say $h = -b^2 - \epsilon$, gives two detached portions of the ellipse semi-axes a, c; viz. each of these portions extends from an umbilicus above the plane of xy, through the extremity of the semi-axis a, to an umbilicus below the plane of xy.

And in like manner $k=$ const. gives the series of curves of curvature one of which is the section by the plane $Z=0$, or ellipse semi-axes a, b; say this is the *major-mean* series. In particular $k=-c^2$ gives the ellipse just referred to; and $k=-b^2$, or say $k=-b^2+\epsilon$, gives the remaining portions of the ellipse semi-axes a, c; viz. these are two portions each extending from an umbilicus above the plane of xy, through the extremity of the semi-axis c, to an umbilicus above the plane of xy.

The ellipse last referred to may be called the umbilicar section, the other two principal sections being the major-mean section and the minor-mean section respectively.

In the limiting case $h=k=-b^2$, we have the umbilici, viz. these are given by

$$\frac{X^2}{a^2}=-\frac{\gamma}{\beta},\quad Y=0,\quad \frac{Z^2}{c^2}=-\frac{\alpha}{\beta}.$$

The two series of curves of curvature cover the whole real surface of the ellipsoid; so that at any real point thereof we have $\xi=h$, $\eta=k$, or else $\xi=k$, $\eta=h$, where h, k are negative real values lying within the foregoing limits $-a^2$, $-b^2$ and $-b^2$, $-c^2$ respectively. But observe that ξ, η taken separately may each extend between the limits $-a^2$, $-c^2$.

6. Suppose $\xi=\eta$, the equation in v will have equal roots, or the condition is

$$(P-X^2-Y^2-Z^2)^2=4\{Q-(b^2+c^2)X^2-(c^2+a^2)Y^2-(a^2+b^2)Z^2\},$$

viz. this surface by its intersection with the ellipsoid determines the envelope of the curves of curvature. This envelope is in fact a system of eight imaginary lines, four of them belonging to one of the systems of right lines on the ellipsoid, the other four to the other of the systems of right lines. For in the values of X^2, Y^2, Z^2 writing $\eta=\xi$, we find

$$\pm\sqrt{-\beta\gamma}\,\frac{X}{a}=a^2+\xi,$$

$$\pm\sqrt{-\gamma\alpha}\,\frac{Y}{b}=b^2+\xi,$$

$$\pm\sqrt{-\alpha\beta}\,\frac{Z}{c}=c^2+\xi,$$

or representing for shortness the left-hand functions by $\pm X'$, $\pm Y'$, $\pm Z'$, the eight lines are

$a^2+\xi=\ \ X'$	$=\ \ X'$	$=-X'$	$=-X'$
$b^2+\xi=\ \ Y'$	$=-Y'$	$=\ \ Y'$	$=-Y'$
$c^2+\xi=\ \ Z'$	$=-Z'$	$=-Z'$	$=\ \ Z'$

$a^2+\xi=-X'$	$=\ \ X'$	$=\ \ X'$	$=-X'$
$b^2+\xi=\ \ Y'$	$=-Y'$	$=\ \ Y'$	$=-Y'$
$c^2+\xi=\ \ Z'$	$=\ \ Z'$	$=-Z'$	$=-Z'$,

so that in the two tetrads each line intersects the four lines of the other tetrad, but it does not intersect the remaining three lines of its own tetrad. The intersections are four points corresponding to $\xi = -a^2$, being the imaginary umbilici in the plane $X = 0$: four to $\xi = -b^2$, being the real umbilici in the plane $Y = 0$: four to $\xi = -c^2$, being the imaginary umbilici in the plane $Z = 0$: and four corresponding to $\xi = \infty$, which may be called the umbilici at infinity [1].

Sequential and Concomitant Centro-curves. Art. No. 7.

7. Consider any particular curve of curvature; the normals at the several points thereof successively intersect each other in a series of points forming a curve; and we have thus, corresponding to the particular curve of curvature, a curve on the centro-surface, which curve may be called the *sequential centro-curve.* Again the same normals, viz. those at the several points of the particular curve of curvature, are intersected, the normal at each point by the consecutive normal belonging to the other curve of curvature through that point; and we have thus, corresponding to the particular curve of curvature, a curve on the centro-surface, which curve may be called the *concomitant centro-curve.* If instead of a single curve of curvature we consider the whole series, say of the major-mean curves of curvature, we have a series of major-mean sequential centro-curves, and also a series of major-mean concomitant centro-curves; and similarly considering the series of the minor-mean curves of curvature we have a series of minor-mean sequential centro-curves and also a series of minor-mean concomitant curves; the configuration of the several curves will be discussed further on, but it may be convenient to remark here that the centro-surface may be considered as consisting of two portions, say,

(A) locus of the major-mean sequential centro-curves; and also of the minor-mean concomitant centro-curves;

(B) locus of the minor-mean sequential centro-curves, and also of the major-mean concomitant centro-curves.

Investigation of expressions for the Coordinates of a point on the Centro-surface. Art. Nos. 8 to 13.

8. Consider the normal at the point (X, Y, Z). Taking in the first instance (x, y, z) as current coordinates, the equations are

$$\frac{x-X}{\frac{X}{a^2}} = \frac{y-Y}{\frac{Y}{b^2}} = \frac{z-Z}{\frac{Z}{c^2}}, \ = \lambda \text{ suppose},$$

[1] According to Salmon, *Solid Geometry*, [2nd Ed. 1865], p. 229, the number of umbilici for a surface of the nth order is $= n(10n^2 - 25n + 16)$; viz. for $n = 2$, this is $= 12$, as in the ordinary theory, not recognizing the umbilici at infinity. But whether properly umbilici or not, the 4 points which I call the umbilici at infinity do in the present theory present themselves in like manner with the 12 umbilici.

or, what is the same thing,

$$x = X\left(1+\frac{\lambda}{a^2}\right),\quad y = Y\left(1+\frac{\lambda}{b^2}\right),\quad z = Z\left(1+\frac{\lambda}{c^2}\right).$$

Suppose now that the normal meets the consecutive normal, or normal at the point $X+dX$, $Y+dY$, $Z+dZ$; and let x, y, z belong to the point of intersection of the two normals; we must have

$$0 = dX\left(1+\frac{\lambda}{a^2}\right)+\frac{X}{a^2}d\lambda,$$

$$0 = dY\left(1+\frac{\lambda}{b^2}\right)+\frac{Y}{b^2}d\lambda,$$

$$0 = dZ\left(1+\frac{\lambda}{c^2}\right)+\frac{Z}{c^2}d\lambda,$$

which determine the direction of the consecutive point; the equations in fact give

$$0 = \begin{vmatrix} dX, & \dfrac{dX}{a^2}, & \dfrac{X}{a^2} \\ dY, & \dfrac{dY}{b^2}, & \dfrac{Y}{b^2} \\ dZ, & \dfrac{dZ}{c^2}, & \dfrac{Z}{c^2} \end{vmatrix},$$

or, what is the same thing,

$$0 = \begin{vmatrix} a^2dX, & dX, & X \\ b^2dY, & dY, & Y \\ c^2dZ, & dZ, & Z \end{vmatrix},$$

which is the differential equation of the curve of curvature. This equation must therefore be satisfied by taking for $X+dX$, $Y+dY$, $Z+dZ$, the coordinates of the consecutive point along either of the curves of curvature,—say along that which is the intersection with the surface

$$\frac{X^2}{a^2+\eta}+\frac{Y^2}{b^2+\eta}+\frac{Z^2}{c^2+\eta}=1.$$

9. To verify this, observe that we then have

$$\frac{XdX}{a^2}+\frac{YdY}{b^2}+\frac{ZdZ}{c^2}=0,$$

$$\frac{XdX}{a^2+\eta}+\frac{YdY}{b^2+\eta}+\frac{ZdZ}{c^2+\eta}=0;$$

or, what is the same thing,

$$XdX : YdY : ZdZ = a^2(a^2+\eta)\,\alpha : b^2(b^2+\eta)\,\beta : c^2(c^2+\eta)\,\gamma.$$

But from the equations $-\beta\gamma X^2 = a^2(a^2+\xi)(a^2+\eta)$ &c., these become

$$XdX : YdY : ZdZ = \frac{X^2}{a^2+\xi} : \frac{Y^2}{b^2+\xi} : \frac{Z^2}{c^2+\xi},$$

or, what is the same thing,

$$dX : dY : dZ = \frac{X}{a^2+\xi} : \frac{Y}{b^2+\xi} : \frac{Z}{c^2+\xi};$$

and, substituting these values in the determinant equation, it becomes

$$0 = \frac{XYZ}{(a^2+\xi)(b^2+\xi)(c^2+\xi)} \begin{vmatrix} a^2, & 1, & a^2+\xi \\ b^2, & 1, & b^2+\xi \\ c^2, & 1, & c^2+\xi \end{vmatrix},$$

which is identically true, since evidently the determinant vanishes.

10. Proceeding with the solution, we have from the three equations

$$XdX + YdY + ZdZ + \lambda\left(\frac{XdX}{a^2} + \frac{YdY}{b^2} + \frac{ZdZ}{c^2}\right) + d\lambda\left(\frac{X^2}{a^2} + \frac{Y^2}{b^2} + \frac{Z^2}{c^2}\right) = 0,$$

and observing that from the equation

$$X^2 + Y^2 + Z^2 = a^2 + b^2 + c^2 + \xi + \eta,$$

considering therein η as constant, we have

$$XdX + YdY + ZdZ = \tfrac{1}{2}d\xi,$$

the equation becomes

$$\tfrac{1}{2}d\xi + d\lambda = 0;$$

and the three equations then are

$$0 = dX\left(1 + \frac{\lambda}{a^2}\right) - \tfrac{1}{2}\frac{X}{a^2}\,d\xi, \text{ \&c.},$$

or say

$$0 = dX\,(a^2+\lambda) - \tfrac{1}{2}X\,d\xi, \text{ \&c.}$$

But from the equation $-\beta\gamma X^2 = a^2(a^2+\xi)(a^2+\eta)$, considering therein η as a constant, we have

$$\frac{dX}{X} = \frac{\tfrac{1}{2}d\xi}{a^2+\xi},$$

and the equations thus become

$$0 = \frac{a^2+\lambda}{a^2+\xi} - 1, \text{ \&c.},$$

viz. these are all satisfied if only $\lambda = \xi$.

11. The coordinates of the point of intersection of the two normals thus are

$$x = X\left(1 + \frac{\xi}{a^2}\right), \quad y = Y\left(1 + \frac{\xi}{b^2}\right), \quad z = Z\left(1 + \frac{\xi}{c^2}\right),$$

or squaring, and substituting for X^2, &c., their values as given by

$$-\beta\gamma X^2 = a^2(a^2+\xi)(a^2+\eta), \text{ \&c.,}$$

the equations become

$$\begin{aligned} -\beta\gamma\, a^2x^2 &= (a^2+\xi)^3(a^2+\eta), \\ -\gamma\alpha\, b^2y^2 &= (b^2+\xi)^3(b^2+\eta), \\ -\alpha\beta\, c^2z^2 &= (c^2+\xi)^3(c^2+\eta), \end{aligned}$$

viz. these equations give (x, y, z) the coordinates of a point on the centro-surface, the intersection of the normal at the point (X, Y, Z) of the ellipsoid, (determined by the parameters ξ, η), by the normal at the consecutive point along the curve of curvature

$$\frac{X^2}{a^2+\eta} + \frac{Y^2}{b^2+\eta} + \frac{Z^2}{c^2+\eta} = 1,$$

or say η is the sequential parameter [1].

Of course by interchanging ξ and η we should obtain the coordinates of the point of intersection of the normal at the same point (X, Y, Z) by the normal at the consecutive point along the other curve of curvature: ξ being in this case the sequential parameter.

12. I stop for a moment to consider the foregoing two equations

$$\lambda = \xi, \quad d\lambda = -\tfrac{1}{2}d\xi,$$

which at first sight appear inconsistent. But observe that in the foregoing solution λ is the parameter of the point (x, y, z) of the centro-surface considered as a point on the normal at (X, Y, Z); $\lambda + d\lambda$ is the parameter of the *same point* considered as a point on the normal at the consecutive point $(X+dX,\ Y+dY,\ Z+dZ)$: the value $\lambda + d\lambda = \xi + d\xi$ would belong to a different point, viz. the consecutive point of the centro-surface considered as a point on the consecutive normal—wherefore the $d\lambda$ of the solution ought not to be $= d\xi$. In further explanation, observe that the equations

$$x = X\left(1+\frac{\lambda}{a^2}\right), \text{ \&c. where } \lambda = \xi,$$

if we pass from (x, y, z) to the consecutive point on the centro-surface, give

$$dx = dX\left(1+\frac{\lambda}{a^2}\right) + \frac{X}{a^2}d\xi;$$

but since by what precedes,

$$0 = dX\left(1+\frac{\lambda}{a^2}\right) - \tfrac{1}{2}\frac{X}{a^2}d\xi,$$

this is

$$dx = \tfrac{3}{2}\frac{X}{a^2}d\xi.$$

[1] The expressions are given in effect, but not explicitly, Salmon, p. 143.

Or since

$$a^2x = X(a^2+\xi),$$

this is

$$\frac{dx}{x} = \tfrac{3}{2}\frac{d\xi}{a^2+\xi};$$

and similarly

$$\frac{dy}{y} = \tfrac{3}{2}\frac{d\xi}{b^2+\xi},$$

$$\frac{dz}{z} = \tfrac{3}{2}\frac{d\xi}{c^2+\xi},$$

which are the correct values of dx, dy, dz as derived from the equations

$$-\beta\gamma a^2x^2 = (a^2+\xi)^3(a^2+\eta), \text{ \&c.}$$

13. The equations $-\beta\gamma a^2x^2 = (a^2+\xi)^3(a^2+\eta)$, &c. give expressions for the coordinates (x, y, z) of a point on the centro-surface in terms of the two parameters (ξ, η): the elimination of (ξ, η) from these equations will therefore lead to the equation of the surface; but the discussion of the surface may also be effected by means of these expressions for the coordinates in terms of the two parameters.

Discussion by means of the equations $-\beta\gamma a^2x^2 = (a^2+\xi)^3(a^2+\eta)$, &c.; *Principal Sections, &c.* Art. Nos. 14 to 24 *(several subheadings).*

14. To fix the ideas consider the section of the surface by the plane $z=0$; we have in the surface $z=0$, that is, $\xi=-c^2$, or else $\eta=-c^2$, values which give respectively

$$\begin{array}{ll|ll} -\beta\gamma a^2x^2 = & -\beta^3(a^2+\eta), & -\beta\gamma a^2x^2 = & -\beta(a^2+\xi)^3, \\ -\gamma\alpha b^2y^2 = & \alpha^3(b^2+\eta); & -\gamma\alpha b^2y^2 = & \alpha(b^2+\xi)^3; \end{array}$$

or, what is the same thing,

$$\begin{array}{ll|ll} \dfrac{\gamma}{\beta^2}a^2x^2 = & a^2+\eta, & \gamma a^2x^2 = & (a^2+\xi)^3, \\ \dfrac{\gamma}{\alpha^2}b^2y^2 = & -b^2-\eta; & \gamma b^2y^2 = & -(b^2+\xi)^3. \end{array}$$

The first set of equations gives

$$\frac{a^2x^2}{\beta^2} + \frac{b^2y^2}{\alpha^2} = 1,$$

which is the equation of an ellipse.

The second set gives

$$(ax)^{\frac{2}{3}} + (by)^{\frac{2}{3}} = \gamma^{\frac{1}{3}},$$

or in a rationalised form

$$(a^2x^2+b^2y^2-\gamma^2)^3 + 27a^2b^2\gamma^2x^2y^2 = 0,$$

which is the equation of an evolute of an ellipse.

15. The ellipse $\frac{a^2x^2}{\beta^2}+\frac{b^2y^2}{\alpha^2}=1$ is a cuspidal curve on the surface, and the section by the plane $z=0$ is consequently made up of this ellipse counting three times, and of the evolute; it is therefore of the twelfth order; and the order of the surface is in fact $=12$.

It is clear that the section of the centro-surface arises from the section $\frac{X^2}{a^2}+\frac{Y^2}{b^2}=1$, viz. the normal at any point of this ellipse lies in the plane $Z=0$, and its intersection by a normal at the consecutive point of the ellipse gives a point of the evolute; the evolute being thus the sequential centro-curve of this section: the intersection by the normal at the consecutive point on the other curve of curvature gives a point on the ellipse $\frac{a^2x^2}{\beta^2}+\frac{b^2y^2}{\alpha^2}=1$, which ellipse is therefore the concomitant centro-curve. Observe that this other curve of curvature cuts the ellipse $\frac{X^2}{a^2}+\frac{Y^2}{b^2}=1$ at right angles, and that the normals at the consecutive points above and below the point on the ellipse will meet each other and also the normal at the point of the same ellipse at the point on the ellipse $\frac{a^2x^2}{\beta^2}+\frac{b^2y^2}{\alpha^2}=1$: this shows that the last-mentioned ellipse is a cuspidal curve on the centro-surface.

16. The three principal sections of the centro-surface are consequently

$$x=0,\quad \frac{b^2y^2}{\gamma^2}+\frac{c^2z^2}{\beta^2}=1,\quad\text{and } (by)^{\frac{2}{3}}+(cz)^{\frac{2}{3}}=\alpha^{\frac{2}{3}};$$

$$y=0,\quad \frac{c^2z^2}{\alpha^2}+\frac{a^2x^2}{\gamma^2}=1,\quad\text{and } (cz)^{\frac{2}{3}}+(ax)^{\frac{2}{3}}=\beta^{\frac{2}{3}};$$

$$z=0,\quad \frac{a^2x^2}{\beta^2}+\frac{b^2y^2}{\alpha^2}=1,\quad\text{and } (ax)^{\frac{2}{3}}+(by)^{\frac{2}{3}}=\gamma^{\frac{2}{3}};$$

viz. each section is made up of an ellipse counting three times and of an evolute (of an ellipse). I have for shortness represented the three evolutes by their irrational equations. It will presently appear that the section (imaginary) by the plane infinity is of the like character.

17. Considering only the positive directions of the axes, we have on each axis two points, viz.

$$\text{axis of } x,\quad x=\frac{\gamma}{a},\quad x=-\frac{\beta}{a};$$

$$\text{axis of } y,\quad y=\frac{\alpha}{b},\quad y=\frac{\gamma}{b};$$

$$\text{axis of } z,\quad z=-\frac{\beta}{c},\quad z=\frac{\alpha}{c};$$

through each of which, in the two different planes through the axis respectively, there passes an ellipse and an evolute. In the assumed case $a^2+c^2>2b^2$, the disposition of the points is as shown in the figure.

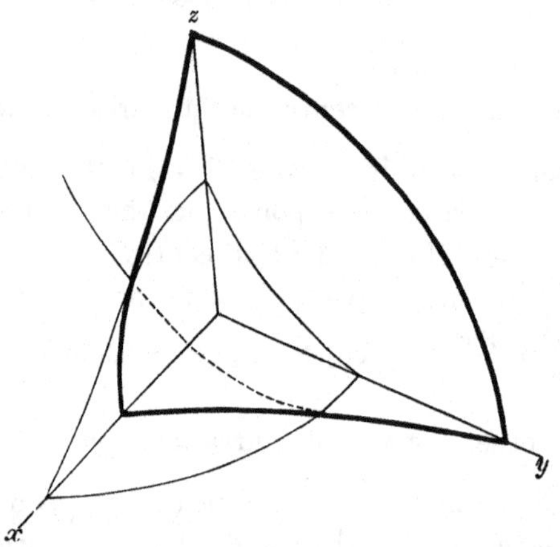

Plane of xz, evolute is outside ellipse,
yz, „ inside „
xy, „ cuts „ ;

but in the contrary case $a^2+c^2<2b^2$, the disposition is

Plane of xz, evolute is outside ellipse,
yz, „ cuts „
xy, „ is inside „ ;

there is no real difference, and to fix the ideas I attend exclusively to the first-mentioned case

$$a^2+c^2>2b^2.$$

18. In each of the principal planes, the evolute and ellipse, quà curves of the orders 6 and 2 respectively, intersect in twelve points, 3 in each quadrant; viz. of the 3 points two unite together into a twofold point or point of contact, and the third is a point of simple intersection; assuming for the moment that this is so, the figure at once shows that in the plane of xz or umbilicar plane the contact is real, the intersection imaginary; in the plane of xy, or major-mean plane, the contact is imaginary, the intersection real; but in the plane of yz or minor-mean plane the contact and intersection are each imaginary. The contacts arise, as will appear, from the umbilici of the ellipsoid, and may be termed "umbilicar centres," or "omphaloi;" the simple intersections "points of outcrop," or simply "outcrops." By what precedes there are in the umbilicar plane, four real umbilicar centres (in each quadrant one); and in the major-mean plane four real outcrops (in each quadrant one); the other umbilicar centres and outcrops are respectively imaginary.

19. The surface consists of two sheets intersecting in a *nodal curve* connecting the outcrop with the umbilicar centre. As to the form of this curve there is a cusp at the outcrop; and the curve does not terminate at the umbilicar centre but, on passing it, from crunodal becomes acnodal (viz. there is no longer through the curve any real sheet of the surface): moreover the curve is not at the umbilicar centre

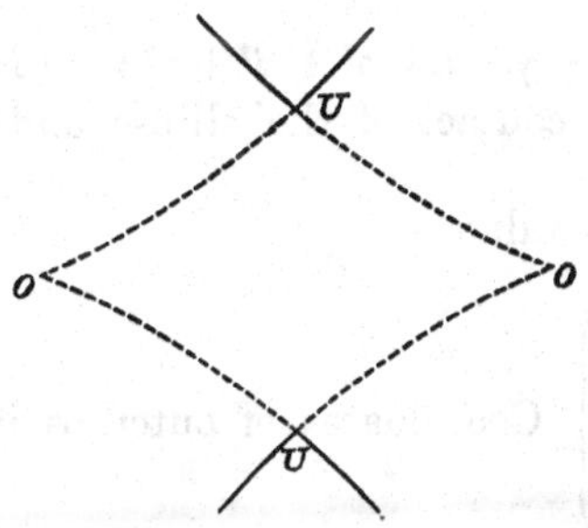

perpendicular to the plane of xz, and there is consequently on the opposite side of the plane a symmetrically situate branch of the curve, viz. the umbilicar centre is a node on the nodal curve. Completing the curve, the nodal curve consists of two distinct portions, one on the positive side of the plane of yz or minor-mean plane consisting of two cuspidal branches as shown in the figure; the other a symmetrically situate portion on the negative side of the minor-mean plane.

Intersections of Evolute and Ellipse.

20. Consider in the plane of xy the ellipse and evolute,

$$\frac{a^2x^2}{\beta^2}+\frac{b^2y^2}{\alpha^2}=1,\quad (a^2x^2+b^2y^2-\gamma^2)^3+27\gamma^2a^2b^2x^2y^2=0.$$

First, these are satisfied by

$$\left.\begin{aligned} a^2x^2 &= -\frac{\beta^3}{\gamma},\\ b^2y^2 &= -\frac{\alpha^3}{\gamma},\end{aligned}\right\}\ \text{Coordinates of Umbilicar centres in plane of } xy \text{ (imaginary)},$$

viz. the equations respectively become

$$-\frac{\beta}{\gamma}-\frac{\alpha}{\gamma}=1,\quad \left(-\frac{\beta^3+\alpha^3}{\gamma}-\gamma^2\right)^3+27\alpha^3\beta^3=0,$$

the first of which is $\alpha+\beta+\gamma=0$, and the second is $(\alpha^3+\beta^3+\gamma^3)^3-27\alpha^3\beta^3\gamma^3=0$. But the equation $\alpha+\beta+\gamma=0$ gives $\alpha^3+\beta^3+\gamma^3=3\alpha\beta\gamma$, and the two equations are thus identically satisfied. Moreover the condition for a contact is at once found to be

$$\beta^2\left[(a^2x^2+b^2y^2-\gamma^2)^2+9\gamma^2b^2y^2\right]=\alpha^2\left[(a^2x^2+b^2y^2-\gamma^2)^2+9\gamma^2a^2x^2\right],$$

or, what is the same thing,

$$(\alpha^2-\beta^2)(a^2x^2+b^2y^2-\gamma^2)^2+9\gamma^2(\alpha^2a^2x^2-\beta^2b^2y^2)=0;$$

and substituting the foregoing values, this is

$$(\alpha^2-\beta^2)\left(-\frac{\alpha^3+\beta^3}{\gamma}-\gamma^2\right)^2+9\gamma^2\frac{-\alpha^2\beta^3+\alpha^3\beta^2}{\gamma}=0,$$

that is,

$$\frac{\alpha^2-\beta^2}{\gamma^2}(\alpha^3+\beta^3+\gamma^3)^2+9\gamma\alpha^2\beta^2(\alpha-\beta)=0,$$

which, putting therein $\alpha+\beta=-\gamma$, and $\alpha^3+\beta^3+\gamma^3=3\alpha\beta\gamma$, is also satisfied; that is, the points in question are points of contact of the ellipse and evolute.

21. Secondly, consider the values

$$\left.\begin{aligned} a^2x^2 &= -\frac{\beta^3}{\gamma}\left(\frac{\gamma-\alpha}{\alpha-\beta}\right)^3, \\ b^2y^2 &= -\frac{\alpha^3}{\gamma}\left(\frac{\beta-\gamma}{\alpha-\beta}\right)^3, \end{aligned}\right\} \text{ Coordinates of outcrops in plane of } xy \text{ (real).}$$

Substituting in the equation of the ellipse, we have

$$\alpha(\beta-\gamma)^3+\beta(\gamma-\alpha)^3+\gamma(\alpha-\beta)^3=0,$$

which is

$$(\beta-\gamma)(\gamma-\alpha)(\alpha-\beta)(\alpha+\beta+\gamma)=0,$$

or the equation is satisfied identically: and substituting in the equation of the evolute, we have first

$$a^2x^2+b^2y^2-\gamma^2=-\frac{\alpha^3(\beta-\gamma)^3+\beta^3(\gamma-\alpha)^3+\gamma^3(\alpha-\beta)^3}{\gamma(\alpha-\beta)^3};$$

which in virtue of $\alpha(\beta-\gamma)+\beta(\gamma-\alpha)+\gamma(\alpha-\beta)=0$ becomes

$$a^2x^2+b^2y^2-\gamma^2=-\frac{3\alpha\beta\gamma(\beta-\gamma)(\gamma-\alpha)(\alpha-\beta)}{\gamma(\alpha-\beta)^3},$$

$$=-\frac{3\alpha\beta(\beta-\gamma)(\gamma-\alpha)}{(\alpha-\beta)^2},$$

and then, completing the substitution, it is seen that the equation of the evolute is also satisfied. The points last considered are simple intersections, and we have thus the complete number $(8+4, =12)$ of the intersections of the evolute and ellipse.

22. We have α, γ positive, β negative; whence $\alpha-\beta$ is positive, $\beta-\gamma$ negative; $\gamma-\alpha\,(=a^2+c^2-2b^2)$ is positive, and hence, the outcrops in the plane of xy are real; the umbilicar centres are imaginary for this plane, but real for the plane of zx, the coordinates being

$$\left.\begin{aligned} c^2z^2 &= -\frac{\alpha^3}{\beta}, \\ a^2x^2 &= -\frac{\gamma^3}{\beta}, \end{aligned}\right\} \text{ Coordinates of Umbilicar centres in plane of } xz \text{ (real).}$$

Nodes of the Evolute.

23. The Evolute is a curve with four nodes, all of them imaginary; viz. for the evolute in the plane of xy, the equation of which is

$$(a^2x^2+b^2y^2-\gamma^2)^3+27\gamma^2a^2b^2x^2y^2=0,$$

these are

$$\left.\begin{array}{l} a^2x^2=-\gamma^2, \\ b^2y^2=-\gamma^2, \end{array}\right\} \text{ Coordinates of Nodes of evolute in plane of } xy \text{ (imaginary)},$$

in fact these values satisfy as well the equation of the evolute, as the two derived equations

$$6a^2x\,[(a^2x^2+b^2y^2-\gamma^2)^2+9\gamma^2b^2y^2]=0,$$
$$6b^2y\,[(a^2x^2+b^2y^2-\gamma^2)^2+9\gamma^2a^2x^2]=0,$$

or the points in question are nodes of the evolute.

The evolute has the four cusps on the axes and two cusps at infinity, in all 6 cusps as just mentioned; it has 4 nodes: and the order being 6, the class is

$$30-2.4-3.6,\ =4.$$

Section by the plane infinity.

24. The surface itself is finite, and the section by the plane infinity is therefore imaginary; but by what precedes the nodal curve must have real points at infinity, viz. there must be real acnodal points on this imaginary section. The section by the plane infinity resembles in fact the principal sections; viz. writing successively $\xi=\infty$, and $\eta=\infty$, we have

$$-\beta\gamma a^2x^2 : -\gamma\alpha b^2y^2 : -\alpha\beta c^2z^2 = a^2+\eta : b^2+\eta : c^2+\eta$$

or

$$=(a^2+\xi)^3 : (b^2+\xi)^3 : (c^2+\xi)^3,$$

giving respectively

$$a^2x^2+b^2y^2+c^2z^2=0, \quad \text{and} \quad (a\alpha x)^{\frac{2}{3}}+(b\beta y)^{\frac{2}{3}}+(c\gamma z)^{\frac{2}{3}}=0,$$

where the first equation represents an imaginary conic which counts three times; and the second equation, the rationalised form of which is

$$(a^2\alpha^2x^2+b^2\beta^2y^2+c^2\gamma^2z^2)^3-27a^2b^2c^2\alpha^2\beta^2\gamma^2x^2y^2z^2=0,$$

an imaginary evolute. The conic and evolute have four contacts and four simple intersections (in all $4.2+4=12$ intersections) which are all of them imaginary. But the evolute has four real nodes (acnodes) $a^2\alpha^2x^2=b^2\beta^2y^2=c^2\gamma^2z^2$; or, what is the same thing, there are four real lines $a^2\alpha^2x^2=b^2\beta^2y^2=c^2\gamma^2z^2$, which are respectively asymptotes of the nodal curve: viz. inasmuch as the equation of the surface contains only the squares

x^2, y^2, z^2, the lines in question will be not merely parallel to, but will be, the asymptotes of the nodal curve.

The plane infinity may be reckoned as a principal plane, and we may say that in each of the four principal planes there are four umbilicar centres, four outcrops, and four evolute-nodes.

The generation of the surface considered geometrically. Art. Nos. 25 to 28.

25. I have deferred until this point the discussion of the generation of the centro-surface by means of the centro-curves, for the reason that it can be carried on more precisely now that we know the forms of the principal sections and of the nodal

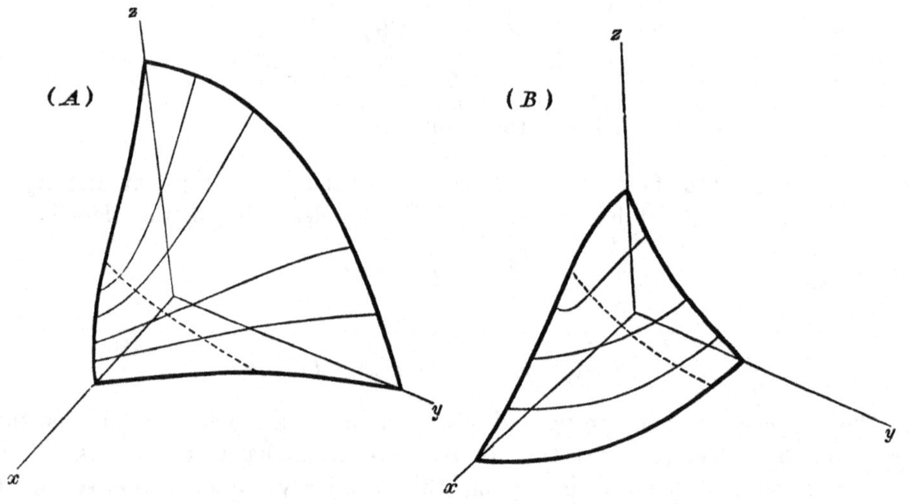

curve. The two figures exhibit (as regards one octant of the surface) the portions already distinguished as (A), and (B): they intersect each other in the nodal curve, shown in each of the figures.

26. Consider first the generation of the portion (A) by means of the major-mean sequential centro-curves. The major-mean curves of curvature (attending to those below the plane of xy) commence with a portion (extending from umbilicus to umbilicus) of the ellipse $\frac{X^2}{a^2}+\frac{Z^2}{c^2}=1$, this may be termed the vertical curve, and they end with the whole ellipse $\frac{X^2}{a^2}+\frac{Y^2}{b^2}=1$, which may be termed the horizontal curve. The normals at the several points of the vertical curve successively intersect along a portion (terminated each way at an umbilicar centre) of the evolute in the plane of xz or umbilicar plane; viz. this portion of the evolute, shown fig. (a), is the sequential centro-curve belonging to the vertical curve of curvature. The curve of curvature is at first a narrow oval surrounding the vertical curve; the corresponding form of the sequential centro-curve is at once seen to be a four-cusped curve as in fig. (b), and which we may imagine as derived from the curve (a) by first doubling this curve and then

opening out the two component parts thereof: the two upper cusps of the curve (*b*) are situate on the *yz*-ellipse of the centro-surface, and the two lower cusps upon two detached portions respectively of the *xz*-ellipse of the centro-surface. And as the curve

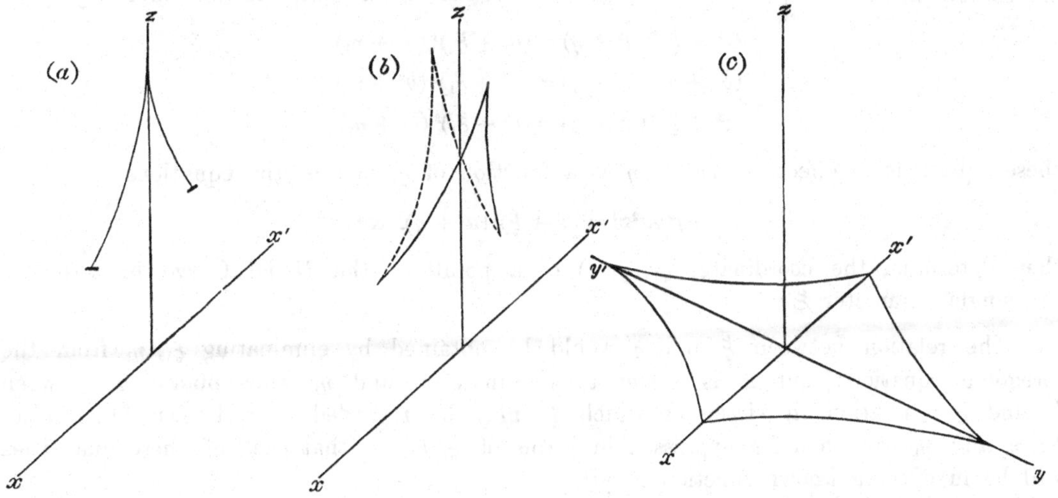

of curvature gradually broadens out and ultimately coincides with the *XY*-section of the ellipsoid, the four-cusped curve continues to open itself out, and ultimately coincides as shown figure (*c*) with the *xy*-evolute of the centro-surface, viz. this evolute is the sequential centro-curve belonging to the horizontal curve of curvature or *XY*-section of the ellipsoid. The successive sequential curves are also shown (so far as regards an octant of the surface) in the figure (A).

27. We consider next the generation of the portion (B) by means of the major-mean concomitant centro-curves. Starting as before with the vertical curve of curvature, the concomitant centro-curve is a finite portion (terminated each way at an umbilicar centre) of the *xz*-ellipse of the centro-surface. As the curve of curvature opens itself out into an oval, the concomitant centro-curve in like manner opens itself out into an oval, the two further vertices thereof situate on two detached portions of the *xz*-evolute of the centro-surface, and the two nearer vertices on the *yz*-evolute of the central surface. And as the curve of curvature continues to open itself out, and ultimately coincides with the horizontal curve or *XY*-section of the ellipsoid, so the concomitant centro-curve continues to open itself out and ultimately coincides with the *xy*-ellipse of the centro-surface. The successive forms (so far as relates to an octant of the surface) are shown in the figure (B). We have in each case attended only to the curves of curvature below the plane of *xy*, and the corresponding centro-curves above the plane of *xy*, but of course everything is symmetrical as regards the two sides of the plane.

28. There is a precisely similar generation of the portion (A) by the minor-mean concomitant centro-curves, and of the portion (B) by means of the minor-mean sequential centro-curves.

The Nodal Curve. Art. Nos. 29 to 60.

29. If two different points on the ellipsoid correspond to the same point on the centro-surface, this will be a point on the Nodal Curve: the conditions for this if (ξ, η), (ξ_1, η_1) are the parameters for the two points on the ellipsoid, are obviously

$$(a^2+\xi)^3(a^2+\eta) = (a^2+\xi_1)^3(a^2+\eta_1),$$
$$(b^2+\xi)^3(b^2+\eta) = (b^2+\xi_1)^3(b^2+\eta_1),$$
$$(c^2+\xi)^3(c^2+\eta) = (c^2+\xi_1)^3(c^2+\eta_1);$$

these equations in effect determine η as a function of ξ, so that the equations

$$-\beta\gamma a^2x^2 = (a^2+\xi)^3(a^2+\eta), \text{ \&c.}$$

then determine the coordinates (x, y, z) of a point on the Nodal Curve in terms of the single parameter ξ.

The relation between ξ and η would be obtained by eliminating ξ_1, η_1 from the foregoing equation: but it is easier to eliminate η and η_1, thus obtaining between ξ_1 and ξ a relation in virtue of which ξ_1 may be regarded as a known function of ξ; η and η_1 can then be expressed in terms of ξ, ξ_1, so that each of these quantities will be in effect a known function of ξ[1].

30. The relation between ξ, ξ_1 is in the first instance given in the form

$$\begin{vmatrix} a^2[(a^2+\xi)^3-(a^2+\xi_1)^3], & (a^2+\xi)^3, & (a^2+\xi_1)^3 \\ b^2[(b^2+\xi)^3-(b^2+\xi_1)^3], & (b^2+\xi)^3, & (b^2+\xi_1)^3 \\ c^2[(c^2+\xi)^3-(c^2+\xi_1)^3], & (c^2+\xi)^3, & (c^2+\xi_1)^3 \end{vmatrix} = 0.$$

Throwing out a factor $(\xi-\xi_1)^2$, this becomes

$$\Sigma\,[a^2\{3a^4+3a^2(\xi+\xi_1)+\xi^2+\xi\xi_1+\xi_1^2\}$$
$$\times(b^2-c^2).(1,\ 1,\ 1\between(b^2+\xi)(c^2+\xi_1),\ (b^2+\xi_1)(c^2+\xi))^2]=0,$$

where the left-hand side is a symmetrical function of ξ, ξ_1 vanishing for $\xi=\xi_1$, and therefore divisible by $(\xi-\xi_1)^2$; it is also divisible by Δ, $=(b^2-c^2)(c^2-a^2)(a^2-b^2)\,(=\alpha\beta\gamma)$. To work this out, write $\xi+\xi_1=p$, $\xi\xi_1=q$, the equation may be written

$$\Sigma\,\{(b^2-c^2)\,a^2 \begin{vmatrix} 3a^4 \\ +3a^2p \\ +p^2-q \end{vmatrix} \begin{vmatrix} 3b^4c^4 \\ +3b^2c^2(b^2+c^2)\,p \\ +(b^4+c^4)(p^2-q) \\ +b^2c^2(p^2+8q) \\ +3(b^2+c^2)\,pq \\ +3q^2 \end{vmatrix}\}=0,$$

where the left-hand side divides by $\Delta\,(p^2-4q)$.

[1] This was my first method of solution; and I have thought the results quite interesting enough to retain them—but it will appear in the sequel that I have succeeded in expressing ξ, η, ξ_1, η_1, in terms of a single parameter σ.

31. Developing and reducing, and omitting this factor, the final result is

$$6R + 3Qp + P(p^2 + 2q) + 3pq = 0,$$

where as before P, Q, R denote $a^2 + b^2 + c^2$, $b^2c^2 + c^2a^2 + a^2b^2$, $a^2b^2c^2$, respectively; that is,

$$6R + 3Q(\xi + \xi_1) + P(\xi^2 + 4\xi\xi_1 + \xi_1^2) + 3(\xi + \xi_1)\xi\xi_1 = 0,$$

or, as this may be written,

$$\begin{aligned} &6R + 3Q\xi + P\xi^2 \\ &+ \xi_1(3Q + 4P\xi + 3\xi^2) \\ &+ \xi_1^2(P + 3\xi) = 0, \end{aligned}$$

viz. the parameters ξ, ξ_1 have a symmetrical (2, 2) correspondence.

32. From the equations $(a^2 + \xi)^3(a^2 + \eta) = (a^2 + \xi_1)^3(a^2 + \eta_1)$, &c., we have

$$\Sigma(b^2 - c^2)\{(a^2 + \xi)^3(a^2 + \eta) - (a^2 + \xi_1)^3(a^2 + \eta_1)\} = 0,$$
$$\Sigma b^2c^2(b^2 - c^2)\{(a^2 + \xi)^3(a^2 + \eta) - (a^2 + \xi_1)^3(a^2 + \eta_1)\} = 0;$$

and observing that the term in { } is

$$\begin{aligned} &a^6(3\xi + \eta - 3\xi_1 - \eta_1) \\ &+ a^4(3\xi^2 + 3\xi\eta - 3\xi_1^2 - 3\xi_1\eta_1) \\ &+ a^2(\xi^3 + 3\xi^2\eta - \xi_1^3 - 3\xi_1^2\eta_1) \\ &+ (\xi^3\eta - \xi_1^3\eta_1), \end{aligned}$$

these are readily reduced to

$$(3\xi + \eta - 3\xi_1 - \eta_1)P + (3\xi^2 + 3\xi\eta - 3\xi_1^2 - 3\xi_1\eta_1) = 0,$$
$$(3\xi + \eta - 3\xi_1 - \eta_1)R + \xi^3\eta - \xi_1^3\eta_1 = 0,$$

or, what is the same thing,

$$3(\xi - \xi_1)(P + \xi + \xi_1) + \eta(P + 3\xi) - \eta_1(P + 3\xi_1) = 0,$$
$$3(\xi - \xi_1)R + \eta(R + \xi^3) - \eta_1(R + \xi_1^3) = 0,$$

and if we hence determine the ratios $3(\xi - \xi_1) : \eta : \eta_1$, the first of the resulting terms divides by $\xi - \xi_1$, and we have

$$\begin{aligned} 3 : \eta : \eta_1 = &- P(\xi^2 + \xi\xi_1 + \xi_1^2) + 3R - 3\xi\xi_1(\xi + \xi_1) \\ &: R(2\xi_1 - \xi) - \xi_1^3(P + \xi + \xi_1) \\ &: R(2\xi - \xi_1) - \xi^3(P + \xi + \xi_1). \end{aligned}$$

Hence observing that by the relation between ξ, ξ_1 the first term is

$$= 3\{P\xi\xi_1 + Q(\xi + \xi_1) + 3R\},$$

the equations become

$$\begin{aligned} 1 : \eta : \eta_1 = &P\xi\xi_1 + Q(\xi + \xi_1) + 3R \\ &: R(2\xi_1 - \xi) - \xi_1^3(P + \xi + \xi_1) \\ &: R(2\xi - \xi_1) - \xi^3(P + \xi + \xi_1); \end{aligned}$$

and we thus have

$$\eta = \frac{R(2\xi_1 - \xi) - \xi_1^3(P + \xi + \xi_1)}{P\xi\xi_1 + Q(\xi + \xi_1) + 3R},$$

which, considering ξ_1 as a given function of ξ, gives η as a function of ξ.

33. I write $\xi + \xi_1 = 2\text{x}$, $\xi - \xi_1 = 2\text{y}$, so that $p = 2\text{x}$, $q = \text{x}^2 - \text{y}^2$. The relation between ξ, ξ_1 takes the form

$$6(R + Q\text{x} + P\text{x}^2 - \text{x}^3) - (6\text{x} + 2P)\,\text{y}^2 = 0,$$

or, what is the same thing,

$$\text{y}^2 = \frac{(\text{x} + a^2)(\text{x} + b^2)(\text{x} + c^2)}{\text{x} + \frac{1}{3}(a^2 + b^2 + c^2)};$$

so that taking x at pleasure and considering y as denoting this function of x, the values of ξ, ξ_1 belonging to a point on the nodal curve are $\xi = (\text{x} + \text{y})$, $\xi_1 = (\text{x} - \text{y})$; and the value of η is then given as before.

34. The form just given is analytically the most convenient, but there is some advantage in writing $\frac{1}{\sqrt{2}}\text{x}$, $\frac{1}{\sqrt{2}}\text{y}$, in the place of x, y respectively; viz. we then have

$$\text{y}^2 = \frac{(\text{x} + a^2\sqrt{2})(\text{x} + b^2\sqrt{2})(\text{x} + c^2\sqrt{2})}{\text{x} + \frac{1}{3}\sqrt{2}(a^2 + b^2 + c^2)},$$

where $\xi = \frac{1}{\sqrt{2}}(\text{x} + \text{y})$, $\xi_1 = \frac{1}{\sqrt{2}}(\text{x} - \text{y})$, so that if (ξ, ξ_1) be taken as rectangular coordinates of a point in a plane, (x, y) will be the rectangular coordinates of the same point referred to axes inclined at angles of 45° to the first-mentioned axes respectively.

35. The curve is a cubic curve symmetrical in regard to the axis of x, and having the three asymptotes,

$$\text{x} = -\tfrac{1}{3}(a^2 + b^2 + c^2)\sqrt{2}, \quad \text{y} = \pm\{\text{x} + \tfrac{1}{3}(a^2 + b^2 + c^2)\sqrt{2}\},$$

viz. these all meet in the point P the coordinates of which are

$$\text{x} = -\tfrac{1}{3}(a^2 + b^2 + c^2)\sqrt{2}, \quad \text{y} = 0:$$

moreover we have $\text{y} = 0$ for the values $\text{x} = -a^2\sqrt{2}$, $-b^2\sqrt{2}$, $-c^2\sqrt{2}$, that is, the curve meets the axis of x in the points A, B, C; the order in the direction of $-\text{x}$ being C, B, P, A as shown in the figure: and with these data it is easy to draw the curve: the portion which gives the crunodal part of the nodal curve is that extending from B to the points Ω; viz. at B we have $\xi = \xi_1 = -b^2$, corresponding to the umbilicar centre; and at Ω, Ω we have ξ or $\xi_1 = -c^2$, ξ_1 or $\xi = -c^2 + \frac{3\alpha\beta}{\alpha - \beta}$, corresponding to the outcrop.

36. The nodal curve passes through (I) the umbilicar centres, (II) the outcrops, (III) the nodes of the evolute. The geometrical construction led to the conclusion that the real umbilicar centre was a node on the nodal curve, and that the real outcrop was a cusp (the tangent lying in the principal plane). It will presently appear generally, as regards the several points real or imaginary, that the umbilicar centre is a node on the nodal curve, and the outcrop a cusp—the tangent at the outcrop being in the principal plane: as regards the node on the evolute this is a simple point on the nodal curve, and by reason of the symmetry in regard to the principal

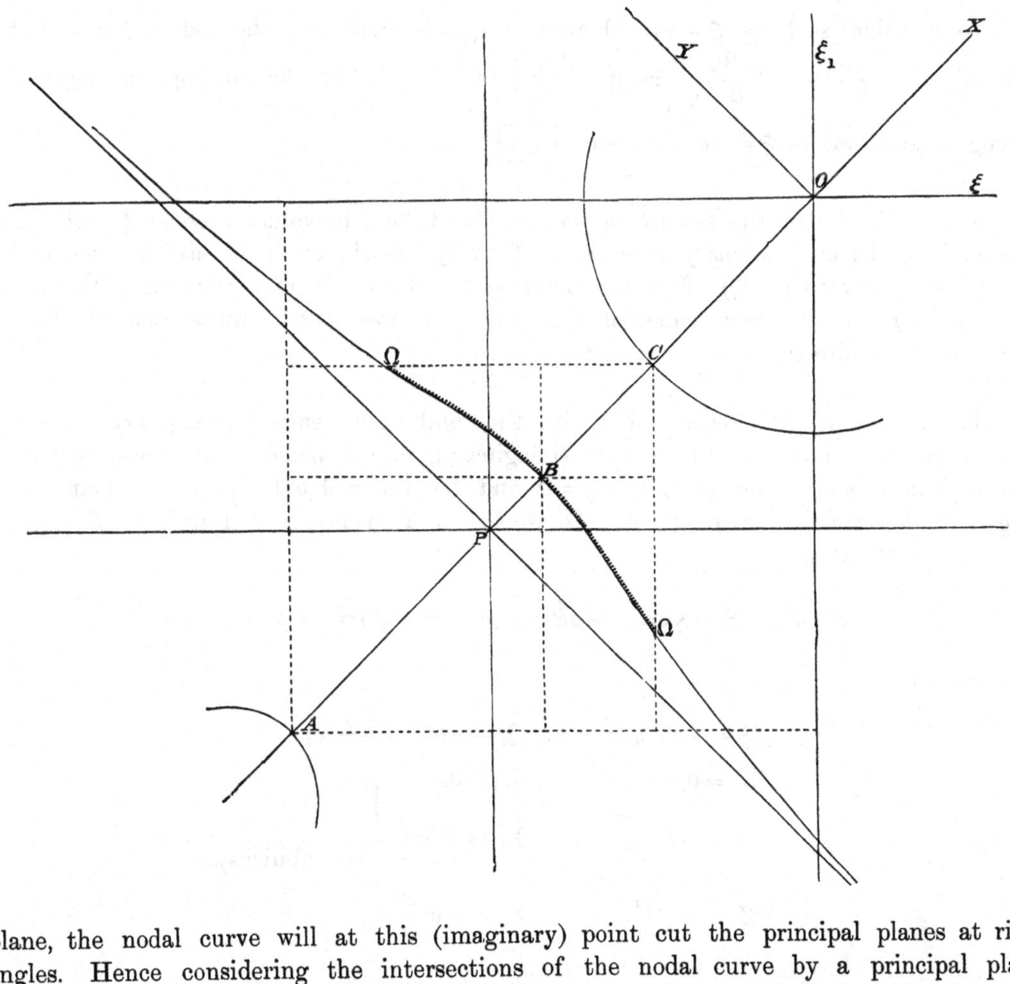

plane, the nodal curve will at this (imaginary) point cut the principal planes at right angles. Hence considering the intersections of the nodal curve by a principal plane, the umbilicar centre, outcrop and node of the evolute count respectively as 2 points, 3 points and 1 point, and as for each kind the number is 4, the whole number of intersections is $4(2+3+1)$, $=24$. It may be shown that these are the only intersections of the nodal curve with the principal plane; and this being so, it follows that the order of the nodal curve is $=24$; which agrees with the result of a subsequent analytical investigation.

37. The umbilicar centres or points (I) belong to values such as $\xi=\xi_1=-a^2$ which are the *united values* in the equation between (ξ, ξ_1), viz. writing herein $\xi_1=\xi$ the equation becomes

$$(\xi+a^2)(\xi+b^2)(\xi+c^2)=0,$$

so that the united values are $\xi=\xi_1=-a^2,\ -b^2$ or $-c^2$. (It may be remarked, that treating this cubic as a degenerate quartic, a united value would be $\xi=\xi_1=\infty$, corresponding to the umbilicar centres at infinity.)

To a value such as $\xi=-a^2$ there corresponds (not only the value $\xi_1=-a^2$, but also) a value $\xi_1=-a^2+\dfrac{3\beta\gamma}{\beta-\gamma}$, as it is easy to verify. And the outcrops or points (II) belong to such values $\xi=-a^2$, $\xi_1=-a^2+\dfrac{3\beta\gamma}{\beta-\gamma}$.

And the nodes of the evolute or points (III) belong to values such as $\xi=\omega b^2+\omega^2c^2$, $\xi_1=\omega^2b^2+\omega c^2$ (ω an imaginary cube root of unity) which, as it is easy to see, satisfy the relation between (ξ, ξ_1). But to complete the theory we require to have the values of η, η_1 and also the coordinates of the points on the centro-surface, and of the two points on the ellipsoid.

38. I exhibit the results first for the umbilicar centres (imaginary), outcrops (imaginary), and nodes of the evolute (imaginary), in the plane $x=0$; secondly for the real umbilicar centres in the plane $y=0$ and for the real outcrops in the plane $z=0$. The formulæ contain an expression Ω which is a symmetrical function of α, β, γ (or a, b, c), viz. it is

$$\Omega=\alpha^2-\beta\gamma=\beta^2-\gamma\alpha=\gamma^2-\alpha\beta=\tfrac{1}{2}(\alpha^2+\beta^2+\gamma^2)=-(\beta\gamma+\gamma\alpha+\alpha\beta).$$

We have

I. $\qquad \xi=-a^2,\ \eta=-a^2;\quad \xi_1=-a^2,\ \eta_1=-a^2.$

$$\left.\begin{array}{ll} X=0, & X_1=0, \\ Y^2=-b^2\dfrac{\gamma}{\alpha}, & Y_1^2=-b^2\dfrac{\gamma}{\alpha}, \\ Z^2=-c^2\dfrac{\beta}{\alpha}, & Z_1^2=-c^2\dfrac{\beta}{\alpha}, \end{array}\right\}\text{(Umbilicus).}$$

$$\left.\begin{array}{l} x=0, \\ b^2y^2=-\dfrac{\gamma^3}{\alpha}, \\ c^2z^2=-\dfrac{\beta^3}{\alpha}, \end{array}\right\}\text{(Umbilicar centre).}$$

II. $\xi = -a^2,\qquad \eta = -a^2 + \frac{9\beta\gamma\Omega}{(\beta-\gamma)^3},$

$$\left(\text{or } \eta + b^2 = -\gamma\frac{(\alpha-\beta)^3}{(\beta-\gamma)^3},\quad \eta + c^2 = \frac{\beta(\gamma-\alpha)^3}{(\beta-\gamma)^3}\right).$$

$$\xi_1 = -a^2 + \frac{3\beta\gamma}{\beta-\gamma},\quad \eta_1 = -a^2.$$

$$\left.\begin{aligned} X &= 0, & X_1 &= 0,\\ Y^2 &= -b^2\frac{\gamma}{\alpha}\frac{(\alpha-\beta)^3}{(\beta-\gamma)^3}, & Y_1^2 &= -b^2\frac{\gamma}{\alpha}\frac{\alpha-\beta}{\beta-\gamma},\\ Z^2 &= -c^2\frac{\beta}{\alpha}\frac{(\gamma-\alpha)^3}{(\beta-\gamma)^3}, & Z_1^2 &= -c^2\frac{\beta}{\alpha}\frac{\gamma-\alpha}{\beta-\gamma}, \end{aligned}\right\}$$

Ellipse, concomitant. Ellipse, sequential.

$$\left.\begin{aligned} x &= 0,\\ b^2y^2 &= -\frac{\gamma^3(\alpha-\beta)^3}{\alpha(\beta-\gamma)^3},\\ c^2z^2 &= -\frac{\beta^3(\gamma-\alpha)^3}{\alpha(\beta-\gamma)^3}, \end{aligned}\right\}\ \text{(Outcrop).}$$

(Observe that at point Y_1, Z_1 of ellipse $\frac{Y^2}{b^2} + \frac{Z^2}{c^2} = 1$, the coordinates of the centre of curvature are $y = \frac{\alpha Y_1^3}{b^4}$, $z = -\frac{\alpha Z_1^3}{c^4}$, and it thence appears that this is the point in regard to which the ellipse is sequential.)

III. $\xi = \omega b^2 + \omega^2 c^2,\quad \eta = -a^2;\qquad \xi_1 = \omega^2 b^2 + \omega c^2,\quad \eta_1 = -a^2.$

$$\begin{aligned} X &= 0, & X_1 &= 0,\\ Y^2 &= -b^2\omega^2, & Y_1^2 &= -b^2\omega,\\ Z^2 &= -c^2\omega, & Z_1^2 &= -c^2\omega^2, \end{aligned}$$

$$\left.\begin{aligned} x &= 0,\\ b^2y^2 &= -a^2,\\ c^2z^2 &= -a^2, \end{aligned}\right\}\ \text{(Node of evolute).}$$

39. Observe that these are the only ways in which it is possible to satisfy the equations

$$0 = (a^2+\xi)^3(a^2+\eta) = (a^2+\xi_1)^3(a^2+\eta_1),$$

viz. starting from this equation we have

I. $\qquad a^2 + \xi = 0,\quad a^2 + \xi_1 = 0,$

whence in the equations for η, η_1, substituting the values $\xi = \xi_1 = -a^2$, we have

$$1 : \eta : \eta_1 = a^4 P - 2a^2 Q + 3R,$$
$$: -a^2 R + a^6 (P - 2a^2),$$
$$: -a^2 R + a^6 (P - 2a^2),$$

that is,

$$1 : \eta : \eta_1 = -a^2\beta\gamma : a^4\beta\gamma : a^4\beta\gamma,$$

or

$$\eta = \eta_1 = -a^2.$$

40. II. $a^2 + \xi = 0$ without $a^2 + \xi_1 = 0$, consequently $a^2 + \eta_1 = 0$; writing $\xi = -a^2$, in the relation between (ξ, ξ_1), this is

$$6R + 3Q(\xi_1 - a^2) + P(\xi_1^2 - 4a^2\xi_1 + a^4) - 3a^2\xi_1(\xi_1 + a^2) = 0,$$

viz. this is

$$\xi_1^2(b^2 + c^2 - 2a^2) + \xi_1(-a^4 - a^2b^2 - a^2c^2 + 3b^2c^2) + a^2(a^4 - 2a^2b^2 - 2a^2c^2 + 3b^2c^2) = 0,$$

where the left-hand side should divide by $\xi_1 + a^2$; the equation in fact is

$$(\xi_1 + a^2)\{\xi_1(b^2 + c^2 - 2a^2) + a^4 - 2a^2b^2 - 2a^2c^2 + 3b^2c^2\} = 0;$$

or, what is the same thing,

$$(\xi_1 + a^2)\{(\xi_1 + a^2)(\beta - \gamma) - 3\beta\gamma\} = 0,$$

whence

$$\xi_1 = -a^2 + \frac{3\beta\gamma}{\beta - \gamma}.$$

41. Considering these values of ξ, ξ_1 as given, the verification of the value $\eta_1 = -a^2$, and determination of $\eta = -a^2 + \frac{9\beta\gamma\Omega}{(\beta - \gamma)^3}$ is somewhat complex.

Writing for a moment $\Lambda = -\frac{3\beta\gamma}{\beta - \gamma}$, we have

$$1 : \eta : \eta_1 = P(a^4 + a^2\Lambda) - Q(2a^2 + \Lambda) + 3R$$
$$: -R(a^2 + 2\Lambda) - (a^2 + \Lambda)^3(2a^2 - P + \Lambda$$
$$: -R(a^2 - \Lambda) - a^6(2a^2 - P + \Lambda).$$

The first term is

$$a^4P - 2a^2Q + 3R + \Lambda(a^2P - Q),$$

which is

$$= -a^2\beta\gamma + \Lambda(a^4 - b^2c^2);$$

and for the value of η_1, proceeding to the third term, this is

$$-a^2R - a^6(2a^2 - P) + \Lambda(R - a^6),$$

which is

$$= a^4\beta\gamma - a^2\Lambda(a^4 - b^2c^2),$$

so that without any further reduction $\eta_1 = -a^2$.

42. We have then

$$\eta = \frac{-R(a^2+2\Lambda)-(a^2+\Lambda)^3(a^2-b^2-c^2+\Lambda)}{-\alpha^2\beta\gamma+\Lambda(a^4-b^2c^2)},$$

and I assume

$$\eta = -a^2 + \frac{9\beta\gamma}{(\beta-\gamma)^3}\,\Omega,$$

and investigate the value of Ω.

We have

$$-R(a^2+2\Lambda)-(a^2+\Lambda)^3(a^2-b^2-c^2+\Lambda),\ =\alpha^4\beta\gamma+\Lambda\odot,\ \text{suppose.}$$

The equation therefore is

$$\frac{\alpha^4\beta\gamma+\Lambda\odot}{-\alpha^2\beta\gamma+\Lambda(a^4-b^2c^2)} = -a^2+\frac{9\beta\gamma}{(\beta-\gamma)^3}\,\Omega,$$

that is,

$$\Lambda\odot = -a^2\Lambda(a^4-b^2c^2)+\frac{9\beta\gamma}{(\beta-\gamma)^3}\,\Omega\,\{-\alpha^2\beta\gamma+\Lambda(a^4-b^2c^2)\}=0,$$

or writing $\dfrac{9\beta\gamma}{(\beta-\gamma)^3} = -\dfrac{3\Lambda}{(\beta-\gamma)^2}$, omitting the factor Λ, and multiplying by $(\beta-\gamma)^2$, this is

$$(\beta-\gamma)^2\{\odot+a^2(a^4-b^2c^2)\}+3\Omega\,\{-\alpha^2\beta\gamma+\Lambda(a^4b^2c^2)\}=0,$$

in which equation

$$\odot = -2R - a^6 - (3a^4+3a^2\Lambda+\Lambda^2)(a^2-b^2-c^2+\Lambda),$$

and thence

$$\begin{aligned}\odot + a^2(a^4-b^2c^2) &= \text{same } + a^2(a^4-b^2c^2),\\ &= -3a^6+3a^4(b^2+c^2)-3a^2b^2c^2\\ &\quad +\Lambda\ \{-6a^4+3a^2(b^2+c^2)\}\\ &\quad +\Lambda^2(-4a^2+\quad b^2+c^2)\\ &\quad -\Lambda^3\\ &= 3a^2\beta\gamma+3a^2\Lambda(\beta-\gamma)+\Lambda^2(\beta-\gamma-2a^2)-\Lambda^3.\end{aligned}$$

43. Hence, substituting for Λ its value and multiplying by $(\beta-\gamma)^3$, we have

$$\begin{aligned}&(\beta-\gamma)^3\{\odot+a^2(a^4-b^2c^2)\}\\ &= 3\alpha^2\beta\gamma(\beta-\gamma)^3-9\alpha^2\beta\gamma(\beta-\gamma)^3+9\beta^2\gamma^2(\beta-\gamma-2a^2)(\beta-\gamma)+27\beta^3\gamma^3,\end{aligned}$$

which is

$$= -6a^2\beta\gamma(\beta-\gamma)^3+9\beta^2\gamma^2(\beta-\gamma)^2-18\alpha^2\beta^2\gamma^2(\beta-\gamma)+27\beta^3\gamma^3;$$

viz. this is

$$\begin{aligned}&= \{-6a^2(\beta-\gamma)+9\beta\gamma\}\{(\beta-\gamma)^2+3\beta\gamma\}\beta\gamma,\\ &= \{-6a^2(\beta-\gamma)+9\beta\gamma\}(\beta^2+\beta\gamma+\gamma^2)\beta\gamma,\end{aligned}$$

and the equation thus is

$$\{-2a^2(\beta-\gamma)+3\beta\gamma\}(\beta^2+\beta\gamma+\gamma^2)\beta\gamma+\Omega\left\{-a^2\beta\gamma-\frac{3\beta\gamma}{\beta-\gamma}(a^4-b^2c^2)\right\}(\beta-\gamma)=0,$$

or finally

$$\Omega\{a^2(\beta-\gamma)+3(a^4-b^2c^2)\}=(-2a^2(\beta-\gamma)+3\beta\gamma)(\beta^2+\beta\gamma+\gamma^2).$$

But $c^2=a^2+\beta$, $b^2=a^2-\gamma$, and hence $a^4-b^2c^2=-a^2(\beta-\gamma)+\beta\gamma$, and therefore

$$a^2(\beta-\gamma)+3(a^4-b^2c^2)=-2a^2(\beta-\gamma)+3\beta\gamma;$$

the equation thus divides by $-2a^2(\beta-\gamma)+3\beta\gamma$ and we have

$$\Omega=\beta^2+\beta\gamma+\gamma^2,$$

or, as this may also be written, $\Omega=\alpha^2-\beta\gamma$, $=\beta^2-\gamma\alpha$, $=\gamma^2-\alpha\beta$. So that Ω has the value originally so denoted, and we have then

$$\eta=-a^2+\frac{9\beta\gamma}{(\beta-\gamma)^3}\Omega.$$

44. III. Lastly the equation $0=(a^2+\xi)^3(a^2+\eta)=(a^2+\xi_1)^3(a^2+\eta_1)$ is satisfied if $a^2+\eta=0$, $a^2+\eta_1=0$: the equations

$$(b^2+\xi)^3(b^2+\eta)=(b^2+\xi_1)^3(b^2+\eta_1),$$
$$(c^2+\xi)^3(c^2+\eta)=(c^2+\xi_1)^3(c^2+\eta_1),$$

then give

$$(b^2+\xi)^3=(b^2+\xi_1)^3,$$
$$(c^2+\xi)^3=(c^2+\xi_1)^3,$$

which can be satisfied by $\xi=\xi_1$, leading to $\xi=\xi_1=-a^2$, which is the case I., or else by

$$b^2+\xi=\omega\,(b^2+\xi_1),$$
$$c^2+\xi=\omega^2(c^2+\xi_1),$$

that is,

$$\xi=\omega b^2+\omega^2c^2,\quad \xi_1=\omega^2b^2+\omega c^2.$$

To show that these values satisfy the relation between ξ, ξ_1, observe that they give

$$\xi+\xi_1=-b^2-c^2,\quad \xi\xi_1=b^4-b^2c^2+c^4,$$

whence also

$$\xi^2+4\xi\xi_1+\xi_1^2=3(b^4+c^4),$$

and the relation becomes

$$6a^2b^2c^2-3[a^2(b^2+c^2)+b^2c^2](b^2+c^2)$$
$$+[a^2+(b^2+c^2)].3(b^4+c^4)-3(b^2+c^2)(b^4-b^2c^2+c^4)=0,$$

which is an identity

45. I will show that these values of ξ, ξ_1 give the foregoing values $\eta = \eta_1 = -a^2$. We have

$$\begin{aligned} 1 : \eta - \eta_1 : \eta + \eta_1 = {} & P\xi\xi_1 + Q(\xi + \xi_1) + 3R \\ & : (\xi_1 - \xi)\{3R - (\xi^2 + \xi\xi_1 + \xi_1^2)(P + \xi + \xi_1)\} \\ & : (\xi_1 + \xi)\{\ R - (\xi^2 - \xi\xi_1 + \xi_1^2)(P + \xi + \xi_1)\}, \end{aligned}$$

this is

$$1 : \eta - \eta_1 : \eta + \eta_1 = \alpha^2(b^2 + c^2) : 0\,(\xi_1 - \xi) : -2a^2\alpha^2(b^2 + c^2),$$

or

$$\eta - \eta_1 = 0,\ \eta + \eta_1 = -2a^2;\ \text{that is, } \eta = \eta_1 = -a^2.$$

46. For the real umbilicar centres and outcrops we have

I. $$\xi = -b^2,\ \eta = -b^2,\ \xi_1 = -b^2,\ \eta_1 = -b^2.$$

$$X^2 = -a^2\frac{\gamma}{\beta}, \qquad X_1^2 = -a^2\frac{\gamma}{\beta},$$

$$Y = 0, \qquad Y_1 = 0,$$

$$Z^2 = -c^2\frac{\alpha}{\beta}, \qquad Z_1^2 = -c^2\frac{\alpha}{\beta},$$

$$\left.\begin{aligned} a^2x^2 &= -\frac{\gamma^3}{\beta}, \\ y &= 0, \\ c^2z^2 &= -\frac{\alpha^3}{\beta}, \end{aligned}\right\} \text{(real umbilicar centre).}$$

II. $\xi = -c^2,$ $$\eta = -c^2 + \frac{9\alpha\beta}{(\alpha - \beta)^3}$$

$$\left(\text{or } \eta + a^2 = -\beta\frac{(\gamma - \alpha)^3}{(\alpha - \beta)^3},\ \eta + b^2 = \frac{\alpha(\beta - \gamma)^3}{(\alpha - \beta)^3}\right).$$

$$\xi_1 = -c^2 + \frac{3\alpha\beta}{\alpha - \beta},\ \eta_1 = -c^2,$$

$$\left.\begin{aligned} X^2 &= -a^2\frac{\beta}{\gamma}\frac{(\gamma - \alpha)^3}{(\alpha - \beta)^3}, & X_1^2 &= -a^2\frac{\beta}{\gamma}\frac{\gamma - \alpha}{\alpha - \beta}, \\ Y^2 &= -b^2\frac{\alpha}{\gamma}\frac{(\beta - \gamma)^3}{(\alpha - \beta)^3}, & Y_1^2 &= -b^2\frac{\alpha}{\gamma}\frac{\beta - \gamma}{\alpha - \beta}, \\ Z &= 0, & Z_1 &= 0, \end{aligned}\right\}$$

ellipse concomitant. ellipse sequential.

$$\left.\begin{aligned} a^2x^2 &= -\frac{\beta^3}{\gamma}\frac{(\gamma - \alpha)^3}{(\alpha - \beta)^3} \\ b^2y^2 &= -\frac{\alpha^3}{\gamma}\frac{(\beta - \gamma)^3}{(\alpha - \beta)^3} \\ z &= 0. \end{aligned}\right\} \text{(real outcrop).}$$

Nodal curve in vicinity of umbilicar centre, $a^2x^2 = -\frac{\gamma^3}{\beta}$, $y=0$, $c^2z^2 = -\frac{\alpha^3}{\beta}$. Art. Nos. 47 to 49.

47. Write

$$\xi = -b^2 + q, \quad \eta = -b^2 + r,$$
$$\xi_1 = -b^2 + q_1, \quad \eta_1 = -b^2 + r_1,$$

we have to find the relation between q, q_1, r, r_1; first for q, q_1, the equation of correspondence gives

$$\begin{aligned} &6R \\ &+ 3Q(-2b^2+q+q_1) \\ &+ \;P\{6b^4 - 6b^2(q+q_1) + q^2 + 4qq_1 + q_1^2\} \\ &+ 3\;\{-2b^6 + 3b^4(q+q_1) - b^2(q^2+4qq_1+q_1^2) + qq_1(q+q_1)\} = 0, \end{aligned}$$

that is,

$$\begin{aligned} &3(q+q_1)(3b^4 - 2b^2P + Q) \\ &+ (q^2 + qq_1 + q_1^2)(-3b^2 + P) \\ &+ 3qq_1(q+q_1) = 0, \end{aligned}$$

viz. this is

$$\begin{aligned} &-3(q+q_1)\alpha\gamma \\ &+ (q^2 + 4qq_1 + q_1^2)(\gamma - \alpha) \\ &+ 3qq_1(q+q_1) = 0, \end{aligned}$$

whence approximately $q + q_1 = 0$; but it will appear that the value is required to the second order; we have therefore

$$\begin{aligned} q + q_1 &= \tfrac{1}{3}\frac{\gamma-\alpha}{\gamma\alpha}(q^2 + 4qq_1 + q_1^2) \\ &= -\tfrac{2}{3}\frac{\gamma-\alpha}{\gamma\alpha}q^2. \end{aligned}$$

48. Now the equations

$$(a^2+\xi)^3(a^2+\eta) = (a^2+\xi_1)^3(a^2+\eta_1), \text{ and } (c^2+\xi)^3(c^2+\eta) = (c^2+\xi_1)^3(c^2+\eta_1),$$

putting therein for ξ, η, ξ_1, η_1, their values, give the first of them

$$\log\left(1+\frac{r}{\gamma}\right) + 3\log\left(1+\frac{q}{\gamma}\right) = \log\left(1+\frac{r_1}{\gamma}\right) + 3\log\left(1+\frac{q_1}{\gamma}\right),$$

that is,

$$r + 3q - \frac{1}{2\gamma}(r^2+3q^2) + \frac{1}{3\gamma^2}(r^3+3q^3) = r_1 + 3q_1 - \frac{1}{2\gamma}(r_1^2+3q_1^2) + \frac{1}{3\gamma^2}(r_1^3+3q_1^3);$$

and similarly the second equation

$$r + 3q + \frac{1}{2\alpha}(r^2+3q^2) + \frac{1}{3\alpha^2}(r^3+3q^3) = r_1 + 3q_1 + \frac{1}{2\alpha}(r_1^2+3q_1^2) + \frac{1}{3\alpha^2}(r_1^3+3q_1^3);$$

whence multiplying by γ, α, and adding,

$$(\gamma+\alpha)\left\{r+3q+\frac{1}{3\alpha\gamma}(r^3+3q^3)\right\}=(\gamma+\alpha)\left\{r_1+3q_1+\frac{1}{3\alpha\gamma}(r_1^3+3q_1^3)\right\},$$

which, neglecting terms of the third order, is

$$r+3q=r_1+3q_1.$$

Subtracting the two equations we have

$$\tfrac{1}{2}\left(\frac{1}{\alpha}+\frac{1}{\gamma}\right)(r^2+3q^2)+\tfrac{1}{3}\left(\frac{1}{\alpha^2}-\frac{1}{\gamma^2}\right)(r^3+3q^3)=\tfrac{1}{2}\left(\frac{1}{\alpha}+\frac{1}{\gamma}\right)(r_1^2+3q_1^2)+\tfrac{1}{3}\left(\frac{1}{\alpha^2}-\frac{1}{\gamma^2}\right)(r_1^3+3q_1^3),$$

viz. this is

$$r^2+3q^2+\tfrac{2}{3}\frac{\gamma-\alpha}{\gamma\alpha}(r^3+3q^3)=r_1^2+3q_1^2+\tfrac{2}{3}\frac{\gamma-\alpha}{\gamma\alpha}(r_1^3+3q_1^3),$$

or, what is the same thing,

$$r^2-r_1^2+3(q^2-q_1^2)+\tfrac{2}{3}\frac{\gamma-\alpha}{\gamma\alpha}\{r^3-r_1^3+3(q^3-q_1^3)\}=0,$$

which, putting therein $r-r_1=-3(q-q_1)$, is

$$-r-r_1+q+q_1+\tfrac{2}{3}\frac{\gamma-\alpha}{\gamma\alpha}(-r^2-rr_1-r_1^2+q^2+qq_1+q_1^2)=0,$$

say this is

$$-r-r_1+q+q_1+2\Delta=0;$$

combining herewith

$$r-r_1+3q-3q_1 \quad =0,$$

we have

$$r+q-2q_1-\Delta=0,$$

and

$$r_1-2q+q_1-\Delta=0,$$

where

$$\Delta=\tfrac{1}{3}\frac{\gamma-\alpha}{\gamma\alpha}(-r^2-rr_1-r_1^2+q^2+qq_1+q_1^2).$$

But substituting herein the values $r=-q+2q_1$, $r_1=2q-q_1$, this becomes

$$\Delta=\tfrac{1}{3}\frac{\gamma-\alpha}{\gamma\alpha}(-2q^2+4qq_1-2q_1^2),\ =-\tfrac{8}{3}\frac{\gamma-\alpha}{\gamma\alpha}q^2,$$

and then

$$r=-q+2q_1+\Delta,$$

that is,

$$r+3q=2(q+q_1)+\Delta,\ =-\frac{4(\gamma-\alpha)}{\gamma\alpha}q^2.$$

49. We have then

$$a^2x^2 = -\frac{\gamma^3}{\beta}\left(1+\frac{r}{\gamma}\right)\left(1+\frac{q}{\gamma}\right)^3,$$

$$= -\frac{\gamma^3}{\beta}\left(1+\frac{r+3q}{\gamma}+\frac{3q(r+q)}{\gamma^2}\right),$$

$$= -\frac{\gamma^3}{\beta}\left(1+\frac{r+3q}{\gamma^2}-\frac{6}{\gamma^2}q^2\right)$$

$$= -\frac{\gamma^3}{\beta}\left\{1+q^2\left(-\frac{4(\gamma-\alpha)}{\gamma^2\alpha}-\frac{6\alpha}{\gamma^2\alpha}\right)\right\}$$

$$= -\frac{\gamma^3}{\beta}\left\{1+q^2\frac{2(\beta-\gamma)}{\gamma^2\alpha}\right\};$$

and in the same way from $c^2z^2 = -\frac{\alpha^3}{\beta}\left(1-\frac{r}{\alpha}\right)\left(1-\frac{r}{\alpha}\right)^3$, we have

$$c^2z^2 = -\frac{\alpha^3}{\beta}\left\{1-q^2\frac{2(\alpha-\beta)}{\gamma\alpha^2}\right\}:$$

moreover we have at once

$$b^2y^2 = -\frac{q^3r}{\gamma\alpha} = \frac{3q^4}{\gamma\alpha}.$$

Hence, writing $x+\delta x$, $0+\delta y$, $z+\delta z$ for x, y, z, we find

$$\delta x = \tfrac{1}{2}x\,.\,\frac{2(\beta-\gamma)}{\gamma^2\alpha}\,.\,q^2,$$

$$\delta y = \pm\frac{1}{b}\sqrt{\frac{3}{\gamma\alpha}}\,.\,q^2,$$

$$\delta z = \tfrac{1}{2}z\,.\,\frac{-2(\alpha-\beta)}{\gamma\alpha^2}\,.\,q^2,$$

or, what is the same thing,

$$\delta x : \delta y : \delta z = x\frac{2(\beta-\gamma)}{\gamma^2\alpha} : \pm\frac{2}{b}\sqrt{\frac{3}{\gamma\alpha}} : z\frac{-2(\alpha-\beta)}{\gamma\alpha^2},$$

where x, z denote the values at the umbilicar centre.

Nodal curve in vicinity of real outcrop, viz.

$$a^2x^2 = -\frac{\beta^3}{\gamma}\frac{(\gamma-\alpha)^3}{(\alpha-\beta)^3},\quad b^2y^2 = -\frac{\alpha^3}{\gamma}\frac{(\alpha-\gamma)^3}{(\alpha-\beta)^3},\quad z=0.$$ Art. Nos. 50 to 52.

50. Write

$$\xi = -c^2+q, \qquad \eta = -c^2+\frac{q\alpha\beta\Omega}{(\alpha-\beta)^3}+\theta,$$

$$\xi_1 = -c^2+\frac{3\alpha\beta}{\alpha-\beta}+q_1, \qquad \eta_1 = -c^2+\theta_1;$$

and first for the relation between q and q_1, writing for a moment $\frac{3\alpha\beta}{\alpha-\beta}+q_1=Q_1$, and therefore $\xi_1=-c^2+Q_1$, the equation of correspondence gives

$$-3\alpha\beta\,(q+Q_1)+(q^2+4qQ_1+Q_1^2)(\alpha-\beta)+3qQ_1(q+Q_1)=0,$$

which, putting for Q_1 its value, is

$$-3\alpha\beta\left(q+q_1+\frac{3\alpha\beta}{\alpha-\beta}\right)$$

$$+(\alpha-\beta)\left(q^2+4qq_1+q_1^2+(4q+2q_1)\frac{3\alpha\beta}{\alpha-\beta}+\frac{9\alpha^2\beta^2}{(\alpha-\beta)^2}\right)$$

$$+3q\left\{q(q+q_1)+(q+2q_1)\frac{3\alpha\beta}{\alpha-\beta}+\frac{9\alpha^2\beta^2}{(\alpha-\beta)^2}\right\}=0;$$

that is,

$$-3\alpha\beta\,(q+q_1)$$

$$+3\alpha\beta\,(4q+2q_1)+(\alpha-\beta)(q^2+4qq_1+q_1^2)$$

$$+\frac{27\alpha^2\beta^2}{(\alpha-\beta)^2}q+\frac{9\alpha\beta}{\alpha-\beta}(q^2+2qq_1)+3qq_1(q+q_1)=0,$$

or, what is the same thing,

$$\left(9\alpha\beta+\frac{27\alpha^2\beta^2}{(\alpha-\beta)^2}\right)q+3\alpha\beta q_1$$

$$+q^2\left\{\frac{\alpha^2+7\alpha\beta+\beta^2}{\alpha-\beta}+qq_1\frac{\alpha^2+16\alpha\beta+\beta^2}{\alpha-\beta}+q_1^2(\alpha-\beta)\right\}$$

$$+3qq_1(q+q_1)=0,$$

or, for small values,

$$\left(3+\frac{9\alpha\beta}{(\alpha-\beta)^2}\right)q+q_1=0,\ \text{that is,}\ \frac{3\Omega}{(\alpha-\beta)^2}q+q_1=0.$$

51. Moreover, from the equation $(c^2+\xi)^3(c^2+\eta)=(c^2+\xi_1)^3(c^2+\eta_1)$, we have

$$q^3\frac{9\alpha\beta\Omega}{(\alpha-\beta)^3}=\left(\frac{3\alpha\beta}{\alpha-\beta}\right)^3.\theta_1,\ \text{that is,}\ \theta_1=\tfrac{1}{3}\frac{\Omega}{\alpha^2\beta^2}\cdot q^3,$$

or, since q and q_1 are of the same order, θ_1 is of the order q_1^3. Hence, starting from the equations $-\beta\gamma a^2x^2=(a^2+\xi_1)^3(a^2+\eta_1)$ &c., the terms of x, y arising from the variation of η_1 are indefinitely small in regard to those arising from the variation of ξ_1; and we have

$$\frac{2\delta x}{x}=\frac{3q_1}{-\beta+\dfrac{3\alpha\beta}{\alpha-\beta}},\ =-3q_1\frac{(\alpha-\beta)}{\beta(\gamma-\alpha)},$$

$$\frac{2\delta y}{y}=\frac{3q_1}{\alpha+\dfrac{3\alpha\beta}{\alpha-\beta}},\ =\ 3q_1\frac{\alpha-\beta}{\alpha(\beta-\gamma)},$$

and for δz $(= z)$ we have

$$c^2(\delta z)^2 = -\frac{1}{\alpha\beta}\left(\frac{3\alpha\beta}{\alpha-\beta}\right)^3 \theta_1, \quad = \frac{-9\Omega q^3}{(\alpha-\beta)^3}, \quad = \frac{(\alpha-\beta)^3}{3\Omega^2} q_1^3,$$

so that writing for greater simplicity, $(\alpha-\beta)q_1 = -\alpha\beta\varpi$, the formulæ become

$$\frac{2\delta x}{x} = \frac{3\alpha}{\gamma-\alpha}\varpi,$$

$$\frac{2\delta y}{y} = -\frac{3\beta}{\beta-\gamma}\varpi,$$

$$c\delta z = \frac{(-\alpha\beta\varpi)^{\frac{3}{2}}}{\Omega\sqrt{3}}.$$

52. This shows that there is at the outcrop a cusp, the cuspidal tangent being in the plane of xy. It appears moreover that this tangent coincides with the tangent of the evolute. In fact, from the equation $(ax)^{\frac{2}{3}} + (by)^{\frac{2}{3}} - \gamma^2 = 0$ of the evolute we have

$$\frac{(ax)^{\frac{2}{3}}.\,dx}{x} + \frac{(by)^{\frac{2}{3}}.\,dy}{y} = 0,$$

or substituting for (x, y) their values at the outcrop,

$$\frac{\beta(\gamma-\alpha)}{\gamma^{\frac{4}{3}}(\alpha-\beta)}\frac{dx}{x} + \frac{\alpha(\beta-\gamma)}{\gamma^{\frac{4}{3}}(\alpha-\beta)}\frac{dy}{y} = 0;$$

that is,

$$\beta(\gamma-\alpha)\frac{dx}{x} + \alpha(\beta-\gamma)\frac{dy}{y} = 0,$$

which is satisfied by the foregoing values of $\frac{\delta x}{x}$, and $\frac{\delta y}{y}$, and the two tangents therefore coincide.

We have

$$4\{(\delta x)^2 + (\delta y)^2\} = \frac{-9\varpi^2\alpha^2\beta^2}{\gamma(\alpha-\beta)^3}\cdot\left\{\frac{\beta(\gamma-\alpha)}{a^2} + \frac{\alpha(\beta-\gamma)}{b^2}\right\},$$

which in virtue of

$$a^2\alpha(\beta-\gamma) + b^2\beta(\gamma-\alpha) + c^2\gamma(\alpha-\beta) = 3\alpha\beta\gamma,$$

is

$$4\{(\delta x)^2 + (\delta y)^2\} = \frac{-9\varpi^2\alpha^2\beta^2}{a^2b^2(\alpha-\beta)^3}\{3\alpha\beta - c^2(\alpha-\beta)\}$$

(observe $3\alpha\beta - c^2(\alpha-\beta)$, $= -c^2(\gamma-\alpha) - 3a^2\alpha$, is negative)

$$= \frac{-9\varpi^2\alpha^2\beta^2}{a^2b^2(\alpha-\beta)^2}\xi_1,$$

if ξ_1 be the value at the outcrop. Writing δs for the element of the arc we have

$$\delta s = -\tfrac{3}{2}\frac{\alpha\beta}{ab(\alpha-\beta)}\sqrt{-\xi_1}\,.\,\varpi,$$

$$\delta z = \frac{(-\alpha\beta\varpi)^{\frac{3}{2}}}{c\,\Omega\sqrt{3}},$$

which exhibit the form at the outcrop.

The Nodal Curve; expressions for the coordinates in terms of a single parameter σ.
Art. Nos. 53 to 60.

53. After the foregoing investigation of the nodal curve, I was led to perceive that it is possible to express ξ, η, ξ_1, η_1 in terms of a single variable σ, and thus to obtain expressions for the coordinates of a point of the nodal curve in terms of the single variable σ. The result was obtained by the consideration that the acnodal portion of the nodal curve could only arise from imaginary values of ξ, η; the question thus was, what imaginary values of these quantities give real values for the coordinates x, y, z. To make y real we may assume

$$\xi = -b^2 - p(\theta-\phi i),$$
$$\eta = -b^2 + p(\theta+\phi i)^3,$$

($i=\sqrt{-1}$ as usual): this being so, if Δ denote one or other of the quantities

$$\gamma,\ -\alpha\ (=a^2-b^2,\ c^2-b^2),$$

the expressions for $-\beta\gamma a^2x^2$, $-\gamma\alpha b^2y^2$ will be

$$=\{\Delta - p(\theta-\phi i)\}^3\{\Delta + p(\theta+\phi i)^3\},$$

and we have therefore the condition that this shall be real (for the two values $\Delta=\gamma$, $\Delta=-\alpha$): being real, it will in certain cases be positive, and we shall then have real values for the remaining coordinates x, z.

54. The condition of reality is easily found to be

$$\Delta^2(3\theta^2-\phi^2+3) - 6\theta p\Delta(\theta^2+\phi^2+1) + p^2\{3(\theta^2+\phi^2)^2+3\theta^2-\phi^2\}=0,$$

viz. this equation in Δ must have the roots γ, $-\alpha$, or the expression on the left hand must be

$$=(3\theta^2-\phi^2+3)\{\Delta^2-(\gamma-\alpha)\Delta-\alpha\gamma\}:$$

we have therefore

$$\frac{(\gamma-\alpha)^2}{-\gamma\alpha} = \frac{36(\theta^2+\phi^2+1)^2}{(3\theta^2-\phi^2+3)\{3(\theta^2+\phi^2)^2+3\theta^2-\phi^2\}},$$

$$\gamma-\alpha = \frac{6\theta p(\theta^2+\phi^2+1)}{3\theta^2-\phi^2+3};$$

and writing $\theta^2+\phi^2=X$, $3\theta^2-\phi^2=Y$, the first of these is

$$\frac{(\gamma-\alpha)^2}{-\gamma\alpha}=\frac{9(X+Y)(X+1)^2}{(Y+3)\{3(X^2-1)+Y+3\}},$$

which regarding X, Y as the coordinates of a point in a plane is a cubic curve having the point $X+1=0$, $Y+3=0$ as a node: hence writing $Y+3=3\sigma(X+1)$, X and Y will be each of them a rational function of σ. The second equation is

$$\frac{6\theta p(X+1)}{Y+3}=\gamma-\alpha, \text{ that is, } p=\frac{(\gamma-\alpha)\sigma}{2\theta}, =\frac{(\gamma-\alpha)\sigma}{\sqrt{X+Y}};$$

and we have also

$$2\theta=\sqrt{X+Y},\quad 2\phi=\sqrt{3X-Y};$$

the equations thus become

$$\xi=-b^2-\frac{(\gamma-\alpha)\sigma}{2}\left\{1-i\sqrt{\frac{3X-Y}{X+Y}}\right\},$$

$$\eta=-b^2+\frac{(\gamma-\alpha)\sigma(X+Y)}{8}\left\{1+i\sqrt{\frac{3X-Y}{X+Y}}\right\}^3,$$

which are better written in the form

$$\xi=-b^2-\tfrac{1}{2}(\gamma-\alpha)\sigma\left\{1-\sqrt{\frac{-3X+Y}{X+Y}}\right\},$$

$$\eta=-b^2+\tfrac{1}{8}(\gamma-\alpha)\sigma(X+Y)\left\{1+\sqrt{\frac{-3X+Y}{X+Y}}\right\}^3,$$

where X, Y are given functions of σ. We in fact thus obtain an analytical expression of the nodal curve, quite independent of the considerations as to real and imaginary which suggested the process: the foregoing values substituted for ξ, η will give $-\beta\gamma a^2x^2$, &c. equal to rational functions of σ, so that taking for ξ_1, η_1 the same expressions, only changing therein the sign of the radical $\sqrt{\dfrac{-3X+Y}{X+Y}}$, these values of ξ_1, η_1 give the very same values of $-\beta\gamma a^2x^2$, &c., or the values of ξ, η, ξ_1, η_1 satisfy the conditions

$$(a^2+\xi)^3(a^2+\eta)=(a^2+\xi_1)^3(a^2+\eta_1),\ \&c.$$

for a point on the nodal curve.

55. To complete the investigation, writing as above $Y+3=3\sigma(X+1)$, we obtain

$$\frac{(\gamma-\alpha)^2}{-\gamma\alpha}=\frac{(3\sigma+1)X+3\sigma-3}{\sigma(X+\sigma-1)};$$

or putting for a moment

$$\frac{(\gamma-\alpha)^2\sigma}{-\gamma\alpha}=K,$$

we have

$$X=\frac{(K-3)(\sigma-1)}{3\sigma+1-K}, \qquad X+1=\frac{K(\sigma-2)+4}{3\sigma+1-K};$$

$$Y+3=\frac{3K\sigma(\sigma-2)+12\sigma}{3\sigma+1-K}, \qquad Y=\frac{3(\sigma-1)\{K(\sigma-1)+1\}}{3\sigma+1-K};$$

$$X+Y=\frac{(\sigma-1)(3\sigma-2)K}{3\sigma+1-K}, \quad -3X+Y=\frac{3(\sigma-1)\{K(\sigma-2)+4\}}{3\sigma+1-K};$$

or substituting for K its value we have

$$K(\sigma-2)+4=-\frac{(\gamma-\alpha)^2}{\gamma\alpha}\left\{\sigma^2-2\sigma-\frac{4\gamma\alpha}{(\gamma-\alpha)^2}\right\}$$

$$=-\frac{(\gamma-\alpha)^2}{\gamma\alpha}\left(\sigma+\frac{2\alpha}{\gamma-\alpha}\right)\left(\sigma-\frac{2\gamma}{\gamma-\alpha}\right),$$

$$3\sigma+1-K=\frac{1}{\gamma\alpha}\{(3\sigma+1)\gamma\alpha+(\gamma-\alpha)^2\sigma\}, \quad =\frac{1}{\gamma\alpha}(\Omega\sigma+\gamma\alpha),$$

if as before $\Omega=\beta^2-\gamma\alpha$; and the result is

$$\xi=-b^2-\tfrac{1}{2}(\gamma-\alpha)\sigma\left\{1-\sqrt{\frac{3\left(\sigma+\frac{2\alpha}{\gamma-\alpha}\right)\left(\sigma-\frac{2\gamma}{\gamma-\alpha}\right)}{\sigma(3\sigma-2)}}\right\},$$

$$\eta=-b^2-\tfrac{1}{8}(\gamma-\alpha)^3\frac{\sigma^2(\sigma-1)(3\sigma-2)}{\Omega\sigma+\gamma\alpha}\left\{1+\sqrt{\frac{3\left(\sigma+\frac{2\alpha}{\gamma-\alpha}\right)\left(\sigma-\frac{2\gamma}{\gamma-\alpha}\right)}{\sigma(3\sigma-2)}}\right\}^3,$$

and changing the sign of the radical we have the values of ξ_1, η_1.

56. Write for a moment

$$\left[\Delta-\tfrac{1}{2}(\gamma-\alpha)\sigma\left\{1-\sqrt{\frac{3\left(\sigma+\frac{2\alpha}{\gamma-\alpha}\right)\left(\sigma-\frac{2\gamma}{\gamma-\alpha}\right)}{\sigma(3\sigma-2)}}\right\}\right]^3=(\Delta-\mathrm{a}+\mathrm{a}\sqrt{S})^3=A+B\sqrt{S},$$

$$\Delta-\tfrac{1}{8}(\gamma-\alpha)^3\frac{\sigma^2(\sigma-1)(3\sigma-2)}{\Omega\sigma+\gamma\alpha}\left\{1+\sqrt{\frac{3\left(\sigma+\frac{2\alpha}{\gamma-\alpha}\right)\left(\sigma-\frac{2\gamma}{\gamma-\alpha}\right)}{\sigma(3\sigma-2)}}\right\}^3=A'+B'\sqrt{S};$$

then in the product of these two expressions the rational part is $=AA'+BB'S$; but from the manner in which they were arrived at we have $0=AB'+A'B$, and the rational part is thus

$$=-\frac{B'}{B}(A^2-B^2S).$$

We have

$$A^2 - B^2 S = \{(\Delta - \mathrm{a})^2 - \mathrm{a}^2 S\}^3,$$

$$B' = -\tfrac{1}{8}(\gamma - \alpha)^3 \frac{\sigma^2(\sigma - 1)(3\sigma - 2)}{\Omega\sigma + \gamma\alpha}(3 + S),$$

$$B = \tfrac{1}{2}(\gamma - \alpha)\{3(\Delta - \mathrm{a})^2 + \mathrm{a}^2 S\};$$

hence the rational part in question is

$$= \tfrac{1}{4}\frac{(\gamma - \alpha)^2}{\Omega\sigma + \gamma\alpha}\frac{\sigma(\sigma - 1)(3\sigma - 2)(3 + S)}{3(\Delta - \mathrm{a})^2 + \mathrm{a}^2 S}\{(\Delta - \mathrm{a})^2 + \mathrm{a}^2 S\}^3,$$

which putting therein $\Delta = 0$ gives the value of $-\gamma\alpha b^2 y^2$; and putting $\Delta = \gamma$, or $\Delta = -\alpha$, gives the value of $-\beta\gamma a^2 x^2$ or $-\alpha\beta c^2 z^2$.

57. We have

$$1 - S = \frac{1}{\sigma(3\sigma - 2)}\left[3\sigma^2 - 2\sigma - 3\left\{\sigma^2 - 2\sigma - \frac{4\gamma\alpha}{(\gamma - \alpha)^2}\right\}\right]$$

$$= \frac{4\left\{\sigma + \dfrac{3\alpha\gamma}{(\gamma - \alpha)^2}\right\}}{\sigma(3\sigma - 2)},$$

$$3 + S = \frac{3}{\sigma(3\sigma - 2)}\left[3\sigma^2 - 2\sigma + \left\{\sigma^2 - 2\sigma - \frac{4\alpha\gamma}{(\gamma - \alpha)^2}\right\}\right]$$

$$= \frac{12}{\sigma(3\sigma - 2)}\left(\sigma + \frac{\alpha}{\gamma - \alpha}\right)\left(\sigma - \frac{\gamma}{\gamma - \alpha}\right).$$

Hence we have at once the value of

$$-\gamma\alpha b^2 y^2, \; = \tfrac{1}{4}\frac{(\gamma - \alpha)^2}{\Omega\sigma + \gamma\alpha}\cdot\frac{\sigma(\sigma - 1)(3\sigma - 2)}{\mathrm{a}^2}\mathrm{a}^6(1 - S)^3,$$

where

$$\mathrm{a} = \tfrac{1}{2}(\beta - \gamma)\sigma.$$

58. Moreover

$$(\Delta - \mathrm{a})^2 - \mathrm{a}^2 S = \Delta^2 - 2\mathrm{a}\Delta + \mathrm{a}^2(1 - S)$$

$$= \frac{1}{3\sigma - 2}[(3\sigma - 2)\{-(\gamma - \alpha)\Delta\sigma + \Delta^2\} + (\gamma - \alpha)^2\sigma^2 + 3\alpha\gamma],$$

where the term in [ ] is

$$\sigma^2(\gamma - \alpha)(\gamma - \alpha - 3\Delta) + \sigma\{3\Delta^2 + 2(\gamma - \alpha)\Delta + 3\alpha\gamma\} - 2\Delta^2,$$

and since $\Delta = \gamma$ or $-\alpha$, that is, $\Delta^2 - (\gamma - \alpha)\Delta - \alpha\gamma = 0$, the coefficient of σ is

$$= \Delta\{6\Delta - (\gamma - \alpha)\},$$

or the term is the product of two linear functions of σ; and we have

$$(\Delta - \mathrm{a})^2 - \mathrm{a}^2 S = \frac{1}{3\sigma - 2}\{\sigma(\gamma - \alpha) - 2\Delta\}\{\sigma(\gamma - \alpha - 3\Delta) + \Delta\}.$$

Similarly

$$3(\Delta - \mathrm{a})^2 + \mathrm{a}^2 S = 3(\Delta^2 - 2\mathrm{a}\Delta) + \mathrm{a}^2(3 + S)$$

$$= \frac{3}{3\sigma - 2}[(3\sigma - 2)\{-(\gamma - \alpha)\sigma\Delta + \Delta^2\} + \sigma\{(\gamma - \alpha)\sigma + \alpha\}\{(\gamma - \alpha)\sigma - \gamma\}],$$

where the term in [ ] is

$$(\gamma - \alpha)^2\sigma^3 - (\gamma - \alpha)(\gamma - \alpha + 3\Delta)\sigma^2 + \{3\Delta^2 + 2(\gamma - \alpha)\Delta - \alpha\gamma\}\sigma - 2\Delta^2,$$

in which the coefficient of σ is $= \Delta\{2\Delta + 3(\gamma - \alpha)\}$, and the term is a product of three linear functions: hence

$$3(\Delta - \mathrm{a})^2 + \mathrm{a}^2 S = \frac{3(\sigma - 1)}{3\sigma - 2}\{(\gamma - \alpha)\sigma - \Delta\}\{(\gamma - \alpha)\sigma - 2\Delta\}.$$

59. Substituting these values we have the expression

$$\frac{1}{\Omega\sigma + \gamma\alpha}\,\frac{\{(\gamma - \alpha)\sigma + \alpha\}\{(\gamma - \alpha)\sigma - \gamma\}\{(\gamma - \alpha)\sigma - 2\Delta\}^2\{(\gamma - \alpha - 3\Delta)\sigma + \Delta\}^3}{\{(\gamma - \alpha)\sigma - \Delta\}(3\sigma - 2)^2};$$

which writing therein $\Delta = \gamma$ gives $-\beta\gamma a^2x^2$, and writing $\Delta = -\alpha$ gives $-\alpha\beta c^2z^2$; we have above an expression for $-\gamma\alpha b^2y^2$ requiring only a simple reduction, and the final results are

$$-\beta\gamma a^2x^2 = \frac{\{(\gamma - \alpha)\sigma + \alpha\}\{(\gamma - \alpha)\sigma - 2\gamma\}^2\{(\beta - \gamma)\sigma + \gamma\}^3}{(\Omega\sigma + \gamma\alpha)(3\sigma - 2)^2},$$

$$-\gamma\alpha b^2y^2 = \frac{(\sigma - 1)\sigma^2\{(\gamma - \alpha)^2\sigma + 3\alpha\gamma\}^3}{(\Omega\sigma + \gamma\alpha)(3\sigma - 2)^2},$$

$$-\alpha\beta c^2z^2 = \frac{\{(\gamma - \alpha)\sigma - \gamma\}\{(\gamma - \alpha)\sigma + 2\alpha\}^2\{(\alpha - \beta)\sigma - \alpha\}^3}{(\Omega\sigma + \gamma\alpha)(3\sigma - 2)^2},$$

where it is to be observed that, equating the denominator to 0, we have a triple root $\sigma = \infty$; to indicate this, we may insert in the denominator the factor $(1 - 0\sigma)^3$.

60. We see here the meaning of all the factors, viz.

Planes.

	$x = 0$	$y = 0$	$z = 0$	∞
Evolute nodes	$\sigma = -\frac{\alpha}{\gamma - \alpha}$	$\sigma = 1$	$\sigma = \frac{\gamma}{\gamma - \alpha}$	$\sigma = \frac{-\gamma\alpha}{\Omega}$
Umbilicar centres	$\sigma = \frac{2\gamma}{\gamma - \alpha}$	$\sigma = 0$	$\sigma = \frac{-2\alpha}{\gamma - \alpha}$	$\sigma = \frac{2}{3}$
Outcrops	$\sigma = \frac{-\gamma}{\beta - \gamma}$	$\sigma = \frac{-3\gamma\alpha}{(\gamma - \alpha)^2}$	$\sigma = \frac{\alpha}{\alpha - \beta}$	$\sigma = \infty$

For the real curve σ extends from $\frac{\alpha}{\alpha-\beta}$ through 0 to $-\frac{\gamma\alpha}{\Omega}$, viz.

$$\sigma = \frac{\alpha}{\alpha-\beta} \text{ gives outcrop in plane } z=0,$$

$$\sigma = \quad 0 \quad \text{,, umbilicar centre in plane } y=0,$$

$$\sigma = \frac{-\gamma\alpha}{\Omega} \quad \text{,, evolute-node in plane } \infty.$$

It is to be noticed that the order of magnitude of the terms in the table is

$$\infty,\ \frac{2\gamma}{\gamma-\alpha},\ \frac{\gamma}{\gamma-\alpha},\ 1,\ \tfrac{2}{3},\ \frac{-\gamma}{\beta-\gamma},\ \frac{\alpha}{\alpha-\beta},\ 0,\ \frac{-\gamma\alpha}{\Omega},\ \frac{-\alpha}{\gamma-\alpha},\ \frac{-2\alpha}{\gamma-\alpha},\ \frac{-3\gamma\alpha}{(\gamma-\alpha)^2},\ -\infty,$$

so that the values $\frac{\alpha}{\alpha-\beta}$, 0, $\frac{-\gamma\alpha}{\Omega}$ which belong to the real curve are contiguous; this is as it should be. Several of the preceding investigations conducted by means of the quantities ξ, η, ξ_1, η_1 might have been conducted more simply by means of the formulæ involving σ.

The Eight Cuspidal Conics. Art. Nos. 61 to 71.

61. The centro-surface is the envelope of the quadric

$$\frac{a^2x^2}{(a^2+\xi)^2} + \frac{b^2y^2}{(b^2+\xi)^2} + \frac{c^2z^2}{(c^2+\xi)^2} - 1 = 0.$$

Hence it has a cuspidal curve given by means of this equation and the first and second derived equations

$$\frac{a^2x^2}{(a^2+\xi)^3} + \frac{b^2y^2}{(b^2+\xi)^3} + \frac{c^2z^2}{(c^2+\xi)^3} = 0,$$

$$\frac{a^2x^2}{(a^2+\xi)^4} + \frac{b^2y^2}{(b^2+\xi)^4} + \frac{c^2z^2}{(c^2+\xi)^4} = 0,$$

which equations determine a^2x^2, b^2y^2, c^2z^2 in terms of ξ, viz. we have

$$-\beta\gamma\, a^2x^2 = (a^2+\xi)^4,$$

$$-\gamma\alpha\, b^2y^2 = (b^2+\xi)^4,$$

$$-\alpha\beta\, c^2z^2 = (c^2+\xi)^4;$$

so that, comparing with the equations $-\beta\gamma\, a^2x^2 = (a^2+\xi)^3(a^2+\eta)$ &c. which give the centro-surface, we see that for the cuspidal curve $\xi=\eta$; or the cuspidal curve now in question arises from the eight lines on the ellipsoid, which lines are the envelope of the curves of curvature: it is clear that the curve is imaginary.

62. From the foregoing equations we have:

$$\sqrt{\alpha}\, ax + \sqrt{\beta}\, by + \sqrt{\gamma}\, cz = \sqrt{-\alpha\beta\gamma},$$

$$\alpha^{\frac{3}{4}} \sqrt{ax} + \beta^{\frac{3}{4}} \sqrt{by} + \gamma^{\frac{3}{4}} \sqrt{cz} = 0,$$

the second of which is best written in the rationalised form

$$(1,\ 1,\ 1,\ -1,\ -1,\ -1)\,(\alpha\sqrt{\alpha}\, ax,\ \beta\sqrt{\beta}\, by,\ \gamma\sqrt{\gamma}\, cz)^2 = 0,$$

and combining herewith the equation

$$\sqrt{\alpha}\, ax + \sqrt{\beta}\, by + \sqrt{\gamma}\, cz = \sqrt{-\alpha\beta\gamma},$$

then for any given signs of $\sqrt{\alpha}$, $\sqrt{\beta}$, $\sqrt{\gamma}$ and $\sqrt{-\alpha\beta\gamma}$ the first of these equations represents a quadric surface, the second a plane, or the two equations together represent a conic.

By changing the signs of the radicals (observing that when all the signs are changed simultaneously the curve is unaltered) we obtain in all 8 conics, but only four quadric surfaces; viz. the two conics

$$\sqrt{\alpha}\, ax + \sqrt{\beta}\, by + \sqrt{\gamma}\, cz = \pm\sqrt{-\alpha\beta\gamma}$$

lie on the same quadric surface.

63. The conics form two sets of four, corresponding to the two sets of four lines on the ellipsoid. The analysis seems to establish a correspondence of each conic of the one set to a single conic of the other set; viz. the conics have been obtained in pairs as the intersections of the same quadric surface by a pair of planes: there is a like correspondence of each line of the one set to a single line of the other set, viz. the lines meet in pairs on the umbilici at infinity, but this correspondence is included in a more general property: in fact each line of the one set meets each line of the other set in an umbilicus; and the corresponding conics (not only meet but) touch at the corresponding umbilicar centre; and *quâ* touching conics they have two points of intersection, and consequently lie on the same quadric surface. It is to be added that the two conics touch also, at the umbilicar centre, the cuspidal conic of the principal section.

64. The 8 conics form two tetrads, and the principal conics (reckoning as one of them the conic at infinity) another tetrad: the complete cuspidal curve consists therefore of three tetrads of conics: with these we may form (one conic out of each tetrad) 16 triads; viz. each conic of one tetrad is combined with each conic of either of the other tetrads, and with a determinate conic of the third tetrad, to form a triad. And the conics of each triad, not only meet but touch at an umbilicar centre, the common tangent being also by what precedes, the tangent of the evolute at that point, which point is also a node of the nodal curve.

65. In fact consider the two conics

$$\sqrt{\alpha}\,ax \pm \sqrt{\beta}\,by + \sqrt{\gamma}\,cz = \sqrt{-\alpha\beta\gamma},$$
$$(1,\ 1,\ 1,\ -1,\ -1,\ -1)(\alpha\sqrt{\alpha}\,ax,\ \pm\beta\sqrt{\beta}\,by,\ \gamma\sqrt{\gamma}\,cz)^2 = 0;$$

for the intersections with the plane $y=0$ we have

$$\sqrt{\alpha}\,ax + \sqrt{\gamma}\,cz = \sqrt{-\alpha\beta\gamma},$$
$$(\alpha\sqrt{\alpha}\,ax - \gamma\sqrt{\gamma}\,cz)^2 = 0;$$

so that the two conics each meet the plane in question in the same two coincident points, that is, they each touch the plane $y=0$ at the same point, viz. the point given by the equations

$$\sqrt{\alpha}\,ax + \sqrt{\gamma}\,cz = \sqrt{-\alpha\beta\gamma},$$
$$\alpha\sqrt{\alpha}\,ax - \gamma\sqrt{\gamma}\,cz = 0;$$

viz. this is the point, $ax - \dfrac{\gamma\sqrt{\gamma}}{\sqrt{-\beta}}$, $cz = \dfrac{\alpha\sqrt{\alpha}}{\sqrt{-\beta}}$, which is one of the umbilicar centres

$$\left(a^2x^2 = -\frac{\gamma^3}{\beta},\quad c^2z^2 = -\frac{\alpha^3}{\beta}\right);$$

and the common tangent at this point is

$$\sqrt{\alpha}\,ax + \sqrt{\gamma}\,cz = \sqrt{-\alpha\beta\gamma},$$

which is also the common tangent of the ellipse and evolute in the plane $y=0$.

66. It has been seen that the nodal curve meets each principal conic at four outcrops, which points are cusps of the nodal curve: it is to be further shown that the nodal curve meets each of the 8 cuspidal conics in four points (giving 32 new points, which may be called 'outcrops,' the 16 points heretofore so called being distinguished as the principal outcrops or 16 outcrops, and the points now in question as the 32 outcrops), which points are cusps of the nodal curve.

In fact to obtain the intersections of the nodal curve with the 8 cuspidal conics, we must in the equation of the nodal curve, or (what is the same thing) in the expressions of ξ, η in terms of σ, write $\eta = \xi$.

67. Putting for shortness,

$$\Theta = \tfrac{1}{4}\frac{(\gamma-\alpha)^2\,\sigma(\sigma-1)(3\sigma-2)}{\Omega\sigma+\gamma\alpha},$$

and as before

$$S = \frac{3\left(\sigma + \dfrac{2\alpha}{\gamma-\alpha}\right)\left(\sigma - \dfrac{2\gamma}{\gamma-\alpha}\right)}{\sigma(3\sigma-2)},$$

we have thus

$$\Theta(1+\sqrt{S})^3 = 1 - \sqrt{S},$$

or, what is the same thing,

$$\Theta(1+3S)-1+\sqrt{S}\{\Theta(3+S)+1\}=0:$$

we have without difficulty

$$\Theta(3+S)+1=\frac{(\gamma-\alpha)^2}{\Omega\sigma+\gamma\alpha}\left\{3\sigma^3-6\sigma^2+4\sigma+\frac{4\gamma\alpha}{(\gamma-\alpha)^2}\right\},$$

$$\Theta(1+3\sqrt{S})-1=\frac{(\gamma-\alpha)^2(3\sigma-2)\left(\sigma+\frac{2\alpha}{\gamma-\alpha}\right)\left(\sigma-\frac{2\gamma}{\gamma-\alpha}\right)}{\Omega\sigma+\gamma\alpha},$$

so that the resulting equation contains the factor

$$\sqrt{\left(\sigma+\frac{2\alpha}{\gamma-\alpha}\right)\left(\sigma-\frac{2\gamma}{\gamma-\alpha}\right)},\ =\sqrt{\sigma^2-2\sigma-\frac{4\gamma\alpha}{(\gamma-\alpha)^2}}.$$

Omitting it, the equation becomes

$$\sqrt{\sigma^2-2\sigma-\frac{4\gamma\alpha}{(\gamma-\alpha)^2}}(3\sigma-2)^{\frac{3}{2}}\sqrt{\sigma}+\sqrt{3}\left\{3\sigma^3-6\sigma^2+4\sigma+\frac{4\gamma\alpha}{(\gamma-\alpha)^2}\right\}^2=0,$$

or putting for shortness $\frac{4\gamma\alpha}{(\gamma-\alpha)^2}=M$, and rationalising, this is

$$-(\sigma^2-2\sigma-M)(3\sigma-2)^3\sigma+3(3\sigma^3-6\sigma^2+4\sigma+M)^2=0,$$

and, working this out, the terms in σ^6, σ^5 disappear, and the result is

$$(36+27M)\sigma^4-(64+36M)\sigma^3+32\sigma^2+16M\sigma+3M^2=0,$$

or, as this may also be written,

$$3M^2+M(27\sigma^4-36\sigma^3+16\sigma)+4(9\sigma^4-16\sigma^3+8\sigma^2)=0,$$

a quartic equation in σ: to each of the 4 roots there correspond 8 intersections, viz. there will be in all 32 intersections, lying in 4's upon the 8 cuspidal conics.

68. To show that these points are cusps, or stationary points on the nodal curve, starting from the expressions of $-\beta\gamma a^2x^2$ &c. in terms of σ we have, first for dy,

$$-\frac{2dy}{y}=d\sigma\left\{\frac{1}{\sigma-1}+\frac{2}{\sigma}+\frac{3(\gamma-\alpha)^2}{(\gamma-\alpha)^2\sigma+3\alpha\gamma}-\frac{\Omega}{\Omega\sigma+\gamma\alpha}-\frac{6}{3\sigma-2}\right\},$$

or, as this may be written,

$$-\frac{2dy}{y}=d\sigma\left\{\frac{1}{\sigma-1}+\frac{2}{\sigma}+\frac{12}{4\sigma+3M}-\frac{4+3M}{(4+3M)\sigma+M}-\frac{6}{3\sigma-2}\right\},$$

$$=d\sigma\left\{\frac{3\sigma^2-6\sigma+4}{3\sigma^3-5\sigma^2+2\sigma}+\frac{(32+24M)\sigma-9M^2}{(16+12M)\sigma^2+(16M+9M^2)\sigma+3M^2}\right\},$$

$$=d\sigma\frac{4\{(36+27M)\sigma^4-(64+36M)\sigma^3+32\sigma^2+16M\sigma+3M^2\}}{\sigma(\sigma-1)(3\sigma-2)(4\sigma+3M)\{(4+3M)\sigma+M\}},$$

viz. the numerator vanishes when σ is a root of the quartic equation.

69. We have next

$$-\frac{2dx}{x}=d\sigma\left\{\frac{\gamma-\alpha}{(\gamma-\alpha)\sigma+\alpha}+\frac{2(\gamma-\alpha)}{(\gamma-\alpha)\sigma-2\gamma}+\frac{3(\beta-\gamma)}{(\beta-\gamma)\sigma+\gamma}-\frac{\Omega}{\Omega\sigma+\gamma\alpha}-\frac{6}{3\sigma-2}\right\},$$

which, putting $\frac{\gamma}{\gamma-\alpha}=B$, and therefore $\frac{\alpha}{\gamma-\alpha}=B-1$, and $\frac{\gamma}{\beta-\gamma}=C$, is

$$=d\sigma\left\{\frac{1}{\sigma+B-1}+\frac{2}{\sigma-2B}+\frac{3}{\sigma+C}-\frac{4+3M}{(4+3M)\sigma+M}-\frac{6}{3\sigma-2}\right\},$$

and adding the fractions except $\frac{3}{\sigma+C}$, the numerator is

$$\begin{aligned}&\sigma^2(27MB+36B-4)\\&+\sigma\{54(B^2-B)M+72B^2-80B+8\}\\&+\quad 4M-16B^2+16B,\end{aligned}$$

which, observing that $B^2-B=\frac{1}{4}M$, is

$$\begin{aligned}=\quad&\sigma^2(27MB+36B-4)\\&+\sigma(\tfrac{27}{2}M^2+18M-8B+8),\end{aligned}$$

and, substituting for M and B their values, this is found to be

$$=\frac{4(2\gamma+\alpha)^3}{(\gamma-\alpha)^3}\sigma^2+\frac{8(2\gamma+\alpha)^3\alpha}{(\gamma-\alpha)^4}\sigma,$$

$$=\frac{4(2\gamma+\alpha)^3}{(\gamma-\alpha)^3}\sigma\left(\sigma+\frac{2\alpha}{\gamma-\alpha}\right).$$

70. Hence observing that $C=\frac{\gamma}{\beta-\gamma}=\frac{-\gamma}{2\gamma+\alpha}$, the whole coefficient of $d\sigma$ is

$$=\frac{\frac{4(2\gamma+\alpha)^3}{(\gamma-\alpha)^3}\left(\sigma^2+\frac{2\alpha}{\gamma-\alpha}\sigma\right)}{(3\sigma-2)(\sigma+B-1)(\sigma-2B)[(3M+4)\sigma+M]}+\frac{3}{\sigma+C},$$

and the numerator of this expressed as a single fraction is

$$\begin{aligned}&=\frac{4(2\gamma+\alpha)^2}{(\gamma-\alpha)^3}\sigma\left(\sigma+\frac{2\alpha}{\gamma-\alpha}\right)[(2\gamma+\alpha)\sigma-\gamma]\\&+3(3\sigma-2)(\sigma^2-\sigma-\tfrac{1}{2}M-B\sigma)\{(3M+4)\sigma+M\},\end{aligned}$$

which is

$$\begin{aligned}&=3(3\sigma-2)(\sigma^2-\sigma-\tfrac{1}{2}M)\{(3M+4)\sigma+M\}\\&+\sigma\Bigg[-3B(3\sigma-2)\{(3M+4)\sigma+M\}\\&\qquad+\frac{4(2\gamma+\alpha)^2}{(\gamma-\alpha)^2}\left(\sigma+\frac{2\alpha}{\gamma-\alpha}\right)\{(2\gamma+\alpha)\sigma-\gamma\}\Bigg]:\end{aligned}$$

the term in [ ] is

$$= \sigma^2 \left[-(27M+36)B + \frac{4(2\gamma+\alpha)^3}{(\gamma-\alpha)^3} \right]$$

$$+ \sigma \left[3B(3M+8) + \frac{4(2\gamma+\alpha)^2}{(\gamma-\alpha)^3} \left(-\gamma + \frac{2\alpha(2\gamma+\alpha)}{\gamma-\alpha} \right) \right]$$

$$+ \left[6BM - \frac{8(2\gamma+\alpha)^2 \gamma x}{(\gamma-\alpha)^4} \right],$$

which is found to be

$$= -4\sigma^2 + \sigma(8 + 15M + \tfrac{27}{2}M^2) - 2M - \tfrac{9}{2}M^2,$$

and the whole numerator is thus

$$3(3\sigma-2)(\sigma^2 - \sigma - \tfrac{1}{2}M)[(3M+4)\sigma + M]$$
$$-4\sigma^3 + \sigma^2(8 + 15M + \tfrac{27}{2}M^2) + \sigma(-2M - \tfrac{9}{7}M^2),$$

which is

$$= (36+27M)\sigma^4 - (64+36M)\sigma^3 + 32\sigma^2 + 16M\sigma + 3M^2.$$

71. We have thus

$$-\frac{2dx}{x} = d\sigma \frac{(36+27M)\sigma^4 - (64+36M)\sigma^3 + 32\sigma^2 + 16M\sigma + 3M^2}{\sigma(\sigma-1)(3\sigma-2)(4\sigma+3M)\{(4+3M)\sigma+M\}\left(\sigma + \frac{\gamma}{\beta-\gamma}\right)},$$

and thence also

$$-\frac{2dz}{z} = d\sigma \frac{(36+27M)\sigma^4 - (64+36M)\sigma^3 + 32\sigma^2 + 16M\sigma + 3M^2}{\sigma(\sigma-1)(3\sigma-2)(4\sigma+3M)\{(4+3M)\sigma+M\}\left(\sigma - \frac{\alpha}{\alpha-\beta}\right)},$$

so that dx and dz also vanish when σ is a root of the quartic equation: the points in question are therefore cusps of the nodal curve.

Centro-surface as the envelope of the quadric $\Sigma a^2x^2(a^2+\xi)^{-2} = 1$. Art. Nos. 72 to 76.

72. The equations $-\beta\gamma a^2x^2 = (a^2+\xi)^3(a^2+\eta)$, &c. considering therein ξ, η as variable give the centro-surface; considering η as a given constant but ξ as variable they give the sequential centro-curve; and considering ξ as a given constant but η as variable they give the concomitant centro-curve.

73. Suppose first that η is a given constant; to eliminate ξ we may write the equations in the form

$$-(\beta\gamma)^{\frac{1}{3}}(ax)^{\frac{2}{3}}(a^2+\eta)^{-\frac{1}{3}} = (a^2+\xi), \text{ \&c.,}$$

and then multiplying first by $\alpha(a^2+\eta)$, &c. and adding, and secondly by α, &c., and adding (observing that $\Sigma\alpha(a^2+\xi)(a^2+\eta) = -\alpha\beta\gamma$, $\Sigma\alpha(a^2+\xi) = 0$); we have

$$\Sigma(\alpha ax)^{\frac{2}{3}}(a^2+\eta)^{\frac{2}{3}} = (\alpha\beta\gamma)^{\frac{2}{3}},$$
$$\Sigma(\alpha ax)^{\frac{2}{3}}(a^2+\eta)^{-\frac{1}{3}} = 0,$$

which equations, considering therein η as a given constant, are the equations of a sequential centro-curve.

If from the two equations we eliminate η we should have the equation of the centro-surface; the second equation is the derivative of the first in regard to η; and it thus appears that the equation of the centro-surface might be obtained by equating to zero the discriminant of the rationalised function

$$\text{norm. } [\{\Sigma (\alpha a x)^{\frac{2}{3}} (a^2+\eta)^{\frac{1}{3}}\} - (\alpha\beta\gamma)^{\frac{1}{3}}];$$

but the form is too inconvenient to be of any use.

74. Taking next ξ as a given constant, and writing the equations in the form

$$-\beta\gamma a^2 x^2 (a^2+\xi)^{-3} = (a^2+\eta), \text{ \&c.};$$

then multiplying by $\alpha(a^2+\xi)$, &c. and adding, and again multiplying by α, &c. and adding, we have

$$\Sigma a^2x^2 (a^2+\xi)^{-2} = 1,$$
$$\Sigma a^2x^2 (a^2+\xi)^{-3} = 0;$$

or writing these equations at full length,

$$\frac{a^2x^2}{(a^2+\xi)^2} + \frac{b^2y^2}{(b^2+\xi)^2} + \frac{c^2z^2}{(c^2+\xi)^2} - 1 = 0,$$
$$\frac{a^2x^2}{(a^2+\xi)^3} + \frac{b^2y^2}{(b^2+\xi)^3} + \frac{c^2z^2}{(c^2+\xi)^3} = 0,$$

which equations, considering therein ξ as a constant, are the equations of any concomitant centro-curve: since the equations are each of the second order it thus appears that the concomitant centro-curves are quadriquadrics.

75. If from the two equations we eliminate ξ, we have the equation of the centro-surface; the second equation is the derivative of the first in regard to ξ; and it thus appears that the equation of the centro-surface is obtained by equating to zero the discriminant in regard to ξ of the integralised function

$$(a^2+\xi)^2 (b^2+\xi)^2 (c^2+\xi)^2 \{(\Sigma a^2x^2 (a^2+\xi)^{-2} - 1\},$$

or, what is the same thing, the discriminant of the sextic function

$$(a^2+\xi)^2 (b^2+\xi)^2 (c^2+\xi)^2 - \Sigma a^2x^2 (b^2+\xi)^2 (c^2+\xi)^2.$$

76. If instead hereof we consider the homogeneous function

$$w^2 (a^2+\xi)^2 (b^2+\xi)^2 (c^2+\xi)^2 - \Sigma a^2x^2 (b^2+\xi)^2 (c^2+\xi)^2,$$

then the coefficients are of the second order in (x, y, z, w), and the discriminant, being of the tenth order in the coefficients, is of the order 20 in (x, y, z, w). But the sextic function has a twofold factor $\left(1+\dfrac{\xi}{\infty}\right)^2$ if $w^2=0$, and it has evidently a twofold factor if $x^2=0$ or $y^2=0$ or $z^2=0$, that is, the discriminant contains the factor $x^2y^2z^2w^2$; or, omitting this factor, it will be of the order 12 in (x, y, z, w); whence writing $w=1$, the centro-surface is of the order 12. I have in this manner actually obtained the equation of the centro-surface: see the memoir "On a certain Sextic Torse," *Camb. Phil. Trans.* t. XI. (1871), pp. 507—523, [436].

Another generation of the Centro-surface. Art. Nos. 77 to 83.

77. By what precedes, the equation of the centro-surface is obtained as the condition in order that the equation

$$\{\Sigma a^2x^2(a^2+\xi)^{-2}\}-1=0$$

may have two equal roots. But taking m an arbitrary constant, this is the derived equation of

$$\{\Sigma a^2x^2(a^2+\xi)^{-1}\}+\xi+m=0,$$

and as such it will have two equal roots if the last-mentioned equation has three equal roots; and conversely, we have thus the equation of the centro-surface by expressing that the last-mentioned equation, or, what is the same thing, the quartic equation

$$(\xi+m)(\xi+a^2)(\xi+b^2)(\xi+c^2)-\Sigma a^2x^2(\xi+b^2)(\xi+c^2)=0$$

has three equal roots. The conditions for this are that the quadrinvariant and the cubinvariant shall each of them vanish; the two invariants are respectively a quadric and a cubic function of m; viz. the equations are

$$(\mathrm{a},\ \mathrm{b},\ \mathrm{c})(m,\ 1)^2=0,\quad (\mathrm{a}',\ \mathrm{b}',\ \mathrm{c}',\ \mathrm{d}')(m,\ 1)^3=0;$$

where the degrees in $(x,\ y,\ z)$ of a, b, c are 0, 2, 4 and those of a′, b′, c′, d′ are 0, 2, 4, 6 respectively: the equation of the centro-surface then is

$$\begin{vmatrix} & & \mathrm{a}, & \mathrm{b}, & \mathrm{c} \\ & \mathrm{a}, & \mathrm{b}, & \mathrm{c} & \\ \mathrm{a}, & \mathrm{b}, & \mathrm{c} & & \\ & \mathrm{a}', & \mathrm{b}', & \mathrm{c}', & \mathrm{d}' \\ \mathrm{a}', & \mathrm{b}', & \mathrm{c}', & \mathrm{d}' & \end{vmatrix}=0,$$

which is of the right order 12; but it would be difficult to obtain thereby the developed equation.

78. For the nodal curve the cubic equation must be satisfied by each root of the quadric equation, or, what is the same thing, the quadric function must completely divide the cubic function; the conditions are

$$\left\|\begin{matrix} & \mathrm{a}, & \mathrm{b}, & \mathrm{c} \\ \mathrm{a}, & \mathrm{b}, & \mathrm{c} & \\ \mathrm{a}', & \mathrm{b}', & \mathrm{c}', & \mathrm{d}' \end{matrix}\right\|=0,$$

where the degrees may be taken to be

$$\begin{vmatrix} 0, & 0, & 2, & 4 \\ 0, & 2, & 4, & 0 \\ 0, & 2, & 4, & 6 \end{vmatrix},$$

and the order of the nodal curve is thus $=24$: two of the equations in fact are

$$\left|\begin{matrix} & \mathrm{a}, & \mathrm{b} \\ \mathrm{a}, & \mathrm{b}, & \mathrm{c} \\ \mathrm{a}', & \mathrm{b}', & \mathrm{c}' \end{matrix}\right| = 0, \qquad \left|\begin{matrix} & \mathrm{b}, & \mathrm{c} \\ \mathrm{a}, & \mathrm{c} & \\ \mathrm{a}', & \mathrm{c}', & \mathrm{d}' \end{matrix}\right| = 0,$$

which are surfaces of the orders 4, 6; or the nodal curve is a complete intersection 4×6. By the results above obtained as to the nodal curve, it appears that the two surfaces must have an ordinary contact at each of the 16 umbilicar centres, and a stationary or singular contact at each of 48 outcrops.

79. The derivation of the centro-surface from the surface

$$\frac{a^2x^2}{a^2+\xi}+\frac{b^2y^2}{b^2+\xi}+\frac{c^2z^2}{c^2+\xi}+\xi-m=0$$

requires to be further explained. The surface, say $V=0$, is a quadric surface depending on the two parameters ξ, m; the axes coincide in direction with those of the ellipsoid, and their relative magnitudes are as

$$\frac{1}{a}\sqrt{a^2+\xi} : \frac{1}{b}\sqrt{b^2+\xi} : \frac{1}{c}\sqrt{c^2+\xi},$$

viz. these are as the axes of the confocal surface

$$\frac{x^2}{a^2+\xi}+\frac{y^2}{b^2+\xi}+\frac{z^2}{c^2+\xi}-1=0$$

divided by a, b, c respectively; to fix the absolute magnitudes observe that the equation may be written

$$x^2+y^2+z^2-m-\xi\left(\frac{x^2}{a^2+\xi}+\frac{y^2}{b^2+\xi}+\frac{z^2}{c^2+\xi}-1\right)=0,$$

viz. the surface $V=0$ is a surface through the spheroconic which is the intersection of the confocal surface by the arbitrary sphere $x^2+y^2+z^2-m=0$; but, while the surface is hereby and by the preceding condition as to the axes completely determined, the geometrical significance is anything but clear.

80. Considering then the quadric surface $V=0$, depending on the parameters ξ, m; suppose that m remains constant while ξ alone varies; we have thus three consecutive surfaces $V=0$, $V'=0$, $V''=0$; and these I say intersect in a point of the centro-surface; the point in question will depend on the two parameters (ξ, m), and if these vary simultaneously we have the whole system of points on the centro-surface; but if only one of them varies, the other being constant, we have a curve on the centro-surface.

The three equations may be replaced by $V=0$, $\delta_\xi V=0$, $\delta_\xi{}^2 V=0$; of which the first alone contains m; and it thus appears that if m be the variable parameter, the equations of the curve are $\delta_\xi V=0$, $\delta_\xi{}^2 V=0$, viz. the curve is then the quadriquadric curve which is the concomitant centro-curve of the curve of curvature for the parameter ξ. But if the variable parameter be ξ, then this is a curve on the 12-thic surface $\Omega=0$ obtained by the elimination of ξ from the equations $V=0$, $\delta_\xi V=0$; viz. we have $\Omega=S^3-T^2=0$, where $S=(\text{a, b, c})(m, 1)^2$, $T=(\text{a}', \text{b}', \text{c}', \text{d}')(m, 1)^3$, and the curve in question is the curve $S=0$, $T=0$, which is the cuspidal curve on the surface $\Omega=0$; the elimination of m from the two equations $S=0$, $T=0$ gives as above the equation of the centro-surface.

81. The surface $\Omega=S^3-T^2=0$ obtained as above by the elimination of ξ from the equations $V=0$, $\delta_\xi V=0$, (or, what is the same thing, by equating to zero the discriminant of V in regard to ξ,) may be termed the sociate-surface: we have then the quartic and sextic surfaces $S=0$, $T=0$ intersecting in the before-mentioned curve, which may be called the sociate-edge; and the locus of these sociate-edges is the centro-surface.

82. We may if we please, changing the parameter in one of the functions, consider the two series of surfaces $S=0$, $T=0$ depending on the parameters m, m' respectively; a surface of the first series will correspond to one of the second series when the parameters are equal, $m=m'$, and we have then a sociate-edge. Taking a point anywhere in space, through this point there pass two surfaces $S=0$, and three surfaces $T=0$; but there is no pair of corresponding surfaces, or sociate-edge. If however the point be taken anywhere on the centro-surface, then there is a pair of corresponding surfaces $S=0$, $T=0$; that is, through each point of the centro-surface there passes a single sociate-edge; and if the point be taken anywhere on the nodal curve of the centro-surface, then there are two pairs of corresponding surfaces; that is, through each point of the nodal curve there are two sociate-edges: this explains the method above made use of for finding the equations of the nodal curve, by giving to the equations $S=0$, $T=0$, considered as equations in m, two equal roots.

83. The *à posteriori* verification that the surfaces $V=0$, $V'=0$, $V''=0$ intersect in a point of the centro-surface, is not without interest; the parameters ξ_1, η_1 of the point of intersection are found to be $\xi_1=\xi$, $\eta_1=m-a^2-b^2-c^2-3\xi$; whence in the equation $V=0$, writing $-\beta\gamma a^2x^2=(a^2+\xi_1)^3(a^2+\eta_1)$ and $m=a^2+b^2+c^2+3\xi_1+\eta_1$, the resulting equation considered as an equation in ξ should have three roots $\xi=\xi_1$: the fourth root is at once seen to be $\xi=\eta_1$, and we ought therefore to have identically

$$\frac{-\alpha(a^2+\xi_1)^3(a^2+\eta_1)}{a^2+\xi}-\&\text{c.}+\alpha\beta\gamma(\xi-3\xi_1-\eta_1-a^2-b^2-c^2)$$
$$=\frac{\alpha\beta\gamma(\xi-\xi_1)^3(\xi-\eta_1)}{(a^2+\xi)(b^2+\xi)(c^2+\xi)};$$

and by decomposing the right-hand side into its component fractions this is at once seen to be true.

Third generation of the Centro-surface. Art. Nos. 84 and 85.

84. Instead of the foregoing equation $V=0$, consider the equation

$$W=\left(\frac{a^2x^2}{a^2+\xi}+\frac{b^2y^2}{b^2+\xi}+\frac{c^2z^2}{c^2+\xi}+\xi\right)-\left(\frac{a^2x^2}{a^2+\eta}+\frac{b^2y^2}{b^2+\eta}+\frac{c^2z^2}{c^2+\eta}+\eta\right)=0.$$

The equations $d_\xi W=0$, $d_\xi^2 W=0$ contain only ξ, and are in fact identically the same as the equations $d_\xi V=0$, $d_\xi^2 V=0$; the elimination of ξ from the equations $d_\xi V=0$, $d_\xi^2 W=0$ would therefore lead to the equation of the centro-surface: and the centro-surface is connected with the surfaces $W=0$, $d_\xi W=0$, $d_\xi^2 W=0$ and the parameters ξ, η in the same way as it is with the surfaces $V=0$, $d_\xi V=0$, $d_\xi^2 V=0$ and the parameters ξ, m. That is, if from the equations $W=0$, $d_\xi W=0$ we eliminate ξ we have a surface $\Omega=0$, depending upon η and having a cuspidal curve; and the locus of the cuspidal curve (as η varies) is the centro-surface. But the equation $W=0$ divides by $\xi-\eta$, and throwing out this factor it becomes

$$\frac{a^2x^2}{(a^2+\xi)(a^2+\eta)}+\frac{b^2y^2}{(b^2+\xi)(b^2+\eta)}+\frac{c^2z^2}{(c^2+\xi)(c^2+\eta)}-1=0,$$

so that the surface $\Omega=0$ is obtained by eliminating ξ from this equation and the derived equation in regard to ξ; or, what is the same thing, by equating to zero the discriminant in regard to ξ of the cubic function

$$(a^2+\xi)(b^2+\xi)(c^2+\xi)-\Sigma\frac{a^2x^2}{a^2+\eta}(b^2+\xi)(c^2+\xi).$$

This surface is in fact the torse generated by the normals at the several points of the curve of curvature belonging to the parameter η; the cuspidal curve is the edge of regression of this torse, that is, it is the sequential centro-curve of the curve of curvature; and we thus fall back upon the original investigation for the centro-surface.

85. In verification I remark that if X, Y, Z be the coordinates of a point on the curve of curvature in question, and (x, y, z) current coordinates, then the tangent plane of the torse, or plane through the normal and the tangent of the curve of curvature, has for its equation

$$\frac{Xx}{a^2+\eta}+\frac{Yy}{b^2+\eta}+\frac{Zz}{c^2+\eta}-1=0,$$

and if in this equation we consider the point (X, Y, Z) to be the point belonging to the parameters (η, ξ), viz. if we have $-\beta\gamma X^2=a^2(a^2+\xi)(a^2+\eta)$, &c., then this plane will be always touched by the before-mentioned ellipsoid,

$$\frac{a^2x^2}{(a^2+\xi)(a^2+\eta)}+\frac{b^2y^2}{(b^2+\xi)(b^2+\eta)}+\frac{c^2z^2}{(c^2+\xi)(c^2+\eta)}=1.$$

The condition for the contact in fact is

$$\Sigma\frac{X^2}{(a^2+\eta)^2}\,\frac{(a^2+\xi)(a^2+\eta)}{a^2}=1,$$

viz. substituting for (X, Y, Z) their values, this is

$$-\frac{1}{\alpha\beta\gamma}\Sigma\alpha(a^2+\xi)^2=1,$$

which is true. And this being so, the ellipsoid and the plane have each the same envelope, viz. this is the torse in question.

Reciprocal Surface. Art. No. 86.

86. The centro-surface is the envelope of

$$\frac{a^2x^2}{(a^2+\xi)^2}+\frac{b^2y^2}{(b^2+\xi)^2}+\frac{c^2z^2}{(c^2+\xi)^2}-1=0;$$

hence the reciprocal surface in regard to the sphere $x^2+y^2+z^2-k^2=0$ is the envelope of

$$\frac{(a^2+\xi)^2}{a^2}X^2+\frac{(b^2+\xi)^2}{b^2}Y^2+\frac{(c^2+\xi)^2}{c^2}Z^2-k^2=0,$$

that is,

$$a^2X^2+b^2Y^2+c^2Z^2-k^2+2\xi(X^2+Y^2+Z^2)+\xi^2\left(\frac{X^2}{a^2}+\frac{Y^2}{b^2}+\frac{Z^2}{c^2}\right)=0,$$

viz. the envelope is

$$(a^2X^2+b^2Y^2+c^2Z^2-k^4)\left(\frac{X^2}{a^2}+\frac{Y^2}{b^2}+\frac{Z^2}{c^2}\right)-(X^2+Y^2+Z^2)^2=0,$$

or, expanding and multiplying by $a^2b^2c^2$, this is

$$a^2(b^2-c^2)^2\,Y^2Z^2+b^2(c^2-a^2)^2\,Z^2X^2+c^2(a^2-b^2)^2\,X^2Y^2$$
$$-k^4(b^2c^2X^2+c^2a^2Y^2+a^2b^2Z^2)=0,$$

or, what is the same thing,

$$a^2\alpha^2\,Y^2Z^2+b^2\beta^2\,Z^2X^2+c^2\gamma^2\,X^2Y^2-k^4(b^2c^2X^2+c^2a^2Y^2+a^2b^2Z^2)=0,$$

which may be written

$$\mathrm{a}^2Y^2Z^2+\mathrm{b}^2Z^2X^2+\mathrm{c}^2X^2Y^2+\mathrm{f}^2X^2+\mathrm{g}^2Y^2+\mathrm{h}^2Z^2=0,$$

where

$$(\mathrm{a},\ \mathrm{b},\ \mathrm{c},\ \mathrm{f},\ \mathrm{g},\ \mathrm{h})=(a\alpha,\ b\beta,\ c\gamma,\ 2k^2bc,\ 2k^2ca,\ 2k^2ab),$$

and consequently,

$$\mathrm{af}+\mathrm{bg}+\mathrm{ch}=2k^2abc(\alpha+\beta+\gamma)=0.$$

It would doubtless be interesting to discuss this surface as it here presents itself, and with reference to its geometrical signification as the locus of the pole, in regard to the sphere, of the plane through two intersecting consecutive normals of the ellipsoid: but I abstain from any consideration of the question.

Delineation of the centro-surface for given numerical values of the semiaxes. Art. Nos. 87 and 88.

87. I constructed on a large scale a drawing of the centro-surface for the values

$$a^2=50,\quad b^2=25,\quad c^2=15.$$

(These were chosen so that a, b, c should have approximately the integer values 7, 5, 4, and that a^2+c^2 should be well greater than $2b^2$; they give a good form of surface, though perhaps a better selection might have been made; there is a slight objection to the existence of the relation $a^2 = 2b^2$, as in the xy-section it brings a cusp of the evolute on the ellipse.) We have therefore

$$\alpha = 10, \quad \beta = -35, \quad \gamma = 25;$$

the ellipses in the principal planes of the centro-surface are

$$\frac{y^2}{(5)^2} + \frac{z^2}{(8\cdot937)^2} = 1,$$

$$\frac{z^2}{(2\cdot582)^2} + \frac{x^2}{(3\cdot535)^2} = 1,$$

$$\frac{x^2}{(4\cdot950)^2} + \frac{y^2}{(2)^2} = 1,$$

and these determine on each axis the two points which are the cusps of the evolutes. We have moreover for the umbilicar centre $x = 2\cdot988$, $y = 0$, $z = 1\cdot380$, and for the outcrop $x = 1\cdot127$, $y = 1\cdot947$, $z = 0$.

88. For the delineation of the nodal curve (crunodal portion) we have first to find the values of ξ, ξ_1; these are given in terms of x, y *ante* No. 33 [p. 334], where y is a given function of x, and x extends between the values $\{-b^2$ and $-\frac{1}{3}(a^2+b^2+c^2)\}$ -25 and $-26\frac{2}{3}$. It was thought sufficient to divide the interval into 6 equal parts, that is, the values of x were taken to be -25, $-25\cdot\dot{3}$, ... $-26\cdot\dot{6}$. The values of ξ, ξ_1 being found, those of η, η_1 were obtained from them by means of the original equations $(a^2+\xi)^3(a^2+\eta) = (a^2+\xi_1)(a^2+\eta)$ &c. viz. we have thus for the determination of η, η_1 three simple equations, affording a verification of each other.

For the performance of these calculations (viz. of the values of y, ξ, ξ_1, η, η_1) I am indebted to the kindness of Mr J. W. L. Glaisher, of Trinity College. The results being obtained it is then easy to calculate as well the coordinates (x, y, z) of the point on the nodal curve as also the coordinates (X, Y, Z) and (X_1, Y_1, Z_1) of the corresponding two points on the ellipsoid (these last are of course not required for the delineation of the nodal curve, but it was interesting to obtain them). The whole series of the results is given in the annexed Table, and from them the drawing was constructed.

I find also in the neighbourhood of the umbilicar centre (if $\xi = -25 + q$),

$$\delta x = \cdot02868\, q^2,$$

$$\delta y = \pm \cdot02484\, q^2,$$

$$\delta z = \cdot02191\, q^2,$$

and in the neighbourhood of the outcrop (if $\xi_1 = -38\cdot333 + \frac{70}{9}\varpi$),

$$\delta x = 1\cdot127\, \varpi,$$

$$\delta y = -1\cdot704\, \varpi,$$

$$\delta z = \pm 4\cdot582\, \varpi^{\frac{1}{2}}.$$

x	y	ξ	ξ_1	η	η_1	x	y	z	X	Y	Z	X_1	Y_1	Z_1
−25	0·	−25·	−25·	−25·	−25·	2·988	0·	1·380	5·326	0·	2·070	5·326	0·	2·070
$-25\cdot\dot{3}$	4·2669	−21·0664	−29·6002	−38·3193	−16·6728	2·543	0·360	0·996	4·394	2·289	2·491	6·233	1·957	1·023
$-25\cdot\dot{6}$	6·3191	−19·3475	−31·9858	−42·9911	−15·4693	2·148	0·721	0·662	3·504	3·189	2·283	5·956	2·580	0·584
−26·	8·1240	−17·8760	−34·1240	−45·7879	−15·1047	1·786	1·175	0·373	2·780	3·847	1·948	5·626	3·005	0·293
$-26\cdot\dot{3}$	9·8760	−16·4573	−36·2094	−47·5684	−15·0106	1·448	1·500	0·139	2·159	4·391	1·426	5·251	3·346	0·098
$-26\cdot\dot{6}$	11·6667	−15·0000	−38·3333	−48·7037	−15·0000	1·127	1·947	0·	1·610	4·869	0·	4·829	3·651	0·

The calculations were performed before I had obtained the formulæ in σ, which would have given the results more easily.

521.

ON DR. WIENER'S MODEL OF A CUBIC SURFACE WITH 27 REAL LINES; AND ON THE CONSTRUCTION OF A DOUBLE-SIXER.

[From the *Transactions of the Cambridge Philosophical Society*, vol. XII. Part I. (1873), pp. 366—383. Read May 15, 1871.]

I.

I CALL to mind that a cubic surface has upon it in general 27 lines which may be all of them real. We may out of the 27 lines (and that in 36 different ways) select 12 lines forming a "double-sixer," viz. denoting such a system of lines by

$$a_1,\ a_2,\ a_3,\ a_4,\ a_5,\ a_6,$$
$$b_1,\ b_2,\ b_3,\ b_4,\ b_5,\ b_6;$$

then no two lines a meet each other, nor any two lines b, but each line a meets each line b, except that the two lines of a pair (a_1, b_1), (a_2, b_2), ... (a_6, b_6) do *not* meet each other. And such a system of twelve lines leads at once to the remaining fifteen lines; viz. we have a line c_{12}, the intersection of the planes which contain the pairs of lines (a_1, b_2) and (a_2, b_1) respectively.

The model is formed of plaster, and is contained within a cube, the edge of which is $= 18{\cdot}2$ inches: the lines a, b, c are coloured blue, yellow, and red respectively; the lines a_1, b_2, b_5 being at right angles to each other, in such wise that taking the origin at the centre of the cube, the axes parallel to the edges, and the unit of length $= 1{\cdot}6$ inches, the equations of these three lines are

$$\begin{array}{lll} a_1, & x=0,\ y=0, & \\ b_2, & x=0, & z=1, \\ b_5, & y=0, & z=-1. \end{array}$$

The model is a solid figure bounded by portions of the faces of the cube, and by a portion of the cubic surface, being a surface with three apertures, the collocation of which is not easily explained.

To determine the construction I measured, on the faces of the cube, the coordinates of the two extremities of each of the twelve lines; these were measured in tenths of an inch (taking account of the half division, or twentieth of an inch), and the resulting numbers divided by 16 to reduce them to the before-mentioned unit of 1·6 inches. These reduced values are shewn in the table: knowing then the coordinates of two points on each line, the equations of the several lines became calculable; the true theoretical form of these results—(viz. the form which, but for errors of the model, or of the measurement, they would have assumed)—is

$$
\begin{array}{llll}
b_1, & x = B_1 z + D, & y = B_1' z + D', & \\
b_2, & x = 0, & & z = 1, \\
b_3, & x = B_3 (z + \beta_3), & y = B_3' (z + \beta_3), & \\
b_4, & x = B_4 (z + \beta_4), & y = B_4' (z + \beta_4), & \\
b_5, & & y = 0, & z = -1, \\
b_6, & x = B_6 (z + \beta_6), & y = B_6' (z + \beta_6). & \\
a_1, & x = 0, & y = 0, & \\
a_2, & x = A_2 z + C_2, & y = A_2' (z - 1), & \\
a_3, & x = A_3 (z + 1), & y = A_3' (z - 1), & \\
a_4, & x = A_4 (z + 1), & y = A_4' (z - 1), & \\
a_5, & x = A_5 (z + 1), & y = A_5' z + C_5', & \\
a_6, & x = A_6 (z + 1), & y = A_6' (z - 1); &
\end{array}
$$

but in consequence of such errors, the results are not accurately of the form in question.

The faces of the cube being as in the diagram:

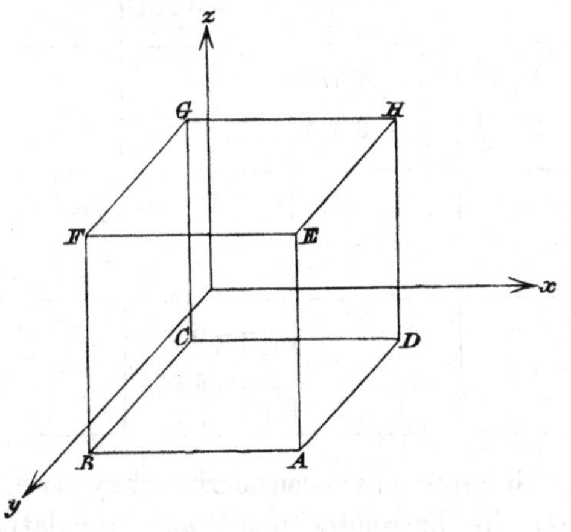

the Table is

Equations calculated from the measurements of the model.		*ABCD* $z=-5{\cdot}688$	*EFGH* $z=+5{\cdot}688$	*AEBF* $y=+5{\cdot}688$	*BCFG* $x=-5{\cdot}688$	*CDGH* $y=-5{\cdot}688$	*AEDH* $x=+5{\cdot}688$
$x=0$ $y=0$	a_1	$x=0$ $y=0$	$x=0$ $y=0$				
$x=-{\cdot}780z-{\cdot}187$ $y=-{\cdot}423z+{\cdot}406$	a_2	$x=4{\cdot}250$ $y=2{\cdot}812$	$x=-4{\cdot}625$ $y=-2{\cdot}000$				
$x=-{\cdot}654z-{\cdot}656$ $y=-{\cdot}588z+{\cdot}531$	a_3	$x=3{\cdot}062$ $y=3{\cdot}875$	$x=-4{\cdot}375$ $y=-2{\cdot}812$				
$x=-2{\cdot}912z-2{\cdot}959$ $y=-{\cdot}736z+{\cdot}752$	a_4	....	....	....	$y={\cdot}0625$ $z={\cdot}9375$	....	$y=2{\cdot}937$ $z=-2{\cdot}969$
$x=1{\cdot}024z+1{\cdot}014$ $y=-1{\cdot}049z-{\cdot}277$	a_5	$x=-4{\cdot}812$ $y=5{\cdot}688$	....	....	....	....	$y=-5{\cdot}063$ $z=4{\cdot}562$
$x={\cdot}264z+{\cdot}187$ $y=-{\cdot}104z+{\cdot}219$	a_6	$x=-1{\cdot}313$ $y={\cdot}8125$	$x=1{\cdot}687$ $y=-{\cdot}375$				
$x=-1{\cdot}611z+{\cdot}151$ $y=-1{\cdot}438z+{\cdot}288$	b_1	....	....	....	$y=5{\cdot}500$ $z=-3{\cdot}625$	....	$y=-4{\cdot}656$ $z=3{\cdot}437$
$x=0$ $z=-1$	b_2	....	....	$x=0$ $z=-1$	....	$x=0$ $z=-1$	
$x=-1{\cdot}352z-{\cdot}685$ $y=-2{\cdot}034z-{\cdot}984$	b_3	....	....	$x=3{\cdot}750$ $z=-3{\cdot}281$	....	$x=-3{\cdot}812$ $z=2{\cdot}313$	
$x=-{\cdot}753z-{\cdot}0315$ $y=-{\cdot}500z-{\cdot}0315$	b_4	$x=4{\cdot}250$ $y=2{\cdot}812$	$x=-4{\cdot}313$ $y=-2{\cdot}875$				
$y=0$ $z=+1$	b_5	....	....	....	$y=0$ $z=1$	....	$y=0$ $z=1$
$x=1{\cdot}123z-{\cdot}702$ $y=-1{\cdot}123z+{\cdot}702$	b_6	....	....	$x=-5{\cdot}688$ $z=-4{\cdot}438$	....	....	$y=-5{\cdot}688$ $z=5{\cdot}688$

I hence calculate the intersections: considering any two lines which ought to intersect, then projecting on the horizontal plane and calculating x, y the coordinates of the point of intersection of the two projections, these values of x, y substituted in the equations should give the same value of z; but if the lines do not accurately intersect, then the values of z will be different.

	b_1	b_2	b_3	b_4	b_5	b_6
a_1	*	0 0 − 1	0 0 − ·495 ± ·011	0 0 − ·052 ∓ ·010	0 0 + 1	0 0 + ·625
a_2	− ·077 + ·455 − ·129 ∓ ·013	*	+ ·381 + ·771 − ·796 ± ·067	+ 4·292 + 2·803 − 5·704 ± ·036	− ·967 − ·008 ∓ ·008 + 1	− ·398 + ·227 + ·346 ± ·076
a_3	− ·423 + ·699 − ·264 ± ·021	− ·001 ∓ ·001 + 1·119 − 1	*	− 4·782 − 3·227 + 6·350 ± ·042	− 1·310 − ·028 ∓ ·028 + 1	− ·673 + ·344 + ·162 ± ·146
a_4	− ·957 + 1·238 − ·674 ± ·014	− ·023 ∓ ·023 + 1·488 − 1	+ 1·286 + 1·736 − 1·398 ± ·060	*	− 5·871 + ·008 ± ·008 + 1	− 1·330 + ·847 − ·344 ± ·215
a_5	+ 2·519 − 1·801 + 1·462 ± ·008	− ·005 ∓ ·005 + ·772 − 1	+ ·282 + ·476 − ·717 ± ·001	+ ·412 + ·194 − ·518 ± ·070	*	+ 18·764 − 14·155 + 15·282 ± 4·052
a_6	+ ·194 + ·214 + ·040 ± ·022	− ·038 ∓ ·038 + ·323 − 1	+ ·045 + ·283 − ·582 ± ·042	− ·131 + ·284 − ·423 ± ·208	+ ·451 + ·057 ± ·057 − 1	*

Starting from the assumed equations of b_2, b_3, b_4, b_5, b_6, a_1, and calculating by the theory the remaining lines, the equations of the b-lines (those of b_1 being calculated) are

$$b_1, \quad x = 1{\cdot}321\,z - {\cdot}310,$$
$$y = -1{\cdot}295\,z + {\cdot}581;$$
$$b_2, \quad x = 0,\ z = -1;$$
$$b_3, \quad x = -1{\cdot}352\,(z + {\cdot}510),$$
$$y = -2{\cdot}034\,(z + {\cdot}510);$$
$$b_4, \quad x = -{\cdot}753\,(z + {\cdot}052),$$
$$y = -{\cdot}500\,(z + {\cdot}052);$$
$$b_5, \quad y = 0,\ z = +1;$$
$$b_6, \quad x = 1{\cdot}123\,(z - {\cdot}624),$$
$$y = -1{\cdot}123\,(z - {\cdot}624);$$

and the equations of the a-lines (those of all but a_1 being calculated) are

$$\begin{aligned}
&a_1, &&x = 0,\ y = 0;\\
&a_2, &&x = -\ \cdot 753\, z - \cdot 091,\\
& &&y = -\ \cdot 498\,(z-1);\\
&a_3, &&x = -\ \cdot 609\,(z+1),\\
& &&y = -\ \cdot 677\,(z-1);\\
&a_4, &&x = -\ 2\cdot 506\,(z+1),\\
& &&y = -\ \cdot 841\,(z-1);\\
&a_5, &&x = \ \cdot 874\,(z+1),\\
& &&y = -\ \cdot 967\, z - \cdot 288;\\
&a_6, &&x = \ \cdot 170\,(z+1),\\
& &&y = -\ \cdot 071\,(z-1);
\end{aligned}$$

and thence for the points of intersection the coordinates are

	b_1	b_2	b_3	b_4	b_5	b_6
a_1	*	0 0 − 1	0 0 − ·510	0 0 − ·052	0 0 + 1	0 0 + ·624
a_2	− ·170 + ·446 + ·105	*	+ ·662 + ·996 − 1	+ 197· + 131· − 262· } i.e. lines a_2, b_4 nearly parallel.	− ·844 0 + 1	− ·336 + ·336 + ·325
a_3	− ·515 + ·782 − ·155	0 + 1·354 − 1	*	− 1·805 − 2·007 + 3·964	− 1·218 0 + 1	− ·641 − ·641 + ·053
a_4	− 1·071 + 1·323 − ·573	0 + 1·682 − 1	+ 1·438 + 2·164 − 1·574	*	− 5·012 0 + 1	− 1·259 + 1·259 − ·497
a_5	+ 3·189 − 2·849 + 2·649	0 + ·679 − 1	+ ·259 + ·291 − ·702	+ ·383 + ·255 − ·561	*	+ 6·410 − 6·410 + 6·333
a_6	+ ·241 + ·041 + ·417	0 + ·142 − 1	− ·074 + ·112 − ·565	+ ·131 + ·087 − ·226	+ ·340 0 + 1	*

II.

I have in a paper "On the double-sixers of a cubic surface," *Quart. Math. Journal*, t. x. (1870), pp. 58—71, [459], obtained analytical expressions for the twelve lines of a double-sixer, and also calculated numerical values, which however (as there remarked) did not come out convenient ones for the construction of a figure. A different mode of treatment since occurred to me, by means of the following equation of the cubic surface

$$\left(\frac{x}{\alpha'}-\frac{y}{\beta'}+\frac{z}{\gamma'}-\frac{w}{\delta'}\right)\left(\frac{xz}{\alpha\gamma}-\frac{yw}{\beta\delta}\right)-k\left(\frac{x}{\alpha}+\frac{y}{\beta}+\frac{z}{\gamma}-\frac{w}{\delta}\right)\left(\frac{xz}{\alpha'\gamma'}-\frac{yw}{\beta'\delta'}\right)=0,$$

which as will appear is a very convenient one for the purpose. We in fact obtain at once eight lines of the double-sixer; viz. these are

1. $x=0,\ w=0,$	2′. $x=0,\ y=0,$
3. $y=0,\ z=0,$	4′. $z=0,\ w=0,$
5. $\frac{x}{\alpha}-\frac{y}{\beta}=0,\ \frac{z}{\gamma}-\frac{w}{\delta}=0,$	5′. $\frac{x}{\alpha'}-\frac{w}{\delta'}=0,\ \frac{y}{\beta'}-\frac{z}{\gamma'}=0,$
6. $\frac{x}{\alpha'}-\frac{y}{\beta'}=0,\ \frac{z}{\gamma'}-\frac{w}{\delta'}=0,$	6′. $\frac{x}{\alpha}-\frac{w}{\delta}=0,\ \frac{y}{\beta}-\frac{z}{\gamma}=0;$

and also five lines not belonging to the double-sixer, viz.

12. $x=0,\quad \left(-\frac{y}{\beta'}+\frac{z}{\gamma'}-\frac{w}{\delta'}\right)\frac{1}{\beta\delta}-k\left(-\frac{y}{\beta}+\frac{z}{\gamma}-\frac{w}{\delta}\right)\frac{1}{\beta'\delta'}=0,$

23. $y=0,\quad \left(\frac{x}{\alpha'}+\frac{z}{\gamma'}-\frac{w}{\delta'}\right)\frac{1}{\alpha\gamma}-k\left(\frac{x}{\alpha}+\frac{z}{\gamma}-\frac{w}{\delta}\right)\frac{1}{\alpha'\gamma'}=0,$

34. $z=0,\quad \left(\frac{x}{\alpha'}-\frac{y}{\beta'}-\frac{w}{\delta'}\right)\frac{1}{\beta\delta}-k\left(\frac{x}{\alpha}-\frac{y}{\beta}-\frac{w}{\delta}\right)\frac{1}{\beta'\delta'}=0,$

41. $w=0,\quad \left(\frac{x}{\alpha'}-\frac{y}{\beta'}+\frac{z}{\gamma'}\right)\frac{1}{\alpha\gamma}-k\left(\frac{x}{\alpha}-\frac{y}{\beta}+\frac{z}{\gamma}\right)\frac{1}{\alpha'\gamma'}=0,$

56. $\frac{x}{\alpha}-\frac{y}{\beta}+\frac{z}{\gamma}-\frac{w}{\delta}=0,\quad \frac{x}{\alpha'}-\frac{y}{\beta'}+\frac{z}{\gamma'}-\frac{w}{\delta'}=0.$

The remaining lines of the double-sixer are then easily determined; viz. the lines 3, 5, 6, and 12 are met by the line 2′, and by a second line 1′; this, as a line meeting 3, 5, 6, will be given by equations of the form

$$x-\frac{\alpha}{\beta}y=\phi\left(\frac{\delta}{\gamma}z-w\right),\quad x-\frac{\alpha'}{\beta'}y=\phi\left(\frac{\delta'}{\gamma'}z-w\right),$$

and observing that these equations, writing therein $x=0$, give

$$\frac{z}{\gamma}-\frac{w}{\delta}=-\frac{\alpha}{\beta\delta\phi}y,\quad \frac{z}{\gamma'}-\frac{w}{\delta'}=-\frac{\alpha'}{\beta'\delta'\phi}y,$$

the condition of intersection with the line 12 gives

$$\phi = -\frac{\alpha' - k\alpha}{\delta' - k\delta},$$

which is the value of ϕ in the foregoing equations: and to these we may join the resulting equation

$$y\gamma\gamma'(\alpha\beta' - \alpha'\beta) = z\phi\beta\beta'(\gamma\delta' - \gamma'\delta).$$

Proceeding in like manner for the lines 3′, 2, 4, the equations for the remaining four lines of the double-sixer are

2. $$\phi = \frac{\alpha' - k\alpha}{\beta' - k\beta},$$

$$x - w\frac{\alpha'}{\delta'} = \phi\left(y - z\frac{\beta'}{\gamma'}\right),$$

$$x - w\frac{\alpha}{\delta} = \phi\left(y - z\frac{\beta}{\gamma}\right),$$

$$w\gamma\gamma'(\alpha\delta' - \alpha'\delta) = z\phi\delta\delta'(\beta\gamma' - \beta'\gamma).$$

4. $$\phi = \frac{\gamma' - k\gamma}{\delta' - k\delta},$$

$$\phi\left(x\frac{\delta'}{\alpha'} - w\right) = y\frac{\gamma'}{\beta'} - z,$$

$$\phi\left(x\frac{\delta}{\alpha} - w\right) = y\frac{\gamma}{\beta} - z,$$

$$x\phi\beta\beta'(\alpha\delta' - \alpha'\delta) = y\alpha\alpha'(\beta\gamma' - \beta'\gamma).$$

1′. $$\phi = \frac{\alpha' - k\alpha}{\delta' - k\delta},$$

$$x - y\frac{\alpha'}{\beta'} = \phi\left(z\frac{\delta'}{\gamma'} - w\right),$$

$$x - y\frac{\alpha}{\beta} = \phi\left(z\frac{\delta}{\gamma} - w\right),$$

$$y\gamma\gamma'(\alpha\beta' - \alpha'\beta) = z\phi\beta\beta'(\gamma\delta' - \gamma'\delta).$$

3′. $$\phi = -\frac{\gamma' - k\gamma}{\beta' - k\beta},$$

$$\phi\left(x\frac{\beta'}{\alpha'} - y\right) = z - w\frac{\gamma'}{\delta'},$$

$$\phi\left(x\frac{\beta}{\alpha} - y\right) = z - w\frac{\gamma}{\delta},$$

$$x\phi\delta\delta'(\alpha\beta' - \alpha'\beta) = w\alpha\alpha'(\gamma\delta' - \gamma'\delta).$$

It may be added that:—

In plane $x = 0$,

intersection of 1′ lies on line $z : w = (\alpha\beta' - \alpha'\beta)\gamma\gamma' : \alpha\gamma\beta'\delta' - \alpha'\gamma'\beta\delta$,

" 2 " $y : z = \alpha\gamma\beta'\delta' - \alpha'\gamma'\beta\delta : (\alpha\delta' - \alpha'\delta)\gamma\gamma'$,

and that the line joining these intersections is the line 12.

In plane $y = 0$,

intersection of 2 lies on line $x : w = \alpha\gamma\beta'\delta' - \alpha'\gamma'\beta\delta : -(\beta\gamma' - \beta'\gamma)\delta\delta'$,

" 3′ " $z : w = \alpha\gamma\beta'\delta' - \alpha'\gamma'\beta\delta : (\alpha\beta' - \alpha'\beta)\delta\delta'$,

and that the line joining these intersections is the line 23.

In plane $z = 0$,

$$\text{intersection of } 3' \text{ lies on line } x : y = (\gamma\delta' - \gamma'\delta)\,\alpha\alpha' : \alpha\gamma\beta'\delta' - \alpha'\gamma'\beta\delta,$$

$$\text{,, } \quad 4 \quad \text{,, } \quad x : w = -(\beta\gamma' - \beta'\gamma)\,\alpha\alpha' : \alpha\gamma\beta'\delta' - \alpha'\gamma'\beta\delta,$$

and that the line joining these intersections is the line 34.

And in plane $w = 0$,

$$\text{intersection of } 4 \text{ is on line } y : z = (\alpha\delta' - \alpha'\delta)\,\beta\beta' : \alpha\gamma\beta'\delta' - \alpha'\gamma'\beta\delta,$$

$$\text{,, } \quad 1' \quad \text{,, } \quad x : y = \alpha\gamma\beta'\delta' - \alpha'\gamma'\beta\delta : (\gamma\delta' - \gamma'\delta)\,\beta\beta',$$

and that the line joining these intersections is the line 14.

The equations of the remaining ten lines of the surface may be obtained without difficulty, and also the forty-five triple planes, but I do not stop to effect this; the planes $x=0$, $y=0$, $z=0$, $w=0$, are, it is clear, triple planes, containing the lines 1, 2′, 12; 2′, 3, 23; 3, 4′, 34; and 4′, 1, 41 respectively.

If, to fix the ideas, the planes $x=0$, $y=0$, $z=0$, $w=0$ are taken to be those of the tetrahedron $ABCD$ ($x = BCD$ &c., as usual), then the edges AB, BC, CD, DA (but not the remaining opposite edges AC, BD) will be lines on the surface. Each plane of the tetrahedron, for instance ABC ($w=0$), is met by the ten lines not contained therein in two vertices A, C, three points on the edge BA, three points on the edge BC, and two other points, viz. these are the intersections of the plane ABC by the lines 4 and 1′. For the construction of a model it is sufficient to determine the three points on each edge, and the two points, say in the plane ABC and in the plane DBC ($x=0$) respectively; for then each of the remaining eight lines will be determined as a line joining two points in these two planes respectively. If in the first instance k is considered as a variable parameter, then the two points in the plane $w=0$ are given as the intersections of two fixed lines by a variable line (14) rotating round the fixed point $\frac{x}{\alpha} - \frac{y}{\beta} + \frac{z}{\gamma} = 0$, $\frac{x}{\alpha'} - \frac{y}{\beta'} + \frac{z}{\gamma'} = 0$; and the like as regards the two points in the plane $x=0$. By making (with assumed values of the other parameters) the proper drawings for the two planes $w=0$, $x=0$, it is easy to fix upon a convenient value of the parameter k; and I have in this manner succeeded in making a string model of the double-sixer; viz. the coordinates x, y, z, w are taken to be as the perpendicular distances of the current point from the faces of a regular tetrahedron (the coordinates being positive for an interior point); the values of α, β, γ, δ were put $= 3, 4, 5, 6$ and those of α', β', γ', $\delta' = 1, 1, 1, 1$; the value of k fixed upon as above was $k = -\frac{1}{8}$; this however brings the lines 2 and 4 too close together (viz. the shortest distance between them is not great enough), and also their apparent intersection too close to their intersections with the line 6′; and it is probable that a slightly different value of k would be better.

The results just obtained may be exhibited in a compendious form as follows:

	x :	y :	z :	w
1 is line BC	0			0
2′ „ CD	0	0		
3 „ DA		0	0	
4′ „ AB			0	0
5 meets CD			γ	δ
„ AB	α	β		
6 „ CD			γ'	δ'
„ AB	α'	β'		
6′ „ BC		β	γ	
„ AD	α			δ
5′ „ BC		β'	γ'	
„ AD	α'			δ'
1′ „ AD	$\alpha'-k\alpha$			$\delta'-k\delta$
„ BCD		$-(\alpha'-k\alpha)(\gamma\delta'-\gamma'\delta)\beta\beta'$	$(\delta'-k\delta)(\alpha\beta'-\alpha'\beta)\gamma\gamma'$	$(\delta'-k\delta)(\alpha\gamma\beta'\delta'-\alpha'\gamma'\beta\delta)$
„ ABC	$(\alpha'-k\alpha)(\alpha\gamma\beta'\delta'-\alpha'\gamma'\beta\delta)$	$(\alpha'-k\alpha)(\gamma\delta'-\gamma'\delta)\beta\beta'$	$-(\delta'-k\delta)(\alpha\beta'-\alpha'\beta)\gamma\gamma'$	
3′ „ BC		$\beta'-k\beta$	$\gamma'-k\gamma$	
„ ACD	$-(\beta'-k\beta)(\gamma\delta'-\gamma'\delta)\alpha\alpha'$		$(\gamma'-k\gamma)(\alpha\gamma\beta'\delta'-\alpha'\gamma'\beta\delta)$	$(\gamma'-k\gamma)(\alpha\beta'-\alpha'\beta)\delta\delta'$
„ ABD	$(\beta'-k\beta)(\gamma\delta'-\gamma'\delta)\alpha\alpha'$	$(\beta'-k\beta)(\alpha\gamma\beta'\delta'-\alpha'\gamma'\beta\delta)$		$-(\gamma'-k\gamma)(\alpha\beta'-\alpha'\beta)\delta\delta'$
2 „ AB	$\alpha'-k\alpha$	$\beta'-k\beta$		
„ BCD		$(\beta'-k\beta)(\alpha\gamma\beta'\delta'-\alpha'\gamma'\beta\delta)$	$(\beta'-k\beta)(\alpha\delta'-\alpha'\delta)\gamma\gamma'$	$(\alpha'-k\alpha)(\beta\gamma'-\beta'\gamma)\delta\delta'$
„ ACD	$(\alpha'-k\alpha)(\alpha\gamma\beta'\delta'-\alpha'\gamma'\beta\delta)$		$-(\beta'-k\beta)(\alpha\delta'-\alpha'\delta)\gamma\gamma'$	$-(\alpha'-k\alpha)(\beta\gamma'-\beta'\gamma)\delta\delta'$
4 „ CD			$\gamma'-k\gamma$	$\delta'-k\delta$
„ ABD	$-(\delta'-k\delta)(\beta\gamma'-\beta'\gamma)\alpha\alpha'$	$-(\gamma'-k\gamma)(\alpha\delta'-\alpha'\delta)\beta\beta'$		$(\delta'-k\delta)(\alpha\gamma\beta'\delta'-\alpha'\gamma'\beta\delta)$
„ ABC	$(\delta'-k\delta)(\beta\gamma'-\beta'\gamma)\alpha\alpha'$	$(\gamma'-k\gamma)(\alpha\delta'-\alpha'\delta)\beta\beta'$	$(\gamma'-k\gamma)(\alpha\gamma\beta'\delta'-\alpha'\gamma'\beta\delta)$	

1′ and 2 meet *BCD* on line $\left(-\frac{y}{\beta'}+\frac{z}{\gamma'}-\frac{w}{\delta'}\right)\frac{1}{\beta\delta}-k\left(-\frac{y}{\beta}+\frac{z}{\gamma}-\frac{w}{\delta}\right)\frac{1}{\beta'\delta'}=0,$

2 and 3′ „ *CDA* „ $\left(\frac{x}{\alpha'}+\frac{z}{\gamma'}-\frac{w}{\delta'}\right)\frac{1}{\alpha\gamma}-k\left(\frac{x}{\alpha}+\frac{z}{\gamma}-\frac{w}{\delta}\right)\frac{1}{\alpha'\gamma'}=0,$

3′ and 4 „ *DAB* „ $\left(\frac{x}{\alpha'}-\frac{y}{\beta'}-\frac{w}{\delta'}\right)\frac{1}{\beta\delta}-k\left(\frac{x}{\alpha}-\frac{y}{\beta}-\frac{w}{\delta}\right)\frac{1}{\beta'\delta'}=0,$

4 and 1′ „ *ABC* „ $\left(\frac{x}{\alpha'}-\frac{y}{\beta'}+\frac{z}{\gamma'}\right)\frac{1}{\alpha\gamma}-k\left(\frac{x}{\alpha}-\frac{y}{\beta}+\frac{z}{\gamma}\right)\frac{1}{\alpha'\gamma'}=0;$

or calculating the numerical values from the foregoing assumed data, say

	x	$:y$	$:z$	$:w$	
1 is line *BC*	0			0	
2′ ... *CD*	0	0			
3 ... *DA*		0	0		
4′ ... *AB*			0	0	
5 meets *CD*			5	6	$z=45{\cdot}5,\ w=54{\cdot}5.$
... *AB*	3	4			$x=42{\cdot}9,\ y=57{\cdot}1.$
6 meets *CD*			1	1	$z=50,\ w=50.$
... *AB*	1	1			$x=50,\ y=50.$
6′ meets *BC*		4	5		$y=44{\cdot}4,\ z=55{\cdot}6.$
... *AD*	3			6	$x=33{\cdot}3,\ w=66{\cdot}7.$
5′ meets *BC*		1	1		$y=50,\ z=50.$
... *AD*	1			1	$x=50,\ w=50.$
1′ meets *AD*	11			14	$x=44,\ z=56.$
... *BCD*		−44	70	126	$y=-25{\cdot}1,\ z=39{\cdot}9,\ w=71{\cdot}8.$
... *ABC*	99	44	−70		Not required.
3′ meets *BC*		12	13		$y=48,\ z=52.$
... *ACD*	−36		117	78	Not required.
... *ABD*	36	108		−78	$x=47{\cdot}2,\ y=141{\cdot}7,\ w=-102{\cdot}3.$
2 meets *AB*	11	12			$x=47{\cdot}8,\ y=52{\cdot}2.$
... *BCD*		100	180	66	$y=26{\cdot}4,\ z=44,\ w=16{\cdot}2.$
... *ACD*	−99		189	66	Not required.
4 meets *CD*			13	14	$z=48{\cdot}1,\ w=51{\cdot}9.$
... *ABD*	42	156		−126	$x=50{\cdot}5,\ y=187{\cdot}6,\ w=-151{\cdot}5.$
... *ABC*	42	156	117		Not required.

$$
\begin{array}{llll}
1' \text{ and } 2 & \text{meet } BCD & \text{on line} & 35y-32z+30w=0, \\
2 \text{ and } 3' & \ldots\ CDA & \ldots\ldots & 26x+22z-21w=0, \\
3' \text{ and } 4 & \ldots\ DAB & \ldots\ldots & 8x-\ 7y-\ 6w=0, \\
4' \text{ and } 1 & \ldots\ ABC & \ldots\ldots & 52x-47y-44z=0,
\end{array}
$$

which last four equations serve as a verification.

The outside numerical values are given in the manner most convenient for the construction of a drawing; viz. when the coordinates refer to a point on an edge of the tetrahedron, or say on the side of an equilateral triangle, then taking the length of this edge (or side) to be $=100$, the numerical values are fixed so that the sum of the two coordinates may be $=100$, and the two coordinates thus denote the distances from the extremities of the edge or side: but when the three coordinates belong to a point in the face of the tetrahedron, or say in the plane of an equilateral triangle, then the sum of the coordinates is made $=86{\cdot}6$, and the three coordinates thus denote the perpendicular distances from the sides of the triangle.

III.

It is possible to find on a cubic curve a double-sixer of points 1, 2, 3, 4, 5, 6 and 1′, 2′, 3′, 4′, 5′, 6′ such that any six points such as 1, 2, 3, 4′, 5′, 6′ lie in a conic. In fact considering a cubic surface having upon it the double-sixer of lines 1, 2, 3, 4, 5, 6 and 1′, 2′, 3′, 4′, 5′, 6′, the section by any plane is a cubic curve meeting the lines, say in the points 1, 2, 3, 4, 5, 6, 1′, 2′, 3′, 4′, 5′, 6′: each of the lines 1, 2, 3 meets each of the lines 4′, 5′, 6′, and consequently the six lines lie in a quadric surface: therefore the points 1, 2, 3, 4′, 5′, 6′ lie in a conic: and so in the other cases; the number of the conics is of course $=60$.

The cubic curve may be a given curve, and six of the points upon it (not being points on a conic) may also be taken to be given; for instance the points 1, 2, 3, 1′, 4′, 5′. For take through the points 2, 3 respectively any two lines 1, 2; through 1′, 4′, 5′ respectively the lines 1′, 4′, 5′ each meeting each of the lines 2, 3: and through 1 a line meeting each of the lines 4′, 5′. It is easy to see that a cubic surface may be drawn through the cubic curve and the lines 1, 2, 3, 1′, 4′, 5′: for the passage through the cubic curve requires 9 conditions; the surface then passes through the point 2 and to make it pass through the line 2 requires 3 conditions; similarly the surface passes through the point 3, and to make it pass through the line 3 requires 3 conditions. The surface now passes through 1′ and through the points of intersection of the line 1′ with the lines 2, 3: to make it pass through the line 1′ requires 1 condition; similarly to make it pass through the lines 4′, 5′, 1 requires in each case 1 condition; or there are in all 19 conditions, so that the cubic surface is completely determined. Take now through the points 1, 2, 3, 4′, 5′, a conic meeting the cubic in the point 6′: then through the lines 1, 2, 3, 4′, 5′ we have a quadric surface passing through this conic, and therefore through 6′: hence through 6′ we may draw a line 6′ meeting each of the lines 1, 2, 3; and since the cubic surface passes through the point 6′ and also through the intersections of the line 6′ with the lines 1, 2, 3, it passes through the line 6′. We complete in this manner by constructions in the plane of the cubic the system of the twelve points, viz. each new point is given as the intersection of the cubic curve by a conic drawn through five points of the cubic curve. It is then shown as for the point 6′ and the line 6′ through it, that through each new point there can be drawn a line denoted by the same number and meeting each of the lines which it ought to meet, and hence lying on the cubic surface: the twelve points are thus the intersections of the plane of the cubic curve by the twelve lines of the double-sixer; and it follows that the six points which ought to lie in a conic (in every case where such conic has not been used in the plane construction) do actually lie in a conic.

I was anxious to construct such a double-sixer of points on a cubic curve; for this purpose I take the equation of the curve to be $y^2 = \left(1 - \frac{x}{a}\right)\left(1 - \frac{x}{b}\right)\left(1 - \frac{x}{c}\right)$, or say for shortness $y^2 = X$; where, to fix the ideas, a, b are supposed to be positive, a greater than b; and c to be negative.

The cubic curve is thus a parabola symmetrical in regard to the axis of x, and consisting of a loop and infinite branch; and I take upon it the points 1, 2, 3, 1′, 4′, 5′

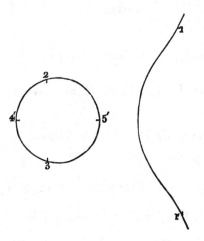

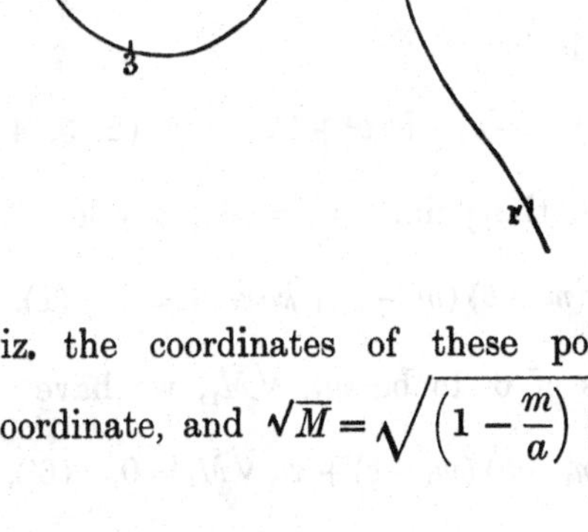

as shown in the figure, viz. the coordinates of these points are as stated in the Table, where m is the x coordinate, and $\sqrt{M} = \sqrt{\left(1 - \frac{m}{a}\right)\left(1 - \frac{m}{b}\right)\left(1 - \frac{m}{c}\right)}$ and so in other cases, $\sqrt{14} = 3{\cdot}74165$.

	x	y	x	y
1	m	$\sqrt{M}$	6	$\sqrt{14}=3{\cdot}742$
2	0	1	0	1
3	0	-1	0	-1
4	θ	$\sqrt{\Theta}$	$2-\frac{75}{1369}=1{\cdot}945$	$-\frac{2280}{(37)^3}\sqrt{14}=-{\cdot}168$
5	ϕ	$\sqrt{\Phi}$	$-1+\frac{75}{361}=-0{\cdot}792$	$\frac{1110}{(19)^3}\sqrt{14}={\cdot}606$
6	m_1	$-\sqrt{M_1}$	$\frac{13}{3}=4{\cdot}333$	$-\frac{4}{9}\sqrt{14}=-1{\cdot}641$
1′	m	$-\sqrt{M}$	6	$-\sqrt{14}=-3{\cdot}742$
2′	σ	$\sqrt{\Sigma}$	$-\frac{1560\,(14-\sqrt{14})}{(31\sqrt{14}-5)^2}=1{\cdot}299$	.. $=+{\cdot}676$
3′	τ	$\sqrt{\mathrm{T}}$	$-\frac{1560\,(14+\sqrt{14})}{(31\sqrt{14}+5)^2}=1{\cdot}887$	.. $=+{\cdot}247$
4′	c	0	-1	0
5′	b	0	2	0
6′	m_1	$\sqrt{M_1}$	$\frac{13}{3}=4{\cdot}333$	$\frac{4}{9}\sqrt{14}=1{\cdot}641.$

The numerical values belong to the curve $y^2=\left(1-\frac{x}{3}\right)\left(1-\frac{x}{2}\right)\left(1+x\right)$ and to $m=6$.

Starting with the points 1, 2, 3, 1′, 4′, 5′ we have to find the remaining points 6′, 6, 4, 5, 2′, 3′.

Point 6′ by means of the conic 1234′5′6′, as follows.

The equation of the conic is

$$(x-b)(x-c)-bc\,y^2+k\,xy=0,\ (2,\ 3,\ 4',\ 5'),$$

and making this pass through the point 1 ($x=m$, $y=\sqrt{M}$) we find

$$(m-b)(m-c)+ka\sqrt{M}=0.\quad (1).$$

Hence taking the coordinates of 6′ to be m_1, $\sqrt{M_1}$, we have

$$(m_1-b)(m_1-c)+ka\sqrt{M_1}=0,\quad (6'),$$

and thence

$$\frac{\sqrt{M_1}}{\sqrt{M}}=\frac{(m_1-b)(m_1-c)}{(m\ -b)(m\ -c)}=\frac{M_1}{M}\frac{(m\ -a)}{(m_1-a)},$$

that is,

$$\frac{\sqrt{M_1}}{\sqrt{M}} = \frac{(m_1 - b)(m_1 - c)}{(m - b)(m - c)} = \frac{m_1 - a}{m - a}.$$

We have thus for m, a quadric equation satisfied by $m = m_1$, so that throwing out the factor $m - m_1$, the equation is a linear one, viz. we find

$$m_1 = \frac{ma - ab - ac + bc}{m - a},$$

or, what is the same thing,

$$m_1 - a = \frac{(a - b)(a - c)}{m - a},$$

and thence also

$$\sqrt{M_1} = \frac{(a - b)(a - c)}{(m - a)^2} \sqrt{M},$$

viz. $\sqrt{M_1}$ is determined rationally in terms of m, $\sqrt{M}$; this is of course as it should be, since the point $6'$ is uniquely determinate.

Point 6 by means of the conic $2361'4'5'$.

In precisely the same manner the coordinates are m_1, $-\sqrt{M_1}$, where m_1, $\sqrt{M_1}$, denote the same quantities as before.

Point 4 by means of the conic $2341'5'6'$.

The equation of the conic is

$$Fx + Gy + H = \frac{1 - y^2}{x}, \qquad (2, 3),$$

where

$$Fb \qquad + H = \frac{1}{b}, \qquad (5')$$

$$Fm_1 + G\sqrt{M_1} + H = \frac{1 - M_1}{m_1}, \qquad (6')$$

$$Fm - G\sqrt{M} + H = \frac{1 - M}{m}, \qquad (1')$$

which give without difficulty

$$abc\,F = -a - c + P,$$

$$\sqrt{M}\,abc\,G = (m - b)(-m + P),$$

$$abc\,H = ab + ac + bc - bP,$$

where $P = 2a - c - \dfrac{2(a - c)(b - c)}{m + m_1 - 2c}$, a quantity which will presently be expressed in terms of m only.

And then

$$F\theta + G\sqrt{\Theta} + H = \frac{1 - \Theta}{\theta},$$

or say

$$F(\theta-b)+G\sqrt{\Theta}=\frac{1-\Theta}{\theta}-\frac{1}{b}$$

$$=-(\theta-b)\left(\frac{1}{ab}+\frac{1}{bc}-\frac{\theta}{abc}\right),$$

that is,

$$(\theta-b)(abc\,F+a+c-\theta)+G\,abc\sqrt{\Theta}=0,$$

viz. that is,

$$(\theta-b)(P-\theta)+(m-b)\frac{\sqrt{\Theta}}{\sqrt{M}}(P-m)=0$$

or, rationalising and throwing out the factor $\theta-b$, this is

$$(\theta-b)(\theta-P)^2-(m-b)(m-P)^2\frac{(\theta-a)(\theta-b)}{(m-a)(m-b)}=0,$$

which is a cubic equation satisfied by $\theta=m$ and $\theta=m_1$; so that throwing out the factors $\theta-m$, $\theta-m_1$ we have for θ a linear equation.

Putting for shortness

$$A=(m-a)^2-(a-b)(a-c),$$

$$B=(m-b)^2-(b-c)(b-a),$$

$$C=(m-c)^2-(c-a)(c-b),$$

the value of θ may be expressed in the forms

$$\theta-a=\frac{B^2}{C^2}(c-a),\quad \theta-b=\frac{A^2}{C^2}(c-b),\quad \theta-c=\frac{4(m-a)(m-b)(m-c)(b-c)(a-c)}{C^2}.$$

We have moreover

$$P-c=\frac{2(a-c)(m-b)(m-c)}{C},\quad P-m=-\frac{(m-c)A}{C},$$

equations which express P in terms of m only; also

$$\theta-P=\frac{-2(a-c)(m-b)(m-c)B}{C^2},$$

and then

$$\sqrt{\Theta}=-\sqrt{M}\,\frac{\theta-b}{m-b}\,\frac{P-\theta}{P-m},$$

whence

$$\sqrt{\Theta}=2\sqrt{M}(b-c)(c-a)\frac{AB}{C^3},$$

so that θ, $\sqrt{\Theta}$ are now determined.

Point 5 by means of the conic 2351′4′6′.

The conic is

$$Fx + Gy + H = \frac{1-y^2}{x}, \qquad (2,\ 3)$$

where

$$Fc. \qquad + H = \frac{1}{c}, \qquad (4')$$

$$Fm_1 + G\sqrt{M_1} + H = \frac{1-M_1}{m_1}, \qquad (6')$$

$$Fm - G\sqrt{M} + H = \frac{1-M}{m}, \qquad (1').$$

Everything is the same as for the point 4 except that b, c are interchanged: hence writing Q instead of P, and using A, B, C to denote as before, we have

$$abc\,F = -a - b + Q,$$

$$\sqrt{M}\,abc\,G = (m-c)(-m+Q),$$

$$abc\,H = ab + ac + bc - cQ,$$

and

$$\phi - a = \frac{C^2(b-a)}{B^2},$$

$$\phi - b = \frac{4(m-a)(m-b)(m-c)(c-b)(a-b)}{B^2},$$

$$\phi - c = \frac{A^2(b-c)}{B^2},$$

$$Q - b = \frac{2(m-b)(m-c)(a-b)}{B},$$

$$Q - m = -\frac{A(m-b)}{B},$$

$$\phi - Q = -\frac{2(m-b)(m-c)\,C\,(a-b)}{B^2},$$

and

$$\sqrt{\Phi} = 2\sqrt{M}(c-b)(b-a)\frac{AC}{B^3},$$

which determine ϕ, $\sqrt{\Phi}$.

Point 3′ by means of the conic 1263′4′5′.

The conic is

$$Fx + Gy + H = \frac{(x-b)(x-c)}{y}, \qquad (4',\ 5')$$

and we have

$$G + H = bc, \qquad (2)$$

$$Fm + G\sqrt{M} + H = \frac{(m-b)(m-c)}{\sqrt{M}}, \qquad (1)$$

$$Fm_1 - G\sqrt{M_1} + H = \frac{(m_1-b)(m_1-c)}{-\sqrt{M_1}}, \qquad (6).$$

Eliminating F, we have

$$G(m_1\sqrt{M} + m\sqrt{M_1}) + H(m - m_1) = \frac{m_1(m-b)(m-c)}{\sqrt{M}} + \frac{m(m_1-b)(m_1-c)}{\sqrt{M_1}},$$

which is easily reduced first to

$$G\frac{2mm_1 - a(m+m_1)}{(m-a)\sqrt{M}} + H(m-m_1) = (m+m_1)\frac{(m-a)(m-b)}{\sqrt{M}},$$

and then to

$$G\{aA + 2m(a-b)(a-c)\} - H\frac{(m-a)A}{\sqrt{M}} + abc\{-A + 2m(m-a)\} = 0;$$

and combining herewith $G + H = bc$, we have

$$H = \frac{2bc\,m[a(m-a) + (a-b)(a-c)]}{aA + 2m(a-b)(a-c) + \dfrac{(m-a)A}{\sqrt{M}}};$$

$$G = bc - H;$$

and we have then

$$F(m+m_1) + G(\sqrt{M} - \sqrt{M_1}) + 2H = 0,$$

that is,

$$F\{2m(m-a) - A\} + G\frac{A\sqrt{M}}{m-a} + 2H(m-a) = 0,$$

or, what is the same thing,

$$F\{2m(m-a) - A\} = -bc\frac{A\sqrt{M}}{m-a} - H\left\{2(m-a) - \frac{A\sqrt{M}}{m-a}\right\}.$$

We then have

$$Fx + H = y\left(-G + \frac{(x-b)(x-c)}{y^2}\right)$$

$$= y\left(-G - \frac{abc}{x-a}\right) = -\frac{y(Ha + Gx)}{x-a},$$

that is,

$$(Fx + H)^2 = -\frac{(x-b)(x-c)}{abc(x-a)}(Ha + Gx)^2,$$

or

$$abc\,(x-a)\,(Fx+H)^2+(x-b)\,(x-c)\,(Gx+Ha)^2=0.$$

Developing and throwing out the factor x, this is

$$\begin{aligned}&G^2x^3\\&+\{2a\,GH-(b+c)\,G^2+abc\,F^2\}\,x^2\\&+\{a^2H^2-2a\,(b+c)\,GH+bc\,G^2+abc\,(2FH-aF^2)\}\,x\\&+\{-(b+c)\,a^2H^2+2abc\,GH+abc\,(H^2-2aFH)\}=0.\end{aligned}$$

This must be satisfied by $x=m$, $x=m_1$; hence the left hand must be $=G^2(x-m)(x-m_1)(x-\sigma)$, or equating the constant terms we have

$$G^2mm_1\sigma=aH\{-2abc\,F+2bc\,G+(bc-ab-ac)\,H\},$$

which gives σ; and we then have

$$\sqrt{\Sigma}=-\frac{\sigma-a}{G\sigma+Ha}(F\sigma+H),$$

but I have not attempted the further reduction of these expressions.

The numerical values for the example are

$$3F=\frac{-140+62\sqrt{14}}{5+21\sqrt{14}},\quad G=\frac{-10+62\sqrt{14}}{5+21\sqrt{14}},\quad H=\frac{-104\sqrt{14}}{5+\sqrt{14}},$$

whence σ as in the Table.

Point $2'$ by means of the conic $1362'4'5'$.

The equation of the conic is

$$Fx+Gy+H=\frac{(x-b)(x-c)}{y},\qquad(4',\ 5')$$

where

$$-G+H=-bc,\qquad(3)$$

$$Fm+G\sqrt{M}+H=\frac{(m-b)(m-c)}{\sqrt{M}},\qquad(1)$$

$$Fm_1-G\sqrt{M_1}+H=\frac{(m_1-b)(m_1-c)}{-\sqrt{M_1}},\qquad(6),$$

which are the same as for point $3'$, if only we reverse the signs of F, H and $\sqrt{M}$, $\sqrt{M_1}$.

Hence the formulæ are

$$H = -\frac{2bc\,m\,[a(m-a)+(a-b)(a-c)]}{aA+2m(a-b)(a-c)-\dfrac{(m-a)A}{\sqrt{M}}},$$

$$G = \quad bc + H,$$

$$F\{2m(m-a)-A\} = -bc\frac{A\sqrt{M}}{m-a} - H\left\{2(m-a)+\frac{A\sqrt{M}}{m-a}\right\},$$

$$G^2\,mm_1\,\tau = aH\{-2abc\,F - 2bc\,G + (bc-ab-ac)\,H\},$$

which gives τ; and then

$$\sqrt{T} = \frac{(\tau-a)}{G\tau - Ha}(F\tau + H),$$

which are also unreduced.

The numerical values are

$$3F = \frac{140+62\sqrt{14}}{5-21\sqrt{14}},\quad G = \frac{-10-62\sqrt{14}}{5-21\sqrt{14}},\quad H = \frac{-104\sqrt{14}}{5-21\sqrt{14}},$$

whence τ as in the Table.

522.

NOTE ON THE THEORY OF INVARIANTS.

[From the *Mathematische Annalen*, vol. III. (1871), pp. 268—271.]

If two binary quantics $(a,..)(x, y)^n$, $(a',..)(x', y')^n$ are linearly transformable the one into the other, and if for the first of them P, Q are any two invariants whatever of the same degree, and P', Q' are the like invariants for the second of them, then we have

$$P : Q = P' : Q',$$

(or, what is the same thing, the absolute invariants have the same values for the two functions respectively); and the entire system of these equations constitutes only a $(n-3)$ fold relation between the two sets of coefficients. But the converse theorem, viz. that if the entire system of equations is satisfied, the two functions are linearly transformable the one into the other, is only true *sub modo.*

For instance, considering the two binary sextics

$$(0, 0, 0, d, e, f, g)(x, y)^6 \text{ and } (0, 0, 0, d', e', f', g')(x', y')^6,$$

or, what is the same thing,

$$(20d, 5e, 2f, g)(x, y)^3 y^3 \text{ and } (20d', 5e', 2f', g')(x', y')^3 y'^3,$$

the invariants of the two functions respectively are each and all of them $=0$, and yet the two functions are not in general linearly transformable the one into the other.

For they can be transformable only by the substitution

$$x = \lambda x' + \mu y', \quad y = \rho y';$$

or, what is the same thing, only if the cubic functions are transformable by the substitution $x = x' + ay'$, $y = y'$; and forming for these the seminvariants $ac - b^2$ and $a^2d - 3abc + 2b^3$ for the cubic $(a, b, c, d)(x, y)^3$, we have as the necessary condition for the transformability

$$(8df - 5e^2)^3 : (8d^2g - 12def + 5e^3)^2 = (8d'f' - 5e'^2)^3 : (8d'^2g' - 12d'e'f' + 5e'^3)^2.$$

To deduce this result from the theory of the sextic function, I observe that denoting by A, B, C, Δ, the values of the quadrinvariant, the sextinvariant, and the discriminant, as given in Salmon's *Higher Algebra*, Ed. 2, pp. 202—211, then in the particular case $a=0$, $b=0$, we have

$$A = -10\,d^2 + 15\,ce, \qquad B = d^4 - 3\,cd^2e + c^2e^2, \qquad C = -8\,d^6 + 36\,cd^4e - 39\,c^2d^2e^2 - 8\,c^3e^3, \qquad \Delta = 0,$$

and hence forming the new invariants

$$\mathrm{B} = 100\,B - A^2,$$
$$\Gamma = 1000\,C - 1200\,AB + 4A^3,$$

the values of these in the same particular case $a=b=0$ are

$$\mathrm{B} = 25\,c^2(8\,df - 5e^2) - 100\,c^3g, \qquad \Gamma = -2500\,c^3(8\,d^2g - 12\,def + 5e^3) + 3000\,c^4(10\,eg - 9f^2).$$

Taking now A, B, C, Δ as the invariants of the sextic, one of the conditions for the transformation is $\mathrm{B}^3 : \Gamma^2 = \mathrm{B}'^3 : \Gamma'^2$.

In the particular case $a=b=c=0$ and $a'=b'=c'=0$, the invariants vanish and the equation is satisfied identically. But if we assume in the first instance only $a=b=0$, $a'=b'=0$, then the terms contain the common factors c^6 and c'^6 respectively; and throwing these out, and then writing $c=0$, $c'=0$, we obtain the condition previously found in a different manner.

It will be observed that the condition is of the original form $P : Q = P' : Q'$, but with the difference that P, Q and the corresponding functions P', Q', are not invariants. As possessing the foregoing property these functions may however be called "imperfect invariants," it being understood that an imperfect invariant is not an invariant, and is not in any case included in the term "invariant" used without qualification.

And we may now establish the general theory as follows: Consider the similarly constituted special forms $(a, ..)(x, y, z, ..)^n$ and $(a', ..)(x', y', z', ..)^n$: to fix the ideas the coefficients $(a, ..)$ may be regarded as homogeneous functions of the elements $(\alpha, \beta, ..)$ which are either independent, or homogeneously connected together in any manner; and then the coefficients $(a', ..)$ will be the like functions of the elements $(\alpha', \beta', ..)$ which are either independent or (as the case may be) homogeneously connected in the like manner.

The entire series of functions $P, Q, \ldots$ of $(\alpha, \beta, ..)$, which are such that P, Q being of the same degree, and P', Q' being the like functions of (α', β'), we have for the linearly transformable functions $(a, ..)(x, y, z, ..)^n$ and $(a', \ldots)(x', y', z', ..)^n$ the relation

$$P : Q = P' : Q',$$

may be called the "perfect and imperfect invariants" of $(a,..)(x, y, z,..)^n$; and the relation in question be briefly referred to by the expression that the perfect and imperfect invariants are proportional.

We have then the theorem that if the two functions $(a,..)(x, y, z,..)^n$ and $(a',..)(x', y', z',..)^n$ are linearly transformable the one into the other, the two functions have their perfect and imperfect invariants proportional; and *conversely* the theorem, that two functions which have their perfect and imperfect invariants proportional, are linearly transformable the one into the other.

There is thus a wide field of inquiry in regard to the imperfect invariants, even of a binary function, but still more so as to those of a ternary or quaternary function representing a curve or surface possessed of singularities.

We have in what precedes the explanation of an error into which I fell in my paper "On the transformation of plane curves," *Proc. Lond. Math. Soc.*, vol. I. No. 3, Oct. 1865, [384], see Arts. Nos. 27—30. Considering a given curve of deficiency D and, by means of a system of $D-3$ points chosen at pleasure on the curve, transforming this into a curve of the order $D+1$ with deficiency D; then for any two of the transformed curves (that is, two curves obtained by means of different systems of the $D-3$ points) I showed that these had the same absolute invariants—or in the language of the present paper, that they had their invariants proportional, and I thence inferred that the two transformed curves were linearly transformable the one into the other—whereas, to sustain this conclusion, it is necessary that the two curves should have their perfect and imperfect invariants proportional; and this was in no wise proved. That the two transformed curves are not in fact linearly transformable the one into the other has since been shown *a posteriori* by Dr Brill in the particular case $D=4$. Riemann's conclusions, with which my own were at variance, are thus correct.

I remark that if a binary function of an odd or even degree $n=2p+1$ or $=2p$, has $p+1$ equal factors, then the invariants all of them vanish; but the equality of the $p+1$ factors implies only a p-fold relation between the coefficients; that is, the vanishing of all the invariants gives only a p-fold relation between the coefficients, viz. the relation is $\frac{1}{2}(n-1)$ fold or $\frac{1}{2}n$-fold according as n is odd or even. Thus for a sextic function the equations $A=0$, $B=0$, $C=0$, $\Delta=0$ constitute only a 3-fold relation between the coefficients.

Similarly if the function has p equal factors, then every invariant is a mere numerical multiple of a power of one and the same function Θ; so that the vanishing invariants can be at once formed. And we have thus only a $(p-1)$ fold relation between the coefficients, viz. the relation is $\frac{1}{2}(n-3)$ fold or $\frac{1}{2}(n-2)$ fold according as n is odd or even.

Cambridge, 4 August, 1870.

523.

ON THE TRANSFORMATION OF UNICURSAL SURFACES.

[From the *Mathematische Annalen*, vol. III. (1871), pp. 469—474.]

I CONSIDER the question of the transformation (Abbildung auf einer Ebene) of unicursal surfaces. Taking (x, y, z, w) for the coordinates of a point on the surface, (x', y', z') for those of the corresponding point on the plane; then if X', Y', Z', W' denote each of them a function $(x', y', z')^{n'}$, the equations of transformation are

$$x : y : z : w = X' : Y' : Z' : W':$$

and assuming that each of the curves

$$X' = 0, \quad Y' = 0, \quad Z' = 0, \quad W' = 0$$

(or, what is the same thing, the general curve

$$aX' + bY' + cZ' + dW' = 0)$$

passes once through each of α_1 points, twice through each of α_2 points, ..., r times through each of α_r points (for convenience I write α_r instead of α_r'); and writing also

$$n = n'^2 - \Sigma r^2 \alpha_r,$$

$$0 = \tfrac{1}{2}n'(n'+3) - 3 - \Theta - \Sigma \tfrac{1}{2}r(r+1)\,\alpha_r,$$

(where Θ is $= 0$ or positive except in the case of special relations between the positions of the fixed points $\alpha_1, \alpha_2, \ldots, \alpha_r$), which equations give

$$-n = 3n' - 6 - 2\Theta - \Sigma r\alpha_r\,;$$

then the order of the surface is $= n$, and the order of the nodal curve is $b = \tfrac{1}{2}(n-2)(n-3) + \Theta$. I assume that the nodal curve has h apparent double points and t actual triple points, but no stationary points, so that q being the class, we have $q = b^2 - b - 2h - 6t$; and I endeavour to find these numbers q, t, h.

For this purpose, imagine through the nodal curve a surface of the order k, which therefore meets the surface besides in a curve of the order $nk-2b$; this curve I call the k-thic residue of the nodal curve, or simply the "residue." The projection (Abbildung) of the complete intersection kn is a curve of the order kn' passing kr times through each of the points α_r: this is made up of the projection of the nodal curve *once*, and of the projection of the residue. But as shown by Dr Clebsch the projection of the nodal curve is of the order $(n-4)n'+3$, and it passes $(n-4)r+1$ times through each of the points α_r; hence the projection of the residue is of the order $(k-n+4)n'-3$, and it passes $(k-n+4)r-1$ times through each of the points α_r. I assume that the projection of the residue is the *general* curve which satisfies the foregoing conditions, viz. that the residue, and its projection as defined by the foregoing conditions, *depend each of them on the same number of constants.* The necessity for this is I confess by no means obvious: but take as an illustration Steiner's quartic surface as transformed by the equations $x : y : z : w = x'^2 : y'^2 : z'^2 : (x'+y'+z')^2$: the nodal curve consists of three lines meeting in a point, the quadric residue is the remaining intersection of the surface by a quadric cone passing through the three lines; and the projection thereof is a line; the quadric cone, and therefore the conic, each depend upon 2 constants; and the line which is the projection of the conic depends upon the same number (2) of constants: at all events I make the assumption provisionally.

Now in the projection of the residue, we have twice the number of constants

$$=[(k-n+4)n'-3](k-n+4)n'-\Sigma\,[(k-n+4)r-1](k-n+4)r\alpha_r,$$

viz. this is

$$=(k-n+4)^2(n'^2-\Sigma r^2\alpha_r)+(k-n+4)(-3n'+\Sigma r\alpha_r),$$

or, what is the same thing, it is

$$=(k-n+4)^2 n+(k-n+4)(n-6-2\Theta),$$

viz. reducing, and replacing Θ by its value $=-\tfrac{1}{2}(n-2)(n-3)+b$, the number in question is

$$=k^2n+k(-n^2+4n-2b)+2(n-4)b.$$

Now k being $=$ or $>n-3$, the curve of intersection of a given surface n by a surface k depends on

$$\tfrac{1}{6}(k+1)(k+2)(k+3)-\tfrac{1}{6}(k-n+1)(k-n+2)(k-n+3)-1$$

constants; and making the surface k to pass through the curve b we have to subtract herefrom $(k+1)b-\tfrac{1}{2}g-2t$; that is, for the residue, twice the number of constants is

$$=\tfrac{1}{3}(k+1)(k+2)(k+3)-\tfrac{1}{3}(k-n+1)(k-n+2)(k-n+3)-2-2(k+1)b+q+4t,$$

viz. this is

$$=k^2n+k(-n^2+4b-2b)+\tfrac{1}{3}(n-1)(n-2)(n-3)-2b+q+4t.$$

Hence comparing the two expressions in question we have

$$2(n-4)b = \tfrac{1}{3}(n-1)(n-2)(n-3) - 2b + q + 4t,$$

that is,

$$0 = \tfrac{1}{3}(n-1)(n-2)(n-3) - 2(n-3)b + q + 4t,$$

or, as I prefer to write it,

$$0 = \tfrac{1}{6}(n-1)(n-2)(n-3) - (n-3)b + \tfrac{1}{2}q + 2t;$$

which agrees with a more general formula in my "Memoir on the theory of Reciprocal Surfaces," *Phil. Trans.* vol. CLIX. (1869), [411], see p. 227, [*Coll. Math. Papers*, vol. VI. p. 356]. I consider any two residues, a k-thic residue and a l-thic residue; to each intersection of these there corresponds an intersection of their projections: or the number of intersections of the two residues must be equal to that of the two projections. Now the projections being (as above)

order $(k-n+4)n'-3$ passing $(k-n+4)r-1$ times through each point α_r,
„ $(l-n+4)n'-3$ „ $(l-n+4)r-1$ „ „ „

the number of the intersections in question is

$$= [(k-n+4)n'-3][(l-n+4)n'-3] - \Sigma[(k-n+4)r-1][(l-n+4)r-1]\alpha_r + \omega,$$

where for a reason which will be afterwards explained I have added the term ω: this is

$$= (k-n+4)(l-n+4)(n'^2 - \Sigma r^2\alpha_r) + (k+l-2n+8)(-3n' + \Sigma r\alpha_r) + 9 - (\Sigma\alpha_r - \omega),$$

viz. it is

$$= (k-n+4)(l-n+4)n + (k+l-2n+8)(n-6-2\Theta) + 9 - (\Sigma\alpha_r - \omega),$$

viz. substituting for Θ its value, $= -\tfrac{1}{2}(n-2)(n-3) + b$, and reducing, the number is

$$= kln - 2(k+l)b - n^3 + 8n^2 - 16n + 9 + 4(n-4)b - (\Sigma\alpha_r - \omega).$$

But the surfaces n, k, l, having in common the curve b which is a nodal curve on n, besides intersect in

$$kln - b(n + 2k + 2l - 4) + 2q + qt$$

points (Salmon's *Geometry of three Dimensions*, 2nd Ed. p. 283, except that in the formula as there given the singularity t is not taken account of); that is, the number of intersections of the two residues is

$$= kln - 2(k+l)b - (n-4)b + 2q + 9t,$$

which is equal to the number of intersections of the two projections (1): or comparing the numbers in question we have

$$-n^3 + 8n^2 - 16n + 9 + 4(n-4)b - (\Sigma\alpha_r - \omega) = -(n-4)b + 2q + 9t,$$

that is,

$$2q + 9t = 5(n-4)b - n^3 + 8n^2 - 16n + 9 - (\Sigma\alpha_r - \omega).$$

1 I remark that $n+\lambda$ being positive or not less than $n-3$, two $(n+\lambda)$ thic residues meet in $n(\lambda+4)(\lambda+6) - 12\lambda - 39 - 4(\lambda+4) - (\Sigma\alpha_r - \omega)$ points: in particular, two $(n-3)$-thic residues meet in $3n - 3 - 4\Theta - (\Sigma\alpha_r - \omega)$ points; and two $(n-2)$-thic residues meet in $8n - 15 - 8\Theta - (\Sigma\alpha_r - \omega)$ points.

But we have already found

$$2q + 8t = 4(n-3)b - \tfrac{2}{3}n^3 + 4n^2 - \tfrac{22}{3}n + 4,$$

and we have therefore

$$t = (n-8)b - \tfrac{1}{3}n^3 + 4n^2 - \tfrac{26}{3}n + 5 - (\Sigma\alpha_r - \omega),$$

and

$$q = -2(n-13)b + n^3 - 14n^2 + 31n - 18 + 4(\Sigma\alpha_r - \omega).$$

I obtain these results in a different manner by investigating expressions for the deficiency (Geschlecht) of the nodal residue $nk - 2b$ and for that of its projection.

First for the projection, we have

$$\begin{aligned}\text{Twice Deficiency} &= [(k-n+4)n' - 4][(k-n+4)n' - 5] \\ &\quad - \Sigma[(k-n+4)r - 1][(k-n+4)r - 2]\alpha_r + 2\omega,\end{aligned}$$

where I have added the term 2ω, as afterwards explained: this is

$$= (k-n+4)^2(n'^2 - \Sigma r^2\alpha_r) + (k-n+4)(-9n' + 3\Sigma r\alpha_r) + 20 - 2(\Sigma\alpha_r - \omega),$$

viz. it is

$$= (k-n+4)^2 n + (k-n+4)(3n - 18 - 6\Theta) + 20 - 2(\Sigma\alpha_r - \omega),$$

or substituting for Θ its value $-\tfrac{1}{2}(n-2)(n-3) + b$ and reducing, it is

$$= k^2n + k(n^2 - 4n - 6b) - 2n^3 + 16n^2 - 32n + 20 + 6(n-4)b - 2(\Sigma\alpha_r - \omega).$$

Next as regards the residue, the number h' of its apparent double points is obtained in terms of h and t by the formula

$$8h + 6t - 2h' = (kn - 4b)(k-1)(n-1) - 2b(k-1),$$

(Salmon, l. c., p. 284, except that the singularity t is not there taken account of); and we thence have

$$\begin{aligned}\text{Twice Deficiency} &= (kn-1)(kn-2) - 2h' \\ &= kn(k+n-4) + 4b^2 + b(-4n - 6k + 12) + 2 - 8h - 6t,\end{aligned}$$

or introducing q instead of h by the formula $4b^2 - 8h = 4q + 4b + 24t$, this is

$$= kn(k+n-4) + b(-4n - 6k + 12) + 4q + 4b + 2 + 18t,$$

viz. it is

$$= k^2n + k(n^2 - 4n - 6b) - 4(n-4)b + 4q + 18t.$$

So that comparing with the deficiency of the projection we have

$$-2n^3 + 16n^2 - 32n + 20 + 6(n-4)b - 2(\Sigma\alpha_r - \omega) = -4(n-4)b + 4q + 2 + 18t,$$

that is,

$$2q + 9t = 5(n-4)b - n^3 + 8n^2 - 16n + 9 - (\Sigma\alpha_r - \omega),$$

the same result as before.

The necessity for the term ω appears by the consideration that if we apply to the plane figure a Cremona-transformation, thus obtaining a new transformation of the surface, the value of $\Sigma\alpha_r$ will in general be altered; whereas the expressions for q, t should it is clear remain unaltered; and it arises as follows, viz. for certain transformations of the surface the curve of the order $(k-n+4)n'-3$, passing $(k-n+4)r-1$ times through each point α_r and assumed to be the projection of the residue, is not an indecomposable curve but contains a certain number ω of factors (each belonging to a unicursal curve definable by means of the number of its passages through the several points α_r), which factors are to be rejected in order to obtain the equation of the proper residue. Thus reverting to the transformation

$$x : y : z : w = x'^2 : y'^2 : z'^2 : (x'+y'+z')^2$$

of Steiner's surface, the projection of the quadric residue was (as already remarked) a line; applying to the plane figure the ordinary quadric (or inverse) transformation we introduce three fixed points, $(\alpha_2 = 3)$, say these are A, B, C; viz. in the new transformation of the surface the projection of any plane section is a quartic curve having a node at each of the fixed points: the projection of the residue ought clearly to be a conic through the three points; but according to the general formula it is a quintic having at each of these points a triple point: the quintic is in fact made up of the lines BC, CA, AB and of the conic which is the proper residue; viz. in the case in question there are 3 factors thrown out, or we have $\omega = 3$. To apply this to the second investigation of $2q+9t$, by comparison of the two deficiencies, observe that in general if a curve is made up of $\omega+1$ indecomposable curves, the deficiency of the compound curve is equal to the sum of the deficiencies of the component curves $-\omega$; hence if ω of the curves are unicursal, the deficiency of the compound curve is equal to that of the remaining curve $-\omega$; or, what is the same thing, the deficiency of the remaining curve is = that of the compound curve $+\omega$; and the addition of the term $+\omega$ to the expression for the deficiency is thus accounted for. It is easy to see that a like explanation applies to the first investigation of $2q+9t$.

I further remark, reverting to the equations

$$x : y : z : w = X' : Y' : Z' : W'$$

of the transformation, that the product of the ω factors is given as the common factor (if any) of the Jacobians

$$J(Y', Z', W'),\quad J(Z', W', X'),\quad J(W', X', Y') \text{ and } J(X', Y', Z').$$

Such common factor exists whenever we can by a Cremona-transformation of the plane figure reduce the number of the points α_r upon which the transformation of the surface depends; viz. for any given transformation of the surface, ω is equal to the excess of $\Sigma\alpha_r$ above the minimum value of $\Sigma\alpha_r$, or, what is the same thing, $\Sigma\alpha_r - \omega$ is equal to the minimum value of $\Sigma\alpha_r$, and is thus independent of the particular transformation. And of course if $\Sigma\alpha_r$ has this minimum value, viz. if the transformation is such that the number of the points α_r cannot be reduced by any Cremona-transformation of the plane figure, then we have $\omega = 0$. I presume that for the most simple transformation, that is, when n' has its least value, $\Sigma \alpha_r$ has also its least value, and consequently that ω is $= 0$.

Recapitulating, the results obtained are

$$q = -2(n-13)b + n^3 - 14n^2 + 31n - 18 + 4(\Sigma \alpha_r - \omega),$$
$$t = (n-8)b - \tfrac{1}{3}n^3 + 4n^2 - \tfrac{26}{3}n + 5 - (\Sigma \alpha_r - \omega),$$

where it will be recollected that

$$b = \tfrac{1}{2}(n-2)(n-3) + \Theta;$$

the formulæ are verified in the several cases:

n'	α_1	α_2'	n	Θ	ω	b	q	t	
2	2	0	2	0	1	0	0	0	Quadric surface
2	1	0	3	1	0	1	0	0	Cubic scroll
2	0	0	4	2	0	3	0	1	Steiner's quartic surface
3	6	0	3	0	0	0	0	0	Cubic surface
3	5	0	4	1	0	2	2	0	Quartic with nodal conic
2	8	1	4	0	0	1	0	0	Do. with nodal line
2	7	1	5	1	0	4	8	0	Quintic with nodal quadriquadric
2	11	0	5	0	0	3	4	0	Do. with nodal skew cubic
2	12	2	5	-1	0	2	0	0	Do. with two non-intersecting nodal lines

which are the transformations chiefly as yet examined: but the first-mentioned case (quadric surface, generalised stereographic projection), although as stated the formulæ are verified with the value $\omega = 1$, does not really come under the foregoing theory. It is interesting to see that they are verified in the last-mentioned case, belonging to a negative value of Θ, that is, to a special system of fixed points.

Cambridge, Dec. 5, 1870.

524.

ON THE DEFICIENCY OF CERTAIN SURFACES.

[From the *Mathematische Annalen*, vol. III. (1871), pp. 526—529.]

If a given point or curve is to be an ordinary or singular point or curve on a surface of the order n, this imposes on the surface a certain number of conditions, which number may be termed the "Postulation"; thus "Postulation of a given curve *quà* i-tuple curve on a surface n" will denote the number of conditions to be satisfied by the surface in order that the given curve may be an i-tuple curve on the surface.

The "deficiency" (Flächengeschlecht) of a given surface of the order n is

$=\frac{1}{6}(n-1)(n-2)(n-3)$ *less* deficiency-value of the several singularities; viz. as shown by Dr Noether, if the surface has a given i-tuple curve, the deficiency-value hereof is

$=$ Postulation of the curve *quà* $(i-1)$ tuple curve on a surface $n-4$;

and if the surface has an i-conical point, the deficiency-value hereof is

$=$ Postulation of the point *quà* $(i-2)$ conical point on a surface $n-4$; viz. this is $=\frac{1}{6}i(i-1)(i-2)$, and is thus independent of the order of the surface.

I remark that if the tangent-cone at the i-conical point has δ double lines and κ cuspidal lines, then the deficiency-value is

$$=\frac{1}{6}i(i-1)(i-2)+(i-2)(n-i-1)(\delta+\kappa).$$

In the case of a double or cuspidal curve i is $=2$, and the deficiency-value is

$=$ Postulation of given curve *quà* simple curve on a surface $n-4$;

and so for an ordinary conical point i is $=2$, and the deficiency-value is $=0$: results which were first obtained by Dr Clebsch.

I found in this manner the expression for the deficiency of a surface n having a double and cuspidal curve and the other singularities considered in my "Memoir on the Theory of Reciprocal Surfaces," *Phil. Trans.* vol. CLIX. (1869), [411, *Coll. Math. Papers*, vol. VI. p. 356]; viz. this was

$$D = \tfrac{1}{6}(n-1)(n-2)(n-3) - (n-3)(b+c) + \tfrac{1}{2}(q+r) + 2t + \tfrac{7}{2}\beta + \tfrac{5}{2}y + i - \tfrac{1}{8}\theta,$$

where we have

b, order of double curve,
q, class of Do.,
c, order of cuspidal curve,
r, class of Do.,
β, number of intersections of the two curves, stationary points on b,
y, number of intersections, stationary points on c,
i, number of intersections, not stationary points on either curve,
θ, number of certain singular points on c, the nature of which I do not completely understand; it is here taken to be $=0$.

Before going further I remark that

Postulation of right line *quâ* i-tuple on surface n

$$= \tfrac{1}{2}i(i+1)n - \tfrac{1}{6}i(i+1)(2i-5),$$
$$= \tfrac{1}{6}i(i+1)(3n-2i+5).$$

Whence if a surface n has an i-tuple right line, the deficiency-value hereof is

$$= \tfrac{1}{6}i(i-1)(3n-2i-5),$$

or we have

$$D = \tfrac{1}{6}(n-1)(n-2)(n-3) - \tfrac{1}{6}i(i-1)(3n-2i-5)$$
$$= \tfrac{1}{6}(i-n+1)(i-n+2)(2i+n-3);$$

so that $D=0$ if either $i=n-1$ or $i=n-2$; the former case is that of a scroll (skew surface) with a $(n-1)$ tuple right line, the latter that of a surface with a $(n-2)$ tuple line: whence (as shown by Dr Noether) such surface is rationally transformable into a plane.

For a surface of the order n with an i-conical point where the tangent cone has δ double lines and κ cuspidal lines, we have

$$D = \tfrac{1}{6}(n-1)(n-2)(n-3) - \{\tfrac{1}{6}i(i-1)(i-2) + (i-2)(n-i-1)(\delta+\kappa)\}$$
$$= \tfrac{1}{6}(n-i-1)\{n^2 + n(i-5) + i^2 - 4i + 6 - 6(i-2)(\delta+\kappa)\};$$

viz. for $i=n-1$ this is $D=0$ (in fact, a surface n with a $(n-1)$ conical point is at once seen to be rationally transformable into a plane): and for $i=n$, that is, for a cone of the order n, we have

$$D = -\tfrac{1}{2}(n-1)(n-2) + (n-2)(\delta+\kappa) - (n-3)(\delta+\kappa),$$

where the last term $-(n-3)(\delta+\kappa)$ is added because in the present case the surface has the δ double lines and the κ cuspidal lines.

The formula therefore gives

$$D = -\tfrac{1}{2}(n-1)(n-2) + \delta + \kappa,$$

viz. this is equal to the deficiency of the plane sections *taken negatively.*

I find that the same property exists *first* in the case of a scroll (skew surface) having only a double curve; and *secondly* in the case of a torse (developable surface) having a cuspidal curve with the ordinary singularities; and this being so there can I think be no doubt but that it is true for any scroll or torse whatever—viz. that for any ruled surface whatever the deficiency is equal to that of the plane section taken negatively.

First, for the scroll, we have

$$D = \tfrac{1}{6}(n-1)(n-2)(n-3) - (n-3)b + \tfrac{1}{2}q + 2t,$$

which should be

$$= -\tfrac{1}{2}(n-1)(n-2) + b.$$

Salmon's equations give in the case of a scroll

$$3t = (n-4)\{3b - n(n-2)\},$$

$$q = n(n-2)(n-5) - 2(n-6)b,$$

and with these values the relation is at once verified.

Secondly, for the torse; changing the notation into that used for the singularities of the curve and torse, we have

$$D = \tfrac{1}{6}(r-1)(r-2)(r-3) - (r-3)(x+m) + \tfrac{1}{2}(q+r) + 2t + \tfrac{7}{2}\beta + \tfrac{5}{2}\gamma + \alpha,$$

which should be

$$= -\tfrac{1}{2}(m-1)(m-2) + h + \beta.$$

We have $q = r(n-3) - 3\alpha$, and substituting this value and expressing everything in terms of r, m, n by means of the formulæ

$$x = \tfrac{1}{2}(r^2 - r - n - 3m),$$

$$\alpha = m - 3r + 3n,$$

$$\beta = n - 3r + 3m,$$

$$t = \tfrac{1}{6}\{r^3 - 3r^2 - 58r - 3r(n+3m) + 42n + 78m\},$$

$$\gamma = rm + 12r - 14m - 6n,$$

$$h = \tfrac{1}{2}(m^2 - 10m - 3n + 8r),$$

we have after all reductions

$$D = -\tfrac{1}{2}(m+n) + r - 1 = -\tfrac{1}{2}(m-1)(m-2) + h + \beta.$$

We have thus a class of surfaces of *negative deficiency;* viz. any rational transformation of a cone for which the plane section has a given (positive) deficiency produces such a surface: and I think it may be assumed conversely that a surface of negative deficiency is always the rational transformation of a cone for which the deficiency is equal to that of the surface taken with the reverse sign. As an instance, take a quintic surface having a nodal conic and two 3-conical (cubiconical) points (this of course implies that the line joining the two cubiconical points is a line on the surface); the formula for the deficiency is ($n=5$, $b=2$, $q=2$, $r=0$, $t=0$)

$$D=\tfrac{1}{6}(n-1)(n-2)(n-3)-(n-3)b+\tfrac{1}{2}q+2t-2,$$

(viz. a term -1 for each of the cubiconical points)

$$=4-4+1 \quad -2, \quad =-1.$$

Such a surface can be obtained as the quadric inverse of a cubic cone; viz. taking for the vertex the point $x:y:z:w=\alpha:\beta:\gamma:\delta$ and the cone to pass through the point $x=0$, $y=0$, $z=0$, the equation of the cone is

$$(\delta x-\alpha w,\quad \delta y-\beta w,\quad \delta z-\gamma w)^3=0,$$

where $(\alpha, \beta, \gamma)^3=0$.

Taking Q a quadric function $(x, y, z)^2$, the transformation in question consists in the change of x, y, z, w into xw, yw, zw, Q; viz. the new equation, rejecting the factor w which divides out, is

$$\frac{1}{w}(\delta xw-\alpha Q,\quad \delta yw-\beta Q,\quad \delta zw-\gamma Q)^3=0,$$

which is a quintic surface, having the two cubiconical points $x=0$, $y=0$, $z=0$ and $x:y:z:w=\alpha:\beta:\gamma:\frac{1}{\delta}Q_0$ (where Q_0 is the value of Q on writing therein α, β, γ in place of x, y, z): and having the nodal conic $w=0$, $Q=0$.

Cambridge, 5 *Jan.* 1871.

525.

AN EXAMPLE OF THE HIGHER TRANSFORMATION OF A BINARY FORM.

[From the *Mathematische Annalen*, vol. IV. (1871), pp. 359—361.]

THE quartic

(1) $$(a, b, c, d, e)(x, y)^4$$

is by means of the two quadrics

(2) $$(\alpha, \beta, \gamma)(x, y)^2 \text{ and } (\alpha', \beta', \gamma')(x, y)^2$$

transformed into

(3) $$(a_1, b_1, c_1, d_1, e_1)(x_1, y_1)^4,$$

that is, eliminating x, y from

$$(a, b, c, d, e)(x, y)^4 = 0,$$

$$x_1(\alpha, \beta, \gamma)(x, y)^2 + y_1(\alpha', \beta', \gamma')(x, y)^2 = 0,$$

we obtain

$$(a_1, b_1, c_1, d_1, e_1)(x_1, y_1)^4 = 0.$$

It is required to express the invariants of (3) in terms of the simultaneous invariants of (1) and (2).

Write

$$P, Q, R = \alpha x_1 + \alpha' y_1, \quad \beta x_1 + \beta' y_1, \quad \gamma x_1 + \gamma' y_1;$$

the equations from which (x, y) have to be eliminated are

$$(a, b, c, d, e)(x, y)^4 = 0, \quad (P, Q, R)(x, y)^2 = 0,$$

and the result of the elimination therefore is

$$\left|\begin{array}{cccccc} & a, & 4b, & 6c, & 4d, & e \\ a, & 4b, & 6c, & 4d, & e & \\ & & & P, & 2Q, & R \\ & & P, & 2Q, & R & \\ & P, & 2Q, & R & & \\ P, & 2Q, & R & & & \end{array}\right| = 0,$$

viz. this determinant is the transformed quartic $(a_1, b_1, c_1, d_1, e_1)(x_1, y_1)^4$.

The developed expression of the determinant is

$$a^2R^4 - 8abQR^3$$

$$+\begin{pmatrix} -12\,ac \\ +16\,b^2 \end{pmatrix} PR^3 - 24\,acQ^2R^2 + \begin{pmatrix} 24\,ad \\ -48\,bc \end{pmatrix} PQR^2 - 32\,adQ^3R$$

$$+\begin{pmatrix} 2\,ae \\ -32\,bd \\ +36\,c^2 \end{pmatrix} P^2R^2 + \begin{pmatrix} -16\,ae \\ +64\,bd \end{pmatrix} PQ^2R + 16\,aeQ^4$$

$$+\begin{pmatrix} 24\,be \\ -48\,cd \end{pmatrix} P^2QR - 32\,bePQ^3 + \begin{pmatrix} -12\,ce \\ +16\,d^2 \end{pmatrix} P^3R \quad + 24\,ceP^2Q^2$$

$$-8\,deP^3Q + e^2P^4,$$

so that writing for P, Q, R their values, we have the transformed function $(a_1, b_1, c_1, d_1, e_1)(x_1, y_1)^4$, the coefficients being of the forms

$$\begin{aligned} a_1 &= (a,\ b,\ c,\ d,\ e)^2 (\alpha,\ \beta,\ \gamma)^4 \\ b_1 &= (\quad ,, \quad)^2 (\alpha,\ \beta,\ \gamma)^3 (\alpha',\ \beta',\ \gamma') \\ &\ \vdots \\ e_1 &= (\quad ,, \quad)^2 \ .\ \ .\ \ .\ (\alpha',\ \beta',\ \gamma')^4. \end{aligned}$$

Writing I, J for the invariants of the quartic (1), and

$$\begin{aligned} A &= 4(\alpha\beta' - \alpha'\beta)(\beta\gamma' - \beta'\gamma) - (\gamma\alpha' - \gamma'\alpha)^2, \\ B &= (e,\ c,\ a,\ b,\ c,\ d)(\alpha\beta' - \alpha'\beta,\ \gamma\alpha' - \gamma'\alpha,\ \beta\gamma' - \beta'\gamma)^2, \end{aligned}$$

we have I, J, A, B simultaneous invariants of the forms (1) and (2). Putting moreover $\nabla = I^3 - 27J^2$, and writing I_1, J_1, ∇_1, for the like invariants of the form (3), I find

$$\begin{aligned} I_1 &= 4\left(4\,IB^2 + 12\,JAB + \tfrac{1}{3}\,I^2A^2\right), \\ J_1 &= 8\left\{8\,JB^3 + \tfrac{4}{3}\,I^2AB^2 + 2\,IJA^2B + \left(2J^2 - \tfrac{1}{27}I^3\right)A^3\right\}, \end{aligned}$$

and thence

$$\nabla_1 = 256\,(4\,B^3 - IA^2B - JA^3)\,\nabla.$$

As a verification, suppose $(a, b, c, d, e)(x, y)^4 = x^4 + y^4$ (whence $I = 1$, $J = 0$). And take $x_1(x+y)^2 - y_1(x-y)^2 = 0$ for the transforming equation, that is, $(\alpha, \beta, \gamma) = (1, 1, 1)$ and $(\alpha', \beta', \gamma') = (-1, 1, -1)$. We have $P = R = x_1 - y_1$ and $Q = x_1 + y_1$, and thence

$$\begin{aligned}\text{Det.} &= (P^2 + R^2)^2 - 16\,PQ^2R + 16\,Q^4 \\ &= (2P^2 - 4Q^2)^2 = (-2x_1^2 - 12\,x_1y_1 - 2y_1^2)^2,\end{aligned}$$

that is,

$$(a_1, b_1, c_1, d_1, e_1)(x_1, y_1)^4 = 4\,(x_1^2 + 6x_1y_1 + y_1^2)^2,$$

whence

$$I_1 = \frac{4096}{3}, \ = \frac{2^{12}}{3}; \quad J_1 = -\frac{262144}{27}, \ = -\frac{2^{18}}{27};$$

also

$$A = -16, \quad B = 8,$$

and the equations for I_1, J_1 become

$$\frac{4096}{3} = 4\,(4\,.\,64 + \tfrac{1}{3}\,256),$$

$$-\frac{262144}{27} = 8\,(\tfrac{4}{3}\,.\,-16\,.\,64 + \tfrac{1}{27}\,4096),$$

which are true. More generally, assuming

$$(a, b, c, d, e)(x, y)^4 = x^4 + 6\Theta x^2y^2 + y^4,$$

(whence $I = 1 + 3\Theta^2$, $J = \Theta - \Theta^3$), and the same transforming equation, we have

$$(a_1, b_1, c_1, d_1, e_1)(x_1, y_1)^4 = 4\,\{(1 + 3\Theta)\,x_1^2 + (3 - 3\Theta)\,2x_1y_1 + (1 + 3\Theta)\,y_1^2\}^2,$$

whence

$$I_1 = \frac{2^{12}}{3}(1 - 3\Theta)^2, \quad J_1 = -\frac{2^{18}}{27}(1 - 3\Theta)^3;$$

also

$$A = -16, \quad B = 8\,(1 - \Theta).$$

Substituting these different values in the equations for I_1, J_1, we obtain

$$16\,(1 - 3\Theta)^2 = 12\,(1 + 3\Theta^2)(1 - \Theta)^2 - 72\,(\Theta - \Theta^3)(1 - \Theta) + 4\,(1 + 3\Theta^2)^2,$$

and

$$\begin{aligned}-8\,(1 - 3\Theta)^3 = 27\,&(\Theta - \Theta^3)(1 - \Theta)^3 - 9\,(1 + 3\Theta^2)^2(1 - \Theta)^2 \\ &+ 27\,(1 + 3\Theta^2)(\Theta - \Theta^3)(1 - \Theta) - 54\,(\Theta - \Theta^3)^2 + (1 + 3\Theta^2)^3,\end{aligned}$$

which are in fact satisfied identically.

Cambridge, 26 July, 1871.

526.

ON A SURFACE OF THE EIGHTH ORDER.

[From the *Mathematische Annalen*, vol. IV. (1871), pp. 558—560.]

I REPRODUCE in an altered form, so as to exhibit the application thereto of the theory of the six coordinates of a line, the analysis by which Dr Hierholzer obtained the equation of the surface of the eighth order, the locus of the vertex of a quadricone which touches six given lines.

I call to mind that if $(\alpha, \beta, \gamma, \delta)$, $(\alpha', \beta', \gamma', \delta')$ are the coordinates of any two points on a line, then the quantities (a, b, c, f, g, h), which denote respectively

$$(\beta\gamma' - \beta'\gamma,\ \gamma\alpha' - \gamma'\alpha,\ \alpha\beta' - \alpha'\beta,\ \alpha\delta' - \alpha'\delta,\ \beta\delta' - \beta'\delta,\ \gamma\delta' - \gamma'\delta),$$

and which are such that $af + bg + ch = 0$, are the six coordinates of the line[1].

Consider the given point (x, y, z, w) and the given line (a, b, c, f, g, h), and write for shortness

$$\begin{aligned} P &= hy - gz + aw, \\ Q &= -hx + fz + bw, \\ R &= gx - fy + cw, \\ S &= -ax - by - cz , \end{aligned}$$

then taking (X, Y, Z, W) as current coordinates, the equation of the plane through the given point and line is

$$PX + QY + RZ + SW = 0.$$

Considering in like manner the given point (x, y, z, w) and the three given lines $(a_1, b_1, c_1, f_1, g_1, h_1)$, $(a_2, \ldots)$, $(a_3, \ldots)$, then we have the three planes

$$\begin{aligned} P_1X + Q_1Y + R_1Z + S_1W &= 0, \\ P_2X + Q_2Y + R_2Z + S_2W &= 0, \\ P_3X + Q_3Y + R_3Z + S_3W &= 0, \end{aligned}$$

[1] Cayley, "On the six coordinates of a line," *Camb. Phil. Trans.* vol. XI. (1869), [435], pp. 290—323.

and if these planes have a common line, the point (x, y, z, w) is in a line meeting each of the three given lines; that is, the locus of the point is the hyperboloid through the three given lines. It follows that the equations

$$\left\|\begin{matrix} P_1, & Q_1, & R_1, & S_1 \\ P_2, & Q_2, & R_2, & S_2 \\ P_3, & Q_3, & R_3, & S_3 \end{matrix}\right\| = 0$$

reduce themselves to a single equation, that of the hyperboloid in question.

I write for shortness

$$\begin{aligned}(000) = (agh)\,x^2 + (bhf\,)\,y^2 + (cfg)\,z^2 + (abc)\,w^2 \\ + [(abg) - (cah)]\,xw \\ + [(bch\,) - (abf)]\,yw \\ + [(caf\,) - (bcg\,)]\,zw \\ + [(bfg\,) + (chf\,)]\,yz \\ + [(cgh\,) + (afg\,)]\,zx \\ + [(ahf) + (bgh)]\,xy,\end{aligned}$$

viz. (123) will mean $(a_1g_2h_3)\,x^2 +$ etc. where $(a_1g_2h_3)$ etc. denote as usual the determinants

$$\left|\begin{matrix} a_1, & g_1, & h_1 \\ a_2, & g_2, & h_2 \\ a_3, & g_3, & h_3 \end{matrix}\right| \text{ etc.};$$

then the equations in question are found to be $x\,(123) = 0$, $y\,(123) = 0$, $z\,(123) = 0$, $w\,(123) = 0$, reducing themselves to the single equation $(123) = 0$, which is accordingly that of the hyperboloid through the three lines[1].

Proceeding now to the above-mentioned problem, we have the point (x, y, z, w), and the six lines $(a_1, b_1, c_1, f_1, g_1, h_1)$, $(a_2, \ldots)$ etc., say the lines 1, 2, 3, 4, 5, 6: the six planes

$$P_1X + Q_1Y + R_1Z + S_1W = 0, \text{ etc.}$$

must be tangents to the same quadricone; that is, considering the sections by the plane $W = 0$, the six lines

$$P_1X + Q_1Y + R_1Z = 0, \text{ etc.}$$

must be tangents to the same conic, and the condition for this is

$$[1\,2\,3\,4\,5\,6] = 0,$$

[1] This equation is given in the paper above referred to, § 54.

where the symbol stands for the determinant

$$\begin{vmatrix} P_1^2, & Q_1^2, & R_1^2, & Q_1R_1, & R_1P_1, & P_1Q_1 \\ P_2^2 \ldots & & & & & \\ \vdots & & & & & \end{vmatrix}.$$

But as is well known this equation may be written

$$(*) \qquad (126)(346)(145)(235) - (146)(236)(125)(345) = 0,$$

where (126) etc. denote the determinants

$$\begin{vmatrix} P_1, & Q_1, & R_1 \\ P_2, & Q_2, & R_2 \\ P_3, & Q_3, & R_3 \end{vmatrix} \text{ etc.};$$

or, what is the same thing, they denote the functions above represented by the like symbols $(126) = (a_1 g_2 h_6) x^2 + \text{etc.}$ The equation (*) just obtained is Hierholzer's equation for the surface of the eighth order, the locus of the vertex of a quadricone which touches six given lines.

I remark that in my "Memoir on Quartic Surfaces," *Proc. Lond. Math. Soc.* vol. III. (1870), [445], pp. 19—69, I obtained the equation of the surface under the foregoing form $[1\,2\,3\,4\,5\,6] = 0$ or say $[(P,\ Q,\ R)^2] = 0$, noticing that there was a factor w^4, so that the order of the surface is $= 8$; and further that the equation might be written

$$w^8 \exp. \frac{1}{w} \{x (g\partial_c - h\partial_b) + y (h\partial_a - f\partial_c) + z (f\partial_b - g\partial_a)\} [(a,\ b,\ c)^2] = 0,$$

where exp. Θ (read exponential) denotes e^Θ, and $[(a,\ b,\ c)^2]$ denotes

$$\begin{vmatrix} a_1^2, & b_1^2, & c_1^2, & b_1c_1, & c_1a_1, & a_1b_1 \\ a_2^2, \ldots & & & & & \\ \vdots & & & & & \end{vmatrix}.$$

Also that the equation contains the four terms

$$x^8 [(a,\ -h,\ g)^2] + y^8 [(h,\ b,\ -f)^2] + z^8 [(-g,\ f,\ c)^2] + w^8 [(a,\ b,\ c)^2] = 0.$$

Cambridge, 12 September, 1871.

527.

ON A THEOREM IN COVARIANTS.

[From the *Mathematische Annalen*, vol. v. (1872), pp. 625—629.]

THE proof given in Clebsch "Theorie der binären algebraischen Formen" (Leipzig, 1872) of the finite number of the covariants of a binary form depends upon a subsidiary proposition which is deserving of attention for its own sake.

I use my own hyperdeterminant notation, which is as follows: Considering a function $U=(a,..)(x, y)^n$, (viz. $U_1=(a,..)(x_1, y_1)^n$ &c.), and writing $\overline{12}=\partial_{x_1}\partial_{y_2}-\partial_{y_1}\partial_{x_2}$ &c., then the general form of a covariant of the degree m is

$$k(\overline{12}^\alpha\, \overline{13}^\beta\, \overline{23}^\gamma \ldots)\, U_1U_2..\, U_m,$$

where k is a merely numerical factor, the indices $\alpha, \beta, \gamma, ..$ are positive integers, and after the differentiations each set of variables $(x_1, y_1), .., (x_m, y_m)$ is replaced by (x, y). I say that the general form of a covariant is as above; viz. a covariant is equal to a single term of the above form, or a sum of such terms.

Attending to a single term: the sum of the indices of all the duads which contain a particular number 1, 2,.. as the case may be is called an index-sum; each index-sum is at most $=n$; so that, calling the index-sums $\sigma_1, \sigma_2, \ldots, \sigma_m$ respectively, we have $n-\sigma_1, n-\sigma_2, \ldots, n-\sigma_m$ each of them zero or positive: the term, before the several sets of variables are each replaced by (x, y), is of the orders $n-\sigma_1, n-\sigma_2, \ldots, n-\sigma_m$ in the several sets of variables respectively.

The term may be expressed somewhat differently: for writing $\nabla_1=x\partial_{x_1}+y\partial_{y_1}$, $\nabla_2=x\partial_{x_2}+y\partial_{y_2}$ &c.—then (except as to a numerical factor) it is for a function $(*)(x_1, y_1)^p$ the same thing whether we change (x_1, y_1) into (x, y), or operate on this function with $\nabla_1{}^p$, and so for the other sets: the term may therefore be written

$$\nabla_1{}^{n-\sigma_1}\ldots\nabla_m{}^{n-\sigma_m}\, k(\overline{12}^\alpha\, \overline{13}^\beta\, \overline{23}^\gamma \ldots)\, U_1U_2\ldots U_m,$$

being now in the first instance a function of the single set (x, y) of variables.

We may omit the operand $U_1U_2 \dots U_m$, and consider only the symbol

$$k(\overline{12}^\alpha\,\overline{13}^\beta\,\overline{23}^\gamma \dots) \text{ or } \nabla_1^{n-\sigma_1} \dots \nabla_m^{n-\sigma_m} k(\overline{12}^\alpha\,\overline{13}^\beta\,\overline{23}^\gamma \dots),$$

which, under either of the two forms, I represent for shortness by $[12 \dots m]$: observe that this is considered as a symbol involving the m symbolic numbers 1, 2, $3 \dots m$, even although in particular cases one or more of these numbers may be wanting from the actual expression of the symbol: thus [123] may denote $\overline{12}^\alpha$, but the operand to be supplied thereto is always $U_1U_2U_3$.

A sum of symbols is not in general equal to a single symbol: but a single symbol can be expressed in a variety of ways as a sum of symbols: the most simple transformation-formulæ relate to three or four symbolic numbers; viz. for three such numbers, say 1, 2, 3, we have

$$\nabla_1 . 23 + \nabla_2 . 31 + \nabla_3 . 12 = 0,$$

showing that in a symbol, which written with the ∇'s involves $\nabla_1 . 23$, this may be replaced by its value $\nabla_2 . 13 - \nabla_3 . 12$; and so in other cases.

For the four numbers 1, 2, 3, 4 we have a group of the like formulae

$$\begin{array}{lllll} . & - \nabla_2 . 34 & + \nabla_3 . 24 & - \nabla_4 . 23 & = 0, \\ \nabla_1 . 34 & . & - \nabla_3 . 14 & - \nabla_4 . 31 & = 0, \\ - \nabla_1 . 24 & + \nabla_2 . 14 & . & - \nabla_4 . 12 & = 0, \\ \nabla_1 . 23 & + \nabla_2 . 23 & + \nabla_3 . 31 & . & = 0, \end{array}$$

leading to

$$23 . 14 + 31 . 24 + 12 . 34 = 0,$$

which is a form not involving the ∇'s and consequently is applicable to the transformation of invariant-symbols where the numbers

$$n - \sigma_1,\ n - \sigma_2,\ \dots,\ n - \sigma_m$$

are all $= 0$.

I establish the following definitions:

A symbol $[12 \dots m]$ is *proximate* when each index-sum is $< n$; otherwise it is *ultimate;* viz. this is the case when any one or more of the index-sums is or are $= n$. We may say that the symbol is ultimate as to 1 if $\sigma_1 = n$; and that it is ultimate as to 1, 2 if σ_1 and σ_2 are each $= n$: and so in other cases.

A proximate symbol which has any one index-sum thereof $< \tfrac{1}{2}n$ is said to be *inferior:* thus if $\sigma_1 < \tfrac{1}{2}n$ the symbol is inferior in regard to 1; and so if σ_1 and σ_2 are each $< \tfrac{1}{2}n$, it is inferior in regard to 1 and 2: and the like in other cases.

Observe that if a symbol is inferior then in the covariant the order exceeds the degree by a number which is greater than $\frac{1}{2}n-1$: in fact, suppose it inferior in regard to 1, then the order is

$$(n-\sigma_1)+(n-\sigma_2)+\ldots+(n-\sigma_m),$$

where each term after the first is at least $=1$, that is, the order is at least $=n-\sigma_1+m-1$; hence order $-$ degree is at least $=n-\sigma_1-1$; viz. σ_1 being less than $\frac{1}{2}n$, this is greater than $\frac{1}{2}n-1$.

Conversely, if for any symbol order-degree is $\overline{\gtrless}\,\frac{1}{2}n-1$, then the symbol is not inferior.

A symbol $[12\ldots m]$ is *sharp* when any index is $\overline{\gtrless}\,\frac{1}{2}n$; otherwise it is *flat;* viz. this is so when each index is $<\frac{1}{2}n$. The symbol is sharp as to any particular duad or duads when the index or indices thereof is or are each of them $\overline{\gtrless}\,\frac{1}{2}n$.

The subsidiary theorem is now as follows: "A symbol is inferior or sharp: or it can be expressed as a sum of symbols each of which is inferior or sharp"—or what is the same thing, the only symbols which need to be considered are those which are either inferior or sharp.

Thus for the degree 1 the symbol is [1] (which is simply unity) $\sigma_1=0$, and the symbol is inferior.

For the degree 2 the symbol is [2], $=\overline{12}^k$; if $k<\frac{1}{2}n$ the symbol is inferior, if $k\,\overline{\gtrless}\,\frac{1}{2}n$ then it is sharp.

A proof is first required for the degree 3, here $[123]=\overline{12}^\alpha\,\overline{13}^\beta\,\overline{23}^\gamma$ ($\beta+\gamma$, $\gamma+\alpha$, $\alpha+\beta$ each $=$ or $<n$) which may very well be neither inferior nor sharp; for instance, if $n=5$, we have $\overline{12}^2\,\overline{13}^2\,\overline{23}^2$, where each index being $=2$, the symbol is not sharp; and each index-sum being $=4$ the symbol is not inferior. But writing the symbol in the form $\nabla_1\nabla_2\nabla_3\,\overline{12}^2\,\overline{13}^2\,\overline{23}^2$, then by means of the relation

$$\nabla_1 . 23+\nabla_2 . 31+\nabla_3 . 12=0,$$

(or, what is the same thing, $\nabla_1 . 23=\nabla_2 . 13-\nabla_3 . 12$), the symbol becomes

$$\nabla_2\nabla_3\,\overline{12}^2\,\overline{13}^2\,\overline{23}^2\,(\nabla_2 . 13-\nabla_3 . 12),$$

$$=\nabla_2{}^2\nabla_3\,\overline{12}^2\,\overline{13}^3\,23-\nabla_2\nabla_3{}^2\,\overline{13}^3\,\overline{12}^2\,\overline{23},$$

where each term, as containing an index 3, is sharp. To complete the reduction, observe that calling the expression $\mathfrak{A}-\mathfrak{B}$, then in the term $\mathfrak{A}$ interchanging the numbers 2 and 3 we obtain $\mathfrak{A}=-\mathfrak{B}$, and thence $\mathfrak{A}-\mathfrak{B}=2\mathfrak{A}$; so that the whole is $2\,\nabla_2{}^2\nabla_3\,\overline{12}^2\,\overline{13}^3\,\overline{23}$, viz. it is a multiple of $\overline{12}^2\,\overline{13}^3\,\overline{23}$.

I prove the general case, substantially in the manner used by Dr Clebsch, as follows. We assume that the theorem is proved up to a particular degree m: that is, we assume that every symbol belonging to a degree not exceeding m can be expressed as a sum of terms each of which is sharp or inferior: and we have to prove this for the next following degree $m+1$, or writing for convenience p in place of $m+1$, (say for the degree p); that is, for a symbol

$$[12 \ldots mp], = \overline{p1}^{\lambda_1}\, \overline{p2}^{\lambda_2} \ldots \overline{pm}^{\lambda_m} [12 \ldots m]$$
$$= P\,[12 \ldots m] \text{ suppose.}$$

I write as before $\sigma_1, \sigma_2, \ldots, \sigma_m$ for the index-sums of $[12 \ldots m]$: those of $[12 \ldots mp]$ are therefore $\sigma_1+\lambda_1, \sigma_2+\lambda_2, \ldots, \sigma_m+\lambda_m$, and (for the duads involving p) $\sigma_p = \lambda_1+\lambda_2 \ldots +\lambda_m$.

If $[12 \ldots m]$ is sharp, then $[12 \ldots mp]$ is sharp, and the theorem is true.

If $\sigma_p < \tfrac{1}{2}n$, then $[12 \ldots mp]$ is inferior in regard to p; and the theorem is true.

The only case requiring a proof is when $[12 \ldots m]$ is not sharp (being therefore inferior) and when σ_p is $\overline{\gtrless}\, \tfrac{1}{2}n$. And in this case if any one of the indices $\lambda_1, \ldots \lambda_m$ is $\overline{\gtrless}\, \tfrac{1}{2}n$ (or say if P is sharp) then the theorem is true.

Consider the expression

$$\overline{p1}^{\lambda_1}\, \overline{p2}^{\lambda_2} \ldots \overline{pm}^{\lambda_m} [12 \ldots m],$$

where $\sigma_1, \sigma_2, \ldots, \sigma_m$ are as before the index-sums for $[12 \ldots m]$ and therefore the numbers

$$n-\sigma_1-\lambda_1, \ldots, n-\sigma_m-\lambda_m$$

are none of them negative.

Assume that when $[12 \ldots m]$ is inferior, and when $\lambda_1 \ldots \lambda_m$ have any values such that their sum is not greater than a given value $\sigma_p - 1$, the expression is a sum of terms each of which is inferior or sharp: we wish to show that when $\lambda_1+\lambda_2 \ldots +\lambda_m$ has the next succeeding value, $=\sigma_p$, the case is still the same.

For this purpose, introducing the ∇'s I write

$$Q = \nabla_1^{\,n-\sigma_1-\lambda_1} \ldots \nabla_m^{\,n-\sigma_m-\lambda_m}\, \nabla_p^{\,n-\sigma_p}\, \overline{p1}^{\lambda_1}\, \overline{p2}^{\lambda_2} \ldots \overline{pm}^{\lambda_m} [12 \ldots m];$$

then supposing for a moment that λ_1 is not $= n-\sigma_1$ and λ_2 not $=0$, the expression contains the factor $\nabla_1 . p2$, which is equal to and may be replaced by $-\nabla_2 . p1 + \nabla_p . 12$: we have thus

$$Q = Q' + \Omega,$$

where omitting the ∇'s

$$Q' = j\, \overline{p1}^{\lambda_1+1}\, \overline{p2}^{\lambda_2-1}\, \overline{p3}^{\lambda_3} \ldots \overline{pm}^{\lambda_m} [12 \ldots m],$$

$$\Omega = k\, \overline{p1}^{\lambda_1}\, \overline{p2}^{\lambda_2-1}\, \overline{p3}^{\lambda_3} \ldots \overline{pm}^{\lambda_m}\, \overline{12}\, [12 \ldots m].$$

Now for Ω the sum of the indices $\lambda_1, \lambda_2-1, \lambda_3 \ldots \lambda_m$ is $\sigma_p - 1$, so that by hypothesis Ω is inferior or sharp: that is, the difference $Q-Q'$ is inferior or sharp: so that to prove that Q is inferior or sharp, we have only to prove this of Q', where Q' is derived from Q by increasing by unity the index of $p1$, at the expense of that of $p2$

which is diminished by unity. Such change is possible so long as the index λ_1 has not attained its maximum value, $n-\sigma_1$ or σ_p as the case may be, and there is any other index $\lambda_2, \ldots, \lambda_m$ which is not $=0$: that is, we may pass from Q to Q', from Q' to Q'' and so on; and it will be sufficient to show that the last term of the series is inferior or sharp. We thus pass from Q to R, where

$$R = \overline{p1}^{\,n-\sigma_1}\,\overline{p2}^{\,\lambda_2-\alpha_2}\ldots\overline{pm}^{\,\lambda_m-\alpha_m}\,[12\ldots m]$$

and $\alpha_2+\alpha_3\ldots+\alpha_m = n-\sigma_1-\lambda_1$; or else to

$$R = \overline{p1}^{\,\sigma_p}\,[12\ldots m],$$

according as $n-\sigma_1$ is not greater or is greater than σ_p.

Now let $[12\ldots m]$ be inferior; suppose it to be so in regard to 1, that is, let σ_1 be less than $\frac{1}{2}n$ or $n-\sigma_1$ greater than $\frac{1}{2}n$. Then if σ_p be less than $\frac{1}{2}n$ it is less than $n-\sigma_1$, that is, we have for R the last-mentioned form which is inferior in regard to p, viz. R is inferior; if σ_p is equal to or greater than $\frac{1}{2}n$, then R, whichever its form may be, is sharp as to $p1$, viz. R is sharp. Hence in either case Q is a sum of terms which are inferior or sharp; that is, assuming the theorem for a form for which $\lambda_1+\lambda_2\ldots+\lambda_m$ does not exceed a given value σ_p-1, the theorem is true for the next succeeding value σ_p; or being true for the case $\sigma_p-1=0$, it is true generally.

Cambridge, 24 April, 1872.

528.

ON THE NON-EUCLIDIAN GEOMETRY.

[From the *Mathematische Annalen*, vol. v. (1872), pp. 630—634.]

THE theory of the Non-Euclidian Geometry as developed in Dr Klein's paper "Ueber die Nicht-Euklidische Geometrie" may be illustrated by showing how in such a system we actually measure a distance and an angle and by establishing the trigonometry of such a system. I confine myself to the "hyperbolic" case of plane geometry; viz. the absolute is here a real conic, which for simplicity I take to be a circle; and I attend to the points *within* the circle.

I use the simple letters $a, A, ..$ to denote (linear or angular) distances measured in the ordinary manner; and the same letters, with a superscript stroke, $\bar{a}, \bar{A}, ..$ to

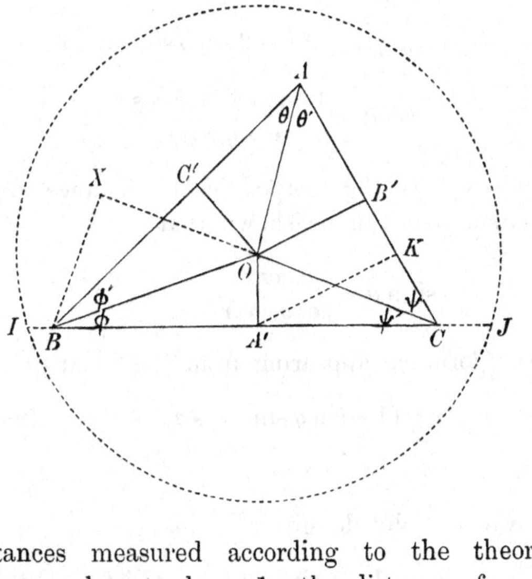

denote the same distances measured according to the theory. The radius of the absolute is for convenience taken to be $=1$; the distance of any point from the centre can therefore be represented as the sine of an angle.

The distance BC, or say $\bar{a}$, of any two points B, C is by definition as follows:

Radius of circle $= 1$:

In $\triangle\ ABC$, sides are a, b, c:

angles „ A, B, C:

OA, OB, OC are $= \sin p$, $\sin q$, $\sin r$:

OA', OB', OC' „ „ $\sin a$, $\sin b$, $\sin c$:

$\measuredangle BOC$, COA, AOB „ „ α, β, γ.

$$\bar{a} = \tfrac{1}{2} \log \frac{BI \,.\, CJ}{BJ \,.\, CI},$$

(where I, J are the intersections of the line BC with the circle); that is,

$$e^{\bar{a}} + e^{-\bar{a}}, \text{ or } 2 \cosh \bar{a} = \sqrt{\frac{BI \,.\, CJ}{BJ \,.\, CI}} + \sqrt{\frac{BJ \,.\, CI}{BI \,.\, CJ}}, \quad = \frac{BI \,.\, CJ + BJ \,.\, CI}{\sqrt{BI \,.\, BJ}\sqrt{CI \,.\, CJ}},$$

where the numerator is

$$\begin{aligned} BI\,(BJ - BC) + CI\,(BC + CJ), \quad &= BI \,.\, BJ + CI \,.\, CJ + BC\,(CI - BI), \\ &= BI \,.\, BJ + CI \,.\, CJ + BC^2. \end{aligned}$$

Hence taking a for the distance BC, and $\sin q$, $\sin r$, for the distances OB, OC respectively, we have $BI \,.\, BJ = \cos^2 q$, $CI \,.\, CJ = \cos^2 r$; and the formula is

$$\cosh \bar{a} = \frac{\cos^2 q + \cos^2 r + a^2}{2 \cos q \cos r},$$

or, what is the same thing, taking α for the angle BOC, and therefore

$$a^2 = \sin^2 q + \sin^2 r - 2 \sin q \sin r \cos \alpha,$$

we have

$$\cosh \bar{a} = \frac{1 - \sin q \sin r \cos \alpha}{\cos q \cos r}.$$

In a similar manner, if $\sin \mathfrak{a}$ is the perpendicular distance from O on the line BC (that is, $a \sin \mathfrak{a} = \sin q \sin r \sin \alpha$) it can be shown that

$$\sinh \bar{a} = \frac{a \cos \mathfrak{a}}{\cos q \cos r},$$

the equivalence of the two formulæ appearing from the identity

$$\cos^2 q \cos^2 r = (1 - \sin q \sin r \cos \alpha)^2 - a^2 + a^2 \sin^2 \mathfrak{a},$$

which is at once verified.

Next for an angle; we have by definition

$$\bar{A} = \frac{1}{2i} \log \frac{\sin BAI \,.\, \sin CAJ}{\sin CAI \,.\, \sin BAJ},$$

where AI, AJ are the (imaginary) tangents from A to the circle; or writing for shortness BI &c. instead of BAI, &c. (the angular point being always at A),

$$\bar{A} = \frac{1}{2i} \log \frac{\sin BI \,.\, \sin CJ}{\sin CI \,.\, \sin BJ},$$

consequently

$$e^{i\bar{A}} - e^{-i\bar{A}} = 2i \sin \bar{A}$$

$$= \sqrt{\frac{\sin BI \,.\, \sin CJ}{\sin CI \,.\, \sin BJ}} - \sqrt{\frac{\sin CI \,.\, \sin BJ}{\sin BI \,.\, \sin CJ}}, = \frac{\sin BI \sin CJ - \sin BJ \,.\, \sin CI}{\sqrt{\sin BI \,.\, \sin BJ}\sqrt{\sin CI \,.\, \sin CJ}},$$

where the numerator is

$$\sin BI \sin (BJ - BC) - \sin BJ \sin (BI + BC) = \sin BC \sin IJ,$$

or say $= \sin A \sin IJ$. Moreover taking the distance OA to be $= \sin p$, and the perpendicular distances from O on the lines AB, AC to be $\sin \mathfrak{c}$ and $\sin \mathfrak{b}$ respectively, then if for a moment the angle IJ is put $= 2\omega$, we have $\sin p \sin \omega = 1$: moreover

$$\sin BI \sin BJ = \sin (\omega - BO) \sin (\omega + BO) = \sin^2 \omega - \sin^2 BO:$$

and $\sin p \sin BO = \sin \mathfrak{c}$; that is, $\sin BI \sin BJ = \dfrac{1 - \sin^2 \mathfrak{c}}{\sin^2 p}, = \dfrac{\cos^2 \mathfrak{c}}{\sin^2 p}$: and similarly $\sin CI \sin CJ = \dfrac{\cos^2 \mathfrak{b}}{\sin^2 p}$; also

$$\sin IJ = - \sin 2\omega = 2 \sin \omega \cos \omega = \frac{2}{\sin p} \frac{i \cos p}{\sin p};$$

whence the required formula

$$\sin \bar{A} = \frac{\cos p \sin A}{\cos \mathfrak{b} \cos \mathfrak{c}}.$$

In the same way, or analytically from this value, we have

$$\cos \bar{A} = \frac{\cos A + \sin \mathfrak{b} \sin \mathfrak{c}}{\cos \mathfrak{b} \cos \mathfrak{c}},$$

and thence also

$$\tan \bar{A} = \frac{\cos p \sin A}{\cos A + \sin \mathfrak{b} \sin \mathfrak{c}}.$$

In particular, taking the line AC to pass through O, or writing in the formula $\mathfrak{b} = 0$, we have $\tan \bar{BO} = \cos p \tan BO = \cos p \tan \theta$; that is, $\bar{BO} = \tan^{-1} . \cos p \tan \theta$; and similarly $\bar{CO} = \tan^{-1} \cos p \tan \theta'$; we ought to have $\bar{A} = \bar{BO} + \bar{CO}$, that is,

$$\bar{A} = \tan^{-1} \cos p \tan \theta + \tan^{-1} \cos p \tan \theta'$$

which, observing that $\sin p \sin \theta = \sin \mathfrak{c}$ and $\sin p \sin \theta' = \sin \mathfrak{b}$, also $A = \theta + \theta'$, is in fact equivalent to the above formula for $\tan \bar{A}$.

Observe in particular that when A is at the centre, p is $= 0$, and the formula becomes $\bar{A} = \theta + \theta'$, $= A$, or say for an angle at the centre, $\bar{O} = O$.

I return to the expression for $\cosh \overline{a}$; in explanation of its meaning, let the distances $\overline{OB}$, $\overline{OC}$ be $\overline{q}$, $\overline{r}$ respectively and let the angle $\overline{BOC}$ be $\overline{\alpha}$; to find $\overline{q}$ we have only to take C at O, that is, in the formula for $\cosh \overline{a}$ to write $r=0$, we thus find $\cosh \overline{q} = \frac{1}{\cos q}$: and similarly $\cosh \overline{r} = \frac{1}{\cos r}$, whence also

$$\cos q = \text{sech}\, \overline{q}, \quad \sin q = i \tanh \overline{q},$$

$$\cos r = \text{sech}\, \overline{r}, \quad \sin r = i \tanh \overline{r},$$

also, as seen above, $\overline{\alpha} = \alpha$; the formula thus is

$$\cosh \overline{a} = \frac{1 + \tanh \overline{q} \tanh \overline{r} \cos \overline{\alpha}}{\text{sech}\, \overline{q}\, \text{sech}\, \overline{r}}$$

$$= \cosh \overline{q} \cosh \overline{r} + \sinh \overline{q} \sinh \overline{r} \cos \alpha,$$

or, what is the same thing, it is

$$\cos \overline{\alpha} = \frac{\cosh \overline{a} - \cosh \overline{q} \cosh \overline{r}}{\sinh \overline{q} \sinh \overline{r}},$$

viz. as will presently appear, this is the formula for $\cos \overline{BOC}$ in the triangle BOC.

From the above formulæ

$$\cosh \overline{a} = \frac{1 - \sin q \sin r \cos \alpha}{\cos q \cos r},$$

and

$$\sin \overline{A} = \frac{\cos p \sin A}{\cos \mathfrak{b} \cos \mathfrak{c}}, \quad \cos \overline{A} = \frac{\cos A + \sin \mathfrak{b} \sin \mathfrak{c}}{\cos \mathfrak{b} \cos \mathfrak{c}},$$

and the like formulæ for $\overline{b}$, $\overline{c}$, $\overline{B}$, $\overline{C}$, it may be shown that in the triangle ABC we have

$$\cosh \overline{a} = \frac{\cos \overline{A} + \cos \overline{B} \cos \overline{C}}{\sin \overline{B} \sin \overline{C}}.$$

In fact, substituting the foregoing values, this equation becomes

$$\frac{(1 - \sin^2 \mathfrak{a})(\cos A + \sin \mathfrak{b} \sin \mathfrak{c}) + (\cos B + \sin \mathfrak{c} \sin \mathfrak{a})(\cos C + \sin \mathfrak{a} \sin \mathfrak{b})}{\sin B \sin C \cos \mathfrak{b} \cos \mathfrak{c}} = \frac{1 - \sin q \sin r \cos \alpha}{\cos q \cos r},$$

that is,

$$\cos A + \cos B \cos C - \sin^2 \mathfrak{a} \cos A + \sin \mathfrak{a} \sin \mathfrak{b} \cos B + \sin \mathfrak{a} \sin \mathfrak{c} \cos C + \sin \mathfrak{b} \sin \mathfrak{c}$$
$$= \sin B \sin C (1 - \sin q \sin r \cos \alpha),$$

or, what is the same thing,

$$\sin^2 \mathfrak{a} (\cos B \cos C - \sin B \sin C) + \sin \mathfrak{a} \sin \mathfrak{b} \cos B + \sin \mathfrak{a} \sin \mathfrak{c} \cos C + \sin \mathfrak{b} \sin \mathfrak{c}$$
$$= - \sin B \sin C \sin q \sin r \cos \alpha,$$

that is,

$$(\sin \mathfrak{a} \cos B + \sin \mathfrak{c})(\sin \mathfrak{a} \cos C + \sin \mathfrak{b}) = \sin B \sin C (\sin^2 \mathfrak{a} - \sin q \sin r \cos \alpha),$$

a relation which I proceed to verify.

We may, from the formulæ

$$a^2 = \sin^2 q + \sin^2 r - 2 \sin q \sin r \cos \alpha, \quad a \sin \mathfrak{a} = \sin q \sin r \sin \alpha, \text{ \&c.},$$

but, more simply, geometrically as presently shown, deduce

$$\sin \mathfrak{a} \cos B + \sin \mathfrak{c} = \frac{1}{a} \sin B \sin q \, (\sin q - \sin r \cos \alpha),$$

$$\sin \mathfrak{a} \cos C + \sin \mathfrak{b} = \frac{1}{a} \sin C \sin r \, (\sin r - \sin q \cos \alpha),$$

and thence

$$(\sin \mathfrak{a} \cos B + \sin \mathfrak{c})(\sin \mathfrak{a} \cos C + \sin \mathfrak{b}) = \frac{1}{a^2} \sin B \sin C \sin q \sin r \begin{Bmatrix} \sin q \sin r \, (1 + \cos^2 \alpha) \\ - \cos \alpha \, (\sin^2 q + \sin^2 r) \end{Bmatrix}$$

$$= \frac{1}{a^2} \sin B \sin C \sin q \sin r \, (\sin q \sin r \sin^2 \alpha - a^2 \cos \alpha)$$

$$= \sin B \sin C \, (\sin^2 \mathfrak{a} - \sin q \sin r \cos \alpha),$$

which is the equation in question. For the subsidiary equations used in the demonstration, observe that the four points O, X, A', B lie in a circle, and consequently that $CO \,.\, CX = CA' \,.\, CB$; or multiplying each side by $\sin C$, then $CO \,.\, CX \,.\, \sin C = A'K \,.\, CB$, that is,

$$\sin r \, (\sin r - \sin q \cos \alpha) \sin C = a \, (\sin \mathfrak{a} \cos C + \sin \mathfrak{b}),$$

and the other of the equations in question is proved in the same manner.

From the formula for $\cosh \bar{a}$ we find

$$\sinh \bar{a} = \frac{1}{\sin \bar{B} \sin \bar{C}} \, \Delta,$$

where

$$\Delta^2 = -(1 - \cos^2 \bar{A} - \cos^2 \bar{B} - \cos^2 \bar{C} - 2 \cos \bar{A} \cos \bar{B} \cos \bar{C}),$$

whence also

$$\sinh \bar{a} : \sinh \bar{b} : \sinh \bar{c} = \sin \bar{A} : \sin \bar{B} : \sin \bar{C};$$

and we can also obtain

$$\cos \bar{A} = \frac{\cosh \bar{a} - \cosh \bar{b} \cosh \bar{c}}{\sinh \bar{b} \sinh \bar{c}} \text{ \&c.}$$

So that the formulæ are in fact similar to those of spherical trigonometry with only $\cosh \bar{a}$, $\sinh \bar{a}$ &c. instead of $\cos a$, $\sin a$ &c. The before-mentioned formula for $\cos \bar{\alpha}$ in terms of $\bar{a}$, $\bar{q}$, $\bar{r}$ is obviously a particular case of the last-mentioned formula for $\cos \bar{A}$.

Cambridge, 11 May, 1872.

529.

A "SMITH'S PRIZE" PAPER[1]; SOLUTIONS.

[From the *Oxford, Cambridge and Dublin Messenger of Mathematics*, vol. IV. (1868), pp. 201—226.]

1. *Find the form of a function of a given number of letters, which has two and only two values.*

It is required to find the general form of a function $\phi(a, b, c, \ldots, k)$, rational and integral, which for all permutations whatever of the letters has two and only two values.

Suppose that any particular permutation of the letters changes $\phi(a, b, c, \ldots, k)$ into $\phi_1(a, b, c, \ldots, k)$; then any permutation of the letters will either leave the functions ϕ, ϕ_1 each of them unaltered, or it will change ϕ into ϕ_1, and ϕ_1 into ϕ. Hence $\phi + \phi_1$ is a symmetrical function of all the letters, say

$$\phi + \phi_1 = 2L;$$

$\phi - \phi_1$ is a function which by any permutation of the letters is either unaltered, or simply changes its sign; and it is to be shown that, writing for shortness V to denote the product $(a-b)(a-c)\ldots(b-c)\ldots$ of the differences of the letters, and denoting by $2M$ a symmetrical function of the letters, we have

$$\phi - \phi_1 = 2VM.$$

These equations give

$$\phi = L + VM,$$

which is the general form required; viz. the function ϕ has then only the two values $L + VM$, $L - VM$.

To prove the subsidiary theorem, observe that there is at least one interchange of two letters which changes $\phi - \phi_1$ (for otherwise $\phi - \phi_1$ would be a symmetrical function);

[1] Set by me, for the Master of Trinity, Thursday, January 30, 1868.

let this be (a, b). Then $\phi - \phi_1$, changing its sign for the interchange in question, must vanish for $a = b$, that is, $\phi - \phi_1$ must contain the factor $(a - b)$; let it contain it in the power $(a - b)^s$; then the quotient $\phi - \phi_1 \div (a - b)^s$, not vanishing for $a = b$, cannot change its sign by the interchange in question, and as by supposition $\phi - \phi_1$ does change its sign, it appears that the exponent s must be odd. But $\phi - \phi_1$, containing the factor $(a - b)^s$, must contain every other like factor $(a - c)^s$ or $(c - d)^s$; in fact, writing $\phi - \phi_1 = K(a - b)^s$, then if the interchange (a, c) alters K into K_1, we have $K(a - b)^s = \pm K_1 (a - c)^s$, and K consequently contains the factor $(a - c)^s$; it does not contain any higher power $(a - c)^{s+s'}$, for if it did, by reversing the process it would appear that $\phi - \phi_1$ contained (contrary to supposition) the factor $(a - b)^{s+s'}$. Similarly $\phi - \phi_1$ contains the factor $(c - d)^s$, but no higher power $(c - d)^{s+s'}$. Hence $\phi - \phi_1$ contains the product of all the factors $(a - b)^s$, that is, it contains V^s, and writing $\phi - \phi_1 = 2V^s M$, the quotient M does not contain any such factor as $(a - b)$; it therefore does not change its sign for any interchange whatever (a, b); and in consequence it remains unaltered for any such interchange, that is, M is a symmetrical function. Observing that any even power of V is a symmetrical function, we may without loss of generality include V^{s-1} in the symmetrical function M, and write therefore

$$\phi - \phi_1 = 2VM,$$

which is the subsidiary theorem in question.

2. *Express* $\dfrac{x^7 - 1}{x - 1}$ *as the product of two cubic factors; and show generally that, for any prime exponent* p, *the function* $\dfrac{x^p - 1}{x - 1}$ *may be broken up into two factors each of the order* $\tfrac{1}{2}(p - 1)$, *by means of a quadratic equation.*

Let r be any root of the given equation, then

$$r^6 + r^5 + r^4 + r^3 + r^2 + r + 1 = 0;$$

the roots of this equation are $r^1, r^2, r^3, r^4, r^5, r^6$.

Hence

$$x^6 + x^5 + x^4 + x^3 + x^2 + x + 1 = (x - r^1)(x - r^2)(x - r^4) . (x - r^3)(x - r^5)(x - r^6),$$

and denoting the two cubic factors by y_1, y_2, or writing

$$y_1 = (x - r^1)(x - r^2)(x - r^4),$$
$$y_2 = (x - r^3)(x - r^5)(x - r^6),$$

we have

$$y_1 = x^3 - x^2(r^1 + r^2 + r^4) + x(r^3 + r^5 + r^6) - 1,$$
$$y_2 = x^3 - x^2(r^3 + r^5 + r^6) + x(r^1 + r^2 + r^4) - 1,$$

and thence

$$y_1 + y_2 = 2x^3 + x^2 - x - 2.$$

Hence y_1, y_2 are the roots of the quadratic equation

$$y^2 - y(2x^3 + x^2 - x - 2) + x^6 + x^5 + x^4 + x^3 + x^2 + x + 1 = 0.$$

Solving the equation, observing that the quantity under the radical sign must of necessity be a square, and that its value is

$$(2x^3+x^2-x-2)^2-4(x^6+x^5+x^4+x^3+x^2+x+1),$$

which is in fact

$$=-7(x^2+x)^2,$$

we find that the roots y_1, y_2, that is, the required cubic factors, are

$$\tfrac{1}{2}\{2x^3+x^2-x-2\pm\sqrt{(-7)}\,(x^2+x)\}.$$

Generally for any prime exponent p, denoting any root by r, and writing

$$y_1=(x-r^{a_1})(x-r^{a_2})\ldots,$$

$$y_2=(x-r^{b_1})(x-r^{b_2})\ldots,$$

where a_1, $a_2\ldots$ are the quadratic residues, and b_1, $b_2\ldots$ the quadratic non-residues of p, we find

$$y^2-yP+Q=0,$$

where P is a function of x of the order $\frac{1}{2}(p-1)$, and Q is $=x^{p-1}+x^{p-2}\ldots+x+1$. Hence

$$y=\tfrac{1}{2}\{P\pm\sqrt{(P^2-4Q)}\},$$

where P^2-4Q is a perfect square, $=\epsilon pZ^2$, Z a rational function of a degree less than $\frac{1}{2}(p-1)$, and $\epsilon=+$ or $-$ according as p is $\equiv 1$ or $\equiv 3$ (mod. 4). Hence the required factors are

$$\tfrac{1}{2}\{P\pm\sqrt{(\epsilon p)}\,Z\}.$$

3. *In a Map of the World, wherein the meridians are projected into right lines meeting in a point, the inclination of any two of the lines being equal to that of the two meridians, and the parallels into circles about the point as centre, the radius of the circle being a given function of the colatitude: compare on the sphere and in the map* (1) *the corresponding elements of area,* (2) *the azimuths of corresponding linear elements; and explain what conveniences may be obtained by proper determinations of the above-mentioned function of the colatitude.*

Let the longitude and colatitude be

on the sphere l, c,

in the map l', c',

then the projection is such, that $l'=l$, $c'=f(c)$, a given function of c.

The lengths of corresponding linear elements in the direction of a meridian, and perpendicular to it, are

$$dc,\ \sin c\,dl,$$

$$dc',\quad c'dl:$$

whence, elements of area are

$$dc \, . \sin c \, dl,$$
$$dc' . \quad c' dl;$$

tangents of azimuth are

$$dc \div \sin c \, dl,$$
$$dc' \div \quad c' dl;$$

and substituting the values $l' = l$, $c' = f(c)$, we have $dl' = dl$, $dc' = f'(c)\, dc$; and thence

$$\text{elements of area are as } \sin c : f(c) f'(c),$$

$$\text{tangents of azimuth as } \frac{1}{\sin c} : \frac{f'(c)}{f(c)}.$$

By proper determinations of the function $f(c)$, we can make

(1) The ratio of the elements of area to be constant; this will be the case if

$$f(c) f'(c) = k \sin c, \text{ that is, } f^2(c) = \text{const.} - 2k \cos c$$
$$= 2k (1 - \cos c),$$

since $f(c)$, $= c'$, must vanish for $c = 0$; that is,

$$f(c) = 2 \sqrt{(k)} \sin \tfrac{1}{2} c:$$

(2) corresponding azimuths to be equal; this will be the case if

$$\frac{f'(c)}{f(c)} = \frac{1}{\sin c}, \text{ that is, } \log f(c) = \text{const.} + \log \tan \tfrac{1}{2} c,$$

or say

$$f(c) = k \tan \tfrac{1}{2} c.$$

This is in fact the stereographic projection, in which (as is known) any indefinitely small figure is in the map represented without distortion.

4. *Explain the general configuration of the contour and slope lines in a tract of Lake and Mountain country.*

A contour line is the locus of points having a given altitude.

To fix the ideas, consider an island forming a two-headed mountain. The contour line at the sea level is a closed curve; at a sufficiently great altitude the contour line consists of two closed curves surrounding the two summits respectively; the transition from one form to the other takes place at the altitude of the pass between the two summits, the contour line then having a node at the top of the pass, and being in form a figure of eight. At the altitude of the lower summit one of the closed curves is reduced to a point, and for greater altitudes it disappears, the contour line being then a single closed curve surrounding the higher summit; and at the altitude of the higher summit this reduces itself to a point. (See fig. 1.) If there is on the breast of the mountain a lake, the contour line at the lake level is (as in the case of the pass) a curve with a node, being however here a closed curve with an interior loop as shown in the figure. The contour line for an altitude below the lake level includes as part of itself a closed curve lying within the contour of the lake,

Fig. 1.

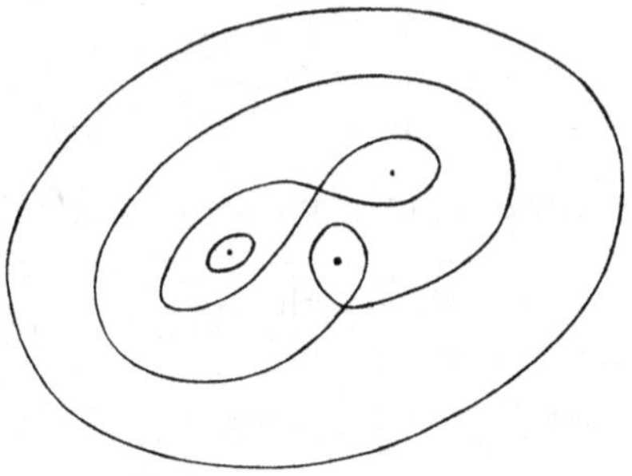

and which for the altitude of the lowest point (or "imit") of the lake reduces itself to a point, and for smaller altitudes disappears.

The contour lines in the immediate neighbourhood of a summit are in general ellipses (geometrically, the indicatrix is an ellipse); they may however be circles (viz. if the summit be an umbilicus).

A slope line is a line of greatest inclination, and it is consequently an orthogonal trajectory to the series of contour lines. A slope line may be considered as always terminated in a summit or an imit; there is through each summit (or imit) an infinity of slope lines, and in general these all *touch* there; if, however, the contour lines in the immediate neighbourhood are circles, then the slope lines, instead of touching, pass from the point in all directions. Through the node at the top of a pass, or outlet of a lake, there are two intersecting slope lines, one ascending each way from the node, and being in general a "ridge-line"; the other descending each way, and being in general a "course-line," viz. in the case of a pass, it is in each direction the course of the principal stream of the valley; and in the case of a lake-outlet, it is in the direction away from the lake, the course of the out-flowing stream. The slope lines which thus pass through the several nodes mark out distinct regions, and so facilitate the tracing of the intermediate slope lines.

5. *Find the differential equations corresponding to the three integral equations respectively* (i) $(y+c)^2 = x(x-1)(x-2)$; (ii) $(y+c)^2 = x^2(x-1)$; *and* (iii) $(y+c)^2 = x^3$: *and discuss geometrically the singular solutions.*

Generally, for the equation $(y+c)^2 - X = 0$, the derived equation is $4X\left(\frac{dy}{dx}\right)^2 - X'^2 = 0$. If from the integral equation, differentiating in regard to c, we attempt to find the singular solution, we obtain $(y+c)=0$; and thence $X=0$ for the singular solution. It is however to be observed that, if X contain single and multiple factors,

$$X = (x+\alpha)(x+\beta)\dots(x+\gamma)^m\dots,$$

then it is only the single factors $x+\alpha=0$, $x+\beta=0$, ..., which are solutions of the differential equation when expressed in its proper form free from extraneous factors.

To explain how this is, write the differential equation in the form

$$X'^2\left(\frac{dx}{dy}\right)^2-4X=0;$$

for a single factor $x+\alpha$, X' does not contain the factor $x+\alpha$, and there is no division by $x+\alpha$. The equation $x+\alpha=0$ gives $\frac{dx}{dy}=0$, $X=0$, and the equation is thus satisfied; and by what precedes $x+\alpha=0$ is a singular solution. Contrariwise, for the multiple factor $(x+\gamma)^m$, X' will contain $(x+\gamma)^{m-1}$, and the equation $X'^2\left(\frac{dx}{dy}\right)^2-4X=0$, will divide by $(x+\gamma)^m$, and divested of this factor it will be of the form

$$\frac{X'^2}{(x+\gamma)^m}\left(\frac{dx}{dy}\right)^2-\frac{4X}{(x+\gamma)^m}=0,$$

where $\frac{X'}{(x+\gamma)^m}$ contains the factor $(x+\gamma)^{m-2}$ (index is 0 or positive) but $\frac{X}{(x+\gamma)^m}$ does not contain $x+\gamma$; hence the equation $x+\gamma=0$, gives $\frac{dx}{dy}=0$, and therefore $\frac{X'}{(x+\gamma)^m}\left(\frac{dx}{dy}\right)^2=0$, but it does not give $\frac{X}{(x+\gamma)^m}=0$, and consequently fails to satisfy the differential equation. Reverting to the integral equation $(y+c)^2=(x+\alpha)(x+\beta)\ldots(x+\gamma)^m\ldots$, we see that $x+\alpha=0$ touches each of the series of curves, and is thus an envelope thereof: that $x+\gamma=0$ is not an envelope, but is the locus of a singular point on the series of curves.

Applying the foregoing considerations to the proposed question,

(i) The differential equation is (fig. 2),

Fig. 2.

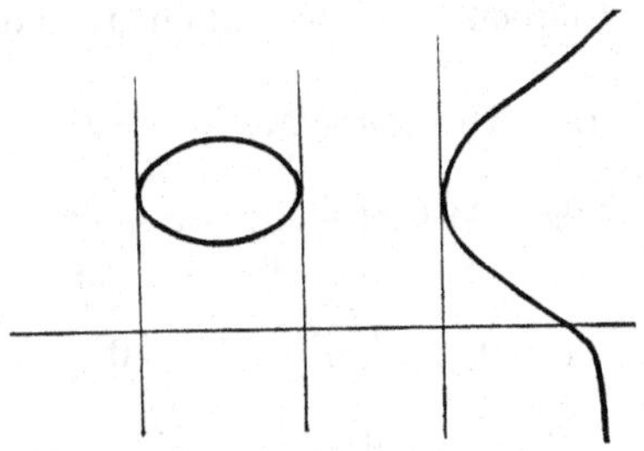

$$(3x^2-6x+2)^2\left(\frac{dx}{dy}\right)^2-4x(x-1)(x-2)=0,$$

having the singular solutions $x=0$, $x-1=0$, $x-2=0$.

(ii) The differential equation is (fig. 3),

$$(3x-2)^2\left(\frac{dx}{dy}\right)^2 - 4(x-1) = 0,$$

Fig. 3.

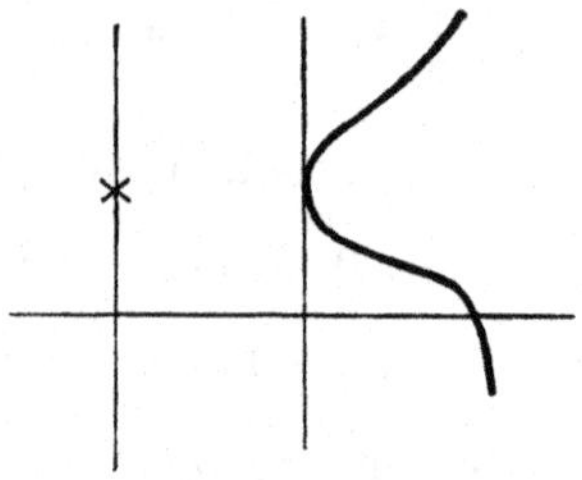

having the singular solution $x-1=0$. But $x=0$ is not a solution; it is the locus of the series of conjugate points of the curves $(y+c)^2 = x^2(x-1)$.

(iii) The differential equation is (fig. 4)

$$9x\left(\frac{dx}{dy}\right)^2 - 4 = 0,$$

which has no singular solution. But $x=0$ is the locus of the cusp of the curves $(y+c)^2 = x^3$.

Fig. 4.

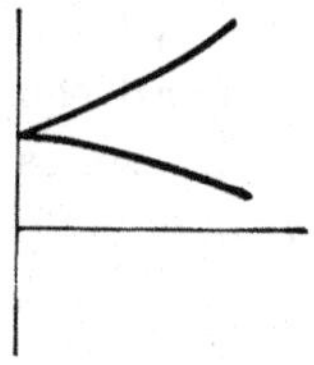

6. *Show that the curve parallel to the parabola is of the order* 6 *and class* 4; *explain the reduction of class; and trace the system of parallel curves.*

The curve parallel to the parabola is the envelope of a circle of constant radius, having its centre on a parabola; taking the equation of the parabola to be $y^2=4x$, the coordinates of any point thereon may be taken to be α^2, 2α, and the equation of the variable circle is

$$(x-\alpha^2)^2 + (y-2\alpha)^2 - r^2 = 0,$$

that is,

$$\alpha^4 + \alpha^2(-2x+4) + \alpha(-4y) + x^2 + y^2 - r^2 = 0;$$

or multiplying by 6, this is

$$(a,\ 0,\ c,\ d,\ e \between \alpha,\ 1)^4 = 0,$$

where

$$a = 6,$$

$$c = -2x+4,$$

$$d = -6y,$$

$$e = 6(x^2+y^2-r^2).$$

The equation of the envelope is obtained by eliminating α between the equation in α and the derived equation; or, what is the same thing, by equating to zero the discriminant of the equation in α; we thus obtain

$$(ae + 3c^2)^3 - 27(ace - ad^2 - c^3)^2 = 0,$$

which, substituting for a, c, d, e their values, is an equation in (x, y) of the order 6: the order of the curve is thus $= 6$.

The class is most easily obtained by geometrical considerations. Seeking the tangents which can be drawn from a given point, it at once appears that, describing about this point a circle radius r, then to each tangent through the point to the parallel curve, there corresponds a common tangent of the circle and the parabola, and reciprocally; whence the class of the curve is equal to the number of common tangents of the circle and the parabola, that is, it is $= 4$.

To the order 6 corresponds in general a class $= 30$, and the reduction from 30 to 4, $= 26$, is caused by the cusps and nodes of the curve.

The cusps are given as the points of intersection of the curves $ae + 3c^2 = 0$, $ace - ad^2 - c^3 = 0$, which being respectively of the orders 2 and 3, give 6 cusps; it may be added that these cusps (2 real and 4 imaginary, or else all 6 imaginary) are, in regard to the parabola, the centres of curvature for those points at which the radius of curvature is $= r$.

There is on the axis a single point (always real) whose normal distance from the parabola is $= r$. Such point is a node of the parallel curve, viz., according to the value of r, either an acnode (conjugate point) or a crunode: we have thus 1 node. The parallel curve at infinity coincides with the two parabolas obtained by the displacement of the given parabola parallel to itself through the distances $+r$, $-r$ along the axis of y. Two such parabolas have at infinity on the axis of x a contact of the second order, equivalent to three coincident intersections; and the parallel curve has thus at infinity on the axis of x, a singular point equivalent to 3 nodes; the number of nodes is thus $= 4$; and the 4 nodes and 6 cusps give the required reduction $2 . 4 + 3 . 6, = 26$.

The parallel curve is most easily traced geometrically; there are two branches, one outside, the other inside the parabola, equidistant from it. The outside branch is a curve of continuous curvature, the form of which requires no explanation. As regards the inside branch, when r is small, this is also of continuous curvature, but there is on the axis of x inside the branch a real acnodal point (the node above referred to): when r becomes = radius of curvature at vertex (or twice the focal distance), the acnode coincides with the branch; and the point on the axis, although presenting no visible singularity, is really in the nature of a triple point composed of two cusps and a node; when r is greater than this limiting value, instead of the acnode we have a crunode;

and two real cusps present themselves; viz. the form of the inside branch is as shown in fig. 5.

Fig. 5.

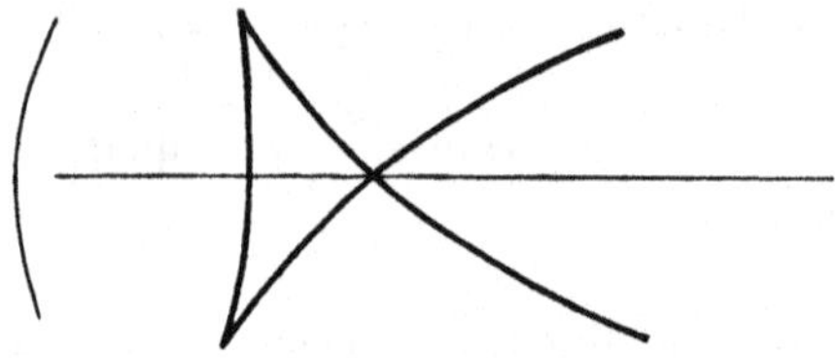

7. *Show that a curve of the order n has at most* $\frac{1}{2}(n-1)(n-2)$ *double points; and that in any curve having this maximum number of double points, the coordinates* (x, y, z) *may be taken to be proportional to rational and integral functions of a variable parameter.*

A curve of the order n cannot have more than $\frac{1}{2}(n-1)(n-2)$ double points; for suppose it had one more, say

$$\tfrac{1}{2}(n-1)(n-2)+1, = \tfrac{1}{2}(n^2-3n)+2, \text{ double points};$$

then since a curve of the order $n-2$ can be drawn through

$$\tfrac{1}{2}(n-2)(n+1), \quad = \tfrac{1}{2}(n^2-n)-1 \text{ points},$$

suppose it drawn through the

	$\frac{1}{2}(n^2-3n)+2$ double points;
and besides through	$n-3$ points on the curve,
together	$\frac{1}{2}(n^2-n)-1$ points,

it will cut the curve of the order n in the double points considered as

	n^2-3n+4 points,
and besides in	$n-3$ points,
together	n^2-2n+1 points,

that is, we should have a curve of order $n-2$ meeting the curve of the order n in $n(n-2)+1$ points, which is of course impossible.

Consider now a curve of the order n with $\frac{1}{2}(n-1)(n-2)$ double points; draw a curve of the order $(n-2)$ through the double points

	$\frac{1}{2}(n^2-3n)+1$ points,
and besides through	$n-3$ points on the curve,
together	$\frac{1}{2}(n^2-n)-2$ points;

this imposes on the curve $\frac{1}{2}(n^2-n)-2$ conditions; but as the curve might have been determined to satisfy $\frac{1}{2}(n^2-n)-1$ conditions, its equation will contain an indeterminate parameter λ: it meets the curve n in the double points considered as

	n^2-3n+2 points,
in the	$n-3$ points,
and besides in	1 point,
altogether	n^2-2n points.

Hence since for any value whatever of λ, there is only one remaining intersection, the coordinates of this point must be rationally determinable in terms of the parameter λ; that is, the coordinates of *any* point on the given curve will be rational functions of λ; or using trilinear coordinates, the coordinates x, y, z of any point on the given curve will be proportional to rational and integral functions of λ.

8. *Show that three bodies attracting each other according to the law of gravitation may move in a line in such wise that the mutual distances are in a constant ratio the value of which depends on the masses.*

Taking the masses m_1, m_2, m_3, to be arranged in this order at distances x_1, x_2, x_3 from a fixed origin, the equations of motion are

$$\frac{d^2x_1}{dt^2}=\frac{m_2}{(x_2-x_1)^2}+\frac{m_3}{(x_3-x_1)^2},$$

$$\frac{d^2x_2}{dt^2}=-\frac{m_1}{(x_1-x_2)^2}+\frac{m_3}{(x_3-x_2)^2},$$

$$\frac{d^2x_3}{dt^2}=-\frac{m_1}{(x_1-x_3)^2}-\frac{m_2}{(x_2-x_3)^2},$$

whence

$$\frac{d^2(x_2-x_1)}{dt^2}=-\frac{m_1+m_2}{(x_2-x_1)^2}+\frac{m_3}{(x_3-x_2)^2}-\frac{m_3}{(x_3-x_1)^2},$$

$$\frac{d^2(x_3-x_1)}{dt^2}=-\frac{m_1+m_3}{(x_3-x_1)^2}-\frac{m_2}{(x_3-x_2)^2}-\frac{m_2}{(x_2-x_1)^2}.$$

Assume

$$x_3-x_1=\alpha(x_2-x_1),$$

and therefore

$$x_3-x_2=x_3-x_1-(x_2-x_1)=(\alpha-1)(x_2-x_1);$$

then the equations become

$$\frac{d^2(x_2-x_1)}{dt^2}=\left\{-m_1-m_2+\frac{m_3}{(\alpha-1)^2}-\frac{m_3}{\alpha^2}\right\}\frac{1}{(x_2-x_1)^2},$$

and

$$\frac{d^2(x_3-x_1)}{dt^2}=\left\{-\frac{m_1+m_3}{\alpha^2}-\frac{m_2}{(\alpha-1)^2}-m_2\right\}\frac{1}{(x_2-x_1)^2},$$

which equations will be one and the same equation if only

$$\alpha\left\{-m_1-m_2+\frac{m_3}{(\alpha-1)^2}-\frac{m_3}{\alpha^2}\right\}=-\frac{m_1+m_3}{\alpha^2}-\frac{m_2}{(\alpha-1)^2}-m_2,$$

an equation which (multiplying out) is of the fifth order, and gives at least one real value of α: and α being thus determined, we may from either equation obtain an equation of the form $\frac{d^2(x_2-x_1)}{dt^2}=\frac{C}{(x_2-x_1)^2}$, which determines x_2-x_1 in terms of t, and then x_3-x_1, $=\alpha(x_2-x_1)$, is also known; that is, the relative motions of the three bodies are determined.

9. *Write down the integral equations for the elliptic motion of a planet; and if r, s, v_1 denote the radius vector, tangent of the latitude, and reduced longitude respectively, show that*

$$\frac{\sqrt{(1+s^2)}}{r}=\frac{1}{a_1(1-e_1^2)}\{\sqrt{(1+s^2)}+e_1\cos(v_1-\varpi_1)\},$$

explaining the significations of the constants a_1, e_1, ϖ_1.

Write

a, the mean distance,

n, the mean motion $=\frac{\sqrt{(\mu)}}{a^{\frac{3}{2}}}$,

e, the eccentricity,

θ, longitude of node,

ϖ, longitude of pericentre in orbit,

ϕ, the inclination,

v, the longitude in orbit.

The position in the orbit is determined in terms of the time by means of an auxiliary quantity u, viz. writing

$$nt+c=u-e\sin u.$$

We then have the radius vector r, and the true anomaly f, given as functions of the time by the equations

$$\left.\begin{aligned}\cos f&=\frac{\cos u-e}{1-e\cos u},\\ \sin f&=\frac{\sqrt{(1-e^2)}\sin u}{1-e\cos u},\end{aligned}\right\}$$

$$r=\frac{a(1-e^2)}{1+e\cos f},\quad =a(1-e\cos u).$$

Take z, x, y the hypothenuse, base, and perpendicular of the right-angled triangle NPP', base angle $=\phi$ (see fig. 6), in which NP is the orbit, N the node, K the pericentre, P the planet; then

Fig. 6.

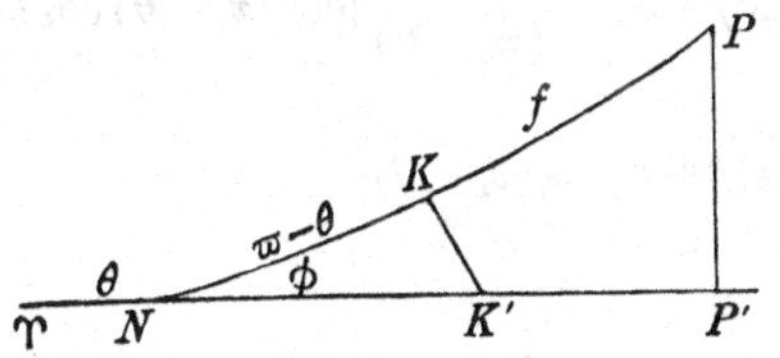

$z=\varpi-\theta+f$ gives z; and x, y are given in terms of z, ϕ, by the formulæ relating to the spherical triangle, so that we have

$$\begin{aligned}\text{longitude in orbit} &= z+\theta\,(=\varpi+f),\\ \text{reduced longitude} &= x+\theta,\\ \text{latitude} &= y.\end{aligned}$$

The expression for r, writing for f its value, $=v-\varpi$, gives

$$\frac{1}{r}=\frac{1}{a(1-e^2)}\{1+e\cos(v-\varpi)\},$$

and we have thence

$$\frac{\sqrt{(1+s^2)}}{r}=\frac{1}{a(1-e^2)}\{\sqrt{(1+s^2)}+\sqrt{(1+s^2)}\,e\cos(v-\varpi)\}.$$

But we have

$$v-\varpi=v-\theta-(\varpi-\theta),$$

and thence

$$\cos(v-\varpi)=\cos(v-\theta)\cos(\varpi-\theta)+\sin(v-\theta)\sin(\varpi-\theta).$$

Consequently

$$\begin{aligned}\sqrt{(1+s^2)}\cos(v-\varpi)=&\sqrt{(1+s^2)}\cos(\varpi-\theta)\cos(v-\theta)\\ &+\sqrt{(1+s^2)}\sin(\varpi-\theta)\sin(v-\theta);\end{aligned}$$

but by the right-angled triangle, we have

$$\begin{aligned}\sin(v-\theta)\sin\phi=\sin y &= \frac{s}{\sqrt{(1+s^2)}},\\ \sin(v_1-\theta) &= \cot\phi\tan y=s\cot\phi,\end{aligned}$$

and thence

$$\sqrt{(1+s^2)}\sin(v-\theta)=\frac{s}{\sin\phi}=\frac{1}{\cos\phi}s\cot\phi=\frac{\sin(v_1-\theta)}{\cos\phi},$$

$$\cos(v-\theta)=\cos(v_1-\theta)\cos y=\frac{\cos(v_1-\theta)}{\sqrt{(1+s^2)}},$$

that is,

$$\sqrt{(1+s^2)}\cos(v-\theta)=\cos(v_1-\theta).$$

The formula thus becomes

$$\frac{\sqrt{(1+s^2)}}{r} = \frac{1}{a(1-e^2)}\left[\sqrt{(1+s^2)} + e\left\{\cos(v_1-\theta)\cos(\varpi-\theta) + \sin(v_1-\theta)\frac{\sin(\varpi-\theta)}{\cos\phi}\right\}\right]$$

$$= \frac{1}{a\sqrt{(1-e^2)}}\left[\sqrt{(1+s^2)} + e\frac{\cos(\varpi-\theta)}{\cos(\varpi_1-\theta)}\{\cos(v_1-\theta)\cos(\varpi-\theta) + \sin(v_1-\theta)\sin(\varpi-\theta)\right]$$

$$= \frac{1}{a_1(1-e_1^2)}\{\sqrt{(1+s^2)} + e_1\cos(v_1-\theta)\},$$

where

$$\frac{1}{\cos\phi}\tan(\varpi-\theta) = \tan(\varpi_1-\theta), \text{ gives } \varpi_1,$$

$$e_1 = e\frac{\cos(\varpi-\theta)}{\cos(\varpi_1-\theta)}, \text{ gives } e_1,$$

$$a_1(1-e_1^2) = a(1-e^2), \text{ gives } a_1.$$

The first equation shows that, drawing the arc KK' at right angles to NP, then

$$\varpi_1 - \theta = NK',$$

and therefore

$$\varpi_1 = \theta + (\varpi_1 - \theta) = \Upsilon K'.$$

The other two equations then give e_1 and a_1: it may be added, that from the right-angled triangle NKK' we have $\cos(\varpi_1-\theta) = \cos(\varpi-\theta)\cos KK'$, and consequently that e_1 is $= e \div \cos KK'$.

10. *Find the differential equations for the motion of a material line acted upon by any forces and moving in a given ruled surface.*

If a, b are the coordinates of the point of intersection of any line of the ruled surface with the plane of xy; α, β, γ the cosines of the inclinations of the line to the three axes respectively; then writing

$$\frac{x-a}{\alpha} = \frac{y-b}{\beta} = \frac{z}{\gamma} (= r),$$

and considering a, b, α, β, γ as given functions of a variable parameter θ, where α, β, γ satisfy the relation $\alpha^2+\beta^2+\gamma^2=1$, the equations in question, exclusive of the equation $(=r)$, determine the particular line on the surface; and taking account of the equation $(=r)$, they determine in terms of the parameters r, θ, the coordinates of a particular point in this line.

The motion of the material line is such that it comes successively to coincide with the several lines on the ruled surface; consider on the material line a fixed point. say its centre of gravity G, and imagine that in the course of the motion the material line comes to coincide with the line determined as above by the parameter θ, and the point G with the point determined as above by the parameters r, θ; consider on the material line any other point P whose distance from G is $=s$; the parameters of P will be $r+s$, θ; and the coordinates will consequently be given in terms of the variable parameters r, θ by the equations

$$\frac{x-a}{\alpha}=\frac{y-b}{\beta}=\frac{z}{\gamma}=r+s,$$

that is, we have

$$x=a+(r+s)\,\alpha,\quad y=b+(r+s)\,\beta,\quad z=(r+s)\,\gamma,$$

where a, b, α, β, γ are given functions of θ; r, θ are parameters varying with the time, and s is a constant in regard to the time, but varies with the point under consideration.

Hence, writing as usual T to denote the *Vis Viva* function

$$=\tfrac{1}{2}\Sigma dm\,(x'^2+y'^2+z'^2),$$

where dm is the element of the material line, and x', y', z' are the velocities of the element, it only remains to express T as a function of r, θ; and the equations of motion will be

$$\frac{d}{dt}\frac{\partial T}{\partial r'}-\frac{\partial T}{\partial r}=0,$$

$$\frac{d}{dt}\frac{\partial T}{\partial \theta'}-\frac{\partial T}{\partial \theta}=0.$$

Using a', b', α', β', γ' to denote the differential coefficients $\frac{\partial a}{\partial \theta}$, &c., but, as above, x', y', z', r', θ' to denote differential equations in regard to t, we have

$$x'=[a'+(r+s)\,\alpha'\,]\,\theta'+\alpha r',$$

$$y'=[b'+(r+s)\,\beta'\,]\,\theta'+\beta r',$$

$$z'=[\quad (r+s)\,\gamma'\,]\,\theta'+\gamma r',$$

and thence

$$x'^2+y'^2+z'^2=A\theta'^2+2Br'\theta'+r'^2,$$

if for shortness

$$A=a'^2+b'^2+2(r+s)\,(\alpha'a'+\beta'b')+(r+s)^2\,(\alpha'^2+\beta'^2+\gamma'^2),$$

$$B=a'\alpha+b'\beta+(r+s)\,(\alpha\,\alpha'+\beta\,\beta'+\gamma\gamma'),$$

$$=a'\alpha+b'\beta,$$

since

$$\alpha\alpha'+\beta\beta'+\gamma\gamma'=0.$$

Hence we have

$$T=\tfrac{1}{2}\theta'^2\Sigma A dm+r'\theta'\Sigma B dm+\tfrac{1}{2}r'^2\Sigma dm,$$

and, observing that in the sum $\Sigma A dm$, the terms involving $\Sigma s dm$ vanish in consequence of G being the centre of gravity, we have

$$\Sigma A dm=[a'^2+b'^2+2r\,(\alpha'a'+\beta'b')+r^2\,(\alpha'^2+\beta'^2+\gamma'^2)]\,\Sigma dm+(\alpha'^2+\beta'^2+\gamma'^2)\,\Sigma s^2 dm.$$

Or writing $\Sigma dm = M$, $\Sigma s^2 dm = Mk^2$, (M being therefore the mass of the line, and Mk^2 its moment about the centre of gravity), we have

$$\Sigma A dm = \tfrac{1}{2}M[a'^2 + b'^2 + 2r(\alpha a' + \beta' b') + (r^2 + k^2)(\alpha'^2 + \beta'^2 + \gamma'^2)].$$

Moreover

$$\Sigma B dm = M(a'\alpha + b'\beta),$$

and hence we have

$$T = \tfrac{1}{2}M\left\{\theta'^2[a'^2 + b'^2 + 2r(\alpha a' + \beta' b') + (r^2 + k^2)(\alpha'^2 + \beta'^2 + \gamma'^2)] + 2\theta' r'(a'\alpha + b'\beta) + r'^2\right\},$$

a given function of r, θ, r', θ'; and the differential equations are therefore given as above.

11. *Explain the mutual connexion of the three theorems in conics: (1) the theorem* ad quatuor lineas; (2) *Pascal's theorem; (3) the theorem of the anharmonic relation of four points.*

The theorem *ad quatuor lineas* is that the locus of a point, such that the product of its distances from two given lines is always in a given ratio to the product of its distances from two other given lines, is a conic.

Consider the lines as forming a quadrilateral $ABCD$. Then A, B, C, D are points on the conic, and writing PAB to denote the perpendicular distance of a point P from the line AB, and so in other cases, the theorem is that the expression

$$\frac{PAB \,.\, PCD}{PAC \,.\, PBD}$$

has a constant value for any point P whatever on the conic. Now the perpendicular distance PAB is $= 2\Delta PAB \div AB$, or, what is the same thing, it is $= PA \,.\, PB \sin PAB \div AB$; transforming the other perpendicular distances in the same manner, the distances PA, PB, PC, PD divide out and the foregoing expression becomes

$$= \frac{AC \,.\, BD \sin PAB \,.\, \sin PCD}{AB \,.\, CD \sin PAC \,.\, \sin PBD},$$

viz. omitting the constant factor, it appears that the expression $\dfrac{\sin PAB \,.\, \sin PCD}{\sin PAC \,.\, \sin PBD}$ is constant for all points P on the conic; or, what is the same thing, that the anharmonic ratio of the pencil $P(A, B, C, D)$ is constant for all points P on the conic. This is the anharmonic property of the points of a conic.

Pascal's theorem is that for any six points 1, 2, 3, 4, 5, 6 (fig. 7) on a conic, the intersections of the lines 12 and 45, the lines 23 and 56, the lines 34 and 61 lie in a line. Marking the points α, β, γ, δ as shown in the figure, it appears by the theorem that we have the two lines $(65\beta\delta)$, $(\alpha 54\gamma)$ meeting in 5, and such that the lines through the corresponding points 6 and α, β and 4, δ and γ meet in a point. Hence the ranges $(65\beta\delta)$ and $(\alpha 54\gamma)$ have the same anharmonic ratio; or, what is the same thing, the pencils $3(65\beta\delta)$ and $1(\alpha 54\gamma)$ have the same anharmonic ratio; that is, the pencils $3(6542)$ and $1(6542)$ have the same anharmonic ratio; or, considering 1 as a variable point, but 2, 3, 4, 5, 6 as fixed points on the conic, the pencil 1 (6542) of lines from the point 1 to the four points 6, 5, 4, 2 has the same anharmonic ratio for all

Fig. 7.

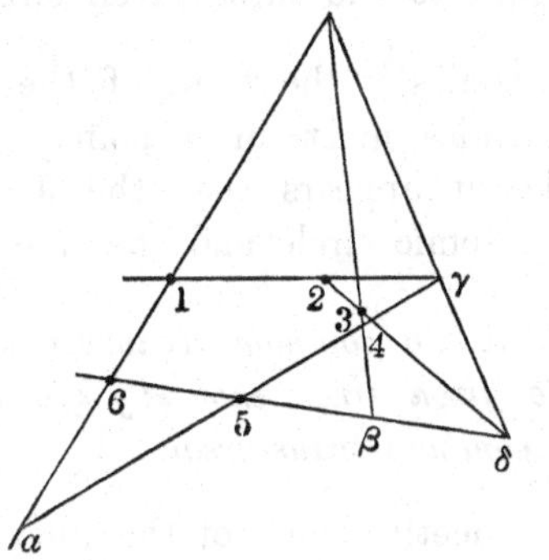

positions whatever of the variable point 1; which is the above-mentioned anharmonic property of the points of a conic.

12. *Show that any line through the centre of either of two orthotomic circles cuts the two circles harmonically; and connect this result with the theorem that the Jacobian of three circles is made up of the line infinity and the orthotomic circle.*

Drawing through the centre of the circle A, (fig. 8) the line $\alpha\beta\alpha'\beta'$, it appears by the figure that we have

$$A\beta \,.\, A\beta' = \text{square of tangential distance of } A \text{ from the circle } B,$$
$$= A\alpha'^2 \text{ since the circles cut at right angles,}$$
$$= A\alpha \,.\, A\alpha';$$

Fig. 8.

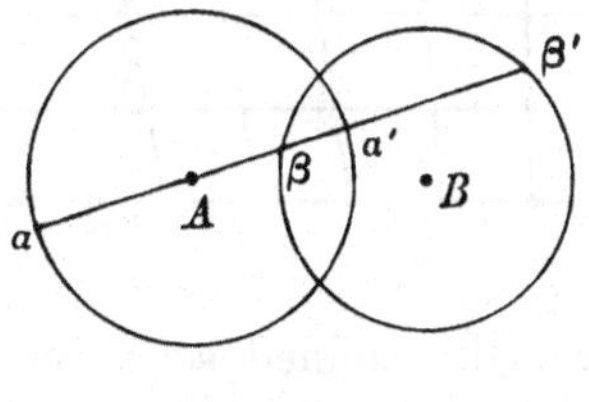

that is, the points β, β' are inverse points in regard to the circle on the diameter $\alpha\alpha'$; and the points α, α'; β, β' are thus harmonically related to each other.

Hence the polar of α, in regard to the circle B, passes through the point α', which is the opposite of α in regard to the circle A; and is consequently a fixed point for all the circles B orthotomic to the circle A. That is, considering any three circles orthotomic to the circle A, the circle A is a locus of points α such that the polars of α in regard to the three circles meet in a point.

Moreover, all circles meet the line at infinity in two fixed points I, J, the circular points at infinity: hence for any point α on the line infinity, the polar of α in regard to any circle whatever meets the line infinity in a fixed point, the harmonic of α in regard to the two points I, J; whence the line infinity is also a locus of points α such that the polars of α in regard to the three given circles meet in a point.

The Jacobian of any three conics is the locus of the points α, such that the polars of α in regard to the three conics meet in a point; and it is in general a cubic curve. Hence by what precedes it appears that the Jacobian of three circles is the cubic curve made up of the orthotomic circle and the line infinity.

13. *If five given lines have a common transversal, then taking the remaining transversal of each four of the given lines, show by statical considerations or otherwise that the five transversals have a common transversal.*

Consider the line 6′ (fig. 9) meeting each of the lines 1, 2, 3, 4, 5, and take

1′	the remaining	transversal of	(2, 3, 4, 5),
2′	„	„	(3, 4, 5, 1),
3′	„	„	(4, 5, 1, 2),
4′	„	„	(5, 1, 2, 3),
5′	„	„	(1, 2, 3, 4);

it is to be shown that the lines 1′, 2′, 3′, 4′, 5′ have a common transversal 6.

Fig. 9.

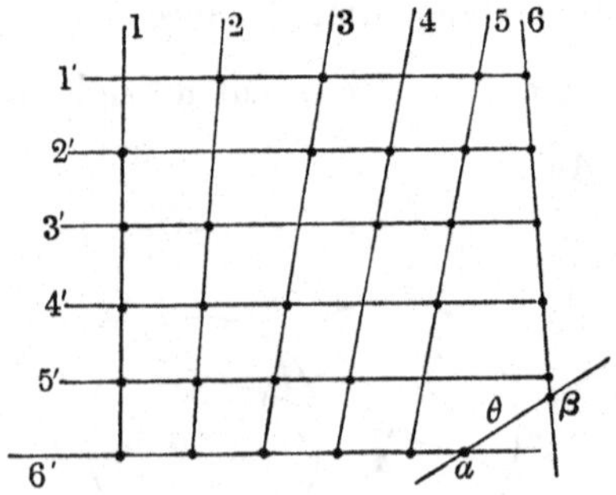

Consider the line 6 as a line determined so as to meet the lines 2′, 3′, 4′ and 5′; and take any line θ meeting each of the lines 6, 6′; since the six lines 1, 2, 3, 4, 5, θ have a common transversal 6′, then considering the lines as fixed lines in a solid body, it is possible to find along the lines 1, 2, 3, 4, 5, θ forces which will be in equilibrium. Suppose for a moment that the line 6, determined as above, does not meet the line 1′; then we have the lines 1′, 2′, 3′, 4′, 5′, θ and 6 such that the line 6 meeting the lines 2′, 3′, 4′, 5′, and θ, does not meet the line 1. The six lines 1′, 2′, 3′, 4′, 5′, and θ would be independent lines, such that there do not exist along them forces in equilibrium, and a force acting in any line whatever may be resolved into forces acting along the lines 1′, 2′, 3′, 4′, 5′, and θ. Hence the original forces along the lines 1, 2, 3, 4, 5 can be each of them so resolved; and combining with the resolved forces the original force along θ, we have a system of forces along the lines 1′, 2′, 3′, 4′, 5′, θ in equilibrium with each other; which is impossible if the lines in question are related as above; that is, the line 6 meeting the lines 2′, 3′, 4′, 5′, θ must also meet the line 1; which is the required theorem.

The statical principles assumed in the above demonstration are as follows:

(1) Six lines may be such that there exist along them forces in equilibrium; or say, for shortness, the six lines may be in involution.

(2) For any seven lines, no six or any less number of which are in involution, there exist along the lines forces in equilibrium; or, what is the same thing, given any six lines not in involution, and a seventh line; then a force along the seventh line may be resolved into forces along the six lines.

(3) Six lines which have a common transversal are in involution (remark in passing that it is not conversely true that if the six lines are in involution they have a common transversal), but if five of the six lines have a common transversal not meeting the sixth line, then the six lines are *not* in involution.

14. *In the theory of the variation of the arbitrary constants of a mechanical problem, state and explain the results obtained by Lagrange and Poisson respectively; and point out the peculiar advantage of Poisson's theory, in regard to the consequences which follow from the coefficients of his formulæ being independent of the time.*

In the theory of the variation of the arbitrary constants of a mechanical problem, it is assumed that the forces consist of principal and disturbing forces, each of them depending on a force function, say there is a principal force function and a disturbing function; (the assumption of a force function however in regard to the disturbing forces, though usual and convenient, is not essential); and that, when the disturbing forces are neglected, or say in the undisturbed problem, the equations of motion can be completely integrated; the theory consists herein that the same integral equations, taking the arbitrary constants to be variable, may be made to satisfy the disturbed equations of motion.

Suppose that the number of the coordinates $x, y, \ldots$ is $=p$, then since the differential equations are of the second order, the number of arbitrary constants $(a, b, c, \ldots)$ will be $=2p$. In Lagrange's solution it is assumed that the coordinates $x, y, \ldots$ (and consequently also the derived functions $x', y', \ldots$) are each of them given in terms of t and the $2p$ constants; the disturbing function Ω is given in the same form; and the expressions for the variations $\frac{da}{dt}$, &c. are obtained by the solution of a system of linear equations

$$\frac{dR}{da} = (a, b)\frac{db}{dt} + (a, c)\frac{dc}{dt} + \&c.,$$

where the coefficients (a, b) are functions involving $\frac{dx}{da}, \frac{dx'}{da}$, &c.; these coefficients are thus *primâ facie* functions of the $2p$ constants *and of the time;* but it is a result of the theory that they are in fact functions of the constants only, the time disappearing of itself.

In Poisson's theory it is assumed that the integrals of the undisturbed problem are obtained in the form

$$a = \phi(x, y, \ldots x', y', \ldots, t), \text{ \&c.},$$

viz. that each constant is given in terms of the coordinates, their derived functions, and the time (so that, if all the integrals are known, this is equivalent to Lagrange's assumption). The expressions for the variations of the constants are given in the form

$$\frac{da}{dt} = [a, b] \frac{dR}{db} + [a, c] \frac{dR}{dc} + \text{\&c.},$$

where each coefficient $[a, b]$ is given as a function of $\frac{da}{dx}, \frac{da}{dx'}$, &c., $\frac{db}{dx}, \frac{db}{dx'}$, &c.: these coefficients are thus in the first instance expressed as functions of the coordinates $x, y, \ldots$, the derived functions $x', y', \ldots$, and the time; but it is a result of the theory that the coefficients $[a, b]$ are really constant; viz. that, if $x, y, \ldots, x', y', \ldots$ were expressed in terms of the $2p$ constants and the time, then that the time would disappear of itself and the coefficients $[a, b]$, &c. would be found to be functions of the constants only.

The formation of the value of any coefficient $[a, b]$ requires only the knowledge of the expressions of a, b in terms of $x, y, \ldots, x', y', \ldots, t$, or say the knowledge of the two integrals a, b. We thence obtain the expression of $[a, b]$ as a function of $x, y, \ldots, x', y', \ldots, t$, say $[a, b] = f(x, y, \ldots, x', y', \ldots, t)$. But, as already mentioned, $[a, b]$ is in fact a constant; calling it c, we have $c = f(x, y, \ldots, x', y', \ldots, t)$; that is, we have an integral of the equations of motion of the undisturbed problem. It may happen that the value of $[a, b]$ is found to be $= 0$; or to be a function of $x, y, \ldots, x', y', \ldots, t$, which in virtue of the given values of a, b in terms of these same quantities reduces itself to a function of (a, b); in either of these cases we obtain no new integral; but if (as may be) neither of the foregoing cases happen, then the equation $c = f(x, y, \ldots, x', y', \ldots, t)$ *is actually a new integral of the equations of motion* (in the undisturbed problem) *obtained by mere differentiations from the two given integrals* a, b. There is nothing analogous to this in Lagrange's theory.

15. *Write a short dissertation on the transformation of coordinates (rectangular in space of three dimensions); and in particular explain under what restriction it is true that two sets of rectangular axes about the same origin may be made to coincide by means of a rotation of either set about a certain axis; and from the formulae of transformation obtain expressions for the position of this axis and the amount of the rotation.*

Two sets of rectangular axes about the same origin (each axis considered, not as a line extending in two opposite senses, but as drawn from the origin in one sense only) are or are not displacements the one of the other; viz., making the axis of x_1 to coincide with that of x, and the axis of y_1 to coincide with that of y, then either the axis of z_1 will coincide with that of z, or it will be in the opposite direction; in the former case the two sets are, in the latter case they are not, displacements one of the other. The restriction referred to in the question, is that the two sets shall be displacements one of the other

In the problem of transformation, the two sets of axes, if not displacements, can always be made so by simply reversing the direction of one of the axes (writing for example $-z$ for z); and there is thus no real loss of generality in considering the two sets as displacements the one of the other; and it is in general convenient to make this assumption.

The transformation between two sets of rectangular axes is at once given by the diagram

	x	y	z
x_1	α	β	γ
y_1	α'	β'	γ'
z_1	α''	β''	γ''

where α is the cosine of the inclination of the axes x, x_1; and so for the rest of the nine quantities. The relation between the two sets of axes is obtained at pleasure by reading the diagram horizontally $x_1 = \alpha x + \beta y + \gamma z$, &c.; or by reading it vertically $x = \alpha x_1 + \alpha' y_1 + \alpha'' z_1$, &c.

We must have identically $x^2 + y^2 + z^2 = x_1^2 + y_1^2 + z_1^2$; and we thus obtain the two equivalent sets of equations

$$\begin{array}{l|l}
\alpha^2 + \beta^2 + \gamma^2 = 1, & \alpha^2 + \alpha'^2 + \alpha''^2 = 1, \\
\alpha'^2 + \beta'^2 + \gamma'^2 = 1, & \beta^2 + \beta'^2 + \beta''^2 = 1, \\
\alpha''^2 + \beta''^2 + \gamma''^2 = 1, & \gamma^2 + \gamma'^2 + \gamma''^2 = 1, \\
\alpha'\alpha'' + \beta'\beta'' + \gamma'\gamma'' = 0, & \beta\gamma + \beta'\gamma' + \beta''\gamma'' = 0, \\
\alpha''\alpha + \beta''\beta + \gamma''\gamma = 0, & \gamma\alpha + \gamma'\alpha' + \gamma''\alpha'' = 0, \\
\alpha\alpha' + \beta\beta' + \gamma\gamma' = 0, & \alpha\beta + \alpha'\beta' + \alpha''\beta'' = 0.
\end{array}$$

Either set leads to the relation

$$\begin{vmatrix} \alpha, & \beta, & \gamma \\ \alpha', & \beta', & \gamma' \\ \alpha'', & \beta'', & \gamma'' \end{vmatrix}^2 = 1, \text{ consequently } \begin{vmatrix} \alpha, & \beta, & \gamma \\ \alpha', & \beta', & \gamma' \\ \alpha'', & \beta'', & \gamma'' \end{vmatrix} = \pm 1,$$

the distinction between the two cases $+1$, -1 being that above explained; viz. if the axes are not displacements the one of the other, the sign is $-$; if they are, (and in all that follows this is assumed to be the case) then the sign is $+$. The equations give further $\alpha = \beta'\gamma'' - \beta''\gamma'$, &c. (nine equations).

It is easy to see geometrically that, as stated in the question, two sets of axes (being as above mentioned displacements the one of the other) can be made to coincide by means of a rotation of either set about a certain axis; inasmuch as the position of the axis itself is not altered by the rotation, it is clear that for any point of the axis, the coordinates x, y, z and x_1, y_1, z_1 must be respectively equal; we thus have

$$\begin{aligned}(\alpha-1)\,x + \beta y + \gamma z &= 0,\\ \alpha' x + (\beta'-1)\,y + \gamma' z &= 0,\\ \alpha'' x + \beta'' y + (\gamma''-1)\,z &= 0,\end{aligned}$$

equations which must be equivalent to two equations; we in fact have

$$\begin{vmatrix} \alpha-1, & \beta\ , & \gamma \\ \alpha'\ , & \beta'-1, & \gamma' \\ \alpha''\ , & \beta''\ , & \gamma''-1 \end{vmatrix} = 0,$$

as is easily verified by means of the foregoing relations between the coefficients. Any two of the three equations will then determine the ratios $x:y:z$; taking the second and third, we have

$$x : y : z = (\beta'-1)(\gamma''-1) - \beta''\gamma' : \gamma'\alpha'' - \alpha'(\gamma''-1) : \alpha'\beta'' - \alpha''(\beta'-1),$$

reducible to

$$x : y : z = 1 + \alpha - \beta' - \gamma'' : \beta + \alpha' : \gamma + \alpha'',$$

and treating in the same way the third and first, and the first and second equations the system of formulæ is

$$\begin{aligned} x : y : z &= 1+\alpha-\beta'-\gamma'' : \beta+\alpha' : \gamma+\alpha''\\ &= \alpha'+\beta : 1-\alpha+\beta'-\gamma'' : \gamma'+\beta''\\ &= \alpha''+\gamma : \beta''+\gamma' : 1-\alpha-\beta'-\gamma'',\end{aligned}$$

equations equivalent to each other; they determine the position of the axis in question, or resultant axis. The foregoing equations may also be written

$$x^2 : y^2 : z^2 : x^2+y^2+z^2 = 1+\alpha-\beta'-\gamma'' : 1-\alpha+\beta'-\gamma'' : 1-\alpha-\beta'+\gamma'' : 3-\alpha-\beta'-\gamma'',$$

and hence, if A, B, C be the inclinations of the resultant axis to the axes of x and x_1, y and y_1, z and z_1 respectively, we have

$$\cos^2 A = \frac{1+\alpha-\beta'-\gamma''}{3-\alpha-\beta'-\gamma''},$$

and thence also

$$\sin^2 A = \frac{2(1-\alpha)}{3-\alpha-\beta'-\gamma''},$$

$$\cos 2A = \frac{-1+3\alpha-\beta'-\gamma''}{3-\alpha-\beta'-\gamma''},$$

$$\alpha - \cos 2A = \frac{(1-\alpha)(1+\alpha+\beta'+\gamma'')}{3-\alpha-\beta'-\gamma''}.$$

Let θ be the amount of the rotation about the resultant **axis**; we have a spherical triangle the two sides whereof are A, A, the included angle θ, and the opposite side $\cos^{-1}\alpha$; that is, we have

$$\cos\theta = \frac{\alpha - \cos^2 A}{\sin^2 A},$$

and thence

$$4\cos^2\tfrac{1}{2}\theta = 2(1+\cos\theta) = \frac{2(\alpha - \cos 2A)}{\sin^2 A}$$

$$= 1 + \alpha + \beta' + \gamma'',$$

that is, the amount of the rotation is given by the formula

$$4\cos^2\tfrac{1}{2}\theta = 1 + \alpha + \beta' + \gamma''.$$

The nine cosines α, β, γ, &c. may be expressed in terms of the inclinations A, B, C, and the rotation θ, or putting λ, μ, ν equal to $\tan\frac{1}{2}\theta\cos A$, $\tan\frac{1}{2}\theta\cos B$, $\tan\frac{1}{2}\theta\cos C$, in terms of the three quantities λ, μ, ν; but the resulting formulæ for the transformation of coordinates can be more readily obtained by other methods.

530.

SOLUTION OF A SENATE-HOUSE PROBLEM.

[From the *Oxford, Cambridge and Dublin Messenger of Mathematics*, vol. v. (1869), pp. 24—27.]

THE Problem, proposed January 7, 1869, is "If θ_1 and θ_2 are two values of θ which satisfy the equation

$$1+\frac{\cos\theta\cos\phi}{\cos^2\alpha}+\frac{\sin\theta\sin\phi}{\sin^2\alpha}=0,$$

show that θ_1 and θ_2, if substituted for θ and ϕ in this equation, will satisfy it."

That is, writing

$$\frac{\cos\theta_1\cos\phi}{a^2}+\frac{\sin\theta_1\sin\phi}{b^2}+1=0,$$

$$\frac{\cos\theta_2\cos\phi}{a^2}+\frac{\sin\theta_2\sin\phi}{b^2}+1=0,$$

where $a^2+b^2=1$, it is to be shown that

$$\frac{\cos\theta_1\cos\theta_2}{a^2}+\frac{\sin\theta_1\sin\theta_2}{b^2}+1=0.$$

From the given equations, we have

$$\frac{\cos\phi}{a^2}:\frac{\sin\phi}{b^2}:1=\sin\theta_1-\sin\theta_2:\cos\theta_2-\cos\theta_1:\sin(\theta_2-\theta_1),$$

which are

$$=\cos\tfrac{1}{2}(\theta_1+\theta_2):\sin\tfrac{1}{2}(\theta_1+\theta_2):-\cos\tfrac{1}{2}(\theta_1-\theta_2).$$

Whence eliminating ϕ, we have

$$a^4 \cos^2 \tfrac{1}{2}(\theta_1 + \theta_2) + b^4 \sin^2 \tfrac{1}{2}(\theta_1 + \theta_2) - \cos^2 \tfrac{1}{2}(\theta_1 - \theta_2) = 0,$$

that is,

$$a^4 \{1 + \cos(\theta_1 + \theta_2)\} + b^4 \{1 - \cos(\theta_1 + \theta_2)\} - \{1 + \cos(\theta_1 - \theta_2)\} = 0,$$

or, what is the same thing,

$$a^4 + b^4 - 1 + (a^4 - b^4 - 1)\cos\theta_1 \cos\theta_2 + (-a^4 + b^4 - 1)\sin\theta_1 \sin\theta_2 = 0.$$

But from the equation $a^2 + b^2 = 1$, we have

$$a^4 + b^4 - 1 = -2a^2b^2,$$
$$a^4 - b^4 - 1 = -2b^2,$$
$$-a^4 + b^4 - 1 = -2a^2,$$

and the equation is thus

$$\frac{\cos\theta_1 \cos\theta_2}{a^2} + \frac{\sin\theta_1 \sin\theta_2}{b^2} + 1 = 0,$$

which is the required equation.

Stopping at the result obtained previous to the use of the relation $a^2 + b^2 = 1$, but making some obvious substitutions in the formulæ, the theorem may be presented in a more general form as follows; viz.:

If we have

$$\frac{xx_1}{a^2} + \frac{yy_1}{b^2} - 1 = 0,$$

$$\frac{xx_2}{a^2} + \frac{yy_2}{b^2} - 1 = 0,$$

where

$$\frac{x^2}{\alpha^2} + \frac{y^2}{\beta^2} - 1 = 0, \quad \frac{x_1^2}{\alpha^2} + \frac{y_1^2}{\beta^2} - 1 = 0, \quad \frac{x_2^2}{\alpha^2} + \frac{y_2^2}{\beta^2} - 1 = 0,$$

then

$$\left(\frac{a^4}{\alpha^4} + \frac{b^4}{\beta^4} - 1\right) + \left(\frac{a^4}{\alpha^4} - \frac{b^4}{\beta^4} - 1\right)\frac{x_1x_2}{\alpha^2} + \left(-\frac{a^4}{\alpha^4} + \frac{b^4}{\beta^4} - 1\right)\frac{y_1y_2}{\beta^2} = 0,$$

a relation, the geometrical signification of which is: If (x, y) be a point on the conic $\frac{x^2}{\alpha^2} + \frac{y^2}{\beta^2} - 1 = 0$, and if the polar hereof in regard to the conic $\frac{x^2}{a^2} + \frac{y^2}{b^2} - 1 = 0$ meet the first-mentioned conic in the points (x_1, y_1) and (x_2, y_2), then these points are harmonics in regard to the conic

$$\left(\frac{a^4}{\alpha^4} + \frac{b^4}{\beta^4} - 1\right) + \left(\frac{a^4}{\alpha^4} - \frac{b^4}{\beta^4} - 1\right)\frac{x^2}{\alpha^2} + \left(-\frac{a^4}{\alpha^4} + \frac{b^4}{\beta^4} - 1\right)\frac{y^2}{\beta^2} = 0;$$

and since the theorem is projective, it is seen that the first two conics may be any conics whatever, the third conic being a conic having with the other two a common system of conjugate points.

If to fix the ideas we write $\alpha=\beta=c$; then the theorem is, if the polar of a point on the circle $x^2+y^2=c^2$ in regard to the conic $\frac{x^2}{a^2}+\frac{y^2}{b^2}-1=0$, meet the circle in two points, these are harmonics in regard to the conic

$$(a^4+b^4-c^4)+(a^4-b^4-c^4)\frac{x^2}{c^2}+(-a^4+b^4-c^4)\frac{y^2}{c^2}=0.$$

This last conic will be similar to the conic $\frac{x^2}{a^2}+\frac{y^2}{b^2}-1=0$, if $c^4=(a^2+b^2)^2$; viz. if $c^2=a^2+b^2$, then the conic is $\frac{x^2}{a^2}+\frac{y^2}{b^2}+1=0$; but if $c^2=-a^2-b^2$, the conic is not only similar to, but is the conic $\frac{x^2}{a^2}+\frac{y^2}{b^2}-1=0$. Considering the given conic $\frac{x^2}{a^2}+\frac{y^2}{b^3}-1=0$ to be an ellipse, the first case $(c^2=a^2+b^2)$ gives the two points (x_1, y_1), (x_2, y_2) harmonics in regard to the imaginary conic $\frac{x^2}{a^2}+\frac{y^2}{b^2}+1=0$, but this is at once transformed into a real theorem, for we have $(-x_1, -y_1)$ and (x_2, y_2), or, what is the same thing, (x_1, y_1), $(-x_2, -y_2)$ harmonics in regard to $\frac{x^2}{a^2}+\frac{y^2}{b^2}-1=0$; and the theorem is: "Given the ellipse $\frac{x^2}{a^2}+\frac{y^2}{b^2}-1=0$, and the circle $x^2+y^2=a^2+b^2$ (which is the locus of the intersection of a pair of orthotomic tangents of the ellipse), if the polar in regard to the ellipse of a point on the circle meet the circle in the points Q, R, and if the *opposite* points to these be Q_1, R_1, then (Q, R_1), or what is the same thing (Q_1, R) are harmonics in regard to the ellipse."

The second case $(c^2=-a^2-b^2)$ gives a real theorem if b^2 be negative; viz. writing $-b^2$ for b^2 we have the hyperbola $\frac{x^2}{a^2}-\frac{y^2}{b^2}=1$, $b^2=a^2+c^2$, an obtuse-angled hyperbola; the circle $x^2+y^2=a^2-b^2$, which is the locus of the intersection of a pair of orthotomic tangents of the hyperbola, is consequently imaginary; but the concentric orthotomic circle hereof, viz. the circle of the theorem, $x^2+y^2=b^2-a^2$, is a real circle; and the theorem is: "Given the hyperbola $\frac{x^2}{a^2}-\frac{y^2}{b^2}-1=0\,(b^2>a^2)$ and the circle $x^2+y^2=b^2-a^2$ (the concentric orthotomic circle of the imaginary circle which is the locus of the intersection of a pair of orthotomic tangents of the hyperbola), if the polar in regard to the hyperbola of a point on the circle meet the circle in two points Q, R, then these are harmonics in regard to the hyperbola."

Of course, if reality be disregarded, the two theorems may each of them be stated of a conic generally; and observe, that in the first theorem the circle is the locus of the intersection of orthotomic tangents, and we have the opposites of the points Q, R; in the second theorem the circle is the concentric orthotomic circle of the circle, which is the locus of the intersection of orthotomic tangents, but we have the points Q, R themselves.

531.

A "SMITH'S PRIZE" PAPER[1]; SOLUTIONS.

[From the *Oxford, Cambridge and Dublin Messenger of Mathematics*, vol. V. (1869), pp. 40—64.]

1. *If* $\sqrt{\{(x-a)^2+(y-b)^2\}}$, $\sqrt{\{(x-a')^2+(y-b')^2\}}$, c, *are respectively real and positive, show that in the equation* $\pm\sqrt{\{(x-a)^2+(y-b)^2\}}\pm\sqrt{\{(x-a')^2+(y-b')^2\}}=c$, *considered as representing a curve, the signs cannot either of them be assumed at pleasure to be* + *or to be* −: *and distinguish the cases of the ellipse and the hyperbola.*

Writing the equation in the form $\pm\sqrt{(S)}\pm\sqrt{(S')}=c$, we find

$$\pm 2c\sqrt{(S)}=c^2+S-S',$$

$$\pm 2c\sqrt{(S')}=c^2-S+S',$$

and thence

$$4c^2S=(c^2+S-S')^2,$$

$$4c^2S'=(c^2-S+S')^2,$$

either of which equations is the rational equation of the curve (the equation being of the second order, inasmuch as $S-S'$ is a linear function of the coordinates). But writing the rational equation under these two forms respectively and passing back to the last preceding forms, it is clear that for any given point of the curve, c^2+S-S' and c^2-S+S', *quà* rational functions of the coordinates, have each of them a completely determinate value; the ambiguous sign therefore cannot be assumed at pleasure, but in the equation $\pm 2c\sqrt{(S)}=c^2+S-S'$ it must be taken to be + or to be − according as the value of c^2+S-S' is positive or is negative; and the like for the other equation. It is with the signs so determined that the two irrational equations hold good, and inasmuch as from them we deduce the original equation $\pm\sqrt{(S)}\pm\sqrt{(S')}=c$, the signs in this equation are not arbitrary, but each of them has, at a given point of the curve, a determinate value, fixed as above.

[1] Set by me for the Master of Trinity, Feb. 3, 1869.

The condition for an ellipse is that the given length c shall be greater than the distance $\surd\{(a-a')^2+(b-b')^2\}$ between the two foci; for a hyperbola that it shall be less. In the former case, for any real point of the curve, it is obvious that the relation *must be* $\surd(S)+\surd(S')=c$; in the latter case, for any real point of the curve it *must be* either $\surd(S)-\surd(S')=c$ or $-\surd(S)+\surd(S')=c$, viz. one of these equations holds for one branch, the other for the other branch of the hyperbola. To see this *à posteriori*, observe that in the ellipse, starting from the equation

$$\pm\surd\{(x-ae)^2+y^2\}\pm\surd\{(x+ae)^2+y^2\}=2a,$$

we find

$$\pm\surd\{(x-ae)^2+y^2\}=a-ex,$$

$$\pm\surd\{(x+ae)^2+y^2\}=a+ex,$$

but here, e being less than 1, and for every real point of the curve x being less in absolute magnitude than a, we have $a-ex$, $a+ex$ each positive for any real point whatever of the curve; the two signs are therefore each of them $+$, or we have $\surd\{(x-ae)^2+y^2\}+\surd\{(x+ae)^2+y^2\}=2a$; and a like verification applies to the hyperbola.

2. *In a system of curves defined by an equation containing a variable parameter, investigate at any point the normal distance between two consecutive curves; and determine the form of the equation for a system of parallel curves.*

Consider the system of curves $f(x, y, c)=0$; then if the point x, y belongs to the curve $f(x, y, c)=0$, and the point $x+\delta x$, $y+\delta y$ to the curve $f(x, y, c+\delta c)=0$, we have

$$\frac{df}{dx}\delta x+\frac{df}{dy}\delta y+\frac{df}{dc}\delta c=0,$$

and if the point $x+\delta x$, $y+\delta y$ be on the normal at (x, y) to the curve $f(x, y, c)=0$, we have

$$\delta x\div\frac{df}{dx}=\delta y\div\frac{df}{dy},$$

or writing

$$\delta x=k\frac{df}{dx},\quad \delta y=k\frac{df}{dy},$$

we have

$$k\left\{\left(\frac{df}{dx}\right)^2+\left(\frac{df}{dy}\right)^2\right\}+\frac{df}{dc}\delta c=0,$$

wherefore

$$k=-\frac{df}{dc}\delta c\div\left\{\left(\frac{df}{dx}\right)^2+\left(\frac{df}{dy}\right)^2\right\},$$

and the normal distance at (x, y) of the curves c and $c+\delta c$ is $=\surd(\delta x^2+\delta y^2)$, viz., it is $=\pm k\sqrt{\left\{\left(\frac{df}{dx}\right)^2+\left(\frac{df}{dy}\right)^2\right\}}$, or finally it is

$$=\pm\frac{df}{dc}\delta c\div\sqrt{\left\{\left(\frac{df}{dx}\right)^2+\left(\frac{df}{dy}\right)^2\right\}},$$

where, if the distance in question be regarded as positive (that is, if we attend only to its absolute value), the sign is to be taken so that $\pm\frac{df}{dc}\delta c$ shall be positive.

If the system be given in the form $V-c=0$ (V a function of (x, y)), we have the normal distance

$$= \pm\, \delta c \div \sqrt{\left\{\left(\frac{dV}{dx}\right)^2 + \left(\frac{dV}{dy}\right)^2\right\}}.$$

For parallel curves, the normal distance is everywhere the same, that is, $\left(\frac{dV}{dx}\right)^2 + \left(\frac{dV}{dy}\right)^2$ must have a constant value for all values of (x, y) satisfying the relation $V=c$, viz. we must have identically $\left(\frac{dV}{dx}\right)^2 + \left(\frac{dV}{dy}\right)^2 = \phi(V)$, ϕ arbitrary, a partial differential equation to be satisfied by V in order that $V=c$ may be the equation of a system of parallel curves. Assuming the equation to be satisfied for any particular form of ϕ, we may, it is clear, find U a function of V, such that $\left(\frac{dU}{dx}\right)^2 + \left(\frac{dU}{dy}\right)^2 = 1$, and inasmuch as the equation $f(V) = Const.$ is the same thing as $V = Const.$, it follows that the equation of the system of parallel curves may be taken to be $V=c$, where

$$\left(\frac{dV}{dx}\right)^2 + \left(\frac{dV}{dy}\right)^2 = 1.$$

3. *Two cannons (each free to recoil) differ only in weight and in the weight of the ball; and it is assumed that at any instant during the explosion the explosive force depends only on the space occupied by the vapour of the gunpowder: compare the emerging velocities of the balls; and also the emerging velocities of balls fired from the same cannon when it is free to recoil, and when it is absolutely fixed.*

Consider a single cannon.

Take M for its mass, m for that of the ball, S, s for the spaces described, backwards by the cannon and forwards by the ball, at the time t during the explosion.

Then by hypothesis the explosive force, forwards on the ball and backwards on the cannon, is a function of $s+S$, $=\phi(s+S)$ suppose, or we have

$$m\frac{d^2s}{dt^2} = \phi(s+S),$$

$$M\frac{d^2S}{dt^2} = \phi(s+S),$$

and thence

$$\frac{d^2(s+S)}{dt^2} = \left(\frac{1}{m}+\frac{1}{M}\right)\phi(s+S).$$

Multiplying each side by $2\frac{d(s+S)}{dt}$, integrating from $s+S=0$, to $s+S=a$, where a is the length of the tube, and observing that the initial velocities are each $=0$, we find

$$\left\{\frac{d(s+S)}{dt}\right\}^2 = 2\left(\frac{1}{m}+\frac{1}{M}\right)\int_0^a \phi(s+S)\,(ds+dS),$$

where $\frac{ds}{dt}$, $\frac{dS}{dt}$ are the velocities of the ball and cannon respectively at the instant of emergence; $\int_0^a \phi(s+S)(ds+dS)$, is a constant (that is a quantity independent of the weights of the ball and cannon), and putting it $=\frac{1}{2}C$, the equation may be written

$$(v+V)^2 = \left(\frac{1}{m}+\frac{1}{M}\right)C,$$

where v, V are the velocities at the instant of emergence.

But from the equation

$$m\frac{d^2s}{dt^2} = M\frac{d^2S}{dt^2},$$

(the initial values of the velocities being each $=0$), we have

$$mv = MV;$$

and hence from the foregoing equation we obtain

$$v = \frac{M}{M+m}(V+v), = \frac{M}{M+m}\,\frac{\sqrt{(M+m)}}{\sqrt{(Mm)}}\sqrt{(C)} = \frac{\sqrt{(M)}}{\sqrt{(m)}\sqrt{(M+m)}}\sqrt{(C)} = \frac{\sqrt{(C)}}{\sqrt{(m)}\sqrt{\left(1+\frac{m}{M}\right)}},$$

or say

$$v\sqrt{(m)} = \frac{\sqrt{(C)}}{\sqrt{\left(1+\frac{m}{M}\right)}}.$$

And similarly for the other cannon, if m_1, M_1 be the mass of the ball and cannon, v_1 the velocity of the ball, we have

$$v_1\sqrt{(m_1)} = \frac{\sqrt{(C)}}{\sqrt{\left(1+\frac{m_1}{M_1}\right)}},$$

whence

$$v\sqrt{(m)} : v_1\sqrt{(m_1)} = \sqrt{\left(1+\frac{m_1}{M_1}\right)} : \sqrt{\left(1+\frac{m}{M}\right)}.$$

If $m_1 = m$, $M_1 = \infty$, then v, v_1 may be taken to be the velocities of the same ball fired from the cannon, mass M, when it is free to recoil and when it is absolutely fixed; and we have

$$v : v_1 = 1 : \sqrt{\left(1+\frac{m}{M}\right)},$$

viz. the velocity in the former case is equal to that in the latter case divided by

$$\sqrt{\left(1+\frac{m}{M}\right)}.$$

4. *If* $X : Y : Z = \eta\zeta' - \eta'\zeta : \zeta\xi' - \zeta'\xi : \xi\eta' - \xi'\eta$, *where* $\xi^2+\eta^2+\zeta^2=0$, $\xi'^2+\eta'^2+\zeta'^2=0$, *show that* $\xi\xi'$, $\eta\eta'$, $\zeta\zeta'$, $\eta\zeta'+\eta'\zeta$, $\zeta\xi'+\zeta'\xi$, $\xi\eta'+\xi'\eta$ *are proportional to quadric functions of* X, Y, Z.

We have

$$Y^2+Z^2 = (\zeta\xi'-\zeta'\xi)^2 + (\xi\eta'-\xi'\eta)^2$$

$$= \xi'^2(\eta^2+\zeta^2) + \xi^2(\eta'^2+\zeta'^2) - 2\xi\xi'(\eta\eta'+\zeta\zeta'),$$

which, in virtue of the equations

$$\xi^2+\eta^2+\zeta^2=0,\quad \xi'^2+\eta'^2+\zeta'^2=0,$$

is

$$= -2\xi^2\xi'^2 - 2\xi\xi'(\eta\eta'+\zeta\zeta')$$

$$= -2\xi\xi'(\xi\xi'+\eta\eta'+\zeta\zeta').$$

And again

$$YZ = (\zeta\xi'-\zeta'\xi)(\xi\eta'-\xi'\eta)$$

$$= \xi\xi'(\eta\zeta'+\eta'\zeta) - \xi'^2\eta\zeta - \xi^2\eta'\zeta',$$

which, in virtue of the same equations, is

$$= \xi\xi'(\eta\zeta'+\eta'\zeta) + \eta\zeta(\eta'^2+\zeta'^2) + \eta'\zeta'(\eta^2+\zeta^2)$$

$$= (\eta\zeta'+\eta'\zeta)(\xi\xi'+\eta\eta'+\zeta\zeta'),$$

and, forming the analogous equations by symmetry, the factor $(\xi\xi'+\eta\eta'+\zeta\zeta')$ divides out, and we have

$$Y^2+Z^2 : Z^2+X^2 : X^2+Y^2 : -2YZ : -2ZX : -2XY$$

$$= \xi\xi' : \eta\eta' : \zeta\zeta' : \eta\zeta'+\eta'\zeta : \zeta\xi'+\zeta'\xi : \xi\eta'+\xi'\eta,$$

which is the required theorem.

5. *Two tangents of a conic are harmonically related to a second conic: find the locus of the intersection of the two tangents.*

In plane geometry the angle in a semicircle is, in spherical geometry it is not, a right angle: show how these conclusions follow from the solution of the above problem.

Let the equation of the first conic be $x^2+y^2+z^2=0$; its equation in line coordinates is therefore $u^2+v^2+w^2=0$; or what is the same thing, the line $ux+vy+wz=0$ will be a tangent of the conic if only $u^2+v^2+w^2=0$. Hence the equations of the two tangents being

$$\xi x + \eta y + \zeta z = 0,$$

$$\xi' x + \eta' y + \zeta' z = 0,$$

we have $\xi^2+\eta^2+\zeta^2=0$, $\xi'^2+\eta'^2+\zeta'^2=0$; and using X, Y, Z for the coordinates of the point of intersection of these tangents, we have

$$X : Y : Z = \eta\zeta'-\eta'\zeta : \zeta\xi'-\zeta'\xi : \xi\eta'-\xi'\eta,$$

and thence, by the last question,

$$\xi\xi' : \eta\eta' : \zeta\zeta' : \eta\zeta' + \eta'\zeta : \zeta\xi' + \zeta'\xi : \xi\eta' + \xi'\eta$$
$$= Y^2 + Z^2 : Z^2 + X^2 : X^2 + Y^2 : -2YZ : -2ZX : -2XY.$$

Suppose that the equation of the second conic in line coordinates is

$$(a, b, c, f, g, h)(u, v, w)^2 = 0;$$

the condition in order that the two lines $\xi x + \eta y + \zeta z = 0$, $\xi' x + \eta' y + \zeta' z = 0$, may be harmonically related to this conic is

$$(a, b, c, f, g, h)(\xi, \eta, \zeta)(\xi', \eta', \zeta') = 0;$$

or, what is the same thing, it is

$$a\xi\xi' + b\eta\eta' + c\zeta\zeta' + f(\eta\zeta' + \eta'\zeta) + g(\zeta\xi' + \zeta'\xi) + h(\xi\eta' + \xi'\eta) = 0,$$

and by what precedes we have

$$a(Y^2 + Z^2) + b(Z^2 + X^2) + c(X^2 + Y^2) - 2fYZ - 2gZX - 2hXY = 0,$$

or, what is the same thing,

$$(a + b + c)(X^2 + Y^2 + Z^2) - (a, b, c, f, g, h)(X, Y, Z)^2 = 0,$$

as the locus of the point of intersection of the two conics; the required locus is therefore a conic.

It is to be observed that the locus is that of a point which is such that the pairs of tangents from it to the two given conics respectively form a harmonic pencil; viz. if the equations of the given conics (in line coordinates) are

$$u^2 + v^2 + w^2 = 0,$$
$$(a, b, c, f, g, h)(u, v, w)^2 = 0;$$

then the equation of the required locus, or say the equation of the harmonic conic, is (in point coordinates)

$$(a + b + c)(x^2 + y^2 + z^2) - (a, b, c, f, g, h)(x, y, z)^2 = 0.$$

In particular, if the second conic be a point-pair or, say, if its equation be

$$(\alpha u + \beta v + \gamma w)(\alpha' u + \beta' v + \gamma' w) = 0,$$

then the equation of the harmonic conic is

$$(\alpha\alpha' + \beta\beta' + \gamma\gamma')(x^2 + y^2 + z^2) - (\alpha x + \beta y + \gamma z)(\alpha' x + \beta' y + \gamma' z) = 0,$$

which is satisfied by writing $(x, y, z) = (\alpha, \beta, \gamma)$, or $= (\alpha', \beta', \gamma')$; viz. the harmonic conic passes through the two points of the point-pair. And so, if the first conic is also a point-pair, the harmonic conic passes through the four points of the two point-pairs.

The equation of the first conic in point coordinates is $x^2+y^2+z^2=0$; hence, supposing as above, that the second conic is a point-pair, the intersections of the first conic with the harmonic conic are given by

$$x^2+y^2+z^2=0, \quad (\alpha x+\beta y+\gamma z)(\alpha' x+\beta' y+\gamma' z)=0;$$

whence the harmonic conic has double contact with the first conic, only if each of the lines $\alpha x+\beta y+\gamma z$, $\alpha' x+\beta' y+\gamma' z=0$, touches the first conic; or, what is the same thing, if each of the points (α, β, γ), $(\alpha', \beta', \gamma')$ lies on the first conic.

In plane geometry, we have (in the plane) a point-pair, the two circular points at infinity; any conic through these points is a circle: the two lines harmonic in regard hereto are at right angles. Hence, by what precedes, taking one of the given conics to be the circular points at infinity, and the other conic to be any two points P, Q; the locus of the intersection of lines through P, Q, cutting each other at right angles, is a circle; this is evidently the circle standing on PQ as diameter, or the angle in a semicircle is a right angle.

In spherical geometry we have (on the sphere) an imaginary conic $x^2+y^2+z^2=0$, called the *absolute;* any conic having double contact herewith is a circle (small circle of the sphere); two lines (arcs of great circles), harmonic in regard hereto, are at right angles. Hence, taking one of the conics to be the conic $x^2+y^2+z^2=0$ and the other to be the two points P, Q; the locus of the intersections of the lines (arcs of great circles) through P, Q, which cut at right angles, is a spherical conic; but it is *not a circle* unless the points P, Q, are each of them on the conic $x^2+y^2+z^2=0$, viz. it is not a circle for any two real points whatever: that is, in spherical geometry, the angle in a semicircle is not a right angle.

6. *A mass M attached to the end A of a chain AC is placed (with the chain) on a horizontal plane, in such wise that a portion AB of the chain forms a straight line, the remaining portion BC being heaped up at B: the mass M is then set in motion in the direction B to A with a given velocity, and so moves in a straight line, dragging the chain: determine the motion; and explain the peculiarity of the dynamical problem.*

The mass attached to the end of the chain is taken to be M, and the mass of a unit of length of the chain to be $=m$; suppose also that at the commencement of the motion the distance CA is $=a$, and that at the end of the time t this distance is $=a+x$, so that the length of chain then in motion is $=a+x$, ($x<l-a$, if l be the whole length of the chain). Suppose also that the velocity at the time t is $=v$; then we have a mass $=M+m(a+x)$ moving with a velocity v; and, in the element of time dt, this sets in motion with the velocity v a length of chain $dx=vdt$, or mass of chain $=mvdt$; if then the impulse backwards on the mass $M+m(a+x)$, and forwards on the element $mvdt$ be $=R$, we have

$$\{M+m(a+x)\}\,dv=-R,$$
$$mvdt\,.\,v=\quad R,$$

that is,

$$\{M+m(a+x)\}\,dv+mv^2dt=0\,;$$

or, observing that $vdt=dx$, this may be written

$$(M+ma)\,dv+md\,.\,xv=0,$$

that is,

$$\{M+m\,(a+x)\}\,v=\text{constant},$$
$$=(M+ma)\,V,$$

if V be the velocity at the commencement of the motion. The equation gives, in terms of the space x, the velocity v at any time t before the whole chain is set in motion: writing it in the form

$$\{M+m\,(a+x)\}\,\frac{dx}{dt}=(M+ma)\,V,$$

we have

$$(M+ma)\,x+\tfrac{1}{2}mx^2=(M+ma)\,Vt,$$

and putting herein $x=l-a$, the equation gives the value of t at the instant when the whole chain is set in motion: after this epoch, the mass and chain will (it is clear) move on with a uniform velocity

$$=\frac{(M+ma)\,V}{M+ml}.$$

The foregoing equation

$$\{M+m\,(a+x)\}\,v=(M+ma)\,V$$

might have been obtained at once by the consideration that the momentum is constant throughout the motion; but the method employed puts more clearly in evidence the peculiarity of the dynamical problem; viz. it is, so to speak, a problem of continuous impulse: in each element of time dt an infinitesimal element of mass has its velocity abruptly altered (in the present problem from 0 to v), but, for the very reason that it is an infinitesimal element of mass which undergoes this abrupt change of velocity, the effect is a continuous, not an abrupt, change of velocity of the whole finite mass which is then in motion.

7. *Show how an ellipse may be constructed as the envelope of a variable circle having its centre upon either of the axes; and examine the geometrical peculiarities which occur according as the major or the minor axis is made use of.*

Considering the ellipse

$$\frac{x^2}{a^2}+\frac{y^2}{b^2}=1,$$

the equation

$$(x-\alpha)^2+y^2=\gamma^2,$$

where α and γ are in the first instance arbitrary parameters, represents a circle having its centre on the major axis: in order that this may touch the ellipse, a relation must be established between α and γ. When this is done we obtain a variable circle (the equation of which contains a single arbitrary parameter), and this circle has the ellipse for its envelope. (On account of the symmetry in regard to the axis of x, the circle will, it is clear, touch the ellipse in *two* points, situate symmetrically in regard to this axis.) Now to make the circle touch the ellipse, eliminating y, we have

$$(x-\alpha)^2 + b^2\left(1-\frac{x^2}{a^2}\right) - \gamma^2 = 0,$$

that is,

$$x^2\left(1-\frac{b^2}{a^2}\right) - 2\alpha x + \alpha^2 - \gamma^2 + b^2 = 0,$$

which equation, considered as an equation in x, must have equal roots; that is, we must have

$$\left(1-\frac{b^2}{a^2}\right)(\alpha^2 - \gamma^2 + b^2) - \alpha^2 = 0;$$

or, what is the same thing,

$$-b^2\alpha^2 + (a^2-b^2)(b^2-\gamma^2) = 0;$$

consequently

$$\gamma^2 = b^2 - \frac{b^2}{a^2-b^2}\alpha^2;$$

or the required equation of the circle is

$$(x-\alpha)^2 + y^2 = b^2 - \frac{b^2}{a^2-b^2}\alpha^2;$$

or, as this may also be written

$$(x-\alpha)^2 + y^2 = (1-e^2)\left(a^2 - \frac{\alpha^2}{e^2}\right).$$

It is to be observed that at the points of contact of the ellipse and circle, we have

$$x = \frac{a^2\alpha}{a^2-b^2} = \frac{\alpha}{e^2}.$$

By simply interchanging a and b, it appears that when the centre is on the minor axis, the equation of the variable circle is

$$x^2 + (y-\beta)^2 = a^2 + \frac{a^2}{a^2-b^2}\beta^2,$$

and that at the points of contact of the ellipse and circle we have

$$y = -\frac{b^2\beta}{a^2-b^2}.$$

In the case where the centre is on the major axis, then attending to positive values of α, the circle remains real so long as α is $\not> ae$, but if α is $= ae$, that is, if the centre be at the focus, the radius of the circle is $=0$. But from the formula $x = \frac{\alpha}{e^2}$, we have $x = a$ for $\alpha = ae^2$, and when α is greater than ae^2, $x > a$, and the points of contact are imaginary. That is, as α passes from 0 to ae^2, the circle is real, and has real contact with the ellipse; for $\alpha = ae^2$ the circle becomes the circle of maximum curvature, touching the ellipse at the extremity of the major axis; as α increases from ae^2 to ae, the circle is still real but its contact with the ellipse is imaginary; for $\alpha = ae$, the circle reduces itself to the focus considered as an evanescent circle; and for greater values of α, the circle is imaginary.

When the centre is on the minor axis it appears in like manner that the circle is always real; but, attending to negative values of β, the circle has real contact with the ellipse only so long as $\beta \not> b - \frac{a^2}{b}$; for this value of β, the circle touches the ellipse at the extremity of the minor axis, being in fact the circle of minimum curvature; and for greater negative values of β, the contact is imaginary.

8. *Show that the number of ways in which n things can be arranged so that no one of them occupies its original position is of the form* $(n-1)A_n$; *and that we have* $A_1 = 0$, $A_2 = 1$, $A_n = (n-2)A_{n-1} + (n-3)A_{n-2}$: *show also that*

$$A_n = (n-1)A_{n-1} - \frac{1}{n-1}\{A_{n-1} + (-)^{n-1}1\}.$$

Supposing the n things arranged as required, we may by a single transposition of two things bring a given thing, say n, into its original place (the last place): and conversely every arrangement of the required form can be obtained from an arrangement in which n occupies the last place, by a transposition of n with another of the things. Now in the arrangement which thus gives rise to an arrangement of the required form, either all the things $1, 2, 3, \ldots, (n-1)$ are out of their original places; and we can then transpose n with any one of the things $1, 2, 3, \ldots, n-1$: or else only one thing is in its original place; say 1 is in its original place, and we can then transpose 1 and n; or 2 is in its original place, and we can transpose 2 and n; ... or $n-1$ is in its original place, and we can transpose $n-1$ and n. And these are the only ways in which an arrangement of the required form is obtained. Hence if for n things we denote by U_n the number of arrangements in which no one thing occupies its original place, we have, by what precedes,

$$U_n = (n-1)(U_{n-1} + U_{n-2}),$$

and it thus appears that U_n is of the form $(n-1)A_n$; and writing accordingly $U_n = (n-1)A_n$, and therefore

$$U_{n-1} = (n-2)A_{n-1}, \quad U_{n-2} = (n-3)A_{n-2},$$

we have

$$A_n = (n-2)A_{n-1} + (n-3)A_{n-2};$$

which is the required equation for A_n; we have, it is clear, $A_1 = 0$, $A_2 = 1$, and the successive values $A_3 = 1$, $A_4 = 3$, &c. can then be calculated. But the equation of differences of the second order may be integrated into one of the first order, viz writing the equation in the form

$$\{(n-1)A_n - n(n-2)A_{n-1}\} + \{(n-2)A_{n-1} - (n-1)(n-3)A_{n-2}\} = 0,$$

the integral is

$$(n-1)A_n - n(n-2)A_{n-1} = (-)^{n-1}C,$$

and writing $n=2$, we have $C=-1$, wherefore

$$(n-1)A_n = n(n-2)A_{n-1} - (-)^{n-1}1,$$

or, what is the same thing,

$$A_n = (n-1)A_{n-1} - \frac{1}{n-1}\{A_{n-1} + (-)^{n-1}1\},$$

which is the other equation for A_n.

The foregoing very elegant proof of the equation

$$U_n = (n-1)(U_{n-1} + U_{n-2}) = (n-1)A_n$$

was unknown to me; but was given in the Examination; my own proof of the equation $U_n = (n-1)A_n$ was derived from the well-known formula

$$U_n = 1.2\ldots n\left\{1 - \frac{1}{1} + \frac{1}{1.2}\ldots + (-)^n \frac{1}{1.2.3\ldots n}\right\}.$$

The theorem itself, and the two equations of differences for A_n are due to Euler, see his memoir, "Sur une espèce particulière de Carrés Magiques," *Comm. Arith. Coll.*, t. I., p. 359.

9. *If in any covariant of a binary form* $(a, b, c, \ldots, k)(x, y)^n$ *the coefficients* $a, b, \ldots, k$ *are replaced by* $ax+by$, $bx+cy$, $\ldots$, $kx+ly$ *respectively, show that the result is a covariant of the next superior form* $(a, b, c, \ldots, l)(x, y)^{n+1}$: *and determine the covariant obtained by thus operating on the discriminant of the cubic form* $(a, b, c, d)(x, y)^3$.

A covariant of the binary form $U, =(a, b, \ldots, k)(x, y)^n$, is either given by a single symbolical expression of the form $\overline{12}^\alpha\,\overline{13}^\beta \ldots \overline{23}^\gamma \ldots U_1U_2U_3\ldots$, or it is the sum of a number of such expressions, each multiplied by a constant (numerical factor): in the latter case each of the expressions in question is a covariant of U. Any such expression is at once seen to be a function of U and of its differential coefficients (i.e. it may contain the differential coefficients of each or any of the orders $0, 1, 2, \ldots, n$), homogeneous as regards the differential coefficients of the same order. But writing $U' = (a, b, c, \ldots, l)(x, y)^{n+1}$, the differential coefficients of any order of U are by the change of $a, b, \ldots, k$ into $ax+by$, $bx+cy$, $\ldots$, $kx+ly$, converted into the differential coefficients of the same order of U', each multiplied into the same merely numerical coefficient; the result (disregarding numerical factors) is thus the same derivative $\overline{12}^\alpha\,\overline{13}^\beta \ldots \overline{23}^\gamma \ldots U_1'U_2'U_3'\ldots$ of U'; viz. it is a covariant of U'; and making the like change in any sum of such expressions (each into a numerical factor), the result is a like sum of expressions referring to U'; that is, making the change in any covariant of U, the result is a covariant of U'.

Or the same thing may be otherwise proved thus; if to fix the ideas we write $U = (a, b, c, d)(x, y)^3$; any covariant Θ of U is reduced to zero by each of the operations $a\delta_b + 2b\delta_c + 3c\delta_d - y\delta_x$; $3b\delta_a + 2c\delta_b + d\delta_c - x\delta_y$; and, conversely, any function Θ which is thus reduced to zero is a covariant of U. Attending to the first operator, we have for any covariant Θ of U,

$$(a\delta_b + 2b\delta_c + 3c\delta_d - y\delta_x)\,\Theta = 0;$$

write $ax + by = a'$, $bx + cy = b'$, $cx + dy = c'$, $dx + ey = d'$, and let Θ' be the function obtained from Θ by changing a, b, c, d into a', b', c', d'. We ought to have

$$(a\delta_b + 2b\delta_c + 3c\delta_d + 4d\delta_e - y\delta_x)\,\Theta' = 0,$$

or considering Θ' as a function of a', b', c', d', x, y this is

$$\{a(y\delta_{a'} + x\delta_{b'}) + 2b(y\delta_{b'} + x\delta_{c'}) + 3c(y\delta_{c'} + x\delta_{d'}) + 4dy\delta_{d'} - y(a\delta_{a'} + b\delta_{b'} + c\delta_{c'} + d\delta_{d'}) - y\delta_x\}\,\Theta' = 0,$$

or, what is the same thing,

$$\{(ax + by)\,\delta_{b'} + 2(bx + cy)\,\delta_{c'} + 3(cx + dy)\,\delta_{d'} - y\delta_x\}\,\Theta' = 0,$$

that is,

$$\{a'\delta_{b'} + 2b'\delta_{c'} + 3c'\delta_{d'} - y\delta_x\}\,\Theta' = 0,$$

an equation which, Θ' being the same function of a', b', c', d', x, y that Θ is of a, b, c, d, x, y, is satisfied identically; and thus Θ' is reduced to zero by the operation $a\delta_b + 2b\delta_c + 3c\delta_d + 4d\delta_e - y\delta_x$; and similarly it is reduced to zero by the operation $4b\delta_a + 3c\delta_b + 2d\delta_c + e\delta_d - x\delta_y$; and it is thus a covariant of $(a, b, c, d, e \between x, y)^4$.

The discriminant of $(a, b, c, d \between x, y)^3$ is

$$= a^2d^2 + 4ac^3 + 4b^3d - 6abcd - 3b^2c^2;$$

making the substitution in question, it is

$$= (ax + by)^2 (dx + ey)^2 + \&c.,$$

viz. it is

$$= (a^2d^2 + 4ac^3 + 4b^3d - 6abcd - 3b^2c^2)\,x^4 + \&c.$$

But there is no such irreducible covariant of the quartic function $(a, b, c, d, e \between x, y)^4$; and observing that we have identically

$$(ae - 4bd + 3c^2)(ac - b^2) - (ace - ad^2 - b^2e - c^3 + 2bcd)\,a = a^2d^2 + 4ac^3 + 4b^3d - 6abcd - 3b^2c^2,$$

the covariant in question must be

$$= (ae - 4bd + 3c^2)\,[(ac - b^2)\,x^4 + \&c.] - (ace - ad^2 - b^2e - c^3 + 2bcd)\,(a, b, c, d, e \between x, y)^4,$$

where $(ac - b^2)\,x^4 + \&c.$ is the Hessian of the quartic function $(a, b, c, d, e \between x, y)^4$.

10. *From the equation of the curves of curvature of an ellipsoid, or otherwise, determine the curves of curvature of a paraboloid: show also that for the paraboloid* $xy = cz$ *(the parallel sections* $z = \text{Const.}$ *being rectangular hyperbolas) the curves of curvature are the intersections of the paraboloid by the system of surfaces*

$$h = \sqrt{(x^2 + z^2)} \pm \sqrt{(y^2 + z^2)}.$$

The curves of curvature of the ellipsoid

$$\frac{x^2}{a^2}+\frac{y^2}{b^2}+\frac{z^2}{c^2}=1$$

are given as the intersection of the surface with the confocal surface

$$\frac{x^2}{a^2+\lambda}+\frac{y^2}{b^2+\lambda}+\frac{z^2}{c^2+\lambda}=1;$$

transforming to the vertex, or writing $z-c$ in place of z, these become

$$\frac{x^2}{a^2}+\frac{y^2}{b^2}+\frac{z^2-2cz}{c^2}=0,$$

$$\frac{x^2}{a^2+\lambda}+\frac{y^2}{b^2+\lambda}+\frac{z^2-2cz-\lambda}{c^2+\lambda}=0,$$

or multiplying by c and writing $\frac{a^2}{c}=l$, $\frac{b^2}{c}=m$, $\frac{\lambda}{c}=\theta$, the equations are

$$\frac{x^2}{l}+\frac{y^2}{m}+\frac{z^2}{c}-2z=0,$$

$$\frac{x^2}{l+\theta}+\frac{y^2}{m+\theta}+\frac{z^2}{c+\theta}-\frac{2z+\theta}{1+\frac{\theta}{c}}=0,$$

or, putting herein $c=\infty$, the equations are

$$\frac{x^2}{l}+\frac{y^2}{m}-2z=0,$$

$$\frac{x^2}{l+\theta}+\frac{y^2}{m+\theta}-2z-\theta=0,$$

viz. these equations, where θ is a variable parameter, determine the curves of curvature of the paraboloid $\frac{x^2}{l}+\frac{y^2}{m}-2z=0$. The second equation may be replaced by

$$\frac{x^2}{l(l+\theta)}+\frac{y^2}{m(m+\theta)}+1=0.$$

Write in the equations $l=-m=c$; they become

$$x^2-y^2-2cz=0,$$

$$\frac{x^2}{c+\theta}+\frac{y^2}{c-\theta}+c=0;$$

or, substituting herein $\frac{x+y}{\sqrt{2}}$ for x, and $\frac{x-y}{\sqrt{2}}$ for y, the equations are

$$xy-cz=0,$$

$$\frac{(x+y)^2}{c+\theta}+\frac{(x-y)^2}{c-\theta}+2c=0$$

the second of which is

$$c(x^2+y^2)-2\theta xy+c(c^2-\theta^2)=0,$$

which is the equation for determining the curves of curvature of the surface $xy-cz=0$.

Writing in this equation $\theta=\sqrt{(h^2+c^2)}$, it becomes

$$c(x^2+y^2)-2xy\sqrt{(h^2+c^2)}-ch^2=0,$$

or, as this may be written,

$$(c^2+x^2)(c^2+y^2)-x^2y^2-2cxy\sqrt{(h^2+c^2)}-c^2h^2-c^4=0,$$

that is,

$$\pm\sqrt{(c^2+x^2)}\sqrt{(c^2+y^2)}=xy+c\sqrt{(h^2+c^2)},$$

and thence

$$\begin{aligned}c^2h^2&=(c^2+x^2)(c^2+y^2)+x^2y^2-c^4\mp 2xy\sqrt{(c^2+x^2)}\sqrt{(c^2+y^2)}\\&=x^2(c^2+y^2)+y^2(c^2+x^2)\mp 2xy\sqrt{(c^2+x^2)}\sqrt{(c^2+y^2)},\end{aligned}$$

that is,

$$ch=x\sqrt{(c^2+y^2)}\pm y\sqrt{(c^2+x^2)};$$

or combining with the equation $xy-cz=0$ of the paraboloid, this is

$$h=\sqrt{(x^2+z^2)}\mp\sqrt{(y^2+z^2)},$$

or the curves of curvature are given as the intersections of the paraboloid by the series of surfaces represented by this equation.

11. *Explain for a surface such as the ellipsoid the form of the curve of given constant slope; and in an ellipsoid having one of its principal planes horizontal determine the limits within which the curve is situate.*

At any point of a surface, the direction of the line of greatest slope is evidently at right angles to the level curve, or, what is the same thing, to the trace of the tangent plane on the horizontal plane; and the inclination of the element of the line of greatest slope is equal to that of the tangent plane to the horizontal plane. The line of given constant slope must therefore lie entirely on that portion of the surface for which the inclination of the tangent plane has a value not less than the given constant slope of the curve; viz. on a closed surface such as the ellipsoid, it will lie upon a certain zone of the surface, included between two boundaries, which boundaries are the curves such that at any point thereof the inclination of the tangent plane is *equal* to the given constant slope; and by what precedes, the direction of the line of given constant slope, at any point where it meets the boundary, is coincident with the direction of greatest slope; viz. it will in general meet the boundary at a finite angle; this implies that the point is a cusp on the curve of given constant slope, and the curve in question will be a curve as shown in the figure, passing continually from one boundary to the other, and when it reaches either boundary, turning back cusp-wise to the other boundary; the curve may in particular cases, after making a circuit, or any number of circuits of the zone, re-enter upon itself, forming a closed curve; but in general this is not the case, and the curve will go on as above between the two boundaries, *ad infinitum*.

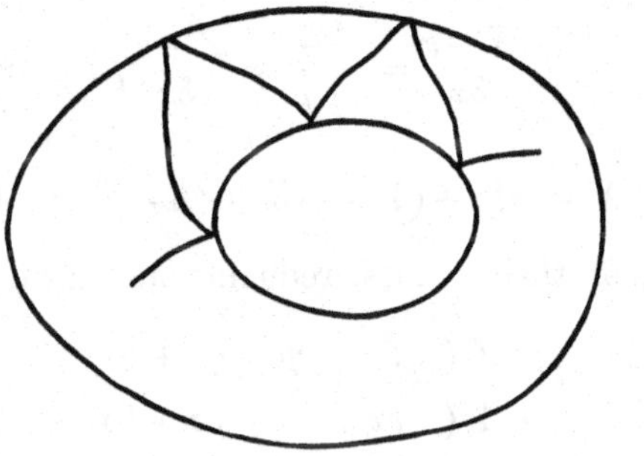

In the case of the ellipsoid, the plane of xy being as usual horizontal, and the equation being

$$\frac{x^2}{a^2} + \frac{y^2}{b^2} + \frac{z^2}{c^2} = 1,$$

then if λ be the given constant inclination, the boundaries are the series of points at which the inclination of the normal to the axes of z is $= \lambda$; that is, we have

$$\frac{z^2}{c^4} = \left(\frac{x^2}{a^4} + \frac{y^2}{b^4} + \frac{z^2}{c^4}\right) \cos^2 \lambda,$$

or, what is the same thing,

$$\left(\frac{x^2}{a^4} + \frac{y^2}{b^4}\right) \cos^2 \lambda - \frac{z^2}{c^4} \sin^2 \lambda = 0,$$

the boundaries are thus two detached ovals, meeting the principal sections $x = 0$, $y = 0$, in the points given by

$$\frac{z}{y} = \pm \frac{c^2}{b^2} \cot \lambda; \quad \frac{z}{x} = \pm \frac{c^2}{a^2} \cot \lambda,$$

and the general form is thus at once perceived.

12. *To every point of space there corresponds a plane, viz. considering the several points as belonging to a solid body which is infinitesimally displaced in any manner, the plane which corresponds to a point is the plane drawn through the point at right angles to the direction of its motion; determine the plane which corresponds to a given point: and connect the result with any general geometrical theory.*

The displacements, in the directions of the axes, of a point whose coordinates are (x, y, z) are given by the ordinary formulæ

$$\delta x = a + qz - ry,$$
$$\delta y = b + rx - pz,$$
$$\delta z = c + py - qx,$$

where a, b, c, p, q, r are constants which determine the particular displacement; hence if X, Y, Z are current coordinates, the plane corresponding to the point (x, y, z) is the plane through this point at right angles to the line

$$\frac{X-x}{\delta x} = \frac{Y-y}{\delta y} = \frac{Z-z}{\delta z},$$

viz. it is the plane

$$(X-x)\,\delta x + (Y-y)\,\delta y + (Z-z)\,\delta z = 0,$$

or, substituting for $\delta x, \delta y, \delta z$ their values, reducing and arranging, this is

$$\begin{aligned} &X(\ \ .\ \ -ry+qz+a)\\ +\,&Y(\ \ rx\ \ .\ \ -pz+b)\\ +\,&Z(-qx+py\ \ .\ \ +c)\\ +\,&(-ax-by-cz\ \ .\ \ )=0; \end{aligned}$$

that is, the equation of the plane corresponding to the point (x, y, z) is an equation linear in regard to the coordinates (x, y, z) of this point; and such that arranging as above in the form of a square, the coefficients which are symmetrically situate in regard to the dexter diagonal are equal and of opposite signs; or say that the coefficients form a skew symmetrical matrix.

More generally, corresponding to the point (x, y, z) we may have the plane

$$AX + BY + CZ + D = 0,$$

where A, B, C, D are any linear functions whatever $(=\alpha x+\beta y+\gamma z+\delta$, &c.) of the coordinates (x, y, z); this is the general relation which is the analytical basis of the theory of duality; it in fact appears that if the point be a variable point situate in a line or a plane, then the corresponding plane is a variable plane passing through a line or a point, &c., &c. In the general case, to a given point there corresponds a plane not in general passing through the point; the case above considered is distinguished by the circumstance that for any point whatever the corresponding plane does pass through the point.

13. *Write down the Lagrangian equations of motion, explaining the notation, and mode of applying them; and by way of illustration deduce the equations of motion of three particles, connected so as to form an equilateral triangle (of variable magnitude), moving in a plane under the action of any forces.*

The Lagrangian equations of motion are

$$\frac{d}{dt}\cdot\frac{dT}{d\xi'} - \frac{dT}{d\xi} = \frac{dU}{d\xi},$$

$$\frac{d}{dt}\cdot\frac{dT}{d\eta'} - \frac{dT}{d\eta} = \frac{dU}{d\eta},$$

&c.,

$\xi, \eta, \ldots$ are here any independent coordinates (in the most general sense of the word) which serve to determine the position of the system at the time t; T is the *vis-viva* function, or half-sum of the mass of each particle of the system into the square of its velocity, expressed in terms of the coordinates $\xi, \eta, \ldots$ and of their differential coefficients $\xi', =\frac{d\xi}{dt}$; $\eta', =\frac{d\eta}{dt}$, &c.; T is thus a *given* function of $\xi, \eta, \ldots \xi', \eta', \ldots$, homogeneous of the second order as regards the quantities $\xi', \eta', \ldots$; $\frac{dT}{d\xi}, \frac{dT}{d\eta}$, &c., $\frac{dT}{d\xi'}, \frac{dT}{d\eta'}$, &c. are the partial differential coefficients of T in regard to $\xi, \eta, \ldots \xi', \eta', \ldots$ respectively, and they are thus given functions of $\xi, \eta, \ldots \xi', \eta', \ldots$; U is the force-function or sum

$$\Sigma dm \int (Xdx + Ydy + Zdz)$$

(it being assumed that $Xdx + Ydy + Zdz$ is a complete differential, that is, that there exists a force-function U) expressed in terms of the coordinates $\xi, \eta, \ldots$ and being thus a *given* function of these coordinates; $\frac{dU}{d\xi}, \frac{dU}{d\eta}$, &c. are the partial differential coefficients of U in regard to $\xi, \eta, \ldots$ respectively; and are thus given functions of these coordinates. The terms $\frac{d}{dt}\frac{dT}{d\xi'}$, contain, it is clear, (and that linearly) the second differential coefficients $\xi'', =\frac{d^2\xi}{dt^2}$, &c. ..., the equations thus establish between the coordinates $\xi, \eta, \ldots$, and their first and second differential coefficients in regard to the time a system of relations, the number of which is equal to that of the coordinates, and which therefore would by integration lead to the expression of the coordinates $\xi, \eta, \ldots$ in terms of the time.

It has been tacitly assumed that T, U were functions of $\xi, \eta, \ldots \xi', \eta', \ldots$, and of $\xi, \eta, \ldots$, not containing the time t; this is the ordinary case, but there are cases in which T and U or either of them may also contain the time t.

In the proposed case of the three particles, if m_1, m_2, m_3 be their masses, r the side of the equilateral triangle, x_1, y_1 the coordinates of m_1, θ, $\theta + 60°$ (or write for convenience $\theta + \alpha$) the inclinations of m_1m_2, m_1m_3 to the axis of x, then the coordinates of m_2, m_3 will be

$$x + r\cos\theta, \quad x + r\cos(\theta+\alpha),$$
$$y + r\sin\theta, \quad y + r\sin(\theta+\alpha).$$

We have

$$\begin{aligned}
T = {} & \tfrac{1}{2} m_1 (x'^2 + y'^2) \\
& + \tfrac{1}{2} m_2 [(x' + r'\cos\theta - r\sin\theta \,.\, \theta')^2 + (y' + r'\sin\theta + r\cos\theta \,.\, \theta')^2] \\
& + \tfrac{1}{2} m_3 [\{x' + r'\cos(\theta+\alpha) - r\sin(\theta+\alpha) \,.\, \theta'\}^2 + \{y' + r'\sin(\theta+\alpha) + r\cos(\theta+\alpha) \,.\, \theta'\}^2] \\
= {} & \tfrac{1}{2}(m_1 + m_2 + m_3)(x'^2 + y'^2) \\
& + m_2 [x'(r'\cos\theta - r\sin\theta \,.\, \theta') + y'(r'\sin\theta + r\cos\theta \,.\, \theta')] \\
& + m_3 [x'\{r'\cos(\theta+\alpha) - r\sin(\theta+\alpha) \,.\, \theta'\} + y'\{r'\sin(\theta+\alpha) + r\cos(\theta+\alpha) \,.\, \theta'\}] \\
& + \tfrac{1}{2}(m_2 + m_3)(r'^2 + r^2\theta'^2),
\end{aligned}$$

a given function of x, y, r, θ, x', y', r', θ'; also

$$U = m_1 \int (Xdx + Ydy) + \&c.$$

will be a given function of x, y, r, θ; and the equations of motion will be

$$\frac{d}{dt}\frac{dT}{dx'} - \frac{dT}{dx} = \frac{dU}{dx},$$

$$\frac{d}{dt}\frac{dT}{dy'} - \frac{dT}{dy} = \frac{dU}{dy},$$

$$\frac{d}{dt}\frac{dT}{dr'} - \frac{dT}{dr} = \frac{dU}{dr},$$

$$\frac{d}{dt}\frac{dT}{d\theta'} - \frac{dT}{d\theta} = \frac{dU}{d\theta}.$$

14. *From the integrals* $x = a\cos t + a'\sin t$, $y = b\cos t + b'\sin t$, *of the dynamical equations* $\frac{d^2x}{dt^2} = -x$, $\frac{d^2y}{dt^2} = -y$, *deduce a simultaneous solution of the two partial differential equations*

$$\frac{dS}{dt} + \tfrac{1}{2}\left\{\left(\frac{dS}{dx}\right)^2 + \left(\frac{dS}{dy}\right)^2\right\} = -\tfrac{1}{2}(x^2 + y^2),$$

$$\frac{dS}{dt} + \tfrac{1}{2}\left\{\left(\frac{dS}{da}\right)^2 + \left(\frac{dS}{db}\right)^2\right\} = -\tfrac{1}{2}(a^2 + b^2).$$

Writing $U = -\tfrac{1}{2}(x^2 + y^2)$, we have

$$x = a\cos t + a'\sin t,$$

$$y = b\cos t + b'\sin t,$$

as the integrals of the equations of motion

$$\frac{d^2x}{dt^2} = \frac{dU}{dx}, \quad \frac{d^2y}{dt^2} = \frac{dU}{dy};$$

moreover

$$x' = -a\sin t + a'\cos t,$$

$$y' = -b\sin t + b'\cos t,$$

where a, b, a', b' are the initial values of x, y, x', y' respectively $\left(x' = \frac{dx}{dt}, y' = \frac{dy}{dt}, \text{ as usual}\right)$.

Hence the Principal Function

$$S = \int_0^t (T + U)\,dt$$

$$= \tfrac{1}{2}\int_0^t (x'^2 + y'^2 - x^2 - y^2)\,dt,$$

expressed as a function of x, y, a, b, t, will satisfy simultaneously the proposed partial differential equations.

We have

$$x'^2 + y'^2 - x^2 - y^2 = (a'^2 + b'^2 - a^2 - b^2)\cos 2t - 2(aa' + bb')\sin 2t.$$

Hence

$$\begin{aligned}4S &= (a'^2+b'^2-a^2-b^2)\sin 2t + 2(aa'+bb')(\cos 2t - 1)\\ &= (a'^2+b'^2)\sin 2t - (a^2+b^2)\sin 2t - 4(aa'+bb')\sin^2 t.\end{aligned}$$

But

$$\begin{aligned}a' \sin t &= x - a\cos t,\\ b'\sin t &= y - b\cos t.\end{aligned}$$

Hence

$$\begin{aligned}(aa'+bb')\sin t &= ax + by - (a^2+b^2)\cos t,\\ (a'^2+b'^2)\sin^2 t &= x^2+y^2-2(ax+by)\cos t + (a^2+b^2)\cos^2 t,\end{aligned}$$

and substituting these values

$$\begin{aligned}4S = \frac{2\sin t\cos t}{\sin^2 t}&\{x^2+y^2-2(ax+by)\cos t + (a^2+b^2)\cos^2 t\}\\ &\qquad - (a^2+b^2)\,2\sin t\cos t\\ -4\sin t\ &\{\qquad ax+by \qquad - (a^2+b^2)\cos t\}\\ = 2(x^2+y^2)\frac{\cos t}{\sin t} &- \frac{4}{\sin t}(ax+by) \quad + 2(a^2+b^2)\frac{\cos t}{\sin t};\end{aligned}$$

or, what is the same thing,

$$S = \tfrac{1}{2}(x^2+y^2+a^2+b^2)\cot t - 2(ax+by)\operatorname{cosec} t,$$

which value of S satisfies the two partial differential equations.

More generally, the two equations are satisfied by

$$S = c + \text{foregoing value},$$

(c an arbitrary constant) which new value, considered as a solution of the first equation, contains the three arbitrary constants c, a, b, and is thus a *complete* solution; and similarly considered as a solution of the second equation, it contains the three arbitrary constants c, x, y, and is thus a *complete* solution.

I venture to add a few remarks in illustration of what is required in the papers sent up in an Examination.

In the latter part of question (2) (form of the equation for a system of parallel curves) it is worse than useless to say $\frac{df}{dc} \div \left\{\left(\frac{df}{dx}\right)^2 + \left(\frac{df}{dy}\right)^2\right\}^{\frac{1}{2}}$ [p. 441] must be constant: a good and sufficient answer would be that it must be constant *in virtue of the given equation* $f(x, y, c) = 0$. So in question (13) (the Lagrangian equations of motion), it is *quite essential* to explain [p. 455] that T, U are *given* functions of $\xi, \eta, \ldots, \xi', \eta', \ldots$ and of $\xi, \eta, \ldots$ respectively; but for this the equations might be partial differential equations for the determination of T, U, or nobody knows what: it is natural and proper to explain further that T is homogeneous of the second order in regard to the derived functions $\xi', \eta', \ldots$. In question (14) the answer [p. 456] that $S = \frac{1}{2}\int_0^t (x'^2+y'^2-x^2-y^2)\,dt$ expressed as a function of x, y, a, b, t will satisfy simultaneously the proposed equations—would be, not of course a complete answer, but a good and creditable one; without the words "expressed as a function of x, y, a, b, t" it would be altogether worthless. A clear and precise indication of a process of solution is very much better than a detailed solution incorrectly worked out.

532.

NOTE ON THE INTEGRATION OF CERTAIN DIFFERENTIAL EQUATIONS BY SERIES.

[From the *Oxford, Cambridge, and Dublin Messenger of Mathematics*, vol. v. (1869), pp. 77—82.]

THERE is a speciality in the integration of certain differential equations by series, which (though evidently quite familiar to those who have written on the subject—Ellis, Boole, Hargreave) has not, it appears to me, been exhibited in the clearest form. To fix the ideas, consider a linear differential equation of the second order integrable in the form $y = Ax^\lambda + Bx^{\lambda+1} + \ldots$; λ is determined by a quadratic equation, and for each value of λ the coefficients $B, C, \ldots$ are given multiples of A; we have thus the general solution

$$y = A\left(x^\alpha + \frac{B}{A}x^{\alpha+1} \ldots\right) + K\left(x^\beta + \frac{L}{K}x^{\beta+1} + \&c. \ldots\right).$$

The speciality referred to is, when the two roots differ by an integer number; suppose α is the smaller root, and $\beta = \alpha + k$ (k a positive integer) the larger. Then, inasmuch as the series

$$y = Ax^\alpha + Bx^{\alpha+1} + \&c.$$

is identical in form with the general solution as above written down, it is clear that, starting with the root $\lambda = \alpha$, the coefficients beginning with that of x^β, $= x^{\alpha+k}$, ought not to be any longer determinate multiples of A, but should contain a new arbitrary constant K, and thus that the series derived from the root $\lambda = \alpha$ should be the general solution containing two arbitrary constants. The most simple case is when the substitution of the series in the differential equation leads to a relation between *two* consecutive coefficients of the series. Here the values $\frac{B}{A}$, $\frac{C}{A}$, &c. are fractions the numerators and denominators of which are factorial functions of α such that, for some coefficient $\frac{F}{A}$ preceding $\frac{K}{A}$ (if K is the coefficient of $x^{\alpha+k}$), and for all the succeeding coefficients $\frac{G}{A}$, &c. there is in the *numerators* one and the same evanescent factor; this being so, it is allowable to write $F=0$, $G=0$, &c. giving for the differential equation the finite solution

$$y = A\left(x^\alpha + \frac{B}{A}x^{\alpha+1}\ldots + \frac{E}{A}x^{\alpha+e}\right):$$

but if, notwithstanding the evanescent factor, we carry on the series, then in the coefficient of $x^{\alpha+k}$ there occurs in the *denominator* the same evanescent factor, so that the coefficient of this term presents itself in the form $A\,\frac{P}{Q}\cdot\frac{0}{0}$, = an arbitrary constant K (since the $\frac{0}{0}$ is essentially indeterminate), and the solution is thus obtained in the form

$$y = A\left(x^\alpha + \frac{B}{A}x^{\alpha+1}\ldots + \frac{E}{A}x^{\alpha+e}\right) + K\left(x^{\alpha+k} + \frac{L}{K}x^{\alpha+k+1} + \&c.\right),$$

viz. there is one particular solution which is finite.

Take for example the equation

$$\frac{d^2y}{dx^2} + q\frac{dy}{dx} - \frac{2}{x^2}y = 0 \qquad \text{(I)},$$

mentioned *Cambridge Math. Journal*, t. II. p. 176 (1840). If the integral is assumed to be

$$y = Ax^\alpha + Bx^{\alpha+1} + Cx^{\alpha+2} + \&c.,$$

then we find

$$\begin{aligned}
(\alpha+1)(\alpha-2)A &= 0,\\
(\alpha-1)(\alpha+2)B + q\alpha A &= 0,\\
\alpha(\alpha+3)C + q(\alpha+1)B &= 0,\\
(\alpha+1)(\alpha+4)D + q(\alpha+2)C &= 0,\\
\&c.
\end{aligned}$$

Hence $\alpha = -1$, or else $\alpha = 2$;

$$B = \frac{-q\alpha}{(\alpha-1)(\alpha+2)}A,$$

$$C = \frac{q^2.\alpha(\alpha+1)}{(\alpha-1)\alpha(\alpha+2)(\alpha+3)}A,$$

$$D = \frac{-q^3\alpha(\alpha+1)(\alpha+2)}{(\alpha-1)\alpha(\alpha+1)(\alpha+2)(\alpha+3)(\alpha+4)}A,$$

&c.

Here taking $\alpha = -1$, we are at liberty to make C and all the following coefficients $= 0$: in fact, if we commence by assuming

$$y = Ax^{\alpha} + Bx^{\alpha+1},$$

the equations

$$\begin{aligned} (\alpha+1)(\alpha-2)\,A &= 0, \\ (\alpha-1)(\alpha+2)\,B + \qquad q\alpha A &= 0, \\ q(\alpha+1)\,B &= 0, \end{aligned}$$

are all satisfied if only $\alpha = -1$, $B = \dfrac{-q\alpha A}{(\alpha-1)(\alpha+2)}$; and we have thus the finite solution $y = A\left(\dfrac{1}{x} - \dfrac{q}{2}\right)$; but if we continue the series, retaining D to represent the indeterminate quantity $\dfrac{-q^3\alpha(\alpha+1)(\alpha+2)}{(\alpha-1)\alpha(\alpha+1)(\alpha+2)(\alpha+3)(\alpha+4)}A$, we have the solution

$$y = A\left(\frac{1}{x} - \tfrac{1}{2}q\right) + D\left(x^2 - \tfrac{1}{2}qx^3 + \&c.\right),$$

the second member of which is in fact the series derived from the root $\alpha = 2$. This series is expressible by means of an exponential, viz. we have

$$x^2 - \tfrac{1}{2}qx^3 + \&c. = \frac{12}{q^3}\left\{\left(\frac{1}{x} + \tfrac{1}{2}q\right)e^{-qx} - \left(\frac{1}{x} - \tfrac{1}{2}q\right)\right\},$$

and the complete integral is thus

$$y = A\left(\frac{1}{x} - \tfrac{1}{2}q\right) + B\left(\frac{1}{x} + \tfrac{1}{2}q\right)e^{-qx},$$

but this result is not immediately connected with the investigation. It may however be noticed that, writing $z = ye^{qx}$, the equation in z is

$$\frac{d^2z}{dx^2} - q\frac{dz}{dx} - \frac{2}{x^2}z = 0,$$

which only differs from the original equation in that it has $-q$ in place of q: there is consequently the particular solution $z = \dfrac{1}{x} + \tfrac{1}{2}q$, giving for y the particular solution $y = \left(\dfrac{1}{x} + \tfrac{1}{2}q\right)e^{-qx}$, and we have thence the complete solution as above.

Consider, secondly, the differential equation

$$\frac{d^2y}{dx^2} - \left(\tfrac{1}{4}q^2 + \frac{2}{x^2}\right)y = 0 \qquad \text{(II)},$$

(derived from the equation (I) by writing therein ye^{-qx} in place of y); this equation is satisfied by the series

$$y = Ax^{\alpha} + Bx^{\alpha+2} + Cx^{\alpha+4} + \&c.$$

Then $\alpha = -1$ or $\alpha = +2$, but here the series belonging to $\alpha = -1$ contains only odd powers, the other contains only even powers of x; hence the two series do not coalesce as in the former case, and the first series is obtained without the indeterminate symbol $\frac{0}{0}$ in any of the coefficients. We have in fact

$$(\alpha+1)(\alpha-2)A = 0,$$
$$(\alpha+3)(\alpha \quad)B - \tfrac{1}{4}q^2A = 0,$$
$$(\alpha+5)(\alpha+2)C - \tfrac{1}{4}q^2B = 0,$$
$$\vdots$$

and there is not in the series in question (or in the other series) any evanescent factor, either in the numerators or in the denominators.

But consider, thirdly, the differential equation

$$\frac{d^2y}{dx^2} + (q-2\theta)\frac{dy}{dx} + \left(\theta^2 - q\theta - \frac{2}{x^2}\right)y = 0 \qquad \text{(III)},$$

(derived from (I) by writing therein $ye^{-\theta x}$ in place of y). This is satisfied by the series

$$y = Ax^\alpha + Bx^{\alpha+1} + Cx^{\alpha+2} + Dx^{\alpha+3} + \&\text{c.},$$

where $\alpha = -1$ or $\alpha = +2$ as before; in the series belonging to $\alpha = -1$, the coefficient D should become indeterminate. The relation between the coefficients is here a relation between three consecutive coefficients, viz. we have

$$(\alpha+1)(\alpha-2)A = 0,$$
$$(\alpha+2)(\alpha-1)B + (q-2\theta)(\ \alpha\ )A = 0,$$
$$(\alpha+3)(\ \alpha\ )C + (q-2\theta)(\alpha+1)B + (\theta^2 - q\theta)A = 0,$$
$$(\alpha+4)(\alpha+1)D + (q-2\theta)(\alpha+2)C + (\theta^2 - q\theta)B = 0,$$
$$(\alpha+5)(\alpha+2)E + (q-2\theta)(\alpha+3)D + (\theta^2 - q\theta)C = 0,$$
&c.

It is to be shown that in the series for $\alpha = -1$, the expression $(q-2\theta)(\alpha+2)C + (\theta^2 - q\theta)B$ contains the evanescent factor $(\alpha+1)$, and consequently that D is indeterminate; we have in fact

$$B = -\frac{(q-2\theta)\alpha}{(\alpha+2)(\alpha-1)}A,$$

$$C = \frac{1}{\alpha(\alpha+3)}\left\{\frac{(q-2\theta)^2\alpha(\alpha+1)}{(\alpha+2)(\alpha-1)} - (\theta^2 - q\theta)\right\},$$

and thence

$$(q-2\theta)(\alpha+2)C + (\theta^2 - q\theta)B$$

$$= \frac{1}{\alpha(\alpha+3)}\left\{-\frac{(q-2\theta)^3\alpha(\alpha+1)(\alpha+2)}{(\alpha+2)(\alpha+1)} - (q-2\theta)(\theta^2 - q\theta)(\alpha+2)\right\}A - \frac{(\theta^2 - q\theta)(q-2\theta)\alpha}{(\alpha+2)(\alpha-1)},$$

and then

$$\frac{\alpha+2}{\alpha(\alpha+3)}+\frac{\alpha}{(\alpha+2)(\alpha-1)}=\frac{2\alpha^3+6\alpha^2-4}{(\alpha-1)\alpha(\alpha+2)(\alpha+3)}=\frac{2(\alpha+1)(\alpha^2+2\alpha-2)}{(\alpha-1)\alpha(\alpha+2)(\alpha+3)},$$

so that the whole expression contains the factor $\alpha+1$. But observe that in the present case, if (as is allowable) we write $D=0$, the next coefficient E (depending not on D only, but on D and C) will not vanish; so that the solution obtained on the assumption $D=0$ will go on to infinity: and if instead of assuming $D=0$, we assume $D=$ an arbitrary quantity D', then E and the subsequent coefficients will contain terms depending on D'; and the complete form of the series belonging to $\alpha=-1$ will be

$$y=A\left(x^{-1}+\frac{B}{A}+\frac{C}{A}x+\frac{D}{A}x^2+\frac{E}{A}x^3+\&\text{c.}\right)+D'\left(x^2+\frac{E}{D'}x^3+\&\text{c.}\right),$$

where the second member is in fact the series belonging to $\alpha=2$. It is hardly necessary to remark that the solution thus obtained can be expressed by means of exponentials, viz. that the solution is

$$y=A\left(\frac{1}{x}-\tfrac{1}{2}q\right)e^{qx}+D'\frac{12}{q^3}\left\{\left(\frac{1}{x}+\tfrac{1}{2}q\right)e^{-qx}-\left(\frac{1}{x}-\tfrac{1}{2}q\right)\right\}e^{qx}.$$

533.

ON THE BINOMIAL THEOREM, FACTORIALS, AND DERIVATIONS.

[From the *Oxford, Cambridge, and Dublin Messenger of Mathematics*, vol. v. (1869), pp. 102—114.]

THE following was part of my course of lectures in the year 1867.

The proof commonly called "Euler's" of the binomial theorem is as follows: the theorem is assumed to be true for positive integer indices; that is, it is assumed that for any positive integer m we have

$$(1+x)^m = 1 + mx + \frac{m(m-1)}{1.2}x^2 + \&c.$$

This being so, since $(1+x)^m.(1+x)^n = (1+x)^{m+n}$, the equation

$$\left\{1 + mx + \frac{m(m-1)}{1.2}x^2 + \&c.\right\}\left\{1 + nx + \frac{n(n-1)}{1.2}x^2 + \&c.\right\}$$
$$= 1 + \frac{m+n}{1}x + \frac{(m+n)(m+n-1)}{1.2}x^2 + \&c.$$

is true for any positive integer values whatever of the indices m, n; the equation is therefore true identically; and it is consequently true for all values whatever of the indices m, n. But any function ϕm of m, satisfying the functional equation $\phi m.\phi n = \phi(m+n)$, is an mth power, $= C^m$ suppose: that is, we have

$$C^m = 1 + mx + \frac{m(m-1)}{1.2}x^2 + \&c.,$$

where C is a constant, viz. it is independent of m; but the value of C will of course depend upon x; and if, in order to determine it, we write $m=1$, the equation gives $C = 1+x$; that is, we have

$$(1+x)^m = 1 + mx + \frac{m(m-1)}{1.2}x^2 + \&c.,$$

which is the binomial theorem in its general form.

It is to be observed that there is not in the demonstration any employment of the so called "principle of equivalent forms," we do not from the truth of an equation for positive integer values of m, n, infer the truth of it for any values whatever of m, n; there is the intermediate step, that being true for integers, it is true identically; and this identical truth of the equation depends on the circumstance that comparing on the two sides of the equation the coefficients of the successive powers x^0, x^1, x^2, &c., these coefficients are in every case finite, rational, and integral functions of m, n.

For instance, comparing the coefficients of x^2, we have

$$(m+n)(m+n-1) = m(m-1) + 2mn + n(n-1),$$

and any such equation, being true for all positive integer values of m, n, will be true identically; developing the two sides, the equation is in fact

$$m^2 + 2mn + n^2 - m - n = m^2 - m + 2mn + n^2 - n.$$

The reasoning is thus perfectly good; but I remark that it is quite as easy to prove the general equation of which the last mentioned equation is an example, as it is to prove the binomial theorem for positive integer indices; and consequently that we can *without the aid of the binomial theorem for positive integer indices* prove the fundamental equation

$$\left\{1 + mx + \frac{m(m-1)}{1.2}x^2 + \&c.\right\}\left\{1 + nx + \frac{n(n-1)}{1.2}x^2 + \&c.\right\}$$
$$= 1 + \frac{m+n}{1}x + \frac{(m+n)(m+n-1)}{1.2}x^2 + \&c.$$

To show this I introduce the factorial notation and write

$$m(m-1)\ldots(m-r+1) = [m]^r;$$

this being so, the equation obtained by comparing the coefficients of x^r is readily found to be

$$[m+n]^r = [m]^r + \frac{r}{1}[m]^{r-1}[n]^1 + \frac{r(r-1)}{1.2}[m]^{r-2}[n]^2 + \&c.,$$

and I say that this, *the factorial binomial theorem for a positive integer index* r, is proved as easily and in the same manner as the binomial theorem for a like value of the index; or say as the equation

$$(m+n)^r = m^r + \frac{r}{1}m^{r-1}n^1 + \frac{r(r-1)}{1.2}m^{r-2}n^2 + \&c.$$

To show how this is, I form the values of $[m+n]^1$, $[m+n]^2$, $[m+n]^3$, &c. successively, by what may be termed the process of *varied multiplication:* we have

$$[m+n]^1 = m+n;$$

to obtain $[m+n]^2$ we have to multiply this by $m+n-1$; in regard to the first term m, I write the multiplier under the form $(m-1)+n$, and in regard to the second term n, under the form $m+(n-1)$; the process then stands

$$\begin{array}{rllll} [m+n]^1 & = m & & +n & \\ m+n-1 & = (m-1)+n & & m+(n-1) & \\ \hline & \quad m(m-1) & +mn & & \\ & & +mn & & +n(n-1) \\ \hline [m+n]^2 & = m(m-1) & +2mn & & +n(n-1) \\ & = [m]^2 & +2[m]^1[n]^1 & +[n]^2. & \end{array}$$

To form $[m+n]^3$ we have to multiply by $m+n-2$; in regard to the first term, this is written under the form $(m-2)+n$; in regard to the second term under the form $(m-1)+(n-1)$; and in regard to the third term under the form $m+(n-2)$; the product is thus obtained in the form

$$\begin{array}{l} [m]^3 + [m]^2[n]^1 \\ \qquad +2[m]^2[n]^1 + 2[m]^1[n]^2 \\ \qquad\qquad + [m]^1[n]^2 + [n]^3 \\ \hline = [m]^3 + 3[m]^2[n]^1 + 3[m]^1[n]^2 + [n]^3, \end{array}$$

and so on; the law of the terms is obvious; and the numerical coefficients are in effect obtained as follows:

$$\begin{array}{rl} & 1,\ 1 \\ \times \text{ by} & \underline{1,\ 1} \\ & 1,\ 1 \\ & \quad\ \ 1,\ 1 \\ \hline & 1,\ 2,\ 1 \\ \times \text{ by} & \underline{1,\ 1} \\ & 1,\ 2,\ 1 \\ & \quad\ \ 1,\ 2,\ 1 \\ \hline & 1,\ 3,\ 3,\ 1 \\ \times \text{ by} & \underline{1,\ 1} \\ & \&\text{c.}, \end{array}$$

viz. by precisely the same process as is used in finding the numerical coefficients in the powers $(m+n)^1$, $(m+n)^2$, $(m+n)^3$, &c.; and we thus see that for any integer value r of the index, we have a factorial binomial theorem, wherein the numerical coefficients are the same as in the binomial theorem for the same index.

But the method of varied multiplication may be applied to the demonstration of a much more general theorem; viz. we may use it to develope a product such as

$$(h-a)(h-b)(h-c)(h-d)(h-e),$$

according to a series of products

$$\begin{aligned}&(h-\alpha)(h-\beta)(h-\gamma)(h-\delta)(h-\epsilon),\\&(h-\alpha)(h-\beta)(h-\gamma)(h-\delta),\\&(h-\alpha)(h-\beta)(h-\gamma),\\&(h-\alpha)(h-\beta),\\&(h-\alpha),\\&1.\end{aligned}$$

For this purpose, starting with

$$h-a=h-\alpha+\alpha-a,$$

we multiply by $h-b$, written first under the form $h-\beta+\beta-b$, and then under the form $h-\alpha+(\alpha-b)$; we have thus

$$\begin{aligned}(h-a)(h-b)=&(h-\alpha)(h-\beta)\\&+(h-\alpha)\quad\{(\alpha-a)+(\beta-b)\}\\&+1\quad(\alpha-a)(\alpha-b),\end{aligned}$$

and so on. It is easy to see that we may for instance write

$$\begin{aligned}&(h-a)(h-b)(h-c)(h-d)(h-e)\\&=(h-\alpha)(h-\beta)(h-\gamma)(h-\delta)(h-\epsilon)\\&+(h-\alpha)(h-\beta)(h-\gamma)(h-\delta)\quad\begin{pmatrix}\alpha, & \beta, & \gamma, & \delta, & \epsilon\\ a, & b, & c, & d, & e\end{pmatrix}_1\\&+(h-\alpha)(h-\beta)(h-\gamma)\quad\begin{pmatrix}\alpha, & \beta, & \gamma, & \delta & \\ a, & b, & c, & d, & e\end{pmatrix}_2\\&+(h-\alpha)(h-\beta)\quad\begin{pmatrix}\alpha, & \beta, & \gamma & & \\ a, & b, & c, & d, & e\end{pmatrix}_3\\&+h-\alpha\quad\begin{pmatrix}\alpha, & \beta & & & \\ a, & b, & c, & d, & e\end{pmatrix}_4\\&+1\quad\begin{pmatrix}\alpha & & & & \\ a, & b, & c, & d, & e\end{pmatrix}_5,\end{aligned}$$

where the symbols $(\;)_\theta$ denote sums of products of the differences of an upper letter and a lower letter, θ factors in each product; and the several sums being formed as follows:

For $\begin{pmatrix}\alpha, & \beta, & \gamma, & \delta, & \epsilon \\ a, & b, & c, & d, & e\end{pmatrix}_1$ write

α	a
β	b
γ	c
δ	d
ϵ	e;

the expression is

$$(\alpha - a) + (\beta - b) + (\gamma - c) + (\delta - d) + (\epsilon - e).$$

For $\begin{pmatrix}\alpha, & \beta, & \gamma, & \delta & \\ a, & b, & c, & d, & e\end{pmatrix}_2$ write

α, α	a, b
α, β	a, c
α, γ	a, d
β, β	b, c
α, δ	a, e
β, γ	b, d
β, δ	b, e
γ, γ	c, d
γ, δ	c, e
δ, δ	d, e;

viz. the expression is

$$(\alpha - a)(\alpha - b) + (\alpha - a)(\beta - c) + \ldots + (\delta - d)(\delta - e).$$

For $\begin{pmatrix}\alpha, & \beta, & \gamma & & \\ a, & b, & c, & d, & e\end{pmatrix}_3$ write

α, α, α	a, b, c
α, α, β	a, b, d
α, α, γ	a, b, e
α, β, β	a, c, d
α, β, γ	a, c, e
β, β, β	b, c, d
α, γ, γ	a, d, e
β, β, γ	b, c, e
β, γ, γ	b, d, e
γ, γ, γ	c, d, e;

viz. the expression is

$$(\alpha - a)(\alpha - b)(\alpha - c) + \ldots + (\gamma - c)(\gamma - d)(\gamma - e).$$

For $\begin{pmatrix}\alpha, \beta \\ a, b, c, d, e\end{pmatrix}_4$ write

$$\begin{array}{l|l} \alpha, \alpha, \alpha, \alpha & a, b, c, d \\ \alpha, \alpha, \alpha, \beta & a, b, c, e \\ \alpha, \alpha, \beta, \beta & a, b, d, e \\ \alpha, \beta, \beta, \beta & a, c, d, e \\ \beta, \beta, \beta, \beta & b, c, d, e; \end{array}$$

viz. the expression is

$$(\alpha-a)(\alpha-b)(\alpha-c)(\alpha-d) \ldots + (\beta-b)(\beta-c)(\beta-d)(\beta-e).$$

Finally for

$$\begin{pmatrix}\alpha \\ a, b, c, d, e\end{pmatrix}_5 \text{ write } \alpha, \alpha, \alpha, \alpha, \alpha \mid a, b, c, d, e;$$

viz. the expression is

$$(\alpha-a)(\alpha-b)(\alpha-c)(\alpha-d)(\alpha-e),$$

which explains the law of the formation of the several coefficients. It is to be observed that in forming the development of any symbol, for instance $\begin{pmatrix}\alpha, \beta, \gamma \\ a, b, c, d, e\end{pmatrix}_3$, the first column contains the homogeneous products, 3 together, of α, β, γ; the second column the combinations (that is, combinations without repetitions) 3 together of a, b, c, d, e: the top line is $\alpha, \alpha, \alpha \mid a, b, c$ and to form the subsequent lines we must for any advance α into β, &c. of a greek letter make the like advance a into b, b into c, or c into d, of the corresponding latin letter.

Two particular cases of the theorem may be noticed: if the latin letters all vanish, we have, for example,

$$\begin{aligned} h^5 = (h-\alpha)(h-\beta)(h-\gamma)(h-\delta)(h-\epsilon)\,. & \\ +(h-\alpha)(h-\beta)(h-\gamma)(h-\delta) \quad & .\,H_1(\alpha, \beta, \gamma, \delta, \epsilon) \\ +(h-\alpha)(h-\beta)(h-\gamma) \quad & .\,H_2(\alpha, \beta, \gamma, \delta) \\ +(h-\alpha)(h-\beta) \quad & .\,H_3(\alpha, \beta, \gamma) \\ +(h-\alpha) \quad & .\,H_4(\alpha, \beta) \\ +\ 1 \quad & .\,H_5(\alpha), \end{aligned}$$

where the symbols H denote the sum of the homogeneous products of the annexed letters, taken together according to the suffix number: the last coefficient $H_5(\alpha)$ is of course $=\alpha^5$. And if the greek letters all vanish, then we have in like manner

$$\begin{aligned} (h-a)(h-b)(h-c)(h-d)(h-e) = h^5 & \\ -h^4C_1(a, b, c, d, e) & \\ +h^3C_2(a, b, c, d, e) & \\ -h^2C_3(a, b, c, d, e) & \\ +h\,C_4(a, b, c, d, e) & \\ -\ \ C_5(a, b, c, d, e), & \end{aligned}$$

where the symbols C denote the combinations of the annexed letters taken together according to the suffix number; the last coefficient $C_5(a, b, c, d, e)$ is of course $= abcde$. This is the ordinary theorem giving the expression of an equation in terms of its roots.

Combining the two theorems, if in the first theorem we express the products $(h-\alpha)(h-\beta)(h-\gamma)(h-\delta)(h-\epsilon)$, &c. in powers of h by means of the second theorem; or if in the second theorem we express the powers h^5, h^4, &c. in terms of the products $(h-a)(h-b)(h-c)(h-d)(h-e)$, &c. by means of the first theorem; then in either case we obtain certain identical relations connecting the C, H of $(\alpha, \beta, \ldots)$ or of $(a, b, \ldots)$.

I have mentioned the factorial notation

$$[m]^r = m(m-1)\ldots(m-r+1),$$

where r is a positive integer; a consequence of this is

$$[m]^{r+s} = [m]^r [m-r]^s,$$

where r and s are positive integers; or as this may also be written

$$[m+r]^{r+s} = [m+r]^r [m]^s.$$

Assuming this to subsist for $s=0$, or a negative integer; first for $s=0$, we have $[m+r]^r = [m+r]^r [m]^0$; that is, $[m]^0$ is $=1$; and then for $s=-r$, we have $1 = [m+r]^r [m]^{-r}$; that is,

$$[m]^{-r} = \frac{1}{[m+r]^r},$$

and in particular, $r = 1, 2$, &c., we have

$$[m]^{-1} = \frac{1}{m+1},$$

$$[m]^{-2} = \frac{1}{(m+1)(m+2)},$$

&c.,

which explains the extension of the factorial notation to negative integer values of the index.

But the equation

$$[m]^{r+s} = [m]^r [m-r]^s$$

does not *in any determinate manner* lead to an extension of the factorial notation to fractional or other values of the index. In fact, assuming $[m]^r = \dfrac{\Pi m}{\Pi(m-r)}$, where Π is an arbitrary functional symbol, the equation in question becomes

$$\frac{\Pi m}{\Pi(m-r-s)} = \frac{\Pi m}{\Pi(m-r)} \frac{\Pi(m-r)}{\Pi(m-r-s)},$$

viz. the original equation is identically satisfied, without any condition whatever being imposed upon the function Π, and on this account we have not, in the notation of the factorial with an integer index, *any sufficient basis for a theory of general differentiation.*

A product

$$m(m-\alpha)\ldots\{m-(r-1)\alpha\}$$

can of course be expressed in the factorial notation, viz. it is

$$=\alpha^r\left[\frac{m}{\alpha}\right]^r,$$

and on this account it is not in general necessary to employ a notation such as $[m,\ \alpha]^r$ to denote such a factorial wherein the difference of the successive factors instead of being $=-1$ is $=-\alpha$; in particular cases where factorials of the kind in question are used, it may be convenient to employ such a notation. In particular it is sometimes convenient to use the notation $[m,\ -1]^r$ or better $\{m\}^r$ to denote the product

$$m(m+1)\ldots(m+r-1),$$

where the successive factors instead of being diminished, are increased by unity. It may be noticed, that reversing in this last product the order of the factors, we find

$$\{m\}^r=[m+r-1]^r;$$

a somewhat similar formula, but employing only the ordinary factorial notation, is obtained from the equation

$$[m]^r=m(m-1)\ldots(m-r+1),$$

by first changing the sign of m and then reversing the order of the factors; viz. we have

$$[-m]^r=(-)^r[m+r-1]^r.$$

Reverting to the process used for the development of the expressions $(\ )_\theta$, where there are two columns, the one of greek, the other of latin letters; it is to be remarked that although the order in which the successive lines are evolved is not material for the purpose of the theorem, yet that a certain definite order of evolution has been made use of; thus in regard to $\begin{pmatrix}\alpha, & \beta, & \gamma, & \delta & \\ a, & b, & c, & d, & e\end{pmatrix}_2$, the column of greek letters, giving the homogeneous products of the second order in $(\alpha,\ \beta,\ \gamma,\ \delta)$, was

$$\begin{array}{cc}
\alpha & \alpha \\ \hline
\alpha & \beta \\ \hline
\alpha & \gamma \\
\beta & \beta \\ \hline
\alpha & \delta \\
\beta & \gamma \\ \hline
\beta & \delta \\
\gamma & \gamma \\ \hline
\gamma & \delta \\ \hline
\delta & \delta
\end{array}$$

this is evolved from the top term (α, α) by a process given implicitly in Arbogast's Calculus of Derivations, and which may be termed the rule *of the last and the last but one.* Let the direction "operate on any letter," be understood to mean that the letter in question is to be changed into that which immediately follows it, but in such wise that when the letter occurs more than once, e.g. as in α, α the operation affects only the letter in the right-hand place. Then operate on the α, α in regard to α, we obtain α, β; operate on this in regard to β, we obtain α, γ; and in regard to α, we obtain β, β. Again we operate on α, γ in regard to γ, and obtain α, δ; we do not operate on it in regard to α for the reason that α is not the letter immediately preceding γ. Operate on β, β in regard to β, we obtain β, γ. The next step, if the series extended to ϵ would be to operate on α, δ in regard to δ, giving α, ϵ; do not operate on it in regard to α, for the reason that α is not the letter immediately preceding δ. But in the example, since the series does not extend to ϵ, there is no operation on α, δ. Passing then to the next term β, γ, we operate in regard to γ, obtaining β, δ, and since β is the letter immediately preceding γ, we also operate in regard to β, obtaining γ, γ. Similarly if ϵ were admissible, β, δ would give β, ϵ, but it in fact gives nothing; γ, γ gives γ, δ; thus if ϵ were admissible would give γ, ϵ and δ, δ, but it in fact gives only δ, δ, and, ϵ being inadmissible, the process is here concluded. The rule is, operate on the last letter, and when the last but one letter is that which, in alphabetical order, immediately precedes the last letter (but in this case only) operate on the last but one letter.

Taking another example, but with numbers instead of letters, and supposing the highest admissible number to be 5, then from 111 we derive as follows:

111	112	113	114	115	125	135	145	155	255	355	455	555
		122	123	124	134	144	235	245	345	445		
			222	133	224	225	244	335	444			
				223	233	234	334	344				
						333,						

the original single column being here for greater convenience broken up into distinct columns; but the order of the terms, when the columns are taken one after the other in order, each being read from the top to the bottom, being the same as before; it will be noticed that the successive divisions are the partitions into 3 parts (no part exceeding 5) of the numbers 3, 4, ..., 15 respectively; the partitions being in each case obtained without repetition, and those of the same number being given, say in their numerical order (corresponding with the alphabetical order when letters are employed). It is necessary to show that the partitions will be obtained without repetitions; and that all the partitions will be obtained; for this purpose consider, for example, the partitions of 9; any one of these is either a partition 135 where the last number 5 is not a repeated number; and in this case there is a partition of 8, viz. 134, from which operating on the last we obtain 135; but there is no other partition of 8 which would give 135, the only such partition would be 125, but here, as 2 is not the number which immediately precedes 5, there is no operation on the last but one, and we do not from it obtain 135. Or else a partition of 9 is of the form 144 where the last letter is repeated; there exists in this case no partition of 8, such that operating on the last we obtain from it 144, but there does exist a partition of 8, viz. 134, such that operating on the last but one we obtain from it 144. That is, for any partition whatever of 9 there exists one (and only one) partition of 8, such that operating on the last or the last but one, we obtain from it the partition of 9; that is, taking the entire system of the partitions of 8, and operating on the last and the last but one, we obtain, and that without repetitions, the entire series of the partitions of 9; and so in general.

Translating the example into letters, but using for greater convenience a^2, &c. instead of a, a, &c. the process will be precisely the same; taking the letters to be a, b, c, d, e, we have

a^3	a^2b	a^2c	a^2d	a^2e	abe	ace	ade	ae^2	be^2	ce^2	de^2	e^3
		ab^2	abc	abd	acd	ad^2	bce	bde	cde	d^2e		
			b^3	ac^2	b^2d	b^2e	bd^2	c^2e	d^3			
				b^2c	bc^2	bcd	c^2d	cd^2				
						c^3.						

Attributing *weights* to the several letters, viz. to a, b, c, d, e the weights 1, 2, 3, 4, 5 respectively, the several columns show the terms of the weights 3, 4, ... 15 respectively.

I have said that the foregoing rule is given implicitly in Arbogast's Calculus of Derivations; this calculus includes in fact a process for the expansion of a function

$$\phi\,(a + bx + cx^2 + dx^3 + \&c.)$$

in powers of x; the expansion in question may be obtained by means of Taylor's theorem, viz.

$$\begin{aligned}&\phi\,(a + bx + cx^2 + dx^3 + \&c.)\\ &= \phi a\\ &+ \frac{\phi' a}{1}\quad (bx + cx^2 + dx^3 + \&c.)\\ &+ \frac{\phi'' a}{1\,.\,2}\quad (bx + cx^2 + dx^3 + \&c.)^2\\ &+ \frac{\phi''' a}{1\,.\,2\,.\,3}(bx + cx^2 + dx^3 + \&c.)^3,\end{aligned}$$

viz. expanding the several powers of the polynomial increment, and arranging in powers of x, this is

$$\begin{aligned}&= \phi a\\ &+ x\ (\phi' a\,.\,b)\\ &+ x^2 \left(\phi' a\,.\,c + \phi'' a\,.\,\frac{b^2}{2}\right)\\ &+ x^3 \left(\phi' a\,.\,d + \phi'' a\,.\,bc + \qquad \phi''' a\,.\,\frac{b^3}{6}\right)\\ &+ x^4 \left\{\phi' a\,.\,e + \phi'' a\,.\,(bd + \tfrac{1}{2}c^2) + \phi''' a\,.\,\frac{b^2c}{2} + \phi'''' a\,.\,\frac{b^4}{24}\right\}\\ &+ \&c.\end{aligned}$$

but the object of the rule is to obtain this last-mentioned result directly, and understanding "operate in regard to a letter" to mean differentiate with regard to this letter and integrate with respect to the next succeeding letter, then the coefficients of the successive powers of x are all obtained from the first coefficient ϕa, by operating thereon according to the rule of the last and the last but one; thus ϕa, operating on a gives $\phi' a \,.\, b$; this operating in regard to b gives $\phi' a \,.\, c$, and in regard to a gives $\phi'' a \,.\, \frac{b^2}{2}$; the term $\phi' a \,.\, c$ is to be operated upon in regard to c only, and it gives $\phi' a \,.\, d$; the other term $\phi'' a \,.\, \frac{b^2}{2}$ operated on in regard to b gives $\phi'' a \,.\, bc$, and in regard to a it gives $\phi''' a \,.\, \frac{b^3}{6}$; and so on. But attending only to the literal parts, the terms, for instance b^2, bc, bd, &c., which present themselves in the formula, are the homogeneous terms derived from $\overset{c^2}{b^2}$, by the rule, as originally stated, with a view to the derivation of such terms.

534.

A "SMITH'S PRIZE" PAPER[1]; SOLUTIONS.

[From the *Oxford, Cambridge and Dublin Messenger of Mathematics*, vol v. (1870), pp. 182—203.]

1. *Mention what form of given relation $\phi(a, b, c, \ldots)=0$ between the roots of a given equation will in general serve for the rational determination of the roots; explain the case of failure; and state what information as to the roots is furnished by a given relation not of the form in question.*

In the given relation, $\phi(a, b, c, \ldots)$ must be a wholly unsymmetrical function of the roots; that is, a function altered by any permutation whatever of the roots; or, what is the same thing, by any interchange whatever of two roots.

For this being so, if $\alpha, \beta, \gamma, \ldots$ be the values of the roots, then for some one order, say $\alpha, \beta, \gamma, \ldots$, of these values the given relation $\phi(a, b, c, \ldots)=0$ will be satisfied by writing therein $a=\alpha$, $b=\beta$, $c=\gamma$, &c.; but it will in general be satisfied for this order only, and not for any other order whatever (viz. it will not be satisfied by writing $a=\beta$, $b=\alpha$, $c=\gamma$, &c., or by any other such system). The given equation determines that the roots are equal to $\alpha, \beta, \gamma, \ldots$ in some order or other, but the given equation combined with the given relation $\phi(a, b, c, \ldots)=0$, determines that a is $=\alpha$ and not equal to any other value, $b=\beta$ and not equal to any other value, &c.; and it thus appears *a priori*, that the two together must rationally determine each of the roots $a, b, c, \ldots$; the *a posteriori* verification, and actual rational determination of the values of $a, b, c, \ldots$ respectively, is a separate question which is not here considered.

The function $\phi(a, b, c, \ldots)$ may be of the proper form, and yet the particular values $\alpha, \beta, \gamma, \ldots$ be such that the given relation $\phi(a, b, c, \ldots)=0$ is satisfied, not only for the single arrangement $a=\alpha$, $b=\beta$, $c=\gamma$, &c., but for some other arrangement,

[1] Set by me for the Master of Trinity, Feb. 3, 1870.

$a=\delta$, $b=\gamma$, $c=\beta,\ldots$ or for more than one such other arrangement. (For instance, if the given relation be $a+2b+3c-32=0$, and the roots are 3, 5, 7; the relation is satisfied by $a=5$, $b=3$, $c=7$, and also by $a=3$, $b=7$, $c=5$.) Here the given equation and relation do not completely determine each root, they only determine that a is $=\alpha$ or $=\delta$ (or as the case may be $=$ some other one value); and similarly that b is $=\beta$ or $=\gamma$ (or as the case may be $=$ some other one value), and so for the other roots $c, d,\ldots$; and it thus appears *a priori*, that in such a case each root is determined, not rationally, but by means of an equation, the order of which is equal to the number of the values of such root; we have here the case of failure of the general theorem.

When the given relation $\phi(a, b, c,\ldots)=0$ is not of the required form; that is, when $\phi(a, b, c,\ldots)$ is a partially symmetrical function, there will be in general several arrangements of $\alpha, \beta, \gamma,\ldots$, such that equating $a, b, c,\ldots$ to $\alpha, \beta, \gamma,\ldots$ according to each of these arrangements, the given relation $\phi(a, b, c,\ldots)=0$ will be satisfied; and it follows that each of the roots $a, b, c,\ldots$ is determined not rationally, but by means of an equation of a certain order (not necessarily the same order for each of the roots). Thus, if the relation be symmetrical as regards a pair of roots a and b; then if it be satisfied, suppose by $a=\alpha$, $b=\beta$, $c=\gamma,\ldots$, it will also be satisfied by $a=\beta$, $b=\alpha$, $c=\gamma,\ldots$, but not in general in any other manner; each of the roots a, b has here either of the values α, β, and the two roots a, b in question will be given, not rationally, but by means of the same quadratic equation. And observe, moreover, that any other function $\psi(a, b, c,\ldots)$ of the same form as ϕ, that is, symmetrical in regard to the two roots a, b, will for the two arrangements $a=\alpha$, $b=\beta$, $c=\gamma\ldots$, and $a=\beta$, $b=\alpha$, $c=\gamma,\ldots$ acquire not two distinct values, but one and the same value, that is, the value of $\psi(a, b, c,\ldots)$ will be determined *rationally*; and so in general.

There is for the partially symmetrical function $\phi(a, b, c,\ldots)$ a case of failure similar to that which arises for the completely unsymmetrical function, viz. the particular values $\alpha, \beta, \gamma\ldots$ may be such as to give more ways of satisfying the given relation $\phi(a, b, c,\ldots)=0$, than there would be but for such particular values of $\alpha, \beta, \gamma,\ldots$; and there is then a corresponding elevation of the order of the equation for the determination of the roots $a, b, c,\ldots$ or some of them.

2. *If the roots* $(\alpha, \beta, \gamma, \delta)$ *of the equation*

$$(a, b, c, d, e)(u, 1)^4=0$$

are no two of them equal; and if there exist unequal magnitudes θ, ϕ *such that*

$$(\theta+\alpha)^4 : (\theta+\beta)^4 : (\theta+\gamma)^4 : (\theta+\delta)^4 = (\phi+\alpha)^4 : (\phi+\beta)^4 : (\phi+\gamma)^4 : (\phi+\delta)^4;$$

show that the cubinvariant $ace-ad^2-b^2e-c^3+2bcd$ *is* $=0$; *and find the values of* θ, ϕ.

We have

$$\left(\frac{\theta+\alpha}{\phi+\alpha}\right)^4=\left(\frac{\theta+\beta}{\phi+\beta}\right)^4=\left(\frac{\theta+\gamma}{\phi+\gamma}\right)^4=\left(\frac{\theta+\delta}{\phi+\delta}\right)^4;$$

and we cannot have any two of the fourth roots, say $\dfrac{\theta+\alpha}{\phi+\alpha}$ and $\dfrac{\theta+\beta}{\phi+\beta}$ equal to each other; for this would imply $(\theta-\phi)(\alpha-\beta)=0$, that is, $\theta=\phi$, or else $\alpha=\beta$.

Hence assuming $\frac{\theta+\alpha}{\phi+\alpha}=\lambda$, we may write

$$\frac{\theta+\alpha}{\phi+\alpha}=\lambda,\quad \frac{\theta+\beta}{\phi+\beta}=-\lambda,\quad \frac{\theta+\gamma}{\phi+\gamma}=i\lambda,\quad \frac{\theta+\delta}{\phi+\delta}=-i\lambda,$$

$$\{i=\surd(-1) \text{ as usual}\},$$

viz. this is one of three systems of equations; the other two may be obtained therefrom by writing γ, δ, β and δ, β, γ successively in place of β, γ, δ. Hence assuming

$$v=\frac{\theta+u}{\phi+u},$$

the four values of u are α, β, γ, δ, and the corresponding four values of v are λ, $-\lambda$, $i\lambda$, $-i\lambda$; and v, u are linearly related to each other; the anharmonic ratio of $(\alpha, \beta, \gamma, \delta)$ is therefore equal to that of $(1, -1, i, -i)$, viz. we have

$$\frac{(\alpha-\gamma)(\beta-\delta)}{(\alpha-\delta)(\beta-\gamma)}=\frac{(1-i)(-1+i)}{(1+i)(-1-i)},\quad =\frac{(1-i)^2}{(1+i)^2},\quad =-1,$$

that is,

$$(\alpha-\gamma)(\beta-\delta)+(\alpha-\delta)(\beta-\gamma)=0,$$

or, what is the same thing,

$$2(\alpha\beta+\gamma\delta)-(\alpha+\beta)(\gamma+\delta)=0,$$

viz. we have this relation, or else one of the like relations

$$2(\alpha\gamma+\delta\beta)-(\alpha+\gamma)(\delta+\beta)=0,$$

$$2(\alpha\delta+\beta\gamma)-(\alpha+\delta)(\beta+\gamma)=0,$$

that is, the product of the three functions $2(\alpha\beta+\gamma\delta)-(\alpha+\beta)(\gamma+\delta)$

is $=0$.

But the product in question is (save as to a numerical factor) the cubinvariant J of the quartic function; or the equation in question is the required equation $J=0$. More simply, the linear transformation $v=\frac{\theta+u}{\phi+u}$, gives for v the equation $v^4-\lambda^4=0$; which is $(1, 0, 0, 0, -\lambda^4 \between v, 1)^4=0$; the cubinvariant hereof is $=0$, and therefore also the cubinvariant of the original function $(a, b, c, d, e \between u, 1)^4$.

Reverting to the equations

$$\frac{\theta+\alpha}{\phi+\alpha}=\lambda,\quad \frac{\theta+\beta}{\phi+\beta}=-\lambda,\quad \frac{\theta+\gamma}{\phi+\gamma}=i\lambda,\quad \frac{\theta+\delta}{\phi+\delta}=-i\lambda,$$

(which, as we have seen, give $2(\alpha\beta+\gamma\delta)=(\alpha+\beta)(\gamma+\delta)$), the same equations give

$$\frac{\theta+\alpha}{\phi+\alpha}+\frac{\theta+\beta}{\phi+\beta}=0;\quad \frac{\theta+\gamma}{\phi+\gamma}+\frac{\theta+\delta}{\phi+\delta}=0,$$

that is,

$$2\theta\phi + 2\alpha\beta - (\theta+\phi)(\alpha+\beta) = 0,$$
$$2\theta\phi + 2\gamma\delta - (\theta+\phi)(\gamma+\delta) = 0,$$

or, what is the same thing,

$$\begin{array}{rcl} 2\theta\phi : 2 : \theta+\phi = & -\alpha\beta(\gamma+\delta) + \gamma\delta(\alpha+\beta) & \\ : & \gamma+\delta \quad - \alpha-\beta & \\ : & \gamma\delta \quad - \quad \alpha\beta, & \end{array}$$

viz. we have thus the values of $\theta\phi$, $\theta+\phi$ (and thence of θ, ϕ) corresponding to the relation $2(\alpha\beta+\gamma\delta) = (\alpha+\beta)(\gamma+\delta)$ of the roots. And by cyclically permuting β, γ, δ as before, we have the values of $\theta\phi$, $\theta+\phi$ corresponding to the other two forms respectively of the relation between the roots.

3. *If in a plane A, B, C, D are fixed points and P a variable point, find the linear relation*

$$\alpha . PAB + \beta . PBC + \gamma . PCD + \delta . PDA = 0$$

which connects the areas of the triangles PAB, &c.

Taking $(x, y, 1)$, $(x_1, y_1, 1)$, &c. for the coordinates of P, A, B, C, D respectively, we have

$$PAB = \begin{vmatrix} x, & y, & 1 \\ x_1, & y_1, & 1 \\ x_2, & x_2, & 1 \end{vmatrix}, = 012, \text{ suppose},$$

$$PBC = 023, \text{ \&c.}$$

(where the values of the several determinants fix the signs of the several triangles). The identical equation then is

$$\alpha . 012 + \beta . 023 + \gamma . 034 + \delta . 041 = 0;$$

(that such an equation exists appears at once by the consideration that α, β, γ, δ can be determined so that the coefficients of x, y, and the constant term shall severally vanish); and in order actually to find the values we may make P coincide with the points A, B, C, D successively. We thus have

$$\beta . 123 + \gamma . 134 = 0,$$
$$\gamma . 234 + \delta . 241 = 0,$$
$$\delta . 341 + \alpha . 312 = 0,$$
$$\alpha . 412 + \beta . 423 = 0,$$

or, what is the same thing,

$$\beta . 123 + \gamma . 341 = 0,$$
$$\gamma . 234 + \delta . 412 = 0,$$
$$\delta . 341 + \alpha . 123 = 0,$$
$$\alpha . 412 + \beta . 234 = 0,$$

and these are at once seen to give

$$\alpha : \beta : \gamma : \delta = 234.341 : -341.412 : 412.123 : -123.341,$$

so that the required identical relation is

$$012.234.341 - 023.341.412 + 034.412.123 - 041.123.341 = 0,$$

in which 012, 023, 034, 041 stand for the triangles *PAB*, *PBC*, *PCD*, *PDA*, and 234, 341, 412, 123 for the triangles *BCD*, *CDA*, *DAB*, *ABC* respectively.

4. *Find at any point of a plane curve the angle between the normal and the line drawn from the point to the centre of the chord parallel and indefinitely near to the tangent at the point.*

Examine whether a like question applies to a point on a surface and the indicatrix section at such point.

Taking the origin at the point on the curve, the axis of x coinciding with the tangent and that of y with the normal; the equation of the curve taken to terms of the third order in x will by

$$y = bx^2 + cx^3,$$

and if, considering x as a small quantity of the first order, and therefore y as a small quantity of the second order, we regard y as given, and find the two values x_1, x_2, each of the order $\sqrt{(y)}$, which satisfy the equation, then, as will appear, $x_1 + x_2$ is a small quantity of the order x^2, and consequently $\frac{x_1 + x_2}{y}$ will have a finite value. And if ϕ be the required angle, then obviously $\tan\phi = \frac{\frac{1}{2}(x_1 + x_2)}{y}$.

We have as a first approximation $bx^2 = y$, or say $x = \frac{y^{\frac{1}{2}}}{b^{\frac{1}{2}}}$, whence to a second approximation $bx^2 = y - \frac{cy^{\frac{3}{2}}}{b^{\frac{3}{2}}}$, $x^2 = \frac{y}{b}\left(1 - \frac{cy^{\frac{1}{2}}}{b^{\frac{3}{2}}}\right)$, whence $x = \frac{y^{\frac{1}{2}}}{b^{\frac{1}{2}}}\left(1 - \frac{cy^{\frac{1}{2}}}{2b^{\frac{3}{2}}}\right)$, $= \frac{y^{\frac{1}{2}}}{b^{\frac{1}{2}}} - \frac{cy}{2b^2}$; say we have

$$x_1 = \frac{y^{\frac{1}{2}}}{b^{\frac{1}{2}}} - \frac{cy}{2b^2},$$

$$x_2 = -\frac{y^{\frac{1}{2}}}{b^{\frac{1}{2}}} - \frac{cy}{2b^2},$$

and thence

$$\tfrac{1}{2}(x_1 + x_2) = -\frac{cy}{2b^2};$$

whence

$$\tan\phi = -\frac{c}{2b^2},$$

which gives the value of the angle ϕ; it would be easy to express b, c in terms of the differential coefficients

$$d_x y,\ d_x^2 y,\ d_x^3 y.$$

It would at first sight appear that a like question might be asked as to a surface; viz. that it might be proposed to determine the angle between the normal and a line drawn from the point to the centre of the indicatrix conic. But this is not so; in fact, taking the origin at a point on the surface, the axes of x, y being in the tangent plane, and the axis of z coinciding with the normal: then to the third order we have

$$z=(A,\ B,\ C\between x,\ y)^2+(a,\ b,\ c,\ d\between x,\ y)^3;$$

but here, regarding z as a given constant, if we take account of the terms of the third order, the section is not a conic but a cubic; and it has not in general any centre; and if (as in the ordinary theory) we neglect the terms of the third order, thus obtaining an indicatrix conic, the centre of this conic lies *on* the normal, and there is no angle corresponding to the angle ϕ of the plane problem.

The only case where there is such an angle is when the cubic terms $(a,\ b,\ c,\ d\between x,\ y)^3$ contain as a factor the quadric terms $(A,\ B,\ C\between x,\ y)^2$ (one relation between the coefficients A, B, C, a, b, c, d). For then we have

$$z=(A,\ B,\ C\between x,\ y)^2(1+2lx+2my), \text{ viz.}$$

$$z=(A,\ B,\ C\between x,\ y)^2+2(lxz+myz),$$

approximately to the third order; and then regarding z as a given constant, this last equation represents a conic having for the coordinates of its centre, say $x=\alpha z$, $y=\beta z$, and there is an angle $\phi=\tan^{-1}\sqrt{(\alpha^2+\beta^2)}$; this is, in fact, what happens in the case of a quadric surface, for the section by a plane parallel and indefinitely near to the tangent plane is then a conic, the centre of which is not on the normal; and the angle ϕ (in the case of a central surface) is in fact the inclination of the normal to the radius from the centre.

I take the opportunity of adding a remark that the indicatrix is never a parabola, but in the separating case between the ellipse and the hyperbola it is a pair of parallel lines. The indicatrix, a parabola, is commonly obtained as follows: viz. taking the axes as before, but starting from an equation $U=0$, the equation presents itself in the form

$$z=(A,\ B,\ C,\ F,\ G,\ H\between x,\ y,\ z)^2,$$

which, considering z as a given constant, represents a conic which, it is said, *may be a parabola*. But observe that z is of the order $(x,\ y)^2$, the terms $2Fyz+Gzx$, are consequently of the order $(x,\ y)^3$, but they are not all the terms of this order which would be obtained by the expansion of z as a function of $(x,\ y)$; there is consequently no meaning in retaining them, and they ought to be rejected; similarly the term in z^2 which is of the order $(x,\ y)^4$ ought to be rejected; the equation is thus reduced to

$$z=Ax^2+2Hxy+By^2,$$

which, when $AB-H^2=0$, represents not a parabola but a pair of parallel lines. On referring to Dupin's *Développements de Géométrie, &c.* (see p. 49) I find that he is quite accurate; his expression is, "elle peut cependant être une parabole; alors elle se présente sous la forme de deux droites parallèles *équidistantes de leur centre*": and he afterwards examines in particular "ce cas remarquable."

5. *Shew that a cubic surface has at most four conical points; and a quartic surface at most sixteen conical points.*

If a cubic surface has two conical points, then the line joining these has with the surface two intersections at each of the conical points, and therefore lies wholly in the surface. Hence, for a cubic surface with three conical points A, B, C, the lines AB, BC, CA lie wholly in the surface, and these three lines form the complete section of the surface by the plane ABC: it is clear that there cannot be in this plane a fourth conical point: but there may be, not in this plane, a fourth conical point D. Suppose that this is so, there cannot be a fifth conical point E; for if there were, the line DE would lie wholly in the surface, and would therefore meet the plane ABC at some point in the section of the surface by this plane; that is, at some point in one of the lines AB, AC, BC; say at a point in AB: but then the lines AB, DE would intersect, or the four conical points A, B, D, E would lie in a plane. Hence there cannot be any fifth conical point E.

For a quartic surface; suppose this has k conical points, and let any one of these be made the vertex of a cone circumscribing the surface; each generating line is a tangent of the surface; and considering any section by a plane through the vertex, and observing that from a double point of a quartic curve we may draw six tangents to the curve, it appears that the order of the cone is $=6$. It is easy to see that the lines from the vertex to the remaining $(k-1)$ conical points are each of them a double line of the cone, and that the cone has not any other double lines; the cone is therefore a cone of the order 6, with $(k-1)$ double lines. A proper cone of the order 6 has at most 10 double lines, but the cone need not be a proper one; it may, in fact, break up into 6 planes, and in this case the double lines are the 15 lines of intersections of the several pairs of planes. Hence $k-1$ is $=15$ at most: or k is $=16$ at most.

6. *Find the differential equation of the parallel surfaces of an ellipsoid.*

Let (x, y, z) be the coordinates of a point on the ellipsoid $\frac{x^2}{a^2}+\frac{y^2}{b^2}+\frac{z^2}{c^2}=1$; (X, Y, Z) the coordinates of a point on the normal at a distance $=k$ from the first-mentioned point. We have

$$\frac{X-x}{\frac{x}{a^2}}=\frac{Y-y}{\frac{y}{b^2}}=\frac{Z-z}{\frac{z}{c^2}}, =\rho \text{ suppose};$$

that is,

$$X=x\left(1+\frac{\rho}{a^2}\right), \quad Y=y\left(1+\frac{\rho}{b^2}\right), \quad Z=z\left(1+\frac{\rho}{c^2}\right),$$

and thence

$$k^2=\rho^2\left(\frac{x^2}{a^4}+\frac{y^2}{b^4}+\frac{z^2}{c^4}\right).$$

Moreover

$$x=\frac{a^2X}{a^2+\rho}, \quad y=\frac{b^2Y}{b^2+\rho}, \quad z=\frac{c^2Z}{c^2+\rho},$$

substituting these values in the equation of the ellipsoid, we have

$$1 = \frac{a^2X^2}{(a^2+\rho)^2} + \frac{b^2Y^2}{(b^2+\rho)^2} + \frac{c^2Z^2}{(c^2+\rho)^2},$$

which determines ρ as a function of X, Y, Z. The tangent plane of the ellipsoid at the point (x, y, z) and of the parallel surface at the point (X, Y, Z), are parallel to each other (or what is the same thing, the parallel surface cuts at right angles the normal of the ellipsoid), we have therefore

$$\frac{x}{a^2}dX + \frac{y}{b^2}dY + \frac{z}{c^2}dZ = 0,$$

or substituting for x, y, z their values, this is

$$\frac{XdX}{a^2+\rho} + \frac{YdY}{b^2+\rho} + \frac{ZdZ}{c^2+\rho} = 0,$$

where ρ denotes as above a function of (X, Y, Z) given by the equation

$$1 = \frac{a^2X^2}{(a^2+\rho)^2} + \frac{b^2Y^2}{(b^2+\rho)^2} + \frac{c^2Z^2}{(c^2+\rho)^2}.$$

We have thus the differential equation of the parallel surfaces. It may be remarked, that the integral equation (involving k as the constant of integration), is found by the elimination of x, y, z, ρ from the foregoing equations

$$x = \frac{a^2X}{a^2+\rho}, \quad y = \frac{b^2Y}{b^2+\rho}, \quad z = \frac{c^2Z}{c^2+\rho},$$

$$\frac{x^2}{a^2} + \frac{y^2}{b^2} + \frac{z^2}{c^2} = 1, \quad k^2 = \rho^2\left(\frac{x^2}{a^4} + \frac{y^2}{b^4} + \frac{z^2}{c^4}\right),$$

or, what is the same thing, by the elimination of ρ from the equations

$$\frac{k^2}{\rho^2} = \frac{X^2}{(a^2+\rho)^2} + \frac{Y^2}{(b^2+\rho)^2} + \frac{Z^2}{(c^2+\rho)^2},$$

$$1 = \frac{a^2X^2}{(a^2+\rho)^2} + \frac{b^2Y^2}{(b^2+\rho)^2} + \frac{c^2Z^2}{(c^2+\rho)^2};$$

these may be replaced by

$$\frac{X^2}{a^2+\rho} + \frac{Y^2}{b^2+\rho} + \frac{Z^2}{c^2+\rho} - \frac{k^2}{\rho} - 1 = 0,$$

$$\frac{X^2}{(a^2+\rho)^2} + \frac{Y^2}{(b^2+\rho)^2} + \frac{Z^2}{(c^2+\rho)^2} - \frac{k^2}{\rho^2} = 0,$$

or, since here the second equation is the derived equation of the first in regard to the parameter ρ, the parallel surface is the envelope of the quadric surface

$$\frac{X^2}{a^2+\rho} + \frac{Y^2}{b^2+\rho} + \frac{Z^2}{c^2+\rho} - \frac{k^2}{\rho} = 0,$$

where ρ is the variable parameter. Or analytically, we find the equation by equating to zero the discriminant in regard to ρ, of the quartic function

$$\rho(a^2+\rho)(b^2+\rho)(c^2+\rho)\left(1 + \frac{k^2}{\rho} - \frac{X^2}{a^2+\rho} - \frac{Y^2}{b^2+\rho} - \frac{Z^2}{c^2+\rho}\right).$$

7. *Explain wherein consists the peculiarity of the following problem, and solve it by geometrical considerations:—*

Determine the least circle inclosing three given points.

The peculiarity of the problem is that the variable parameters upon which the circle depends, (say α, β the coordinates of the centre and k the radius), are not subject to any equations, but only to the inequalities

$$k^2 > (\alpha - \alpha_1)^2 + (\beta - \beta_1)^2,$$
$$k^2 > (\alpha - \alpha_2)^2 + (\beta - \beta_2)^2,$$
$$k^2 > (\alpha - \alpha_3)^2 + (\beta - \beta_3)^2,$$

(α_1, β_1; α_2, β_2; α_3, β_3, the coordinates of the given points, and the sign $>$ including $=$). The problem therefore cannot be solved by the ordinary analytical method, but it is easily solved geometrically as follows: Let A, B, C be the three points; consider all the circles inclosing the three points, viz. O a circle not passing through any of them; A a circle through the point A, B a circle through the point B, AB a circle through the points A and B, &c. Then for any circle O, if the centre be fixed and the radius gradually diminish, the circle will at last pass through one of the points ABC; that is, every circle O is greater than some circle A, B, or C; and the circle O is therefore not a minimum. Taking next a circle A, we may imagine the centre to move from its original position in a straight line towards the point A, the circle thus gradually diminishing until it passes through one of the points B or C; that is, every circle A is greater than some circle AB or AC, and therefore no circle A is a minimum; and in like manner no circle B or C is a minimum. There remain the circles AB, AC, BC; if the triangle ABC is acute-angled, then in each series, the least circle is the circle ABC circumscribed about the triangle; and this is then the minimum circle inclosing the three points. But if the triangle is obtuse-angled, say at C, then the least circle CA or CB is the circle ABC circumscribed about the triangle; but this is not the least circle AB, viz. the circle AB, being diminished to ABC, may be further diminished until it becomes the circle on the diameter AB; but below this it cannot be diminished; and consequently the minimum circle inclosing the three points is in this case the circle on the diameter AB.

8. *A particle describes an ellipse under the simultaneous action of given central forces, each varying as (distance)$^{-2}$, at the two foci respectively: find the differential relation between the time and the eccentric anomaly.*

Taking the equation of the ellipse to be $\frac{x^2}{a^2} + \frac{y^2}{b^2} = 1$, and the absolute forces at the two foci $(ae, 0)$, $(-ae, 0)$ to be μ, μ' respectively, the differential equations of motion will be

$$\frac{d^2x}{dt^2} = -\mu \frac{x - ae}{(a - ex)^3} - \mu' \frac{(x + ae)}{(a + ex)^3},$$

$$\frac{d^2y}{dt^2} = -\mu \frac{y}{(a - ex)^3} - \mu' \frac{y}{(a + ex)^3}.$$

But if u be the eccentric anomaly, then

$$x = a\cos u, \quad y = b\sin u, \quad = a\sqrt{(1-e^2)}\sin u,$$

and the equations become

$$-\sin u\frac{d^2u}{dt^2} - \cos u\left(\frac{du}{dt}\right)^2 = -\frac{\mu}{a^3}\frac{\cos u - e}{(1-e\cos u)^3} - \frac{\mu'}{a^3}\frac{\cos u + e}{(1+e\cos u)^3}$$

$$+\cos u\frac{d^2u}{dt^2} - \sin u\left(\frac{du}{dt}\right)^2 = -\frac{\mu}{a^3}\frac{\sin u}{(1-e\cos u)^3} - \frac{\mu'}{a^3}\frac{\sin u}{(1+e\cos u)^3},$$

and multiplying by $-\cos u$, $-\sin u$ respectively, and adding, we have

$$\left(\frac{du}{dt}\right)^2 = \frac{\mu}{a^3}\frac{1}{(1-e\cos u)^2} + \frac{\mu'}{a^3}\frac{1}{(1+e\cos u)^2},$$

which is the required differential relation.

9. *Show that the attraction of an indefinitely thin double-convex lens on a point at the centre of one of its faces is equal to that of the infinite plate included between the tangent plane at the point and the parallel tangent plane of the other face of the lens.*

The figure represents the upper half only of the lens, but in speaking of any portion thereof, such as PRQ, we include the symmetrically situate portion of the under-half of the lens.

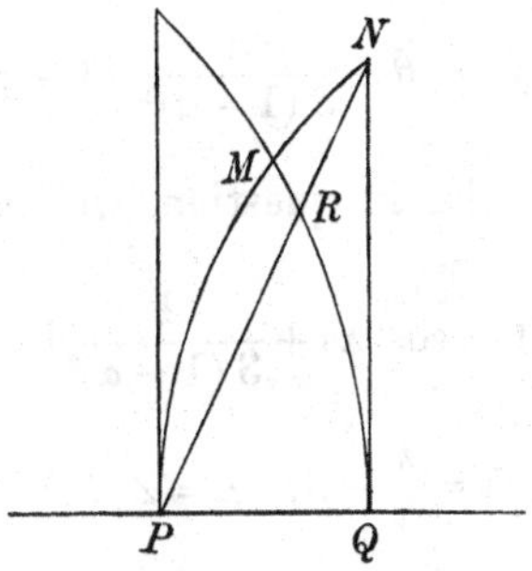

Let α, $= PQ$, be the thickness of the lens, $\angle NPQ = \lambda$, which angle is ultimately $= \frac{\pi}{2}$. Then it is at once seen that the attraction of the cone NPQ is $= 2\pi\alpha(1-\cos\lambda)$: and from this it follows that the attraction of the infinite plate is $= 2\pi\alpha$. The attraction of the whole infinite plate except the cone NPQ is $= 2\pi\alpha\cos\lambda$, which is indefinitely small in regard to $2\pi\alpha$; and, *a fortiori*, the attraction of the portion MPR of the lens is indefinitely small in regard to $2\pi\alpha$. We have then only to show that the attraction of the solid NRQ is indefinitely small in regard to $2\pi\alpha$; for, this being so, the attraction of the lens may be taken to be equal to that of the cone NPQ, and will therefore ultimately be $= 2\pi\alpha$, the attraction of the infinite plate.

Let the position of an element of the solid in question be determined by r its distance from P, θ the inclination of r to the axis PQ, and ϕ the azimuth in regard to any fixed plane through the axis; then $dm = r^2 \sin\theta\, dr\, d\theta\, d\phi$, and the attraction in the direction PQ is $= \int \sin\theta\cos\theta\, dr\, d\theta\, d\phi$, $= 2\pi \int \left(\frac{\alpha}{\cos\theta} - r\right) \sin\theta\cos\theta\, d\theta$, the integral in regard to ϕ having been taken from $\phi = 0$ to $\phi = 2\pi$, and that in regard to r from $r = r$ (value at the face MQ of the lens) to $r = \frac{\alpha}{\cos\theta}$ (value at the tangent plane QN). Taking the radius of the surface QM of the lens to be $= 1$, we have

$$(1 - \alpha + r\cos\theta)^2 + r^2\sin^2\theta = 1,$$

that is,

$$r^2 + 2r\cos\theta\,(1-\alpha) = 2\alpha - \alpha^2,$$

$$\{r + (1-\alpha)\cos\theta\}^2 = (1-\alpha)^2\cos^2\theta + 2\alpha - \alpha^2,$$

or

$$r = -(1-\alpha)\cos\theta + \sqrt{\{(1-\alpha)^2\cos^2\theta + 2\alpha - \alpha^2\}},$$

which is the value of r to be substituted in the formula

$$\frac{1}{2\pi} A = \int (\alpha\sin\theta - r\sin\theta\cos\theta)\, d\theta,$$

and the integral is to be taken from $\theta = 0$ to $\theta = \lambda$; viz. this is

$$\int [\alpha\sin\theta + (1-\alpha)\sin\theta\cos^2\theta - \sin\theta\cos\theta\sqrt{\{(1-\alpha)^2\cos^2\theta + 2\alpha - \alpha^2\}}]\, d\theta,$$

$$= -\alpha\cos\theta - \tfrac{1}{3}(1-\alpha)\cos^3\theta + \frac{1}{3(1-\alpha)^2}\{(1-\alpha)^2\cos^2\theta + 2\alpha - \alpha^2\}^{\frac{3}{2}};$$

so that taking this between the limits in question, we have

$$\frac{1}{2\pi} A = \alpha(1-\cos\lambda) + \tfrac{1}{3}(1-\alpha)(1-\cos^3\lambda) + \frac{1}{3(1-\alpha)^2}[\{(1-\alpha)^2\cos^2\lambda + 2\alpha - \alpha^2\}^{\frac{3}{2}} - 1]$$

or writing for greater convenience $\lambda = \frac{\pi}{2} - \mu$, ($\mu = \angle\, PNQ$), this is

$$\frac{1}{2\pi} A = \alpha(1-\sin\mu) + \tfrac{1}{3}(1-\alpha)(1-\sin^3\mu) + \frac{1}{3(1-\alpha)^2}[\{(1-\alpha)^2\sin^2\mu + 2\alpha - \alpha^2\}^{\frac{3}{2}} - 1]$$

$$= \alpha + \tfrac{1}{3}\left\{1 - \alpha - \frac{1}{(1-\alpha)^2}\right\} - \alpha\sin\mu$$

$$+ \frac{1}{3(1-\alpha)^2}[\{(1-\alpha)^2\sin^2\mu + 2\alpha - \alpha^2\}^{\frac{3}{2}} - (1-\alpha)^3\sin^3\mu]$$

$$= \frac{1}{3(1-\alpha)^2}(-3\alpha^2 + 2\alpha^3) - \alpha\sin\mu$$

$$+ \frac{1}{3(1-\alpha)^2}[\{(1-\alpha)^2\sin^2\mu + 2\alpha - \alpha^2\}^{\frac{3}{2}} - (1-\alpha)^3\sin^3\mu];$$

$\sin\mu$ is here an indefinitely small quantity of the order $\alpha^{\frac{1}{2}}$, all the terms are therefore at least of the order $\alpha^{\frac{3}{2}}$, and are to be neglected in comparison with α; or neglecting such terms we have $A=0$ (that is, the attraction of the solid NRQ is indefinitely small in regard to α); and the theorem is thus proved.

10. *Indicate in what manner the Lagrangian equations of motion*

$$\frac{d}{dt}\frac{dT}{d\xi'}-\frac{dT}{d\xi}=\frac{dV}{d\xi},\ \&c.$$

lead to the equations

$$A\frac{dp}{dt}+(C-B)\,qr=0,\ \&c.$$

for the motion of a solid body about a fixed point.

The expression of the *vis viva* function T is

$$T=\tfrac{1}{2}(Ap^2+Bq^2+Cr^2),$$

but this expression will not by itself lead to the equations of motion; we require to know also the expressions of p, q, r in terms of certain coordinates λ, μ, ν, which determine the position of the body in regard to axes fixed in space, and of the differential coefficients λ', μ', ν' of these coordinates in regard to the time; each of the quantities p, q, r will be a linear function of λ', μ', ν' ($p=a\lambda'+b\mu'+c\nu'$, &c.), containing in any manner whatever the coordinates λ, μ, ν. This being so, the equations of motion will be

$$\frac{d}{dt}\frac{dT}{d\lambda'}-\frac{dT}{d\lambda}=0,\ \&c.;\quad \frac{dT}{d\lambda'}=Ap\frac{dp}{d\lambda'}+Bq\frac{dq}{d\lambda'}+Cr\frac{dr}{d\lambda'},$$

where $\frac{dp}{d\lambda'}$, $\frac{dq}{d\lambda'}$, $\frac{dr}{d\lambda'}$ are each independent of λ', μ', ν'; hence, in the equation, the only terms containing the differential coefficients of p, q, r, are the terms

$$\frac{dp}{d\lambda'}\cdot A\frac{dp}{dt}+\frac{dq}{d\lambda'}\cdot B\frac{dq}{dt}+\frac{dr}{d\lambda'}\cdot C\frac{dr}{dt}$$

of $\frac{d}{dt}\cdot\frac{dT}{d\lambda'}$; and hence, assuming that the equations of motion are the known equations $A\frac{dp}{dt}+(C-B)\,qr=0$, it appears that the equation $\frac{d}{dt}\cdot\frac{dT}{d\lambda'}-\frac{dT}{d\lambda}=0$ will assume the form

$$\frac{dp}{d\lambda'}\left\{A\frac{dp}{dt}+(C-B)\,qr\right\}+\frac{dq}{d\lambda'}\left\{B\frac{dq}{dt}+(A-C)\,rp\right\}+\frac{dr}{d\lambda'}\left\{C\frac{dr}{dt}+(B-A)\,pq\right\}=0\,;$$

there are of course two other equations only differing from this in that in place of λ', they contain μ' and ν' respectively; and since p, q, r regarded as functions of λ', μ', ν' are independent functions, the determinant formed with the differential coefficients

$\frac{dp}{d\lambda'}, \frac{dq}{d\lambda'}, \frac{dr}{d\lambda'}$, &c. is not $=0$; and the three equations are therefore equivalent (as they should be) to the equations

$$A\frac{dp}{dt} + (C-B)\,qr = 0, \text{ \&c.}$$

What precedes is a complete answer to the question, but in regard to the actual expressions of p, q, r, it may be remarked, that these quantities may be expressed very symmetrically in terms of the quantities

$$\lambda,\ \mu,\ \nu = \tan\tfrac{1}{2}\theta\cos f,\quad \tan\tfrac{1}{2}\theta\cos g,\quad \tan\tfrac{1}{2}\theta\cos h,$$

which determine the positions of the principal axes in regard to the axes fixed in space, by means of the angles of position $(\cos f, \cos g, \cos h)$ of the resultant axis, and the rotation θ about this axis; viz. writing $\kappa = 1 + \lambda^2 + \mu^2 + \nu^2$, we then have

$$\begin{aligned}\kappa p &= 2\,(\quad \lambda' + \nu\mu' - \mu\nu'),\\ \kappa q &= 2\,(-\nu\lambda' + \quad \mu' + \lambda\nu'),\\ \kappa r &= 2\,(\ \ \mu\lambda' - \lambda\mu' + \quad \nu'),\end{aligned}$$

and the above result may be verified *a posteriori* without any difficulty. See *Camb. Math. Jour.*, vol. III. (1843), [6], p. 224, [*Coll. Math. Papers*, vol. I. p. 33].

11. *Find in the Hamiltonian form,*

$$\frac{d\eta}{dt} = \frac{dH}{d\varpi},\quad \frac{d\varpi}{dt} = -\frac{dH}{d\eta},\ \text{\&c.}$$

the equations for the motion of a particle acted on by a central force.

Taking as coordinates r the radius vector, v the longitude, y the latitude, the equation of the *vis viva* function is

$$T = \tfrac{1}{2}\{r'^2 + r^2(\cos^2 y\,.\,v'^2 + y'^2)\},$$

hence

$$\begin{aligned}\frac{dT}{dr'} &= \quad r' \quad = \mathrm{r} \text{ suppose},\\ \frac{dT}{dv'} &= r^2\cos^2 y\,.\,v' = \mathrm{v} \quad ,, \quad ,\\ \frac{dT}{dy'} &= \quad r^2 y' \quad = \mathrm{y} \quad ,, \quad ,\end{aligned}$$

and the expression of T in terms of r, v, y, and of the new coordinates r, v, y is

$$T = \tfrac{1}{2}\left(\mathrm{r}^2 + \frac{\mathrm{v}^2}{r^2\cos^2 y} + \frac{\mathrm{y}^2}{r^2}\right);$$

whence writing

$$H = \tfrac{1}{2}\left(\mathrm{r}^2 + \frac{\mathrm{v}^2}{r^2\cos^2 y} + \frac{\mathrm{y}^2}{r^2}\right) - V,$$

the equations are

$$\frac{dH}{d\mathrm{r}}=\frac{dr}{dt},\quad \frac{dH}{d\mathrm{v}}=\frac{dv}{dt},\quad \frac{dH}{d\mathrm{y}}=\frac{dy}{dt},$$

$$\frac{dH}{dr}=-\frac{d\mathrm{r}}{dt},\quad \frac{dH}{dv}=-\frac{d\mathrm{v}}{dt},\quad \frac{dH}{dy}=-\frac{d\mathrm{y}}{dt}.$$

We have

$$\frac{dH}{d\mathrm{r}}=\mathrm{r},\quad \frac{dH}{d\mathrm{v}}=\frac{\mathrm{v}}{r^2\cos^2 y},\quad \frac{dH}{d\mathrm{y}}=\frac{\mathrm{y}}{r^2},$$

$$\frac{dH}{dr}=-\frac{1}{r^3}\left(\frac{\mathrm{v}^2}{\cos^2 y}+\mathrm{y}^2\right)-\frac{dV}{dr},\quad \frac{dH}{dv}=0,\quad \frac{dH}{dy}=-\frac{\mathrm{v}^2\sin y}{r^2\cos^3 y};$$

and, substituting these values, the equations of motion present themselves as six equations of the first order between r, v, y, r, v, y, and t in the form

$$dt=\frac{dr}{\mathrm{r}}=\frac{dv}{\dfrac{\mathrm{v}}{r^2\cos^2 y}}=\frac{dy}{\dfrac{\mathrm{y}}{r^2}}=\frac{d\mathrm{r}}{-\dfrac{1}{r^3}\left(\dfrac{\mathrm{v}^2}{\cos^2 y}+\mathrm{y}^2\right)-\dfrac{dV}{dr}}=\frac{d\mathrm{v}}{0}=\frac{d\mathrm{y}}{-\dfrac{\mathrm{v}^2\sin^2 y}{r^2\cos^3 y}}.$$

12. *An unclosed polygon of* $(m+1)$ *vertices is constructed as follows: viz. the abscissæ of the several vertices are* $0, 1, 2 \ldots m$, *and, corresponding to the abscissa* k, *the ordinate is equal to the chance of* $(m+k)$ *heads in* $2m$ *tosses of a coin; and* m *then continually increases up to any very large value: what information in regard to the successive polygons, and to the areas of any portions thereof, is afforded by the general results of the Theory of Probabilities?*

It is somewhat more convenient to take account also of the abscissæ $-1, -2, \ldots, -m$, thereby obtaining a polygon of $2m+1$ vertices, symmetrical in regard to the axis of y. In such a polygon, the sum of the $2m+1$ ordinates is $=1$; the central ordinate is the largest, and the ordinates continually diminish as k increases: moreover for any large value of m the area of the whole polygon is very nearly, and may be regarded as being, $=1$; and the area between the ordinates corresponding to the abscissæ $+k$, $-k$ as being equal to the probability of a number of heads between $m+k$, $m-k$, in the $2m$ tosses of the coin. A general result of the Theory of Probabilities is that in a great number of trials the several events tend to happen in the proportion of their respective probabilities; viz. in the case of the $2m$ tosses there is a tendency to an equal number of heads and tails. But observe that this does not mean that the probability of m heads and m tails increases with the number $2m$ of the trials; or even that, α being any given number, the probability of a number of heads between $m+\alpha$ and $m-\alpha$ increases with the number $2m$ of trials; on the contrary, it diminishes; what it does mean is that taking the limit of deviation to vary with m, say a number of heads between $m+\alpha m$, $m-\alpha m$, the probability of such a number increases with m; viz. that taking α a fraction however small, m can be taken so large that the probability of a number of heads between $m+\alpha m$, $m-\alpha m$ in the $2m$ trials, shall be as nearly as we please $=1$.

The conclusion in regard to the areas of the polygons is that, taking k any given value whatever, however large, the ratio (m being of course $>k$) which the area between the ordinates to the abscissæ $m+k$, $m-k$ bears to the area of the whole polygon (or to unity) continually decreases as $2m$ increases, and ultimately vanishes; but contrarywise, taking α any given fraction whatever, however small, the ratio which the area between the ordinates to the abscissæ $m+\alpha m$, $m-\alpha m$ bears to the area of the whole polygon (or to unity) continually increases as $2m$ increases, and ultimately becomes $=1$.

13. *Show that for the quadric cones which pass through six given points the locus of the vertices is a quartic surface having upon it twenty-five right lines; and, thence or otherwise, that for the quadric cones passing through seven given points the locus of the vertices is a sextic curve.*

Suppose $U=0$, $V=0$, $W=0$, $S=0$ are any particular four quadric surfaces passing through the six points, say

$$\left(U=(a, \ldots)(x, y, z, w)^2, \quad V=(b, \ldots)(x, y, z, w)^2, \text{ \&c.}\right);$$

then the equation of the general quadric surface through the six points will be

$$\alpha U+\beta V+\gamma W+\delta S=0,$$

and this surface will be a cone, having (x, y, z, w) for the coordinates of its vertex, if only we have simultaneously

$$\begin{aligned}
&\alpha\frac{dU}{dx}+\beta\frac{dV}{dx}+\gamma\frac{dW}{dx}+\delta\frac{dS}{dx}=0,\\
&\alpha\frac{dU}{dy}+\text{\&c.} \qquad\qquad\qquad\quad =0,\\
&\alpha\frac{dU}{dz}+\text{\&c.} \qquad\qquad\qquad\quad =0,\\
&\alpha\frac{dU}{dw}+\text{\&c.} \qquad\qquad\qquad\quad =0.
\end{aligned}$$

Eliminating $(\alpha, \beta, \gamma, \delta)$ we have an equation $\nabla=0$, where ∇ is the Jacobian or functional determinant $\dfrac{d(U, V, W, S)}{d(x, y, z, w)}$ formed with the differential coefficients of the four functions (U, V, W, S): the locus of the vertex is thus a quartic surface.

Calling the six points 1, 2, 3, 4, 5, 6, then taking as vertex any point in the line 12, the lines from such point to the points 1 and 2 coincide with the line 12, and we can through this line and the lines to the remaining points 3, 4, 5, 6 describe a quadric cone; the quartic surface therefore passes through the line 12; and similarly it passes through each of the fifteen lines 12, 13, ..., 56.

Again, taking the vertex anywhere in the line of intersection of the planes 123 and 456, we have an improper quadric cone, viz. the plane-pair formed by these two planes; the line in question is therefore a line of the quartic surface; and similarly the quartic surface contains each of the ten lines 123.456, 124.356,..., 156.234. We have thus in all 25 lines on the quartic surface.

In the case of seven points 1, 2, 3, 4, 5, 6, 7, the locus is the curve of intersection of the quartic surfaces which correspond to the points 1, 2, 3, 4, 5, 6 and the points 1, 2, 3, 4, 5, 7 respectively: these have in common the ten lines 12, 13, 14, 15, 23, 24, 25, 34, 35, 45 (which it is easy to see do not form part of the required locus), and they have therefore, as a residual intersection, a curve of the order $16-10$, $=6$, or sextic curve, which is the locus of the vertices of the cones which pass through the seven given points.

14. *Show that the envelope of a variable circle having its centre on a given conic and cutting at right angles a given circle is a bicircular quartic; which, when the given conic and circle have double contact, becomes a pair of circles; and, by means of the last-mentioned particular case of the theorem, connect together the porisms arising out of the two problems:*

(1) *given two conics, to find a polygon of n sides inscribed in the one and circumscribed about the other;*

(2) *given two circles, to find a closed series of n circles each touching the two given circles and the two adjacent circles of the series.*

The equation of the given circle is taken to be

$$(x-\alpha)^2+(y-\beta)^2=\gamma^2,$$

and that of the conic $\dfrac{x^2}{a^2}+\dfrac{y^2}{b^2}=1$. This being so, we have $a\cos\theta$, $b\sin\theta$ as the coordinates of a point on the conic, which point may be taken to be the centre of the variable circle, and introducing the condition that the two circles cut at right angles, the equation of the variable circle is

$$(x-a\cos\theta)^2+(y-b\cos\theta)^2=(\alpha-a\cos\theta)^2+(\beta-b\sin\theta)^2-\gamma^2,$$

that is,

$$x^2+y^2-\alpha^2-\beta^2+\gamma^2-2ax\cos\theta-2by\sin\theta=0,$$

where θ is the variable parameter; and the equation of the envelope therefore is

$$(x^2+y^2-\alpha^2-\beta^2+\gamma^2)^2-4a^2x^2-4b^2y^2=0,$$

which is a quartic curve; and writing herein $\dfrac{x}{z}$, $\dfrac{y}{z}$ in place of x, y the equation would be of the second order in regard to x^2+y^2, z, and it thus appears that the curve has double points at each of the points $x^2+y^2=0$, $z=0$, viz. that the envelope is a bicircular quartic.

If the fixed circle touches the conic, then by a consideration of the figure it at once appears that the point of contact is a double point on the curve; and so if there is a double contact, then each of the points of contact is a double point on the curve. But in this case the curve is a bicircular quartic with *four* double points; viz. it is a pair of circles.

The porism in regard to the two conics is, that in general it is not possible to find any polygon of n sides satisfying the conditions; but that the conics may be such that there exists an infinity of polygons; viz. any point whatever of the one conic may then be taken as a vertex of the polygon, and then constructing the figure, the $(n+1)^{\text{th}}$ vertex will coincide with the first vertex, and there will be a polygon of n sides.

Now imagine that the conic touched by the sides is a circle having double contact with the other conic. Describe any one of the polygons, and with each vertex as centre describe the orthotomic circle, which will, it is clear, be a circle passing through the points of contact with the fixed circle of the sides through the vertex. We have thus a closed series of n circles, each touching the two adjacent circles of the series. And by considering any other polygon, we have a like series of n circles: and by what precedes the envelope of all the circles of the several series is a pair of circles; that is, the circles of every series touch these two circles. We have consequently two circles, such that there exists an infinity of closed series of n circles, each circle touching the two fixed circles, and also the two adjacent circles of the series; which is the porism arising out of the second problem.

535.

NOTE ON THE PROBLEM OF ENVELOPES.

[From the *Messenger of Mathematics*, vol. I. (1872), pp. 3, 4.]

THERE is a mode of looking at the problem of Envelopes, which, so far as I am aware, has not been explicitly noticed. Let $U=(x, y, z)^m$ be a function of the coordinates (x, y, z), $\Theta=\Theta'=(x, y, z)^a (x', y', z')^{a'}$ a function of the two sets of coordinates (x, y, z) and (x', y', z'); it being understood that when we write Θ we regard (x, y, z) as the current coordinates, when Θ' we regard (x', y', z') as the current coordinates. Suppose that we have $U=0$; the curve $\Theta'=0$ is then a curve the equation whereof contains as parameters the coordinates (x, y, z) of a point P on the curve $U=0$; and we may seek for the envelope of the curve $\Theta'=0$ as P describes the curve $U=0$; the required envelope is of course obtained as an equation in (x', y', z') given by the elimination of x, y, z, λ from the equations (equivalent to four equations only)

$$U=0,\quad \Theta'=0,$$
$$d_x\Theta' + \lambda d_x U = 0,$$
$$d_y\Theta' + \lambda d_y U = 0,$$
$$d_z\Theta' + \lambda d_z U = 0.$$

But, observe that the required envelope is the locus of the points of intersection of the curve $\Theta'=0$ belonging to a particular point (x, y, z) of the curve $U=0$, by the curve $\Theta'=0$ which belongs to a consecutive point of U. The curve $\Theta=0$, considering therein (x', y', z') as the coordinates of a given point of the plane, determines by its intersection with $U=0$ those points (x, y, z) on the curve $U=0$, to each of which belongs a curve $\Theta'=0$ passing through the point in question (x', y', z'). Hence, if the curve $\Theta=0$ touch the curve $U=0$, the point of contact, coordinates (x, y, z), is a point such that to it and to the consecutive point there belong curves, each of them passing through the given point (x', y', z'). Hence expressing that the curves

$\Theta=0$, $U=0$ touch each other, we have a relation in (x', y', z') which is the locus of the point of intersection of the curves $\Theta'=0$ belonging to two consecutive points of the curve $U=0$; that is, the equation of the required envelope is obtained as the condition that the curves $U=0$, $\Theta=0$ shall touch each other. But when the curves touch each other, they have at the point of contact their derived functions proportional, or we have simultaneously

$$U=0, \quad \Theta=0,$$
$$d_x\Theta+\lambda d_x U=0,$$
$$d_y\Theta+\lambda d_y U=0,$$
$$d_z\Theta+\lambda d_z U=0,$$

the same equations as before, since Θ and Θ' denote the same function.

It is to be added that, when $a=m$, the equations

$$d_x\Theta+\lambda d_x U=0,$$
$$d_y\Theta+\lambda d_y U=0,$$
$$d_z\Theta+\lambda d_z U=0,$$

are homogeneous in (x, y, z), and we may by the elimination of (x, y, z) from these equations obtain an equation Disct. $(\Theta+\lambda U)=0$, say for shortness $\Lambda=0$, involving λ and also the coordinates (x', y', z'). Now it is a known theorem that the condition for the contact of the two curves $U=0$, $\Theta=0$ can be obtained by expressing that the equation $\Lambda=0$ shall have a pair of equal roots, or, what is the same thing, by equating to zero the discriminant of the function Λ; this last-mentioned process leads therefore to the equation of the envelope of the curve $\Theta'=0$, viz. (a being $=m$ as above) the equation of the envelope of the curve $\Theta'=0$, is in fact

$$\text{Disct.}_\lambda\ \text{Disct.}_{(x,\,y,\,z)}\,(\Theta+\lambda U)=0,$$

viz. we first take the discriminant of the function $\Theta+\lambda U$ in regard to the coordinates (x, y, z), and then taking the discriminant in regard to λ of this discriminant we equate it to zero. This is in many cases a more simple process than that of the direct elimination of x, y, z, λ from the five equations.

536.

NOTE ON LAGRANGE'S DEMONSTRATION OF TAYLOR'S THEOREM.

[From the *Messenger of Mathematics*, vol. I. (1872), pp. 22—24.]

I TAKE the occasion of the publication of the last edition of Mr Todhunter's *Treatise on the Differential Calculus* to make some remarks on the demonstration in question. Mr Todhunter proposes to himself to exhibit a comprehensive view of the Differential Calculus *on the method of Limits;* but he very properly introduces in some cases demonstrations founded upon other views of the subject, pointing out that this is the case, and explaining or indicating his objections. Thus (Chapter VI.) upon Taylor's Theorem, he remarks "Before we offer a strict demonstration of the theorem in question, we shall notice the method which it was usual to adopt in treatises on the Differential Calculus not based on the doctrine of limits," and then, after giving a demonstration depending on the relation $\frac{d}{dx}f(x+h) = \frac{d}{dh}f(x+h)$,[1] he goes on "There are numerous objections to the method of the preceding articles, and especially the use of an infinite series, without ascertaining that it is convergent, is inadmissible; we proceed then to a rigorous investigation," which investigation (after Mr Homersham Cox) is a demonstration of the equation

$$f(x+h) = f(x) + hf'(x) \ldots + \frac{h^n}{\underline{|n}} f^n(x) + \frac{h^{n+1}}{\underline{|n+1}} f^{n+1}(x+\theta h),$$

(θ between 0 and 1) whence "if the function $f^{n+1}(x+\theta h)$ is such that by making n sufficiently great the term $\frac{h^{n+1}}{\underline{|n+1}} f^{n+1}(x+\theta h)$ can be made as small as we please, then by carrying on the series

$$f(x) + hf'(x) + \frac{h^2}{\underline{|2}} f''(x) + \frac{h^3}{\underline{|3}} f'''(x) + \ldots$$

[1] This demonstration is similar in principle to Lagrange's but I think his is preferable; viz. the principle made use of by Lagrange is that the series has the same value whether x is changed into $x+k$, or h into $h+k$.

to as many terms as we please we obtain a result differing as little as we please from $f(x+h)$. Under these circumstances then we may assert the truth of Taylor's theorem."

I share Abel's horror of divergent series([1]), and I maintain the validity of Lagrange's demonstration. When by an algebraic process we expand a function in a series, for instance the function $\frac{1}{1-x}$, by division

$$\begin{array}{lll} 1-x)\ 1 & & (1+x+x^2+\&c. \\ & 1-x & \\ \hline & \quad x & \\ & \quad x-x^2 & \\ \hline & \qquad x^2\ \&c. & \end{array}$$

in the series $1+x+x^2+$&c., and write accordingly

$$\frac{1}{1-x}=1+x+x^2+\&c.$$

all that is (or ought to be) meant is that the algebraical operations continued as far as we please will give the series of terms $1, x, x^2, \ldots$ or say the series of coefficients $1, 1, 1, \ldots$ And of course with this meaning of the equation, the objection "*non constat* that the series is convergent" would be wholly irrelevant, we do not say that it is, we do not care whether it is so or not. In further illustration, remark that we frequently use such an equation merely as the means of expressing the law of a series of numbers $a_0, a_1, a_2, \ldots$, say $a_n =$ coeff. x^n in $f(x)$, where the function is assumed to be by a definite process expansible in the form $a_0+a_1x+a_2x^2+$&c. in question. Any objection that the series is not convergent would be simply irrelevant. Now any rational or irrational algebraic function $f(x+h)$ can by ordinary algebraical processes be expanded in the form $f(x)+$ terms in h, h^2 &c.... And if in regard to a function $f(x)$ we make the *single assumption* that $f(x+h)$ is *expansible in a form containing powers of h and reducing itself to $f(x)$ when h is put* $=0$, then Lagrange's demonstration shows that the powers of h are h, h^2, h^3, &c.... and that the expansion in fact is

$$f(x+h)=f(x)+hf'(x)+\frac{h^2}{1\,.\,2}f''(x)+\&c.;$$

viz. $f(x+h)$ acquires the same value $f(x+h+k)$ whether we change therein x into $x+k$ or h into $h+k$; and the expression on the right-hand side is the only series in h possessed of the same property. It is to be remarked that the equation contains in itself the definition of the operation of derivation, viz. the equation being true, $f'(x)$ can only denote the coefficient of h in the expansion of $f(x+h)$; and what

[1] Peut-on imaginer rien de plus horrible que de débiter

$$0=1^n-2^n+3^n-4^n+\text{etc.},$$

n étant un nombre entier positif?—*Œuvres*, t. II., p. 266; [Nouv. Éd., 1881, t. II., p. 257].

really is shown is that admitting such an operation to be possible in regard not only to $f(x)$, but to $f'(x)$, &c., then the coefficients $f'(x)$, $\frac{f''(x)}{1.2}$, &c., are obtained from $f(x)$ by the successive repetitions of this operation and by dividing by the proper numerical denominator.

By what precedes, any objection in regard to convergency, I regard as irrelevant; and if it is said that the above-mentioned single assumption is not granted, I would either ask "What is a function"—or I would content myself with the hypothetical statement—*if* $f(x)$ be such that $f(x+h)$ is expansible *ut suprà*, then Taylor's theorem.

In regard to the demonstration given by Mr Todhunter, it implicitly assumes that x and h are both real, and (although doubtless possible) it would be considerably more difficult to find an analogous demonstration of the formula involving $f^{n+1}(x+\theta h)$ in the case of x and h imaginary. But the formula *with the term in question* is not (nor does Mr Todhunter consider it as being) Taylor's theorem; to obtain from it Taylor's theorem, we require (in the foregoing point of view) the property that $h^{n+1}f(x+\theta h)$ is expansible in a series involving h^{n+1} and the higher powers of h, that is, the very property that $f(x+h)$ is expansible in positive powers of h.

Moreover admitting that the formula with the term $f^{n+1}(x+\theta h)$ is demonstrable for imaginary values of x, h, the formula is *meaningless* in the case where x, h are one or both a symbol or symbols of operation: θ would certainly have no definable numerical magnitude, and if it is considered as meaning anything, then the equation in question is a mere definition of what it does mean, and ceases to be a theorem in regard to $f(x+h)$. It is impossible, in a quantitative algebra such as is presupposed in the method of limits, to put any meaning on the equation

$$f\left(\frac{d}{dx}+h\right)=f\left(\frac{d}{dx}\right)+hf'\left(\frac{d}{dx}\right)+\text{\&c.},$$

which however I regard as a legitimate particular form of Taylor's theorem.

537.

SOLUTIONS OF A SMITH'S PRIZE PAPER FOR 1871.

[From the *Messenger of Mathematics*, vol. I. (1872), pp. 37—47, 71—77, 89—95.]

1. *A point moves in a plane with a given velocity, and also with a given velocity about a fixed point in the plane: show that the locus is either a circle passing through the fixed point, or else a circle having the fixed point for its centre; and explain the relation between the two solutions.*

We have in general

$$v^2 = \left(\frac{dr}{dt}\right)^2 + r^2\left(\frac{d\theta}{dt}\right)^2,$$

and in the present question, taking the fixed point as the origin, and measuring θ from any fixed line through this point,

$$\frac{d\theta}{dt} = \omega,\quad V^2 = \left(\frac{dr}{dt}\right)^2 + r^2\omega^2,$$

where V, ω are given constants. Hence

$$\left(\frac{dr}{d\theta}\right)^2 = \left(\frac{dr}{dt}\right)^2 \div \left(\frac{d\theta}{dt}\right)^2 = \frac{V^2}{\omega^2} - r^2,$$

or, writing $V = a\omega$,

$$\left(\frac{dr}{d\theta}\right)^2 = a^2 - r^2,$$

therefore

$$d\theta = \frac{dr}{\sqrt{(a^2 - r^2)}},$$

or

$$\theta + \beta = \sin^{-1}\frac{r}{a},\ (\beta \text{ the constant of integration}),$$

that is,

$$r = a \sin(\theta + \beta),$$

which is the equation of a circle (radius $= \frac{1}{2}a$) passing through the fixed point. In fact, the point moving in such a circle with a constant velocity, moves about the centre with a constant angular velocity, and about any fixed point in the circumference with an angular velocity which is one-half of that about the centre, and is therefore also constant.

Treating β as a variable parameter, to obtain the envelope we have

$$0 = a \cos(\theta + \beta),$$

that is, $\theta + \beta = \frac{\pi}{2}$ and therefore $r = a$, which is the equation of a circle (radius $= a$) having the fixed point for its centre. This is consequently the singular solution.

2. *Determine the system of curves which satisfy the differential equation*

$$dx\,\{\sqrt{(1+x^2)} + ny\} + dy\,\{\sqrt{(1+y^2)} + nx\} = 0\,;$$

and show that the curve which passes through the point $x = 0$, $y = n$ *contains as part of itself the conic*

$$x^2 + y^2 + 2xy\sqrt{(1+n^2)} - n^2 = 0.$$

The equation is integrable *per se*, viz. we have

$$x\sqrt{(1+x^2)} + \log\{x + \sqrt{(1+x^2)}\} + y\sqrt{(1+y^2)} + \log\{y + \sqrt{(1+y^2)}\} + 2nxy = C,$$

or, determining the constant so that for $x = 0$, y may be $= n$,

$$C = n\sqrt{(1+n^2)} + \log\{n + \sqrt{(1+n^2)}\},$$

and the equation may be written

$$x\sqrt{(1+x^2)} + y\sqrt{(1+y^2)} + 2nxy - n\sqrt{(1+n^2)} + \log\frac{\{x+\sqrt{(1+x^2)}\}\{y+\sqrt{(1+y^2)}\}}{n+\sqrt{(1+n^2)}} = 0,$$

which is evidently a transcendental curve; it may however be shown that, if

$$x^2 + y^2 + 2xy\sqrt{(1+n^2)} - n^2 = 0,$$

then we have

$$x\sqrt{(1+x^2)} + y\sqrt{(1+y^2)} + 2nxy - n\sqrt{(1+n^2)} = 0,$$

and

$$\{x + \sqrt{(1+x^2)}\}\{y + \sqrt{(1+y^2)}\} = n + \sqrt{(1+n^2)},$$

so that the equation of the curve is thus satisfied; wherefore the transcendental curve contains as part of itself the conic $x^2 + y^2 + 2xy\sqrt{(1+n^2)} - n^2 = 0$.

[As a simple illustration as to how this may happen, take the transcendental curve $y - \sin xy = 0$, which it is clear contains as part of itself the line $y = 0$.]

We have, from the equation of the conic

$$\{x + y\sqrt{(1+n^2)}\}^2 = n^2(1+y^2),$$

that is,

$$x + y\sqrt{(1+n^2)} = \pm n\sqrt{(1+y^2)},$$

but considering the radicals as positive, the sign must be taken so that we have simultaneously $x = 0$, $y = n$. We have therefore

$$x + y\sqrt{(1+n^2)} = n\sqrt{(1+y^2)},$$

and similarly

$$y + x\sqrt{(1+n^2)} = n\sqrt{(1+x^2)}.$$

Then

$$\begin{aligned} n\{x\sqrt{(1+x^2)} + y\sqrt{(1+y^2)}\} &= 2xy + (x^2+y^2)\sqrt{(1+n^2)} \\ &= 2xy + \sqrt{(1+n^2)}\{n^2 - 2xy\sqrt{(1+n^2)}\} \\ &= n^2\{\sqrt{(1+n^2)} - 2xy\}, \end{aligned}$$

which is the *first* of the relations in question; and

$$\begin{aligned} &n^2\{x+\sqrt{(1+x^2)}\}\{y+\sqrt{(1+y^2)}\} \\ &\quad= n^2xy + nx\{x + y\sqrt{(1+n^2)}\} + ny\{y + x\sqrt{(1+n^2)}\} \\ &\qquad + xy + (x^2+y^2)\sqrt{(1+n^2)} + xy(1+n^2) \\ &\quad= \{n+\sqrt{(1+n^2)}\}\{x^2+y^2+2xy\sqrt{(1+n^2)}\} \\ &\quad= \{n+\sqrt{(1+n^2)}\}n^2, \end{aligned}$$

which is the *second* of the two relations. And the theorem is thus proved.

[The foregoing is the easiest and most obvious solution, but it is interesting to consider the question differently, as follows:

Write

$$Q = \frac{\{x+\sqrt{(1+x^2)}\}\{y+\sqrt{(1+y^2)}\}}{n+\sqrt{(1+n^2)}},$$

we have

$$\begin{aligned} Q\ \{\sqrt{(n^2+1)}+n\} &= \{\sqrt{(1+x^2)}+x\}\{\sqrt{(1+y^2)}+y\} = A + B, \\ Q^{-1}\{\sqrt{(n^2+1)}-n\} &= \{\sqrt{(1+x^2)}-x\}\{\sqrt{(1+y^2)}-y\} = A - B, \end{aligned}$$

if

$$\begin{aligned} A &= \sqrt{(1+x^2)}\sqrt{(1+y^2)} + xy, \\ B &= x\sqrt{(1+y^2)} + y\sqrt{(1+x^2)}; \end{aligned}$$

and then

$$\begin{aligned} AB &= x^2y\sqrt{(1+y^2)} + xy^2\sqrt{(1+x^2)} \\ &\qquad + y(1+x^2)\sqrt{(1+y^2)} + x(1+y^2)\sqrt{(1+x^2)} \\ &= x\sqrt{(1+x^2)} + y\sqrt{(1+y^2)} + 2xyB, \end{aligned}$$

that is,

$$Q^2 \{2n^2 + 1 + 2n \sqrt{(1 + n^2)}\} - \frac{1}{Q^2} \{2n^2 + 1 - 2n \sqrt{(1 + n^2)}\}$$
$$= 4 \{x \sqrt{(1 + x^2)} + y \sqrt{(1 + y^2)}\}$$
$$+ 4xy \left[Q \{\sqrt{(1 + n^2)} + n\} - \frac{1}{Q} \{\sqrt{(1 + n^2)} - n\} \right],$$

whence

$$4 \{x \sqrt{(1 + x^2)} + y \sqrt{(1 + y^2)} + 2nxy - n \sqrt{(1 + n^2)}\}$$
$$= Q^2 \{2n^2 + 1 + 2n \sqrt{(1 + n^2)}\} - \frac{1}{Q^2} \{2n^2 + 1 - 2n \sqrt{(1 + n^2)}\}$$
$$+ 8nxy - 4n \sqrt{(1 + n^2)}$$
$$- 4xy \left[Q \{\sqrt{(1 + n^2)} + n\} - \frac{1}{Q} \{\sqrt{(1 + n^2)} - n\} \right]$$
$$= \left(Q^2 - \frac{1}{Q^2}\right) (2n^2 + 1) + \left(Q^2 + \frac{1}{Q^2} - 2\right) 2n \sqrt{(1 + n^2)}$$
$$- \left(Q - \frac{1}{Q}\right) 4xy \sqrt{(1 + n^2)} - 4 \left(Q + \frac{1}{Q} - 2\right) nxy$$
$$= (Q - 1) \left\{ \frac{(Q + 1)(Q^2 + 1)}{Q^2} (2n^2 + 1) \right.$$
$$+ \frac{(Q - 1)(Q + 1)^2}{Q^2} 2n \sqrt{(1 + n^2)}$$
$$- \frac{4(Q + 1)}{Q} xy \sqrt{(1 + n^2)}$$
$$\left. - 4 \frac{(Q - 1)}{Q} nxy \right\}$$
$$= (Q - 1)\, \Omega \text{ suppose,}$$

that is,

$$x \sqrt{(1 + x^2)} + y \sqrt{(1 + y^2)} + 2nxy - n \sqrt{(1 + n^2)} = \tfrac{1}{4} (Q - 1)\, \Omega.$$

And the integral equation is

$$\tfrac{1}{4} (Q - 1)\, \Omega + \log Q = C,$$

which, for $C = 0$, is satisfied by $Q = 1$.

Now starting from

$$Q = \frac{\{x + \sqrt{(1 + x^2)}\} \{y + \sqrt{(1 + y^2)}\}}{n + \sqrt{(1 + n^2)}},$$

we have

$$\sqrt{(1 + x^2)} + x = Q \{\sqrt{(1 + n^2)} + n\} \{\sqrt{(1 + y^2)} - y\},$$
$$\sqrt{(1 + x^2)} - x = \frac{1}{Q} \{\sqrt{(1 + n^2)} - n\} \{\sqrt{(1 + y^2)} + y\},$$

and thence

$$2x = K\sqrt{(1+y^2)} - Ly,$$

if

$$K = Q\{\sqrt{(1+n^2)}+n\} - \frac{1}{Q}\{\sqrt{(1+n^2)}-n\},$$

$$L = Q\{\sqrt{(1+n^2)}+n\} + \frac{1}{Q}\{\sqrt{(1+n^2)}-n\},$$

wherefore

$$L^2 - K^2 = 4.$$

Moreover

$$(2x+Ly)^2 = K^2(1+y^2),$$

that is,

$$4x^2 + (L^2-K^2)y^2 + 4Lxy = K^2,$$

or, what is the same thing,

$$x^2+y^2+Lxy = \tfrac{1}{4}(L^2-4),$$

which is the rationalised form of

$$Q = \frac{\{x+\sqrt{(1+x^2)}\}\{y+\sqrt{(1+y^2)}\}}{n+\sqrt{(1+n^2)}}.$$

And if $Q=1$ then $L=2\sqrt{(1+n^2)}$, $\tfrac{1}{4}(L^2-4)=n^2$, so that this equation is

$$x^2+y^2+2xy\sqrt{(1+n^2)}-n^2=0;$$

or, when $C=0$, the complete integral is satisfied by

$$\frac{\{x+\sqrt{(1+x^2)}\}\{y+\sqrt{(1+y^2)}\}}{n+\sqrt{(1+n^2)}} = 1,$$

that is, by

$$x^2+y^2+2xy\sqrt{(1+n^2)}-n^2=0.$$

We may without difficulty rationalise, and present the result as follows: the equation

$$\left\{2\left(x+\frac{1}{x}\right)+\left(n-\frac{1}{n}\right)\left(y-\frac{1}{y}\right)\right\}\left(1+\frac{1}{x^2}\right)dx$$

$$+\left\{2\left(y+\frac{1}{y}\right)+\left(n-\frac{1}{n}\right)\left(x-\frac{1}{x}\right)\right\}\left(1+\frac{1}{y^2}\right)dy=0,$$

has the complete integral

$$x^2-\frac{1}{x^2}+y^2-\frac{1}{y^2}+\left(x-\frac{1}{x}\right)\left(y-\frac{1}{y}\right)\left(n-\frac{1}{n}\right)-\left(n^2-\frac{1}{n^2}\right)=C+4\log\frac{xy}{n},$$

and a particular integral $xy-n=0$: the complete integral is in fact

$$(n-xy)\{-n^3x^2y^2+n^2xy(-x^2-y^2+1)+n(x^2y^2-x^2-y^2)-xy\} = x^2y^2n^2\left(C+4\log\frac{xy}{n}\right),$$

satisfied, for $C=0$, by $xy-n=0$.]

3. *Write* $\alpha = b - c$, $\beta = c - a$, $\gamma = a - b$; *then considering the three circles and the three conics*

$$(x-a)^2 + y^2 = -\beta\gamma, \quad \frac{x^2}{bc} + \frac{y^2}{K+bc} = 1,$$

$$(x-b)^2 + y^2 = -\gamma\alpha, \quad \frac{x^2}{ca} + \frac{y^2}{K+ca} = 1,$$

$$(x-c)^2 + y^2 = -\alpha\beta, \quad \frac{x^2}{ab} + \frac{y^2}{K+ab} = 1,$$

where K is arbitrary; it is required to show that if a variable circle having its centre on one of the conics cuts at right angles the corresponding circle, the envelope is in each of the three cases one and the same bicircular quartic.

Consider the circle $(x-a)^2 + y^2 = -\beta\gamma$ and the conic $\frac{x^2}{bc} + \frac{y^2}{K+bc} = 1$, the coordinates of a point on the conic are $\cos\theta\sqrt{(bc)}$, $\sin\theta\sqrt{(K+bc)}$, where θ is a variable parameter; say for a moment these values are p and q. The equation of the variable circle is

$$(x-p)^2 + (y-q)^2 = r^2,$$

and in order that this may cut at right angles the circle

$$(x-a)^2 + y^2 = -\beta\gamma,$$

we must have

$$(p-a)^2 + q^2 = r^2 - \beta\gamma,$$

or, substituting for r^2 its value from this equation, the equation of the variable circle is

$$(x-p)^2 + (y-q)^2 = (a-p)^2 + q^2 + \beta\gamma,$$

that is,

$$x^2 + y^2 - a^2 - \beta\gamma - 2p(x-a) - 2qy = 0,$$

viz. this is

$$(x^2 + y^2 - a^2 - \beta\gamma) - 2(x-a)\sqrt{(bc)}\cos\theta - 2y\sqrt{(K+bc)}\sin\theta = 0.$$

Hence taking the envelope in regard to θ, the equation is

$$(x^2 + y^2 - a^2 - \beta\gamma)^2 - 4(x-a)^2 bc - 4y^2(K+bc) = 0,$$

that is,

$$(x^2 + y^2 - ab - ac + bc)^2 - 4(x-a)^2 bc - 4y^2(K+bc) = 0,$$

or, what is the same thing,

$$(x^2+y^2)^2 - 2(bc+ca+ab)(x^2+y^2) - 4Ky^2 + 8abcx$$
$$+ b^2c^2 + c^2a^2 + a^2b^2 - 2a^2bc - 2b^2ca - 2c^2ab = 0,$$

viz. this equation, being symmetrical in regard to a, b, c, is the same equation as would have been obtained from either of the other conics and the corresponding circle; and from the form of the equation it is clear that the curve is a bicircular quartic.

4. *Show that the caustic by refraction for parallel rays of a circle, radius c, index of refraction μ, is the same curve as the caustic by refraction for parallel rays of the concentric circle, radius $\frac{c}{\mu}$, index of refraction $\frac{1}{\mu}$.*

Take as usual $\mu > 1$. Imagine the ray AP (fig. 1) parallel to the axis of x, incident at P on the circle radius c, and let the refracted ray after cutting the circle radius $\frac{c}{\mu}$, cut it again in Q, and then cut the axis in R. Take ϕ, ϕ' for the angles of incidence and refraction; $\sin\phi = \mu\sin\phi'$.

Fig. 1.

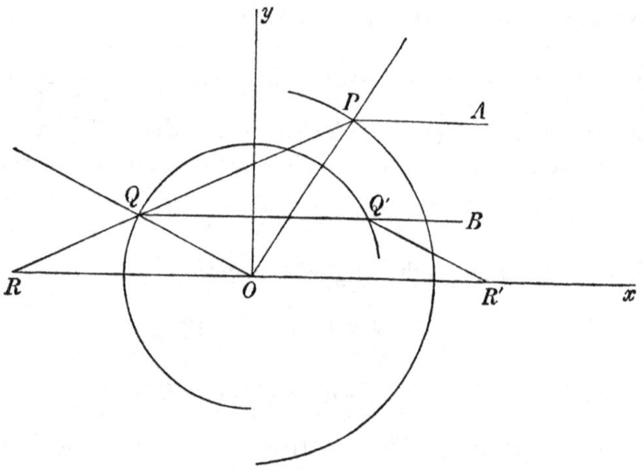

Moreover in the triangle PQO, we have $\sin Q : \sin P = c : \frac{c}{\mu}$; that is, $\sin Q = \mu \sin P$, $= \mu\sin\phi', = \sin\phi$; or $\angle Q = \phi$. And then in the triangle RQO, $\angle R = \phi - \phi'$, $\angle Q = 180° - \phi$, whence $\angle O = \phi'$, that is, $\angle QOR = \phi'$.

Consider now a ray BQ incident at Q and refracted in the direction QR; the index of refraction being $\frac{1}{\mu}$, that is, the denser medium being on the outside of the small circle. Taking θ, θ' for the angles of incidence and refraction, we have $\sin\theta = \frac{1}{\mu}\sin\theta'$; but, the refracted ray being by hypothesis QR, we have by what precedes $\theta' = \phi$, hence $\sin\theta = \frac{1}{\mu}\sin\phi = \sin\phi'$, that is, $\theta = \phi'$, or $\angle BQO = \angle QOR$, that is, the incident ray BQ is parallel to the axis of x. We have thus two pencils of rays each parallel to the axis, such that for any ray AP of the first pencil there is a corresponding ray BQ of the second pencil, the rays AP and BQ each giving rise to the same refracted ray PQR; hence the two pencils have the same caustic.

[It is proper to remark that for the ray BQ it has been assumed that the refraction takes place not at Q' where it first meets the small circle, but at Q; if we consider the refraction at Q', then the index of refraction is still to be $=\frac{1}{\mu}$, that is, the denser medium must now be *inside* the small circle; the refracted ray is in the direction $R'Q'$ situate symmetrically with RQ on the opposite side of the axis of y; and it would at first sight appear that the caustic was a curve equal and similar to the original caustic, but situate on the opposite side of the axis of y. But geometrically the complete caustic consists of two equal and similar portions situate on opposite sides of the axis of y; so that we really obtain, not an equal and opposite caustic, but in each case one and the same caustic.

I originally obtained the theorem in a different manner; viz. the equation for the caustic for the first pencil of rays was found to be

$$(1-\mu^2)\frac{x}{c} = \left\{1-\mu^{\frac{4}{3}}\left(\frac{y}{c}\right)^{\frac{2}{3}}\right\}^{\frac{2}{3}} + \mu\left\{1-\mu^{-\frac{2}{3}}\left(\frac{y}{c}\right)^{\frac{2}{3}}\right\}^{\frac{3}{2}},$$

which equation (as is easily seen) remains unaltered when c, μ are changed into $\frac{c}{\mu}$, $\frac{1}{\mu}$ respectively.—See my "Memoir on Caustics," *Phil. Trans.*, t. CXLVII. (1857), [145], p. 285.]

5. *Given at each point of space the direction-cosines* (α, β, γ) *of a line through that point: it is required to find the conditions in order that the lines may be not a triple but a double system.*

For any given point P the values of the quantities α, β, γ which determine the direction of the line through that point are given as functions of the coordinates (x, y, z) of the point P. Hence passing from a point P to a consecutive point P' *on the line*, the coordinates of P' will be $x+\rho\alpha$, $y+\rho\beta$, $z+\rho\gamma$; and the values of α, β, γ for the point P' will be

$$\alpha+\rho\left(\alpha\frac{d\alpha}{dx}+\beta\frac{d\alpha}{dy}+\gamma\frac{d\alpha}{dz}\right),$$

$$\beta+\rho\left(\alpha\frac{d\beta}{dx}+\beta\frac{d\beta}{dy}+\gamma\frac{d\beta}{dz}\right),$$

$$\gamma+\rho\left(\alpha\frac{d\gamma}{dx}+\beta\frac{d\gamma}{dy}+\gamma\frac{d\gamma}{dz}\right).$$

But if the lines form a double system, we must have the same line for the point P, and for any other point P' on the line; and in particular the same line for the point P, and for the consecutive point P'. Hence as conditions for the double system we obtain

$$\alpha\frac{d\alpha}{dx}+\beta\frac{d\alpha}{dy}+\gamma\frac{d\alpha}{dz}=0,$$

$$\alpha\frac{d\beta}{dx}+\beta\frac{d\beta}{dy}+\gamma\frac{d\beta}{dz}=0,$$

$$\alpha\frac{d\gamma}{dx}+\beta\frac{d\gamma}{dy}+\gamma\frac{d\gamma}{dz}=0.$$

But in virtue of the relation $\alpha^2+\beta^2+\gamma^2=1$, we have

$$\alpha\frac{d\alpha}{dx}+\beta\frac{d\beta}{dx}+\gamma\frac{d\gamma}{dx}=0,$$

$$\alpha\frac{d\alpha}{dy}+\beta\frac{d\beta}{dy}+\gamma\frac{d\gamma}{dy}=0,$$

$$\alpha\frac{d\alpha}{dz}+\beta\frac{d\beta}{dz}+\gamma\frac{d\gamma}{dz}=0.$$

Hence subtracting the corresponding equations we have three equations, which are at once seen to be equivalent to the two equations

$$\frac{d\beta}{dz}-\frac{d\gamma}{dy}:\frac{d\gamma}{dx}-\frac{d\alpha}{dz}:\frac{d\alpha}{dy}-\frac{d\beta}{dx}=\alpha:\beta:\gamma,$$

equations which must be satisfied identically, whatever are the values of (x, y, z). The equations have been obtained as necessary conditions; they are, in fact, the sufficient conditions for a double system; for the line being unaltered in passing from P to P', it remains unaltered when we pass to the following point P'', and so on; that is, for the passage to any point Q whatever on the line.

COR. If the equation $\alpha dx+\beta dy+\gamma dz=0$ be integrable by a factor, it must be integrable *per se*: in fact, the condition that it may be integrable by a factor is

$$\alpha\left(\frac{d\beta}{dz}-\frac{d\gamma}{dy}\right)+\beta\left(\frac{d\gamma}{dx}-\frac{d\alpha}{dz}\right)+\gamma\left(\frac{d\alpha}{dy}-\frac{d\beta}{dx}\right)=0.$$

But we have

$$\frac{d\beta}{dz}-\frac{d\gamma}{dy}=k\alpha, \text{ \&c.},$$

and the equation thus becomes

$$k(\alpha^2+\beta^2+\gamma^2)=0,$$

that is, $k=0$, and therefore

$$\frac{d\beta}{dz}-\frac{d\gamma}{dy}=0,\quad \frac{d\gamma}{dx}-\frac{d\alpha}{dz}=0,\quad \frac{d\alpha}{dy}-\frac{d\beta}{dx}=0.$$

Hence, also, if the lines cut at right angles a surface, we must have $\alpha dx+\beta dy+\gamma dz$ a complete differential.

The foregoing theory is given in Sir W. R. Hamilton's "Memoir on Ray-Systems."

6. *If* $X=0$, $Y=0$, $Z=0$, $W=0$ *are four given conics in the same plane and having a common point: show that, in the system of conics* $aX+bY+cZ+dW=0$, *there are in general four (improper) conics the equations of which may be taken to be* $x^2=0$, $y^2=0$, $xz=0$, $yz=0$.

Taking the conics to pass through the point $x=0$, $y=0$; their equations will be of the form

$$X = a_1x^2 + 2h_1xy + b_1y^2 + 2f_1yz + 2g_1zx = 0,$$
$$Y = a_2x^2 + 2h_2xy + b_2y^2 + 2f_2yz + 2g_2zx = 0,$$
$$Z = a_3x^2 + 2h_3xy + b_3y^2 + 2f_3yz + 2g_3zx = 0,$$
$$W = a_4x^2 + 2h_4xy + b_4y^2 + 2f_4yz + 2g_4zx = 0.$$

Now multiplying by the indeterminate quantities α, β, γ, δ, the three ratios $\alpha : \beta : \gamma : \delta$ may be determined so that the terms in yz, zx shall vanish, and the terms in x^2, xy, y^2 be a perfect square: we thus arrive at a quadric equation for any one of the ratios, say $\alpha : \beta$, the remaining ratios being then linearly determined; viz. there are two sets of values of α, β, γ, δ: and changing the coordinates (x, y), the two resulting forms may be represented by $x^2 = 0$, $y^2 = 0$.

And it is clear that we thus have in the system of conics $\alpha X + \beta Y + \gamma Z + \delta W = 0$, four conics the equations of which may be represented by

$$X' = x^2 = 0,$$
$$Y' = y^2 = 0,$$
$$Z' = h_3xy + f_3yz + g_3zx = 0,$$
$$W' = h_4xy + f_4yz + g_4zx = 0,$$

where of course the coefficients f, g, h have new values.

We may then form the equations

$$\alpha X' + f_4Z' - f_3W' = x\{\alpha x + (f_4h_3 - f_3h_4)\, y + (f_4g_3 - f_3g_4)\, z\},$$
$$\beta Y' - g_4Z' + g_3W' = y\{(g_3h_4 - g_4h_3)\, x + \beta y + (f_4g_3 - f_3g_4)\, z\},$$

so that, by writing $\alpha = g_3h_4 - g_4h_3$ and $\beta = f_4h_3 - f_3h_4$, the terms in $\{\ \ \}$ will be one and the same linear function of (x, y, z); that is, changing the z so as to denote the linear function in question by z, we have as conics of the series $xz = 0$, and $yz = 0$, that is, we have in the series the four conics $x^2 = 0$, $y^2 = 0$, $xz = 0$, $yz = 0$; whence also any other conic of the series, and consequently each of the original four conics, may be represented by an equation of the form

$$ax^2 + by^2 + 2fyz + 2gzx = 0.$$

7. *The coordinates (x, y, z, w) of a point P in space are connected with the coordinates (x', y', z') of a point P' in a plane by the equations*

$$x : y : z : w = X' : Y' : Z' : W',$$

where X', Y', Z', W' are quadric functions of (x', y', z') such that $X' = 0$, $Y' = 0$, $Z' = 0$, $W' = 0$ represent conics having a common point: show that the locus of P is a cubic scroll (skew surface of the third order): and find the curves in the plane which correspond to the generating lines of the scroll.

The equations $x : y : z : w = X' : Y' : Z' : W'$ are three equations containing the indeterminate parameters $x' : z'$ and $y' : z'$, so that eliminating these we have between (x, y, z, w) a single (homogeneous) equation representing a surface. To each point (x', y', z') of the plane, there corresponds a single point of the surface, and to each point (x, y, z, w) of the surface a single point (x', y', z') of the plane. The only exception is that for the common point of the four conics, the ratios $x : y : z : w$ are essentially indeterminate, and there is not corresponding hereto any determinate point of the surface.

To find the order of the surface, consider its intersection with any arbitrary line

$$ax + by + cz + dw = 0,$$
$$a_1x + b_1y + c_1z + d_1w = 0.$$

We have corresponding hereto in the plane the points of intersection of the conics

$$aX' + bY' + cZ' + dW' = 0,$$
$$a_1X' + b_1Y' + c_1Z' + d_1W' = 0,$$

viz. these are conics each of them passing through the common point of the four conics, and therefore they intersect besides in three points: that is, the order of the surface is $= 3$.

To show that the common point ought to be (as above) excluded, some further explanation is desirable. To the section of the surface by the plane $ax + by + cz + dw = 0$, corresponds the conic $aX' + bY' + cZ' + dW' = 0$; and similarly to the section by the plane $a_1x + b_1y + c_1z + d_1w = 0$, corresponds the conic $a_1X' + b_1Y' + c_1Z' + d_1W' = 0$. Now to the common point *considered as belonging to the first conic* there corresponds a determinate point of the surface; and to the common point *considered as belonging to the second conic* there corresponds a determinate point of the surface; but these are two distinct points on the surface: so that corresponding to the common point of the four conics, there is not on the surface any point of intersection of the two plane sections; but these intersect in only three points of the surface; viz. the line of intersection of the two planes meets the surface in three points: or the surface is a cubic surface.

The same result may be obtained, and it may be further shown that the surface is a scroll, by means of the property in the foregoing question 6; viz. it thereby appears that each of the functions X', Y', Z', W' may be taken to be of the form $ax'^2 + by'^2 + fy'z' + gz'x'$; hence replacing the original coordinates x, y, z, w, by properly selected linear functions of these coordinates, the given relations may be presented in the form

$$x : y : z : w = x'^2 : y'^2 : x'z' : y'z',$$

whence eliminating, we have

$$xw^2 - yz^2 = 0$$

the equation of a cubic scroll, having the line $z = 0$, $w = 0$ for a double line, and the line $x = 0$, $y = 0$ for a directrix line. The equations of a generating line of the scroll are, it is clear, $z - \theta w = 0$, $x - \theta^2 y = 0$, where θ is a variable parameter; and corresponding hereto in the plane we have the line $x' - \theta y' = 0$, viz. this is any line through the common intersection of the four conics.

8. *If U, V are binary functions of the form $(a, b, \ldots)(x, y)^m$ with arbitrary coefficients, and if the equations $U = 0$, $V = 0$ have a common root, show how this can be determined in terms of the derived functions of the Resultant in regard to the coefficients of either function.*

Show what results in regard to the common root can be obtained when the coefficients are not all of them arbitrary but (1) each or either of the functions depends in any manner whatever on a set of arbitrary coefficients not entering into the other function, (2) the two functions depend in any manner whatever on one and the same set of arbitrary coefficients.

How is the theory modified when, instead of the two equations, there is a single equation $U = 0$ having a double root?

Suppose

$$U = (a, b, \ldots)(x, y)^m \left(= a x^m + \frac{m}{1} b x^{m-1} y + \&c.\right),$$

$$V = (a', b', \ldots)(x, y)^{m'} \left(= a' x^{m'} + \frac{m'}{1} b x^{m'-1} y + \&c.\right).$$

Then if R is the resultant, the equation $R = 0$ is the relation which must exist between the coefficients $(a, b, \ldots)$ and $(a', b', \ldots)$ in order that the equations $U = 0$ and $V = 0$ may have a common root (that is, in order that the functions U, V may have a common factor $x - \alpha y$). Imagine the relation subsisting, and that x, y are the values belonging to the common root, or (what is the same thing) that we have $x - \alpha y = 0$; we have then simultaneously $U = 0$, $V = 0$, $R = 0$. Now suppose the coefficients $a, b, \ldots$ to be infinitesimally varied in such manner that U, V have still a common root; say the new values are $a + \delta a$, $b + \delta b, \ldots$: this implies between $\delta a, \delta b, \ldots$ the relation

$$\frac{dR}{da}\delta a + \frac{dR}{db}\delta b + \ldots = 0.$$

But the common factor $x - \alpha y$ is a factor of the *unaltered* equation $V = 0$; and the values of (x, y) are thus unaltered, viz. the equation $U = 0$ is satisfied with the original values of (x, y); so that we have

$$\frac{dU}{da}\delta a + \frac{dU}{db}\delta b + \ldots = 0,$$

or, what is the same thing,

$$x^m \delta a + m x^{m-1} y \delta b + \ldots = 0,$$

an equation which must agree with the former one, that is, we have

$$x^m : m x^{m-1} y : \&c. = \frac{dR}{da} : \frac{dR}{db} : \&c.,$$

a series of equations giving the value of the common root $\frac{x}{y}(=\alpha)$ in the several forms

$$\frac{1}{m}\frac{x}{y}=\frac{dR}{da}\div\frac{dR}{db},\quad \frac{2}{m-1}\frac{x}{y}=\frac{dR}{db}\div\frac{dR}{dc},\ \&\text{c.}$$

And it is clear that we have in like manner

$$x^{m'} : m'x^{m'-1}y : \&\text{c.} = \frac{dR}{da'} : \frac{dR}{db'} : \&\text{c.}$$

It is clear that if U involves, *in any manner whatever*, the coefficients $a, b, \ldots$ which do not enter into the function V, then we have in precisely the same manner

$$\frac{dU}{da} : \frac{dU}{db} : \&\text{c.} = \frac{dR}{da} : \frac{dR}{db} : \&\text{c.},$$

a system of equations satisfied by the values x, y which belong to the common root.

But if the coefficients $a, b, \ldots$ are contained in any manner whatever in both of the functions U, V; then by altering $a, b, \ldots$ we alter the common root; say that $x+\delta x$, $y+\delta y$ belong to its new value; then we have

$$\frac{dU}{dx}\delta x+\frac{dU}{dy}\delta y+\frac{dU}{da}\delta a+\frac{dU}{db}\delta b+\ldots=0,$$

$$\frac{dV}{dx}\delta x+\frac{dV}{dy}\delta y+\frac{dV}{da}\delta a+\frac{dV}{db}\delta b+\ldots=0.$$

Now the values of x, y which satisfy $U=0$, $V=0$ also satisfy

$$\frac{dU}{dx}\frac{dV}{dy}-\frac{dU}{dy}\frac{dV}{dx}=0;$$

hence from the foregoing two equations eliminating δx or δy, the other of these two quantities will disappear of itself, and we thus obtain an equation

$$A\delta a+B\delta b+\ldots=0,$$

which must agree with the above equation

$$\frac{dR}{da}\delta a+\frac{dR}{db}\delta b+\ldots=0,$$

or we have

$$\frac{dR}{da} : \frac{dR}{db} : \&\text{c.} = A : B : \&\text{c.},$$

a system of equations satisfied by the values x, y which belong to the common root.

In the case of a single equation $U=0$ having a double root, the condition for this is $\Delta=0$, where Δ is the discriminant of the function U; and the like reasoning shows that for the values x, y which belong to the double root we have

$$\frac{dU}{da} : \frac{dU}{db} : \&\text{c.} = \frac{d\Delta}{da} : \frac{d\Delta}{db} : \ldots;$$

viz. if U is of the form $(a, b, \ldots)(x, y)^m$ with arbitrary coefficients, then we have thus a series of equations giving the required value of $\frac{x}{y}$; but if $(a, b, \ldots)$ are arbitrary coefficients contained in any manner whatever in the function U, then we have a series of equations satisfied by the values x, y which belong to the double root.

9. *The normal at each point of a principal section of an ellipsoid is intersected by the normal at a consecutive point not on the principal section: show that the locus of the point of intersection is an ellipse having four (real or imaginary) contacts with the evolute of the principal section.*

The principal section is for convenience taken to be that in the plane of zx; the coordinates of any point thereof are therefore X, 0, Z where

$$\frac{X^2}{a^2}+\frac{Z^2}{c^2}=1.$$

Consider the normal at a point X, Y, Z of the ellipsoid; taking x, y, z as current coordinates, the equations of the normal are

$$\frac{x-X}{\frac{X}{a^2}}=\frac{y-Y}{\frac{Y}{b^2}}=\frac{z-Z}{\frac{Z}{c^2}}.$$

Writing herein $y=0$, we have

$$x=X\left(1-\frac{b^2}{a^2}\right),\quad z=Z\left(1-\frac{b^2}{c^2}\right);$$

viz. x, z are here the coordinates of the point where the normal meets the plane of xz; and observing that the point in question lies on the normal at the point X, 0, Z, it is clear that x, y, z will be the coordinates of the intersection of the last-mentioned normal by the normal at the consecutive point not on the principal section.

Writing for shortness

$$\alpha=b^2-c^2,\quad \beta=c^2-a^2,\quad \gamma=a^2-b^2,$$

($\alpha+\beta+\gamma=0$, α and γ positive, β negative) the values are

$$x=\frac{\gamma X}{a^2},\quad z=-\frac{\alpha Z}{c^2},$$

wherefore

$$\frac{X}{a}=\frac{ax}{\gamma},\quad \frac{Z}{c}=-\frac{cz}{\alpha};$$

or, substituting in

$$\frac{X^2}{a^2}+\frac{Z^2}{c^2}=1,$$

we have

$$\frac{a^2x^2}{\gamma^2}+\frac{c^2z^2}{\alpha^2}=1,$$

the required locus, which is thus an ellipse.

If the point $(X, 0, Z)$ is an umbilicus, it is clear that the corresponding point of the locus will be a point of the evolute of the principal section; and to prove that the locus touches the evolute, it is only necessary to show that the tangent of the locus is also the tangent of the evolute; or what is the same thing, that the tangent of the locus passes through the umbilicus.

Now for the umbilicus we have

$$X^2=-a^2\frac{\gamma}{\beta},\quad Z^2=-c^2\frac{\alpha}{\beta};$$

the corresponding values of x, z being

$$x=\frac{\gamma X}{a^2},\quad z=-\frac{\alpha Z}{c^2}.$$

Take ξ, ζ as the current coordinates of a point on the tangent of the locus, we have

$$\frac{a^2x\xi}{\gamma^2}+\frac{c^2z\zeta}{\alpha^2}=1,$$

or, substituting for x, z the foregoing values,

$$\frac{X\xi}{\gamma}-\frac{Z\zeta}{\alpha}=1,$$

and these should be satisfied by $\xi, \zeta = X, Z$; viz. we ought to have

$$\frac{X^2}{\gamma}-\frac{Z^2}{\alpha}=1,$$

and this equation is in fact true for the values of X, Z at the umbilicus; viz. for these values we have

$$-\frac{a^2}{\beta}+\frac{c^2}{\beta}=1,$$

that is, $\beta=c^2-a^2$, which is in fact the value of β.

There is obviously a contact in each quadrant, that is, there are four contacts (in the present case all real) of the locus with the evolute.

The same theorem holds good in regard to the other principal sections; only for these, the umbilici being imaginary, the points of contact of the locus with the evolute are also imaginary.

Remark. There is a great convenience in questions relating to the ellipsoid, in the use of the foregoing notations α, β, γ.

10. *An endless heavy chain of given length is suspended from two fixed points in the same horizontal plane: show that (subject to a condition as to the length) the figure of equilibrium may consist of portions of two distinct catenaries.*

The two parts of the chain will each of them be a portion of a catenary, viz. they will either coincide with each other, forming a twice repeated portion of a catenary (which is always a possible position of equilibrium), or they will form portions of two distinct catenaries. That the latter form is in some cases possible, appears from the case of a very long chain. It is then clear that there is a position of equilibrium in which the upper catenary is nearly a straight line. It may be added, that, as the length of the chain diminishes, the two distinct catenaries approach more and more, and for a certain value of the length become coincident; for any smaller value of the length, the only position is that consisting of a twice repeated portion of a catenary. But to obtain the solution in a regular manner, observe that, in order to the existence of such a form of equilibrium, the necessary condition is, that the tension at A (or B) must be equal in the two catenaries. Now the tension at any point of a catenary is proportional to the height above the directrix of the catenary; hence the condition is, that there shall be through the points A, B two catenaries having the same directrix, and such that the sum of the lengths is equal to the given length of the chain.

Take $AB = 2a$, the length of the chain $= 2l$. Take β for the distance of the directrix below the points A, B; c for the parameter of the catenary (or distance of its lowest point above the directrix), β, c being of course unknown. Then taking the origin at the mid-point of the directrix, and the axis of y vertically upwards, the equation of the catenary is

$$y = \frac{c}{2}(e^{\frac{x}{c}} + e^{-\frac{x}{c}}),$$

whence for the point A or B,

$$\beta = \frac{c}{2}(e^{\frac{a}{c}} + e^{-\frac{a}{c}}),$$

and the arc measured from the lowest point is

$$s = \frac{c}{2}(e^{-\frac{x}{c}} - e^{-\frac{x}{c}}).$$

Hence, assuming that there are two distinct catenaries, if the parameters are c, c', we have

$$\tfrac{1}{2}c(e^{\frac{a}{c}} + e^{-\frac{a}{c}}) = \tfrac{1}{2}c'(e^{\frac{a}{c'}} + e^{-\frac{a}{c'}}),$$

$$\tfrac{1}{2}c(e^{\frac{a}{c}} - e^{-\frac{a}{c}}) + \tfrac{1}{2}c'(e^{\frac{a}{c'}} - e^{-\frac{a}{c'}}) = l,$$

which are the conditions for the determination of c, c'; and it is to be shown that these can be satisfied otherwise than by taking $c = c'$.

Trace the two curves

$$y = \frac{x}{2}\left(e^{\frac{a}{x}} + e^{-\frac{a}{x}}\right),$$

$$y' = \frac{x}{2}\left(e^{\frac{a}{x}} - e^{-\frac{a}{x}}\right),$$

shown respectively by the black line and the dotted line in fig. 2. Draw any line parallel to the axis of x, meeting the first curve in the points P, P' respectively, and let the ordinates MP, $M'P'$ meet the second curve in the points Q, Q' respectively;

Fig. 2.

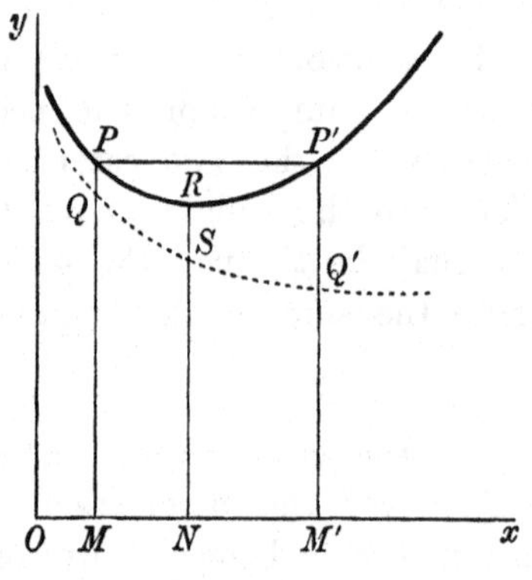

then it is clear, that if for a given value of l the line PP' can be drawn in suchwise that $MQ + M'Q' = l$, there will be in fact the required two values $c = OM$ and $c' = OM'$.

And since for MP very large we have MQ, and therefore also $MQ + M'Q'$, very large, and as MP diminishes, $MQ + M'Q'$ also diminishes until it attains a certain minimum value, say $= \lambda$, it is clear that if l has any value greater than this minimum value, PP' can be so drawn that $QM + Q'M' = l$.

[The above remarkably elegant investigation in regard to the two values c, c' was given in the Examination; it seems to be the case that as PP' moves downwards, $MQ + M'Q'$ continually decreases (viz. MQ decreases more rapidly than $M'Q'$ increases), its value being least, and $= 2NS$ when PP' becomes a tangent to the first curve at its lowest point R; but it is not by any means easy to prove that this is so. The question depends on the form of the curve defined by the equations

$$X = \tfrac{1}{2}x_1\left(e^{\frac{a}{x_1}} - e^{-\frac{a}{x_1}}\right) + \tfrac{1}{2}x_2\left(e^{\frac{a}{x_2}} - e^{-\frac{a}{x_2}}\right),$$

$$Y = \tfrac{1}{2}x_1\left(e^{\frac{a}{x_2}} + e^{-\frac{a}{x_2}}\right) = \tfrac{1}{2}x_2\left(e^{\frac{a}{x_2}} + e^{-\frac{a}{x_2}}\right)$$

where X and Y are the current coordinates.]

11. *A particle is attracted to two centres of force, one of them at the origin, the other revolving about the origin in a circle in the plane of xy with a uniform angular velocity n': find the equations of motion; and writing v for the velocity of the particle and A for the resolved area (about the fixed centre) in the plane of xy, show that there is a first integral giving the value of $v^2 - 4n' \frac{dA}{dt}$ in terms of the coordinates of the particle and of the revolving centre.*

Take ξ, η, 0 for the coordinates of the moving centre, so that

$$\xi = a \cos n't, \quad \eta = a \sin n't;$$

the equations of motion are

$$\frac{d^2x}{dt^2} = -\phi r \frac{x}{r} - \psi\rho \frac{x-\xi}{\rho},$$

$$\frac{d^2y}{dt^2} = -\phi r \frac{y}{r} - \psi\rho \frac{y-\eta}{\rho},$$

$$\frac{d^2z}{dt^2} = -\phi r \frac{z}{r} - \psi\rho \frac{z}{\rho},$$

where

$$r^2 = x^2 + y^2 + z^2,$$

$$\rho^2 = (x-\xi)^2 + (y-\eta)^2 + z^2.$$

We have

$$r\,dr = x\,dx + y\,dy + z\,dz,$$

$$\rho\,d\rho = (x-\xi)(dx - d\xi) + (y-\eta)(dy - d\eta) + z\,dz.$$

But

$$d\xi = -n'a \sin n't\,dt = -n'\eta\,dt,$$

$$d\eta = \quad n'a \cos n't\,dt = \quad n'\xi\,dt,$$

whence

$$\begin{aligned}\rho\,d\rho &= (x-\xi)\,dx + (y-\eta)\,dy + z\,dz \\ &\quad + n'[\eta(x-\xi) - \xi(y-\eta)]\,dt \\ &= (x-\xi)\,dx + (y-\eta)\,dy + z\,dz \\ &\quad - n'[x(y-\eta) - y(x-\xi)]\,dt.\end{aligned}$$

Hence from the equations of motion

$$\begin{aligned}&2\left(\frac{dx}{dt}\frac{d^2x}{dt^2} + \frac{dy}{dt}\frac{d^2y}{dt^2} + \frac{dz}{dt}\frac{d^2z}{dt^2}\right) \\ &- 2n'\left(x\frac{d^2y}{dt^2} - y\frac{d^2x}{dt^2}\right) \\ &= -\frac{\phi r}{r}\, 2\left(x\frac{dx}{dt} + y\frac{dy}{dt} + z\frac{dz}{dt}\right) \\ &\quad - \frac{\psi\rho}{\rho}\left[2\left\{(x-\xi)\frac{dx}{dt} + (y-\eta)\frac{dy}{dt} + z\frac{dz}{dt}\right\}\right. \\ &\quad \left. - 2n'\{x(y-\eta) - y(x-\xi)\}\right].\end{aligned}$$

But we have

$$v^2 = \left(\frac{dx}{dt}\right)^2 + \left(\frac{dy}{dt}\right)^2 + \left(\frac{dz}{dt}\right)^2,$$

$$x\frac{d^2y}{dt} - y\frac{d^2x}{dt^2} = \frac{d}{dt}r^2\frac{d\theta}{dt} = 2\frac{dA}{dt};$$

the foregoing equation may be written

$$\frac{d\,.\,v^2}{dt} - 4n'\frac{d^2A}{dt^2} = -2\phi r\frac{dr}{dt} - 2\psi\rho\frac{d\rho}{dt},$$

whence

$$v^2 - 4n'\frac{dA}{dt} = C - 2\int\phi r dr - 2\int\psi\rho d\rho,$$

the required result.

12. *If x, y are the coordinates of a particle moving* in plano *under the action of a central force varying as* (*distance*)$^{-2}$: *write down the expressions of the coordinates x, y in terms of the time t and of four arbitrary constants: and* (*in case of disturbed motion*) *starting from the equations*

$$\delta x = 0,\ \delta y = 0,\ \delta x' = \frac{d\Omega}{dx}dt,\ \delta y' = \frac{d\Omega}{dy}dt,$$

(*the notation to be explained*), *indicate the process of finding the variations of the constants in terms of* (1) $\frac{d\Omega}{dx}, \frac{d\Omega}{dy}$, (2) *the derived functions of Ω in regard to the constants.*

We have

$$x = a\left\{\frac{\cos u - e}{1 - e\cos u}\cos\varpi + \frac{\sqrt{(1-e^2)}\sin u}{1 - e\cos u}\sin\varpi\right\},$$

$$y = a\left\{\frac{\sqrt{(1-e^2)}\sin u}{1 - e\cos u}\cos\varpi - \frac{\cos u - e}{1 - e\cos u}\sin\varpi\right\},$$

where

$$n - e\sin u = t\sqrt{\left(\frac{\mu}{a^3}\right)} + c,$$

an equation serving to express u in terms of t and the constants a, e, c; the foregoing equations, therefore, in effect give x, y in terms of t and the four constants a, e, c, ϖ.

In the second part of the question, Ω is a given function of x, y, t, the differential coefficients $\frac{d\Omega}{dx}, \frac{d\Omega}{dy}$ being the partial ones in regard to x, y respectively. The equation $\delta x = 0$ signifies that the variation of x, in so far as it arises from the variation of the constants, is $= 0$; it in fact *means*

$$\frac{dx}{da}\frac{da}{dt} + \frac{dx}{de}\frac{de}{dt} + \frac{dx}{dc}\frac{dc}{dt} + \frac{dx}{d\varpi}\frac{d\varpi}{dt} = 0.$$

The value of $x'\left(=\frac{dx}{dt}\right)$ is therefore obtained from that of x by differentiating in regard to t alone, as if a, e, c, ϖ were constants: viz. x' will be a given function of t, a, e, c, ϖ; $\delta x'$ then denotes the variation of x' in so far as it arises from the variation of the constants, viz. the equation $\delta x' = \frac{d\Omega}{da}dt$ *means*

$$\frac{dx'}{da}\frac{da}{dt}+\frac{dx'}{de}\frac{de}{dt}+\frac{dx'}{dc}\frac{dc}{dt}+\frac{dx'}{d\varpi}\frac{d\varpi}{dt}=\frac{d\Omega}{dx}.$$

There are the like equations in regard to y, y', viz. in all, four equations which are linear in regard to $\frac{da}{dt}, \frac{de}{dt}, \frac{dc}{dt}, \frac{d\varpi}{dt}$; and which serve to determine these quantities in terms of $\frac{d\Omega}{dx}, \frac{d\Omega}{dy}$.

Now considering the x, y as expressed in terms of a, e, c, ϖ, t, then Ω becomes a function of these quantities; the differential coefficients $\frac{d\Omega}{da}$, &c., being connected with the original differential coefficients $\frac{d\Omega}{dx}, \frac{d\Omega}{dy}$ by the equations

$$\frac{d\Omega}{da}=\frac{d\Omega}{dx}\frac{dx}{da}+\frac{d\Omega}{dy}\frac{dy}{da},$$

$$\frac{d\Omega}{de}=\frac{d\Omega}{dx}\frac{dx}{de}+\frac{d\Omega}{dy}\frac{dy}{de},$$

&c.

As there are four equations, $\frac{d\Omega}{dx}, \frac{d\Omega}{dy}$ can be expressed in an infinity of ways in terms of $\frac{d\Omega}{da}, \frac{d\Omega}{de}, \frac{d\Omega}{dc}, \frac{d\Omega}{d\varpi}$, and considering $\frac{da}{dt}$, &c., as given in terms of $\frac{d\Omega}{dx}, \frac{d\Omega}{dy}$, we can in an infinity of ways express $\frac{da}{dt}$, &c., as linear functions of $\frac{d\Omega}{da}, \frac{d\Omega}{de}, \frac{d\Omega}{dc}, \frac{d\Omega}{d\varpi}$. But there is one form (obtained by combining the equations in a particular manner) wherein the coefficients of the last-mentioned quantities are functions of a, e, c, ϖ *without* t; and this is the form actually employed for the expression of $\frac{da}{dt}, \frac{de}{dt}, \frac{dc}{dt}, \frac{d\varpi}{dt}$ in terms of $\frac{d\Omega}{da}, \frac{d\Omega}{de}, \frac{d\Omega}{dc}, \frac{d\Omega}{d\varpi}$, in the method wherein these quantities are made use of.

I remark upon the present question, that the answer *ought* to be *in substance perfectly familiar to every student in Physical Astronomy;* and that a student *ought* to be able to present it *in a clear and logical form:* the question being in fact intended as a test of ability *in this respect.*

13. *Explain the course of the geodesic lines on a spheroid of revolution: and in particular show that the condition is satisfied in virtue of which any geodesic line, considered as starting from a given point, ceases at some point of its course to be a shortest line.*

From each point on the surface a geodesic line may be drawn in any direction whatever along the surface, that is, through each point of the surface there is a singly infinite series of geodesic lines. A geodesic line undulates (in the manner of a sinusoid) between two parallels equidistant from the equator on opposite sides thereof; viz. considering it as starting from a point A on the equator, it arrives at a point V on the upper parallel (there touching the parallel), and passes downwards to cut the equator at A', and thence arrives at a point V' on the lower parallel (there touching the parallel), and again passes upwards to meet the equator at A'', and so on; the arcs AV, VA', $A'V'$, $V'A''$, &c., being similar and equal to each other (differing only in position). The equatoreal arc $AA'(=A'A''=\&c.)$ or difference of the longitudes A, A', is always less than 180°, its value increasing with the inclination at which the geodesic line cuts the equator, viz. when this angle is indefinitely small, the arc is $=\frac{c}{a}180°$ (c, a the polar and equatoreal axes respectively), and as the inclination becomes indefinitely near to 90°, the value of the arc becomes indefinitely near 180°. If the arc in question is commensurable with 180°, the geodesic line will be, it is clear, a closed curve; but if not, then it is not a closed curve, but proceeds undulating for ever between the two parallels. In the limiting case where the inclination is $=90°$, the geodesic line is obviously a meridian.

Considering a geodesic line starting in a given direction from a point A, and the geodesic line from the same point A in the consecutive direction, it appears from the foregoing account of the configuration of the lines, that the two lines will intersect each other in general an indefinite number of times: supposing that they first intersect in a point K, then by a general theorem of Jacobi's, the geodesic line AK is a shortest line from A to any point nearer than K, but it is not a shortest line from A to any point beyond K.

538.

EXTRACT FROM A LETTER TO MR. C. W. MERRIFIELD.

[From the *Messenger of Mathematics*, vol. I. (1872), pp. 87, 88.]

THE general integral of the equations

$$\frac{\alpha}{\beta}=\frac{\beta}{\gamma}=\frac{\gamma}{\delta},$$

$\left[\text{where } \alpha, \beta, \gamma, \delta=\frac{d^3z}{dx^3}, \frac{d^3z}{dx^2dy}, \frac{d^3z}{dx\,dy^2}, \frac{d^3z}{dy^3}\right]$, can, I think, be found, viz. $\frac{\alpha}{\beta}=\frac{\beta}{\gamma}$ gives $r=$ function s, and $\frac{\beta}{\gamma}=\frac{\gamma}{\delta}$ gives $s=$ function t. But $r=$ function s, is integrated as the equation of a developable surface (p instead of z), viz. we have

$$\left.\begin{array}{l} p=ax+hy+g\\ 0=a'x+\ \ y+g'\end{array}\right\},$$

a and g functions of h, and

$$\left(a'=\frac{da}{dh},\ \ g'=\frac{dg}{dh}\right);$$

similarly, $s=$ function t, gives

$$\begin{array}{l} q=hx+by+f,\\ 0=\ \ x+b'y+f',\ \ \left(b'=\frac{db}{dh},\ \ f'=\frac{df}{dh}\right).\end{array}$$

Observe that the constants have been so taken, that $\frac{dp}{dy}=h$, $\frac{dq}{dx}=h$; but in order that h may, in the two pairs of equations, mean the same function of (x, y), we must have

$$a'=\frac{1}{b'}=\frac{g'}{f'},$$

that is,

$$b = \int \frac{dh}{a'}, \quad f = \int \frac{g'dh}{a'};$$

or, writing $a = \phi h$, $g = \chi h$, we have

$$p = x\phi h + yh + \chi h,$$

$$q = hx + y\int \frac{dh}{\phi' h} + \int \frac{\chi' h \,.\, dh}{\phi' h},$$

where

$$x\phi' h + y + \chi' h = 0.$$

The last equation gives h as a function of (x, y), and the values of p, q are then such that $dz = pdx + qdy$ is a complete differential, so that we obtain z by the integration of that equation.

A simple example is

$$p = \tfrac{1}{2}h^2 x - hq, \quad q = -hx + y\log h, \quad hx - y = 0,$$

that is,

$$p = -\tfrac{1}{2}\frac{y^2}{x}, \quad q = -y + y\log\frac{y}{x},$$

whence

$$z = \tfrac{1}{2}y^2\log\frac{y}{x} - \tfrac{3}{4}y^2,$$

we have

$$r = \tfrac{1}{2}\frac{y^2}{x^2}, \quad s = -\frac{y}{x}, \quad t = \log\frac{y}{x},$$

$$\alpha = -\frac{y^2}{x^3}, \quad \beta = \frac{y}{x^3}, \quad \gamma = -\frac{1}{x}, \quad \delta = \frac{1}{y},$$

or

$$\frac{\alpha}{\beta} = \frac{\beta}{\gamma} = \frac{\gamma}{\delta}\left(= -\frac{y}{x}\right),$$

as it should be.

Cambridge, 28 *July*, 1871.

539.

FURTHER NOTE ON LAGRANGE'S DEMONSTRATION OF TAYLOR'S THEOREM.

[From the *Messenger of Mathematics*, vol. I. (1872), pp. 105, 106.]

THIS refers to a paper, Wilkinson, "Note on Taylor's Theorem," *Messenger of Mathematics*, same volume, pp. 36, 37, discussing the "Note on Lagrange's Demonstration of Taylor's Theorem," [536, ante, pp. 493—495].

540.

ON A PROPERTY OF THE TORSE CIRCUMSCRIBED ABOUT TWO QUADRIC SURFACES.

[From the *Messenger of Mathematics*, vol. I. (1872), pp. 111, 112.]

THE property mentioned by Mr Townsend in his paper[1] in the August No., "On a Property in the Theory of Confocal Quadrics," may be demonstrated in a form which, it appears to me, better exhibits the foundation and significance of the theorem.

Starting with two given quadric surfaces, the torse circumscribed about these touches each of a singly infinite series of quadric surfaces, any two of which may be used (instead of the two given surfaces) to determine the torse; in the series are included four conics, one of them in each of the planes of the self-conjugate tetrahedron of the two given surfaces; and if we attend to only two of these conics, the two conics are in fact any two conics whatever, and the torse is the circumscribed torse of the two conics; or, what is the same thing, it is the envelope of the common tangent-planes of the two conics.

Consider now two conics U, U', the planes of which intersect in a line I; and let I meet U in the points L, M, and meet U' in the points L', M': take A the pole of I in regard to the conic U, and A' the pole of I' in regard to the conic U'.

Take T any point on I, and draw TP touching U in P, and TP' touching U' in P': the points P, P' may be considered as corresponding points on the two conics respectively.

Join AP and produce it to meet the line I in G; the line APG is in fact the polar of T in regard to the conic U (for T being a point on I, the polar of T passes through A; and this polar also passes through P); that is, the points T, G and L, M are harmonics on the line I; whence also, in the plane of the conic U',

[1 *Messenger of Mathematics*, same volume, pp. 49, 50.]

the lines $P'G$, $P'T$ and $P'L$, $P'M$ are harmonic lines through the point P'. It thus appears that in the particular case where the points L, M are the foci of the conic U', the line $P'G$ is the normal at the point P'; and we may say in general that $P'G$ is the quasi-normal at the point P' of the conic U'.

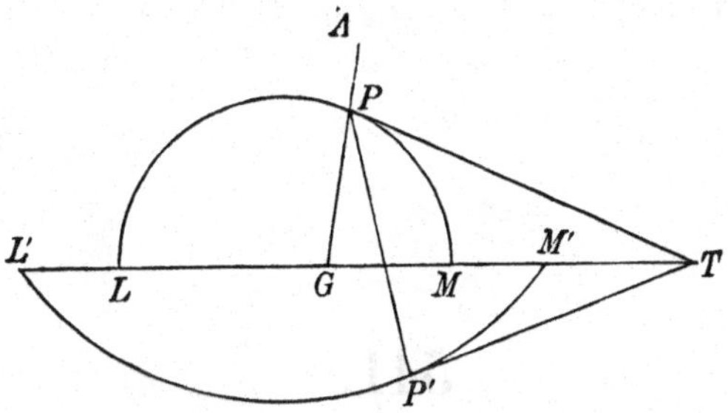

Consider now the torse circumscribed about the conics U, U'; the plane PTP' will represent any plane, and the line PP' any line of this torse: projecting on the plane of U' with the point A as centre of projection, the projection of PP' is the line $P'G$; which, as just seen, is the quasi-normal of the conic U' at the point P'.

The projection of the cuspidal curve is the envelope of line $P'G$, which is the projection of the generating line PP' of the torse—viz. this envelope is the quasi-evolute of the conic U'; which is the theorem in question.

541.

ON THE RECIPROCAL OF A CERTAIN EQUATION OF A CONIC.

[From the *Messenger of Mathematics*, vol. I. (1872), pp. 120, 121.]

THE following formula is useful in various problems relating to conics: the reciprocal equation of the conic

$$\lambda\,(ax+by+cz)\,(a'x+b'y+c'z)-\mu\,(a''x+b''y+c''z)\,(a'''x+b'''y+c'''z)=0$$

may be written indifferent in either of the forms

$$\left\{\lambda\begin{vmatrix}\xi, & \eta, & \zeta\\ a', & b', & c'\\ a, & b, & c\end{vmatrix}+\mu\begin{vmatrix}\xi\,, & \eta\,, & \zeta\\ a'', & b'', & c''\\ a''', & b''', & c'''\end{vmatrix}\right\}^2+4\lambda\mu\begin{vmatrix}\xi\,, & \eta\,, & \zeta\\ a\,, & b\,, & c\\ a''', & b''', & c'''\end{vmatrix}\begin{vmatrix}\xi\,, & \eta\,, & \zeta\\ a'\,, & b'\,, & c'\\ a'', & b'', & c''\end{vmatrix}=0,$$

and

$$\left\{\lambda\begin{vmatrix}\xi, & \eta, & \zeta\\ a', & b', & c'\\ a, & b, & c\end{vmatrix}-\mu\begin{vmatrix}\xi\,, & \eta\,, & \zeta\\ a'', & b'', & c''\\ a''', & b''', & c'''\end{vmatrix}\right\}^2+4\lambda\mu\begin{vmatrix}\xi\,, & \eta\,, & \zeta\\ a\,, & b\,, & c\\ a'', & b'', & c''\end{vmatrix}\begin{vmatrix}\xi\,, & \eta\,, & \zeta\\ a'\,, & b'\,, & c'\\ a''', & b''', & c'''\end{vmatrix}=0.$$

In fact, in the reciprocal equation, seeking for the coefficient of ξ^2, it is

$$\{\lambda\,(bc'+b'c)-\mu\,(b''c'''+b'''c'')\}^2-(2\lambda bb'-2\mu b''b''')\,(2\lambda cc'-2\mu c''c'''),$$

viz. this is

$$\lambda^2\,(bc'-b'c)^2+\mu^2\,(b''c'''-b'''c'')^2+2\lambda\mu\begin{Bmatrix}2bb'c''c'''+2b''b'''cc'\\ -(bc'+b'c)(b''c'''+b'''c'')\end{Bmatrix},$$

or, as it may be written,

$$\{\lambda(bc' - b'c) \pm \mu(b''c''' - b'''c'')\}^2 + 2\lambda\mu \left\{\begin{array}{c} 2bb'c''c''' + 2b''b'''cc' \\ -(bc' + b'c)(b''c''' + b'''c'') \\ \mp(bc' - b'c)(b''c''' - b'''c'') \end{array}\right\}.$$

Taking the upper signs, this is

$$\{\lambda(bc' - b'c) + \mu(b''c''' - b'''c'')\}^2 + 4\lambda\mu \left(\begin{array}{c} bb'c''c''' + b''b'''cc' \\ - bc'b''c''' - b'cb'''c'' \end{array}\right),$$

viz. the term in $\lambda\mu$ is

$$+ 4\lambda\mu(bc''' - b'''c)(b'c'' - b''c').$$

Taking the lower signs, it is

$$\{\lambda(bc' - b'c) - \mu(b''c''' - b'''c'')\}^2 + 4\lambda\mu \left(\begin{array}{c} bb'c''c''' + b''b'''cc' \\ - bc'b'''c'' - b'cb''c''' \end{array}\right),$$

viz. the term in $\lambda\mu$ is

$$+ 4\lambda\mu(bc'' - b''c)(b'c''' - b'''c').$$

And it is thence easy to infer the forms of the other coefficients, and to arrive at the foregoing result.

542.

FURTHER NOTE ON TAYLOR'S THEOREM.

[From the *Messenger of Mathematics*, vol. I. (1872), p. 137.]

THIS paper refers to a "Further Note on Taylor's Theorem," *Messenger of Mathematics*, same volume, pp. 135—137.

543.

ON AN IDENTITY IN SPHERICAL TRIGONOMETRY.

[From the *Messenger of Mathematics*, vol. I. (1872), p. 145.]

In a spherical triangle, writing for shortness α, β, γ for the cosines and α', β', γ' for the sines, of the sides: also

$$\Delta^2 = 1 - \alpha^2 - \beta^2 - \gamma^2 + 2\alpha\beta\gamma;$$

we have

$$\cos A = \frac{\alpha - \beta\gamma}{\beta'\gamma'}, \quad \sin A = \frac{\Delta}{\beta'\gamma'},$$

with the like expressions in regard to the other two angles B, C respectively.

Hence

$$\begin{aligned}\cos(A+B+C) &= \cos A \cos B \cos C - \cos A \sin B \sin C - \&c.\\ &= \frac{(\alpha-\beta\gamma)(\beta-\gamma\alpha)(\gamma-\alpha\beta) - \Delta^2(\alpha+\beta+\gamma-\beta\gamma-\gamma\alpha-\alpha\beta)}{(1-\alpha^2)(1-\beta^2)(1-\gamma^2)}.\end{aligned}$$

The numerator is identically

$$= (1-\alpha)(1-\beta)(1-\gamma)[\Delta^2 - (1+\alpha)(1+\beta)(1+\gamma)],$$

viz. comparing the two expressions, we have

$$\begin{aligned}&(1-\alpha)(1-\beta)(1-\gamma)\Delta^2 - (1-\alpha^2)(1-\beta^2)(1-\gamma^2)\\ &\quad = (\alpha-\beta\gamma)(\beta-\gamma\alpha)(\gamma-\alpha\beta) + \Delta^2(-\alpha-\beta-\gamma+\beta\gamma+\gamma\alpha+\alpha\beta);\end{aligned}$$

or, what is the same thing,

$$(1-\alpha\beta\gamma)\Delta^2 = (1-\alpha^2)(1-\beta^2)(1-\gamma^2) + (\alpha-\beta\gamma)(\beta-\gamma\alpha)(\gamma-\alpha\beta),$$

which is the identity in question and can be immediately verified. We have thus

$$\cos(A+B+C) = \frac{\Delta^2 - (1+\alpha)(1+\beta)(1+\gamma)}{(1+\alpha)(1+\beta)(1+\gamma)},$$

and thence

$$1 + \cos(A+B+C) = \frac{\Delta^2}{(1+\alpha)(1+\beta)(1+\gamma)},$$

$$1 - \cos(A+B+C) = \frac{2(1+\alpha)(1+\beta)(1+\gamma) - \Delta^2}{(1+\alpha)(1+\beta)(1+\gamma)},$$

giving at once the values of $\cos^2\frac{1}{2}(A+B+C)$, $\sin^2\frac{1}{2}(A+B+C)$, $\sin(A+B+C)$, and $\tan^2\frac{1}{2}(A+B+C)$: these are known expressions in regard to the spherical excess.

544.

ON A PENULTIMATE QUARTIC CURVE.

[From the *Messenger of Mathematics*, vol. I. (1872), pp. 178—180.]

I HAVE had occasion to consider with some particularity the form of a curve about to degenerate into a system of multiple curves; a simple instance is a trinodal quartic curve about to degenerate into the form $x^2y^2=0$, or say a "penultimate" of $x^2y^2=0$. To fix the ideas, take x, y, z to denote the perpendiculars on the sides of an equilateral triangle, altitude $=1$ (so that $x+y+z=1$), and let the curve be symmetrical in regard to the coordinates x, y, its equation being thus

$$(a,\ a,\ 1,\ f,\ f,\ h)(x,\ y,\ z)^2=0,$$

where a, f, h are ultimately all indefinitely small in regard to unity: to diminish the number of cases I further assume

$$a=+,\quad f \text{ and } h=-,$$
$$h^2>a^2,\quad \text{that is, } a+h=-,$$
$$f>a,\quad ,,\quad ,,\quad \sqrt{(a)}+f=-,$$

but I do not in the first instance take a, f, h to be indefinitely small. Then if $-f$ is not too large, the curve is as shown in the figure[1], viz. it is a triloop curve, with two horizontal double tangents, 3 touching the curve in two real points, 4 touching it in two imaginary points. Imagine $-f$ increased: the new curve will have the same general form, intersecting the first curve at A and B but touching it at C, viz. it will pass inside the loop C but outside the loops A, B; and outside the remainder of the curve; and the 4 will also move downwards as shown. The new position of 4 will be below the first position.

Supposing that a, h have given values, and that $-f$ continually is increased in regard to $\sqrt{(a)}$; two things may happen. First, the double tangent 3 may move down

[1] The figure is drawn with very small values of a, f, h, in order to exhibit as nearly as may be one of the penultimate forms of the curve; but this is not in anywise assumed in the reasoning of the text. Observe in the figure that the points A, B are ordinary double points, and that there are at each of them two distinct tangents inclined at a small angle to each other.

to $z=-\infty$, the lower loops lengthening out and finally becoming each of them a pair of parabolic branches parallel at infinity; and then reappearing at $z=+\infty$, again move downwards, each loop becoming in this case a pair of hyperbolic branches touching two asymptotes at $z=-\infty$, and then again on the opposite sides thereof at $z=+\infty$,

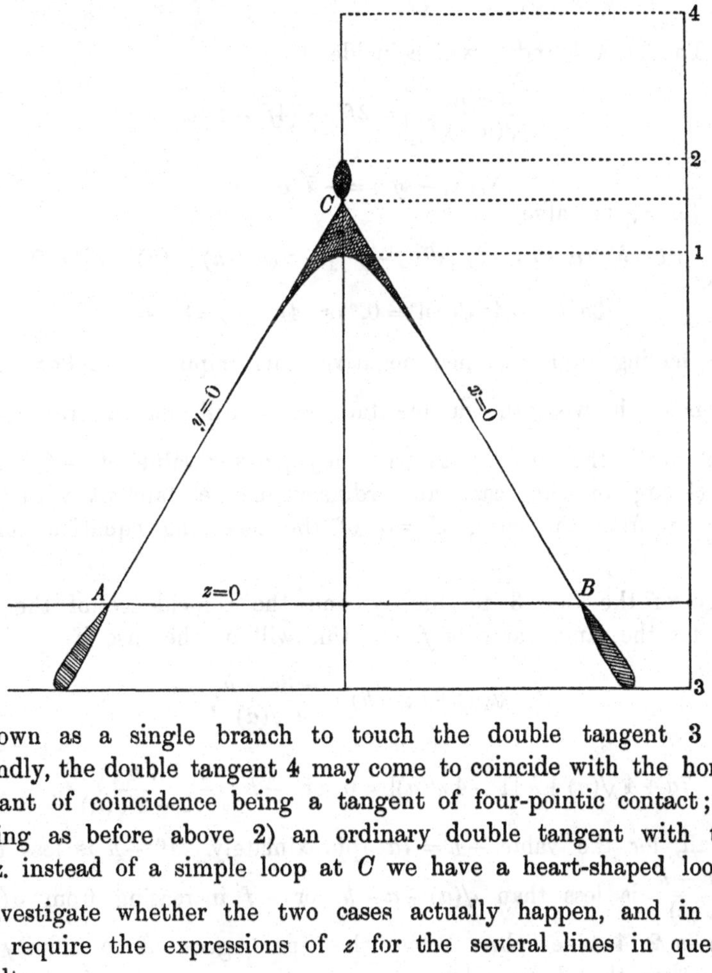

and coming down as a single branch to touch the double tangent 3 which is now above 4. Secondly, the double tangent 4 may come to coincide with the horizontal tangent 2, at the instant of coincidence being a tangent of four-pointic contact; and becoming afterwards (being as before above 2) an ordinary double tangent with two real points of contact; viz. instead of a simple loop at C we have a heart-shaped loop.

But to investigate whether the two cases actually happen, and in what order of succession, we require the expressions of z for the several lines in question; we find, without difficulty,

$$\text{for line } 1,\quad z_1=\frac{1}{1+2\lambda_1},\quad \text{where } \lambda_1=-2f+\sqrt{\{4f^2-2(a+h)\}},$$

$$\text{,,}\quad 2,\quad z_2=\frac{1}{1-2\lambda_2},\quad \text{,,}\quad \lambda_2=\ \ 2f+\sqrt{\{4f^2-2(a+h)\}},$$

$$\text{,,}\quad 3,\quad z_3=\frac{1}{1-2\lambda_3},\quad \text{,,}\quad \lambda_3=\frac{a-h}{2\{-\sqrt{(a)}-f\}},$$

$$\text{,,}\quad 4,\quad z_4=\frac{1}{1-2\lambda_4},\quad \text{,,}\quad \lambda_4=\frac{a-h}{2\{\sqrt{(a)}-f\}},$$

where λ_1, λ_2, λ_3, λ_4 are all positive. Observe that in the limiting case $-f=\sqrt{(a)}$, where, instead of the loops at A, B, we have cusps; z_1, z_2, and z_4 are (in general) positive; $\lambda_3 = \infty$, and therefore $z_4 = 0$; that is, the line 3 coincides with AB, ceasing to be a double tangent; there is in this case the one double tangent 4.

First. z_3 becomes infinite for $1-2\lambda_3=0$; that is, $a-h=-\surd(a)-f$, or $-f=\surd(a)+(a-h)$; viz. for $-f=\surd(a)+(a-h)-\epsilon$, we have $z_3=-\infty$, and for $-f=\surd(a)+(a-h)+\epsilon$, we have $z_3=+\infty$.

Secondly. The lines 4 and 2 will coincide if

$$\lambda_4\left[=\frac{a-h}{2\{\surd(a)-f\}}\right]=2f+\surd\{4f^2-2(a+h)\},$$

that is, if

$$\lambda_4(\lambda_4-4f)=-2(a+h),$$

or, substituting for λ_4 its value,

$$(a-h)[(a-h)-8f\{\surd(a)-f\}]+8(a+h)\{\surd(a)-f\}^2=0,$$

the condition is

$$\{3a+h-4f\surd(a)\}^2=0, \text{ or } 4f\surd(a)=3a+h,$$

(observe that, f having been assumed negative, this implies $-h>3a$). That is, $3a+h$ being $=-$ but not otherwise, the double tangent 4 will, for the value $-f=\dfrac{-3a-h}{4\surd(a)}$, come to coincide with the line 2; and for any greater value of $-f$ will be as before above line 2, (being in this case an ordinary double tangent with real points of contact) as appears from the form, $U^2=0$, of the foregoing equation for the determination of f.

The passage of the line 3 to infinity, and the coincidence of the lines 4 and 2 may take place for the same value of f, viz. this will be the case if

$$\surd(a)+(a-h)=\frac{-3a-h}{4\surd(a)},$$

that is, if

$$7a+4\surd(a)+h\{1-4\surd(a)\}=0 \text{ or } -h=\frac{a\{7+4\surd(a)\}}{1-4\surd(a)},$$

or, a being small, for the value $-h=7a$ approximately. If $-h$ is less than the above value, then $\dfrac{-3a-h}{4\surd(a)}$ is less than $\surd(a)+a-h$, or $-f$ increasing from $\surd(a)$, the coincidence of the lines 2, 4 takes place before the line 3 goes off to infinity: contrarywise, if $-h$ is greater than the above value.

In any form of the curve (i.e. whatever be the value of f in regard to a, h), if we imagine a, h indefinitely diminished, the lines 1, 2 and 4 will continually approach C, and the curve will gather itself up into certain definite portions of the lines $x=0$, $y=0$. Thus any secant through A (not being indefinitely near to the line AC), which meets the curve in real points, will meet it in two points tending to coincide at the intersection of the secant with the line $x=0$; analytically there are always two intersections real or imaginary which (the secant not being indefinitely near the line AC) tend to coincide at the intersection of the secant with the line $x=0$; and we thus see how we ultimately arrive at the line $x=0$ twice repeated; and similarly for the line $y=0$.

545.

ON THE THEORY OF THE SINGULAR SOLUTIONS OF DIFFERENTIAL EQUATIONS OF THE FIRST ORDER.

[From the *Messenger of Mathematics*, vol. II. (1873), pp. 6—12.]

I CONSIDER a differential equation under the form

$$\phi(x, y, p) = 0,$$

where

1°. ϕ is as to p, rational and integral of the degree n;

2°. it is, or is taken to be, one-valued in regard to (x, y);

3°. it has no mere (x, y) factor;

4°. it is indecomposable as regards p.

Considering (x, y) as the coordinates of a point *in plano*, the differential equation determines a system of curves, in general indecomposable, the system depending on a single variable parameter, and such that through each point of the plane there pass n curves.

Such a system is represented by an integral equation

$$f(x, y, c_1, c_2 \dots c_m) = 0,$$

where

5°. f is rational and integral in regard to the m constants, which constants are connected by an algebraic $(m-1)$ fold relation;

6°. it is, or is taken to be, one-valued in regard to (x, y);

7°. it has no mere (x, y) factor;

8°. it is indecomposable as regards (x, y);

9°. Considering (x, y) as given, the equation $f=0$, together with the $(m-1)$ fold relation between the constants, must constitute a m-fold relation of the order n, that is, must give for the constants n sets of values. We may, if we please, take f to be linear in regard to the constants $c_1, c_2, \ldots, c_m$, and then the condition simply is, that the $(m-1)$ fold relation shall be of the order n.

I give in regard to these definitions such explanations as seem necessary.

2°. A one-valued function of (x, y) is either a rational function, or a function such as $e^x \sin y$, &c., which for any given values whatever, real or imaginary, of the variables, has only one value. A function is taken to be one-valued when, either for any values whatever of the variables, or for any class of values (e.g. all real values), we select for any given values of the variables one value, and attend exclusively to such one value, of the function.

Thus, U a rational function of (x, y), $\phi=p^2-U=0$, ϕ is one-valued in regard to (x, y). But if, U not being the square of a rational function, we take $\surd(U)$ to be a one-valued function (consider it as denoting, say for all real values of x, y for which U is positive, the positive square-root of U), then, $\phi=p+\surd(U)=0$, ϕ is taken to be one-valued in regard to (x, y).

3°. The meaning is, that the equation $\phi=0$ is not satisfied irrespectively of the value of p, by any relation between the variables x, y.

4°. The meaning is that ϕ is not the product of two factors, each rational and integral in regard to p, and being or being taken to be one-valued in regard to (x, y). Thus, if as before U is a rational function of (x, y) but is not the square of a rational function, and if we do not take any one-valued function, then the equation $\phi=p^2-U=0$ is indecomposable; but if we take $\surd(U)$ as one-valued, then we have $\phi=\{p-\surd(U)\}\{p+\surd(U)\}=0$, and the equation breaks up into the two equations $p-\surd(U)=0$ and $p+\surd(U)=0$. I assume as an axiom, that the curves represented by the indecomposable differential equation are in general indecomposable; for supposing the differential equation satisfied in regard to a system of curves, the general curve breaking into two curves, each depending on the arbitrary parameter, then we have two distinct systems of curves; either the differential equation is satisfied in regard to each system separately, and in this case they are the same system twice repeated; or the differential equation is satisfied in regard to one system only, and in this case the other system is not part of the solution, and it is to be rejected. As an instance, take the equation $\phi=px+y=0$, $(x\,dy+y\,dx=0)$; if we choose to integrate this in the form $x^2y^2-c=0$, this equation represents the two hyperbolas $xy+\surd(c)=0$, $xy-\surd(c)=0$, but considering each of these separately, and giving to the constant theory any value whatever, we have simply the system of hyperbolas $xy-c=0$ twice repeated. But if by any (faulty) process of integration the solution had been obtained in a form such as $(c+x)(c-xy)=0$, then the differential equation is not satisfied in regard to the system $c+x=0$; and the factor $c+x$ is to be rejected. Observe that it is said, that the curves are *in general* indecomposable; particular curves of the system may very well be decomposable; thus in the foregoing example, where the system of curves is $xy-c=0$, in the particular case $c=0$, the hyperbola breaks up into the two lines $x=0$, $y=0$.

5°. It is *necessary* to consider a form $f=0$ involving the m constants connected by the $(m-1)$ fold relation. For taking such a system of constants, imagine an equation $f(x, y, c_1, c_2, \ldots, c_m)=0$ rational and integral in regard to (x, y), and representing an indecomposable curve; such an integral equation leads to a differential equation of the form $\phi(x, y, p)=0$, rational in regard to (x, y); whence, conversely, a differential equation of the form last referred to *may have* for its integral the equation $f(x, y, c_1, c_2, \ldots, c_m)=0$. And we cannot *in a proper form* exhibit this integral in terms of a single constant. For first consider for a moment $c_1, c_2, \ldots, c_m$ as the coordinates of a point in m-dimensional space; the curve is not in general unicursal, and unless it be so, we cannot express the quantities $c_1, c_2, \ldots, c_m$ rationally in terms of a parameter; that is, we cannot in general express $c_1, c_2, \ldots, c_m$ rationally in terms of a parameter. Secondly, if by means of the $(m-1)$ fold relation we sought to eliminate from the equation $f=0$ all but one of the m constants, we should indeed arrive at an equation $F(x, y, c)=0$ rational and integral as regards x and y, and also as regards c; but this equation *would not represent an indecomposable curve.*

6°. It is important to remark that, even in the case where $\phi(x, y, p)$ is one-valued in regard to (x, y) (2°), there is not in every case a form $f=0$ one-valued in regard to (x, y). A simple example shows this; let α, β be incommensurable (e.g. $\alpha=e$, $\beta=\pi$), then the equation $\phi=\beta xp+\alpha y=0\,(\alpha y\,dx+\beta x\,dy=0)$ has for its integral $c=x^\alpha y^\beta$, where $x^\alpha y^\beta$ is not a one-valued function of x, y, and we cannot in any way whatever transform the integral so as to express it in terms of one-valued functions of x and y. But taking $x^\alpha y^\beta$ to be a one-valued function—if e.g. for all real values of x, y we consider $x^\alpha y^\beta$ as representing the real value of $(\pm x)^\alpha(\pm y)^\beta$—, we shall have, without any loss of generality, the integral of the differential equation; the *whole* system of curves $c=x^\alpha y^\beta$, c any value whatever, is the same whether we attribute to $x^\alpha y^\beta$ its infinite series of values, or only one of these values.

7°. The meaning is, that the equation $f=0$ is not satisfied irrespectively of the values of $c_1, c_2, \ldots, c_m$ by any relation between x, y only.

8°. The meaning is, that the function f is not the product of two factors, rational or irrational in regard to $c_1, c_2, \ldots, c_m$, but each of them one-valued or taken to be one-valued in regard to x, y. Thus the function

$$f=x^2y^2-c, \quad =\{xy+\sqrt{(c)}\}\{xy-\sqrt{(c)}\},$$

is decomposable; but if we do not take any one-valued function, then $f=xy-c$ is indecomposable; if we take $\sqrt{(xy)}$ to be one-valued, then it is decomposable.

The case $f=x^\alpha y^\beta-c$ (α and β incommensurable) is to be noticed; starting from the differential equation $\alpha y\,dx+\beta x\,dy=0$, there is no reason for writing the integral $f=x^\alpha y^\beta-c=0$ rather than in either of the forms $f=x^{m\alpha}y^{m\beta}-c=0$, $f=x^{\frac{\alpha}{m}}y^{\frac{\beta}{m}}-c=0$, ($m$ an integer), (α, β), $(m\alpha, m\beta)$, $\left(\frac{\alpha}{m}, \frac{\beta}{m}\right)$ are, each pair as well as the others, two incommensurable magnitudes. If we choose to take $x^\alpha y^\beta$ but not $x^{\frac{\alpha}{m}}y^{\frac{\beta}{m}}$ as one-valued, then $f = x^{\alpha}y^{\beta} - c$ and $f = x^{m\alpha}y^{m\beta} - c$ are each one-valued, and the former is, the latter is not, indecomposable; they would be each decomposable if we chose to take $x^{\frac{\alpha}{m}}y^{\frac{\beta}{m}}$ as one-valued; and if only $x^{m\alpha}y^{m\beta}$ were taken to be one-valued, then $f = x^{m\alpha}y^{m\beta} - c$ would be indecomposable.

9°. This is a mere statement of the condition in order that the system of curves represented by the integral equation may be such that, through a given point of the plane, there may pass n of these curves. Since the number of constants is unlimited, there is clearly no loss of generality in assuming that the equation is linear in regard to the several constants.

I consider now the differential equation

$$\phi(x, y, p) = 0,$$

(as already stated of the degree n as regards p), and its integral equation

$$f(x, y, c_1, c_2, \ldots, c_m) = 0.$$

I take (x, y) to be the coordinates of a point, say in the horizontal plane, and I use C to refer to the constants $c_1, c_2, \ldots, c_m$ collectively, thus for given values of x, y I speak of the n values of C, meaning thereby the n values of the set $(c_1, c_2, \ldots, c_m)$; of C having a two-fold value, meaning thereby that two of the sets $c_1, c_2, \ldots, c_m$ become identical; and so on.

The case $C = c$, where there is only a single constant c, is interesting as affording an easier geometrical conception; we may take $c = z$ to be a third coordinate; the equation $f(x, y, z) = 0$ thus represents a surface, such that its plane sections $z = c$, or say these sections projected by vertical ordinates on the horizontal plane $z = 0$ are the series of curves $f(x, y, c) = 0$. But the case is not really distinct from the general one.

The theory of singular solutions depends on the following considerations:

To a given point P on the horizontal plane belong n values of C, each determining a curve $f(x, y, C) = 0$ through P; and also n values of p, viz. these give the directions at P of the n curves respectively.

The curve $f(x, y, C) = 0$ may be such as to have in general a certain number of nodes and of cusps (either or each of these numbers being $= 0$): we may imagine C determined, say $C = C_0$, so that the curve shall have one additional node: this node I call a "level point." Take P at the level point, there are n values of C, viz. C_0 and $n-1$ other values; that is, there are through P the nodal curve, and $n-1$ other curves, and therefore $2 + (n-1)$, $= n+1$ directions of the tangent; but the directions are determined by the equation $\phi = 0$ of the order n: and the only way in which we can have more than n values is when this equation becomes an identity $0 = 0$; that is, P at the level point, the function $\phi(x, y, p)$ will vanish identically, irrespectively of the value of p.

The ordinary nodes (if any) on the curves $f(x, y, C) = 0$ form a locus called the "nodal locus," and the cusps (if any) a locus called the "cuspidal locus." Take the point P a given point on the nodal locus, C has a two-fold value answering to the curve in regard to which P is a node, and $n-2$ other values; that is, the n curves through P are the nodal curve reckoned twice and $n-2$ other curves; the directions are the directions at the node, and the $n-2$ other directions, in all n directions, which are the directions given by the equation $\phi = 0$; there is no peculiarity in regard to this equation.

Similarly, take P anywhere on the cuspidal locus: C has a two-fold value, answering to the curve in regard to which P is a cusp, and $n-2$ other values; that is, the n curves through P are the cuspidal curve reckoned twice and $n-2$ other curves; the directions are the direction at the cusp reckoned twice and $n-2$ other directions: in all n directions, which are the directions given by the equation $\phi = 0$; this equation thus gives a two-fold value of p.

There is a locus (distinct from the nodal and cuspidal loci) which may be called the "envelope locus," such that taking P anywhere on this locus C has a two-fold value; for such position of P the n values of C are the value in question reckoned twice and $n-2$ other values; the n curves through P are that belonging to the two-fold value of C, or say the two-fold curve, and $n-2$ other values; and the n directions are the direction along the two-fold curve counted twice, and $n-2$ other directions; these are the n directions given by the equation $\phi=0$, viz. this equation gives a two-fold value of p.

The envelope locus may be an indecomposable curve, or it may break up into two or more curves; and it may happen that either the whole curve or one or more of the component curves may coincide with a particular curve or curves of the system $f(x, y, C) = 0$.

There is a locus (distinct from the cuspidal and envelope loci) which may be called the tac-locus, such that taking P anywhere on this locus p has a two-fold value; for such position of P, there is no peculiarity as regards C, viz. C has n distinct values giving rise to n curves through P; but as the directions are given by the equation $\phi = 0$, two of the curves touch each other, viz. the tac-locus is the locus of points, such that at any one of them two of the curves $f(x, y, C) = 0$ through the point touch each other.

We may by an extension of the received notation write

$$\text{disct}_C f(x, y, C) = 0$$

to denote the equation between (x, y), such that, for any values of (x, y), which satisfy the condition, or say for any position of P on the C-discriminant locus, there is a two-fold value of C. By what precedes it appears that the C-discriminant locus is made up of the nodal, cuspidal, and envelope loci, and without going into the proof I infer that it is in fact made up of the nodal locus twice, the cuspidal locus three times, and the envelope locus once.

Writing moreover

$$\text{disct}_p\, \phi(x,\ y,\ p) = 0$$

to denote the equation between (x, y), such that for any values of (x, y) which satisfy the condition, or say for any position of P on the p-discriminant locus, there is a two-fold value of p. By what precedes, it appears that the p-discriminant locus is made up of the envelope locus, cuspidal locus, and the tac-locus; as I infer, each of them once.

The foregoing are the abstract principles: I consider the singular solution to be that given by the equation which belongs to the envelope-locus (viz. *I do not recognise any singular solution which is not of the envelope species*); and the result of the investigation is, when we seek in the ordinary way to obtain the singular solution, whether from the integral equation or from the differential equation, that we account for the extraneous factors which present themselves in the two processes respectively. I reserve for another communication the discussion of particular examples.

546.

THEOREMS IN RELATION TO CERTAIN SIGN-SYMBOLS.

[From the *Messenger of Mathematics*, vol. II. (1873), pp. 17—20.]

I FIND the following among my papers:

Let the latin letters $a, b, \ldots$ denote *lines* of n signs $\pm$, and the greek letters $\alpha, \beta, \ldots$ *columns* of the same number n of signs $\pm$; two symbols of the same kind are multiplied together by multiplying their corresponding terms, the product being thus a symbol of the same kind; in particular, the product of a symbol by itself, or square of a symbol, is a line (or column as the case may be) of +'s: and the symbol itself is thus a square root of a line (or column) of +'s. Thus n being $=5$, we say that the latin letters denote roots of $+++++$ and the greek letters roots of

$$\begin{matrix}+\\+\\+\\+\\+.\end{matrix}$$

The roots a, b, c, d, e will be *independent* if no one of them is equal to the product of all or any of the others; and, this being so, the 32 roots are the terms of

$$(1+a)(1+b)(1+c)(1+d)(1+e);$$

it follows that, for any other system of independent roots a', b', c', d', e', we have

$$(1+a')(1+b')(1+c')(1+d')(1+e')=(1+a)(1+b)(1+c)(1+d)(1+e):$$

and conversely if either system be independent and this equation is satisfied, then the other system is also independent.

In particular a, b, c, d, e being independent, then a, b, c, bd, e (viz. any term d is replaced by its product by some other term b) is also independent; and by a similar transformation on the new series a, b, c, bd, e, and so on in succession we can pass from a given independent system a, b, c, d, e to *any other independent system whatever.*

A similar but more general theorem is the following: let a, b, c, d, e be independent, and l be equal to the product of all or any of these roots, but so that as regards, suppose (b, c, e), the number of these factors contained in l is even (or may be $=0$), e.g. l is $=a$, or abc, &c., but it is not $=ab$, or bce, &c.

Then a, lb, lc, d, le is an independent system: to show this we must show that

$$\overline{1+a}\,\overline{1+b}\,\overline{1+c}\,\overline{1+d}\,\overline{1+e} = \overline{1+a}\,\overline{1+lb}\,\overline{1+lc}\,\overline{1+d}\,\overline{1+le},$$

that is,

$$\overline{1+a}\,\overline{1+d}\,(\overline{1+lb}\,\overline{1+lc}\,\overline{1+le} - \overline{1+b}\,\overline{1+c}\,\overline{1+e}) = 0,$$

that is,

$$\overline{1+a}\,\overline{1+d}\,(\overline{l-1}\,\overline{b+c+e} + \overline{l^2-1}\,\overline{bc+be+ce} + \overline{l^3-1}\,bce) = 0,$$

or, since $l^2=1$, $l^3=l$, this is

$$(1+a)(1+d)(l-1)(b+c+e+bce) = 0,$$

which is easily verified under the assumed conditions as to l, e.g. $l=abc$,

$$(1+a)\,l = abc + a^2bc = abc + bc = (1+a)\,bc,$$

$$(1+a)(l-1) = (1+a)(bc-1),$$

and the equation is

$$(1+a)(1+d)(bc-1)(b+c+e+bce) = 0;$$

and we in fact have

$$bc\,(b+c+e+bce) = c+b+ebc+e;$$

that is,

$$(bc-1)(b+c+e+bce) = 0.$$

The proof is obviously quite general.

All that precedes applies also to the columns.

Now consider a square of 5×5 signs $\pm$; I say that, if this is *independent as to its lines*, it will be also *independent as to its columns.*

To prove this consider any particular square, say

	α	β	γ	δ	ϵ
a					
b	+	−	−	+	−
c					
d					
e					

independent as to its lines, and also independent as to its columns: I derive from this the square

	α	$\theta\beta$	$\theta\gamma$	δ	$\theta\epsilon$
a					
b					
c					
bd					
e					

viz. in the new square the line d is replaced by bd, the designation of the columns will be presently explained. This new square is, by what precedes, independent as to its lines; we have to show that it is also independent as to its columns.

As regards the columns, any column is either unchanged or it is changed in its fourth place only, according as the sign in b is for that column $+$ or $-$; that is, if we write $\theta = +$, the columns of the new square are (as above written down) α, $\theta\beta$, $\theta\gamma$, δ, $\theta\epsilon$;

$$\begin{matrix} + \\ + \\ - \\ + \end{matrix}$$

and θ is a product of all or some of the original columns α, β, γ, δ, ϵ: but as regards β, γ, ϵ it contains an even number (or it may be 0) of these factors; for otherwise the sign in the second line of θ instead of being $+$ would be $-$. But these are the very conditions that show that the columns α, $\theta\beta$, $\theta\gamma$, δ, $\theta\epsilon$ are independent.

Hence starting from the square

−	+	+	+	+
+	−	+	+	+
+	+	−	+	+
+	+	+	−	+
+	+	+	+	−

which obviously is independent as to its lines and also as to its columns; and transforming as above any number of times in succession, we obtain ultimately a square which has for its lines *any system* whatever of independent roots, and by what precedes each of the new squares is also independent as to its columns; that is, every square independent as to its lines is also independent as to its columns. Q.E.D.

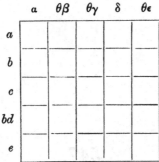

547.

ON THE REPRESENTATION OF A SPHERICAL OR OTHER SURFACE ON A PLANE: A SMITH'S PRIZE DISSERTATION.

[From the *Messenger of Mathematics*, vol. II. (1873), pp. 36, 37.]

IN the Smith's Prize Examination for 1871 I set as the subject for a dissertation:

The representation of a spherical or other surface on a plane.

I give the following as a specimen of the sort of answer required: an answer which, without so much as noticing that projection (in its restricted sense) is only one kind of representation, goes into the details of the constructions for the different projections of the sphere, and even into the demonstrations of these constructions, errs quite as much by excess as by defect, and is worth very little indeed.

The question is understood to refer to Chartography, viz. the kind of representation is taken to be such as that of a hemisphere or other portion of the earth's surface in a map.

An implied condition is that each point of the surface (viz. of the portion thereof comprised in the map) shall be represented by a single point on the map; and conversely, that each point on the map shall represent a single point on the surface. And further, any closed curve on the surface must be represented by a closed curve on the map, and the points within the one by the points within the other. If for shortness the term element is used to denote an infinitesimal area included within a closed curve, we may say that each element of the surface must be represented by an element of the map; and conversely, each element of the map must represent an element of the surface.

A map would be perfect if each element of the surface and the corresponding element of the map were of the same form, and were in a constant ratio as to magnitude; say if it were free from the defects of "distortion" and "inequality" (of scale); the condition as to form, or freedom from distortion, may be otherwise expressed by saying that any two contiguous elements of length on the surface and the corresponding two contiguous elements of length on the map must meet at the same angle (this at once appears by taking the two elements of area to be each of them a triangle). But for a spherical or other non-developable surface, it is not possible to construct a map free from the two defects.

An obvious and usual kind of representation is that by projection: viz. taking any fixed point and plane, the line joining any point P of the surface with the fixed point meets the fixed plane in a point P' which is taken to be the representation of the point P on the surface.

When the surface is a sphere the projection is called orthographic, gnomonic or stereographic according to the positions of the fixed point and plane: the last kind is here alone considered; viz. in the stereographic projection the fixed point is on the surface of the sphere, and the fixed plane is parallel to the tangent plane at that point, and is usually and conveniently taken to pass through the centre of the sphere.

The stereographic projection is one of those which is free from the defect of distortion; it is consequently, and that in a considerable degree, subject to the defect of inequality. It possesses in a high degree the important quality of facility of construction, viz. any great or small circle on the sphere is represented by a circle in the map; and from the general property of the equality of corresponding angles, or otherwise, there arise easy rules for the construction of such circles.

The so-called Mercator's projection is an instance of a representation which is not in the above restricted sense a projection; and which is free from the defect of distortion: viz. the (equidistant) meridians are here represented by a system of (equidistant) parallel lines; and the parallels of latitude by a set of lines at right angles thereto: the distance between consecutive parallels in the map being taken in such wise as is required to obtain freedom from distortion; for this purpose the increments of latitude and longitude must have in the map the same ratio that they have on the sphere, and since in the map the length of a degree of longitude (instead of decreasing with the latitude) remains constant, the lengths of the successive degrees of latitude in the map must increase with the latitude: the scale of the representation thus increases with the latitude, and would for the latitude $\pm 90°$ become infinite.

There is a simple representation of a hemisphere, due to M. Babinet, in which the defect of inequality is avoided, viz. the meridians are represented by ellipses having their major axes coincident with the diameter through the poles and dividing the equator into equal distances, and the parallels by straight lines parallel to the equator.

548.

ON LISTING'S THEOREM.

[From the *Messenger of Mathematics*, vol. II. (1873), pp. 81—89.]

LISTING'S theorem, (established in his Memoir*, *Die Census räumlicher Gestalten*), is a generalisation of Euler's theorem $S+F=E+2$, which connects the number of summits, faces, and edges in a polyhedron; viz. in Listing's theorem we have for a figure of any sort whatever

$$A-B+C-D-(p-1)=0,$$

or, what is the same thing,

$$A+C=B+D+(p-1),$$

where

$$A=a,$$
$$B=b-\kappa',$$
$$C=c-\kappa''+\pi,$$
$$D=d-\kappa''',$$

in which theorem a relates to the points; b, κ' relate to the lines; c, κ'', π to the surfaces; d, κ''' to the spaces; and p relates to the detached parts of the figure, as will be explained.

a is the number of points; there is no question of multiplicity, but a point is always a single point. A point is either detached or situate on a line or surface.

b is the number of lines (straight or curved). A line is always finite, and if not reentrant there must be at each extremity a point: no attention is paid to cusps, inflexions, &c., and if the line cut itself there must be at each intersection a point;

* *Gött. Abh.* t. x. (1862).

and in general a point placed on a line constitutes a termination or boundary of the line. Thus a line is either an oval (that is, a non-intersecting closed curve of any form whatever), a punctate oval (oval with a single point upon it), or a biterminal (line terminated by two distinct points). For instance, a figure of eight is taken to be two punctate ovals; an oval, placing upon it two points, is thereby changed into two biterminals.

κ'. The definition, analogous to the subsequent definitions of κ'' and κ''', would be that κ' is the sum, for all the lines, of the number of circuits for each line; but inasmuch as for an oval the number of circuits is $=1$, and for any other line (punctate oval, or biterminal) it is $=0$, κ' is in fact the number of ovals.

c is the number of surfaces. A surface is always finite, and if not reentrant there must be at every termination thereof a line: no attention is paid to cuspidal lines, &c., and if the surface cut itself there must be at each intersection a point or a line; and in general a point or a line placed on a surface constitutes a termination or boundary thereof. It may be added that if a line intersects a surface there must be at the intersection a point, constituting a termination or boundary as well of the line as of the surface. Thus a surface is either an ovoid (simple closed surface, such as the sphere or the ellipsoid), a ring (surface such as the torus or anchor-ring), or other more complicated form of reentrant surface; or else it is a surface in part bounded by a point or points, line or lines. We may in particular consider a blocked surface having upon it one or more blocks: where by a block is meant a point, line, or connected superficial figure composed of points and lines in any manner whatever, the superficial area (if any) included within the block being disregarded as not belonging to the surface, or being, if we please, cut out from the surface. Thus an ovoid having upon it a point, and a segment or incomplete ovoid bounded by an oval, are each of them to be regarded as a one-blocked ovoid; the boundary being in the first case the point, and in the second case the oval; and so in general the blocked surface is bounded by the boundary or boundaries of the block or blocks. It will be understood from what precedes, and it is almost needless to mention, that for any surface we can pass along the surface from each point to each point thereof; any line which would prevent this would divide the surface into two or more distinct surfaces.

κ'' is the sum, for the several surfaces, of the number of circuits on each surface. The word circuit here signifies a path on the surface from any point to itself: all circuits which can by continuous variation be made to coincide being regarded as identical; and the evanescible circuit reducible to the point itself being throughout disregarded. Moreover, we count only the simple circuits, disregarding circuits which can be obtained by any repetition or combination of these. Thus for an ovoid, or for a one-blocked ovoid, there is only the evanescible circuit, that is, no circuit to be counted; but for a two-blocked ovoid there is besides one circuit, or we count this as one; and so for a n-blocked ovoid we count $n-1$ circuits. For a ring it is easy to see that (besides the evanescible circuit) there are, and we accordingly count, two circuits; and so in other cases.

π. It might be possible to find an analogous definition, but the most simple one is that π denotes the number of ovoids (unblocked ovoids) or other surfaces not bounded by any point or line.

d is the number of spaces, reckoning as one of them infinite space.

κ''' is the sum, for the several spaces, of the number of circuits in each space: the word circuit here signifying a path in the space from a point to itself; all circuits which can by continuous variation be made to coincide being considered as identical, and the evanescible circuit reducible to the point itself being throughout disregarded. Moreover, we count only the simple circuits, disregarding circuits which can be obtained by a repetition or combination of these. Thus for infinite space, or for the space within an ovoid, there is only the evanescible circuit, or there is no circuit to be counted; and the same is the case if within such space we have any number of ovoidal blocks (the term will, I think, be understood without explanation); but if within the space we have an oval, ring, or other ring-block of any kind whatever, then there is (besides the evanescible circuit) a circuit interlacing the ring-block, and we count one circuit; and so if there are n ring-blocks, either separate or interlacing each other in any manner, then there are, and we accordingly count, n circuits. So for the space inside a ring there is (besides the evanescible circuit), and we count, one circuit; and the case is the same if we have within the ring any number of ovoidal blocks whatever; but if there is within the ring an oval ring or other ring-block, then there is one new circuit, and we count in all (for the space in question) two circuits.

p is the number of detached parts of the figure; or, say the number of detached aggregations of points, lines, and surfaces. Observe, that rings interlacing each other in any manner (but not intersecting) are considered as detached; so also two closed surfaces, one within the other, are considered as detached. The figure may be infinite space alone; we have then $p = 0$.

The examples which follow will further illustrate the meaning of the terms and nature of the theorem; and will also indicate in what manner a general demonstration of the theorem might be arrived at.

1. Infinite space.

$$\begin{array}{llll} a = 0, & & & A = 0, \\ b = 0, & \kappa' = 0, & & \qquad B = \ \ 0 \\ c = 0, & \kappa'' = 0, & \pi = 0, & C = 0, \\ d = 1, & \kappa''' = 0, & & \qquad D = \ \ 1 \\ p = 0, & & & \qquad p - 1 = -1 \\ \hline & & & 0 \quad = \ 0. \end{array}$$

2. Spherical surface.

$$\begin{array}{lllll}
a=0, & & & A=0, & \\
b=0, & \kappa'=0, & & & B=0 \\
c=1, & \kappa''=0, & \pi=1, & C=2, & \\
d=2, & \kappa'''=0, & & & D=2 \\
p=1, & & & & p-1=0 \\
\hline
 & & & 2 & =2:
\end{array}$$

viz. the effect is to increase C by 2 and D and $p-1$ each by 1.

3. Spherical surface, with point upon it.

$$\begin{array}{lllll}
a=1, & & & A=1, & \\
b=0, & \kappa'=0, & & & B=0 \\
c=1, & \kappa''=0, & \pi=0, & C=1, & \\
d=2, & \kappa'''=0, & & & D=2 \\
p=1, & & & & p-1=0 \\
\hline
 & & & 2 & =2:
\end{array}$$

viz. the effect is to increase a and diminish π each by 1; that is, A is increased and C diminished each by 1.

4. Spherical surface with two points.

$$\begin{array}{lllll}
a=2, & & & A=2, & \\
b=0, & \kappa=0, & & & B=0 \\
c=1, & \kappa''=1, & \pi=0, & C=0, & \\
d=2, & \kappa'''=0, & & & D=2 \\
p=1, & & & & p-1=0 \\
\hline
 & & & 2 & =2:
\end{array}$$

viz. the second point increases a and κ'' each by 1, that is, it increases A and diminishes C each by 1.

And for each new point on the spherical surface there is this same effect; so that we have, for the next case:

5. Spherical surface with n points ($n \geqq 2$).

$$\begin{array}{lllll}
a=n, & & & A=\ \ n, & \\
b=0, & \kappa=0, & & & B=0 \\
c=1, & \kappa''=n-1, & \pi=0, & C=2-n, & \\
d=2, & \kappa'''=0, & & & D=2 \\
p=1, & & & & p-1=0 \\
\hline
 & & & 2 & =2.
\end{array}$$

Imagine that besides the n points there is an aperture (bounded by a closed curve); the case is:

6. Spherical surface with n points ($n \overline{>} 2$) and aperture.

$$\begin{array}{llll}
a=n, & & & A=\ \ n, \\
b=1, & \kappa\ \ =1\ \ , & & \qquad B=0 \\
c=1, & \kappa''=n\ \ , & \pi=0, & C=1-n, \\
d=1, & \kappa'''=0\ \ , & & \qquad D=1 \\
p=1, & & & \qquad p-1=0 \\
\hline
 & & & 1 \qquad =1:
\end{array}$$

viz. b and κ are each increased by 1, and therefore B is unaltered; κ'' is increased, and therefore C is diminished, by 1; but d is diminished, and therefore D also diminished, by 1.

7. Spherical surface with n points ($n \overline{>} 2$) and two apertures.

$$\begin{array}{llll}
a=n, & & & A=\ \ n, \\
b=2, & \kappa\ \ =2\ \ , & & \qquad B=0 \\
c=1, & \kappa''=n+1, & \pi=0, & C=\ \ -n, \\
d=1, & \kappa'''=1\ \ , & & \qquad D=0 \\
p=1, & & & \qquad p-1=0 \\
\hline
 & & & 0 \qquad =0:
\end{array}$$

viz. b and κ are each increased by 1, and thus B is still unaltered; κ'' is increased, and therefore C diminished, by 1; κ''' is increased, and therefore D diminished, by 1, and each new aperture produces the like effect. Thus we have:

8. Spherical surface with m apertures ($m \overline{>} 2$).

$$\begin{array}{llll}
a=0, & & & A=0, \\
b=m, & \kappa'\ \ =m\ \ , & & \qquad B=0 \\
c=1, & \kappa''=m-1, & \pi=0, & C=2-m, \\
d=1, & \kappa'''=m-1, & & \qquad D=2-m \\
p=1, & & & \qquad p-1=0 \\
\hline
 & & & 2-m \qquad =2-m;
\end{array}$$

where, comparing with case 5, we see the different effects of a point and an aperture.

9. Spherical surface with m apertures ($m \overline{>} 2$) and a point or points on each or any of the bounding curves of the aperture.

If on the bounding curve of any aperture we place a point, this increases a, and therefore A, by 1; the bounding curve is no longer a simple closed curve, and we thus also have κ diminished, and therefore B increased by 1; and the balance still holds.

Placing on the same bounding curve a second point, a, and therefore A, is increased by 1; but the bounding curve is converted into two distinct curves; that is, b, and therefore B, is increased by 1; and the balance still holds. And the like for each new point on the same bounding curve.

10. Spherical surface with n points connected in any manner by lines.

Reverting to the cases 4 and 5, by joining any two points by a line, we increase b, and therefore B, by 1; but as regards κ'' the two united points take effect as a single point; that is, κ'' is diminished, and therefore C increased, by 1; the balance is therefore undisturbed.

The case is the same for each new line, if only we do not thereby produce on the surface a closed polygon, or partition an existing closed polygon; in each of these cases we still increase b, and therefore B, by 1; and instead of diminishing κ'', we increase c, by 1, and therefore still increase C by 1; and the balance continues to subsist.

By continuing to join the several points we at last arrive at a spherical surface partitioned into polygons in any manner whatever; or, what is the same thing, we have:

11. Closed polyhedral surface. Here, if S is the number of summits, F the number of faces, E the number of edges; then

$$\begin{array}{lllll} a=S, & & & A=S, & \\ b=E, & \kappa\ =0, & & & B=E \\ c=F, & \kappa''=0, & \pi=0, & C=\quad F, & \\ d=2, & \kappa'''=0, & & & D=\quad 2 \\ p=1, & & & & p-1=\quad 0 \\ \hline & & & S+F & =E+2, \end{array}$$

so that we have Euler's theorem. Observe that this theorem (Euler's) does not apply to annular polyhedral surfaces, or to polyhedral shells. For instance, consider a shell, the exterior and interior surfaces of which are each of them a closed polyhedral surface; $S=S'+S''$, $F=F'+F''$, $E=E'+E''$, where $S'+F'=E'+2$, $S''+F''=E''+2$, and therefore $S+F=E+4$. Listing's theorem, of course, applies, viz. we have

12.
$$\begin{array}{lll} a=S'+S'', & A=S'+S'', & \\ b=E'+E'', & B= & E'+E'' \\ c=F'+F'', & C=F'+F'', & \\ d=3, & D= & 3 \\ p=2, & p-1= & 1 \\ \hline & S+F & =E'+E''+4. \end{array}$$

As another group of examples, consider a plane rectangle, for instance, a sheet of paper bounded by its four edges; here

13.
$$\begin{array}{llll} a=4, & & & A=4, \\ b=4, & \kappa'=0, & & \quad B=4 \\ c=1, & \kappa''=0, & \pi=0, & C=1, \\ d=1, & \kappa'''=0, & & \quad D=1 \\ p=1, & & & \quad p-1=0 \\ \hline & & & 5 \quad =5. \end{array}$$

Let the paper be formed into a tube by uniting two opposite sides, the suture not being obliterated, but continuing as a line drawn lengthwise from one extremity of the tube to the other: here

14.
$$\begin{array}{llll} a=2, & & & A=2, \\ b=3, & \kappa'=0, & & \quad B=3 \\ c=1, & \kappa''=0, & \pi=0, & C=1, \\ d=1, & \kappa'''=1, & & \quad D=0 \\ p=1, & & & \quad p-1=0 \\ \hline & & & 3 \quad =3. \end{array}$$

Let the suture be obliterated, so that we have simply a tube open at each end; here

15.
$$\begin{array}{llll} a=0, & & & A=0, \\ b=2, & \kappa'=2, & & \quad B=0 \\ c=1, & \kappa''=1, & \pi=0, & C=0, \\ d=1, & \kappa'''=1, & & \quad D=0 \\ p=1, & & & \quad p-1=0 \\ \hline & & & 0 \quad =0. \end{array}$$

Let the tube be formed into an annulus by bending it round and joining the two extremities, the suture not being obliterated, but continuing as a closed curve round the tube; here

16.
$$\begin{array}{llll} a=0, & & & A=0, \\ b=1, & \kappa'=1, & & \quad B=0 \\ c=1, & \kappa''=1, & \pi=0, & C=0, \\ d=2, & \kappa'''=2, & & \quad D=0 \\ p=1, & & & \quad p-1=0 \\ \hline & & & 0 \quad =0. \end{array}$$

Let the suture be obliterated, so that we have simply a tubular annulus; here

17.
$$\begin{array}{llll} a=0, & & & A=0, \\ b=0, & \kappa'=0, & & \qquad B=0 \\ c=1, & \kappa''=2, & \pi=1, & C=0, \\ d=2, & \kappa'''=2, & & \qquad D=0 \\ p=1, & & & \qquad p-1=0 \\ \hline & & & 0 \qquad =0. \end{array}$$

We may compare herewith the case of a simple annulus or closed curve.

18.
$$\begin{array}{llll} a=0, & & & A=0, \\ b=1, & \kappa'=1, & & \qquad B=0 \\ c=0, & \kappa''=0, & \pi=0, & C=0, \\ d=1, & \kappa'''=1, & & \qquad D=0 \\ p=1, & & & \qquad p-1=0 \\ \hline & & & 0 \qquad =0. \end{array}$$

Add to such an annulus, for instance, three radii meeting in the centre; then

19.
$$\begin{array}{llll} a=4, & & & A=4, \\ b=6, & \kappa'=0, & & \qquad B=\ \ 6 \\ c=0, & \kappa''=0, & \pi=0, & C=0, \\ d=1, & \kappa'''=3, & & \qquad D=-2 \\ p=1, & & & \qquad p-1=\ \ 0 \\ \hline & & & 4 \qquad =\ \ 4. \end{array}$$

Let the last-mentioned figure become tubular, all sutures being obliterated; then

20.
$$\begin{array}{llll} a=0, & & & A=\ \ 0, \\ b=0, & \kappa'=0, & & \qquad B=\ \ 0 \\ c=1, & \kappa''=6, & \pi=1, & C=-4, \\ d=2, & \kappa'''=6, & & \qquad D=-4 \\ p=1, & & & \qquad p-1=\ \ 0 \\ \hline & & & -4 \qquad =-4. \end{array}$$

And so if instead of the tubular figure, annulus with three radii, we had a tubular figure, annulus with diameter, then

21.
$$\begin{array}{llll} a=0, & & & A=\ \ 0, \\ b=0, & \kappa'=0, & & \qquad B=\ \ 0 \\ c=1, & \kappa''=4, & \pi=1, & C=-2, \\ d=2, & \kappa'''=4, & & \qquad D=-2 \\ p=1, & & & \qquad p-1=\ \ 0 \\ \hline & & & -2 \qquad =-2; \end{array}$$

and the like in other cases.

549.

NOTE ON THE MAXIMA OF CERTAIN FACTORIAL FUNCTIONS.

[From the *Messenger of Mathematics*, vol. II. (1873), pp. 129, 130.]

I CONSIDER the functions

$$y_1 = x(x-1),$$
$$y_2 = x(x-\tfrac{1}{2})(x-1),$$
$$y_3 = x(x-\tfrac{1}{3})(x-\tfrac{2}{3})(x-1),$$
$$\vdots$$
$$y_n = x\left(x-\frac{1}{n}\right)\left(x-\frac{2}{n}\right)\dots\left(x-\frac{n-1}{n}\right)(x-1).$$

Attending only to the absolute values, disregarding the signs, y_n has n maxima, viz. if n be odd, $=2p+1$ suppose, these are

$$Y_1,\ Y_2,\dots,\ Y_p,\ Y_{p+1},\ Y_p,\dots Y_1,$$

where Y_{p+1} corresponds to the value $x=\frac{1}{2}$, and $Y_1,\ Y_2,\dots,\ Y_p$ to values of x between

$$0 \text{ and } \frac{1}{2p+1},\ \frac{1}{2p+1} \text{ and } \frac{2}{2p+1},\dots,\ \frac{p-1}{2p+1} \text{ and } \frac{p}{2p+1}.$$

But if n be even, $=2p$ suppose, then the maxima are

$$Y_1,\ Y_2,\dots,\ Y_p,\ Y_p,\dots,\ Y_1,$$

where $Y_1,\ Y_2,\dots,\ Y_p$ correspond to values of x between

$$0 \text{ and } \frac{1}{2p},\ \frac{1}{2p} \text{ and } \frac{2}{2p},\dots,\ \frac{p-1}{2p} \text{ and } \tfrac{1}{2}.$$

In every case the maxima decrease from Y_1 which is the greatest, to Y_p or Y_{p+1} which is the least; in particular, $n = 2p+1$, then

$$Y_{p+1} = \tfrac{1}{2}\left(\tfrac{1}{2} - \frac{1}{2p+1}\right) \dots \left(\tfrac{1}{2} - 1\right)$$

$$= \left(\tfrac{1}{2}\,\frac{2p-1}{2\,.\,2p+1} \dots \frac{1}{2\,.\,2p+1}\right)^2$$

$$= \frac{\{1\,.\,3 \dots (2p-1)\}^2}{2^{2p+2}\,.\,(2p+1)^{2p}} = \tfrac{1}{4}\,\frac{\left(\tfrac{1}{2}\,.\,\tfrac{3}{2} \dots p-\tfrac{1}{2}\right)^2}{(2p+1)^{2p}},$$

which is

$$= \tfrac{1}{4}\,\frac{\{\Gamma(p+\tfrac{1}{2}) \div \Gamma\tfrac{1}{2}\}^2}{(2p+1)^{2p}} = \frac{\Gamma^2(p+\tfrac{1}{2})}{4\pi(2p+1)^{2p}}.$$

Suppose p is large; then, as for large values of x,

$$\Gamma x = \sqrt{(2\pi)}\, x^{x-\frac{1}{2}} e^{-x},$$

we have

$$\Gamma(p+\tfrac{1}{2}) = \sqrt{(2\pi)}\,(p+\tfrac{1}{2})^p\, e^{-p-\frac{1}{2}}$$

$$= \sqrt{(2\pi)}\, p^p e^{p\log\left(1+\frac{1}{2p}\right)} e^{-p-\frac{1}{2}} = \sqrt{(2\pi)}\, p^p\, e^{-p},$$

$$(2p+1)^{2p} = (2p)^{2p}\,.\, e^{2p\log\left(1+\frac{1}{2p}\right)} = 2^{2p} p^{2p} e,$$

and so

$$Y_{p+1} = \frac{2\pi p^{2p} e^{-2p}}{4\pi 2^{2p} p^{2p} e} = \frac{p^2 e^{-2p-1}}{2^{2p-1}} = p^2 \left(\frac{1}{2e}\right)^{2p+1}.$$

Also Y_1 corresponds *approximately* to

$$x = \tfrac{1}{2}\,\frac{1}{2p+1} = \frac{1}{2n},$$

$$Y_1 = \frac{1}{2n}\cdot\frac{1}{2n}\cdot\frac{3}{2n}\dots\frac{2n-1}{2n} = \frac{1}{n^{n+1}}\,\tfrac{1}{2}\,.\,\tfrac{1}{2}\,.\,\tfrac{3}{2}\dots\left(n-\tfrac{1}{2}\right)$$

$$= \frac{1}{n^{n+1}}\,\tfrac{1}{2}\,\frac{\Gamma(n+\tfrac{1}{2})}{\Gamma\tfrac{1}{2}} = \frac{1}{2\,(2p+1)^{2p+1}}\,\frac{\Gamma(2p+\tfrac{3}{2})}{\sqrt{(\pi)}}.$$

Now

$$\Gamma(2p+\tfrac{3}{2}) = \sqrt{(2\pi)}\,(2p+\tfrac{3}{2})\, e^{-2p-\frac{3}{2}} = \sqrt{(2\pi)}\,(2p)^{2p+2} e^{(2p+\frac{3}{2})\log\left(1+\frac{3}{4p}\right)}\, e^{-2p-\frac{3}{2}}$$

$$= \sqrt{(2\pi)}\, 2^{2p+2} p^{2p+2} e^{-2p},$$

and

$$(2p+1)^{2p+1} = (2p)^{2p+1}\, e^{(2p+1)\log\left(1+\frac{1}{2p}\right)} = (2p)^{2p+1} e;$$

so that

$$Y_1 = \frac{1}{2^{2p+2} p^{2p+1} e}\cdot\frac{\sqrt{(2\pi)}\,.\,2^{2p+2}\,.\,p^{2p+2} e^{-2p}}{\sqrt{(\pi)}}$$

$$= \frac{p\sqrt{(2)}}{e^{2p+1}},$$

so that, p being large, Y_1 is far larger than Y_{p+1}.

550.

PROBLEM AND HYPOTHETICAL THEOREMS IN REGARD TO TWO QUADRIC SURFACES.

[From the *Messenger of Mathematics*, vol. II. (1873), p. 137.]

Two conics may be circum-and-inscribable to an n-gon; viz. the conics may be such that there exists a singly infinite series of n-gons each inscribed in the first and circumscribed about the second of the conics. In particular they may be circum-and-inscribable to a triangle.

The following problem arises:

Consider two given quadric surfaces and a given line S; to find the planes through S, which cut the surfaces in two conics circum-and-inscribable to a triangle (it is presumed there are two or more such planes).

Let the surfaces be Θ, Θ', and let the line S a tangent to Θ' meet Θ in the points A, B; if through S we draw two planes as above, then in the first plane the tangents from A, B to the section of Θ' will meet in a point C of Θ; and in the second plane the tangents from A, B to the section of Θ' will meet in a point D of Θ. The points C, D being thus determined the lines AB, AC, BC, AD, BD all touch the surface Θ', and it is presumed that the surfaces Θ, Θ' may be such that CD also touches the surface Θ'; viz. in this case we have a tetrahedron $ABCD$, the summits of which lie in the surface Θ, and the edges touch the surface Θ'; and not only so, but it is further presumed that the surfaces may be such that starting from *any* point A of Θ and using either *any* tangent or a properly selected tangent AB of Θ', it shall be possible to complete the figure as above; or, what is the same thing, the surfaces may be such that there exists a *doubly* or a *triply* infinite series of tetrahedra, the summits of each lying in Θ and its edges touching Θ'. It is also presumed that the faces of the tetrahedra all touch one and the same quadric surface Θ''.

551.

TWO SMITH'S PRIZE DISSERTATIONS.

[From the *Messenger of Mathematics*, vol. II. (1873), pp. 145—149.]

WRITE dissertations on the following subjects:

1. The theory of interpolation, with a determination of the limits of error in the value of a function obtained by interpolation.

2. Determinants.

1. The general problem is to find y a function of x having given values for given values of x. The problem thus stated is of course indeterminate; in practice, we assume a certain form for the function y, the coefficients of which form are determined by the given conditions, viz. either y is known to be of the form in question, the actual value being then determined as above, or it is assumed that y is approximately equal to a function of the form in question, and the value is then approximately determined in such wise that, for the given values of x, the function y shall have its given values.

The ordinary case is when we have the values of y corresponding to n given values of x, and y is taken to be a function of the form $A + Bx + \ldots + Kx^{n-1}$.

Suppose to fix the ideas $n = 4$, and that y_1, y_2, y_3, y_4 are the values of y corresponding to the values a, b, c, d of x. We may at once write down the expression

$$\begin{aligned} y = &\frac{(x-b)(x-c)(x-d)}{(a-b)(a-c)(a-d)} y_1 \\ &+ \frac{(x-c)(x-d)(x-a)}{(b-c)(b-d)(b-a)} y_2 \\ &+ \frac{(x-d)(x-a)(x-b)}{(c-d)(c-a)(c-b)} y_3 \\ &+ \frac{(x-a)(x-b)(x-c)}{(d-a)(d-b)(d-c)} y_4, \end{aligned}$$

for clearly this is of the form in question $A+Bx+Cx^2+Dx^3$, and y becomes $=y_1$ for $x=a$; $=y_2$ for $x=b$, &c. And the like for any value of n. This is known as Lagrange's interpolation formula.

The given values of x may be equidistant, say they are 0, 1, 2,..., $n-1$, and the corresponding values of y are $y_0, y_1, \ldots, y_{n-1}$; then writing down the expression

$$y_x = y_0 + \frac{x}{1}\Delta y_0 + \frac{x \cdot \overline{x-1}}{1 \cdot 2}\Delta^2 y_0 + \ldots + \frac{x \cdot \overline{x-1} \ldots \overline{x-n+2}}{1 \cdot 2 \ldots \overline{n-1}}\Delta^{n-1} y_0,$$

where, as usual,

$$\Delta y_0 = y_1 - y_0, \quad \Delta^2 y_0 = y_2 - 2y_1 + y_0, \text{ \&c.};$$

then for $x=0$, 1, 2, &c. the values of y are

$$\begin{aligned} &y_0, \\ &y_0 + \Delta y_0, \quad = y_1, \\ &y_0 + 2\Delta y_0 + \Delta^2 y_0, \quad = y_2, \\ &\text{\&c.}, \end{aligned}$$

or the required conditions are satisfied.

As regards the determination of the limits of error, taking a particular case $n=4$, suppose that we have the values y_0, y_1, y_2, y_3 of y corresponding to the values 0, 1, 2, 3 of x, and that the true value of y is known to be

$$= A + Bx + Cx^2 + Dx^3 + Kx^4,$$

where K is a function of x, which for any value of x within the given values (i.e. from $x=0$ to $x=3$) is known to be at least $=P$ and at most $=Q$, i.e., $K>P<Q$, where to fix the ideas P and Q are each positive, Q being the greater. Here calculating the interpolation value of $y-Kx^4$ (the last term Kx^4 by Lagrange's formula), we have

$$\begin{aligned} y = y_0 + \frac{x}{1}\Delta y_0 + \frac{x \cdot \overline{x-1}}{1 \cdot 2}\Delta^2 y_0 + \frac{x \cdot \overline{x-1}\,\overline{x-2}}{1 \cdot 2 \cdot 3}\Delta^3 y_0 \\ + Kx^4 \qquad\qquad\qquad\qquad \\ - \tfrac{1}{2} K_1 x(x-2)(x-3) \\ + 8\, K_2 x(x-1)(x-3) \\ - \tfrac{27}{2} K_3 x(x-1)(x-2), \end{aligned}$$

viz. this is the true value of y. Hence using the approximate formula as given by the first line, the last four lines give the error, viz. this is

$$= Kx^4 + K_3 \tfrac{81}{2} x^2 + K_2(8x^3+24x) + K_1 \tfrac{5}{2} x^2 - K_3(\tfrac{27}{2}x^3 + 27x) + K_2(32x^2) - K_1(\tfrac{1}{2}x^3 + 3x).$$

But K_1, K_2, K_3 being each $>P$ and $<Q$, this is

$$\begin{aligned} &> P(x^4 + \ \ 8x^3 + 43x^2 + 24x) \\ &- Q(\qquad 14x^3 + 32x^2 + 30x), \end{aligned}$$

and it is

$$\begin{aligned} &< Q\,(x^4 + \ \ 8\,x^3 + 43x^2 + 24x) \\ &- P(\qquad 14x^3 + 33x^2 + 30x), \end{aligned}$$

the difference of these limits being

$$= (Q - P)\,(x^4 + 22x^3 + 75x^2 + 54x).$$

2. A determinant is a function of n^2 letters; viz. arranging these in the form of a square, the determinant

$$\begin{vmatrix} a_1, & b_1, & c_1 \\ a_2, & b_2, & c_2 \\ a_3, & b_3, & c_3 \end{vmatrix}$$

is a function linear in regard to each of the n^2 letters, and such that interchanging any two entire lines, or any two entire columns, the sign of the determinant is reversed, its absolute value being unaltered.

The above definition leads to a rule for calculating the actual value of the determinant, which rule may be taken as a definition, viz. the determinant is the sum of $1 . 2 . 3 \ldots n$ terms obtained as follows: starting from the term

$$+ a_1 b_2 c_3 \ldots,$$

we permute in every possible way the suffixes 1, 2, 3, ..., and give to the term a sign, $\pm$, which is that compounded of as many $-$ signs as there are cases in which an inferior number succeeds a superior number. Or, what is the same thing, any arrangement may be obtained by a succession of interchanges of two letters; and then taking for each interchange the sign $-$, we obtain the sign $\pm$ of the term in question. The positive and the negative terms are each of them $\tfrac{1}{2}(1 . 2 . 3 \ldots n)$ in number.

To show the connexion of the two definitions, it is sufficient to observe that in the second definition, attending for instance only to the first and second columns, to any terms Ma_1b_2, Na_1b_3, &c., there always correspond other terms $-Ma_2b_1$, $-Na_3b_1$, &c., so that taking the pairs together, these are $M(a_1b_2 - a_2b_1)$, $N(a_1b_3 - a_3b_1)$, &c., terms which change their sign, but remain unaltered as to their absolute values by the interchange of the first and second columns.

Among the fundamental properties of determinants are as follows:

The properties are the same as regards lines and columns.

A determinant vanishes if any line vanishes (that is, if each term of the line is $=0$).

A determinant vanishes if two lines are identical.

A determinant is a linear function of its lines.

Whence—

Determinant having a line sA is $= s$ times the determinant having the line A (sA is here used to denote the line each term of which is s times the corresponding term of the line A).

Determinant having a line $A + A'$ = determinant with line A + determinant with line A'.

It follows that, if any line of a determinant is the sum of the other lines, each multiplied by an arbitrary coefficient, or, what is the same thing, if we can with any of the lines, each multiplied by an arbitrary coefficient, compose a line 0, then the determinant is $= 0$.

The same principle leads to a theorem for the product of two determinants of the same order n, viz. it is found that the product is a determinant of the same order n, each term thereof being a sum of the products of the terms of a line of one of the factors into the corresponding terms of a line of the other factor. Starting with this expression of the product, we decompose it into a series of determinants each of which is either $= 0$, or it is a product of a single term of the one factor into the other factor, and the sum of all these products is equal to the product of the two factors.

If we have n quantities $x, y, \dots$ connected by as many linear equations

$$a_1 x + b_1 y + c_1 z + \dots = 0,$$

then the determinant

$$\begin{vmatrix} a_1, & b_1, & c_1, \dots \\ a_2, & b_2, & c_2, \dots \\ a_3, & b_3, & c_3, \dots \\ \vdots & & \end{vmatrix} \text{ is } = 0;$$

and so, if we have n linear equations

$$a_1 x + b_1 y + c_1 z + \dots = u,$$

then each of the quantities $x, y, z, \dots$ is given as the quotient of two determinants, the denominator being in each case

$$\begin{vmatrix} a_1, & b_1, & c_1, \dots \\ a_2, & b_2, & c_2, \dots \\ a_3, & b_3, & c_3, \dots \\ \vdots & & \end{vmatrix},$$

and the numerators being (save as to their signs) that for x

$$\begin{vmatrix} u_1, & b_1, & c_1, \dots \\ u_2, & b_2, & c_2, \dots \\ u_3, & b_3, & c_3, \dots \\ \vdots & & \end{vmatrix},$$

and the like for $y, z, \dots$.

A determinant remains unaltered when the lines and columns are interchanged, the dexter diagonal ($\diagdown$) remaining unaltered.

A determinant

$$\begin{vmatrix} a_1, & b_1, & c_1, & d_1, \dots \\ a_2, & b_2, & c_2, & d_2, \dots \\ a_3, & b_3, & c_3, & d_3, \dots \\ a_4, & b_4, & c_4, & d_4, \dots \\ \vdots & & & \end{vmatrix},$$

is a sum of products of complementary determinants

$$\Sigma \pm \begin{vmatrix} a_1, & b_1 \\ a_2, & b_2 \end{vmatrix} \begin{vmatrix} c_3, & d_3, \dots \\ c_4, & d_4, \dots \\ \vdots & \end{vmatrix},$$

or, say of products of complementary minors.

In particular, it is a sum

$$\Sigma \pm a_1 \begin{vmatrix} b_2, & c_2, & d_2, \dots \\ b_3, & c_3, & d_3, \dots \\ b_4, & c_4, & d_4, \dots \\ \vdots & & \end{vmatrix},$$

of products of first minors into single terms or $(n-1)^{\text{th}}$ minors.

This last theorem affords a convenient rule for the development of a determinant.

552.

ON A DIFFERENTIAL FORMULA CONNECTED WITH THE THEORY OF CONFOCAL CONICS.

[From the *Messenger of Mathematics*, vol. II. (1873), pp. 157, 158.]

THE following transformations present themselves in connexion with the theory of confocal conics.

The coordinates x, y of a point are considered as functions of the parameters h, k where

$$\frac{x^2}{a+h}+\frac{y^2}{b+h}=1,$$

$$\frac{x^2}{a+k}+\frac{y^2}{b+k}=1;$$

and then assuming $\xi=x+iy$, $\eta=x-iy$ $\left(i=\sqrt{(-1)}\right.$ as usual$\left.\right)$, and writing $c=a-b$, we find

$$h=\tfrac{1}{2}(-a-b+\xi\eta)+\tfrac{1}{2}\sqrt{\{(\xi^2-c)(\eta^2-c)\}},$$
$$k=\tfrac{1}{2}(-a-b+\xi\eta)-\tfrac{1}{2}\sqrt{\{(\xi^2-c)(\eta^2-c)\}},$$

whence if

$$H=(a+h)(b+h),\quad K=(a+k)(b+k),$$

we have

$$H=\tfrac{1}{4}\{\xi\sqrt{(\eta^2-c)}+\eta\sqrt{(\xi^2-c)}\}^2,$$
$$K=\tfrac{1}{4}\{\xi\sqrt{(\eta^2-c)}-\eta\sqrt{(\xi^2-c)}\}^2,$$

or, say

$$\sqrt{(H)}=\tfrac{1}{2}\{\xi\sqrt{(\eta^2-c)}+\eta\sqrt{(\xi^2-c)}\},$$
$$\sqrt{(K)}=\tfrac{1}{2}\{\xi\sqrt{(\eta^2-c)}-\eta\sqrt{(\xi^2-c)}\},$$

and thence

$$h + \tfrac{1}{2}(a+b) + \sqrt{(H)} = \tfrac{1}{2}\{\xi + \sqrt{(\xi^2 - c)}\}\{\eta + \sqrt{(\eta^2 - c)}\},$$

$$k + \tfrac{1}{2}(a+b) + \sqrt{(K)} = \tfrac{1}{2}\{\xi + \sqrt{(\xi^2 - c)}\}\{\eta - \sqrt{(\eta^2 - c)}\}$$

$$= \tfrac{1}{2}c\,\frac{\xi + \sqrt{(\xi^2 - c)}}{\eta + \sqrt{(\eta^2 - c)}}.$$

These also follow from the known differential formula

$$4(dx^2 + dy^2) = (h-k)\left(\frac{dh^2}{H} - \frac{dk^2}{K}\right),$$

that is,

$$\frac{4d\xi d\eta}{\sqrt{(\xi^2 - c)}\sqrt{(\eta^2 - c)}} = \frac{dh^2}{H} - \frac{dk^2}{K},$$

implying

$$\frac{2\alpha d\xi}{\sqrt{(\xi^2 - c)}} = \frac{dh}{\sqrt{(H)}} + \frac{dk}{\sqrt{(K)}},$$

$$\frac{2d\eta}{\alpha\sqrt{(\eta^2 - c)}} = \frac{dh}{\sqrt{(H)}} - \frac{dk}{\sqrt{(K)}},$$

where α is a constant. The foregoing integral formulæ give at once

$$\frac{dh}{\sqrt{(H)}} = \frac{d\xi}{\sqrt{(\xi^2 - c)}} + \frac{d\eta}{\sqrt{(\eta^2 - c)}},$$

$$\frac{dk}{\sqrt{(K)}} = \frac{d\xi}{\sqrt{(\xi^2 - c)}} - \frac{d\eta}{\sqrt{(\eta^2 - c)}},$$

and substituting these values we find $\alpha = 1$, and the differential formulæ are then satisfied.

We thence have

$$\text{const.} = \sqrt{\{(a+h)(b+h)\}} \pm \sqrt{\{(a+k)(b+k)\}},$$

as the integral of the differential equation

$$\frac{dh}{\sqrt{(H)}} \pm \frac{dk}{\sqrt{(K)}} = 0.$$

553.

TWO SMITH'S PRIZE DISSERTATIONS.

[From the *Messenger of Mathematics*, vol. II. (1873), pp. 161—166.]

WRITE dissertations:

1. On the condition of the similarity of two dynamical systems.

2. On orthogonal surfaces.

1. We may consider two particles m, m', describing similarly two similar paths (which for convenience may be taken to be similarly situate in regard to two sets of rectangular axes respectively), viz. this means that the times t, t' of passage through corresponding arcs s, s' are proportional. The ratios $\frac{s'}{s}$, $\frac{t'}{t}$, are thus each of them constant; and this must also be the case with the ratio $\frac{v'}{v}$, of the velocities v, v' at corresponding points; since it is clear that we must have $\frac{s'}{s}=\frac{v'}{v}\cdot\frac{t'}{t}$.

Now in order that the two particles may move as above under the action of any forces upon the two particles respectively, it is clearly necessary that the forces F, F' at corresponding points shall act in the same direction, and be in a constant ratio of magnitude. To obtain this ratio, imagine the two particles, masses m, m', moving as above, in the corresponding infinitesimal elements of time τ, τ', with the velocities v, v' through the infinitesimal arcs σ, σ' respectively, $\left(\frac{\tau}{\tau'}=\frac{t}{t'},\ \frac{v}{v'}=\frac{v}{v'},\ \frac{\sigma}{\sigma'}=\frac{s}{s'}\right)$; the deflections from the tangent will be $\frac{1}{2}\frac{F}{m}\tau^2$, $\frac{1}{2}\frac{F'}{m'}\tau'^2$ respectively, and these must be in the ratio of the corresponding arcs σ, σ', viz. we must have

$$\frac{F\tau^2}{m} : \frac{F'\tau'^2}{m'} = \sigma : \sigma',$$

or, what is the same thing,

$$\frac{Ft^2}{m} : \frac{F't'^2}{m'}, = s : s',$$

that is,

$$\frac{F'}{F} = \frac{m'}{m}\frac{s'}{s} \div \left(\frac{t'}{t}\right)^2,$$

and this relation subsisting, and the velocities at the beginnings of the elements of time τ, τ' being in the assumed ratio, it is clear that the velocities at the ends of these elements of time will be in the same ratio; and thus the two particles will go on moving in the manner in question.

All that has been said as to two particles, applies without alteration to any two systems of particles moving under the like geometrical conditions, and we thus arrive at the conclusion; given two similarly constituted systems, which at any instant are in a given magnitude-ratio $\frac{s'}{s}$, their component particles being in a given ratio $\frac{m'}{m}$ (the same for each pair of component particles), then if the particles of the two systems respectively are to move in similar paths of the same magnitude-ratio $\frac{s'}{s}$, the times of describing corresponding arcs being in a given constant ratio $\frac{t'}{t}$ (this implying as above that the ratio of the velocities at corresponding points is $\frac{v'}{v}, = \frac{s'}{s} \div \frac{t'}{t}$), it is necessary that the forces on corresponding particles in corresponding positions shall act in the same directions, and shall be in the constant magnitude-ratio

$$\frac{F'}{F} = \frac{m'}{m} \cdot \frac{s'}{s} \div \left(\frac{t'}{t}\right)^2,$$

and this being so, the motion of the two systems will in fact be similar as above explained.

Taking $\frac{m'}{m}, = \mu$ for the mass-ratio, $\frac{s'}{s}, = \sigma$ for the length-ratio, and $\frac{t'}{t}, = \tau$ for the time-ratio; also $\frac{F'}{F}, = \phi$ for the force-ratio, the condition determining the force-ratio ϕ is thus

$$\phi = \frac{\mu\sigma}{\tau^2}.$$

It is to be observed, that if the forces are entirely internal, and proportional to homogeneous functions of the same order, say $-n$, of the coordinates of all or any of the particles; e.g. if they are central forces varying as the inverse nth power of the distances; then the condition as to the action of the forces in the two systems respectively can always be satisfied by giving a proper constant value to the ratio of the absolute forces (or forces at unity of distance); thus, if in the first system we have two particles m_1, m_2 attracting or repelling each other with a force $\frac{km_1m_2}{r^n}$, and if in the second system the force is $\frac{k'm_1'm_2'}{r'^n}$; then the condition as to the direction of the forces at corresponding positions is satisfied *ipso facto;* and the condition as to magnitude is

$$\frac{k'm_1'm_2'}{r'^n} \times \frac{r^n}{km_1m_2} = \frac{m'}{m}\frac{s'}{s}\frac{t^2}{t'^2},$$

that is,

$$\frac{k'}{k} = \frac{m'}{m_1'm_2'}\frac{m_1m_2}{m}\frac{s'r'^n}{sr^n}\frac{t^2}{t'^2}$$

$$= \frac{m}{m'}\left(\frac{s'}{s}\right)^{n+1}\frac{t^2}{t'^2},$$

or, say

$$\frac{t'}{t} = \left(\frac{s'}{s}\right)^{\frac{n+1}{2}}\left(\frac{km}{k'm'}\right)^{\frac{1}{2}}.$$

In the case $n=2$, the present theorem (applying however only to the case of two elliptic orbits of the same eccentricity) agrees with Kepler's third law, or say with the theorem

$$T = \frac{2\pi a^{\frac{3}{2}}}{\sqrt{(\mu)}},$$

that is,

$$T \propto \frac{a^{\frac{3}{2}}}{\sqrt{(\mu)}},$$

where observe that the μ, or mass in the sense of the formula, is the km, or attractive force on a unit of mass, of the theorem as above written down.

2. In a family of surfaces $F(x, y, z, p) = 0$, containing a single variable parameter p, there is through any given point of space, a surface or surfaces of the family; or (if more than one, confining the attention to one of these surfaces) we may say that there is, through any given point of space, a surface of the family.

Considering now two other families of surfaces, there will be through any given point of space, three surfaces, one of each family; and if (for every given point of space whatever) these intersect each other at right angles, we have a system of orthogonal surfaces.

Supposing the equations of the three families to be

$$F(x, y, z, p) = 0,$$

$$\Phi(x, y, z, q) = 0,$$

$$\Psi(x, y, z, r) = 0,$$

then the requisite conditions are

$$\frac{dF}{dx}\frac{d\Phi}{dx}+\frac{dF}{dy}\frac{d\Phi}{dy}+\frac{dF}{dz}\frac{d\Phi}{dz}=0,$$

$$\frac{dF}{dx}\frac{d\Psi}{dx}+\ldots\ldots\ldots\ldots\ldots\ldots=0,$$

$$\frac{d\Phi}{dx}\frac{d\Psi}{dx}+\ldots\ldots\ldots\ldots\ldots\ldots=0,$$

viz. these equations must be satisfied, not in general identically, but in virtue of the given equations $F=0$, $\Phi=0$, $\Psi=0$.

Or, what is more convenient, we may take the equations of the three families to be

$$p-f(x,\ y,\ z)=0,\quad q-\phi(x,\ y,\ z)=0,\quad r-\psi(x,\ y,\ z)=0;$$

and write the conditions in the form

$$\frac{dp}{dx}\frac{dq}{dx}+\frac{dp}{dy}\frac{dq}{dy}+\frac{dp}{dz}\frac{dq}{dz}=0,$$

$$\frac{dp}{dx}\frac{dr}{dx}+\ldots\ldots\ldots\ldots\ldots=0,$$

$$\frac{dq}{dx}\frac{dr}{dx}+\ldots\ldots\ldots\ldots\ldots=0,$$

where of course p, q, r stand for their given functional values, $p=f(x,\ y,\ z)$, &c.; the equations in this form contain only $(x,\ y,\ z)$, and not the parameters p, q, r; so that, if satisfied at all, they must be satisfied identically; and the required conditions therefore are that the last-mentioned system of equations shall be satisfied identically by the values p, q, r considered as given functions of $(x,\ y,\ z)$.

The last-mentioned conditions lead to the theorem known as Dupin's; viz. it follows from them that the surfaces intersect along their curves of curvature; or more definitely, each surface of one family is intersected by the surfaces of the other two families in its two sets of curves of curvature respectively.

To indicate the geometrical ground of the theorem, consider on a surface of one family a point P, and at this point the normal meeting the consecutive surface in P'; the surfaces through P of the other two families respectively will pass through P', and meet the given surface in two curves PA, PB (viz. PA, PB represent infinitesimal arcs on these two curves respectively), the angle at P being a right angle.

Drawing at A, B normals to the given surface to meet the consecutive surface in the points A', B' respectively, the same two surfaces will meet the consecutive surface in the arcs $P'A'$, $P'B'$ respectively; and (the system being orthogonal), we must have the angle at P' a right angle. This imposes a condition upon the direction (in the tangent plane at P) of the orthogonal directions PA, PB; viz. it is found that these must be such that the normals PP', AA' intersect, or, what is the same

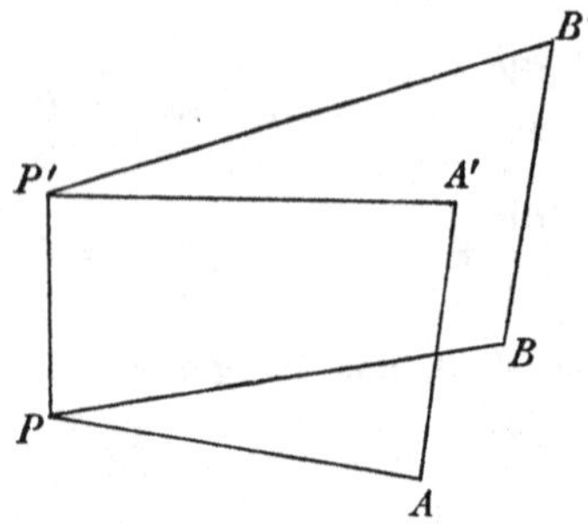

thing, the normals PP', BB' (one of these conditions implying the other); that is, that the lines PA, PB shall be the directions of the two curves of curvature through P on the given surface.

Observe that PP', AA' intersecting each other, the four points P, P', A, A' are in the same plane, that is, PA, $P'A'$ intersect, these lines being the normals at P, P' respectively of the surface through P of one of the other two families; and similarly PP', BB' intersecting each other, the lines PB, $P'B'$ intersect; these being the normals at P, P' respectively of the surface through P of the other two families. We have through PP' two planes at right angles to each other; and these are met by a plane $A'P'B'$, in two lines $A'P'$, $B'P'$, the inclinations of which to the line PP' differ only infinitesimally from a right angle, say they are $90° - a$ and $90° - b$ respectively; hence if the angle $A'P'B'$ is $= 90° - c$, this is the hypotenuse of a right-angled spherical triangle, the sides whereof are $90° - a$, $90° - b$; wherefore $\sin c = \sin a \sin b$, viz. $\sin c$ is an infinitesimal of a higher order which may be neglected, or the angle P' will be $= 90°$; that is, the surfaces through P of the other two families, intersecting the given surface in the directions PA, PB of the two curves of curvature, will intersect the consecutive surface at P' in the two directions $P'A'$, $P'B'$ *at right angles to each other;* which is an *a posteriori* verification of Dupin's theorem.

In what precedes the given surface through P may be regarded as a surface assumed at pleasure; and it in effect appears that taking the consecutive surface also at pleasure (but varying only infinitesimally from the given surface), the condition in order that the two surfaces, which at P intersect each other and the given surface at right angles, shall at P' intersect the consecutive surface in two directions at right angles to each other, is that they shall intersect the given surface in the directions PA, PB of the two curves of curvature. But if we thus take the consecutive surface at pleasure,—or say if we construct it by measuring off along the normal at *each* point P of the given surface an infinitesimal distance PP', $= \rho$, where ρ an arbitrary function of the coordinates of the point P,—then although on the consecutive surface the lines $P'B'$, $P'A'$ are at right angles to each other, there is nothing to show, and it is not in fact the case, that these lines $P'A'$, $P'B'$ are the directions of the curves of curvature on the consecutive surface. In the orthogonal system they must be so; and this imposes upon the infinitesimal normal distance ρ, a condition; viz. it is found that ρ considered as a function of (x, y, z) must satisfy a certain partial differential equation of the second order.

It hence appears that no one of the three families of surfaces can be assumed at pleasure; for taking the equation of a family to be $p - f(x, y, z) = 0$, then p being the value of the parameter for the given surface of the foregoing investigation, and $p + \delta p$ the value of the parameter for the consecutive surface, the normal distance at the point (x, y, z) between the two surfaces is

$$= \delta p \div \sqrt{\left\{\left(\frac{dp}{dx}\right)^2 + \left(\frac{dp}{dy}\right)^2 + \left(\frac{dp}{dz}\right)^2\right\}};$$

viz. δp is here a constant; and we have

$$1 \div \sqrt{\left\{\left(\frac{dp}{dx}\right)^2 + \left(\frac{dp}{dy}\right)^2 + \left(\frac{dp}{dz}\right)^2\right\}},$$

satisfying the foregoing partial differential equation; or, what is the same thing, p considered as a function of x, y, z must satisfy a certain partial differential equation of the third order; viz. this is the condition to be satisfied in order that a family of surfaces $p - f(x, y, z) = 0$ may belong to an orthogonal system.

554.

AN ELLIPTIC-TRANSCENDENT IDENTITY.

[From the *Messenger of Mathematics*, vol. II. (1873), p. 179.]

THE following is a singular identity:

$$\begin{aligned}&(1+q)(1+q^3)(1+q^5)(1+q^7)^2(1+q^9)\ldots\\&-(1-q)(1-q^3)(1-q^5)(1-q^7)^2(1-q^9)\ldots\\&=2q(1+q^2)(1+q^4)(1+q^6)(1+q^8)(1+q^{10})(1+q^{12})(1+q^{14})^2(1+q^{16})\ldots,\end{aligned}$$

where in each of the three terms every factor has the exponent 1 or 2 according as the exponent of q is not, or is, divisible by 7.

555.

NOTICES OF COMMUNICATIONS TO THE BRITISH ASSOCIATION FOR THE ADVANCEMENT OF SCIENCE.

[From the *Reports of the British Association for the Advancement of Science*, 1865 to 1873, *Notices and Abstracts of Miscellaneous Communications to the Sections.*]

1. *On the Problem of the in-and-circumscribed Triangle.* Report, 1870, pp. 9, 10.

I HAVE recently accomplished the solution of this problem, which I spoke of at the Meeting in 1864. The problem is as follows: required the number of the triangles the angles of which are situate in a given curve or curves, and the sides of which touch a given curve or curves. There are in all 52 cases [see 514] of the problem, according as the curves which contain the angles and are touched by the sides are distinct curves, or are any or all of them the same curve. The first and easiest case is when the curves are all of them distinct; the number of triangles is here $= 2aceBDF$, where a, c, e are the *orders* of the curves containing the angles (or, say, of the angle-curves) respectively; and B, D, F are the *classes* of the curves touched by the sides (or, say, of the side-curves) respectively. An interesting case is when the angle-curves are one and the same curve; or, say, $a = c = e$ (where the sign $=$ is used to denote the identity of the curves); the number of triangles is here $= \{2a(a-1)(a-2) + A\}\, BDF$, where a, A are the order and class of the curve $a = c = e$. In the reciprocal case, where the side-curves are one and the same curve, say $B = D = F$, we have of course a like formula, viz. the number of triangles is here $= \{2B(B-1)(B-2) + b\}\, ace$, where B, b are the class and order of the curve $B = D = F$. The last and most difficult case is when the six curves are all of them one and the same curve, say $a=c=e=B=D=F$; the number of triangles is here $=$ one-sixth of

$$\begin{array}{l} A^4(\ .\quad .\quad .\quad .\quad +\ 1), \\ +A^3(\ .\quad 2a^3 - 18a^2 + 52a - 46) \\ +A^2(\ .\ - 18a^3 + 162a^2 - 420a + 221) \\ +A\ (\ .\quad 52a^3 - 420a^2 + 704a + 172) \\ +\quad a^4 - 46a^3 + 221a^2 + 172a \\ +\alpha \left\{\begin{array}{l} A^2(\ .\quad .\quad .\ -\ 9) \\ +A\ (\ .\quad .\ - 12a + 135) \\ \quad - 9a^2 + 135a - 600 \end{array}\right\}, \end{array}$$

where a is the order, A the class of the curve; α is the number, three times the class + the number of cusps, or (what is the same thing) three times the order + the number of inflexions.

---

2. *On a Correspondence of Points and Lines in Space.* Report, 1870, p. 10.

NINE points in a plane may be the intersection of two (and therefore of an infinite series of) cubic curves; say, that the nine points are an "ennead": and similarly nine lines through a point may be the intersection of two (and therefore of an infinite series of) cubic cones; say, the nine lines are an ennead. Now, imagine (in space) any 8 given points; taking a variable point P, and joining this with the 8 points, we have through P 8 lines, and there is through P a ninth line completing the ennead; this is said to be the corresponding line of P. We have thus to any point P a single corresponding line through the point P; this is the correspondence referred to in the heading, and which I would suggest as an interesting subject of investigation to geometers. Observe that, considering the whole system of points in space, the corresponding lines are a triple system of lines, *not* the whole system of lines in space. It is thus, not any line whatever, but only a line of the triple system, which has on it a corresponding point. But as to this some explanation is necessary; for starting with an arbitrary line, and taking upon it a point P, it would seem that P might be so determined that the given line and the lines from P to the eight points should form an ennead,—that is, that the arbitrary line would have upon it a corresponding point or points.

The question of the foregoing species of correspondence was suggested to me by the consideration of a system of 10 points, such that joining any one whatever of them with the remaining nine points, the nine lines thus obtained form an ennead; or, say, that each of the 10 points is the "enneadic centre" of the remaining nine. I have been led to such a system of 10 points by my researches upon Quartic Surfaces; but I do not as yet understand the theory.

---

3. *On the Number of Covariants of a Binary Quantic.* Report, 1871, pp. 9, 10.

THE author remarked [see 462] that it had been shown by Prof. Gordan that the number of the covariants of a binary quantic of any order was finite, and, in particular, that the numbers for the quintic and the sextic were 23 and 26 respectively. But the demonstration is a very complicated one, and it can scarcely be doubted that a more simple demonstration will be found. The question in its most simple form is as follows: viz. instead of the covariants we substitute their leading coefficients, each of which is a "seminvariant" satisfying a certain partial differential equation; say, the quantic is $(a, b, c, \ldots, k \between x, y)^n$, then the differential equation is $(a\partial_b + 2b\partial_c \ldots + nj\partial_k)\, u = 0$, which *quà* equation with $n+1$ variables admits of n independent solutions: for instance, if $n = 3$, the equation is $(a\partial_b + 2b\partial_c + 3c\partial_d)\, u = 0$, and the solutions are a, $ac - b^2$,

$a^2d - 3abc + 2b^3$; the general value of u is u = any function whatever of the last-mentioned three functions. We have to find the rational non-integral functions of these functions which are rational and integral functions of the coefficients; such a function is

$$\frac{1}{a^2}\{(a^2d - 3abc + 2b^3)^2 + 4(ac - b^2)^3\},$$

$$= a^2d^2 + 4ac^3 + 4b^3d - 3b^2c^2 - 6abcd,$$

and the original three solutions, together with the last-mentioned function a^2d^2 + &c., constitute the complete system of the seminvariants of the cubic function; viz. every other seminvariant is a rational *and integral* function of these. And so, in the general case, the problem is to complete the series of the n solutions a, $ac - b^2$, $a^2d - 3abc + 2b^3$, $a^3e - 4a^2bd + 6ab^2c - 3b^4$, &c. by adding thereto the solutions which, being rational but non-integral functions of these, are rational and integral functions of the coefficients; and thus to arrive at a series of solutions such that every other solution is a rational and integral function of these.

---

4. *Note on certain Families of Surfaces.* Report, 1871, pp. 19, 20.

SEE the paper numbered 538, of which this Note is a duplicate.

---

5. *On the Mercator's Projection of a Surface of Revolution.* Report, 1873, p. 9.

THE theory of Mercator's projection is obviously applicable to any surface of revolution; the meridians and parallels are represented by two systems of parallel lines at right angles to each other, in such wise that for the infinitesimal rectangles included between two consecutive arcs of meridian and arcs of parallel the rectangle in the projection is similar to that on the surface. Or, what is the same thing, drawing on the surface the meridians at equal infinitesimal intervals of angular distance, we may draw the parallels at such intervals as to divide the surface into infinitesimal squares; the meridians and parallels are then in the projection represented by two systems of equidistant parallel lines dividing the plane into squares. And if the angular distance between two consecutive meridians instead of being infinitesimal is taken moderately small (5° or even 10°), then it is easy on the surface or *in plano*, using only the curve which is the meridian of the surface, to lay down graphically the series of parallels which are in the projection represented by equidistant parallel lines. The method is, of course, an approximate one, by reason that the angular distance between the two consecutive meridians is finite instead of infinitesimal.

I have in this way constructed the projection of a skew hyperboloid of revolution: viz. in one figure I show the hyperbola, which is the meridian section, and by means of it (taking the interval of the meridians to be = 10°) construct the positions of the successive parallels; I complete the figure by drawing the hyperbolas which are the orthographic projections of the meridians, and the right lines which are the orthographic projections of the parallels; the figure thus exhibits the orthographic projection (on the plane of a meridian) of the hyperboloid divided into squares as above. The other figure, which is the Mercator's projection, is simply two systems of equidistant parallel lines dividing the paper into squares. I remark that in the first figure the projections of the right lines on the surface are the tangents to the bounding hyperbola; in particular, the projection of one of these lines is an asymptote of the hyperbola. This I exhibit in the figure, and by means of it trace the Mercator's projection of the right line on the surface; viz. this is a serpentine curve included between the right lines which represent two opposite meridians and having these lines for asymptotes. It is sufficient to show one of these curves, since obviously for any other line of the surface belonging to the same system the Mercator's projection is at once obtained by merely displacing the curve parallel to itself, and for any line of the other system the projection is a like curve in a reversed position.

A Mercator's projection might be made of a skew hyperboloid not of revolution; viz. the curves of curvature might be drawn so as to divide the surface into squares, and the curves of curvature be then represented by equidistant parallel lines as above; and the construction would be only a little more difficult. The projection presented itself to me as a convenient one for the representation of the geodesic lines on the surface, and for exhibiting them in relation to the right lines of the surface; but I have not yet worked this out. In conclusion, it may be remarked that a surface in general cannot be divided into squares by its curves of curvature, but that it may be in an infinity of ways divided into squares by two systems of curves on the surface, and any such system of curves gives rise to a Mercator's projection of the surface.

NOTES AND REFERENCES.

518. Ribaucour, *C. R.*, t. LXXV. (1872), pp. 533—536, referring to my Note remarks that the condition can be (by means of the imaginary coordinates of M. Ossian Bonnet) expressed in a simple form communicated by him to the Philomathic Society, May, 1870. I reproduce this investigation, although it is not easy to present it in a quite intelligible form. We take $p=f(x, y, z)$ to represent a family of surfaces belonging to a triply orthotomic system, and consider two neighbouring surfaces (A) and (A') corresponding to the values z and $z+dz$; A and A' the two points where they meet the trajectories of the surfaces; AT, $A'T'$ the tangents to the curves of curvature of the same system at A, A' respectively. Then according to the remark of M. Lévy, it is to be expressed that these lines meet, and this is done by expressing that *along the trajectory AA', the variation of the angle of AT with the osculating plane at A is equal to the angle of the osculating planes at A, A' respectively.*

Let B' be the spherical image of A', the plane OBB' is parallel to the osculating plane at A of the trajectory, and the angle of the two osculating planes measures the geodesic curvature of BB': denote this by $d\gamma$.

Let β be the angle of BB' with BX, θ the angle of AT with BX, $\beta-\theta$ is the angle of AT with the osculating plane at A of the trajectory: $d\beta-d\theta=d\gamma$. Introducing the symmetric imaginary coordinates x and y, we write

$$a=\frac{dp}{\lambda^2 dx},\quad b=\frac{dp}{\lambda^2 dy},\quad c=\frac{1}{\lambda^2}\frac{d^2p}{dx\,dy},\quad ds^2=4\lambda^2\frac{da}{dx}\frac{db}{dy}\,dx\,dy.$$

But dx and dy being the increments of x, y corresponding to dz in the passage from A to A', then by a theorem of M. Liouville

$$d\gamma=d\beta-i\left(\frac{d\lambda}{\lambda\,dx}dx-\frac{d\lambda}{\lambda\,dy}dy\right);$$

the condition thus is

$$d\theta=i\left(\frac{d\lambda}{\lambda\,dx}dx-\frac{d\lambda}{\lambda\,dy}dy\right),$$

and the formula

$$e^{-2i\theta} = \pm\sqrt{\frac{da}{dx}} \div \sqrt{\frac{db}{dy}},$$

enables this to be written in the definitive form

$$dx\frac{d}{dx}l\left(\lambda^4\frac{db}{dy}\div\frac{da}{dx}\right) + dy\frac{d}{dy}l\left(\frac{db}{dy}\div\lambda^4\frac{da}{dx}\right) + dz\left\{\frac{d}{dz}\left(l\frac{db}{dy}\right) - \frac{d}{dz}\left(l\frac{da}{dx}\right)\right\} = 0.$$

We have

$$dx\,(\tfrac{1}{2}p + c) + dy\frac{db}{dy} \qquad + dz\frac{db}{dz} = 0,$$

$$dx\frac{da}{dx} \qquad + dy\,(\tfrac{1}{2}p + c) + dz\frac{da}{dz} = 0,$$

and thence eliminating dx, dy, dz, we have

$$\begin{vmatrix} \frac{d}{dx}l\left(\lambda^4\frac{db}{dy}\div\frac{da}{dx}\right), & \frac{d}{dy}l\left(\frac{db}{dy}\div\lambda^4\frac{da}{dx}\right), & \frac{d}{dz}l\left(\frac{db}{dy}\div\frac{da}{dx}\right) \\ \tfrac{1}{2}p + c \quad , & \frac{db}{dy} \quad , & \frac{db}{dz} \\ \frac{da}{dx} \quad , & \tfrac{1}{2}p + c \quad , & \frac{da}{dz} \end{vmatrix} = 0,$$

which defines the triply orthotomic system.

END OF VOL. VIII.

CAMBRIDGE: PRINTED BY J. AND C. F. CLAY, AT THE UNIVERSITY PRESS.